Predictability and Nonlinear Modelling
in Natural Sciences and Economics

Support

The conference has been financially supported by the following institutes and research programmes:

Commission of the European Communities, Directorate for Science, Research and Development, Climatology Programme

Dutch Ministry of Housing, Physical Planning and Environment, National Research Programme "Global Air Pollution and Climate Change" (NOP)

Netherlands Organization for Scientific Research, Priority Program "Nonlinear Systems"

Royal Netherlands Academy of Sciences

Wageningen Agricultural University, as part of its programme to commemorate its 75th Anniversary.

Sponsors

Shell Nederland bv

Unilever

Predictability and Nonlinear Modelling in Natural Sciences and Economics

edited by

J. Grasman

Department of Mathematics,
Agricultural University,
Wageningen, The Netherlands

and

G. van Straten

Department of Agricultural Engineering and Physics,
Agricultural University,
Wageningen, The Netherlands

SPRINGER-SCIENCE+BUSINESS MEDIA, B.V.

A C.I.P. Catalogue record for this book is available from the Library of Congress

ISBN 978-0-7923-2943-5   ISBN 978-94-011-0962-8 (eBook)
DOI 10.1007/978-94-011-0962-8

Printed on acid-free paper

CONTENTS

4. Systems sciences

5. Environmental Sciences

6. Economics

Preface

These proceedings have been composed from selected papers presented at the International Conference on 'Predictability and Non-Linear Modelling in Natural Sciences and Economics', held at the occasion of the 75th anniversary of Wageningen Agricultural University, The Netherlands. Some 160 participants from 18 countries gathered at Wageningen from 5-7 April 1993 for a most challenging and interesting meeting.

The organization of such a meeting is not possible without the help of many people. The members of the National Programme Committee have been very instrumental in the preliminary stages of the organization, in particular in the choice of outstanding plenary speakers. Apart from us, members of the National Programme Committee were: H. Folmer (Wageningen), J. Goudriaan (Wageningen), L. Hordijk (Wageningen), H.A.M. de Kruijf (Bilthoven), J. Leentvaar (Lelystad), S. Parma (Nieuwersluis), C.J.E. Schuurmans (De Bilt), H.N. Weddepohl (Amsterdam) and J.C. Willems (Groningen). We are also greatful for advice of the International Scientific Advisory Committee, consisting of M.B. Beck (London), J.D. Farmer (Sante Fé), J.-M. Grandmont (Paris), T.N. Palmer (Reading), M.L. Parry (Oxford), W.M. Schaffer (Tucson) and L. Somlyódy (Laxenburg). Several members of both committees have been of great help in finding referees, and have played an active role in the reviewing process.

The meeting itself, taking place at the International Agricultural Centre, was excellently organized by the Wageningen University Congress Bureau, an accomplishment that should largely be attributed to Mr. J. Meulenbroek and Mrs. A. van Wijk-van der Leeden.

Finally, the production of a book of this size is not an easy task. The cooperation of the authors and numerous anonymous reviewers has been acknowledged. Special thanks have to be awarded to Mrs. H. Evers-van Holland, Mrs. D.J. Massop-Schreuders and Mrs. M. Slootman-Vermeer of the secretariat of the Department of Mathematics for their excellent job in producing the camera ready text.

We are proud to present this book, and we hope that the reader will find the contents interesting, stimulating and rewarding.

Johan Grasman
Department of Mathematics

Gerrit van Straten
Department of Agricultural Engineering and Physics

Agricultural University
Wageningen, The Netherlands

INTRODUCTION

J. GRASMAN and G. VAN STRATEN

Agricultural University, Wageningen

Predicting the future behaviour of natural and economic processes is the subject of research in various fields of science. After a period of considerable progress by refining models in combination with large scale computer calculations, the scientific community is presently confronted with problems that require a novel approach to further extend the range of forecasts and to improve their quality. It is recognized that nonlinearity of a system may significantly complicate the predictability of future states of the system. A small variation of parameters can drastically change the dynamics, while sensitive dependence on the initial state may severely limit the predictability horizon.

Predictability of dynamic systems has two sides. The notion that non-linear dynamic systems may exhibit chaotic behaviour has contributed enormously to the understanding of phenomena that could not be explained before. Although precise predictions of the state of the system in the far or even nearby future is impossible in the chaotic regime, chaos theory may have the key to open new pathways in predictability theory. Moreover, it answers the question under which circumstances chaotic behaviour can be expected.

On the other hand, not every non-linear system shows chaotic behaviour in the parameter range of interest, and therefore do not have a fundamental prediction problem. Nevertheless, prediction capability is hampered also here, because of uncertainties. This is the other side of the coin of predictability. Since models are abstractions of the real world, they always are no more than approximations. Thus, errors made during model construction and calibration will propagate when forecasts are made. The analysis of uncertainty is therefore the second important issue when dealing with predictability.

The papers brought together in this book have been presented at a conference organized at the occasion of the 75th anniversary of Wageningen Agricultural University in the Netherlands. The conference aimed at bringing together scientists who are modelling the dynamics of natural and economic processes on the one hand, with system analysts and mathematicians who develop methods for quantifying the characteristic features of such models on the other. Quite naturally, this has led to the meeting of the two main methodological tools - the mathematical theory of nonlinear dynamical systems and the analysis of uncertainty in modelling - as is reflected in the various contributions.

Chaos

Non-linearity in the dynamic equations can give rise to irregular dynamics of a system. This so-called chaotic behaviour is characterized by sensitive dependence on the initial state. The degree of divergence of the trajectories in chaotic systems is determined by the Lyaponov exponents. In contrast to stochastic systems, in chaotic systems the irregular trajectories concentrate in state space on the (strange) attractor of the system. Such strange attractors have a fractal structure and are said to have a non-integer dimension.

In the applied sciences chaos was first noticed from the properties of a simple model of the atmospheric circulation (Lorenz, 1965) and in biology from the dynamics of an even more simple model of the changing size of a biological population (May, 1974). Not amazingly, these two fields of applications are represented in this proceedings again, but the questions that are addressed have evolved since then. Knowing that the error in the meteorological forecast increases with time, one may wish to quantify the expected size of the error, because this defines the forecast skill (Royer et al.). The analysis is based on the behaviour of the system near the trajectory starting in the present state that is computed from the observations. In the other field, that of population biology, presently a strong interest exists in the data of childhood diseases, such as measles (Kendall et al., Engbert and Drepper). Large time series of cities like New York are available. The first problem concerning the irregular behaviour of such a time series was to find out whether the dynamics of this epidemical process is really chaotic. Recently the answer could be given and turned out to be confirmative, see Olsen and Schaffer (1990). Presently other properties of these processes, such as the nonuniform information content of data points (Kendall et al.) and the possibility of controlled outbreaks, are being studied (Schwartz and Triandaf).

Chaos as an explanation for irregular behaviour and unpredictability was taken over by other disciplines. At the conference it was made explicit that economic processes can have a nonlinear structure that may give rise to chaos. In these contributions the models are in the form of nonlinear difference equations (Hommes, Weddepohl, Hommes et al.). Large time series of prices at stock markets can be analyzed with the method of rescaled range analysis introduced by Hurst, see Feder (1988). It turns out that this stochastic fractal approach may help to quantify long term dependence in a time series (Jacobsen).

Uncertainty

Uncertainty arises from different sources. In the construction phase of the model, observational noise and errors in input and output sequences lead to possibly erroneous structures and biased parameters. Moreover, lack of fundamental knowledge, simplifications, aggregations of variables, neglect of variables and processes, and approximations to functional relationships all contribute to systems noise, and thus to uncertainty when the model is used for predictions.

In the stage where calibrated and validated models are being used, prediction uncertainty may arise from three sources. First, there can be uncertainty in the expected future input sequences. If the model responds to its inputs in a non-linear fashion, it is not justified to use average values for future inputs (Semenov & Porter). But, if properly handled, input uncertainty can be taken care off in scenario studies, and does not arise from the model itself. Second, the uncertainties in model structure and parameters, as identified in the construction stage, can be propagated into future projections, thus leading to distributions or regions around future state trajectories. The third kind of uncertainty arises when models are used outside their range of validity. This situation is met quite frequently in the environmental field, where the purpose of the model is to forecast the future state of the system under changed pollution load conditions. In all situations where the future is different from the present, there is no absolute guarantee that the model structure will not change. The dilemma here is to use complex, detailed models, which however cannot be fully identified on the basis of present knowledge and data, or to accept smaller well calibrated models, which however may give the wrong predictions under changed conditions (Beck, 1991). A possible remedy to this situation has been indicated with the term 'educated speculation', where relatively simple models are used as frame of reference for expert judgement to speculate about possible modifications of the model parameters in the future (Van Straten & Keesman, 1991). Another idea is to use various models, and study the sensitivity of the final management decisions, possibly with the use of experts via Delphi methods (Komen). These proposals are all elements of what Beck in his contribution calls the quest for 'robust' predictions insensitive to the lack of identifiability.

The analysis of uncertainty already starts in the stage of model formulation, calibration and validation. The two basic approaches to model building - data oriented versus first principles oriented - are also present in this book. System identification by approximate realization (e.g. Heij) is an exponent of the first, while validation applied to a priori formulated model structures is an exponent of the latter. This issue is not an easy matter in large-scale, spatial models, as shown in the study by Fleming.

The bringing together of people from different disciplines have also brought to light a difference in terminology. While systems dynamics people like to restrict the term inputs to time sequences of independent variables, as a distinction to the essentially constant parameters of the model, to the computer science people these are all 'inputs', i.e. quantities to be known, before any computation can be made (see e.g. Kleijnen).

The majority of contributions on uncertainty in this proceedings deal with the methodology of uncertainty assessment, most notably with the uncertainty component that can be quantified in the calibration stage of model development. Several useful measures of uncertainty have been proposed (Janssen). The prevailing technique for uncertainty analysis in non-linear models is Monte Carlo simulation, followed by regression type of analyses of the results. This can be done in a probabilistic framework, using prior distributions of uncertainties in the parameters (e.g. Heuberger & Janssen), in a Bayesian framework, where prior distributions are derived from the data (Kramer et al.), or in a set-theoretic setting (Keesman).

Finally, the developed methods of various kinds have been applied to a variety of rather complex uncertainty assessment problems, for example in environmental risk assessment (e.g. Hommen, Traas & Aldenberg), environmental scenario studies (e.g. Hettelingh & Posch), pest control (Rossing et al.), and macro-economics (Don).

Retrospective

Going over the contributions, we observe a remarkable dichotomy in approach which seems to be related to the dimensionality of the problem. In low dimensional models, the chaos approach is prevailing, while for large dimensional models prediction uncertainty is analyzed with Monte Carlo or other statistical methods. In high dimensional models such as climate models or the models used for pest control of crops, uncertainty in the values of the parameters make it very unattractive to start large scale computer computations for pinning down the values of the largest Lyapunov exponents or the dimension of the attractor. An exceptional position is taken by general atmospheric circulation models, which are being analysed with the chaos approach, despite the fact that the dimension of the state space is extremely large. The reason lies in the fact that the dynamical equations are known to be correct and that the parameter values can also be specified with a sufficient accuracy.

The question that remains untouched in this book is whether the chaos approach in association with low dimensional models can form an alternative to large size models. Although the underlying mechanisms in, for example, childhood epidemics or business cycles can be very complex, and thus would normally give rise to high dimensional models, there are strong indications that the phenomena occurring near the transition to chaos can very well be described by non-linear models of low dimension. In addition, aggregation of variables such as in ecological compartment models, will also lead to models of low order. Even so, in these and similar problems, there is no reason to assume that the deterministic equations used in chaos analysis are 'exact', and not subject to uncertainty, both in structure and in parameters. Also, the questions of calibration and validity still apply. Consequently, it seems that calibration, structure identification and the introduction of intrinsic parametric or system uncertainty in forecasts for chaotic systems can add an additional dimension to the theory and application of non-linear systems analysis.

On the other hand, the possibility that high dimensional non-linear models might exhibit chaotic behaviour has not been given very much attention, with the exception of large scale meteorological models. If chaos can occur, the results of Monte Carlo analysis need to be questioned, or at least reexamined. In this context further probing to discover signs of possible chaotic behaviour in for instance environmental and ecological models is likely to be rewarding.

Another dichotomy that comes into mind when reviewing the papers in the book is between deterministic and stochastic models. Practically all models analyzed in this book belong to the class of deterministic models, even though the analysis of uncertainty is done by Monte Carlo or other statistical methods. This, we guess, is largely related to the tradition in the natural sciences to give mechanistic descriptions of observed phenomena. Nevertheless, since models are approximations to the real world, the mismatch between model and reality may take the form of a stochastic process, which can be modelled accordingly, apart from the possibility that processes are stochastic by nature. The question of stochasticity versus chaos is relevant to the solution of real world problems, see also Casdagli (1992), and also here much is still to be done.

We have observed that most work on chaotic systems is restricted to unforced systems. Apart from analytical tractability there is no good reason for this, and further work in analyzing the behaviour of potentially chaotic systems under forcing inputs is of much practical value. This is so, because much of the work on predictability is motivated not just by scientific curiosity, but also by the desire of controlling systems in order to prevent them from moving into undesirable directions. The idea to study uncertainty in the context of control and controllability, which has created so much interest in robust design methods in the control community, should be promoted vigorously in the world of natural scientists and economists, if the results are going to be used for management purposes.

Looking back at the contents of the book, and from what we have heard from the participants of the conference, we believe that we have succeeded in achieving the main goal: the cross-fertilization of ideas between the different disciplines that share the objective of making forecasts of dynamical processes. We hope that with this book we can convey some of our shared enthousiasm to a larger audience, and that the topics covered will provide a source of inspiration for further work on this intriguing topic.

References

Beck, M.B. (1991), 'Forecasting Environmental Change', *J. Forecasting*, **10**(1&2), 3-19.
Casdagli, M. (1992), 'Chaos and deterministic versus stochastic modelling', *J. Royal Statistical Society* **B54**, 303-328.
Feder, J. (1988), 'Fractals', Plenum Press, New York.

Lorenz, E. (1963), 'Deterministic non-periodic flow', *J. Atmosph. Sci* **20**, 130-141.

May, R.M. (1976), 'Simple mathematical models with very complicated dynamics', *Nature* **261**, 459-467.

Olsen, L.F., and W. Schaffer (1990), 'Chaos versus noisy periodicity: alternative hypotheses for childhood epidemics', *Science* **249**, 499-504.

van Straten, G. and K.J. Keesman (1991), 'Uncertainty propagation and speculation in projective forecasts of environmental change: a lake eutrophication example', *J. Forecasting*, **10**(1&2), 163-190.

KARL POPPER AND THE ACCOUNTABILITY OF SCIENTIFIC MODELS

H. TENNEKES

Royal Netherlands Meteorological Institute, de Bilt

Abstract

Karl Popper has written extensively on the methodology of scientific research. In his "Postscript to the Logic of Scientific Discovery" he formulates the principle of scientific accountability: "Scientific determinism requires the ability to predict every event with any desired degree of precision, provided we are given sufficiently precise intitial conditions. Our theory will have to account for the imprecision of the prediction: given the degree of precision which we require of the prediction, the theory will have to enable us to calculate the degree of precision which we require of the prediction, the theory will have to enable us to calculate the degree of precision in the initial conditions. For systems that exhibit sensitive dependence on initial conditions, this demand is nearly impossible to meet. This paper focuses on the potential consequences of insisting on accountability in the development of mathatical models of chaotic systems.

1. Introduction

On occasion one wonders what certain famous people would have said about a subject one is vitally interested in, provided they had been in a position to study the subject in depth. This thought occurred to me when I read Karl Popper's 1972 essay a few years ago. Karl Popper (born 1902), the outspoken philosopher of science, is the originator of the sobering idea that science progresses by falsification, not verification (the latter being ultimately impossible). He introduced clouds in his essay to represent physical systems "which are highly irregular, disorderly, and more or less unpredictable", and clocks to represent systems "which are regular, orderly and highly predictable in their behavior".

Popper's interest in predictability was fueled by his opposition to the views held by the proponents of determinism. The central thesis of determinism is the "staggering proposition" that all clouds are clocks; in this view the distinction between clocks and clouds is not based on their intrinsic nature, but on our lack of knowledge. If only we knew as much about clouds as we do about clocks, clouds would be just as predictable as clocks. Or, in meteorological terms, a perfect model of the atmosphere, initialized with perfect data from an observation network of infinite resolution, and run on an infinitely powerful computer, should in principle produce a perfect forecast with an unlimited range of validity.

If, at the time he wrote his essay, Popper had been living in 1990 and therefore had been aware of the explosive growth of our knowledge about the chaotic behavior of nonlinear systems since the publication of Ed Lorenz' 1963 paper on deterministic, nonperiodic flow, it would have been a lot easier for him to demolish the determinist position. We now know that the effective forecast range for a complex physical system is finite and limited to the typical life span of its most energetic phenomena. The "prediction horizon" of midlatitude weather is comparable to the average life span of extratropical cyclones: a week, no more. Even satellites, the most clocklike of human artifacts, are not immune: the small deviations in their orbits caused by irregularities in the gravitational field also can be predicted just a few days ahead. In other words: all clocks are clouds, if only one looks closely enough.

My own interest in predictability dates from 1976, when I attempted to reconcile my experience in turbulence research with the error cascade calculations published by Lorenz (1969), and Leith and Kraichnan (1972). By 1986 my thoughts had evolved to the point where I realized that skill forecasting should be one of the central goals of predictability research. At a workshop held in April 1986 I introduced the slogan: "No forecast is complete without a forecast of forecast skill". The admittedly vague motivation for coining this phrase was that it will enhance the credibility of meteorology if we are honest about the intrinsic limitations to forecast skill. Since it is unlikely that the average performance of numerical weather forecasts can be substantially improved, it is prudent to invest in research that might lead to tolerably skilful skill forecasts. Toward the end of the forecast range the quality of a forecast will always be poor, no matter how much that range is extended. A little bit of guidance on the anticipated reliability of weather forecasts is a whole lot better than no guidance at all.

Little did I know at the time that I was dabbling in the periphery of a fundamental problem in the methodology of scientific inquiry. Fortunately, in 1990 I chanced upon a copy of the second volume of Popper's "Postscript to the Logic of Scientific Discovery" (later published as The Open Universe, 1982) and read the astonishing paragraphs on scientific accountability that I shall quote from *in extenso* below. Those texts, based on two lectures Popper gave as early as 1950, made it painfully clear that I had merely scratched the surface of the problem I had pronounced upon. In this essay I am trying to make up for this defect. Better late than never.

2. Accountability

Popper (1982) begins his quest for honesty in scientific inquiry by pointing out that it is necessary to lay down certain ground rules on the precision required of scientific calculations:

> "The fundamental idea underlying scientific determinism is that the structure of the world is such that every future event can in principle be rationally calculated in advance, if only we know the laws of nature and the present state of the world. But if *every* event is to be predictable, it must be predictable *with any desired degree of precision*: for even the most minute difference in measurement may be claimed to distinguish between different events" (The Open Universe, p.6).

In other words, we have to agree in advance on the accuracy we demand. We should not allow ourselves an escape route if things go wrong. A few pages later, Popper explains the issue in more detail:

> "Scientific determinism requires the ability to predict every event with any desired degree of precision, provided we are given sufficiently precise initial conditions. But what does 'sufficiently' mean here? Clearly, we have to explain 'sufficiently' in such a way that we deprive ourselves of the right to plead - every time we *fail* in our predictions - that we were given initial conditions which were not sufficiently precise. In other words, our theory will have to account for the imprecision of the prediction:

given the degree of precision which we require of the prediction, the theory will have to enable us to calculate the degree of precision in the initial conditions that would suffice to give us a prediction of the required degree of precision. I call this demand the principle of accountability" (The Open Universe, p.11).

The thrust of this argument is evident. Whenever they fail in their predictions, scientists tend to blame the poor accuracy of the observations, the lack of computer power and the inadequate parameterization in their numerical models, rather than their own lack of skill in computing the accuracy that can be obtained with present resources. Sloppy reasoning of this kind is responsible for much of the thoughtless expansion and escalation numerical modelers in all branches of science indulge in.

Popper holds researchers accountable for their own work. He feels, and so do I, that it is besides the point to complain about available resources being finite. They will always be. Instead, one should focus one's modeling skills on the creation of algorithms that compute the forecast skill obtainable at the present state of the art. To put it bluntly, a calculation that does not include a calculation of its predictive skill is not a legitimate scientific product.

If calculations of forecast skill were easy, there would be no reason to worry about accountability. Does one satisfy the principle of accountability if one collects a skill climatology by repeated experimentation? Popper quickly plugs that hole:

"We have to demand that we must be able to find out, *before* we test the result of our predictions, whether or not the initial conditions are sufficiently precise. To put it more fully, we must be able to account *in advance* for any failure to predict an event with the desired degree of precision, by pointing out that our initial conditions are not precise enough, and by stating how precise they would have to be for the particular prediction task at hand" (The Open Universe, p.12).

My own hesitation to rely on skill climatology was rather more pragmatic. In weather forecasting, there are large fluctuations in forecast skill from day to day; the average forecast skill is often not a reliable indication of the skill of an individual forecast. In the same way, the effective forecast range or prediction horizon, which currently is about five days on average, may be as little as two or as much as ten days from one event to the next. With that much variability, quoting an average value just does not fit the bill. Popper is a lot tougher: he insists that, since we are interested in predicting individual events (single realizations, in the language of statistics), we have to make *advance* calculations of the predictive skill *for each event*. In other words, if we wish to make a deterministic forecast, it will not do to make anything less than a deterministic skill forecast. Don't retreat into statistics when the going gets tough, that is what Popper says.

3. Infinite regress

It is evident that Popper allows himself little leeway when he thinks about the methodology of research. But once he has defined the core of the problem, he can permit himself to be considerate. He realizes that it is tougher to meet nonnegotiable demands than to formulate them. The production of skill forecasts is likely to be much more difficult than the production of the forecasts themselves. In fact, it is not at all clear whether skill forecasts of sufficient quality to meet the principle of accountability are possible, even in principle. The skill of a skill forecast is likely to be less than the skill of the original forecast, and the entire process demanded by accountability may not even converge. Popper starts his investigation into this problem by pointing out that accountability is concerned not only with the accuracy of the initial conditions, but also with the limitations inherent in the models we employ to describe the evolution of a physical system:

> "The method of science depends on our attempts to describe the world with simple theories. Theories that are complex may become untestable, even if they happen to be true (numerical modelers, please note - H.T.). Science may be described as the art of systematic over-simplification: the art of discerning what we may with advantage omit" (The Open Universe, p.44).

Since every prediction task that may be conceivably accountable operates with a simplifying model, a probably unresolvable paradox arises: if the model contains too many simplifications, it may not be possible to perform adequate calculations of forecast skill. Models have to be simple to be testable (another Popper favourite!), but simplicity and accountability may turn out to be mutually exclusive categories. In Popper's own words:

> "The question arises: how good does the model have to be in order to allow us to calculate the approximation required by accountability? Since the goodness of the model is its degree of approximation or precision, we are threatened by an infinite regress, and this threat will be very serious for systems which are complex (such as the atmosphere - H.T.). But the complexity of the system can also be assessed only if an approximate model is at hand; a consideration which again indicated that we are threatened by an infinite regress" (The Open Universe, p.50).

The matter of "infinite regress" was pointed out to me within days after launching my slogan on forecasts of forecast skill. What about the skill of the skill forecast, my friends would tease me, how do you forecast that? And what about having to repeat the process *ad infinitum*? Is is not true that these are progressively harder questions, and is there any chance the process will converge? Popper would have been sympathetic:

> "It is clear that we must, in advance, limit either the permissible number of demands for an improved model, or else the goodness, i.e. the degree of precision which we may require of the model. But the task of calculating either of these may lead us merely to a problem of accountability of a higher order. And with this, we may be well on our way to an infinite regress. For there is no reason whatever to believe that the higher-order problem is easier to solve than is the lower-order problem, or that a less good model is needed for its solution than for the solution of the lower-order problem. Nor is there any reason to believe that methods of approximation are always capable of improving results indefinitely" (The Open Universe, p.51).

With all of these problems piling up, there is, Popper says, "no reason whatever to believe that Newtonian mechanics is accountable". The results obtained by chaos theory bear him out. Determinism is an untenable proposition, even in the branch of physics in which it seemed most secure.

4. Marching orders

Obviously, science is not an occupation for the weak at heart. If one is afraid of tackling problems that seem insoluble, one would do well to seek another profession. Fortunately, the human spirit is such that the hardest questions often attract the best minds. That bodes well for skill prediction research.

Because of the problem of "infinite regress", I do not think it advisable to hold skill forecasts accountable in the strict sense of the word. For the time being, statistical verifications of the skill of skill forecasts will have to suffice. The conceptual problems associated with higher-order accountability will have to wait. This puts skill forecasting midway between the presumed determinism of numerical weather prediction and the ad-hoc statistics of second-order skill. Skill forecasting deals with the statistical dynamics of the atmosphere; it is in that general area that one may anticipate the best prospects for advancement.

Acknowledgment

This essay appeared first in a slighly different form in Weather, **47**, 343-346. Reprinted by permission from the Royal Meteorological Society, England.

References

Leith, C.E. and R.H. Kraichnan (1972), 'Predictability of turbulent flows', *J.Atmos.Sci.*, **29**, 1041-1058.

Lorenz, E.N. (1963), 'Deterministic, nonperiodic flow', *J.Atmos.Sci.*, **20**, 130-141.

Lorenz, E.N. (1969), 'The predictability of a flow which possesses many scales of motion', *Tellus*, **21**, 289-307.

Popper, K.R. (1972), 'Of clouds and clocks', *Objective Knowledge*, Clarendon Press, Oxford.

Popper, K.R. (1982), *The Open Universe*, Hutchinson, London.

EVALUATION OF FORECASTS

ALLAN H. MURPHY[1] and MARTIN EHRENDORFER[2]

[1] Professor Emeritus, College of Oceanic and Atmospheric Sciences,
Oregon State University, Prediction and Evaluation Systems
Corvallis, Oregon 97330, USA

[2] National Center for Atmospheric Research
Boulder, Colorado 80307, USA
On leave from the Institute for Meteorology and Geophysics,
University of Vienna, Hohe Warte 38, A-1190 Vienna, Austria

Abstract

Evaluation of forecasts encompasses the processes of assessing both forecast quality and forecast value. These processes necessarily play key roles in any effort to improve forecasting performance or to enhance the usefulness of forecasts.

A framework for forecast verification (the process of assessing forecast quality) based on the joint distribution of forecasts and observations - and on the conditional and marginal distributions derived from factorizations of this joint distribution - is described. The joint, conditional, and marginal distributions relate directly to basic aspects of forecast quality, and evaluation methods based on these distributions - and associated statistics and measures - provide a coherent, diagnostic approach to forecast verification. This approach - and its attendant methodology - is illustrated using a sample of probabilistic long-range weather forecasts.

A decision-analytic approach to the problem of assessing the value of forecasts is outlined, and this approach is illustrated by considering the so-called fallowing-planting problem. In addition to providing estimates of the value of state-of-the-art and hypothetically improved long-range weather forecasts, the results of this case study illustrate some of the fundamental properties of quality/value relationships. These properties include the inherent nonlinearity of such relationships and the existence of quality thresholds below which the forecasts are of no value.

The sufficiency relation is used to explore quality/value relationships; this relation embodies the conditions that must exist between the joint distributions of two forecasting systems to ensure that one system's forecasts are better in all respects (i.e., in terms of quality and value) than the other system's forecasts. The applicability of the sufficiency relation is illustrated by comparing forecasting systems that produce prototypical long-range weather forecasts. This application also demonstrates that quality/value reversals can occur when the multifaceted nature of forecast quality is not respected.

Some outstanding problems in forecast evaluation are identified and briefly discussed. Recommendations are made regarding improvements in evaluation methods and practices.

1. Introduction

In the meteorological community, both individuals who formulate numerical and/or statistical forecasting models (modelers) and individuals who actually make forecasts on an operational basis (forecasters) are concerned with the quality of their forecasts. Identification of the characteristics of forecasting performance is an essential first step in the processes of model refinement and forecast improvement. Since the underlying rationale for developing forecasting systems in the first place is to provide information that can enhance the welfare (economic or otherwise) of potential users of the forecasts, modelers and forecasters are - or should be - concerned as well with the value of their forecasts. Forecast evaluation, which encompasses the processes of assessing both forecast quality and forecast value, is thus a problem of considerable importance.

Notwithstanding the importance of this problem, traditional methods of assessing forecast quality and forecast value are seriously deficient and potentially misleading. With regard to quality, forecast verification (the name usually given to the component of the evaluation process concerned with forecast quality) generally consists of the calculation of one or two measures of the overall correspondence between forecasts and observations (e.g., a skill score, a correlation coefficient). Consideration of the dimensionality of most verification problems suggests that this measures-oriented approach is inadequate as a means of characterizing the fundamental strengths and weaknesses in forecasting perform- ance. Forecast value is seldom assessed in anything but the most informal and ad hoc manner. It is often simply assumed that improvements in forecast quality, as reflected by an increase in a skill score or a correlation coefficient, necessarily are accompanied by increases in forecast value. It is relatively easy to show that this assumption generally is unwarranted (see section 4).

Coherent approaches to the problems of assessing forecast quality and forecast value are described in this paper. In the case of forecast quality, considered in section 2, the approach is based on the concept that the joint distribution of forecasts and observations contains all of the (nontime-dependent) information relevant to forecast verification. Section 3 outlines a decision-analytic approach to the assessment of forecast value. Both sections 2 and 3 contain brief descriptions of applications of evaluation methods associated with the respective approaches. The relationship between forecast quality and forecast value is discussed in section 4, with emphasis on the role and usefulness of the sufficiency relation in comparative evaluation. Section 5 identifies some of the outstanding methodo- logical and practical problems in this area and includes some recommendations regarding changes in current practices.

2. Forecast Quality

2.1 Approach and Methodology

The quality of forecasts produced by a model or a forecaster (denoted by F) can be defined as the totality of the statistical characteristics embodied in the joint distribution of forecasts (f) and observations (x), $p(f,x)$ (Murphy and Winkler 1987). Under the assump- tion that the joint probabilities that constitute $p(f,x)$ are stationary parameters, they can be estimated from the joint relative frequencies obtained from a single realization of a verification process that extends over a reasonably long but finite time period. The information contained in $p(f,x)$ becomes more accessible when this distribution is factored into conditional and marginal distributions. Two such factorizations are possible: (1) $p(f,x) = p(x|f) \, p(f)$ and (2) $p(f,x) = p(f|x)p(x)$. These expressions are referred to as the calibra- tion-refinement and likelihood-base rate factorizations of $p(f,x)$, respectively. In effect, these factorizations provide two different but complementary - and obviously related -

descriptions of forecast quality in the context of absolute verification problems (i.e., problems involving the quality of a single forecasting system).

If the dimensionality (d) of an absolute verification problem is defined as the number of parameters (e.g., joint probabilities) that must be specified to determine $p(f,x)$, then it follows that $d = 3$ when both f and x are binary, and $d = 21$ when f can assume any of 11 equally-spaced probability values and x is binary (Murphy 1991). (The inadequacy of the traditional measures-oriented approach to verification problems, described briefly in section 1, should now be clear.) The multidimensional nature of forecast quality also can be appreciated by recognizing that quality possesses several aspects, including bias, accuracy, skill, reliability, resolution, sharpness, and discrimination (Murphy and Winkler 1987; Murphy 1993). These aspects of quality can be related directly to the aforementioned joint, conditional, and marginal distributions (see section 2b).

To accommodate comparative verification, which is concerned with the relative quality of two sets of forecasts (produced by systems, models, or forecasters F and G), the framework for absolute verification must be extended to include two joint distributions, $p(f,x)$ and $p(g,x)$. (For simplicity, we assume throughout this paper that F and G produce forecasts for the same events at the same location on the same occasions.) In this context, it is necessary to compare the conditional and marginal distributions associated with the respective factorizations of these joint distributions. Clearly, comparative verification is considerably more complex than absolute verification, and space limitations preclude detailed treatment of this problem here (however, see section 4 for some relevant discussion).

Within the framework of this distributions-oriented approach to forecast verification, three general classes of verification methods can be identified: (1) the joint, conditional, and marginal distributions themselves; (2) statistics of these distributions (e.g., means, medians, standard deviations, interquartile ranges); and (3) performance measures and terms in decompositions of performance measures. These methods can be used to evaluate - both qualitatively and quantitatively - various aspects of forecast quality, and some of these methods are illustrated in section 2b.

In describing the distributions-oriented approach to verification problems, it has been assumed implicitly that both f and x are scalar quantities, in the sense that they denote forecasts and observations at a specific location. However, this approach can be extended readily to situations in which the arguments of the joint distribution are vectors; for example, these vectors might define two-dimensional arrays of the forecast and observed (or analyzed) values of a meteorological variable at a particular time. Verification of these fields could proceed by applying distributions-oriented methods to the data set obtained by pooling the forecast-observation pairs from all locations. More general methods that take into account the spatial characteristics of the arrays of forecasts and observations would require a further extension of the approach described in this paper.

2.2 Applications

To illustrate the distributions-oriented approach to forecast verification, we describe briefly an application of this methodology to long-range weather forecasts formulated by forecasters at the Climate Analysis Center of the U.S. National Weather Service (Murphy and Huang 1993). The results presented here involve probabilistic forecasts of monthly mean temperatures (MT) and monthly precipitation amounts (MP), which were produced twice each month for approximately 100 locations during the period 1982-1991. These forecasts relate to below-normal, near-normal, and above-normal categories of MT or MP, which are defined at each location such that their historical probabilities of occurrence are equal to 0.30, 0.40, and 0.30, respectively. These probabilities are referred to here as the climatological probabilities. Due to space limitations we will focus primarily on the quality of the forecast probabilities assigned to the below-normal category. These probabilities are denoted here by f. Since the probability assigned to the near-normal category always is fixed at 0.40, the range of values of f is $0 \leq f \leq 0.60$. Since the below-normal category either occurs or does not occur on each occasion (i.e., each month-location combination), the corresponding observation x equals 1 or 0.

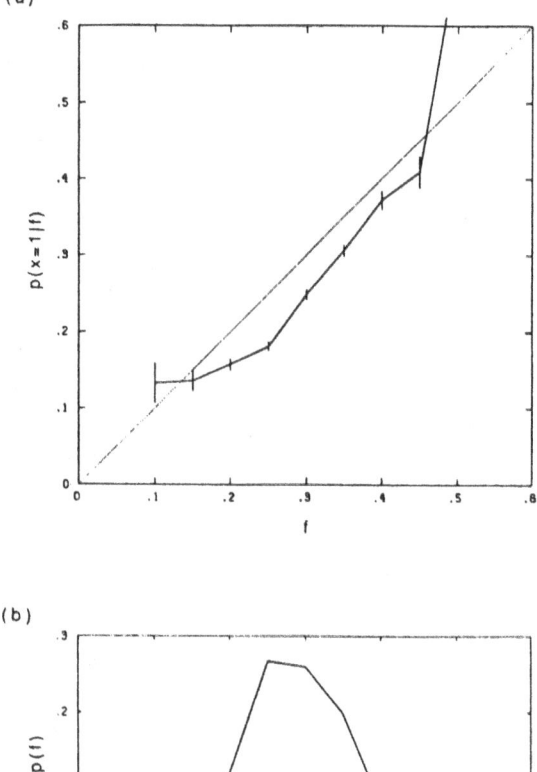

Figure 1. (a) Reliability diagram and (b) sharpness diagram, for monthly temperature forecasts.

Figures 1 and 2 contain reliability diagrams for the below-normal MT and MP forecasts. Probabilistic forecasts are completely reliable (or well-calibrated) if, for all distinct forecasts, the conditional relative frequency of occurrence of the event of interest given a particular forecast is equal to that probability [i.e., $p(x=1 \mid f) = f$ for all f]. The reliability diagram for the MT forecasts (Fig. 1a) indicates that these forecasts exhibit modest but consistent overforecasting [$p(x=1 \mid f) < f$] for $0.20 \le f \le 0.45$ and possibly some underforecasting [$p(x=1 \mid f) > f$] for $f = 0.50$. The MP forecasts (Fig. 2a) exhibit quite good reliability over a limited range of probability values.

To complete the description of forecast quality based on the calibration-refinement factorization of $p(f,x)$, sharpness (refinement) diagrams also are included in Figures 1 and 2. Probabilistic forecasts are sharp if low and high probabilities are used relatively frequently and intermediate probabilities are used relatively infrequently [i.e., if $p(f)$ possesses a u-shaped distribution]. It is evident that the MT and MP forecasts (see Figs. 1b and 2b) are not very sharp, although the former are somewhat sharper than the latter. Most forecast probabilities fall in the range between 0.20 and 0.40; that is, within 0.10 of the climatological value of 0.30.

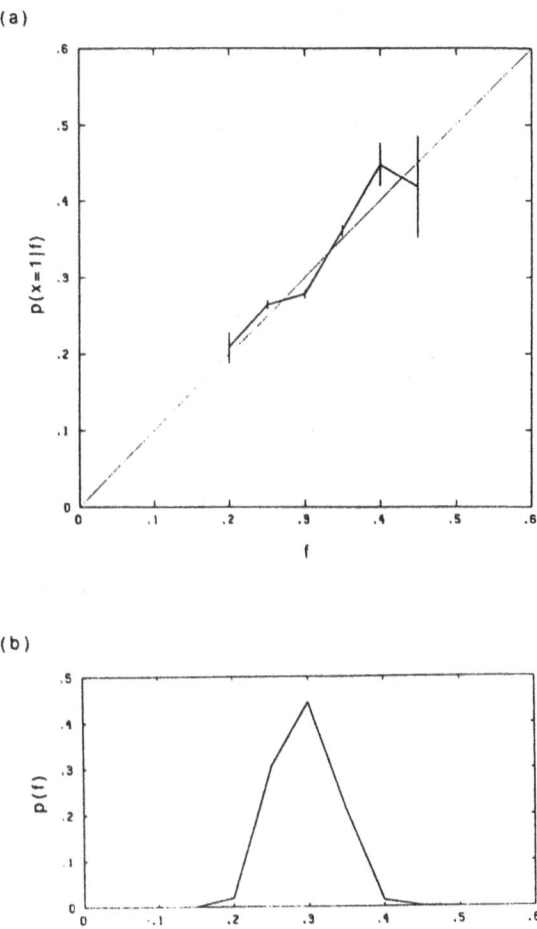

Figure 2. Same as Figure 1, except for monthly precipitation forecasts.

Discrimination (likelihood) diagrams for the below-normal MT and MP forecasts are
included in Figures 3 and 4. These diagrams depict the conditional distributions of the
forecasts given that the event in question does and does not occur; that is, $p(f|x=1)$ and
$p(f|x=0)$, respectively. Perfect discrimination would be represented by conditional distribu-
tions that do not overlap. The greater the overlap between $p(f|x=1)$ and $p(f|x=0)$, the
weaker the discrimination. Figures 3 and 4 reveal only modest discrimination in the case
of the MT forecasts and little if any discrimination in the case of the MP forecasts.
Evidently, it is difficult to achieve strong discrimination for these long lead-time forecasts.
To complete the description of forecast quality based on the likelihood-base rate
factorization of $p(f,x)$, the sample climatological probabilities (i.e., base rates), $p(x=0)$ and
$p(x=1)$, are included as inserts in the discrimination diagrams.

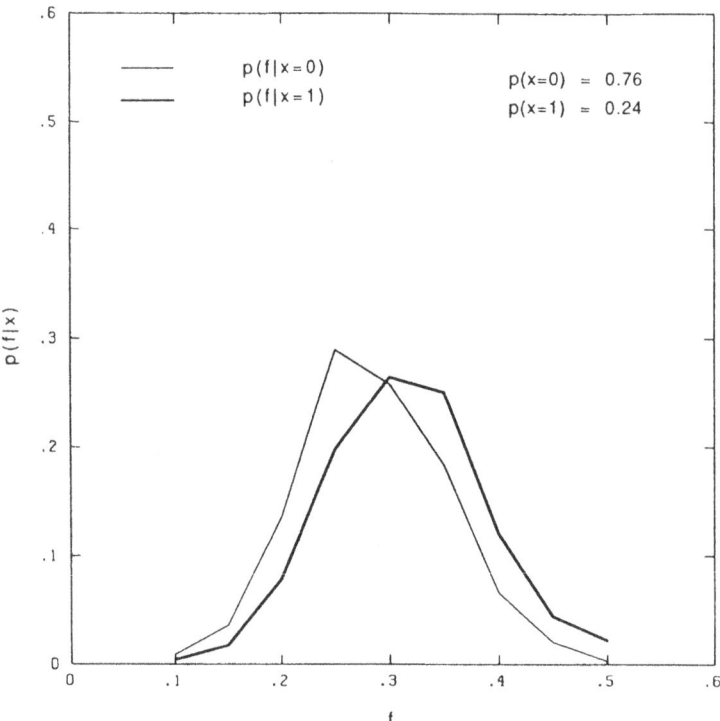

Figure 3. Discrimination diagram for monthly temperature forecasts.

Some statistics related to the joint, conditional, and marginal distributions are included in
Table 1. These statistics (Table 1a) involve only the forecast probabilities assigned to the
below-normal category. Comparison of $\bar{f}$ and $\bar{x}$ reveals that the MT forecasts exhibit some
overall bias (i.e., overforecasting), whereas the MP forecasts exhibit no systematic bias at
all. The standard deviations of both types of forecasts are considerably smaller than the
respective standard deviations of the observations, reflecting in part a basic difference
between these two quantities (i.e., the f's are probabilities whereas the x's are binary
numbers). Note also that s_f(MT) is almost twice as large as s_f(MP). The correlation
between f and x is relatively modest for both the MT and MP forecasts, with the magnitude of r_{fx}(MT) being more than twice that of r_{fx}(MP).

Conditional means of the distributions $p(f|x=1)$ and $p(f|x=0)$ also are included in Table 1a. The difference between these means is an overall one-dimensional measure of discrimination. In the case of both types of forecasts this difference is quite small, indicating relatively little discrimination (cf. Figs. 3 and 4). Since the conditional means of the distributions $p(x|f)$ represent the points that define the empirical curves in the reliability diagrams (see Figs. 1a and 2a), these means are not reproduced here.

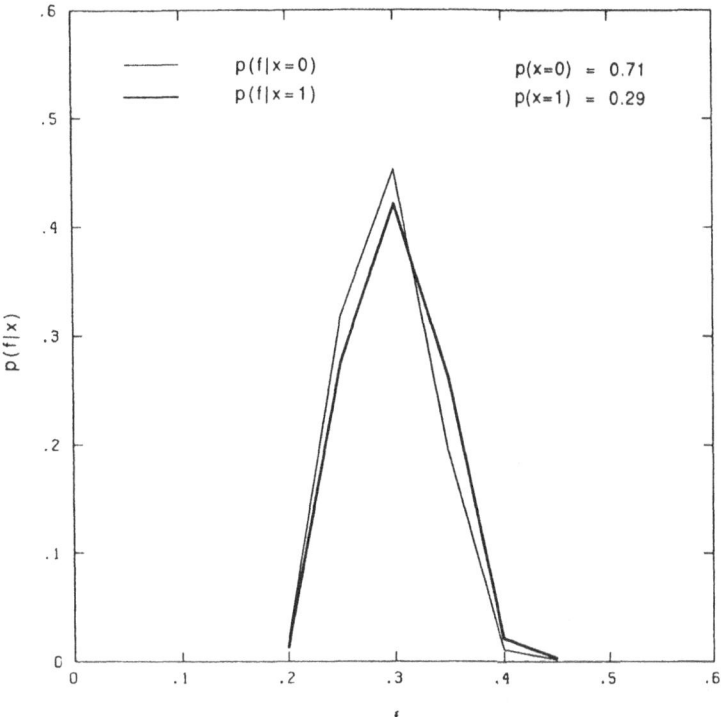

Figure 4. Same as Figure 3, except for monthly precipitation forecasts.

Table 1b contains overall measures of accuracy and skill (these measures involve the forecasts and observations for all three categories). The measure of accuracy is the ranked probability score (RPS) (Epstein 1969; Murphy 1971), a mean-square-error measure defined as the average squared difference between the forecasts and observations after they have been translated into empirical cumulative distribution functions. The skill score (SS) represents the fractional improvement in the RPS of the forecasts over the RPS of constant forecasts based solely on historical climatological probabilities. The latter indicates a 4% (1%) improvement in the case of the MT (MP) forecasts.

Finally, the terms in two decompositions of the RPS (see Murphy and Winkler 1987) are presented in Table 1c. The first decomposition (RPS = s_x^2 + REL - RES) is related to the calibration-refinement factorization of $p(f,x)$. This decomposition indicates that the failure of the MT and MP forecasts to achieve perfect reliability (see Figs. 1a and 2a) contributes very little to the magnitude of this error measure (i.e., REL is small). On

Table 1. Statistics of basic distributions, performance measures, and decompositions of RPS for long-range monthly temperature (MT) and monthly precipitation (MP) forecasts. See section 2b for additional details.

(a) Statistics of joint, conditional, and marginal distributions

Type of forecast	$\bar{f}$	$\bar{x}$	s_f	s_x	r_{fx}	$\overline{f}x = 1$	$\overline{f}x = 0$	n
MT	0.29	0.24	0.07	0.43	0.19	0.31	0.28	22374
MP	0.29	0.29	0.04	0.46	0.09	0.30	0.29	22262

(b) Performance measures

Type of forecast	RPS	SS
MT	0.42	0.04
MP	0.43	0.01

(c) Decompositions of RPS

Type of forecast	RPS	=	s_x^2 (s_f^2)	+	REL(DIS1)	–	RES(DIS2)
MT	0.42		0.42 (0.01)		0.01 (0.41)		0.02 (0.00)
MP	0.43		0.43 (0.00)		0.00 (0.43)		0.00 (0.00

the other hand, both types of forecasts exhibit very little resolution (i.e., RES is small). (For perfectly reliable forecasts, resolution is equivalent to sharpness.) Moreover, the fact that RES is small and only slightly larger than REL indicates that forecast skill, although positive, is low.

The second decomposition (RPS = s_f^2 + DIS1 - DIS2) is related to the likelihood-base rate factorization of $p(f,x)$. This decomposition indicates that the magnitude of the overall error measure is due to the fact that the conditional mean forecasts (i.e., $\bar{f}|x=1$ and $\bar{f}|x=0$) differ very little from each other and from the unconditional mean forecast (i.e., $\bar{f}$). As a result, these conditional means are quite distant from the observations $x = 1$ and $x = 0$ (i.e., DIS2 is small and DIS1 is large; cf. Figs. 3 and 4). The variability of the forecasts (i.e., s_f^2) contributes very little to the magnitude of the RPS.

Space limitations have precluded a more detailed examination and interpretation of the quality of these long-range weather forecasts. For an in-depth diagnostic verification of these forecasts, see Murphy and Huang (1993). Recent applications of these distributions-oriented methods to short-range weather forecasts include Murphy et al. (1989) and Murphy and Winkler (1992).

3. Forecast Value

3.1 Approach and Methodology

Forecasts possess no intrinsic value. Instead, they acquire value through their use by individuals whose decisions are influenced by the information contained in the forecasts. Here we adopt a decision-analytic approach to the problem of assessing the value of forecasts (Raiffa 1968; Bunn 1984; Clemen 1991). In this approach the basic determinants of forecast value are: (1) the courses of action available to the decision maker; (2) the payoff structure associated with the decision-making problem; (3) the quality of the information on which decisions are based in the absence of the forecasts; and (4) the quality of the forecasts themselves (Hilton 1981). Changes in any of these determinants (e.g., the addition or deletion of an action, the reevaluation of a payoff) can lead to changes in forecast value. This prescriptive approach to decision-making problems assumes that decision makers behave in a coherent manner and choose the actions that maximize their expected payoffs (expected payoffs are the probability-weighted averages of the payoffs associated with the outcomes, where the relevant probabilities are derived from the information on which the decisions are based).

In the context of this paper, it is important to note that forecast value depends on both the quality of the forecasts and the quality of the information on which decisions are based in the absence of the forecasts. In particular, if the quality of the forecasts is such that the decision maker makes the same decisions with and without the forecasts, then the forecasts are of no value. To simplify this discussion we will assume that payoffs expressed in monetary terms reflect the true worth of the outcomes to the decision maker. Under this assumption (of a linear utility function), the value of the forecasts is simply the difference between the expressions for expected payoff when the individual's decisions are made with and without the forecasts. The expression for expected payoff with the forecasts involves both the conditional distributions of the observations given the forecasts, $p(x|f)$, and the marginal distribution of the forecasts, $p(f)$. Thus, forecast value generally depends on forecast quality in its full dimensionality.

3.2 Applications

As an example of an application of the decision-analytic approach to forecast-value assessment, we present some results of the so-called fallowing-planting problem (Brown et al. 1986; Katz et al. 1987). This study involved farmers in the drier areas of the northern Great Plains region of the U.S. who must decide each year whether or not to plant a spring wheat crop. These farmers routinely grow a crop every other year, thereby allowing the land to lie fallow in alternate years. The primary reason for this practice is to ensure the availability of sufficient soil moisture at planting time the following year to grow an economically viable crop. However, information in the form of a forecast of growing-season precipitation amount might influence the farmer's decision to let the land lie fallow, especially on those occasions on which the forecast indicated a relatively high probability of above-normal precipitation.

A dynamic decision-analytic model of the fallowing-planting problem was formulated. In this model the farmer was assumed to make the decision to plant or fallow each year for a period of 50 years (the lifetime of the farmer) and to act in such a way as to maximize total expected economic return over this period. The model is dynamic in the sense that next year's soil moisture at planting time is assumed to depend on this year's soil moisture at planting time, the farmer's decision as to whether or not to plant a crop this year, and the precipitation amount in this year's growing season. Soil moisture at the time a crop is planted is the so-called state variable in this dynamic decision-making model (precipitation amount augments soil moisture). The expected yield when a crop is planted is based on a linear regression model in which the independent variables are soil moisture and growing-season precipitation amount. These yields were translated into economic returns to the farmer using 1983 estimates of crop prices, and a discount factor of 0.90 was applied to all future returns (equivalent to an interest rate of about 11%). The method of stochastic-dynamic programming was used to determine the farmer's optimal strategies and to obtain forecast-value estimates.

The forecasts of interest here are growing-season forecasts of precipitation amount. Forecasts of three types were considered: (1) climatological forecasts; (2) imperfect forecasts; and (3) perfect forecasts. Climatological forecasts are forecasts based solely on historical climatological probabilities and represent the zero point on the scale of forecast quality. In the absence of (imperfect) forecasts, the farmers are assumed to base their decisions on climatological forecasts. (In the dynamic decision-analytic model, decisions based on climatological forecasts lead to fallowing and planting in alternate years; see Katz et al. 1987.) Perfect forecasts, although unattainable, provide useful upper bounds on the quality and value of all forecasts. The imperfect forecasts considered here are defined in terms of a simple model of the performance of the seasonal forecasts of precipitation amount produced by the U.S. National Weather Service (these forecasts are similar in format to the monthly forecasts described in section 2b). Specifically, these forecasts are assumed to be completely reliable and to characterize the current level of forecasting performance. Given these assumptions, the variance of the forecasts represents a reasonable one-dimensional measure of forecast quality (a larger variance indicates higher quality).

The decision-analytic model (including the model of growing-season precipitation amount forecasts) was applied to representative farmers in Havre, Montana, and Williston, North Dakota. These two locations were considered to capture some of the natural variability in climatological precipitation amounts over the region. Both soil moisture and climatological precipitation amount are greater at Williston than at Havre. The overall results indicate that the value of imperfect forecasts of current quality is $4 per acre at Havre, whereas these forecasts are of no value at Williston. Perfect forecasts, on the other hand, would be worth $79 per acre at Havre and $47 per acre at Williston. Thus, the value of current forecasts at Havre is about 5% of the value of perfect forecasts.

These results, as well as results related to the value of hypothetical improvements in current forecast quality, are depicted in relative terms in Figure 5. This figure also reveals some general features of the relationships between forecast quality and forecast value (see also section 4). Specifically, quality/value relationships are inherently nonlinear and frequently possess quality thresholds, below which forecasts are of no value. In the fallowing-planting context, considerable improvement in forecast quality would be required before seasonal forecasts are of positive value at Williston. On the other hand, substantial increases in forecast value could be realized by relatively modest improvements in forecast quality at Havre. The "kink" in the quality/value curves corresponding to pseudoperfect forecasts (i.e., forecasts in which a probability of 0.60 is assigned to either the below-normal or above-normal precipitation amount category and the respective category subsequently occurs) may be an artifact of the definition of these forecasts and/or the way in which the forecasts were improved.

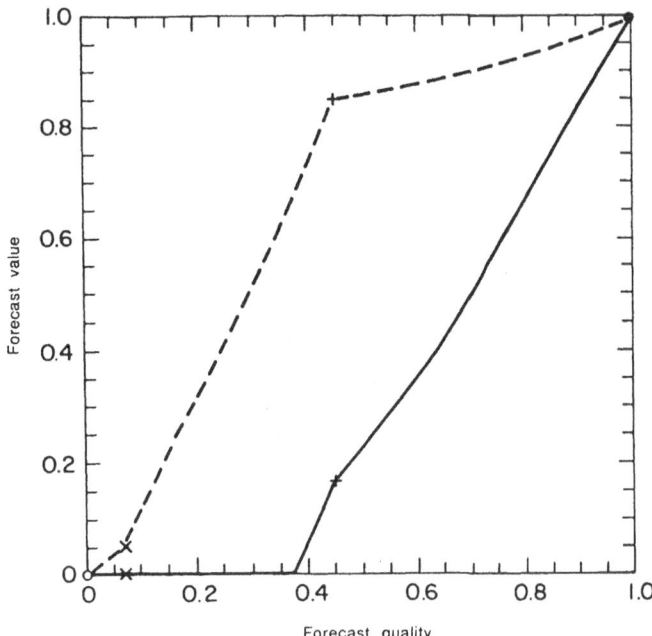

Figure 5. Relationship between forecast quality and forecast value, measured relative to the quality and value of perfect forecasts, in the fallowing-planting problem. Dashed (solid) curve represents Havre (Williston). Point (1,1) denotes perfect forecasts, point (0,0) denotes climatological forecasts, crosses (×) denote forecasts of current quality (in 1983), and pluses (+) denote pseudoperfect forecasts. See section 3b for additional details.

In addition to the fallowing-planting study, several other decision-analytic studies of the value of weather forecasts have been conducted in recent years. These studies include analyses of both prototypical (i.e., idealized) decision-making problems (e.g., Epstein and Murphy 1988; Katz 1993; Katz and Murphy 1990; Murphy 1985; Murphy et al. 1985; Wilks 1991; Winkler et al. 1983) and real-world decision-making problems (e.g., Katz et al. 1982; Mjelde et al. 1988; Sonka et al. 1987; Wilks and Murphy 1986; Wilks et al. 1993). For a more complete list of such studies, see Ehrendorfer and Murphy (1992a).

4. Quality/Value Relationships

4.1 Approach and Methods

This discussion of relationships between forecast quality and forecast value focuses on the following question: What conditions must exist between the joint distributions of forecasting systems F and G, $p(f,x)$ and $p(g,x)$, to ensure that F's forecasts can be judged unambiguously to be better in all respects than G's forecasts (or vice versa)? These conditions are embodied in the sufficiency relation, originally developed by Blackwell

(1953) to compare statistical experiments, further refined by Marschak (1971) in the context of information systems, and introduced into the forecasting literature by DeGroot and Fienberg (1982). According to this relation, F's forecasts are sufficient for G's forecasts when G's likelihoods, $p(g \mid x)$, can be obtained from F's likelihoods, $p(f \mid x)$, by a stochastic transformation. When such a stochastic transformation exists, it consists of a set of conditional probabilities defined over all possible combinations of F's and G's forecasts, and these conditional probabilities are used as weights to transform F's likelihoods into G's likelihoods (e.g., see Ehrendorfer and Murphy 1992b). The existence of such a stochastic transformation possesses two important consequences: (1) F's forecasts are of higher quality than G's forecasts in all relevant aspects and (2) F's forecasts are of greater value than G's forecasts to all decision makers regardless of their payoff structures. These powerful consequences make the sufficiency relation an attractive framework within which to perform comparative evaluation studies and to investigate quality/value relationships. However, the stringent conditions imposed by the sufficiency relation raise questions about its applicability in real-world situations.

4.2 Applications

Relationships between forecast quality and forecast value, in the context of comparative evaluation, are illustrated here by applying the sufficiency relation to prototypical forecasting systems that produce forecasts similar in format to the forecasts considered in sections 2b and 3b. Each forecast specifies the probabilities of three events, which represent below-normal, near-normal, and above-normal weather conditions, respectively. Moreover, the forecasting systems are constrained to use only three distinct forecasts: (1) a climatological forecast, with probabilities equal to the historical climatological probabilities (namely, 0.30, 0.40, and 0.30); (2) a forecast with a higher (lower) probability of below-normal (above-normal) conditions; and (3) a forecast with lower (higher) probability of below-normal (above-normal) conditions. In the cases of these latter two non-climatological forecasts, the probability assigned to near-normal conditions remains fixed at its climatological value (0.40). The quality of the forecasts produced by these prototypical forecasting systems can be described completely by two parameters, denoted here by δ and π. The parameter $\delta(-0.30 \le \delta \le 0.30)$ represents the magnitude of the deviation of the non-climatological forecast from the climatological forecast, whereas the parameter $\pi(0 \le \pi \le 0.50)$ specifies the relative frequency with which each non-climatological forecast is used (for a more detailed description of these forecasting systems, see Ehrendorfer and Murphy 1992b). The implications of the sufficiency relation for the comparative evaluation of these prototypical forecasting systems can be investigated within the framework of a sufficiency diagram. For these forecasting systems, this diagram is two-dimensional with coordinates δ and π. An example of such a sufficiency diagram is presented in Figure 6; in this case, the reference system F is defined by $\delta = -0.10$ and $\pi = 0.15$. Given F, the two-dimensional sufficiency diagram is divided into three regions: (1) a region S containing the systems G for which F is sufficient (denoted by diamonds); (2) regions S' containing the systems G that are sufficient for F (denoted by crosses); and (3) regions I containing the systems G that are insufficient for F (blank). Systems G represented by regions I are neither sufficient for F, nor is F sufficient for these systems.

Inspection of this diagram reveals that systems G that use more extreme non-climatological forecasts more frequently than the reference system F are sufficient for F. However, it also can be seen that systems G that use the non-climatological forecasts less frequently than the reference system F can still be sufficient for F, provided that the non-climatological forecast used by G is extreme enough. On the other hand, a system G that makes more frequent use of a less extreme non-climatological forecast can never be sufficient for F. Since sufficiency is determined by the values of the parameters that define the respective forecasting systems, it provides a coherent approach to the problem of comparative evaluation. Unfortunately, as this example illustrates, the stringent conditions imposed by the sufficiency relation may lead to situations in which the forecasting systems of interest are not comparable, in the sense that they are insufficient for each other. To investigate other features of quality/value relationships in this context, we introduce specific measures of both forecast quality and forecast value. As a one-dimensional measure of quality (in particular that aspect of quality called accuracy), we use the expected ranked probability score (ERPS) (see Ehrendorfer and Murphy 1992b).

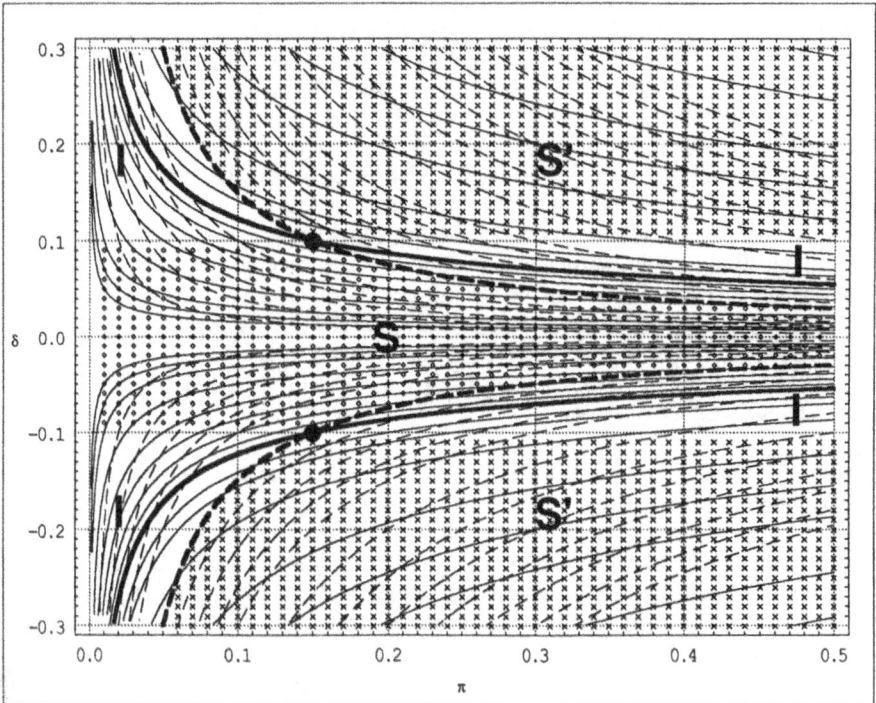

Figure 6. Example of a sufficiency diagram. The large dot represents the reference system $F(\delta = -0.10$ and $\pi = 0.15)$. The regions denoted by S (diamonds), S' (crosses), and I (blank) represent those alternative systems G for which F is sufficient, which are sufficient for F, and which are insufficient for F, respectively. Broken (solid) lines represent VF (ERPS) - contoured at unequal spacing - with decreasing (increasing) numerical values from the lower-right and upper-right corners towards the middle of the diagram. See section 4b for additional details.

To measure the value of these forecasts, it is necessary to consider a specific decision-making problem. Here we consider an extension of the so-called cost-loss ratio problem (see Murphy 1985), in which a decision maker must choose among three actions involving different levels of protection against adverse weather conditions that can occur on three different levels of severity. The payoff structure for this simple problem is described by a single parameter, the cost-loss ratio. The expected value of the forecasts (VF) is calculated as the difference between the expected expense when decisions are based on climatological information [i.e., on $p(x)$] and the expected expense when decisions are based on the forecasts [i.e., on $p(x|f)$ and $p(f)$] (see section 3). Numerical values of ERPS and VF are displayed in Figure 6 in the form of isopleths, where the solid lines are ERPS-contours and the' broken lines are VF-contours (the value of the cost-loss ratio was taken to be 0.3). Examination of these isopleths leads to two important conclusions. First, a one-dimensional measure of quality such as the ERPS is inadequate in this situation because it cannot discriminate between sufficiency and insufficiency. Specifically, isopleths of ERPS pass through regions in which F is sufficient for G (or G is sufficient for F) and regions in which F and G are insufficient for each other. Second, forecast quality, as measured by a one-dimensional measure of accuracy, and forecast value do not possess a one-to-one relationship. A given isopleth of ERPS intersects more than one isopleth of VF (and vice versa), which implies that an increase in forecast accuracy (as measured by ERPS) can be associated with a decrease in forecast value. The existence of such accuracy/value reversals raises serious questions about the use of one-dimensional measures of aspects of forecast quality as surrogates for measures of forecast value. To establish whether system F is better in all respects than system G it is necessary to consider forecast quality in its full dimensionality, as embodied in the sufficiency relation or some other equivalent representation of the relationships between the respective forecasts and observations.

Other applications of the sufficiency relation as a means of investigating (inter alia) various aspects of the relationship between forecast quality and forecast value include Ehrendorfer and Murphy (1988), Krzysztofowicz (1992), Krzysztofowicz and Long (1991a, 1991b), and Murphy and Ye (1990). Many of the forecast-value studies referenced at the end of section 3b also contain results and discussions related to quality/value relationships for weather forecasts.

5. Conclusion

This paper has addressed the problem of forecast evaluation, which includes the assessment of both forecast quality (forecast verification) and forecast value. A general frame work for forecast verification based on the joint distribution of forecasts and observations - and on conditional and marginal distributions associated with factorizations of this joint distribution - has been described. This framework is consistent with the multidimensional nature of verification problems and the multifaceted nature of forecast quality. From the perspective of this framework, forecast verification is seen as a means of obtaining a coherent and complete assessment of forecast quality, identifying basic strengths and weaknesses in forecasting performance, and guiding efforts to improve the systems and models (numerical, statistical, and conceptual) that are used to produce the forecasts. Some results of an application of verification methods consistent with this framework to long-range weather forecasts were presented to illustrate this distributions-oriented approach to verification problems. Since most forecasts in other fields (e.g., business, economics) possess formats similar to those of weather forecasts, these distributions-oriented methods also could be used to verify such forecasts.

A decision-analytic approach to forecast-value assessment was outlined. An application of this approach to a problem involving fallowing-planting decisions and long-range weather forecasts was described and forecast-value estimates were reported. Since these estimates are derived from an approach based on an optimization process (i.e., the decision maker is assumed to choose courses of action that maximize expected payoffs or minimize expected expenses), it follows that the relationship between forecast quality and forecast value is nonlinear. Moreover, this and other decision-analytic case studies have demonstrated that many forecast-sensitive decision-making problems involve quality thresholds, below which the forecasts are of no value.

Relationships between forecast quality and forecast value in the context of comparative evaluation were described with the aid of the sufficiency relation. This relation, which defines a stochastic transformation between the likelihoods associated with the two forecasting systems of interest [i.e., between $p(f \mid x)$ and $p(g \mid x)$; see section 4a], embodies the conditions under which one system can be judged to be better in all respects - that is, in terms of both quality and value - than the other system. In an application involving prototypical long-range weather forecasting systems, it was shown that the stringent conditions imposed by the sufficiency relation imply that it is not always possible to show that one system is sufficient for another system. Moreover, this application was used to demonstrate the possibility of so-called quality/value reversals (i.e., decreases in value associated with increases in an aspect of quality), which can occur when the multifaceted nature of forecast quality is not respected.

Although the evaluation methods described in this paper appear to provide a reasonable framework for assessing both forecast quality and forecast value, as well as for investigating quality/value relationships, a variety of methodological issues and practical problems exist that warrant further attention. With regard to forecast quality (forecast verification), for example, it is not entirely clear what constitutes an adequate or complete assessment of forecast quality. In the context of absolute verification, is it always necessary to examine the conditional and marginal distributions associated with *both* factorizations of the joint distribution of forecasts and observations (in the sense that the respective factorizations contain different information)? This question assumes even greater significance in the context of comparative verification, in view of the substantially greater complexity and dimensionality of these problems. In this regard, it would be useful to explore ways of reducing the dimensionality of verification problems. One possibility would be to fit statistical models to the joint distribution of forecasts and observations, or to the conditional and marginal distributions associated with factorizations of this joint distribution. Given that acceptable fits can be obtained, forecast verification could be based on the parameters of the model(s), thereby reducing dimensionality. Use of statistical models (instead of empirical relative frequencies) in the verification process also should reduce the impact of sampling variability on the results of such assessments.

With regard to forecast value and quality/value relationships, it is clear that additional case studies of forecastsensitive decision-making problems are needed. In addition to yielding forecast-value estimates, such studies would provide further insight into the general and specific characteristics of quality/value relationships as well as potentially useful information regarding the structure of decision makers' payoff functions. Even partial knowledge of the structure and behavior of payoff functions in particular decision-making situations could be used as a basis for developing "tailored" versions of the sufficiency relation (the basic version of the sufficiency relation assumes that nothing is known about the decision makers' payoff functions). Tailored versions of the sufficiency relation generally would impose less stringent conditions on the joint distributions associated with the respective forecasting systems, thereby increasing the likelihood that one system could be judged unambiguously superior (or inferior) to another system in the situations of concern. In view of the strong conclusions that can be drawn when the conditions for sufficiency are met, efforts to enhance the practical applicability of this relation appear to be especially worthwhile.

From a practical point of view, the methods commonly used to evaluate forecasts need to be improved (they are incomplete and potentially misleading; see section 1). In the case of forecast quality, for example, the distributions-oriented approach described here provides the basis for a more coherent and insightful body of verification methods than is generally used in practice. Moreover, in view of recent technological developments, organizations involved in routine forecasting operations should develop :on-line" forecast verification systems that can provide a wide variety of outputs in real time (to satisfy the needs of managers, modelers, and forecasters). Such systems, if properly designed and implemented, would be a valuable source of information for those individuals who are responsible for formulating forecasts as well as for those individuals who are concerned primarily with improving forecasting methods and models.

Finally, the discussion of forecast verification in this paper has focused on the use of the verification process as a means of obtaining insight into the basic strengths and weaknesses in forecasting performance. Clearly, this process is an essential part of any effort to develop and implement state-of-the-art forecasting systems and necessarily plays a vital role in efforts to improve such systems. Notwithstanding the importance of these particular uses of the verification process and its output, it should be noted here that this process also can produce information related to the predictability of the events of concern and the uncertainty inherent in forecasts of these events. Statistical studies of the deterioration of forecast quality (or aspects of quality) as lead time increases can provide empirical estimates of predictability. Moreover, studies of the joint distribution of forecasts and observations, and the associated conditional and marginal distributions, can produce quantitative information regarding the overall uncertainty inherent in the forecasts. In the future it may prove useful to compare empirical estimates of predictability and uncertainty derived from the verification process with theoretical estimates of these quantities obtained from modeling studies.

Acknowledgment

This work was supported in part by the National Science Foundation under Grant SES-9106440.

References

Blackwell, D., (1953), 'Equivalent comparisons of experiments'. *Annals of Mathematical Statistics*, **24**, 265-272.

Brown, B.G., R.W. Katz, and A.H. Murphy, (1986), 'On the economic value of seasonal-precipitation forecasts: the fallowing-planting problem'. *Bulletin of the American Meteorological Society*, **67**, 833-841.

Bunn, D., (1984), *Applied Decision Analysis*. New York, McGraw-Hill, 251 pp.

Clemen, R.T., (1991), *Making Hard Decisions: An Introduction to Decision Analysis*. 4 Boston, PWS-Kent, 557 pp.

DeGroot, M.H., and S.E. Fienberg, (1982), 'Assessing probability assessors: calibration and refinement'. *Statistical Decision Theory and Related Topics III*, **1** (S.S. Gupta and J.O. Burger, Editors). New York, Academic Press, 291-314.

Ehrendorfer, M., and A.H. Murphy, (1988), 'Comparative evaluation of weather forecasting systems: sufficiency, quality, and accuracy'. *Monthly Weather Review*, **116**, 1757-1770.

Ehrendorfer, M., and A.H. Murphy, (1992a), 'On the relationship between the quality and value of weather and climate forecasting systems'. *Idojaras*, **96**, 187-206.

Ehrendorfer, M., and A.H. Murphy, (1992b), 'Evaluation of prototypical climate forecasts: the sufficiency relation'. *Journal of Climate*, **5**, 876-887.

Epstein, E.S., (1969), 'A scoring system for probabilities of ranked categories'. *Journal of Applied Meteorology*, **8**, 985-987.

Epstein, E.S., and A.H. Murphy, (1988), 'Use and value of multiple-period forecasts in a dynamic model of the cost-loss ratio situation'. *Monthly Weather Review* **116**, 746-761.

Hilton, R.W., (1981), 'The determinants of information value: synthesizing some general results'. *Management Science*, **27**, 57-64.

Katz, R.W., (1993), 'Dynamic cost-loss ratio decision-making model with an autocorrelated climate variable'. *Journal of Climate*, **6**, 151-160.

Katz, R.W., B.G. Brown, and A.H. Murphy, (1987), 'Decision-analytic assessment of the economic value of weather forecasts: the fallowing/planting problem'. *Journal of Forecasting*, **6**, 77-89.

Katz, R.W., and A.H. Murphy, (1990), 'Quality/value relationships for imperfect weath\-er forecasts in a prototype multistage decision-making model'. *Journal of Forecasting*, **9**, 75-86.

Katz, R.W., A.H. Murphy, and R.L. Winkler, (1982), 'Assessing the value of frost forecasts to orchardists: a dynamic decision-making approach'. *Journal of Applied Meteorology*, **21**, 518-531.

Krzysztofowicz, R., (1992), 'Bayesian correlation score: a utilitarian measure of forecast skill'. *Monthly Weather Review*, **120**, 208-219.

Krzysztofowicz, R., and D. Long, (1991a), 'Forecast sufficiency characteristic: construction and application'. *International Journal of Forecasting*, **7**, 39-45.

Krzysztofowicz, R., and D. Long, (1991b), 'Beta likelihood models of probabilistic forecasts'. *International Journal of Forecasting*, **7**, 47-55.

Marschak, J., (1971), 'Economics of information systems'. *Journal of the American Statistical Association*, **66**, 191-219.

Mjelde, J.W., S.T. Sonka, B.L. Dixon, and P.J. Lamb, (1988), 'Valuing forecast characteristics in a dynamic production system'. *American Journal of Agricultural Economics*, **70**, 674-684.

Murphy, A.H., (1971), 'A note on the ranked probability score'. *Journal of Applied Meteorology*, **10**, 155-156.

Murphy, A.H., (1985), 'Decision making and the value of forecasts in a generalized model of the cost-loss ratio situation'. *Monthly Weather Review*, **113**, 362-369.

Murphy, A.H., (1991), 'Forecast verification: its complexity and dimensionality'. *Monthly Weather Review*, **119**, 1590-1601.

Murphy, A.H., (1993), 'What is a good forecast? An essay on the nature of goodness in weather forecasting'. *Weather and Forecasting*, **8**, 281-293.

Murphy, A.H., B.G. Brown, and Y.-S. Chen, (1989), 'Diagnostic verification of temperature forecasts'. *Weather and Forecasting*, **4**, 485-501.

Murphy, A.H., and J. Huang, (1993), 'Diagnostic verification of the Climate Analysis Center's probabilistic monthly and seasonal forecasts'. In preparation.

Murphy, A.H., R.W. Katz, R.L. Winkler, and W.-R. Hsu, (1985), 'Repetitive decision-making and the value of forecasts in the cost-loss ratio situation: a dynamic model'. *Monthly Weather Review*, **113**, 801-813.

Murphy, A.H., and R.L. Winkler, (1987), 'A general framework for forecast verification'. *Monthly Weather Review*, **115**, 1330-1338.

Murphy, A.H., and R.L. Winkler, (1992), 'Diagnostic verification of probability forecasts'. *International Journal of Forecasting*, **7**, 435-455.

Murphy, A.H., and Q. Ye, (1990), 'Comparison of objective and subjective precipitation probability forecasts: the sufficiency relation'. *Monthly Weather Review*, **118**, 1783-1792.

Raiffa, H., (1968), *Decision Analysis*. Reading, MA, Addison Wesley, 309 pp.

Sonka, S.T., J.W. Mjelde, P.J. Lamb, S.E. Hollinger, and B.L. Dixon, (1987), 'Valuing climate forecast information'. *Journal of Climate and Applied Meteorology*, **26**, 1080-1091.

Wilks, D.S., (1991), 'Representing serial correlation of meteorological events and forecasts in dynamic decision-analytic models'. *Monthly Weather Review*, **119**, 1640-1662.

Wilks, D.S., and A.H. Murphy, (1986), 'A decision-analytic study of the joint value of seasonal precipitation and temperature forecasts in a choice-of crop problem'. *Atmosphere-Ocean*, **24**, 353-368.

Wilks, D.S., R.E. Pitt, and G.W. Fick, (1993), 'Modeling optimal alfalfa harvest scheduling using short-range weather forecasts'. *Agricultural Systems*, **42**, 277-305.

Winkler, R.L., A.H. Murphy, and R.W. Katz, (1983), 'The value of climate information: a decision-analytic approach'. *Journal of Climatology*, **3**, 187-197.

THE LIOUVILLE EQUATION
AND PREDICTION OF FORECAST SKILL

MARTIN EHRENDORFER

National Center for Atmospheric Research
Boulder, Colorado 80307, USA
e-mail: ehren@ncar.ucar.edu

On leave from: Institute for Meteorology and Geophysics
University of Vienna, A-1190 Vienna, Austria

Abstract

The Liouville equation represents the consistent and comprehensive framework for dealing with uncertainty arising in meteorological forecasts due to uncertainty in the initial condition. This equation expresses the conservation of the phase space integral of the number density of realizations of a dynamical system originating at the same time instant from different initial conditions, in a way completely analogous to the continuity equation for mass in fluid mechanics. Its solution describes the temporal evolution of the probability density function of the state vector of a given dynamical model.

The main purposes of this paper are (i) to review the basic form of the Liouville equation, (ii) to present the explicit general solution of the Liouville equation for a large class of dynamical systems in analytical terms, and (iii) to investigate the potential usefulness of the Liouville equation in the context of the prediction of forecast skill.

As an illustration, the general analytical solution of the Liouville equation is used to obtain the solution of the Liouville equation relevant for a low-dimensional chaotic dynamical system. The information contained in this solution is compared with results obtained by application of the method of ensemble forecasting. It is found that a large number of ensemble integrations is required in order to obtain estimates of statistics, such as means, variances, and covariances, with the accuracy that is obtained by integration of the solution of the LE over phase space, even in the low-dimensional situation considered.

The paper is concluded with a discussion of the fundamental role of the Liouville equation in dealing with initial state uncertainties in dynamical models, and of the problems that arise in this context. Even though some of these problems may be difficult to deal with in situations more realistic than considered here, the argument is made that the Liouville equation must be considered as an extremely valuable and useful guideline in the process of studying, developing and refining methods for the prediction of forecast skill.

1. Introduction

The widely recognized uncertainty inherent in meteorological forecasts has led to various approaches to develop methods suitable to adequately describe and quantify this uncertainty. Such quantification should ideally be expressed in terms of the precise and unambiguous language of probability (see, Lindley 1987).

The natural way to account for the uncertainty and the probabilistic nature of forecasts made by numerical meteorological models that are based on a set of physical conservation laws is to consider the probability density function (pdf) of the model state vector in phase space. A number of sophisticated procedures have been   developed and used both experimentally and semi-operationally to assess certain characteristics of this pdf; among these methods are *stochastic-dynamic prediction* (the integration of the model equations supplemented by equations describing higher moments of the pdf; Epstein 1969), *ensemble forecasting* (Monte Carlo forecasting; Leith 1974, Mureau *et al.* 1993), *lagged average forecasting* (Hoffman and Kalnay 1983), and *statistical schemes* (Molteni and Palmer 1991).

The Liouville equation (LE) represents the fundamental basis of these approaches in the sense that it describes explicitly the dynamical evolution of the full pdf of the model state vector given a particular meteorological model. The potential importance of the LE (and its associated Fokker-Planck equation), as well as methods suitable to solve these equations analytically in general contexts have been discussed by Thompson (1983, 1985). By its nature, the LE expresses a continuity requirement for a large number of realizations in phase space. Thus, its solution contains all the relevant information of the time development of the pdf of the state vector. The LE is a quasi-linear, inhomogeneous, first-order partial differential equation with a single dependent variable (namely, the density, or pdf) and a (possibly) large number of independent variables (namely, time and the coordinates of phase space).

The purposes of this paper are to discuss several issues related to the LE and to indicate its potential importance in the context of forecasting forecast skill. The LE is introduced and discussed in the context of Liouville's theorem in section 2. The general solution of the LE, valid for a large class of dynamical systems, is presented in explicit form in section 3. The structure of the solution of the LE is investigated for a low-dimensional chaotic system in section 4. The relevance and information content of the solution of the LE are contrasted with results obtained through the above-mentioned ensemble forecasting technique. The paper is concluded with a discussion of the results and a critical assessment of the practical utility of the LE in the context of the prediction of forecast skill.

2. The Liouville equation: historical and theoretical background

In January 1838, Joseph Liouville, who lived from 1809 to 1882, published a note on the time-dependence of the Jacobian of the "transformation" exerted by the solution of an ordinary differential equation on its initial condition (see, Lützen 1990). His result, known today as the material derivative of the Jacobian, may be viewed as the common basis for the Liouville equation and for the famous Liouville theorem in classical mechanics on the conservation of phase space volume.

The relationship of both results and their common origin, namely Liouville's original contribution, are most easily demonstrated by considering the transport theorem (2.1), describing the time change of a volume-integrated generic quantity φ within a material region $R(t)$:

$$\frac{dI(t)}{dt} \equiv \frac{d}{dt}\int_{R(t)} \varphi(X,t)dX = \int_{R(0)} \frac{d}{dt}|_{\Xi}(\varphi(X(\Xi,t),t)J(\Xi,t))\, d\Xi =$$

$$= \int_{R(0)}\left(J(\Xi,t)\frac{d\varphi(X(\Xi,t),t)}{dt}|_{\Xi} + \varphi(X(\Xi,t),t)(\sum_{i=1}^{N}\frac{\partial \dot X_i}{\partial X_i}|_{X=X(\Xi,t)})J(\Xi,t)\right) d\Xi = \tag{2.1}$$

$$= \int_{R(t)}\left(\frac{\partial\varphi(X,t)}{\partial t} + \sum_{j=1}^{N}\dot X_j\frac{\partial\varphi(X,t)}{\partial X_j} + \varphi(X,t)\sum_{i=1}^{N}\frac{\partial \dot X_i}{\partial X_i}\right) dX,$$

where Liouville's result has been used to express the time derivative of the Jacobian as the product of divergence and Jacobian in going from the first to the second line. In this context, it is assumed that X is the N-dimensional state vector of a (possibly non-autonomous) dynamical system, given by the following set of ordinary differential equations:

$$\dot X = \Phi(X,t), \tag{2.2}$$

subject to the initial condition:

$$X(t = 0) = \Xi, \tag{2.3}$$

whose solution is of the form:

$$X = X(\Xi,t). \tag{2.4}$$

It is noted that, in a mechanical context, eq. (2.2) may be interpreted as specifying the (Eulerian) velocity for a given time and location in phase space, whereas eq. (2.4) specifies the position that a given particle, which originated at position Ξ at time $t=0$, attains at time t given the velocity field (2.2). In connection with the transport theorem, it is important to note that, in order to be able to transform from the domain of integration $R(t)$ to $R(0)$, it is necessary to assume that the Jacobian of the "transformation" (2.4), given by:

$$J(\Xi,t) \equiv J(\Xi_1,...,\Xi_N,t) \equiv \frac{\partial(X_1,...,X_N)}{\partial(\Xi_1,...,\Xi_N)}, \tag{2.5}$$

is non-vanishing, which is always true for a single-valued mapping (2.4), corresponding to the assumption of *impenetrability of matter* (see, Lin and Segel 1988). A non-vanishing Jacobian also implies that relationship (2.4) may be inverted to express the initial condition Ξ in terms of the state vector X at a given time t in the form:

$$\Xi = \Xi(X,t). \tag{2.6}$$

At this point, the LE is easily obtained as a special case of the transport theorem (2.1), by requiring that the phase space integral over the number density $\rho(X,t)$ of realizations in phase space remains constant in time (set $\varphi \equiv \rho$ in (2.1)); specifically, this constant is chosen to be one, which implies that $\rho(X,t)$ is the multi-variate pdf of the model state vector X at time t. Thus, the LE assumes the following form:

$$\frac{\partial \rho(X,t)}{\partial t} + \sum_{k=1}^{N} \frac{\partial}{\partial X_k}\left[\rho(X,t)\dot{X}_k(X,t)\right] = 0. \tag{2.7}$$

It is immediately obvious that the LE (2.7) is completely analogous to the equation of mass conservation in fluid mechanics (note that the $\dot{X}_k$ are the velocity components in phase space, see above). As such, it simply expresses the fact that the local change of ρ must be balanced by the net flux of realizations across the faces of a small volume surrounding the point under consideration. It is also noted here that, since (2.7) results from a conservation requirement, any solution to (2.7) will have the property to be correctly normalized.

The central role of the LE in the context of the prediction of forecast skill results from the fact that this equation governs the time evolution of the multi-variate pdf of the model state vector X. Thus, given a dynamical model (2.2), and the solution to the LE (2.7), probabilistic statements of any kind concerning the model state can be made on the basis of knowledge of the pdf $\rho(X,t)$, through a ρ-weighted integration over (parts of) phase space. This construction of probabilistic statements includes the computation of moments (or expected values) of the pdf as a special case.

The LE (2.7) is a quasi-linear, inhomogeneous, partial differential equation (quasi-linearity implies that the coefficients multiplying the derivatives of ρ, as well as the inhomogeneous term may depend on ρ, but not on its derivatives). The single dependent variable is the pdf ρ, and the $N+1$ independent variables are the phase space coordinates X and time t. It is also clear from the form of the LE, that the form of the dynamical system considered does determine the behavior of the solution of the LE, due to the direct insertion of the model dynamics.

In concluding this section, the relationship of the LE to Liouville's theorem in classical mechanics is illustrated. This theorem states that Hamiltonian phase flow conserves volume. In order to establish this statement, the transport theorem is used with $\varphi \equiv 1$, which implies the investigation of a material volume $V(t) = I(t)$. Further, the dynamics of a Hamiltonian system are governed - as a special case of (2.2) - by:

$$\begin{pmatrix} \dot{p} \\ \dot{q} \end{pmatrix} = \begin{pmatrix} -\partial H/\partial q \\ \partial H/\partial p \end{pmatrix}, \tag{2.8}$$

with the property of vanishing divergence:

$$\sum_{i=1}^{N} \frac{\partial \dot{X}_i}{\partial X_i} \equiv \frac{\partial}{\partial p}\left(-\frac{\partial H}{\partial q}\right) + \frac{\partial}{\partial q}\left(\frac{\partial H}{\partial p}\right) \equiv 0. \tag{2.9}$$

Inserting these results into the transport theorem proves that $V(t) = const$ for Hamiltonian flow. More generally, as is also immediately clear from the transport theorem, the time change of V is given by the phase-space integrated divergence:

$$\frac{dV(t)}{dt} = \int\limits_{R(t)} \left(\sum_{i=1}^{N} \frac{\partial \dot{X}_i}{\partial X_i} \right) dX, \tag{2.10}$$

Thus, both the LE and Liouville's theorem may be traced back to a common origin, namely the transport theorem, which may be interpreted as a powerful consequence of Liouville's original result of 1838. However, the LE is derived by requiring the constancy of a certain phase space integral, whereas Liouville's theorem assures the constancy of another phase space integral, given that the special class of dynamical systems (2.8) is considered.

3. Solution of the Liouville equation

As a quasi-linear, inhomogeneous, partial differential equation, the LE (2.7), rewritten here - through use of (2.2) - in the form:

$$\frac{\partial \rho(X,t)}{\partial t} + \sum_{k=1}^{N} \Phi_k(X,t)\frac{\partial \rho(X,t)}{\partial X_k} + \rho(X,t)\sum_{k=1}^{N} \frac{\partial \Phi_k(X,t)}{\partial X_k} = 0, \tag{3.1}$$

can, in principle, be solved for an arbitrary initial condition:

$$\rho(X,t=0) = f(X) \tag{3.2}$$

by using the method of characteristics (see, Zwillinger 1989). This solution technique is based on the idea that a partial differential equation can be transformed into a set of ordinary differential equations defining the characteristics of the partial differential equation, and describing how the solution changes along the characteristics. In the case of the LE, a set of $N+2$ ordinary differential equations is obtained (one equation for the time derivative of ρ, one equation for each of the phase space coordinates, and one equation for the inhomogeneous term); in addition, the initial condition (3.2) must be expressed in parametric form. The solution to (3.1) and (3.2) is then found to be:

$$\rho(X,t) = f(\Xi)\exp\left[- \int\limits_{0}^{t}\psi(X(\Xi,t'),t')dt' \right] \tag{3.3}$$

with (see (3.1)):

$$\psi(X,t) \equiv \sum_{k=1}^{N} \frac{\partial \Phi_k(X,t)}{\partial X_k} \, .$$

This solution is to be understood in the context of the dynamical system (2.2) and (2.3) in the following way. For given arguments X and t, it is first necessary to find Ξ, as a function of the model state X at time t according to (2.6), where Ξ and X are related through the dynamical system (2.2). The function f is to be evaluated for this argument Ξ. This argument Ξ is then used to define the model state X through the solution (2.4) of the dynamical system (2.2) within the time interval $(0,t)$; this, in turn, allows for the subsequent computation of the function ψ - which is defined in (3.1) and does explicitly depend on t for non-autonomous systems - followed by the indicated integration over t'. Summarizing, the complete explicit solution to (3.1) and (3.2), given the dynamical system (2.2), is given by (3.3), where it is understood that (2.6) is used to express Ξ in terms of X and t.

The above interpretation of the result (3.3) reflects the method of characteristics for solving a partial differential equation, in the sense that the initial solution (3.2) is carried forward in time, by observing the characteristics of the LE and the change of ρ along these characteristics. Consequently, it is clear that the solution (3.3) satisfies the initial condition (3.2). In order to verify that the solution (3.3) satisfies the differential equation (3.1), it is necessary to take the derivatives of ρ with respect to its arguments while carefully observing the dependencies implicit in eq. (3.3).

It is immediately evident, that the solution described by (3.3) possesses the two essential properties that qualify ρ as a pdf, namely non-negativeness, and normalization to one. Both properties are, of course, dependent on the proper specification of the initial pdf through the function f (see (3.2)). The maintenance of the normalization property, once realized at the initial time, is most easily understood by recalling the conservation requirement that was made in the derivation of the LE (see section 2).

In order to be able to exploit the analytical solution (3.3) of the LE (3.1), given the initial condition (3.2), it is necessary to provide the solution to the dynamical system (2.2) for a given initial condition. It is well known that - under certain regularity conditions - the solution to (2.2) and (2.3) exists, and is unique, and is of the form (2.4) (see, Arnold 1992). For a set of ordinary differential equations (see (2.2)) - that can be viewed as representing a meteorological model - this solution may enter into the solution (3.3) of the LE either in analytical form, or in the form of a consistent and stable numerical solution of the governing equations. In addition, expressions for the "inverse" dynamical model must be available for the computation of the initial state from the current state vector (see eq. (2.6)). Again, due to existence and uniqueness theorems for ordinary differential equations, this inversion is theoretically available (no intersections of integral curves occur in the extended phase space; this is equivalent to the requirement of a non-vanishing Jacobian of the mapping (2.4); see eq. (2.5)). However, in the absence of an analytical solution for a given dynamical system, the numerical solution may become increasingly inaccurate as the system evolves for longer times. The limitations resulting from these considerations for the practical applicability of the solution to the LE are further discussed in section 5.

As an illustration of the foregoing developments, the solution (3.3) is considered for a first-order, non-autonomous, homogeneous, *linear* dynamical system (2.2), following the developments of Tribbia and Baumhefner (1993). In this situation, the function Φ in (2.2) takes on the special form:

$$\Phi(X,t) \equiv A(t)X, \qquad\qquad\qquad\qquad (3.4)$$

where $A(t)$ is a time-dependent matrix. This implies that, in this case, the solution (2.4) can be written in the form:

$$X(\Xi,t) = R(t)\Xi, \tag{3.5}$$

where the resolvent $R(t)$ is defined to link the initial state Ξ to the model state X at time t. Further, use of the inverse resolvent allows to specify relationship (2.6) in the form:

$$\Xi(X,t) = [R(t)]^{-1}X. \tag{3.6}$$

Finally, since in this case the function ψ is independent of its argument X, and simply the trace of the matrix $A(t)$, which is seen from:

$$\psi(X,t) \equiv \sum_{k=1}^{N} \frac{\partial \Phi_k(X,t)}{\partial X_k} = \sum_{k=1}^{N} \frac{\partial (A(t)X)_k}{\partial X_k} = \sum_{k=1}^{N} (A(t))_{k,k} \equiv \sigma(t), \tag{3.7}$$

the general solution (3.3) of the LE (3.1) relevant in this case specializes to:

$$\rho(X,t) = f([R(t)]^{-1}X)\exp[-\int_0^t \sigma(t')dt']. \tag{3.8}$$

Tribbia and Baumhefner (1993) proceed to interpret the state vector in this case as the vector of deviations from a time-dependent reference trajectory that is used to linearize the model dynamics. This, in turn, allows to investigate the form of the pdf as a function of deviatoric coordinates.

4. The Liouville equation for a low-dimensional chaotic system

The applicability of the general solution (3.3) of the LE has been demonstrated for the case of a one-dimensional, analytically solvable, dynamical system given in form of a Riccati equation by Ehrendorfer (1992). Further extensions are discussed in Ehrendorfer (1993). In this section, the structure of the solution of the LE relevant for a chaotic dynamical system is investigated. Further, the usefulness of this solution of the LE in the context of predicting forecast skill is demonstrated by a comparison of statistics of the model state vector derived from the solution of the LE and derived through ensemble integrations.

For the case of the specific dynamical model considered, the solution of the LE is obtained by specializing the general solution (3.3). The dynamical model under consideration here, is the set of three nonlinear, autonomous, ordinary differential equations, introduced by Lorenz (1984) as:

$$\dot{X} \equiv \begin{pmatrix} \dot{X} \\ \dot{Y} \\ \dot{Z} \end{pmatrix} = \begin{pmatrix} -Y^2 - Z^2 - aX + aF \\ XY - bXZ - Y + G \\ bXY + XZ - Z \end{pmatrix} \equiv \begin{pmatrix} \Phi_1(X,Y,Z) \\ \Phi_2(X,Y,Z) \\ \Phi_3(X,Y,Z) \end{pmatrix}, \tag{4.1}$$

(see (2.2)), with the parameters:

$$a = \frac{1}{4}, \quad b = 4, \quad F = 8. \tag{4.2}$$

The typical behavior of this system changes, as the system parameter G is changed. For example, for $G = 0.2$, and $G = 0.8$, solutions are periodic, whereas for $G = 1.25$, the dynamics are chaotic. This behavior may be globally quantified by noting, that, in the latter case, the Lyapunov exponents (see, e.g., Parker and Chua 1989) for system (4.1) are:

$$\lambda_1 = 0.229, \quad \lambda_2 = 0, \quad \lambda_3 = -0.534, \tag{4.3}$$

whereas, for $G = 0.8$, these exponents are:

$$\lambda_1 = 0, \quad \lambda_2 = -0.071, \quad \lambda_3 = -0.186. \tag{4.4}$$

System (4.1) is considered here in the chaotic regime with $G = 1.25$. In the case of system (4.1), the function $\psi(X,Y,Z)$ has no explicit time-dependence due to the autonomous nature of (4.1) and is obtained as:

$$\psi(X,Y,Z) \equiv \frac{\partial \Phi_1(X,Y,Z)}{\partial X} + \frac{\partial \Phi_2(X,Y,Z)}{\partial Y} + \frac{\partial \Phi_3(X,Y,Z)}{\partial Z} =$$

$$= (-a) + (X - 1) + (X - 1) = 2X - 2 - a. \tag{4.5}$$

Therefore, the LE (see (3.1)) relevant for system (4.1) is given by:

$$\frac{\partial \rho(X,Y,Z,t)}{\partial t} + (-Y^2 - Z^2 - aX + aF)\frac{\partial \rho(X,Y,Z,t)}{\partial X} +$$

$$+ (XY - bXZ - Y + G)\frac{\partial \rho(X,Y,Z,t)}{\partial Y} + (bXY + XZ - Z)\frac{\partial \rho(X,Y,Z,t)}{\partial Z} = \tag{4.6}$$

$$= -(2X - 2 - a)\rho(X,Y,Z,t).$$

The solution of the LE (4.6), subject to an initial condition expressed in terms of an arbitrary, appropriately normalized, function f as:

$$\rho(X,Y,Z,t = 0) = f(X,Y,Z), \tag{4.7}$$

is given by specialization of (3.3) as:

$$\rho(X,Y,Z,t) = f\left(\Xi(X,t)\right)\exp[-\int_0^t \left(2X(\Xi(X,t),t') - 2 - a\right)dt']. \tag{4.8}$$

As discussed in the general context of section 3, Ξ is that point in phase space, at time $t=0$, that is taken to point X at time t, under the governing equations (4.1). The point Ξ has been computed for given X and t (see eq. (2.6)) by a numerical backward integration in time of eqs. (4.1) by a predictor-corrector scheme with a time step of 0.01 time units.

Given Ξ, eqs. (4.1) are integrated forward in time to obtain X - as the first component of X - evaluating at the same time the integral with respect to t' by a trapezoidal scheme. Finally, the solution (4.8) has been evaluated for an initial condition specified as a tri-variate normal pdf (assuming initial independence of X, Y, and Z):

$$f(X,Y,Z) = \left(\frac{1}{2\pi\tilde{\sigma}^2}\right)^{3/2} \exp\left[-\frac{1}{2\tilde{\sigma}^2}[(X - \tilde{\mu}_X)^2 + (Y - \tilde{\mu}_Y)^2 + (Z - \tilde{\mu}_Z)^2]\right], \tag{4.9}$$

with

$$\tilde{\mu}_X = 2; \quad \tilde{\mu}_Y = -1; \quad \tilde{\mu}_Z = 0; \quad \tilde{\sigma}^2 = 0.01^2. \tag{4.10}$$

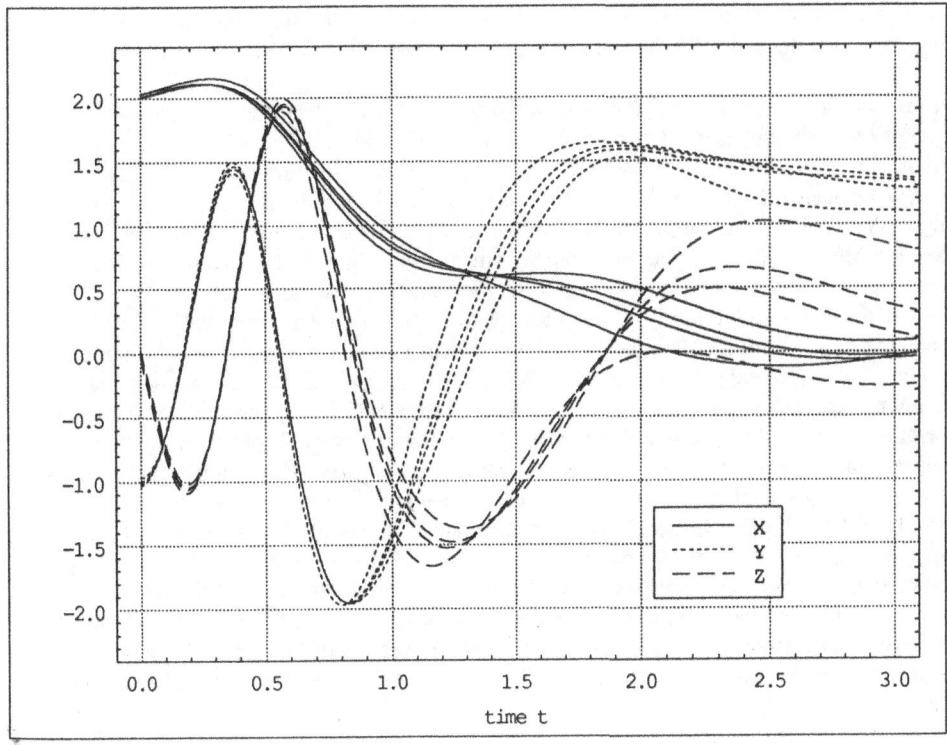

Figure 1. Numerical solution of system (4.1), given the parameter values of section 4, for four slightly different initial conditions. The numerical solution is obtained by a predictor-corrector method with a time step of 0.01 time units.

Fig. 1 shows the time evolution of four realizations of system (4.1) originating at initial points that are close to the mean of the initial pdf, namely $(2,-1,0)$ (see (4.10)). It is first of all evident that, due to the chaotic nature of the system, the temporal behavior of the trajectories is aperiodic and complicated; second, initially close trajectories tend to diverge. It is noted at this point that, due to the chaotic nature of the system, the backward integration required for evaluation of the pdf can accurately be carried out only for rather short times (see also the related comments in section 3, and the discussion in section 5). For this reason, the results presented are restricted to the interval $0 \leq t \leq 3$.

In order to capture the behavior of system (4.1) apparent from Fig. 1 in a probabilistic framework, that is, by means of the solution of the LE given in (4.8), the pdf ρ was evaluated at a regular grid in three-dimensional phase space for the time points $t=0, 1, 2$, and 3. This grid has been placed over the region in phase space with significantly non-zero pdf; the resolution in phase space varied between a grid increment of 10^{-3} and 10^{-2} with a total number of gridpoints of approximately 250^3.

At each time point, the pdf is a cloud in three-dimensional phase space with varying density, evolving in time according to (4.6). Because of the difficulty to visualize such a cloud, only certain aspects can be shown here; specifically, attention is focused on the marginal densities of each variable, each of which is obtained by integrating ρ over the remaining two variables. These marginal densities are shown in Fig. 2 for the four time points given above (note the different scales in these plots). Their accurate representation - as measured, for example, by the integral over the entire phase space - requires at least the resolution of phase space described above. Obviously, the initial marginal densities in Fig. 2 are just the univariate normal densities prescribed by the initial density (4.9).

From Fig. 2, it can be seen that the marginal densities reflect some of the behavior of system (4.1) already evident from Fig. 1, namely, they are concentrated near the mean of the four trajectories shown before. However, the picture shows more details. In particular, it may be seen that the variance of the marginal densities does not simply increase with time; for example, the spread of ρ_Y is much larger at time $t=1$, than at the later time $t=2$ (see Fig. 2b). Further, the marginal densities tend to become slightly skewed at later times (see, e.g., ρ_Y and ρ_Z at $t=2$), which is directly related to the nonlinearities present in the system under consideration. The densities at time $t=3$ appear to be highly localized, and asymmetric, for both the variables X and Y, whereas variable Z exhibits large variance.

Given the full solution of the LE, namely the pdf ρ, it is also straightforward to compute expected values of functions of the variables under consideration by a numerical integration over phase space. Such numerical integrations over phase space have been carried out to compute means (μ_X; μ_Y; μ_Z), variances ($\theta_{X,X}$; $\theta_{Y,Y}$; $\theta_{Z,Z}$), and covariances ($\theta_{X,Y}$; $\theta_{X,Z}$; $\theta_{Y,Z}$) at times $t=0,1,2$, and 3 for the components of the state vector directly from the pdf. In order for these integrations to be accurate it is essential that the pdf is known at least at the resolution described above. The results are presented in Table 1. For the validation of these computations that rely on the solution of the LE, the same statistics have, in addition, been computed from a large number (i.e.,10^5) of Monte Carlo (MC) ensemble integrations of system (4.1), where the initial condition for each integration has been specified by sampling from the initial density (4.9). The results are also included in Table 1.

Table 1. Means, variances, and covariances of the probability density function of the state vector of the low-dimensional chaotic model (4.1) at selected times t. The statistics are derived by numerically integrating the solution (4.8) of the LE over phase space (rows denoted by LE), as well as from Monte Carlo ensemble integrations with an ensemble size of 10^5 (rows denoted by MC).

t	method	μ_X	μ_Y	μ_Z
0	LE	0.2000001E+01	−0.1000001E+01	0.1189315E−09
	MC	0.1999971E+01	−0.1000014E+01	0.1121332E−05
1	LE	0.8669946E+00	−0.1408491E+01	−0.1042182E+01
	MC	0.8667593E+00	−0.1414703E+01	−0.1039724E+01
2	LE	0.2724522E+00	0.1596231E+01	0.2740712E+00
	MC	0.2780187E+00	0.1604498E+01	0.2775604E+00
3	LE	−0.4570496E−01	0.1336498E+01	0.1499279E+00
	MC	−0.4493906E−01	0.1352473E+01	0.1634511E+00

t	method	$\theta_{X,X}$	$\theta_{Y,Y}$	$\theta_{Z,Z}$
0	LE	0.9999991E−04	0.1000001E−03	0.1000001E−03
	MC	0.9986159E−04	0.9965572E−04	0.1004246E−03
1	LE	0.7185004E−03	0.4393051E−02	0.9647106E−02
	MC	0.7140323E−03	0.4337194E−02	0.9677102E−02
2	LE	0.4933610E−02	0.4924354E−03	0.4874722E−02
	MC	0.4958276E−02	0.4137549E−03	0.4843830E−02
3	LE	0.3859935E−03	0.1008074E−02	0.3346903E−01
	MC	0.4359647E−03	0.9038529E−03	0.3480525E−01

t	method	$\theta_{X,Y}$	$\theta_{X,Z}$	$\theta_{Y,Z}$
0	LE	−0.8996339E−12	0.2498075E−13	−0.3043887E−13
	MC	0.1201472E−06	0.7523069E−07	−0.1407540E−06
1	LE	0.1620662E−02	−0.2393321E−02	−0.6459885E−02
	MC	0.1613808E−02	−0.2391009E−02	−0.6440926E−02
2	LE	−0.1360140E−02	0.4744190E−02	−0.1251778E−02
	MC	−0.1400017E−02	0.4739842E−02	−0.1283271E−02
3	LE	−0.4063890E−03	0.3313069E−02	−0.2242368E−02
	MC	−0.4959261E−03	0.3609491E−02	−0.2945630E−02

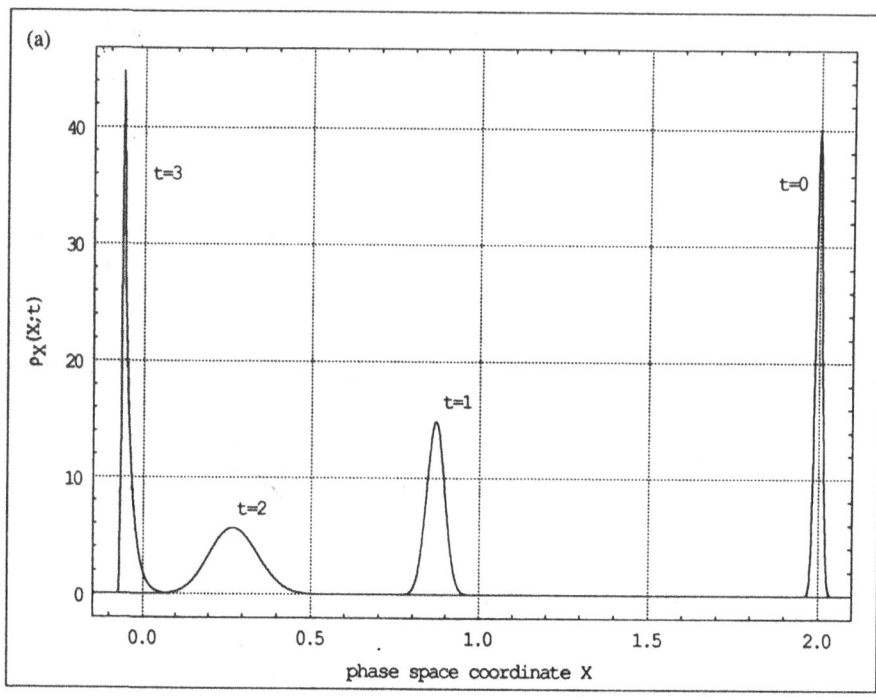

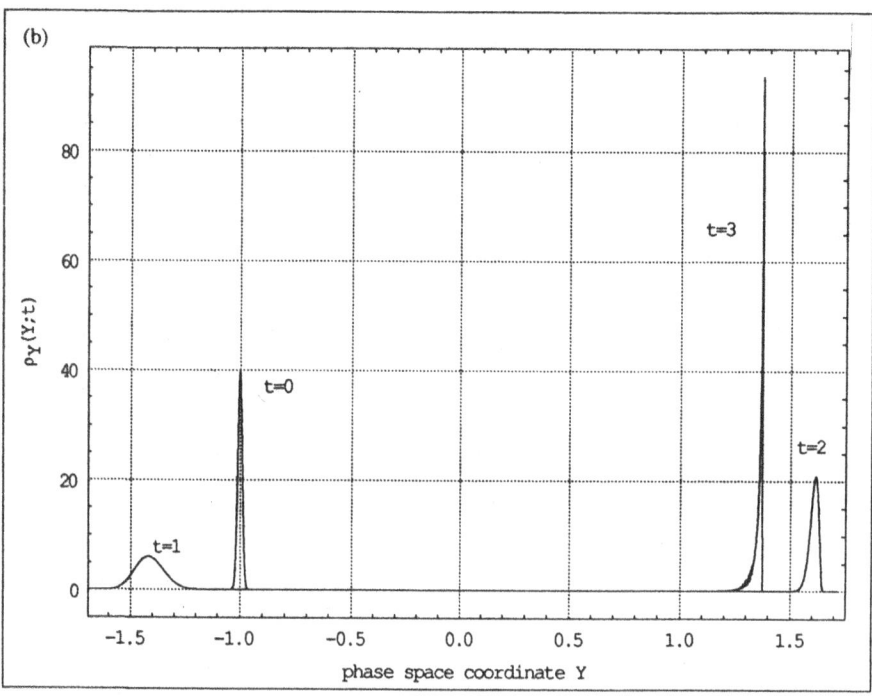

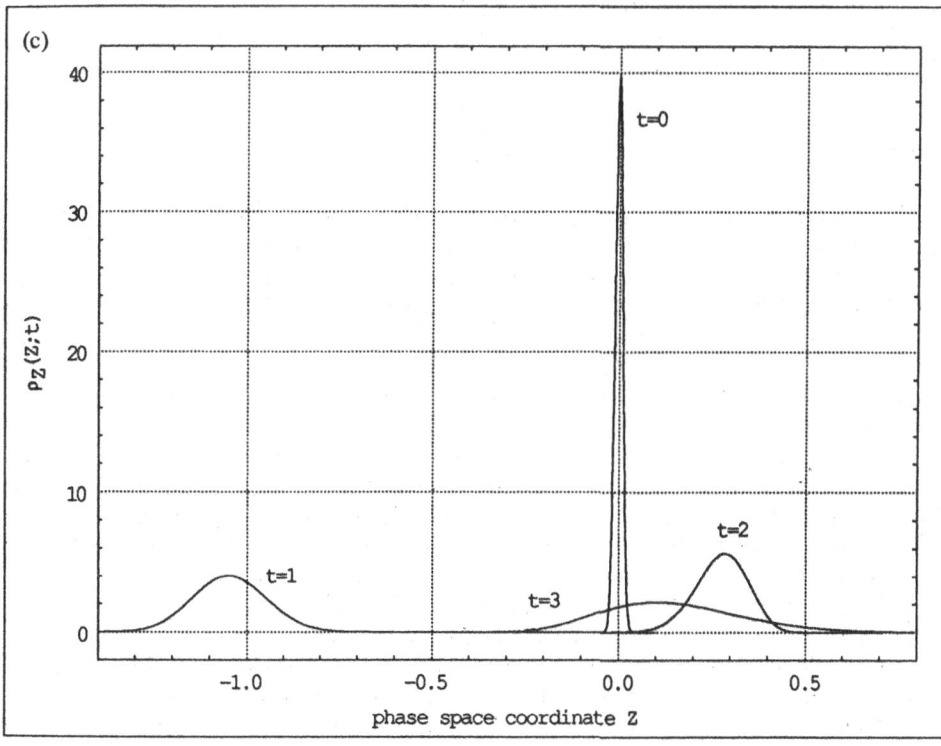

Figure 2. Marginal densities of variable (a) X, (b) Y, and (c) Z of system (4.1) obtained by phase space integration from the joint pdf $\rho(X,Y,Z,t)$ for four different time points. The joint pdf is evaluated according to the analytical solution of the LE given in eq. (4.8) for the initial condition (4.9). Note the different scales on individual panels.

There are two points worth mentioning in this context. First, as can be seen from Table 1, both approaches, namely the LE and the MC integration yield approximately the same numerical values for the expected values; by this fact the consistency of the computations and the correctness of both approaches are ensured. MC ensemble integrations with smaller ensemble sizes (results not shown) indicate the necessity to maintain the ensemble size mentioned above for the accurate estimation of the statistics investigated, even in the low-dimensional situation considered here. These rather large ensemble sizes are specifically required for the accurate computation of variances and covariances. On the other hand, it is necessary to have the numerical values of the pdf available at a large number of grid points (see above) for accurate integration over phase space. However, the evaluation of the density at a specific point in phase space for a given time may be accomplished very cheaply by using (4.8). A further discussion of these issues may be found in the next section.

Second, it is evident from the results presented in Table 1 that the initially uncorrelated variables (by specification) become correlated after short times. This fact - not necessarily related to the nonlinearities of the system considered - implies that the joint pdf ρ is no longer simply the product of the marginal densities, but can only be represented in various ways as factorizations of conditional and marginal densities.

This preliminary investigation of the joint pdf ρ for a chaotic dynamical system is concluded by the remark that it may serve as the starting point for work to get more detailed insight into the shape of the pdf at later times and/or the equilibrium pdf.

5. Discussion

The LE is central to the question of predicting forecast skill in the sense that it governs the temporal evolution of the multi-variate pdf ρ of the model state vector of a dynamical model - specified in terms of a set of ordinary differential equations - given information about the uncertainty of the initial model state in terms of an initial pdf. The LE arises from consideration of a conservation requirement for the phase space integral of the number density of realizations of the model under consideration. It is a quasi-linear, partial differential equation, and, as such, its explicit analytical solution can be found by the method of characteristics.

The direct application of the LE to realistic meteorological models - represented as a set of ordinary differential equations - encounters a number of practical difficulties. These have prevented the wide-spread use of this concept, and have led to the development and refinement of more easily accessible approaches. Some of these difficulties have already become apparent in the main body of the paper; further related remarks are made here.

The first issue relates to the actual evaluation of the explicit solution (3.3) of the LE for given arguments X and t. This evaluation relies on the construction of the characteristics of the LE through the solution of the dynamical system considered. In the absence of analytical solutions, a numerical algorithm can be used to express phase space points at a certain time as functions of initial points in phase space. Such forward integrations are the essence of meteorological models. However, in order to evaluate the solution to the LE, it is also necessary to trace points in phase space back to their origin implying the necessity of a backward integration of the model. Even when restricting attention to meteorological models expressed as a set of ordinary differential equations, which allows to refer to existence and uniqueness theorems for the solution, such a backward integration may be impossible or inaccurate for nonlinear dynamical models at larger times, seemingly representing an inherent limitation in the application of the solution of the LE. This situation could possibly be circumvented by consideration of the evolution of the pdf over short time intervals, while restarting the solution from previous solutions. Alternatively, a fully numerical solution of the LE might be attempted through consideration of an initial value problem. However, the high dimensionality of phase space encountered in the case of realistic meteorological models (see below) seems to prohibit this approach, because the solution over the complete phase space must be considered while advancing the solution forward in time.

The second issue is concerned with the dimensionality of the phase space, which for realistic meteorological models might be of the order of 10^6 to 10^7. As the computation of the value of the pdf at a certain point in phase space at a given time through the solution (3.3) is actually independent of dimensionality, since this evaluation simply requires a backward and a forward model integration, the large number of degrees of freedom of a realistic meteorological model does not, by itself, represent a problem. However, if certain statistics - to be obtained by numerical integration over (parts of) phase space - are actually of interest, large dimensionality leads to enormous increases in computational cost. Efforts to reduce the large cost of numerical integration given high-dimensional phase spaces have been reported by Woźniakowski (1992).

In the context of the issue of dimensionality, it is worth mentioning that some of the information actually contained in the full solution of the LE might be considered as superfluous for the question of predicting forecast skill. Nevertheless, the full solution might be quite valuable for the investigation of certain questions related to the prediction of forecast skill.

This discussion of the LE illustrates that its applicability in contexts more realistic than considered in this work may be subject to considerable problems. Nevertheless, these problems should be judged in light of the general nature of the LE, and in view of its clear advantages, such as avoiding the need of making any closure assumptions (as are necessary for stochastic-dynamic prediction), and avoiding the need to construct explicitly a large number of realizations in ensemble prediction. Therefore, the LE deserves further attention in the future, and its potential in the context of predicting forecast skill should be explored for more complex models and higher-dimensional situations. Finally, it is emphasized that the LE must be considered as a potentially important tool within the area of forecasting forecast skill, presenting valuable guidance in the process of developing new, and/or refining existing methodology suitable for the prediction of forecast skill.

Acknowledgments

The author appreciates discussions with Dr. Joseph J. Tribbia related to the subject treated in this work. The remarks by an anonymous reviewer on an earlier version of this manuscript were helpful in improving the presentation. Part of this research was carried out, while the author was a Visiting Scientist at the National Center for Atmospheric Research. The National Center for Atmospheric Research is sponsored by the National Science Foundation.

References

Arnold, V.I., 1992, *Ordinary differential equations*, Springer, 334 pp.

Ehrendorfer, M., 1992, 'Predicting forecast skill: The Liouville equation'. *Preprint Volume Twelfth Conference on Probability and Statistics in the Atmospheric Sciences*, Toronto, 56-61.

Ehrendorfer, M., 1993, 'The Liouville equation and its potential usefulness for the prediction of forecast skill'. Submitted to *Mon. Wea. Rev.*

Epstein, E.S., 1969, 'Stochastic dynamic prediction'. *Tellus*, **21**, 739-759.

Hoffman, R.N., and E. Kalnay, 1983, 'Lagged average forecasting, an alternative to Monte Carlo forecasting'. *Tellus*, **35A**, 100-118.

Leith, C.E., 1974, 'Theoretical skill of Monte Carlo forecasts'. *Mon. Wea. Rev.*, **102**, 409-418.

Lin, C.C., and L.A. Segel, 1988, *Mathematics applied to deterministic problems in the natural sciences*. Society for Industrial and Applied Mathematics, 609 pp.

Lindley, D.V., 1987, 'The probability approach to the treatment of uncertainty in artificial intelligence and expert systems'. *Stat. Sci.*, **2**, 17-24.

Lorenz, E.N., 1984, 'Irregularity: a fundamental property of the atmosphere'. *Tellus*, **36A**, 98-110.

Lützen, J., 1990, *Joseph Liouville 1809-1882: Master of pure and applied mathematics.* Springer, 884 pp.

Molteni, F., and T.N. Palmer, 1991, 'A real-time scheme for the prediction of forecast skill'. *Mon. Wea. Rev.*, **119**, 1088-1097.

Mureau, R., F. Molteni, and T.N. Palmer, 1993, 'Ensemble prediction using dynamically conditioned perturbations'. *Q. J. R. Meteorol. Soc.*, **119**, 299-323.

Parker, T.S., and L.O. Chua, 1989, *Practical numerical algorithms for chaotic systems.* Springer, 348 pp.

Thompson, P.D., 1983, 'Equilibrium statistics of two-dimensional viscous flows with arbitrary random forcing'. *Phys. Fluids*, **26**, 3461-3470.

Thompson, P.D., 1985, 'A statistical-hydrodynamical approach to problems of climate and its evolution'. *Tellus*, **37A**, 1-13.

Tribbia, J.J., and D.P. Baumhefner, 1993, 'On the problem of prediction beyond the deterministic range'. *Prediction of interannual climate variations*, J. Shukla, Ed., Springer, 251-265.

Woźniakowski, H., 1992, 'Average case complexity of linear multivariate problems. I. Theory'. *Journal of Complexity*, **8**, 337-372.

Zwillinger, D., 1989, *Handbook of differential equations.* Academic Press, 673 pp.

AN IMPROVED FORMULA TO DESCRIBE ERROR
GROWTH IN METEOROLOGICAL MODELS

JEAN-FRANÇOIS ROYER[1], RODICA STROE[1], MICHEL DÉQUÉ[1]
AND STÉPHANE VANNITSEM[2]

[1] Météo-France, Centre National de Recherche Météorologique,
42 Av. G. Coriolis, 31057 Toulouse Cédex, France

[2] Institut Royal Météorologique de Belgique,
Avenue Circulaire 3, B-1180 Bruxelles, Belgique

Summary

In meteorological models, the logistic growth law has been used traditionally to describe the error growth due to sensitivity to the initial conditions. A detailed analysis obtained from long range forecasting experiments using a GCM model, as well as from simulations based on a simple 3-variable model, has revealed significant deviations from the logistic law. A natural generalization is proposed, giving a law that has been used previously for the description of biological growth. A new characteristic parameter, which can be interpreted as a saturation rate for error growth, is identified. Further studies, based on a simple 3-variable model for different magnitudes of the initial error, reveal a more complex behaviour having a transient initial regime that is independent of the error magnitude, a regime of exponential growth, and a "deceleration regime". The deceleration regime as defined here includes both the phases of linear and saturated error growth in time. For the case of large initial errors, the vanishing of the exponential regime, as a result of the coalescence of the initial and deceleration regimes, gives a continuous decrease in error growth rate with time, which can be well represented by the Gompertz growth law.

Keywords error-growth, predictability, dynamical system

1. Presentation and generalization of error growth models

It has long been recognized that the numerical prediction of detailed instantaneous atmospheric states is limited because small uncertainties in the initial state are amplified during the forecasts by the non-linearity and the instabilities of the atmospheric dynamics (Thompson, 1957). Error growth arising from sensitivity to initial conditions can be studied by monitoring the time evolution of the average distance between two initially nearby trajectories of a dynamical system. Estimates of error growth in realistic atmospheric models were made by Lorenz (1982), Dalcher and Kalnay (1987), Brankovic et al.(1990), Chen (1989), Schubert and Suarez (1989) and Toth (1989).

In these studies, idealized models of error growth were fitted to the data in order to describe the main features of the error growth curves with a small number of parameters.

The assumption made by Lorenz (1969) is that the idealized error growth is governed by the first order differential equation:

$$dE/dt = \alpha\, E\, (1 - E/E_\infty)\,. \tag{1a}$$

The parameter α represents the exponential growth rate of small errors and is usually identified with the largest Lyapunov exponent of the system. The factor $(1 - E/E_\infty)$ represents the deceleration of the error growth when the error E asymptotically reaches a limiting value E_∞, which is the mean error between two randomly chosen points on the attractor of the dynamical system.

The solution of this differential equation is the well-known logistic growth law, which has found widespread use in describing biological growth since its introduction by Verhulst in 1838 (Savageau, 1980):

$$E(t) = E_\infty\, \{\, 1 + (E_\infty/\,E_0 - 1)\, e^{-\alpha t}\, \}^{-1}\,, \tag{1b}$$

where E_0 is initial error at time $t = 0$.

Noticing the discrepancy between the studies of Lorenz (1982), which assumes that the logistic growth law (1) applies to the root mean square error R (rmse), and of Dalcher and Kalnay (1987) who apply it to the error variance R^2, Stroe and Royer (1992, 1993) have proposed a possible generalization of the error growth formula by making the assumption that the logistic law applies to some power p of the rmse R. The change of variable $E = R^p$ in (1a) readily gives the differential equation for the relative error growth rate of R:

$$(1/R)\, dR/dt = \alpha/p\, \{\, 1 - (R/R_\infty)^p\, \}\,. \tag{2a}$$

A similar growth equation has been proposed and used as a model of biological growth function by Bertalanffy (1957), and has been studied further by Richards (1959).

It is convenient to introduce as a new independent variable the logarithm of the error: $y = \ln(R)$, for which the relative growth rate $(1/R)dR/dt$ becomes simply the time derivative dy/dt. Then (2a) can be rewritten as:

$$dy/dt = \alpha/p\, \{\, 1 - \exp[\, p\, (y - y_\infty)]\, \}\, = f(y)\,. \tag{2b}$$

For $(y \rightarrow -y)$ the relative growth rate tends asymptotically to a constant value $\sigma = \alpha/p$, which represents therefore the growth rate of small errors. Considering the other limit when y tends to its saturation value y_∞, we have:

$$\lim_{y \rightarrow \infty} \{[d(y_\infty - y)/dt]/(y_\infty - y)\} = [df(y)/dy](y = y_\infty) = -\alpha\,. \tag{2}$$

This shows that the parameter α can be interpreted as a convergence rate to the saturation value, and can be called in short the "saturation rate" (Stroe and Royer, 1993).

The generalized formula thus allows to distinguish two characteristic parameters of error growth: the *growth rate of small errors* σ, which corresponds to the horizontal asymptote of the curve $f(y)$, and the saturation rate α, which is given by the opposite of the slope of $f(y)$ near the accumulation point y_∞. With the logistic equation ($p=1$), these two parameters become equal, and most previous studies of error growth, at least with meteorological models, have not made any distinction between these two different characteristics.

The generalized formula (2a) can be extended by continuity to $p = 0$. It is easy to check, by making a Taylor series expansion of the exponential in (2b), that the limit for vanishing p becomes a linear differential equation for y:

$$dy/dt = -\alpha \ (y - y_\infty) \ . \tag{3}$$

The solution of (3) is a doubly exponential law for R, known in biology as the Gompertz law

$$R(t) = R_\infty \exp[-\ln(R_\infty/R_o) \ e^{-\alpha t}] \ . \tag{4}$$

It has been used in several studies of the growth laws for biological systems and in population studies (Savageau, 1980).

Trevisan et al. (1992) have shown that the error growth between analogs, obtained in a long simulation with a simplified quasigeostrophic model of atmospheric dynamics, is well approximated by a linear relationship in the plane $(y, \ dy/dt)$, which, for their system of equations, has a slope equal to the largest Lyapunov exponent.

Adjusting the generalized model (2a) to the mean square error growth of 500 hPa geopotential in a realistic general circulation model, Stroe and Royer (1992) have found that the parameter p giving the optimal fit was generally rather close to zero. The study was based on a series of 44-day integrations performed as monthly forecast experiments for the months of November, December, January and February of the eight winters from 1983 to 1991. The skill for a 21 case subset of these monthly forecasts has been evaluated (Déqué and Royer, 1992). The distance between two integrations starting from initial conditions 24 hours apart is interpreted as the forecast error of a perfect model arising from uncertainties in the initial conditions.

To illustrate that the Gompertz law gives a good fit to error growth in a numerical model, additional results from these experiments are presented here. The error growth as a function of time for the geopotential height of the 500 hPa pressure surface in 3 different ranges of spatial scales is shown in Fig.1A. The spatial scale is represented by the wavenumber n resulting from a spectral decomposition of the global fields in spherical harmonics. The evolution of the error is characterized by an initial linear growth followed by a progressive saturation, which is faster for the short waves ($n = 11–21$), and slower for the long waves ($n = 1–5$) that appear to be still growing slightly until the end of the 44 days of integration.

Figure 1B shows the values in the plane $(y, \ dy/dt)$, i.e., they show the function $f(y)$ giving the relative error growth rate as a function of the logarithm of the error. The $(y,dy/dt)$ values, (where dy/dt has been computed by centered finite differences of the model y data), appear to be linearly related, which indicates that the error growth can be well represented by the Gompertz law (Eqn.3), except for the first point for the planetary waves (wavenumber 1-5). We can notice that the initial error is smaller for the planetary waves than for the other wavenumber ranges. The success of the linear fit comes in part from the fact that the initial error is relatively large. A possible explanation for the behaviour of the initial values of error for the planetary waves will be given in paragraph 3.

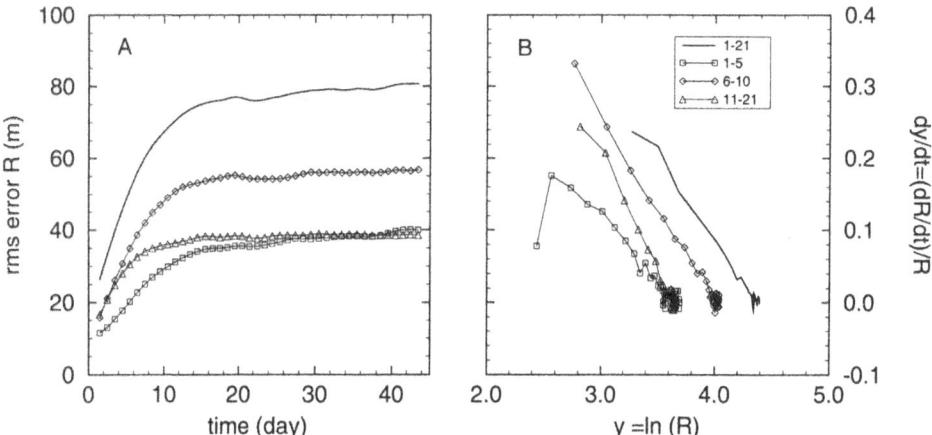

Figure 1. Error growth between a pair of long-range forecasts of the 500 hpa geopotential field using the CNRM model. Figure 1A shows the rms error R (in geopotential meters) as a function of forecast range (in days). The upper curve correspond to total error (solid line), and the 3 other to decomposition of this error in the planetary waves $n = 1$-5 (squares), the synoptic waves $n = 6$-10 (diamonds), and the shorter waves $n = 11$-21 (triangles). Figure 1B shows the same data in the plane $(y, dy/dt)$ with $y=ln(R)$. Note that the Gompertz law (Eqn.3) gives a straight line in this plane.

In the plane $(y, dy/dt)$, the parameters of Gompertz law can be simply determined by a linear regression. The slope of the straight line that fits the data, corresponds to the parameter $-\alpha$, and its intersection with the axis $dy/dt = 0$ corresponds to the saturation value y_∞ of the logarithm of the error. A similar fit has also been made separately for each spectral wavenumber n, in order to examine the spatial scale dependence. The saturation spectrum for the error variance has a maximum at wavenumber $n = 7$ which corresponds to synoptic scale waves (Fig.2A). The linear decrease in log-log coordinates for the short waves $(n > 13)$, corresponds to a power law decrease close to n^{-5} in the high wavenumbers region similar to what has been found for observed 500 hPa geopotential spectra. The values of the parameter α increase from *0.15 per day* for the planetary waves to *0.4 per day* for the short waves (Fig. 2B), confirming that errors in smaller scales saturate more rapidly.

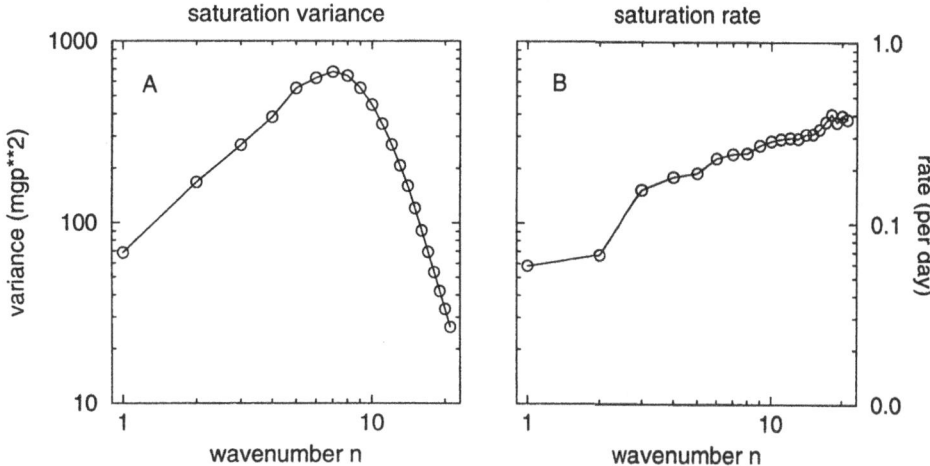

Figure 2. Parameters of Gompertz law fitted by linear regression to the error growth of 500 hPa geopotential for each wavenumber n (horizontal axis): A) saturation variance R_∞^2 (A) in units m2. B) the saturation rate σ in 1/day. Note that the two axes are in logarithmic coordinates where a straight line correspond to a power law.

2. New experiments with the Lorenz model

The fit of error growth using the Gompertz law raises some fundamental questions since its extrapolation to infinitesimal initial errors implies an initial growth rate increasing without bounds. Forecast experiments with real data cannot adress this problem since the initial errors are usually somewhat large, and predictability experiments with general circulation models are computationally very expensive.

Thus it is more efficient to study the problem of error growth using simpler dynamical systems, which can allow a larger number of simulations (Nicolis and Nicolis, 1991). A model that has already been used in several studies is the system of 3 differential equations proposed by Lorenz (1984) as a minimal model of the general circulation.

$$dX/dt = -Y^2 - Z^2 - aX + aF,$$
$$dY/dt = XY - bXZ - Y + G,$$
$$dZ/dt = bXY + XZ - Z.$$

(5)

For the parameter values $a = 1/4$; $b = 4$; $F = 8$; $G = 5/4$, which have been used in several studies (Benzi and Carnevale, 1989; Nicolis, 1992), the model gives rise to chaotic dynamics with a single positive Lyapunov exponent.

Using the simulations of error growth performed by Nicolis (1992) starting from an initial error of $1.E-4$, Royer et al. (1993) have found that the Gompertz law (4) could reproduce the overall error evolution in a more satisfactory manner than the logistic law.

Yet a simulation of the fine structure of error growth has shown that the Gompertz law was not able to reproduce the more complex behaviour of initial error growth.

In order to examine in more detail the initial error growth, several new simulations using Eqn.(5) have been performed using differing initial errors. The technique of twin experiments was used starting from a large number, N, of points lying on the attractor of the model. These points were generated from a long integration of the dynamical system (5) sampled every 10 time units. The differential equations were integrated with the fourth-order Runge-Kutta method, as described in Press et al. (1986), with a time step of 0.01 time units. New trajectories were generated by introducing perturbations of a fixed magnitude R_o but with a random orientation in (X,Y,Z) space. The error was defined as the

Euclidean distance in (X,Y,Z) space between the control and perturbed trajectories at time t:

$$R_n(t) = \{(X_n' - X_n)^2 + (Y_n' - Y_n)^2 + (Z_n' - Z_n)^2\}^{1/2} .$$

Using the notation: $<A> = (1/N) \Sigma A_n$ to represent the ensemble mean over N trajectories, we have investigated the effect of the averaging operator by considering 3 possibilities for defining the mean error:

Arithmetic mean : $<R>_1 = <R>$,
Quadratic mean : $<R>_2 = (<R^2>)^{1/2}$,
Geometric mean : $<R>_0 = \exp(<ln(R)>)$.

The 3 methods of averaging have been used in previous studies; the arithmetic mean by Nicolis (1992); the quadratic mean by Dalcher et Kalnay (1987), and Stroe and Royer (1992); and the geometric mean, which appears naturally if we work with the logarithm of the error, by Lorenz (1969) and Trevisan et al. (1992).

Whilst with the results of Fig. 1, the choice of the averaging operator makes little difference, this is not so for the dynamical system of Lorenz. Due to the large variability of the local error growth rate over the attractor of the Lorenz system, the saturation value of the error depends strongly on the averaging procedure used to define the ensemble mean (Fig. 3). The error growth is faster for the quadratic and arithmetic mean than for the geometric mean. Such results are in agreement with an analytic expression, derived by Benzi and Carnevale (1989), for the variation of the growth rate of different moments of the error, in the case when fluctuations of the growth rate can be approximated by a white noise process. The quadratic mean gives more weight to large deviations, whereas the geometric mean gives more weight to small deviations, and this explains why the saturation values and the error growth rate are higher for the quadratic mean, and smaller for the geometric mean (Fig.3A). All the three curves show a transient initial regime of very rapid error growth which last only 2 or 3 time units (Fig.3B).

The representation of the mean error in the plane $(y, dy/dt)$ reveals additional differences. The values of error growth rate obtained with the quadratic and arithmetic mean are systematically larger and appear also more unstable and show large oscillations, though the averaging has been done with a large number ($N = 100000$) of independent trajectories (Fig. 3C). With the logarithmic mean, a regime with a constant growth rate is clearly apparent. The corresponding exponential growth rate is very close to the value of the largest Lyapunov exponent of the system (0.22) computed with the method of Shimada and Nagashima (1979). On the other hand, with the other averaging operators the values are systematically larger. Apparently, the Lyapunov exponent gives the rate of exponential error growth for this system, only if we define the averaging with the logarithm of the error, as is consistent with the definition of the Lyapunov exponent. With the other averaging procedures the exponential growth rate of small errors, after the period of transient growth is systematically higher than the largest Lyapunov exponent. Due to this increase of the growth rate a straight line fits better, and over a larger range of errors, the curves obtained with the quadratic and arithmetic averaging than the one with the geometric mean that has a horizontal asymptote at 0.22 (Fig. 3D). This shows that part of the success of the Gompertz law in fitting the results of Nicolis (1992) is due to the

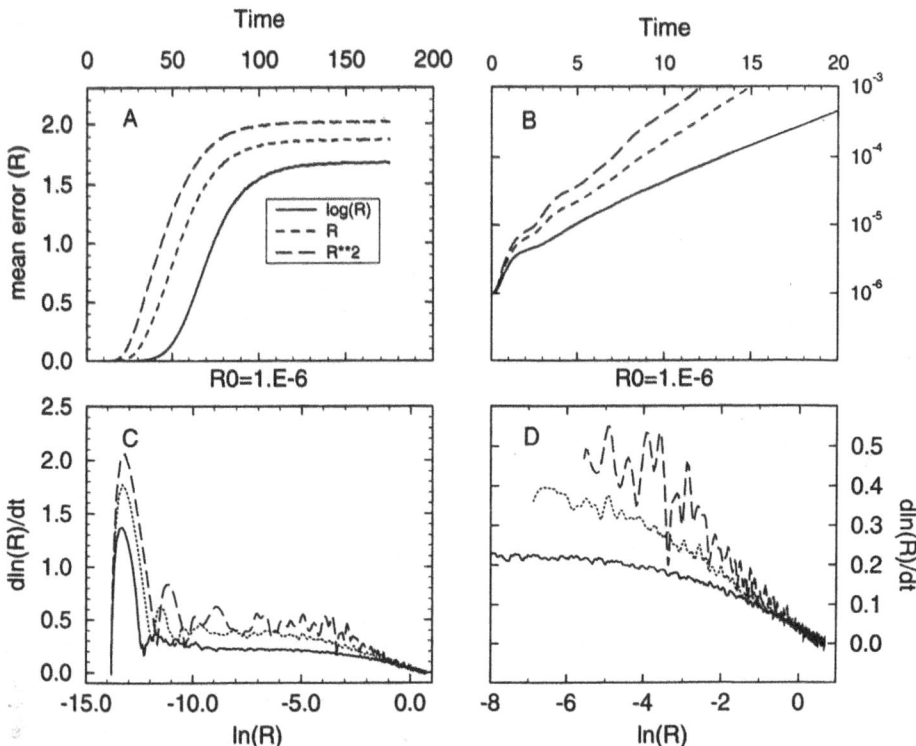

Figure 3. Error growth in the Lorenz model computed with different averaging operator: quadratic mean ($R**2$), arithmetic mean (R), geometric mean ($\log(R)$). A) mean error as a function of time. B) Same data as in A1 for the first 20 time units to show the initial error growth. C) Representation of the data of A1 in the plane (y, dy/dt) with $y = \ln(R)$. D) Enlargement of a portion of C showing the behaviour of the curves near the saturation value. The curves have been computed from 10^5 trajectories perturbed with an initial error of 1.E-6.

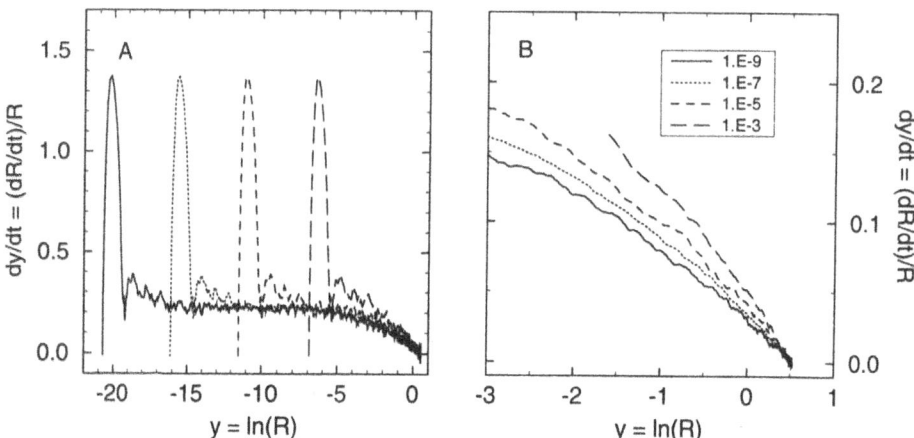

Figure 4. Representation in the plane (y, dy/dt) of the dependence on the initial error values: 1.E-9 (solid line), 1.E-7 (dotted), 1.E-5 (dashed), 1.E-3 (long dashed), on the logarithmic mean error growth computed from 5.10^4 trajectories of the Lorenz system with a sampling of 0.1 time units. In Figure 4B a running mean of 5 time units has been applied, and the first part of the record influenced by transient error growth ($0 < t < 15$) has been removed, so as to show more clearly the differences close to saturation.

choice of an arithmetic averaging for the computation of the mean error.

It may be preferable to use the logarithmic averaging since it minimizes fluctuations and gives a well defined exponential growth at the rate close to the largest Lyapunov exponent. Only this averaging will be considered in what follows.

Several integrations have been made with different initial errors R_o ranging from *1.E-9* to *1.E-3*. The curves of error growth in the plane (y, dy/dt) show a strong peak corresponding to the initial transient rapid error growth (Fig. 4A).

With the smaller initial errors 3 separate regimes of error growth can be easily distinguished: a *transient initial regime* of rapid error growth corresponding to the large peak and subsequent smaller oscillations; a *regime of exponential growth* corresponding to a constant value of dy/dt equal to *0.22*, and finally a "*deceleration regime*" where dy/dt starts to decrease, due to the slowing down of the error growth.

The initial regime of transient error growth has been identified and studied by Trevisan (1993), who has pointed out its important effect on predictability time. The "deceleration regime", as defined here in the plane (y, dy/dt), encompasses both the regime of linear error growth in time, identified by Nicolis and Nicolis (1991), and the regime of saturated growth, which can be distinguished by looking at the error growth curve $R(t) = exp[y(t)]$. For the range of values considered here, the magnitude of the initial error has little influence on the regime of transient initial growth that keeps the same shape in the diagram (Fig. 4A). When the initial error magnitude is increased, the regime of exponential growth becomes shorter, since the non-linear terms, which produce the deceleration of error growth, always start to become significant at similar values of the error magnitude. With an initial error magnitude of $1.E-3$ the exponential regime has almost disappeared and the regimes of transient growth and deceleration become close together. The disappearance of the constant growth rate regime, as the regimes of transient growth and decelerated growth coalesce together, yields a linear relationship between dy/dt and y, in particular for values of initial error between $1.E-5$ and $1.E-3$ (Fig.4A). The coalescence of the transient and decelerated growth regimes contributes, along with the arithmetic averaging, to the linear decrease that was responsible for the success of Gompertz law with the initial error of $1.E-4$ in Royer et al.(1993).

Hence, it becomes apparent that the Gompertz law may have no real physical significance but is more likely due to the contributions of two different factors, arithmetic averaging and the existence of transient initial error growth. Both are capable of increasing the growth rate for small errors, thereby improving the linear fit, for the cases *when the initial error magnitude is sufficiently large*. The deviation from a straight line which happens in Figure 1 for the planetary wavenumbers 1-5 seems consistent with this explanation since this wavenumber range has the smallest initial error , and can be interpreted as a manifestation of oscillatory transient error growth.

The initial transient regime and the exponential regime are governed by the linear tangent system (Lacarra and Talagrand, 1988), and thus are independent of the magnitude of the error while the error remains small enough for non-linear terms to be negligible. This is evident from looking at the curve of the relative error growth rate as a function of time during the first time units of the integration (Fig. 5A), which shows clearly that this is independent of the initial error magnitude. The transient initial regime with a large peak and several smaller oscillations is damped in less than 15 time units. The curves are nearly coincident, and only for the larger error magnitude of $1.E-3$ do the non-linear terms start to decrease the growth rate after $t = 10$.

3. Conclusion and perspectives

The results suggest that the general error growth equation for the transient initial regime is of the form

$$dy/dt = \sigma \, g(t) \quad \text{for } t < t1$$

where $g(t)$ is an oscillating function tending towards 1 when t increases. The function $g(t)$ can be determined from the linear tangent model and depends on the orientation of the initial perturbation.

The usual equations for error growth relating the error growth to the magnitude of the error can be used only after the transient initial growth phase. A more detailed examination of the final evolution (Fig. 4B) reveals that the function $f(y)$ and the saturation rate α are not independent of the initial error. The saturation rate, given by the slope of the curve $f(y)$ close to the axis $dy/dt=0$, is apparently smaller for smaller initial errors. This is because deceleration appears at smaller values of the mean error. Therefore, the mean error is insufficient to characterize the error growth. For the same value of the mean error the variance of the logarithm of the error, which grows approximately linearly in time until the error growth starts to decelerate, has a larger maximum for the trajectories

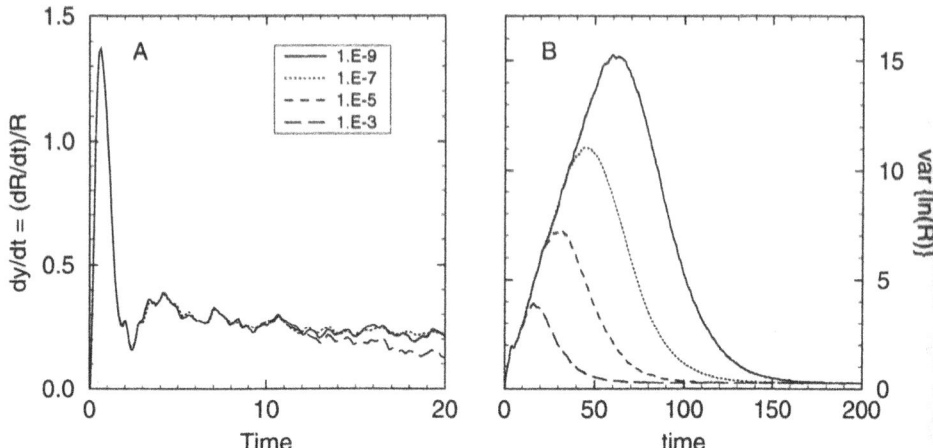

Figure 5. A) Evolution of the relative error growth $dy/dt = (1/R)dR/dt$ during the first 20 time units illustrating the similar transient behaviour for the different initial errors.
B) Evolution of the variance of the logarithm of the error.

starting from the smaller initial error. When starting from smaller initial errors, the error variance reaches a much larger value (Fig. 5B). For the same value of the ensemble mean error, a larger number of trajectories can reach the threshold where non-linear terms start to reduce significantly the error growth rate. The appearance of saturation of error growth for these trajectories, has the effect of spreading the saturation stage over a larger domain of the ensemble mean error, and the saturation rate is thus reduced.

Hence, the saturation rate is not independent of the initial error. The minimal equation needed to describe error growth should incorporate a time-dependent description of the transient regime. If we assume that transient and deceleration regimes correspond to different processes, we could attempt to model error growth by an equation of the following form:

$$dy/dt = g(t)\ f(y, y_0) \tag{6}$$

where $g(t)$ explains the transient regime, and $f(y,y_0)$ explains the deceleration regime.

One of the simplest functions would be the generalized model with p depending on y_0

$$f(y,y_0) = \sigma \{\ 1 - \exp [p(y_0)\ (y - y_\infty)]\ \} \tag{7}$$

Alternatively the error growth equation could be written as a function of the error variance, which would then lead to a closure problem, since an equation to represent the variance growth would then be necessary. Finally, it appears that a general representation of error growth would have to be rather complex, and a parameterization by a stochastic differential equation such as that used by Nicolis (1992) would seem to provide a more detailed and adequate representation.

Error growth dynamics is of fundamental interest in predictability studies (Nicolis and

Nicolis, 1987). The present study has clarified some points concerning the parametric representation of error growth curves. In general the exponential growth rate of small errors and the saturation rate need to be considered as two different characteristics of error growth. Therefore, the logistic law, which implicitly assumes that these two parameters are identical, cannot provide a sufficiently general representation of error growth curves. The generalized formula of Bertalanffy (Eqn. 2a or 2b) is a more satisfactory alternative since it has the possibility of being able to distinguish between these two parameters. For systems with a large variability of local error growth rates, such as the simple dynamical model of Lorenz (1984), the shape of the error growth curve is dependent upon the ensemble averaging operator used to define the mean error. The exponential growth rate for small errors tends to the Lyapunov exponent only after a phase of transient growth, and only with the use of a geometric mean of the error (averaging of the logarithm), whereas the use of arithmetic or quadratic averaging will usually give larger error growth rates. The Gompertz law appears as a limiting case of error growth that cannot provide a generally valid model, and it happens to fit the error growth well in some cases only because of a combination of factors, such as quadratic mean averaging which enhances the growth of small errors, transient error growth, and a large enough initial error magnitude. In conclusion, it seems that error growth is a more complex problem than has generally been recognized, and that the fitting of mathematical formulas to estimate characteristic parameters such as the Lyapunov exponent or the saturation rate has to be done with caution. Finally a probabilistic description of error growth appears as a more suitable framework.

Acknowledgments

We thank Marc Pontaud for programming the Lorenz model, David Stephenson for comments and help with the English, and Catherine Nicolis for helpful suggestions. We acknowledge the support of the Programme National d'Etude de la Dynamique du Climat (PNEDC), *of the Programme Global Change du gouvernement Belge,* and of the Commission of the European Communities (Contract EPOC-0003-C-MB, and EV5V-CT92-0121). This paper is dedicated to the memory of Rodica Stroe, who tragically died on 18 November 1992, whilst preparing part of this work for her research thesis.

References

Benzi, R. and F. C. Carnevale (1989), 'A possible measure of local predictability', *J. Atmos. Sci.*, **46**, 3595-3598.
Bertalanffy, L. von (1957), 'Quantitative laws in metabolism and growth', *Quart. Rev. of Biol.*, **32**, 217-231.
Brankovic, C., T.N. Palmer, F. Molteni, S. Tibaldi and U. Cubasch (1990), 'Extended-range predictions with ECMWF models: Time-lagged ensemble forecasting', *Q. J. R. Meteorol. Soc.*, **116**, 867-912.
Chen, W.Y. (1989), 'Estimate of dynamical predictability from NMC DERF experiments', *Mon. Wea. Rev.*, **117**, 1227-1236.
Dalcher, A. and E. Kalnay (1987), 'Error growth and predictability in operational ECMWF

forecasts', *Tellus*, **39** A, 474-491.

Déqué, M. and J.F. Royer (1992), 'The skill of extended-range extratropical winter dynamical forecasts', *J. of Climate*, **5**, 1346-1356.

Lacarra, J.F. and O. Talagrand (1988), 'Short-range evolution of small perturbations in a barotropic model', *Tellus*, **40A**, 81-95.

Lorenz, E.N. (1969), 'Atmospheric predictability as revealed by naturally occuring analogues', *J. Atmos. Sci.*, **26**, 636-646.

Lorenz, E.N. (1982), 'Atmospheric predictability experiments with a large numerical model', *Tellus*, **34**, 505-513.

Lorenz, E.N. (1984), 'Irregularity: a fundamental property of the atmosphere', *Tellus*, **36A**, 98-110.

Lorenz, E.N. (1987), 'Deterministic and stochastic aspects of atmospheric dynamics', In: *Irreversible Phenomena and Dynamical Systems Analysis in Geosciences*, Nicolis C., and G. Nicolis (éds.), D. Reidel Publishing Company, Dordrecht, Holland, pp. 159-179.

Nicolis, C. (1992), 'Probabilistic aspects of error growth in atmospheric dynamics', *Quart. J. Roy. Meteorol. Soc.*, **118**, 553-568.

Nicolis, C. and G. Nicolis, Eds. (1987), *'Irreversible Phenomena and Dynamical Systems Analysis in Geosciences'*, D. Reidel Publishing Company, Dordrecht, Holland.

Nicolis, C. and G. Nicolis (1991), 'Dynamics of error growth in unstable systems', *Phys. Rev. A*, **43**, 5720-5723.

Press, W.H., B. Flannery, S.A. Teukolsky and W.T. Vetterling (1986), 'Numerical recipes "Fortran version"', Cambridge University Press.

Richards, F.J. (1959), 'A flexible growth function for empirical use', *J. of Experimental Botany*, **10**, 290-300.

Royer, J.F., R. Stroe and C. Nicolis (1993), 'Croissance de l'erreur dans un modèle atmosphérique simple: comparaison entre la loi logistique et la loi de Gompertz', *C. R. Acad. Sci. Paris*, t.**316**, Série II, 193-200.

Savageau, M.A. (1980), 'Growth equations: a general equation and a survey of special cases', *Math. Biosci.*, **48**, 267-278.

Schubert, S.D. and M. Suarez (1989), 'Dynamical predictability in a simple general circulation model: average error growth', *J. Atmos. Sci.*, **46**, 353-370.

Shimada, I. and T. Nagashima (1979), 'A numerical approach to ergodic problem of dissipative dynamical systems', *Progress in Theoretical Physics*, **61**, 1605-1616.

Stroe, R. and J.-F. Royer (1992), 'Ecart entre des intégrations jumelles d'un modèle de circulation générale: Etude comparative de plusieurs lois de croissance', *C. R. Acad. Sci. Paris*, t.**315**, Série II, 445-451.

Stroe, R. and J.F. Royer (1993), 'Comparison of different error growth formulas and predictibility estimation in numerical extended-range forecasts', *Ann. Geophysicae*, **11**, 296-316.

Thompson, P.D. (1957), 'Uncertainty of initial state as a factor in the predictability of large scale atmospheric flow patterns', *Tellus*, **9**, 275-295.

Toth, Z. (1991), 'Estimation of atmospheric predictability by circulation analogs', *Mon. Wea. Rev.*, **119**, 65-72.

Trevisan, A., P. Malguzzi and M. Fantini (1992), 'On Lorenz's law for the growth of large and small errors in the atmosphere', *J. Atmos. Sci.*, **49**, 713-719.

Trevisan, A. (1993), 'Impact of transient error growth on global average predictability measures ', *J. Atmos. Sci.*, **50**, 1016-1028.

SEARCHING FOR PERIODIC MOTIONS IN LONG-TIME SERIES

RUBÉN A. PASMANTER

Royal Netherlands Meteorological Inst. (KNMI)
P.O. Box 201, 3730 AE De Bilt, Netherlands
e-mail: pasmante@knmi.nl

Abstract

We present a method for the detection of periodic motions in long-time series of data. In contrast to other commonly used approaches, e.g., empirical orthogonal functions, principal oscillation patterns, singular spectrum analyses, this method is oriented towards scalar-quantities, therefore, it does not require the introduction of an arbitrary metric in the space of the dynamical variables. Nonlinear effects are included; needless to say, the higher the non-linearities included the longer and more complicated the actual calculations become. We are trying the method on a Lorenz model and on a simple model of the dynamics of the atmosphere.

Keywords: Periodic motion, nonlinear systems, data analysis.

1. Introduction

In a meteorological and climatological context, like in many other of the natural sciences, there is great interest in identifying, characterizing and understanding any periodic motion that may be present in the system under consideration. For the sake of precision:

1) The systems of interest are usually *chaotic*, therefore, by periodic motion we mean actually *unstable* periodic motion; we could also refer to the phenomenon as 'transient periodicity', see Kendall *et al.* (1993).

2) These are nonlinear systems so that, in general, the oscillations cannot be described by one trigonometric function (= Fourier component).

3) Quasi-stationary, non-oscillating patterns, like atmospheric blockings, are a special in stance of periodic motion.

4) In a system with a large number of degrees of freedom, it may happen that only the motion of a fraction of the variables can be approximately described as periodic while the motion of the other variables is clearly chaotic. In meteorology, this corresponds to the fact that blockings, large-scale waves and other weather patterns are, more often than not, localized in space.

Unstable periodic orbits (UPO) are being actively studied also in the general theory of dynamical systems, where they play a central role (Ruelle, 1978; Cvitanović, 1988). A number of techniques have been proposed and developed in order to find UPO's in low-dimensional dynamical systems (Auerbach *et al.*, 1987); unfortunately, these techniques become impractical for systems with more than three degrees of freedom.

I have been developing an algorithm that may succeed in extracting periodic motions in nonlinear dynamical systems if these occur in long series of measurements. It is focused on finding a *scalar* nonlinear function of the dynamical variables such that it approaches a pure oscillation as much as possible. Since one deals with just a scalar, there is no need of introducing an arbitrary metric in phase space, an arbitrariness that is unavoidable in other methods, like the ones based on empirical orthogonal functions (EOF's) (Loéve, 1955) and principal oscillation patterns (POP's) (Hasselmann, 1988), that minimize the "amplitude" of an error vector in this space. The papers in (Auenbach et al., 1987) employ time-delayed coordinates so that the use of an Euclidean metric is justified.

Another important difference with respect to the above-mentioned methods is that ours tries to find a *subset* of variables that participate in a periodic motion while the others search for oscillations of the whole set of variables.

2. Search of oscillations as a variational problem

Suppose we are dealing with an N-dimensional dynamical system whose variables we denote by $v_1(t),...,v_N(t)$. By taking nonlinear combinations of these variables, we form a larger set of zero-mean variables $\{x_i(t), i = 1,...,n\}$ with $n > N$. We search for a complex linear combination $y(t)$ of the larger set

$$y(t) \equiv \sum_i^n c_i x_i(t) \, ,$$

(1)

such that the ratio

$$\langle (\frac{d\tilde{y}}{dt} + i\tilde{v}\tilde{y}) (\frac{dy}{dt} - ivy) \rangle / \langle \tilde{y}y \rangle \geq 0.$$

(2)

is a minimum and where we have introduced the complex 'eigenfrequency' $v \equiv \omega + i\gamma$; the tilde indicates complex conjugate and the pointed brackets stand for time-average. In words: By minimazing (2) we are finding the $y(t)$, a non-linear function of the original variables $\{v_i\}$, that comes as close as possible to a pure oscillation with frequency ω. Notice that it is essential to divide by $\langle \tilde{y}y \rangle$ in (2); in this way, we get rid of the amplitude of the oscillation and no metric is required in the space $\{x_i\}$ or in $\{v_i\}$. The quantity we minimaze is a scalar with dimensions of frequency and it is directly related to the inverse of the life-time of the oscillation.

The minimum of (2) is found by taking independent variations of (the real and imaginary parts of) the coefficients c_i *and* of the 'eigenfrequency' $\omega + i\gamma$. In this way, one obtains

$$\omega = - i\langle \tilde{y}\frac{dy}{dt} - y\frac{d\tilde{y}}{dt} \rangle / 2 \langle \tilde{y}y \rangle,$$

(3)

$$\gamma = - \langle \tilde{y}\frac{dy}{dt} + y\frac{d\tilde{y}}{dt} \rangle / 2\langle \tilde{y}y \rangle,$$

(4)

$$[D + 2\omega iA + 2\gamma R + |v|^2 V]\bar{c} = \mu^2 V\bar{c},$$

(5)

where $\bar{c}$ is a column matrix formed by the coefficients $\{c_i\}$, μ^2 is the value taken by the ratio (2) we minimize and the antisymmetric matrix A and the symmetric matrices D, R

and V are given by

$$A_{ij} = \frac{1}{2} \langle x_i \frac{dx_j}{dt} - x_j \frac{dx_i}{dt} \rangle, \tag{6}$$

$$D_{ij} = \langle \frac{dx_i}{dt} \frac{dx_j}{dt} \rangle, \tag{7}$$

$$R_{ij} = \frac{1}{2} \langle x_i \frac{dx_j}{dt} + x_j \frac{dx_i}{dt} \rangle, \tag{8}$$

$$V_{ij} = \langle x_i x_j \rangle, \qquad i,j = 1,...,n. \tag{9}$$

In a stationary state both the matrix R and γ vanish, therefore equation (5) recudes to

$$[D + 2i\omega A + \omega^2 V] \, \bar{c} = \mu^2 V \bar{c}, \tag{10}$$

while the value of ω associated with a minimal μ^2, confer eq. (3), expressed in terms of the matrices A, V and of the coefficients $\{c_i\}$, is

$$\omega = -i \frac{\bar{c}^* A \bar{c}}{\bar{c}^* V \bar{c}}, \tag{11}$$

where $\bar{c}^*$ is the Hermitian conjugate of $\bar{c}$. A possible solution is $\omega = 0$, in such a case $\bar{c}$ is real, otherwise the solutions occur in pairs, $\bar{c}$ and $\bar{c}$, corresponding to $\pm\omega$.

Recapitualing: the simultaneous solution of equations (11) and (10) leads to a set of complex 'modes' $y_k(t) = \Sigma_i \, c_i^k x_i(t)$ each one associated with a frequency ω_k and an average life-time $\mu_k^{-1} = \{\mu(\omega_k)\}^{-1}$. By taking nonlinear combinations of the original variables, one is able to construct periodic but non-purely-oscillatory motions; these motions are usually unstable, i.e., $\mu \neq 0$, but recurring. If the original variables describe spatial structures then those variables associated with the non-vanishing coefficients $\{c_i\}$ describe the spatial structure of time-dependent, recurring motions.

3. Method of solution

Equation (10) looks like a standard eigenvalue problem, but it is not since the parameter ω is itself a (homogeneous) function of the unknown coefficients $\bar{c} = \{c_i\}$, confer eq. (11). An iteration procedure could be a way of finding solutions since an analytic expression for the gradient of $\mu(\omega)$ can be derived from (10) (Pasmanter, 1993). It turns out to be more convenient and probably essential, to take first ω as a free parameter and to plot the eigenvalues μ_i^2 of (10) as functions of ω. The reason for doing this can be understood by looking at the results in Fig. 1a: Many of the extrema observed in this figure could be spurious in the sense that they may be associated with the crossing of two different eigenvalues, say $\mu_i(\omega) \, \mu_{i+1}(\omega)$, i.e., they may be associated with accidental degneracies of

(10). When such a crossing occurs, small errors may lead to a 'level repulsion' characterized by a close pair of a maximum and a minimum at the same value of ω; one can check whether the extrema are real or due to such a level-repulsion effect by looking at the behaviour of the corresponding *eigenvectors* in a neighbourhood of the crossing. This extrapolation may also be the most accurate way of determining the eigenvectors at the crossing.

4. An example

As a first check on the feasibility of the method, we considered the 'new Lorenz model' (Lorenz, 1984),

$$\frac{dv_1}{dt} = - v_2^2 - v_3^2 - a(v_1 - F), \tag{12}$$

$$\frac{dv_2}{dt} = v_1 v_2 - b v_1 v_3 - v_2 + G, \tag{13}$$

$$\frac{dv_3}{dt} = b v_1 v_2 + v_1 v_3 - v_3, \tag{14}$$

where v_1 is the intensity of the westerly jet which is assumed to be permanently in equilibrium with the poleward temperature gradient and a is its damping rate, v_2, v_3 are the sine and cosine phases of rescaled, large-scale (Rossby) waves, whose damping time has been chosen as the unit of time, F and G represent symmetric and asymmetric thermal forcing and b is related to the dragging of the waves by the jet. In the numerical simulations presented below, the same values used by Lorenz, $a = 1/4$, $G = 1$ and $b = 4$, were employed.

When $F = 6.40$, the motion is periodic and it is sufficient to include, besides the original variables $\{v_1,...,v_3\}$, no more than quadratic combinations and so to form an extended set of variables $\{x_i\}$, $i = 1,...,9$, in order ot find and to characterize with great precision the non-linear oscillation; see Fig. 1. I.e., in this case, $y(t)$ is just a second order polynomial in the original dynamical variables.

At $F = 6.0$ the motion is also purely periodic, however, the oscillation in terms of the original variables is more complicated and in order to achieve a precision comparable to the one in the previous example, we have to include also cubic combinations, i.e., the extended set $\{x_i\}$ contains a total of 19 variables, see Fig. 2.

If we include some quartic combinations, then also the first harmonic is identified with the same high precision. Since in both of these examples the motion is purely periodic, the minimum value of μ^2 can be made arbitrarily small by taking, if necessary, higher non-linear combinations of the original variables.

At $F = 8.00$, the motion is chaotic; taking only quadratic combinations still allows for the identification of periodic motion, however, even including all cubic combinations and some quartic ones to form a set of 23 variables is not enough to find a sinusoidal oscillation to the high precision of the previous examples, see Fig. 3.

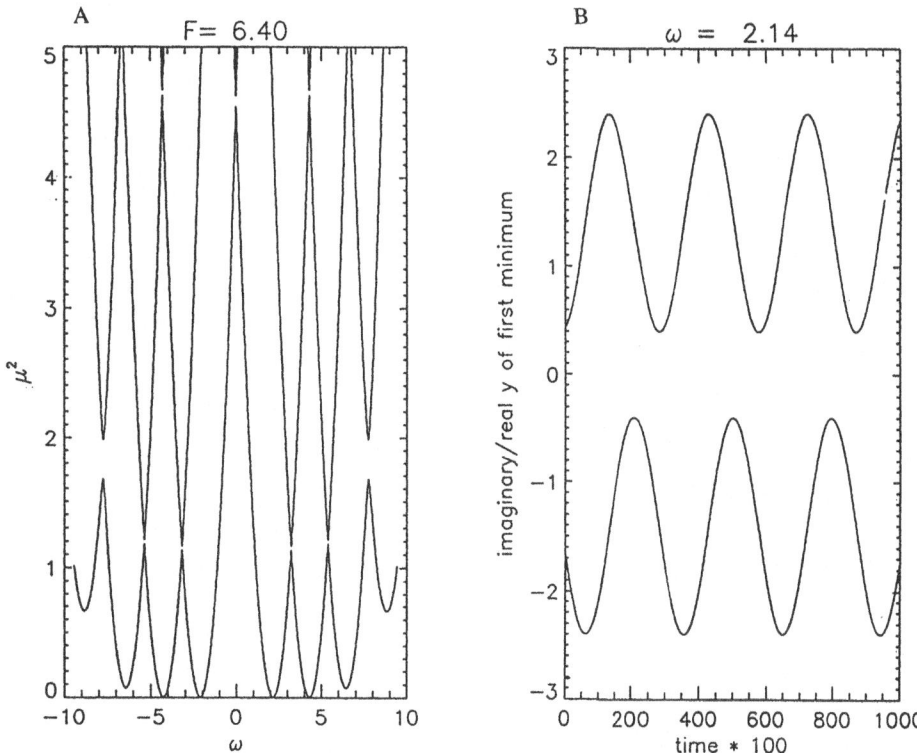

Figure 1. Results for the new Lorenz model (12) – (14) with $F = 6.40$ (periodic regime). The three original variables and their six quadratic combinations have been included in the analysis. a) The lowest eigenvalues μ^2 as a function of ω. b) Time evolution of the real and imaginary parts of $y_1(t)$, the eigenmode corresponding to the smallest value of μ^2.

It should be noticed that, e.g., the value of the 'resonant' frequency ω remains nearly constant when one extends the length of the series of the number of variables included in the extended set $\{x_i\}$. In the chaotic regime the life time of any periodic motion is necessarily finite, therefore, by taking higher order combinations, the minimum value of μ^2 can be reduced but it cannot be made arbitrarily small.

5. Comments and discussion

1) Once an unstable periodic motion has been detected and characterized by the method presented above, how could one make use of this information? The knowledge of the coefficients c_i unables one to develop an "index", i.e., a criterium (actually, a series of tests) in order to check whether the state (instantaneous configuration) of the system falls on the periodic orbit or not; if it does, then knowledge of the corresponding life-time makes some predictions possible. This is especially interesting when the life-time is longer than the inverse of the largest Lyapunov exponent.

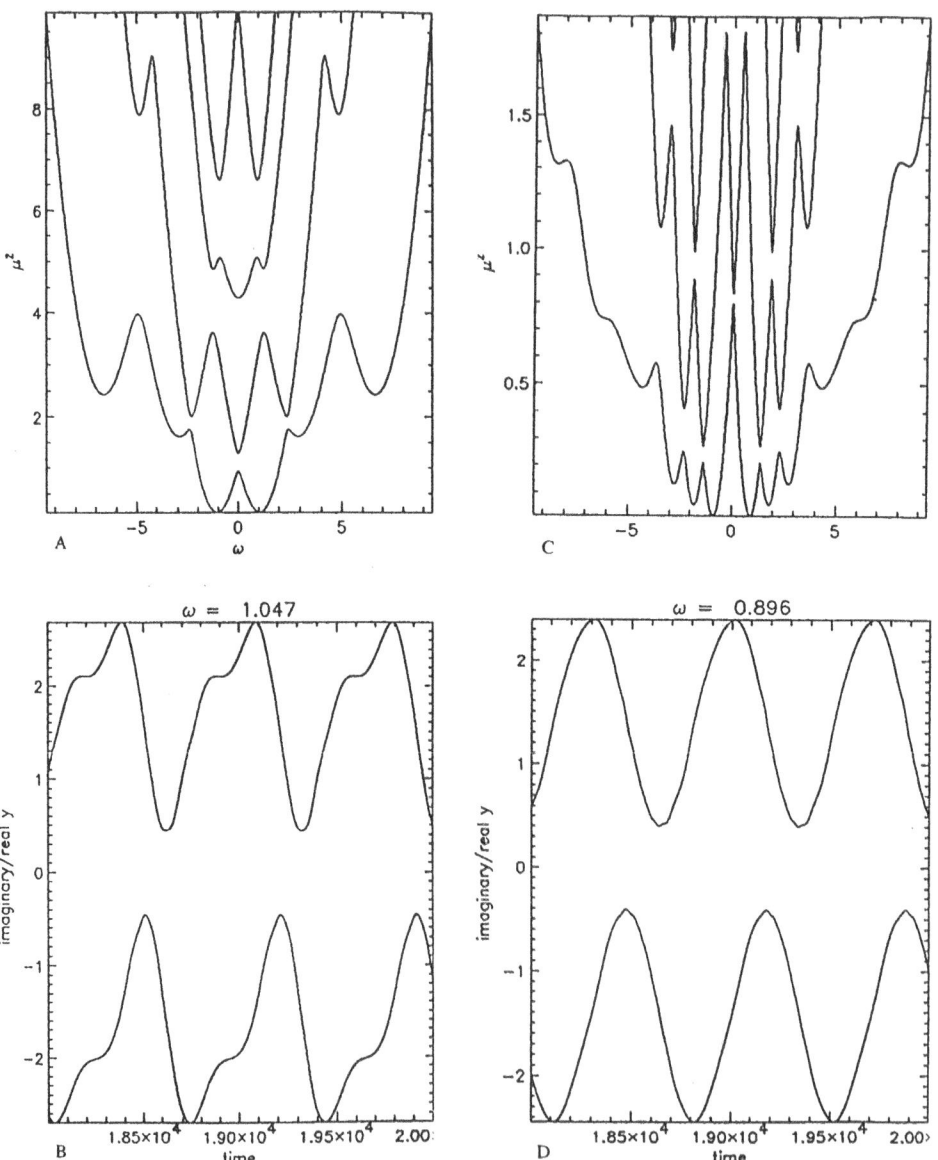

Figure 2. As in Fig. 1 but for $F = 6.00$ (also periodic regime). a) and b) Including only quadratic combinations c) and d) Including also all cubic combinations. Notice the change of scale on the vertical axis of a) and c).

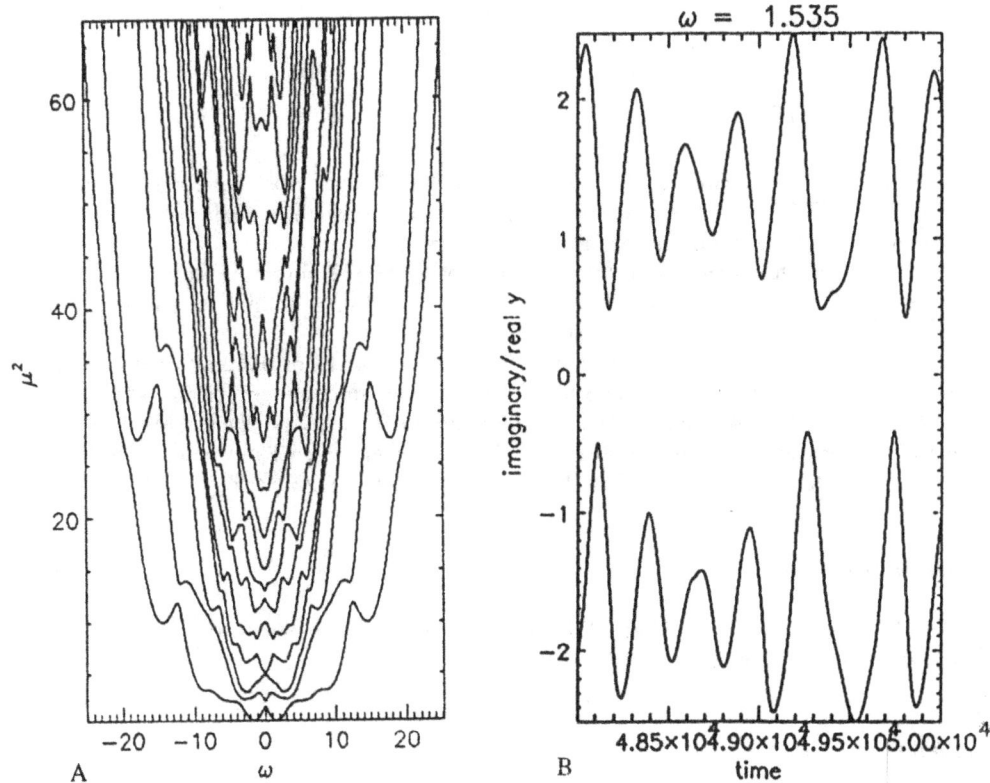

Figure 3. As in Figs. 1 and 2 but for $F = 8.00$, chaotic regime. Besides all cubic combinations, some quartic ones have been included; the total number of variables is 23.

2) The analysis presented above will, in general, lead to 'modes' that differ from those generated by a standard EOF analysis of the same data. This becomes evident when one realizes that, e.g., a low-amplitude EOF mode may correspond to a sharp spectral component or that a combination of EOF modes may be spectrally sharper than any isolated EOF mode, etc. From a practical point of view, when working with a system with a large number of degrees of freedom, one should first perform an EOF analysis and apply our method to the reduced system of the largest EOF-components. The effects that such a truncation has on eq. (10) have been analysed in terms of projection operators (Pasmanter, 1993).

3) In a stationary state, the matrix R, eq. (8), should vanish. In practice, however, it will never vanish exactly due to numerical errors and because of the finite temporal length of the data (remember that one expects the number of unstable periodic orbits with period T to grow, for large T, like $\exp(KT)$, where K is the Kolmogorov-Sinai entropy of the system (Ruelle, 1978). Therefore, it is always worthwhile checking how small the matrix R is in comparison to the matrix A.

4) In contraposition to the standard EOF or POP analysis, the minimum of (2) does not require the introduction of a metric (or inner product) in the space of the variables

$\{x_i(t), i = 1,...,n\}$. As a matter of fact, since the vectors $\{\bar{c}_i(\omega), i = 1,...,n\}$ solving eq. (10) are orthogonal with respect to V, i.e.,

$$\bar{c}_i^*(\omega) \cdot V \cdot \bar{c}_j(\omega) = C_i^2 \delta_{ij} \tag{15}$$

and since V is real, symmetric and positive defined, one could talk of the natural appearance of a metric V^{-1}. (Strictly speaking, this V^{-1} is an acceptable metric as long as we limit ourselves to linear transformations of the variables.)

5) It is worthwhile mentioning that it is also possible to express the sharpness of the minima in $\mu^2(\omega)$, i.e., the dimensionless quantity $d^2\mu/d\omega^2$, in terms of the statistical matrices D, A, V and of $\bar{c}_k$, ω, etc. (Pasmanter, 1993).

At present, we are testing the method in the chaotic regimes of the new Lorenz model and in a two-layer model of the atmosphere with 30 real, spectral coefficients. The results obtained up till now, compare favourably with, e.g., a POP analysis of the same data. The method has been also modified in order to make it focus on the spells of time when period motion is manifest (Pasmanter, 1993); the new version is yet to be tested.

Acknowledgement

I would like to thank Dr. K. Verbeek for the numerical calculations on the Lorenz model and for numerous stimulating conversations.

References

Kendall B.E., Schaffer W.M. and Tidd C.W. (1993), 'Transient periodicity in chaos', *Phys. Lett. A*, **117**, 13-20.

Ruelle D. (1978), 'Statistical Mechanics, Thermodynamic Formalism', Addison-Wesley, Reading MA, (1986), 'Resonances of chaotic dynamical systems', *Phys. Rev. Letters*, **56**, 405 and (1986), 'Locating resonances for axiom A dynamical systems', *J. Statist. Phys.*, **44**, 281.

Cvitanović P. (1988), 'Invariant measurement of strange sets in terms of cycles', *Phys. Rev. Letters*, **61**, 2729; Artuso R., Aurell E. and Cvitanović P. (1990), 'Recycling of strange sets: I (Cycle expansions) and II (Applications)', *Nonlinearity*, **3**, 325--359 and 360--386; Cvitanović P., Gaspard P. and Schreiber T. (1992), 'Investigation of the Lorentz gas in terms of periodic orbits', *Chaos*, **2**, 85.

Auerbach D., Cvitanović P., Eckmann J.-P., Gunaratne G.H. and Procaccia I. (1987), *Phys. Rev. Letters*, **58**, 2387; Biham O. and Wenzel W. (1989), 'Characterization of unstable periodic orbits in chaotic attractors and repellers', *Phys. Rev. Letters*, **63**, 819.

Loéve M.M. (1955), 'Probability Theory', van Nostrand, Princeton, NJ; Lumley J.L.(1970), 'Stochastic Tools in Turbulence', Academic Press, N.Y.

Hasselmann K. (1988), 'PIPs and POPs - A general formalism for the reduction of dynamical systems in terms of Principal Interaction Patterns and Principal Oscillation Patterns', *J. Geophys. Res.*, **93**, 11015-11021.

R.A. Pasmanter, to be published.

Lorenz E.N. (1984), 'Irregularity: a fundamental property of the atmosphere', *Tellus*, **36** A, 98.

COMPARISON STUDY OF THE RESPONSE OF THE CLIMATE SYSTEM TO MAJOR VOLCANIC ERUPTIONS AND EL NIÑO EVENTS

W. BÖHME

Kunersdorfer Str. 16, D-14473 Potsdam, Germany

Summary

This comparative study employs some special phase space representation of the coupled thermal behaviour of subsystems of the atmosphere. The course of the time derivatives or differences (especially the first ones) prove to be rather informative. Major volcanic eruptions (MVEs) are shown to activate a special dynamical state (a series of states) that, having some attractor properties, can be clearly traced up to the sixth year after the eruption with a high statistical significance. The cooling of the troposphere tends to start in the southern hemisphere and covers for the period from the first to the second year after the eruption both hemispheres with a small but very definite amount. From the second to the third year a remarkable warming of both hemispheres follows nearly without exception. There is a quasi-periodic fluctuation of a little less than three years the amplitude of which diminishes up to the sixth year only slowly but afterwards more rapidly. - The behaviour of the atmosphere in the same phase space representation in connection with El Niño events (ENEs) is rather different. The ENEs do not create a starting behaviour of the atmosphere that one might expect in view of the intensified latent heat flux into the atmosphere. There is rather a continuing fluctuation of the tropospheric temperature with a period of about 4 years. So the ENE is not simply the cause of this fluctuation, but is a manifestation of the fluctuation during a special phase. There is some evidence that a global cooling of the troposphere stimulates the fluctuation mentioned. - The method used is a powerful tool to detect weak modes in complex and/or noisy systems. The method will be applicable to all systems that can be thought to be composed of some relevant subsystems. As an example the method is applied to the relation between the global/hemispheric behaviour and the Middle European winters (MEWI) including the consequences of the global volcanic impact.

Keywords Nonlinear systems and subsystems, atmosphere, thermal behaviour, phase space representation, impact of major volcanic eruptions, El Niño events, weak modes in noisy systems

1. Phase space representations of the dynamics of a complex system and between subsystems

The behaviour (dynamics) of a complex system can in principle be derived from the time series of the time derivatives $d^n E/dt^n$ ($n = 0...N$, $N \rightarrow \infty$) or time differences D^n of a property E of the system (at one point) (e.g. Ruelle, 1981, Nicolis and Nicolis, 1984, Tsonis and Elsner, 1989). The same must be valid for diagnoses of interaction between parts of systems. Of course, it is not feasible to visualize an n-dimensional system with a large N. A graphic representation of the behaviour in the phase space is normally only possible in two or three dimensions. Which phase coordinates should preferably be used in diagnostic studies of system dynamics? A diagnosis should start with the frist derivative (n = 1) for the following reasons (Böhme, 1993): This derivative and the higher ones are not as much disturbed by systematic observational or measuring errors and large scale trends as the values of E itself ($n = 0$). But with increasing n the disturbing influence of noise and/or the divergence of the trajectories in systems with deterministic chaos tend to increase. The following sections will show that such a diagnostic approach proves to be rather useful.

2. On the thermal behaviour of the northern and southern hemisphere of the atmosphere

It is understandable, that the courses of the difference of the annual averages of the mean tropospheric temperatures T of the layer 850 to 300 hPa (in the following abbreviated to "tropospheric temperature") from the long-term mean for the northern (NH) and southern (SH) hemisphere are nearly identical (Fig. 1).

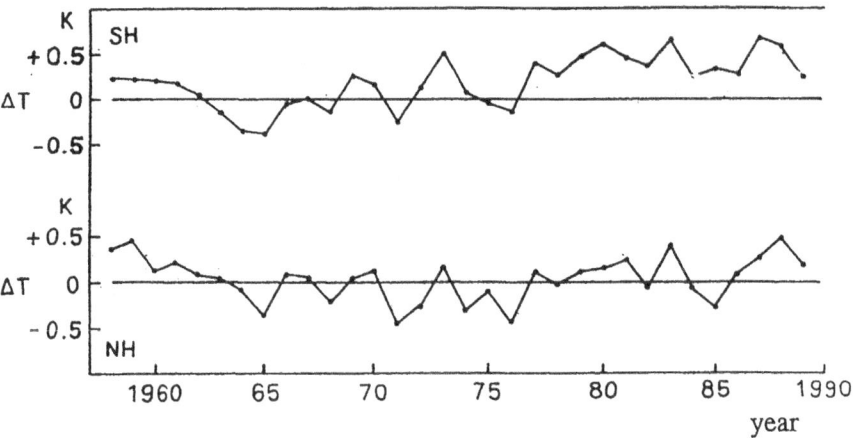

Figure 1. Deviation (ΔT) of the annual values of tropospheric mean temperature (800 - 300 hPa) from the 1958-1977 average (after Boden *et al.*, 1990) - NH resp. SH = northern resp. southern hemisphere.

This was illustrated by the author (Böhme, 1993) using the period (1958-1989) for which sufficiently homogeneous data have been published (Angell, 1988, Boden *et al.*, 1990). The interannual differences DT of the tropospheric (Tr) temperatures should group in the phase space representation DT (NH, Tr), DT (SH, Tr), being an intersection through the phase space, rather closely around the diagonal as a kind of attractor. But the display of the 31 data of the interannual differences DT in this phase plane (Fig. 2; in Fig. 2 to Fig. 4 the abbreviated denotations DT (NH) and DT (SH) are used) seems to show that the points are grouped around the broken curve.

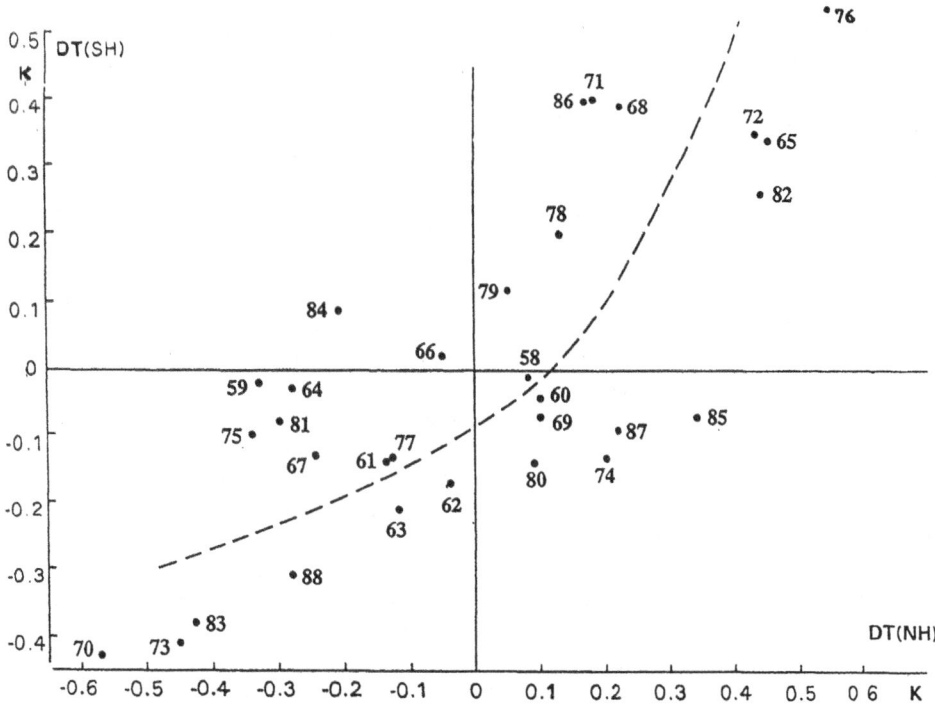

Figure 2. Distribution (phase plane) of the 31 values (*DT*) of interannual changes of mean NH and SH tropospheric temperature from 1958/59 through 1988/89. The distribution forms around the broken curve. 58 means the change from 1958 to 1959, etc.

This distribution is different from the expected diagonal distribution with a statistical significance of 80 %. The low significance is due to the small volume of the time series useable. More details and a discussion of the meaning of trajectories in this phase space are given in Böhme (1993). For instance, a movement along the curve in Fig. 2 from the upper right quadrant through the lower right to the lower left one (and this is a preferred trajectory) does mean, that the interannual change of the temperature of the NH troposphere (in this case a cooling) lags behind the respective change of the SH. The movement in the other direction is more rare. Further, there are jumps from the lower left to the upper right quadrant, meaning simultaneous changes from interannual cooling to interannual warming of both hemispheres (6 cases: 1964/65, 67/68, 70/71, 75/76, 77/78 and 81/82, to which we return in the following section), and from the upper right to the lower left one (3 cases). - A concentration of the phase points under similar conditions (e.g. the same number of years after similar events, as volcano eruptions) in special parts of the phase plane *DT* (NH, Tr), *DT* (SH, Tr) would mean, that similar conditions would create similar phase states of the atmosphere.

3. Impact of major volcanic eruptions

The atmospheric impact of major volcanic eruptions (MVEs) has been studied by many authors (quotations can be found, for instance, in Mass and Portman, 1989, as well as in Asaturov *et al.*, 1986; see also IPCC, 1990). It is quite sure that the eruptions enhance the stratospheric aerosol concentration and enlarge the optical depth of the atmosphere up to 30 %. That may lead to a reduction of the global average of the temperature in the lower troposphere by 0.1 to 0.5 K (maximum 1 K) for one or two years after the eruption. Our question is how the interannual changes of the troposphere of NH and SH arrange in the phase plane representation in connection with MVEs.

The 5 MVEs between 1958 and 1989 have been the eruptions of Agung (March 1963, 8°S, mass of aerosol ejected into the stratosphere about 15.2 Mt), Awu (August 1966, 4°N, 4.4 Mt), Fuego (October 1974, 15°N, 1.3 Mt) and El Chichón (April 1982, 17°N, 9.7 Mt) in the tropics and the eruption of St. Helens in May 1980 in temperate latitude (46°N, 0.59 Mt). The impact of the mentioned eruptions in the tropics has been thoroughly investigated by Mass and Portman (1989), but also the impact of St. Helens' eruption fits well into the picture we found (Fig. 3), although it is probable, that only a small part of the gases and aerosols may have reached the stratosphere, because the matter was mainly ejected in horizontal direction. The masses ejected into the stratosphere have been taken from Khmelevtsov (1987) but the data given by other authors (e.g. Asaturov *et al.*, 1986) show a spread to each side by a factor of 2 to 3.

A close inspection of daily weather maps and monthly reports (e.g. Deutscher Wetter- dienst, Seewetteramt, 1963 - 1982, Hydrometeorological Service of the USSR, 1974 -1982, National Oceanic and Atmospheric Administration, USA, 1963 - 1967, Deutscher Wetterdienst, 1970 - 1982) as well as of climatological results (e.g. Pogosjan, 1972, and Liebmann *et al.*, 1982) has been made with respect to the position of the eruption site relative to the intertropical convergence zone (ITCZ). The result was: The MVEs of Fuego 1974 and El Chichón 1982 clearly happened north of the ITCZ, that is, volcanic matter was ejected into the upper parts of the northern hemispheric Hadley cell and the strato- sphere above it. The same seems to be the case for the 1963 Agung eruption; during this period the ITCZ was situated more south than in the climatological average. In the case of Awu (1996; 3°20'N) the eruption site was far south of the main ITCZ, that was clearly situated north of 15°N. But there was a secondary convergence at about 1°N being the westerly wing of the southern pacific convergence zone, so that it may be well the case, that volcanic aerosol was immediately fed into the circulation of both hemispheres. A splitting of the ITCZ is a frequent behaviour, especially in the Indian and Pacific sector of the tropics between 70°E and 150°W (e.g. Pogosjan, 1972, and Gadgil *et al.*, 1990). Also in the American sector local convergences outside the main ITCZ are sometimes features of the general circulation, and so it was at the time of the eruptions of El Niño and Fuego. Summing up one can state that at least in 3 of the 4 MVEs in tropic latitudes the eruption site was situated north of the ITCZ; in one case the site was situated well south of the main ITCZ but a secondary branch of the ITCZ was situated just south of this site.

Figure 3 displays the parts of the phase plane DT (NH), DT (SH) that are covered by the interannual changes of the tropospheric temperatures (phase states) DT in each of the years $V + i$ ($i = -1$ to $i = +6$) for the volcanic eruptions ($i = 0$ means the year of the volcanic eruption).

The polygon V (afterwards called volcano polygone - for the year V) in Fig. 3a connects (so that we have a maximum area) the phase points for the five eruptions. That means also that this is the volcano composite of the phase state for the years V. The point

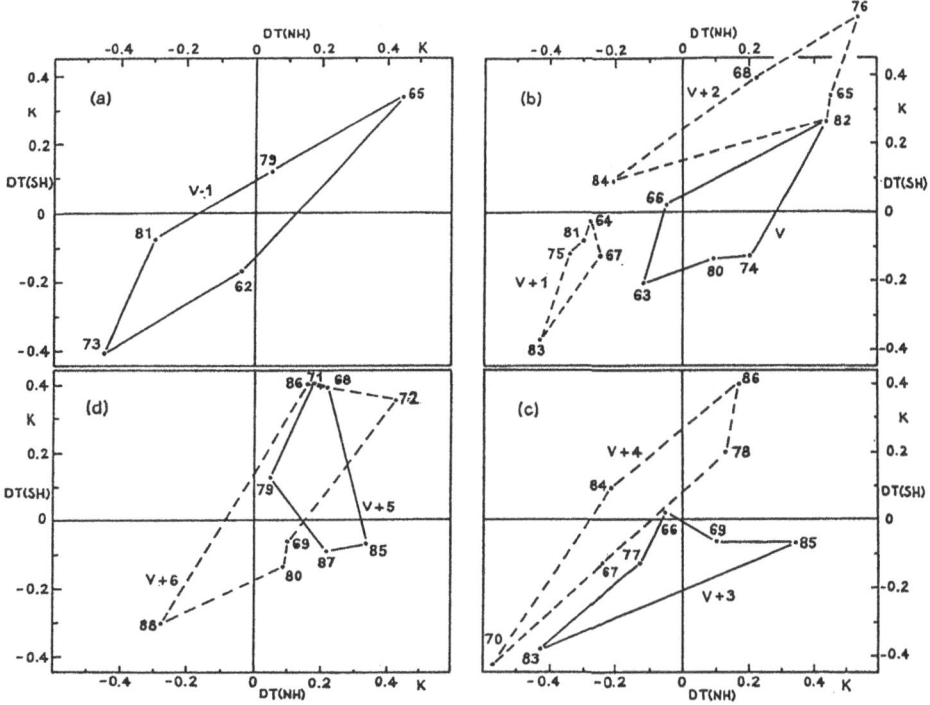

Figure 3. Areas $V + i$ covered from the volcano eruption composite for i from -1 to +6 in the phase plane DT (NH, Tr), DT (SH, Tr); V = year of the volcano eruption with DT (V) = T $(V + 1)$ - T (V). For further detail see text.

with the number 63, for instance, represents the change of the tropospheric temperatures DT (NH), DT (SH) from 1963 (eruption of Agung) to 1964. The polygon V - 1 (Fig. 3a), representing the interannual changes DT from the years before the eruptions $(V - 1)$ to the eruption years covers an area which is similar to the area covered by the 31 points (Fig. 2). That means that the atmospheric system behaves as usual. A similar result we find for almost all polygones $V - i$ $(i = 2$ to $5)$ (not presented here). The changes from the eruption years V to the following years $V + 1$ (polygon V in Fig. 3b) are already concentrated to a smaller area mainly with SH cooling and NH warming (see also Fig. 4). That gets clearer when one takes into account that in the polygon V (Fig. 3b) the phase point (19)82 represents the change of the hemispheric temperatures from the El Chichón eruption year 1982 to 1983 but may at the same time be mainly influenced, indeed, from the St. Helens eruption two years earlier (1980). This concentration process continues for $V + 1$ (Fig. 3b) (for the changes from the year $V + 1$ to the year $V + 2$). That means, the MVEs force the atmosphere to nearly one point of the state in the phase plane used. Only "1983" is a small exception, probably because the already mentioned impact of St. Helens' eruption continues to interact. In all 5 cases we have a cooling of both hemispheres with an average amount of 0.3 (NH) resp. 0.1 K (SH). Such an amount normally vanishes in the strong noise, in the dispersion of the points DT (NH), DT (NS) as given in Fig. 2. But the forcing exerted by the MVE appears quite clearly within this phase space representation. The centre of gravity of the areas covered by the phase states after the major eruptions from V to $V + 1$ obviously follows the broken curve in Fig. 2. The area $V + 2$ (changes from the year $V + 2$ to the year $V + 3$) is mainly situated in the upper left, the global warming quadrant. A jump from global cooling (area $V + 1$) has happened. 4 of the 6 jumps, we find in the period 1958 to 1988 (see section 2), belong to the trajectories from $V + 1$ to $V + 2$ of the impact of MVEs. The area $V + 2$ is larger than $V + 1$, but does not overlap the areas $V + 1$ and V. The enlargment means that in the different cases there are different disturbances or a dispersion of the trajectories in the case of deterministic chaos. The movement continues on a trajectory in a clockwise sense up to $V + 5$ (DT from the year $V + 5$ to $V + 6$) and after a little less than three years a similar position in the phase plane is reached. There is no overlapping of the successive phase areas $V + i$ up to $i = 4$. The amount of the areas has been nearly constant from $V + 2$ up to $V + 4$, but the phase points of $V + 5$ are again more concentrated. This may be a sign of the frequent tendency of a 4 to 5 years period in climate. The area of the polygon $V + 6$ is again larger, covers a great part of $V + 5$ and is similar to the figure of the ensemble of 31 points. That altogether means, that the changes DT from the sixth to the seventh years after the eruption are no longer influenced by the eruption.

Of course, as already mentioned, a quick succession of MVEs (as in the case of the St. Helens' eruption 1980 and that of El Chichón 1982) creates some interference. The interannual hemispheric temperature change from 1982 to 1983 (Fig. 3b) fits better polygone $V + 2$ than polygone V; that means that the interannual change of the hemi-spheric temperatures from 1982 to 1983 is more strongly influenced by the St. Helens' eruption 1980 than by the El Chichón eruption 1982. The extreme position, that 1982 does have in the polygon V, continues with 1983 in polygon $V + 1$ and so on up to 1986 in polygon $V + 4$ (Fig. 3c), but with a gradually diminishing degree as one might expect it. - The interference of two MVEs three years apart (1963 and 1966 in polygon V) is -as one may see from Fig. 3- not as pronounced. This may be also supported by the fact mentioned above that phase states after MVEs have a tendency to be repeated after a little less than 3 years.

There may be also some interaction between impacts of MVE and of El Niño events. The El Niño events within the period 1958 to 1985 are mentioned in section 4 (1) here. One may note, that in two cases an El Niño event happened within one year after an MVE (more specific: in the northern hemisphere winter after the year of the MVE). These MVEs are the Agung eruption 1966 and the El Chichón eruption 1982. In Fig. 3 the interannual hemispheric temperature changes, related to these two MVE (represented by the phase points 63 and 82 at the polygon V, the points 64 and 83 at $V + 1$ and so on up to the points 68 and 87 at $V + 5$) show no distinct, specific behaviour within these polygons besides some already mentioned (see above) extreme positions of the 82...86 series which seem rather connected to the interference of the impact of the El Chichón eruption with that of the St. Helens' eruption 1980.

Thus altogether -and notwithstanding the mentioned interferences of the impacts of MVEs and with impacts of El Niño events- Fig. 3 and its discussion in section 3 (3) shows that the ensemble or composite of the states of the free troposphere (represented by the interannual changes of the hemisphere temperatures) after the MVEs maintains a clear, individual character (despite of the smallness of the signal relative to the noise) up to the sixth year. This is much longer as normally mentioned in the literature (e.g. Mass and Portman, 1989, Asaturov et al., 1986).

The statistical significance of the signal found was tested by the application of a Monte Carlo method (Böhme, 1993). The probability that such a frequent non-overlapping of successive areas happens by chance is much lower than 0.1 %.

I would like to underline, that the MVEs lead to a behaviour of the interannual change of the NH and SH tropospheric temperatures, which is not so much dependent of such properties of the eruption as its intensity (provided it is a MVE), its geographical latitude, and the season of it. This seems also to concern the position of the eruption relative to the ITCZ; yet, we have to keep in mind, that nearly all MVEs in the period 1958 to 1989 happened north of the ITCZ; even in the case of the Awu eruption 1966, when the main ITCZ was north of the eruption site, a share of the volcanic aerosol may have been fed immediately in the northern hemispheric circulation (see 3(2)). Unfortunately, there was after 1958 no MVE in such southern latitudes, that the matter would have been ejected immediately only into the circulation of the southern hemisphere. So, for the time being, the statement could only be made for MVEs by which a substantial share of the volcanic matter is immediately ejected into the northern hemisphere circulation. - The rather uniform behaviour for these cases in the phase plane used feeds the presumption that the state trajectory of the climate system after such MVEs crosses a bifurcation in a way which independently of the special volcanic forcing leads to a rather uniform reaction of the climate system. There seems to be a special attractor in the climate system, which is activated by such MVEs, and which may be named a volcanic (volcanically excited) attractor of the climate system. Further: The attaching of a single case to the ensemble obviously says only, that this case stays within the ensemble, but not, where this case exactly lies one year later in the phase plane used. This is a typical behaviour when a strange attractor exists.

The impact character of the atmospheric response is also clearly displayed by the composite time series DT (SH, Tr) and DT (NH, Tr) (Fig. 4), especially that of the SH (Fig. 4b).

Unfortunately the time series of tropospheric temperatures are rather short and prevent to check whether the behaviour found may be stable over a longer history of the atmosphere. Of course, there are time series (e.g. Boden et al., 1990) of hemispheric and global mean "surface temperatures" (derived from observations of meteorological surface stations) for more than 100 years. But an investigation into the behaviour of the surface (Sf) temperature for the same period as used above (1958-1989) for the troposhere showed, that the signal is weaker; the phenomenon of non-overlapping of successive ensemble states is nearly not present (compare Fig. 5 with Fig. 3b). This is not astonishing: the radiative forcing of the MVEs will mainly happen in the troposphere and stratosphere, and the heterogeneity of the earth surface creates small-scale noise, the importance of which diminishes with increasing elevation. - The presentation (Fig. 6) of the trajectories of the average phase state of the volcano composite in the DT (NH), DT (SH) plane for the tropospheric (Tr) temperature and for the surface (Sf) layer shows, that their is a similar behaviour.

In both regions of the atmosphere we find a clear starting (impact) behaviour followed by a clockwise rotation with a period of about 3 years in the phase plane used; this rotation breaks down after 6 (troposphere) respectively 4 (surface layer) years. So, there remains some hope to exploit longer time series. Some further results are given in section 5(3).

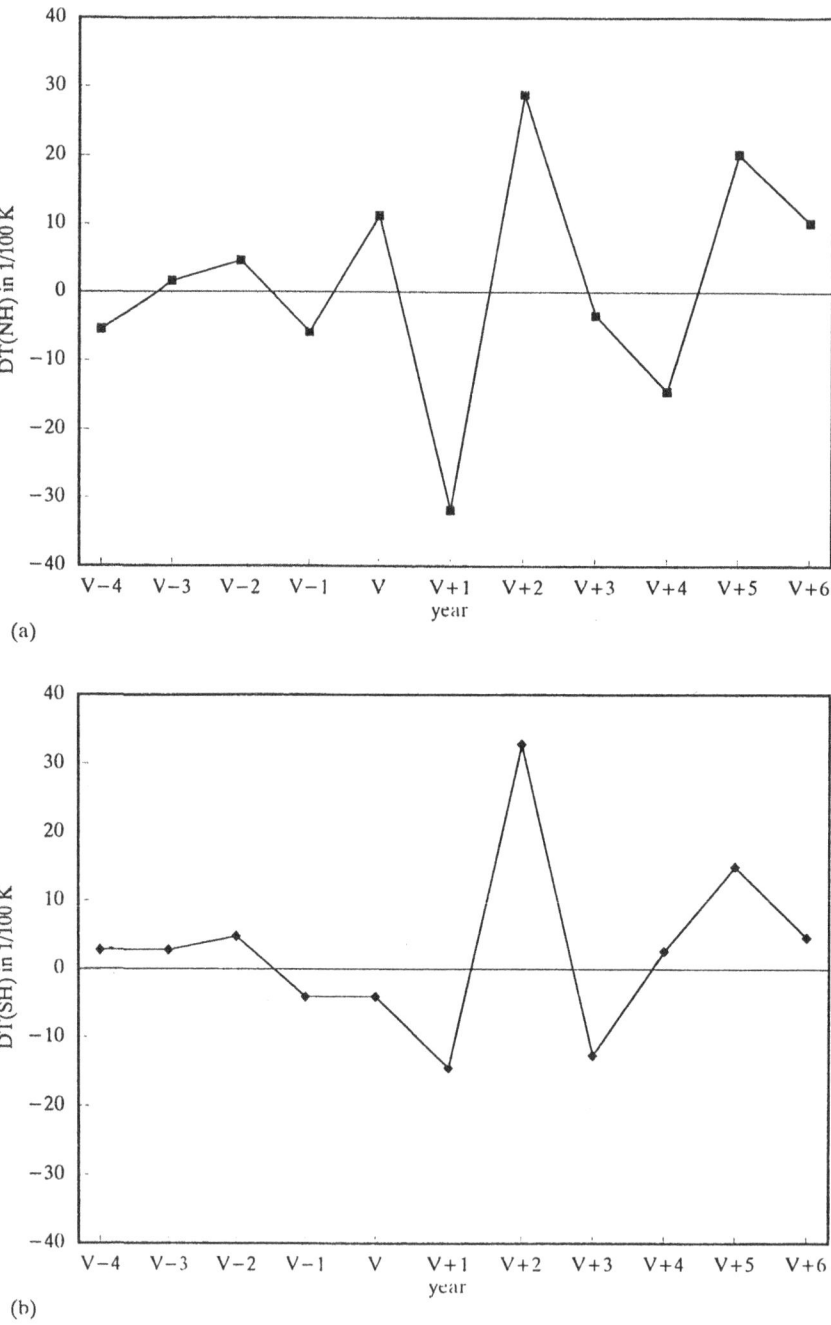

Figure 4. Volcano eruption composite average of interannual change of tropospheric temperature *DT* (a) NH, (b) SH for the years V - 4 to V + 6. (DT (V + 1) = T (V + 2) - T (V + 1), etc.).

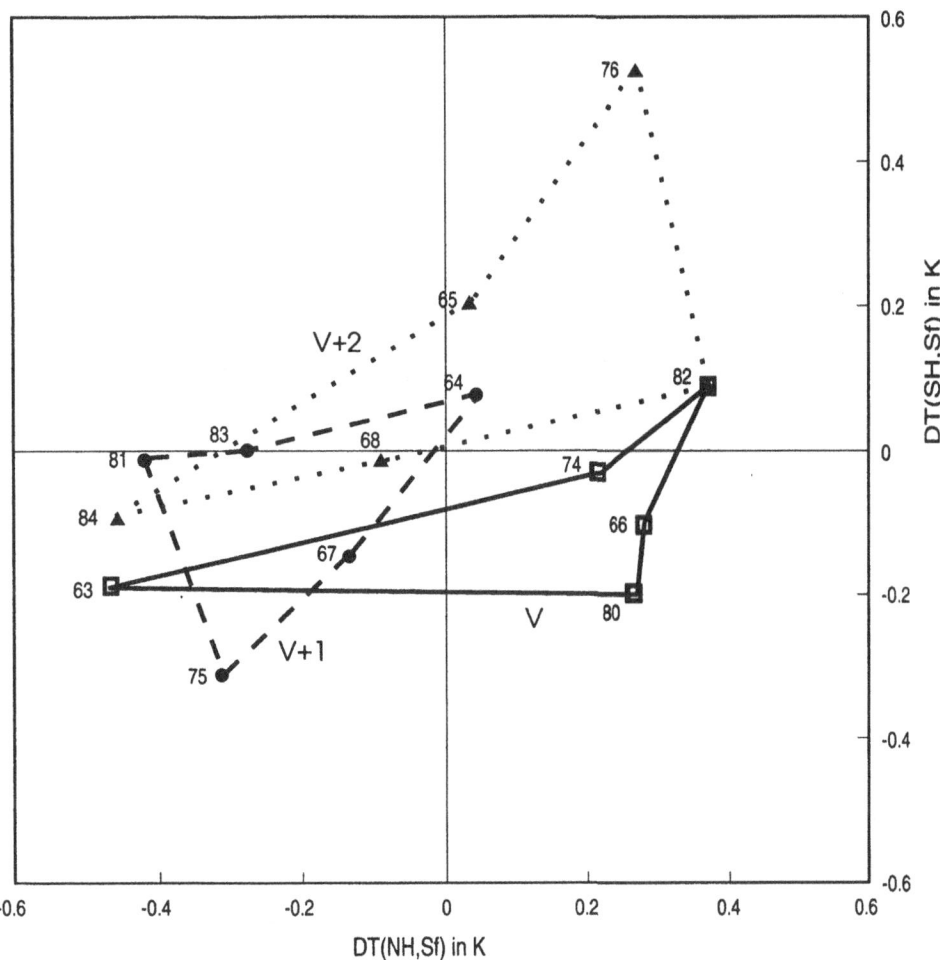

Figure 5. Area $V + i$ covered from the volcano eruption composite for i from 0 to +2 in the phase plane DT (NH, Sf), DT (SH, Sf).

4. El Niño events and global/hemispheric behaviour of the atmosphere

The El Niño event (ENE) is another transient phenomenon related to atmosphere and climate. A multitude of comprehensive studies has been published, and a composite method has found wide application (e.g. Rasmusson and Carpenter, 1982, Philander, 1990). Here we apply the same phase space representation method to the ENEs as used for the study of the impact of MVEs. The same time series for the hemispheric temperatures as above were used to study the relation between climate fluctuations and the ENEs. The El Niño years within the period 1958 to 1985 were taken from Kiladis and Diaz (1989):

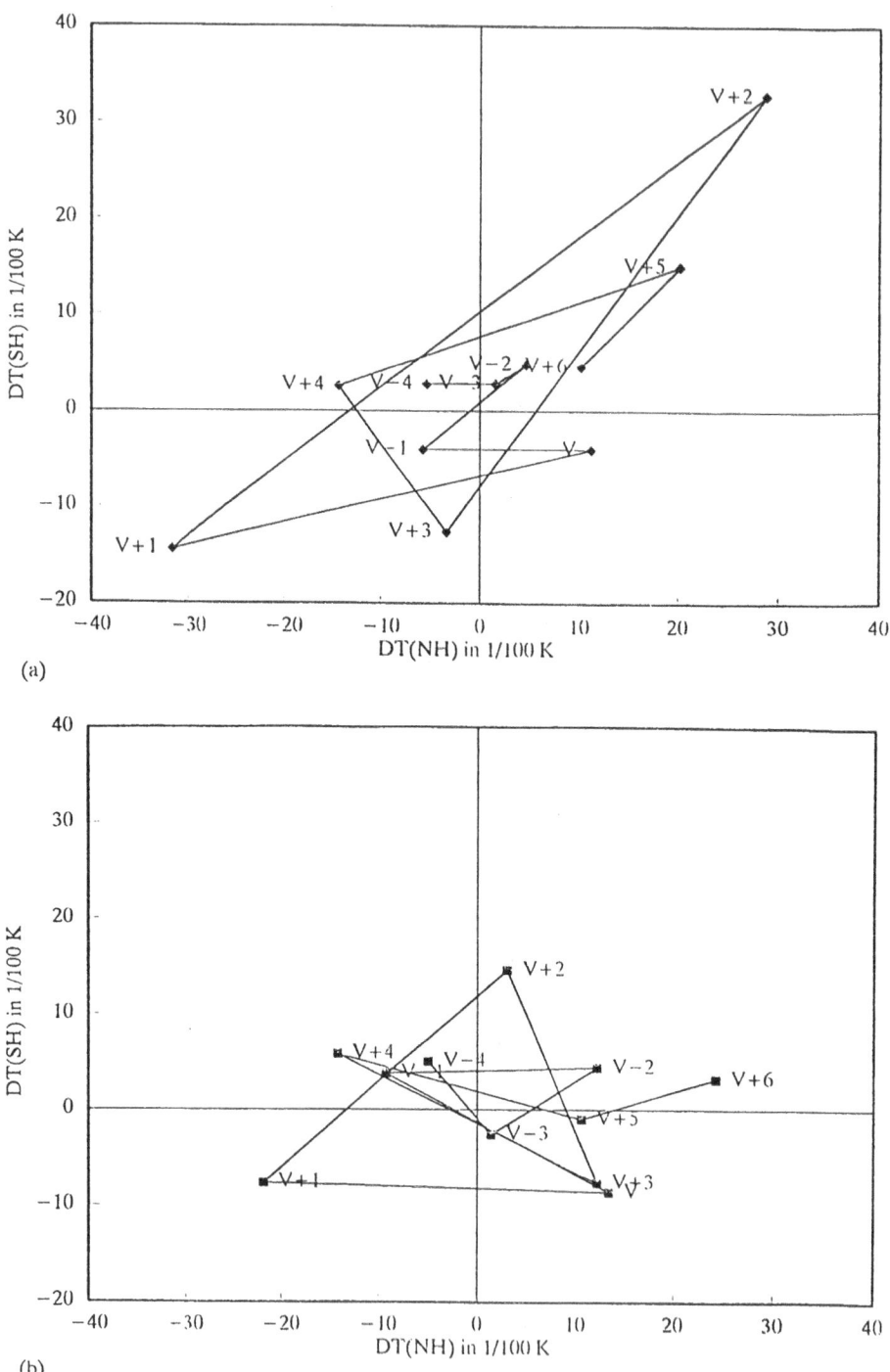

Figure 6. Trajectory of volcano eruption composite average in *DT* (NH), *DT* (SH) plane for (a) tropospheric and (b) surface temperature.

These are 1963, 1965, 1969, 1972, 1976 and 1982. Normally, during an ENE the positive temperature anomaly of the eastern equatorial Pacific culminates around December/January; the years mentioned concern the years, to which the December belongs.

One could suspect, due to the intensified latent heat flux into the atmosphere during El Niño episodes, that one would find a similar dynamic behaviour of the atmosphere as in the case of the volcano impact.

But this is not the case: This is already visible by Fig. 7, which represents the trajectories of the ENE composite averages of the interannual changes of the NH and SH tropospheric temperatures, thus being the counterpart to Fig. 6a. Here we do not display the thermal behaviour of the surface layer: it is, with a lower amplitude, similiar to that of the troposphere but more erratic. The trajectories in Fig. 7 do not show the quasi-circular form as in the case of the MVEs. They take a course along the diagonal of the phase plane, which means that NH and SH interannual temperature changes show nearly an identical course. Furthermore, there are great interannual changes from the fourth to the third year before the event (phase point w - 4) and from the last but one to the last year before theevent (point w - 2). So we do not having a starting behaviour as in the case of the MVE but rather a continuing fluctuation (as it is the Southern Ocsillation to which the El Niño events are related) with a period of about 4 years and a possible amplification around the ENE.

Such a course is also displayed (Fig. 8) by the composite average time series of the interannual changes of the NH tropospheric and surface temperature (the SH time series are very similar).

A pronounced global tropospheric cooling lasting two years (w - 3, w - 2) is remarkable. That may act as a kind of a trigger for the El Niño and the global atmospheric warming event w. Further, this cooling starts first in the troposphere and one year later at the surface. This is in accordance with findings of Graf (1984, 1986, 1989).

Having such a symmetrical and continuing fluctuation it is not astonishing, that there are no signs for non-overlapping of successive phase areas in connection with ENEs. So we can refrain from bringing relevant figures here. Only the overlapping between the areas of polygones w and w + 1 is small. In the contrary, the areas of the cooling (w - 3 and w - 2) are nearly identical, saying that the cooling effect of the global troposphere covers two years. But we will show (Fig. 9) the time series of the amount of the area surrounded from the polygone of the ENE composite, which supports the presumption already made that a global cooling triggers or stimulates the ENE. We see, that the aerea is smallest for w - 3 and w - 2.

In the further course of the El Niño process such low values are not reached again. That means, that there is a two year long lasting process, which forces the phase state of the troposphere in the second and last but one year befor each ENE into a small part of the total available phase plane.

The identification of such a process needs further study. One possibility is, that a warmer than normal troposphere does not favour convection which would continue to bring heat into the atmosphere. So the free troposphere starts to cool. At the same time more solar energy begins to be stored in the tropic oceans. When this process continues for a long enough time (this may be two years) then the stratification of the troposphere against the state of the earth surface becomes less stable and enough energy will be accumulated especially in the (tropic) ocean, so that an El Niño process could start as the probably most efficient process to bring heat stored into the atmosphere.

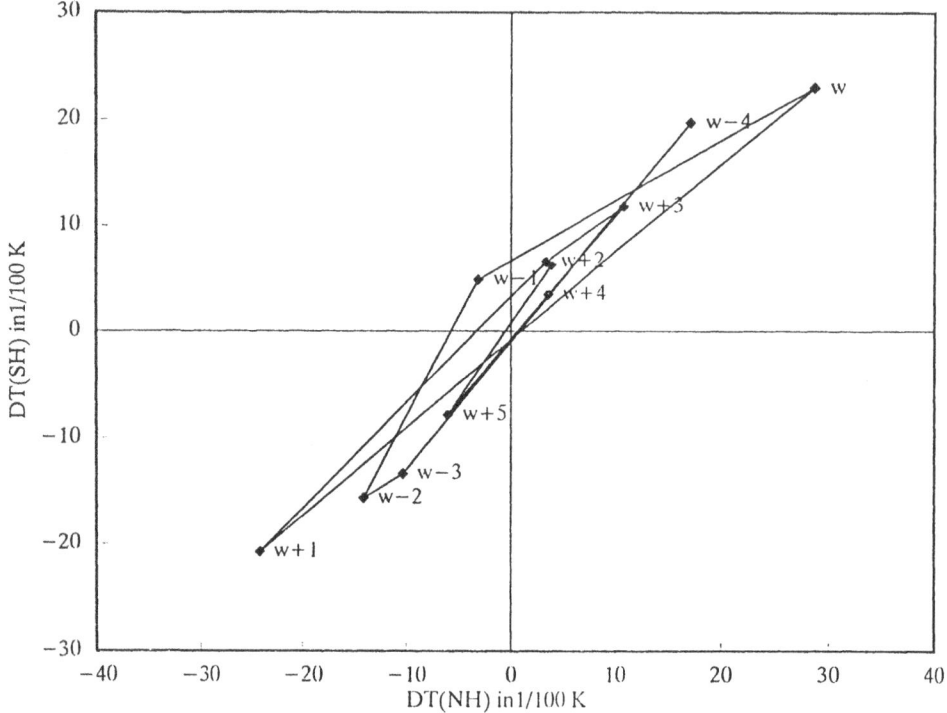

Figure 7. as Fig. 6 (a), but for the El Niño composite.

5. Application of the method to smaller subsystems

The question arises whether the approach used here for the investigation of the global system is also applicable for smaller subsystems. This is really the case. A small selection of interesting results shall illustrate this, the power of this approach, and possible problems. We demonstrate this by results of an investigation of the relation between interannual changes of Middle European winters (MEWI) to those of hemispheric or global mean temperatures and to the MVEs mentioned in section 3. As representative for MEWI the so-called Baur series (Linke and Baur, 1962, 1970; Deutscher Wetterdienst, 1969 and following years) of the deviation of the monthly mean Middle European temperature (as the average of the deviations of the monthly mean temperatures of De Bilt, Potsdam, Basel and Vienna) from the long-term average have been used to derive the deviation of the winter mean temperature (mean of Dezember of the year i and of January and February of year $i + 1$) for the winter i. Further, as a measure for the heterogeneity (HET) of each winter the standard deviation of the temperature of the three months has been applied.

The phase space representation in Fig. 10 using the phase plane DT (NH, Tr), DT (MEWI) shows the distribution of the phase points (representing the interannual changes

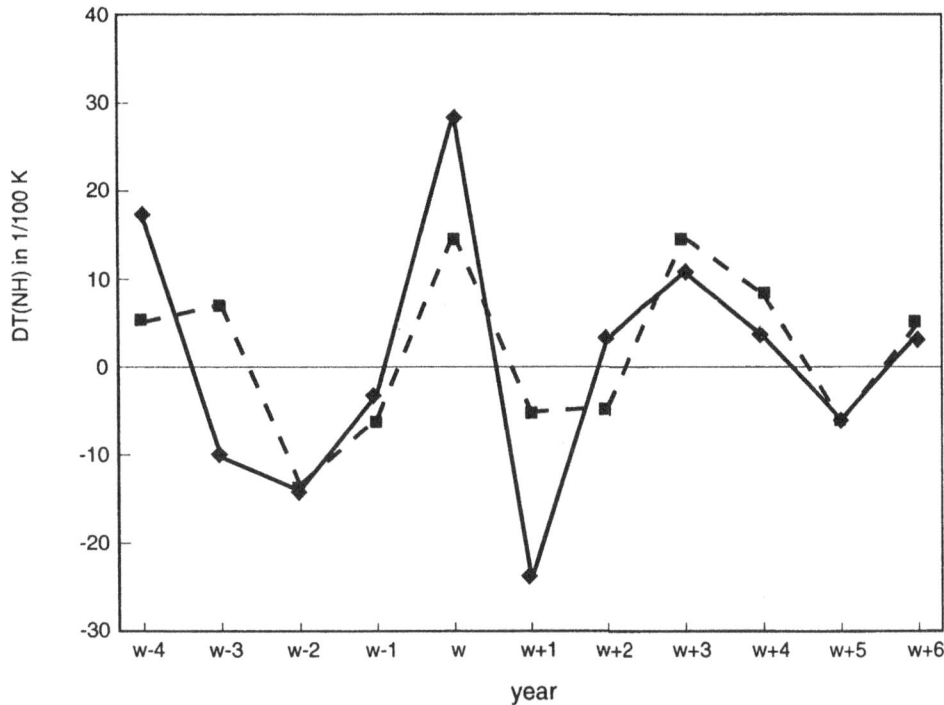

Figure 8. El Niño composite average of interannual change of NH tropospheric (♦) and surface (■) temperature.

of the mean NH tropospheric and the MEWI temperature respectively). This distribution is very peculiar.

Phase points cover two regions (a higher and a lower region or "band") which are seperated by a bandlike gap (in the case of longer time series this gap may be replaced by a band with a very low density of phase points). The lower band is occupied somewhat more frequently than the higher one. The bands seem to be the traces of two attractor branches or sections. This may also mean, that there are in the phase space two quasi-stable regions ("levels"). Trajectories (not given in Fig. 10, yet easily to be visualized between the points) remain for some time in one of the levels (e.g. 1966-67-68, this says that the interannual changes of T (MEWI) for 66 to 67, 67 to 68 and 68 to 69 have continued to be negative). But there are also many jumps from one level to the other, for instance for 68-69, which means that interannual MEWI cooling from 68 to 69 is replaced by MEWI warming from 69 to 70. It is clear, that such jumps must occur, when the MEWI climate shall be stationary, having only transient changes. It is visible that the lower level includes a small region of interannual MEWI warming, but this may not be sufficient to compensate the prevailing cooling, when the state of the systems stays continuously in the lower level. An inspection into the phase plan DT (NH, Sf), DT (MEWI) shows that, when passing over from the troposphere (Tr) to the surface (Sf), the gap becomes smaller and gets bridged near the origin (0, 0) of this phase plan (no figure given). It is clear, that this interesting situation and its explanation needs further study.

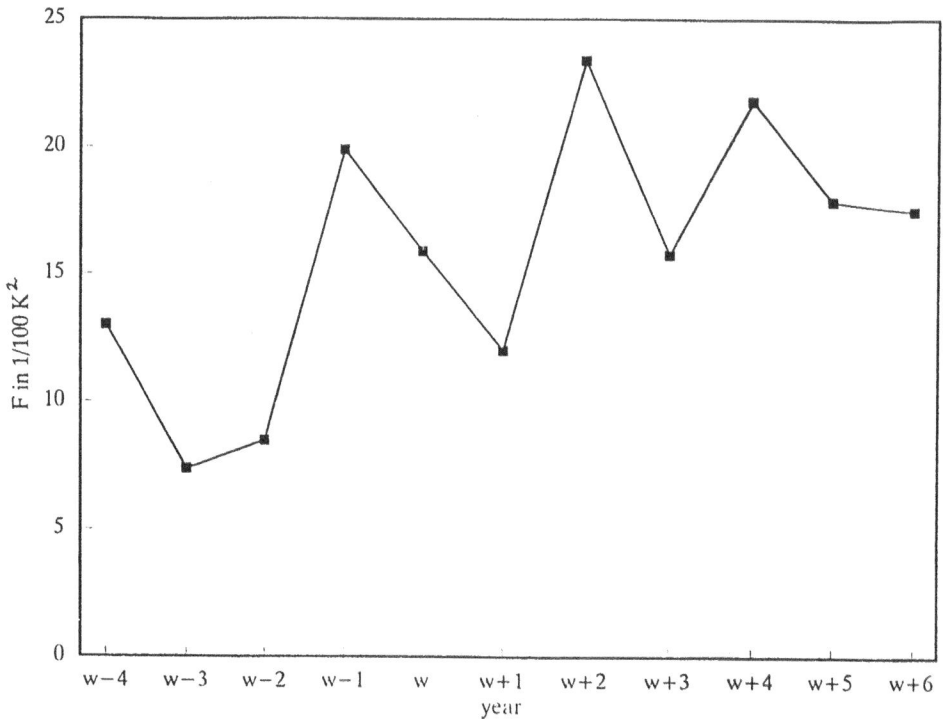

Figure 9. Amount F of El Niño composite polygon area of the interannual changes of NH and SH tropospheric temperature (*w* concerns the change from the year *w* to the year *w* + 1).

There are definite signs that the impact of the MVEs is present also in the behaviour of the Middle European winter (MEWI). In the phase plane DT (NH, Tr), DT (MEWI) the polygons drawn for the different years before and after MVEs show similar properties as the respective polygons in Fig. 3. There are also some non-overlappings of succesive ($V + i$, $V + i + 1$) polygons but not as pronounced as in the case of Fig. 3. Fig. 11a may serve as an example: The area covered by the polygon V does not overlap with that of $V + 1$. The same is true for $V + 1$ and $V + 2$.

The secondary maximum concentration of the phase points for $V + 5$ in the phase plane DT (NH), DT (SH) (Fig. 3d) is greatly enhanced in the phase plane DT (NH, Tr), DT (MEWI) (Fig. 11b) and the polygone $V + 5$ is situated totally in the lower level. This means: From the 5th to the 6th year after the eruption in all 5 cases a small warming of the NH troposphere (between +0.05 and +0.34) is connected with a relatively small change of the MEWI temperature around -0.5 K (between +0.3 K and -1.5 K).

A similar, but less pronounced behaviour is seen, when one uses the interannual temperature changes of the NH surface layer (that means the application of the phase plane DT (NH, Sf), DT (MEWI) instead of the NH tropospheric one. In this case we found, in contrast to the situation mentioned in 3 (7) an access to variables (i.e. hemispheric surface and MEWI temperatures) on which we have long time series. This is underlined by a comparison (Fig. 12) of the course of the composite averages of the impact of the volcanoes on the MEWI interannual temperatur changes (DT (MEWI)) between the MVEs from 1958 to 1988 (5 cases) and the MVEs from 1879 to 1957 (11 cases, i.e. 1883, 1888, 1902, 07, 12, 18, 19, 32, 45, 47 and 1956, taken from a list given by Khmelevtsov (1987) as far as the mass of matter ejected into the stratosphere was estimated to be equal or larger than 2,5 Mt).

There is a very good similarity, which is characterized by a correlation coefficient for the values of DT (MEWI) of 0,799. This means that there is a statistical significance for this correlation with an error probability lower than 0.3 %. This similarity of the volcanic impact for independent data sets underlines the reality of the volcanic impact additional to the result we mentioned in section 3(5). The strongest signal is found with $V - 1$. This does not necessarily mean that the signal happens before the volcanic eruption,

for DT(MEWI) at $V - 1$ gives the change of the Middle European winter temperature from the winter before the eruption ($V - 1$) to the winter (V) after the eruption (average of the temperature of December of year V and temperatures of January and February of year $V +$ 1); however, this means that the impact of a MVE to the MEWI temperature needs in the average half a year. With respect to the Pinatubo eruption (1991) the interannual changes of the winter temperature of Middle Europe from the winter 1990/91 to the winter 91/92 (this corresponds to $V - 1$) and from 91/92 to winter 92/93 (this corresponds to V) with +2.00 K resp. -1.20 K fit very well into the composite values.

However, the high correlation of the two independent composite averages within the 12 winters used in Fig. 12 does not necessarily mean that there is a significant correlation for all shorter periods of this 12 years interval. Of cause, there is, due to the very high significance of the eruption influence to the global atmosphere with its two hemispheres (see section 3(5)), no doubt for the period $V - 1$ (this represents the interannual change from the winter $V - 1$ to V) up to $V + 5$. Indeed, for the 7 winters from $V - 1$ to $V + 5$ we find a correlation coefficient of 0.806; that is a statistical significance for this low number of winters with an error probability of less than 5 %. For the period afterwards (for $V + 6$ to $V + 8$) there may be the possibility, that the fluctuation of the atmospheric behaviour continues at least sometimes with gradually diminishing amplitude. For the period before (for $V - 4$ to $V - 2$) there seems to be also some parallel course of the two independent composite averages. At least in $V - 2$ and $V - 3$ both composite averages have only small amounts, so that the similar course may have happened for this interval by chance. To find a definite conclusion more study of this problem is necessary. If it could be shown, that this similar behaviour for $V - 4$ to $V - 2$ is not caused by chance, it would mean that the atmospheric dynamics or parts of it, which influence the Middle European winter, may be able to trigger volcanic eruptions. This odd result, however, would support a sometimes raised opinion (e.g. Rampino et al., 1979), that the atmosphere and its dynamics may trigger the release of tectonic energy (earthquakes and even volcanic eruptions). Of cause, there is no doubt, that the atmosphere will have certain influence to the "solid" earth including its rotation (e.g. Volland, 1989), but a triggering of volcanic eruptions is a much more difficult and doubtful question.

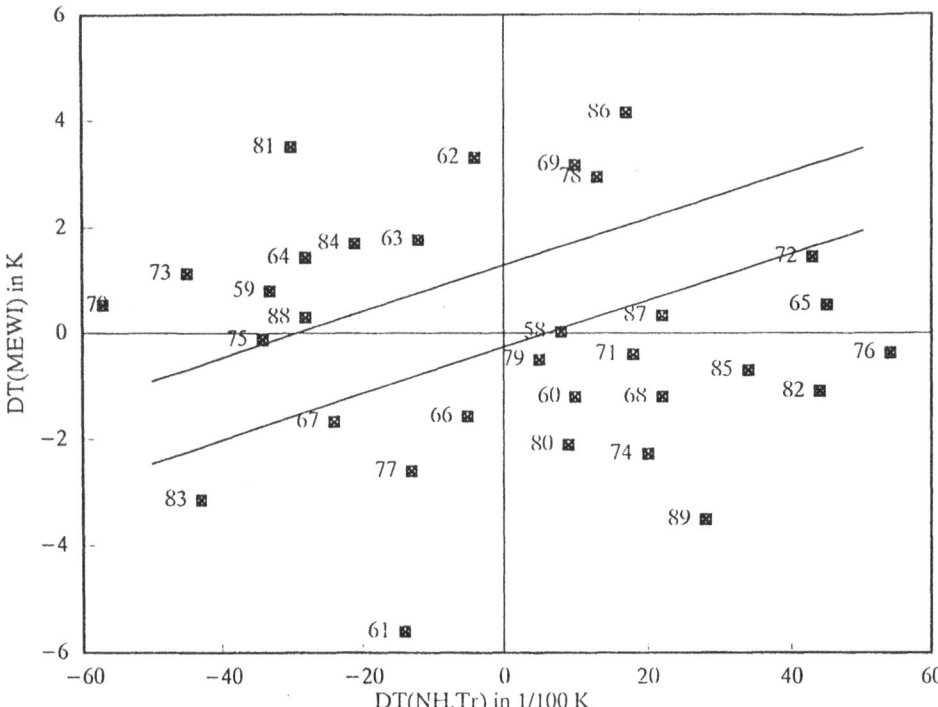

Figure 10. Interannual changes of the Middle European winter temperature *DT* (WEMI) relative to these of the NH tropospheric temperature *DT* (NH, Tr). 58 means the change from the year 1958 to 1959, etc.

When investigating the MEWI temperatures I saw, that the hetereogeneity (HET) of the winters (the standard variation of the three winter month in each winter) undergoes large variations, and this in a way, that a great majority of the interannual changes of HET has the opposite sign of those of the NH tropospheric temperature: warming of the NH troposphere is normally connected with diminishing HET, and cooling nearly in all cases with increase of HET (Fig. 13).

The correlation coefficient has a value of - 0.56. This means a statistical significance of this correlation with an error probability of 0.1 %. The behaviour of the volcano composite for $V + 1$ ($i = 0$, + 1 and + 5) is indicated in Fig. 13, too. With ongoing studies a qualitatively similar behaviour is also being found in the completely "regional" phase space D (HET), DT (MEWI) that uses only data of Middle Europe. This similarity may be mainly an expression of the validity of the first sentence in section 1.

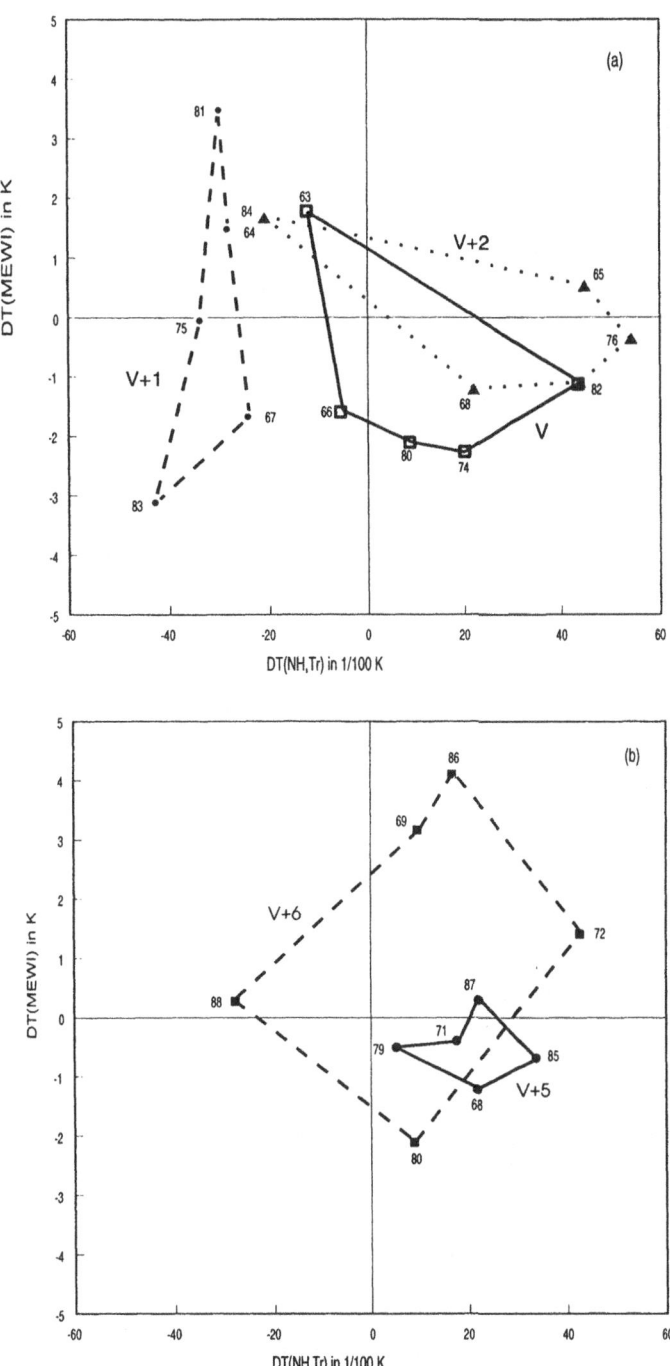

Figure 11. Areas $V + i$ covered from the volcano eruption composite in (a) for $i = 0$, $+1$, $+2$, and in (b) for $i = +5$ and $i = +6$ in phase plane DT (NH, Tr), DT (MEWI).

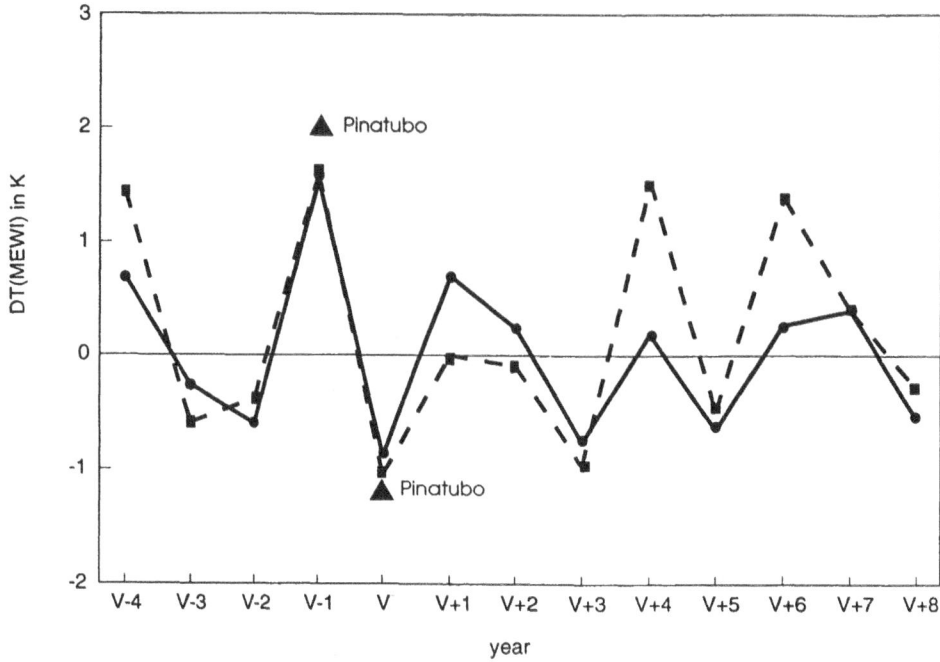

Figure 12. Volcano eruption composite average of interannual change of Middle Europe-an winter temperature for 5 cases 1958-88 (broken line) and for 11 cases 1879-1957 (full line). ▲ = related to Pinatubo eruption 1991. (*DT* (MEWI, *V*) means the winter tempera-ture change from the eruption year *V* to the year *V* + 1, etc.).

6. Conclusions

The use of the first time derivatives or time differences of properties of complex dynamical systems, and especially coupled subsystems, as phase space coordinates proves to be an effective tool for the diagnostic study of such systems. They are appropriate to detect and trace weak modes in complex and noisy systems.

This is demonstrated at the behaviour of the atmosphere as an essential part of the climate system. Although the course of the annual averages of the tropospheric tempera-ture of NH and SH is very similar, one can find characteristic details of the behaviour and the interaction by a differential study employing the phase plane of interannual tempera-ture changes of both hemispheres. The relations between or with the surface temperatures is here and in the following in most cases not so pronounced as that of the tropospheric ones.

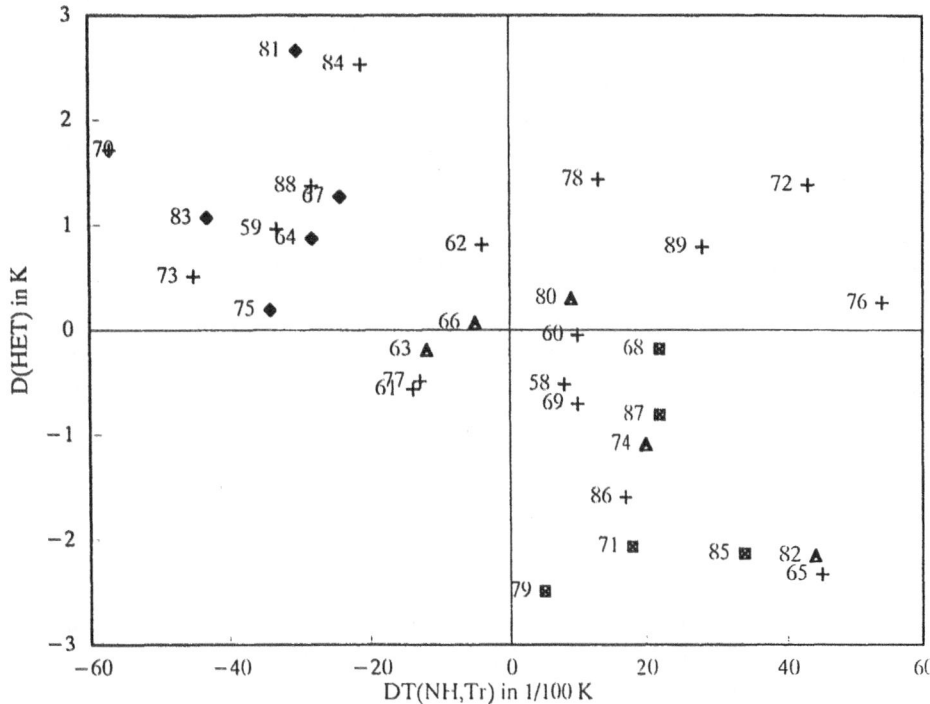

Figure 13. Interannual changes of heterogeneity of Middle European winters D (HET) relative to these of the NH tropospheric temperature DT (NH, Tr). ▲ = volcano eruption year V, ◆ = $V + 1$, ▨ = $V + 5$, + = all other years. (DT at V means the interannual change from the volcano eruption year to the following year, etc.).

The application of such an approach to the coupling of smaller subsystems or special regions to the global or hemispheric behaviour shows also encouraging results: For instance, the phase plane using the interannual changes of the Middle European winter temperature and of NH tropospheric temperature displays unexpected and rather peculiar characteristics: The phase states group in two "levels" clearly separated by a gap. - Further, the interannual change of the heterogeneity of each winter as a measure of its internal variability is closely related to the simultaneous interannual change of the NH tropospheric temperature.

The capacity of the method used is also revealed by studies of the impact of certain disturbances exerted to the atmosphere. A characteristic example is the impact of major volcanic eruptions. A composite study shows that a special phase state of the atmosphere clearly starting with major volcanic eruptiones (in the period between 1958 and 1989) can be traced at least up to the sixth year after the eruption despite of the otherwise present large noise and different possible interferences. They seem to activate a special (time dependent) dynamical state of the global atmosphere having attractor properties. This volcano impact is also reflected in the same strong way in the interannual changes of the Middle European winter temperature (order of magnitude several K) for independent volcano eruption ensembles within the last 110 years, is confirmed by the Pinatubo eruption 1991, and may have some long-term forecast potential. - The additional question whether climatic changes could trigger volcanic events needs further study.

The El Niño events, characterized by a transient warming of the equatorial east Pacific surface water, is one other kind of possible disturbance of the global atmosphere. In contrast to the volcano eruption impact there is, in the phase plane used, no starting behaviour of the global atmosphere as a consequence of El Niño events. The El Niño is confirmed to be a special phase in a continuing fluctuation (Southern Oscillation) with a somewhat variable period. There are some more hints that a global tropospheric cooling stimulates the fluctuation mentioned. This cooling covers two successive years beginning with the temperature change from the third year to the second one before the starting year of the real El Niño phase.

The phase space representation, using as variables with priority the first time derivatives or time differences, does obviously have a wide field of application with respect to diagnostic studies of complex dynamical systems in scientific and practial fields. It needs a further theoretical development. Some experiences show, in accordance with theoretical reason, that the application of higher derivatives or differences gives deeper insight into the behaviour of complex systems.

References

Ruelle, D. (1981), 'Chemical kinetics and differential dynamical systems', In: *Nonlinear Phenomena in Chemical Dynamics*, Springer-Verlag, Berlin.

Nicolis, C. and G. Nicolis (1984), 'Is there a climatic attractor?', *Nature* **311**, 519-532.

Tsonis, A.A. and J.B. Elsner (1989), 'Chaos, strange attractors, and weather', *Bull. Amer. Meteor. Soc.*, **70**, No. 1, 14-23.

Böhme, W. (1993), 'Untersuchungen zur Reaktion des Klimasystems auf große vulkanische Eruptionen mittels Phasenebenen-Darstellung', *Meteorol. Zeitschrift*, N.F. **2**, 76-80.

Angell, J.K. (1988), 'Variations and trends in tropospheric and stratospheric global temperatures, 1958 - 1987', *J. of Climate*, **1**, 1296-1313.

Boden, T.A., P. Kanciruk and M.P. Farrell (1990), 'Trends '90: A compendium of data on global change', Carbon Dioxide Information Analysis Center, Oak Ridge.

Mass, C.F. and D.A. Portman (1989), 'Major volcanic eruptions and climate: a critical evaluation', *J. of Climate*, **2**, 566-593.

Asaturov, M.L., M.I. Budyko, K.J. Vinnikov, P.J. Groisman, A.S. Kabanov, I.L. Karol, M.P. Kolomejew, Z.I. Pivovarova, J.W. Rozanov and S.S. Khmelevtsov (1986), ''Volcaneos, stratospheric aerosol and the climate of the Earth' (in Russian), Gidrometeoizdat, Leningrad.

IPPC (1990), 'Climate change, the IPCC Scientific Assessment', Section 2.3.2., WMO/UNEP Intergovernmental Panel on Climate Change, Cambridge University Press, Cambridge.

Khmelevtsov, S.S. (1987), 'Aerosol and modern climate change' (in Russian), *Meteorologia i Gidrologia*, **11**, 59-64.

Deutscher Wetterdienst, Seewetteramt (1963 - 1982), 'Die Witterung in Übersee', Hamburg.

Hydrometeorological Service of the USSR (1974 - 1882), 'Synoptical Bulletin, Northern

Hemisphere, Part III' (in Russian), Moscow.

National Oceanic Atmospheric Administration (USA) (1963 - 1967), 'Daily Series, Synoptic Weather Maps', Asheville, N.C.

Deutscher Wetterdienst (1970 - 1982), 'Täglicher Wetterbericht' (up to 1975) and 'Europäischer Wetterbericht' (1976 - 1982), Offenbach a.M.

Pogosjan, H.P. (1972), 'General Circulation of the atmosphere' (in Russian), Gidro-meteoizdat, Leningrad.

Liebmann, B. and D.L. Hartmann (1982): 'Interannual variations of outgoing IR associated with tropical circulation changes during 1974 - 1978', *J. Atmos. Sci.*, **39**, 1153-1162.

Gadgil, S. and A. Guruprasad (1990), 'An objective method for the identification of the intertropical convergence zone', *J. of Climate*, **3**, 558-567.

Rasmusson, E. and T. Carpenter (1982), 'Variations in tropical sea surface temperature and surface wind fields associated with the Southern Oscillation/El Niño', *Month. Weath. Rev.*, **110**, 354-384.

Philander, S.G. (1990), 'El Niño, la Niña, and the Southern Oscillation', Academic Press, San Diego.

Kiladis, G.N. and H.F. Diaz (1989), 'Global climatic anomalies associated with extremes of the Southern Oscillation', *J. of Climate*, **2**, 1069-1090.

Graf, H.-F. (1984), 'Eine globale Eigenschwingung des Systems Ozean-Atmosphäre', *Z. Meteorol.*, **35**, 223-226.

Graf, H.-F. (1986), 'Abkühlung der Nordhemisphäre - ein möglicher Trigger für El Niño/Southern Oscillation-Episoden', *Naturwissenschaften*, **73**, 258-263.

Graf, H.-F. (1989), 'Forced cooling of the polar T21 atmosphere and tropical climate variability', Report No. 45, Max-Planck-Institut für Meteorologie, Hamburg.

Linke, F. and F. Baur (1962), 'Meteorologisches Taschenbuch', Neue Ausgabe, I. Band, 2. Aufl., Akad. Verlagsgesellschaft, Geest u. Portig, Leipzig.

Linke, F. and F. Baur (1970), 'Meteorologisches Taschenbuch', Neue Ausgabe, II. Band, 2. Aufl., Akad. Verlagsgesellschaft, Geest u. Portig, Leipzig.

Deutscher Wetterdienst (1969 - 1992), 'Die Großwetterlagen Europas', Offenbach.

Rampino, M.R., S. Self and R.W. Fairbridge (1979), 'Can rapid climatic change cause volcanic eruptions?', *Science*, **206**, 826-829.

Volland, H. (1989), 'On the seasonal components of polar motion', *Geophys. Res. Letters*, **16**, 303-305.

DETECTION OF A PERTURBED EQUATOR-POLE TEMPERATURE GRADIENT IN A SPECTRAL MODEL OF THE ATMOSPHERIC CIRCULATION

SIPKO L.J. MOUS

Dept. of Mathematics, Wageningen Agricultural University
Dreijenlaan 4, 6703 HA, Wageningen, The Netherlands
e-mail: mous@rcl.wau.nl

Summary

This paper deals with the assessment of external perturbations in nonlinear chaotic dynamical processes using an extended Kalman filter treatment. We consider processes that can be modelled by a system of nonlinear ordinary differential equations and we use the extended Kalman filter to estimate the size of external perturbations. As an example to illustrate this approach we have analyzed a lower order spectral model of the atmospheric circulation with a perturbed equator-pole temperature gradient.

1. Introduction

Many physical processes, e.g., the atmospheric circulation, exhibit irregular behaviour. We may model this behaviour by a system of nonlinear differential equations that manifest chaotic behaviour (de Swart,1988a). A characteristic feature of chaotic models is the sensitive dependence on the initial state. In this paper we analyze the difference between the dynamical process as it is observed and the solution of a differential equation for this process. Due to sensitive dependence on the initial state, the observed values and the model values will diverge. This divergence may get worse in case the physical process is not correctly modelled or certain external perturbations are neglected. One would like to have available procedures that account for the propagation of the errors in the initial state, so that the other sources of divergence can be assessed.

In many cases, we can parameterize the external perturbations with a small number of unknown parameters. These parameters may often be interpreted as the size of the external perturbations. The classical least squares method cannot be applied to estimate the unknown parameters. The reason for this is the exponential growth of an error in the initial state, making accurate predictions of the output of the model impossible (Baake et. al., 1992). In this study we describe the uncertainty in the initial state with a Gaussian distribution and we use the extended Kalman filter to approximate the time evolution of the density function. In fact, we then reïnterpret the state vector as a stochastic process, satisfying a stochastic differential equation in the îto sense (Jazwinski, 1970).

The unknown parameters can be simply estimated by regarding them as random variables and augmenting the state vector with these variables. Since the extended Kalman filter yields estimates of the state vector, it also provides an estimate of these unknown parameters (Jazwinski, 1970).

We have applied this method to a low-order spectral model of the atmospheric circulation with a perturbed equator-pole temperature gradient. The idea is that one detects a change in the heat flux (green-house effect) from the observation of systematic deviation in the circulation.

2. Extended Kalman filter for estimating uncertain parameters.

We assume that the physical process with a strange attractor can be modelled by a stochastic differential equation. As mentioned above, the physical process is perturbed by some external force, denoted in the model by the term $\lambda g(t)$. We assume that the time dependency $g(t)$ of this external perturbation is known and that $g(t)$ can be normalized in some way. The size of this perturbation is then denoted by the parameter λ (Grasman and Houtekamer, 1992). All other perturbations and errors in the model are modelled by a white Gaussian noise process $\xi_1(t)$ with mean $E(\xi_1(t)) = 0$ and autocovariance function $E(\xi_1(t)\xi_1'(t+\tau)) = Q_1(t)\delta_D(\tau)$, where $\delta_D(.)$ denotes the Dirac delta function. Furthermore we assume that the initial error in the state vector is Gaussian, with $E(\xi_0) = 0$ and $E(\xi_0\,\xi_0') = P_0$. The model equations are then given by

$$\frac{dx}{dt} = f(x) + \lambda g(t) + \xi_1(t), \quad x(t_0) = x_0 + \xi_0 . \tag{1}$$

In most practical situations the observations are taken at discrete times t_i, $i=1,...n$. We assume that the observations are linear combinations of the state vector and that the observation errors are Gaussian distributed with mean $E(\xi_2(t_k)) = 0$ and covariance matrix $E(\xi_2(t_k)\xi_2'(t_k)) = R(t_k)$. Furthermore we assume that the errors in the initial state, the errors in the observations and the errors in the state equation are mutually independent. The observations are then given by

$$y(t_k) = Mx(t_k) + \xi_2(t_k) . \tag{2}$$

The observations' $y(t_1)$ up to and including $y(t_k)$ and the model description (1,2) can be used to estimate the state at time t_k. This is called filtering. In the early sixties Kalman and Bucy derived the optimal unbiased filter for the state, for a linear dynamical process, in the sense of minimum variance. For nonlinear dynamical systems there are many approximated optimal filters developed (Jazwinski, 1970, Sorenson, 1988). In this study we will use the extended Kalman filter and we apply a procedure described in Jazwinski (1970, pp. 281-282) to estimate an uncertain parameter. For completeness we will give a brief description of the extended Kalman filter for a dynamical system of the form (1,2). Starting from estimates of the mean $\hat{x}(t_k|t_k)$ and the error covariance matrix $P(t_k|t_k)$ of the state vector at time t_k, given the observations $y(t_1),...,y(t_k)$, we predict the state $\hat{x}(t_{k+1}|t_k)$ and the error covariance matrix $P(t_{k+1}|t_k)$ at time t_{k+1} with the first two equations of the extended Kalman filter

$$\hat{x}(t_{k+1}|t_k) = \hat{x}(t_k|t_k) + \int_{t_k}^{t_{k+1}} [f(\hat{x}(t|t_k))+\lambda g(t)]\,dt, \tag{3}$$

$$P(t_{k+1}|t_k) = \Phi(t_{k+1},t_k;\hat{x}(t_k|t_k))P(t_k|t_k)\Phi^T(t_{k+1},t_k;\hat{x}(t_k|t_k))+Q(t_{k+1}), \tag{4}$$

where $\Phi(t_{k+1}|t_k;\hat{x}(t_k|t_k))$ is the state transition matrix of the linearized state equation

$$\frac{d(\delta x)}{dt} = f'(\hat{x}(t|t_k))\delta x \tag{5}$$

with

$$f'(\hat{x}(t|t_k)) = \left[\frac{\partial f(\hat{x}(t|t_k))}{\partial x}\right]_{n \times n}$$

and where

$$Q(t_{k+1}) = \int_{t_k}^{t_{k+1}} \Phi(t_{k+1},t)Q_1(t)\Phi^T(t_{k+1},t)\,dt.$$

The observation $y(t_{k+1})$ is used with the remaining three equations of the Kalman filter to update the estimates of the mean and the error covariance matrix of the state vector,

$$\hat{x}(t_{k+1}|t_{k+1}) = \hat{x}(t_{k+1}|t_k) + K(t_{k+1})(y(t_{k+1}) - M\hat{x}(t_{k+1}|t_k)), \tag{6}$$

$$P(t_{k+1}|t_{k+1}) = (I - K(t_{k+1})M)P(t_{k+1}|t_k)(I - K(t_{k+1})M)^T + K(t_{k+1})R(t_{k+1})K^T(t_{k+1}), \tag{7}$$

$$K(t_{k+1}) = P(t_{k+1}|t_k)M^T(MP(t_{k+1}|t_k)M^T + R(t_{k+1}))^{-1}. \tag{8}$$

The filter is initialized with estimates $\hat{x}(t_0|t_0)$ and $P(t_0|t_0)$. These estimates of the initial state and the covariance matrix are assumed to be known.

We will make use of a method that is described in Jazwinski (1970) to estimate the size λ of the external perturbation of the dynamical system (1). In this method the uncertain parameter λ is regarded as a random variable and the state equation is then augmented with this random variable. The augmented model is then given by

$$\frac{dX}{dt} = \frac{d}{dt}\begin{bmatrix} x \\ \lambda \end{bmatrix} = \begin{bmatrix} f(x) \\ 0 \end{bmatrix} + \begin{bmatrix} \lambda g(t) \\ 0 \end{bmatrix} + \begin{bmatrix} \xi_1(t) \\ 0 \end{bmatrix} \tag{9}$$

with state vector $X(t) = (x(t),\lambda(t))'$ (Mous and Grasman, 1993). This augmented model is still in state space form, so the extended Kalman filter can be applied. The result of this filter gives an on-line estimate of the size of the external perturbation.

3. Numerical experiments with a low-order spectral model of the atmosphericcirculation

We study the use of the extended Kalman filter for a 10-component spectral model of the atmospheric circulation that has been analyzed by De Swart (1988a, 1988b). For barotropic flow the atmospheric circulation is described by a streamfunction. This streamfunction, $\psi(x,y,t)$, satisfies the so-called quasi-geostrophic barotropic potential vorticity equation:

$$\frac{\partial}{\partial t}\nabla^2\psi + J(\psi,\nabla^2\psi + f) + \gamma J(\psi,m) + C\nabla^2(\psi - \psi^*) = 0, \tag{10}$$

where J is the Jacobian operator given by

$$J(a,b) = a_2 b_1 + a_1 b_2, \qquad a=(a_1,a_2)', \ b=(b_1,b_2)'. \tag{11}$$

The gradient of the Coriolis parameter f is taken fixed, meaning that the flow is restricted to a channel in the tangent plane at a given latitude. The term $m(x,y)$ describes the topography (mountains) of the domain and the coefficient γ accounts for the effect of this topography. The coefficient C is a measure for frictional effects and finally the term $\psi^*(x,y,t)$ models the driving force of the atmospheric flow: the equator-pole temperature gradient.

The solution $\psi(x,y,t)$ of (10) is approximated by expanding ψ, ψ^* and m in a series of eigenfunctions $\{ \phi_j \}$ of the Laplace operator for the channel being a domain in the x,y-plane, periodic in x and bounded by lower and upper values of y:

$$\psi(x,y,t) = \sum_{j=1}^{\infty} x_j(t)\phi_j(x,y). \tag{12}$$

In De Swart (1988a) the system of nonlinear differential equations for the coefficients' $x_j(t)$, $j=1,...,10$ of the truncated series $\psi(x,y,t)$ is derived. This truncated system is a state space model of the form

$$\frac{dx}{dt} = f(x) + Cx^*, \tag{13}$$

(see appendix A). This spectral model has a vacillating solution and is a system of the lowest possible order that still exhibits this behaviour (De Swart, 1988a). Vacillation means that the solution visits irregularly domains in state space where it remains for some time. These domains correspond to so-called preferent weather regimes.

The effect of the equator-pole temperature gradient is represented in (13) by the term Cx^*, with $x^*=(x_1^*,0,0,x_4^*,0,...,0)$. We want to study the influence of a perturbation in this temperature gradient on the evolution of this spectral model.

In the numerical examples the "real" process is simulated by integrating equation (13), with the initial conditions contaminated with white noise with variance τ^2 and with the vector x^* perturbed with $\lambda g(t)$. We have taken $g(t)=(1,0,...,0)'$. For simplicity, all state variables are observed. The observations are sampled with a sampling period $T_s = 0.25$. We have also contaminated the observations with white observation noise, with variance τ^2. The augmented model is then given by

$$\frac{dX}{dt} = \frac{d}{dt}\begin{bmatrix} x \\ \lambda \end{bmatrix} = \begin{bmatrix} f(x) \\ 0 \end{bmatrix} + \begin{bmatrix} C(x^*+\lambda g(t)) \\ 0 \end{bmatrix},$$

$$y(t_k) = \begin{bmatrix} I & 0 \end{bmatrix}\begin{bmatrix} x(t_k) \\ \lambda(t_k) \end{bmatrix} + \xi_2(t_k), \tag{14}$$

where $y(t_k)$ are the observations of the state vector at time t_k.

In the extended Kalman filter, we take as the initial estimate for $\hat{x}(t_0|t_0)$ the first observation of the data set. The covariance matrices are set to:

$$P_{i,j}(t_0|t_0) = \tau^2 \text{ if } i=j,$$
$$= 0 \text{ if } i\neq j,$$

$$Q = 0$$

$$R_{i,j} = \tau^2 \text{ if } i=j,$$
$$= 0 \text{ if } i\neq j.$$

We take $\hat{\lambda}(t_0|t_0)=0$ and $P_\lambda(t_0|t_0)=100$ as the initial statistics for λ. This large variance $P_\lambda(t_0|t_0)$ stands for the initial uncertainty in the parameter λ.

We will filter two data sets, where we take as "real" parameter values $\lambda=0.01$, $\tau^2=10^{-6}$ and $\lambda=0.05$, $\tau^2=10^{-6}$ respectively. The filtering results are shown in figures 1a and 1b. In these figures we plotted the estimated values $\hat{\lambda}(t_k|t_k)$. The dotted line represents the error standard deviation, $\lambda\pm\sqrt{P_\lambda(t_k|t_k)}$ (root mean square error curve). We have found as final estimates $\hat{\lambda}(t_{100}|t_{100})=0.010$ for the first data set and $\hat{\lambda}(t_{100}|t_{100})=0.051$ for the second data set.

Figures 1a and 1b show that the error standard deviation decreases fast which means that the data contain much information on λ. Especially for the data set with $\lambda=0.01$ we see that the standard deviations' decreases very fast for samples taken at the interval (t_{30},t_{40}). Apparently, the process is very sensitive for this parameter on this part of the attractor. It is noted that for the data set with $\lambda=0.05$, the estimates $\hat{\lambda}(t_k|t_k)$, for large t_k fall somewhat outside the rms-error curve. This may be due to the particular realizations of the observation errors and the choice of the initial estimate $\hat{\lambda}(t_0|t_0)$. Since the filter is an approximated optimal filter, it may also be that the higher order moments of the probability density functions and numerical errors are not negligible for this perturbation.

In the continuation of this section we consider a completely unknown perturbation and we want to study how well the filter can track a time-varying perturbation. For this purpose we have to change the initial settings of the filter, since otherwise the estimated perturbation will become finally constant. The idea is to use a small artificial random perturbation term in (14) for the state variable λ and to take for Q_λ a small value. The effect of this random forcing term is that the filter retains the possibility to adjust the estimate $\hat{\lambda}$ since the error covariance function P_λ will not tend to 0 for large t_k. In other words, the filter does not learn the perturbation too well.

Since we sample the state variables with a sampling period of $T_s=0.25$, only slowly varying perturbations can be estimated. In two test examples we perturbed the "real" process with a sine-shaped forcing term, $g(t)=(sin\omega t,0,...,0)$. In figure 2 and 3 it is seen that, for low frequency perturbations, the filter can track the "real" perturbation rather well for $Q_\lambda=1.0\times10^{-4}$. For larger values of Q_λ the tracking signal becomes rather noisy. For Q_λ tending to 0 it is observed that we get a phase shift between the real and the estimated perturbation and that the amplitude of the perturbation decreases.

4. Conclusions

In this paper we have used an extended Kalman filter treatment to estimate the size of a perturbed equator-pole temperature gradient in a low order spectral model of the atmospheric circulation. We regarded the size of this perturbation as an uncertain parameter in

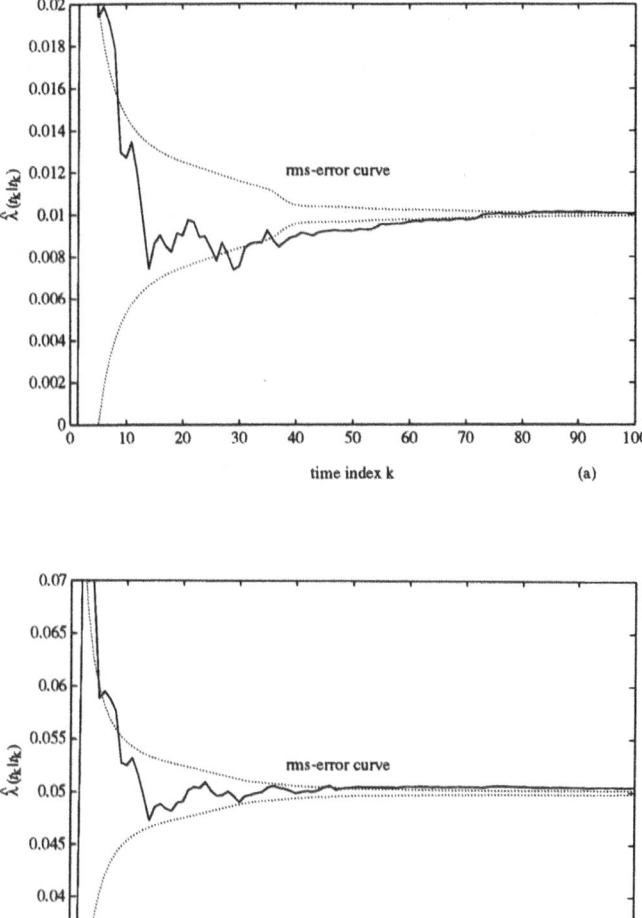

Figure 1. The extended Kalman filter for the atmospheric circulation problem. The solid line denotes the estimated $\hat{\lambda}(t_k|t_k)$ and the dotted line denotes the root mean square error curve, $\lambda \pm \sqrt{P_\lambda(t_k|t_k)}$. (a) The "real" process was simulated with $\lambda=0.01$ and $\tau^2=10^{-6}$, (b) the "real" process was simulated with $\lambda=0.05$ and $\tau^2=10^{-6}$.

the model and we have constructed an estimator for this uncertain parameter based on the extended Kalman filter. In this case study we compared simulations of the perturbed model, representing the "real" process, with simulations of the model. In the two cases that we have analyzed, the errors in the estimates of this perturbation were small, $\varepsilon = |\hat{\lambda} - \lambda| <$ 2%.

From figure 1 it is seen that the root mean square error curve calculated by the extended Kalman filter drops considerably after a few observations are taken. This implies that the data contains much information about the size of the perturbation. This observation is also reported in a paper of Baake et al. (1992), where they concluded that an observed trajectory of a chaotic process may be expected to contain a large amount of information about uncertain parameters.

We also studied the case where the external perturbation was completely unknown. In this situation we have to take a small value for Q_λ so that the filter can adjust the parameter λ. It appeared that the filter tracks the external perturbation rather well for low frequency perturbations and for a well-chosen value of Q_λ.

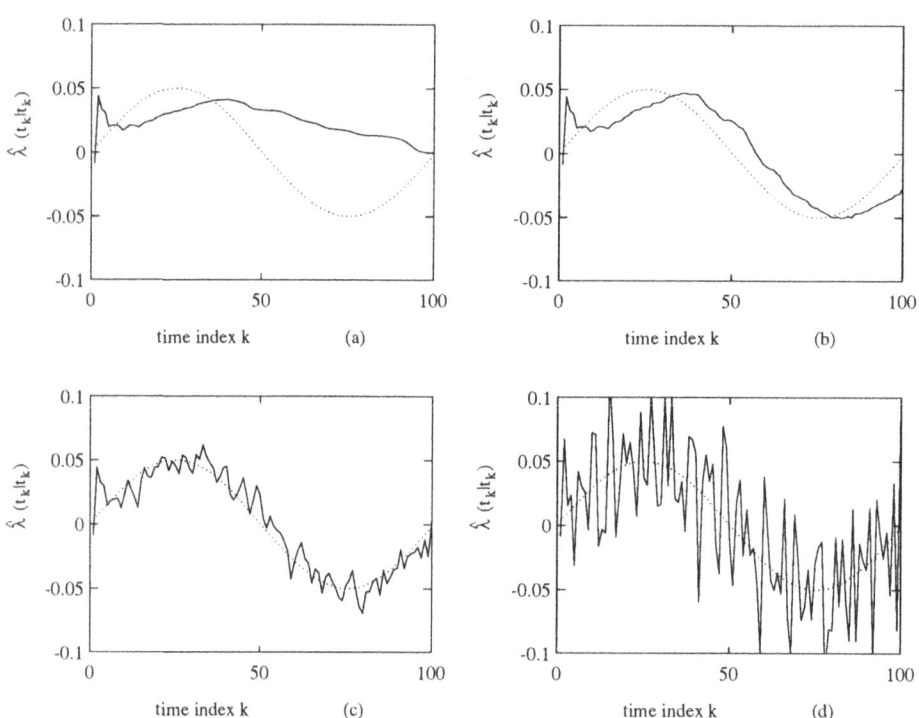

Figure 2. Tracking of an sine-shaped perturbation using the extended Kalman filter. The frequency of the sine-shaped perturbation is equal to $0.25 \times 2\pi$. In (a) $Q_\lambda = 0.0$, in (b) $Q_\lambda = 1.0 \times 10^{-6}$, in (c) $Q_\lambda = 1.0 \times 10^{-4}$, in (d) $Q_\lambda = 1.0 \times 10^{-2}$. The solid line denotes the estimated $\hat{\lambda}(t_k|t_k)$ and the dotted line denotes the "real" perturbation.

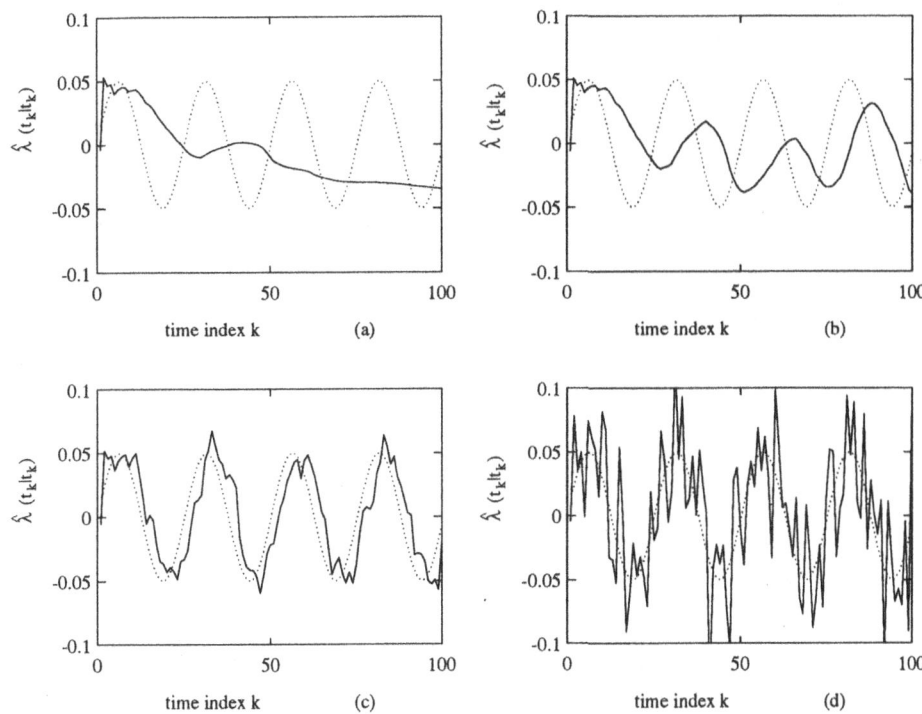

Figure 3. See figure 2 for explenation. Now with the frequency of the sine-shaped perturbation equal to 1.0×2π.

Acknowledgement

I would like to thank Huib de Swart for providing the code of his 10-component spectral model of the atmospheric circulation.

References

Baake, E., M. Baake, H.G. Bock and K.M. Briggs, (1992), 'Fitting ordinary differential equations to chaotic data', *Phys. Review A*, **45(8)**, 5524-5529.

De Swart, H.E., (1988a), *'Vacillation and predictability properties of low-order atmospheric spectral models'*, Ph.D.Thesis, Rijksuniversiteit Utrecht.

De Swart, H.E., (1988b), 'Low-order spectral models of the atmospheric circulation: a survey'. *Acta Appl. Math.*, **11**, 49-96.

Grasman, J., and P. Houtekamer, (1992), 'Methods for improving the prediction of dynamical processes with special reference to the atmospheric circulation', In: *Proc. of IUGG Symp. "Nonlinear Dynamics and Predictability of critical Geophysical Phenomena*, W.I. Newman and A.M. Gabrielov (eds.), Am. Geophysical Union.

1. Geophysics

Jazwinski, A.H., (1970), '*Stochastic Processes and Filtering Theory*', Academic Press, Paris

Mous, S.L.J and J. Grasman, (1993), 'Two methods for assessing the size of external perturbations in chaotic processes', *Math. Models and Methods in appl. sc.*, **3**(4), 577-593.

Sorenson, H.W., (1988), 'Recursive estimation for nonlinear dynamical systems', In: *Bayesian analysis of time series and dynamic models*, J.C. Spall ed., Dekker, New York.

Appendix A

In De Swart (1988a) a 10-component spectral model for the barotropic flow (23) in a rectangular channel is derived. The channel has length 2π in the zonal direction and width πb in the meridional direction. The 10-component model is given by

$$\frac{d}{dt}\begin{bmatrix} x_1 \\ x_2 \\ x_3 \\ x_4 \\ x_5 \\ x_6 \\ x_7 \\ x_8 \\ x_9 \\ x_{10} \end{bmatrix} = \begin{bmatrix} & +\overset{*}{\gamma}_{11}x_3 & -C(x_1-x_1^*) \\ -(\alpha_{11}x_1-\beta_{11})x_3 & -Cx_2 & -\delta 11x_4x_6 & -\rho_{11}(x_5x_8-x_6x_7) \\ +(\alpha_{11}x_1-\beta_{11})x_2 & -\overset{*}{\gamma}_{11}x_1 -Cx_3 & +\delta_{11}x_4x_5 & +\rho_{11}(x_5x_7+x_6x_8) \\ & +\overset{*}{\gamma}_{12}x_6 & -C(x_4-x_4^*) & +\varepsilon_1(x_2x_6-x_3x_5) & +\varepsilon_2(x_7x_{10}-x_8x_9) \\ -(\alpha_{12}x_1-\beta_{12})x_6 & -Cx_5 & -\delta_{12}x_3x_4 & +\rho_{12}(x_2x_8-x_3x_7) & +\overset{\prime}{\gamma}_{12}x_8 \\ +(\alpha_{12}x_1-\beta_{12})x_5 & -Cx_6 & +\delta_{12}x_2x_4 & -\rho_{12}(x_2x_7+x_3x_8) & -\overset{\prime}{\gamma}_{12}x_7 \\ -(\alpha_{21}x_1-\beta_{21})x_8 & -Cx_7 & -\delta_{21}x_4x_{10} & -\rho_{21}(x_2x_6+x_3x_5) & +\overset{\prime}{\gamma}_{21}x_6 \\ +(\alpha_{21}x_1-\beta_{21})x_7 & -Cx_8 & +\delta_{21}x_4x_9 & +\rho_{21}(x_2x_5-x_3x_6) & -\overset{\prime}{\gamma}_{21}x_5 \\ -(\alpha_{22}x_1-\beta_{22})x_{10} & -Cx_9 & -\delta_{22}x_4x_8 \\ +(\alpha_{22}x_1-\beta_{22})x_9 & -Cx_{10} & +\delta_{22}x_4x_7 \end{bmatrix}$$

| advection | topo-graphy | forcing/advection | advection | wave triad | topo-graphy |

where

$$\alpha_{nm} = \frac{8\sqrt{2}\,n}{\pi}\frac{m^2}{4m^2-1}\frac{n^2b^2+m^2-1}{n^2b_2+m^2} \quad , \quad \beta_{nm} = \frac{\beta nb^2}{n^2b^2+m^2} \quad ,$$

$$\delta_{nm} = \frac{64\sqrt{2}\,n}{15\pi}\frac{n^2b^2-(m^2-1)}{n^2b^2+m^2} \quad , \quad \overset{*}{\gamma}_{nm} = \frac{4m}{4m^2-1}\frac{\sqrt{2}\,nb\gamma}{\pi} \quad ,$$

$$\varepsilon_n = \frac{16\sqrt{2}\,n}{5\pi} \quad , \quad \gamma_{nm} = \frac{4m^3}{4m^2-1}\frac{\sqrt{2}\,nb\gamma}{\pi(n^2b^2+m^2)} \quad ,$$

$$\rho_{nm} = \frac{9}{2}\frac{(n-2)^2-(m-2)^2}{n^2b^2+m^2} \quad , \quad \overset{\prime}{\gamma}_{nm} = \frac{3b\gamma}{4(n^2b^2+m^2)} \quad .$$

The β_{nm}-contributions represent planetary vorticity advection, the $\overset{*}{\gamma}_{nm}$, γ_{nm} and $\overset{\prime}{\gamma}_{nm}$-terms the various couplings between flow and topography and Cx_n^* the equator pole-temperature gradient. We have taken $b=1.6$, $\beta=1.25$ $\gamma=1$, $C=0.1$, $x_1^*=4$ and $x_4^*=8$.

A SIMPLE TWO-DIMENSIONAL CLIMATE MODEL WITH OCEAN AND ATMOSPHERE COUPLING

E.J.M. VELING[1] and M.E. WIT[2]

[1] National Institute of Public Health and Environmental Protection
Centre of Mathematical Methods
P.O. Box 1, 3720 BA Bilthoven, The Netherlands
e-mail: cwmedve@rivm.nl

[2] University of Utrecht
Department of Mathematics and Informatics
P.O. Box 80010, 3508 TA Utrecht, The Netherlands

Summary

To study the effect of changes of essential climate parameters (such as the solar flux) a two-dimensional ocean model for the northern hemisphere is constructed which takes into account advection and diffusion in the top layer and upwelling of cold water formed at the ice edge from the deeper parts of the ocean (for the not-frozen part). At the ice-line, which acts as an internal boundary condition, continuity of temperature and heat flux is required. This model is solved by analytical means in terms of hypergeometric functions, which give the possibility to determine the latitude of the ice-line by a non-linear equation and to study the stability in terms of the physical parameters.

Keywords climate model, ocean circulation, ice-line, upwelling, hypergeometric function, stability

1. Introduction

The climate has always got the interest of mankind from historical times on. On large time-scales one is interested in the change of the climate which induces changes in the vegetation zones. On shorter time-scales people like to know how the climate depends e.g. on changes in oceanic currents. Nowadays, with the increased possibility of calculation power, very sophisticated models, so-called General Circulation Models (GCM) simulate quite realistically the climate, let it be with an enormous effort in tuning the software to the hardware. The drawback of these models is that they are less accessible and time-consuming. Furthermore, by the enormous number of degrees of freedom and the amount of output the essential underlying mechanisms of the formation of the climate are hidden. Therefore, there exists a need for more accessible and schematized models, maybe also on pedagogic grounds.

At the National Institute of Public Health and Environmental Protection (RIVM) some years ago one started to study the effects of the emission of some gases (e.g. CO_2) in order to assess an increased Greenhouse effect. This study resulted in the IMAGE-model (Rotmans, 1990), in which the formation and interaction of these gases in the atmosphere has been modelled on a global scale. One of the main driving forces of the climate is the temperature of the ocean. In the IMAGE-model some coupling exists between the atmosphere and the ocean on a very coarse spatial scale. Therefore, to improve upon this matter, in this paper attention is focussed on the ocean and on the effects of the temperature distribution in the ocean as function of the depth and latitude.

The underlying model is extension of a simple model of North (1975), see also North et al. (1981), in such a way that it is still possible to solve the model analytically, which shows the dependency of the temperature with respect to the parameters in a closed form. One extension is the inclusion of the upwelling of colder water. This water has been formed near the ice-line, and sunk into deeper water by its greater density, where it disperses to lower latitudes before it wells up. The effect of this upwelling water is that in the top layer, the so-called mixing layer, a current flows from the equator to the ice-line (Watts and Morantine, 1990). Another extension is the allowance of different values for the parameters in the frozen and unfrozen part, plus another formulation of the time-dependent flux at the ice-line.

We consider a two-dimensional model of the temperature of the ocean on the Northern Hemisphere as a function of latitude and depth, and more in particular, we pay attention to the top layer, and formulate the terms which determine the temperature therein. The ocean is split into two parts, an unfrozen and a frozen part. To keep the calculations tractable, the temperature in the top layer will be averaged over the thickness. Finally, we end up for the stationary case with two differential equations with conditions at the boundaries (i.e. the equator and the pole) and at the internal boundary, the ice-line, where the temperature is fixed at the freezing point of oceanic water and where the outgoing flux from the water-part equals the incoming flux into the frozen part. The location of the ice-line will be determined by these conditions and will not be known in advance (i.e. a Stefan-problem). It turns out that for the set of parameter values chosen in this study, there are three solutions. Attention will be given to the question of stability. Furthermore, some results will be presented with respect to the influence of a few essential parameters (namely A, B). For example, for a doubling of the concentration of CO_2, which means a reduction in A of 4.32 W/m^2 (see IPCC (1990)), the temperature will increase by 2 °C, and the latitude of the ice-line will move at least 2° polewards.

2. Model

Firstly, we consider the general energy balance equation

$$\frac{\partial T}{\partial t} = div(K \cdot gradT - uT)$$

on a sphere from the equator to the ice-line in spherical coordinates (r, ϕ). We assume rotational symmetry, so we do not take into account longitudinal dependency.

$$\frac{\partial T}{\partial t} + \frac{1}{r^2\cos\phi} \frac{\partial}{\partial \phi}(r\cos\phi u_\phi T) + \frac{1}{r^2\cos\phi} \frac{\partial}{\partial r}(r^2\cos\phi u_r T) =$$

$$\frac{1}{r^2\cos\phi} \frac{\partial}{\partial \phi}\left(\cos\phi K_{hw}\frac{\partial}{\partial \phi}T\right) + \frac{1}{r^2\cos\phi} \frac{\partial}{\partial r}\left(r^2\cos\phi K_v\frac{\partial}{\partial r}T\right),$$

(1)

for $t > 0$, $R-D \leq r \leq R$, $0 \leq \phi \leq \phi_i$ with initial and boundary conditions. Here is

t[y(ear)] the time,

r[m] the distance from the centre of the earth,
ϕ[rad] the latitude,
ϕ_i[rad] the latitude of the ice-line,
R[m] the radius of the earth,
D[m] the depth of the ocean,
$T = T(\phi,r,t)[^{\circ}C]$ the yearly averaged temperature at latitude ϕ, radius r and time t,
$u_\phi = u_\phi(\phi,r,t)$[m/y] the horizontal velocity polewards,
$u_r = u_r(\phi,r,t)$[m/y] the vertical velocity upwards,
K_{hw}[m^2/y] the horizontal diffusion in water,
K_v[m^2/y] the vertical diffusion in the "deep ocean".

Because $D \ll R$, we introduce the variable $z = R - r$. In that case eq. (1) can be approximated by:

$$\frac{\partial T}{\partial t} + \frac{1}{R\cos\phi}\frac{\partial}{\partial \phi}(\cos\phi u_\phi T) + \frac{\partial}{\partial z}(u_z T) =$$

$$\frac{1}{R^2\cos\phi}\frac{\partial}{\partial \phi}\left(\cos\phi K_{hw}\frac{\partial}{\partial \phi}T\right) + \frac{\partial}{\partial z}\left(K_v\frac{\partial}{\partial z}T\right), \tag{2}$$

for $t > 0$, $0 \leq z \leq D$, $0 \leq \phi \leq \phi_i$, with initial and boundary conditions.

From here on, there holds

$T = T(\phi,z,t)[^{\circ}C]$ the yearly averaged temperature at latitude ϕ, depth z and time t,
$u_\phi = u_\phi(\phi,z,t)$[m/y] the horizontal velocity polewards,
$u_z = u_z(\phi,z,t)$[m/y] the vertical velocity downwards ($u_z = -u_r$).

Since we have assumed rotational symmetry, we have neglected the land masses. The transport of heat takes place in the top layer from equator to ice-line. At the ice-line creation of ''deep water'' takes place. During the long winter nights the water is not heated by the sun, but it will lose heat by radiation, so ice will be formed. Since ice contains less salt than oceanic water, the upper layer of water will contain more salt than the average, is therefore heavier and sinks by the greater density. This induces a current flowing at the bottom of the ocean to the equator, upwelling over all latitudes and flowing in the top layer to the ice-line, see Fig. 1. Because $\text{div}\mathbf{u} = 0$, with $\mathbf{u}$ the velocity field, we have

$$\frac{1}{r^2\cos\phi}\frac{\partial}{\partial r}(r^2\cos\phi u_r) + \frac{1}{r^2\cos\phi}\frac{\partial}{\partial \phi}(r\cos\phi u_\phi) = 0.$$

By the same approximation with respect to r as before, we find

$$\frac{\partial}{\partial z}u_z + \frac{1}{R\cos\phi}\frac{\partial}{\partial \phi}(\cos\phi u_\phi) = 0.$$

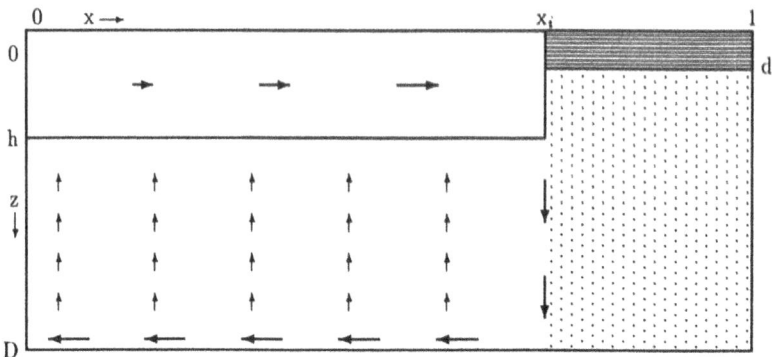

Figure 1. Cross-section ocean.

Integration over the top layer with depth h, with the extra condition that $u_z(0) = 0$, gives

$$u_z(h) = -\frac{h}{R\cos\phi}\frac{\partial}{\partial\phi}(\cos\phi u_\phi). \tag{3}$$

Multiplication of both sides by $\cos\phi$ and integration from $\phi = 0$ to $\phi = \phi$ we obtain (with $u_\phi = 0$ for $\phi = 0$) for u_ϕ in the top layer

$$u_z(h)\sin\phi = -\frac{h}{R}\cos\phi u_\phi.$$

Substitution of this relation in eq. (2) we find in the top layer

$$\frac{\partial T}{\partial t} - \frac{1}{R\cos\phi}\frac{\partial}{\partial\phi}\left(\frac{R}{h}\sin\phi u_z(h)T\right) + \frac{\partial}{\partial z}(u_z T) =$$

$$\frac{1}{R^2\cos\phi}\frac{\partial}{\partial\phi}\left(\cos\phi K_{hw}\frac{\partial}{\partial\phi}T\right) + \frac{\partial}{\partial z}\left(K_v\frac{\partial}{\partial z}T\right). \tag{4}$$

Integration of eq. (4) over the depth of the top layer, we obtain

$$h\frac{\partial T}{\partial t} = \frac{1}{\cos\phi}\frac{\partial}{\partial\phi}(\sin\phi u_z(h)T) + \frac{h}{R^2\cos\phi}\frac{\partial}{\partial\phi}\left(\cos\phi K_{hw}\frac{\partial}{\partial\phi}T\right) +$$

$$\left(K_v\frac{\partial}{\partial z}T - u_z T\right)_{z=h} - \left(K_v\frac{\partial}{\partial z}T - u_z T\right)_{z=0}. \tag{5}$$

The first boundary term becomes F_l, the flux through the level $z = h$. At the transition of the layers we require continuity of the flux, that is

$$F_l = \left(K_v \frac{\partial}{\partial z} T_l - u_z T_l \right)_{z=h},$$

in which T_l is the temperature of the bottom layer. The righthand side becomes $g - u_z(h)T$, by the assumption $T_l(h) = T$, and by the introduction of $g = K_v \frac{\partial}{\partial z} T_l|_{z=h}$. The second boundary term in eq. (5) is the flux at the top of the top layer. This equals hF, with F defined as:

$$F(T, \sin\phi) = \frac{S(1 - \alpha) - (A + BT)}{\rho_w c_w h}.$$

Here is,

$S(\phi)[\text{W/m}^2]$ the yearly averaged incoming radiation of the sun,
$\alpha(\phi)[\text{-}]$ the albedo at latitude ϕ,
$A[\text{W/m}^2]$ and $B[\text{W/m}^2/^\circ\text{C}]$ constants connected with radiation of heat by earth,
$\rho_w[\text{kg/m}^3]$ the density of oceanic water,
$c_w[\text{Wy/kg}/^\circ\text{C}]$ the heat capacity of oceanic water,
$h[\text{m}]$ the thickness of the top layer.

The mean averaged energy flux from the sun to the centre of the earth is denoted by S. The total energy per unit time, received by the earth $\pi R^2 S$. The surface of the earth is $4\pi R^2$. The total amount of energy which the earth per unit time and unit surface receives is thus $Q = S/4$. In view of the oblique position of the earth a correction has to be made for the differences during the seasons in the distribution of the received energy flux. The function $S(\phi)$ can be approximated by (see North (1975) and Watts and Morantine (1990)):

$$S(\phi) = Q\left(1 - \frac{1}{2}P_2(\sin\phi)\right),$$

where

$Q[\text{W/m}^2]$ the solar constant divided by 4, and
$P_2(x) = \frac{1}{2}(3x^2 - 1)$, the second order Legendre polynomial is.

The albedo will be approximated by (see Watts and Hayder (1984)):

$\alpha(\phi) = \alpha_1 + \alpha_2 P_2(\sin\phi)$ $\sin\phi \leq \sin\phi_i$, with ϕ_i the latitude where $T = T_i$,
$\alpha(\phi) = \alpha_3$ elsewhere, with α_i's constants.

It is convenient to make a coordinate transformation for the latitude ϕ to x given by $x = \sin\phi$. From here on, it will be supposed that $T = T(x,t)[^\circ\text{C}]$ the yearly averaged temperature at $\phi = \arcsin x$ and time t in the top layer. There follows

$$\frac{\partial T}{\partial t} = \frac{1}{h}\frac{\partial}{\partial x}(xu_z(h)T) + \frac{1}{R^2}\frac{\partial}{\partial x}\left((1 - x^2)K_{hw}\frac{\partial}{\partial x}T\right) - \frac{u_z(h)T}{h} + \frac{g}{h} + F(T,x). \tag{6}$$

Because we suppose that $u_z(h)$ does not depend on x, we can take together the first and the third term in the righthand side. This leads to the following equation for the water part of our model for the top layer (compare Watts and Morantine (1990), eq. A.7)

$$\frac{\partial T}{\partial t} = \frac{xu_z(h)}{h}\frac{\partial}{\partial x}T + \frac{1}{R^2}\frac{\partial}{\partial x}\left((1 - x^2)K_{hw}\frac{\partial}{\partial x}T\right) + \frac{g}{h} + F(T,x), \tag{7}$$

for $t > 0$, $0 \le x \le \sin\phi_i$,

$$\left((1 - x^2)\frac{\partial}{\partial x}T(x,t)\right)_{x=0} = 0, t > 0, \text{ with an initial condition.}$$

The condition

$$\left((1 - x^2)\frac{\partial}{\partial x}T(x,t)\right)_{x=0} = 0$$

implies that the derivative at the equator equals 0, which is induced by the requirement of symmetry with respect to the equator. It means that there is no transport of heat across the equator.

For the frozen part there holds an equivalent equation. However, no advection term is present and no diffusion takes place from the bottom layer. Furthermore, the thickness of the ice-layer, d, will be used in the function $F(T,x)$ in stead of h, the thickness of the top layer. The equation for the ice part will be

$$\frac{\partial T}{\partial t} = \frac{1}{R^2}\frac{\partial}{\partial x}\left((1 - x^2)K_{hi}\frac{\partial}{\partial x}T\right) + \frac{S(1 - \alpha) - (A + BT)}{\rho_i c_i d}, \tag{8}$$

for $t > 0$, $\sin\phi_i \le x \le 1$,

$$\left((1 - x^2)\frac{\partial}{\partial x}T(x,t)\right)_{x=1} = 0, \quad t > 0, \text{ with an initial condition.}$$

Here is,

$K_{hi}[m^2/y]$ the horizontal diffusion in the ice-layer,
$\rho_i[kg/m^3]$ the density of ice,
$c_i[Wy/kg/^0C]$ the heat capacity of ice,
$d[m]$ the thickness of the ice-layer.

The condition

$$\left((1 - x^2)\frac{\partial}{\partial x}T(x,t)\right)_{x=1} = 0$$

has been put because there is no heat transport across the pole. This condition determines the rate of growth of $T(x,t)$ for $x \uparrow 1$. The total model consists of the eq. (7) and (8), with

the following conditions on the ice-line
$x_i = \sin\phi_i$:

$$T_{water}(x_i,t) = T_i , \tag{9}$$

$$T_{ice}(x_i,t) = T_i , \tag{10}$$

$$-\rho_w c_w h K_{hw}\left((1 - x^2)\frac{\partial T_{water}}{\partial x}\right)_{(x=x_i,t=t)} =$$

$$-\rho_i c_i d K_{hi}\left((1 - x^2)\frac{\partial T_{ice}}{\partial x}\right)_{(x=x_i,t=t)} - \rho_i d R^2 L\left(\frac{\partial}{\partial t}(1 - x)\right)_{(x=x_i,t=t)} \tag{11}$$

Note that in general the ice-line x_i depends on t. Here $L[m^3/kg]$ is the latent heat of ice (i.e. heat of solidification). The third condition describes the heat balance on the ice-layer. The term in the lefthand side is the heat flux from the ocean polewards. The first term in the righthand side is the heat flux from the ice polewards. The second term in the righthand side is the heat necessary to melt the ice during warmer periods. If this term is negative, it represents the heat coming free during formation of ice.

3. Solution

In this section we will specify the solution of the stationary problem as specified above. We will assume that $u_z(h)$ does not depend on h, so we write in the sequel u_z. To summarize, we have the two equations

<u>ice:</u>   for $x_i \leq x < 1$,

$$\frac{1}{R^2}\frac{d}{dx}\left((1 - x^2)K_{hi}\frac{dT}{dx}\right) + \frac{Q(1 - \frac{1}{2}P_2(x))(1 - \alpha_3) - (A + BT)}{\rho_i c_i d} = 0, \tag{12}$$

$$\left((1 - x^2)\frac{\partial}{\partial x}T(x,t)\right)_{x=1} = 0, \quad t > 0.$$

<u>water:</u> for $0 \leq x \leq x_i$,

$$\frac{xu_z}{h}\frac{dT}{dx} + \frac{1}{R^2}\frac{d}{dx}\left((1 - x^2)K_{hw}\frac{dT}{dx}\right) + \frac{g}{h} +$$

$$\frac{Q(1 - \frac{1}{2}P_2(x))(1 - (\alpha_1 + \alpha_2 P_2(x))) - (A + BT)}{\rho_w c_w h} \approx 0 , \tag{13}$$

$$\left((1 - x^2)\frac{\partial}{\partial x}T(x,t)\right)_{x=0} = 0, \quad t > 0.$$

As noted before, $g = (K_v \frac{d}{dz} T_l)_{z=h}$, and furthermore, we have the conditions:

1. $T_{water}(x_i) = T_i$,

2. $T_{ice}(x_i) = T_i$,

3. $-\rho_w c_w h K_{hw} \left((1 - x^2) \frac{\partial T_{water}}{\partial x} \right)_{(x=x_i, t=l)} = -\rho_i c_i d K_{hi} \left((1 - x^2) \frac{\partial T_{ice}}{\partial x} \right)_{(x=x_i, t=l)}$

The values of the parameters are given in Table 1.

We shall not go into the derivation of the solutions of these equations, but just specify the results. Therefore, the following abbreviations are introduced

$$\frac{u_z}{h} = \alpha, \qquad \frac{K_{hw}}{R^2} = \beta_w, \qquad \frac{K_{hi}}{R^2} = \beta_i, \qquad \frac{Q}{\rho_w c_w h} = \gamma_w, \qquad \frac{Q}{\rho_i c_i d} = \gamma_i,$$

$$\frac{A}{\rho_w c_w h} = \delta_w, \qquad \frac{A}{\rho_i c_i d} = \delta_i, \qquad \frac{B}{\rho_w c_w h} = \varepsilon_w, \qquad \frac{B}{\rho_i c_i d} = \varepsilon_i.$$

We shall show that g/h can be approximated by αT. Let $T_l(z)$ denote the temperature in the "deep ocean". Here, in the deep ocean holds approximately $(h < z < D)$:

$$K_v \frac{d^2 T_l}{dz^2} - u_z \frac{dT_l}{dz} = 0, \text{ with } T_l(h) = T_i \text{ and } T_l(D) = T_b.$$

The solution reads

$$T_l(z) = \frac{T_i - T_b}{e^{\frac{u_z}{K_v} h} - e^{\frac{u_z}{K_v} D}} e^{\frac{u_z}{K_v} z} + T_b - \frac{T_i - T_b}{e^{\frac{u_z}{K_v} h} - e^{\frac{u_z}{K_v} D}} e^{\frac{u_z}{K_v} D} .$$

Because $e^{\frac{u_z}{K_v} D} \approx 0$ ($\frac{u_z}{K_v} < 0$ and D large), we have $T_l \approx (T_i - T_b) e^{\frac{u_z}{K_v}(z - h)} + T_b$. Therefore

$$g = \left(K_v \frac{dT_l}{dz} \right)_{z=h}$$

is thus approximately

$$K_v (T_i - T_b) \frac{u_z}{K_v} e^{\frac{u_z}{K_v}(h - h)} = u_z (T_i - T_b).$$

In our model we suppose $T_b = 0$ and T_t is the temperature in the top layer ($= T$), so $g/h \approx u_z T/h = \alpha T$. The solution will be expressed in hypergeometric functions, defined by

$$F(a,b;c;z) = \sum_{n=0}^{\infty} \frac{(a)_n (b)_n}{(c)_n n!} z^n, \quad |z| < 1,$$

with
$$(a)_n = a(a + 1)(a + 2) \ldots (a + n - 1), \quad (a)_0 = 1.$$

Table 1 Physical parameters

Variable		Unit	Value
u_z	velocity upwelling water	m/y	−4
h	depth top layer	m	60
d	depth ice-layer	m	5
R	radius earth	m	$6.3785\ 10^6$
D	depth ocean	m	4000
K_{hw}	horizontal diffusion in the water in the top layer	m²/y	$1.2\ 10^{12}$
K_{hi}	horizontal diffusion in the ice	m²/y	$0.6 K_{hw}$
K_v	vertical diffusion in the "deep ocean"	m²/y	2000
A	constants related with the radiation	W/m²	208
B	of the earth	W/m²/°C	1.8
ρ_w	density oceanic water	kg/m³	1035
ρ_i	density ice	kg/m³	1035
c_w	heat capacity of water	Wy/kg/°C	$1.33\ 10^{-4}$
c_i	heat capacity of ice	Wy/kg/°C	$0.666\ 10^{-4}$
Q	solar constant S divided by 4	W/m²	340
α_1	constants related with the albedo	-	0.31
α_2		-	0.12
α_3		-	0.6
L	latent heat of ice	Wy/kg	0.0106

The total solution has the following form:

$$T(x) = C_1 F(a,b;1;1 - x^2) + \bar{a}_0 + \bar{a}_z x^2, \quad x_i \leq x \leq 1, \tag{14}$$

with

$$a_i = \frac{1}{4} + \frac{1}{4}\sqrt{1 - 4\frac{\varepsilon_i}{\beta_i}} \quad \text{and} \quad b_i = \frac{1}{4} - \frac{1}{4}\sqrt{1 - 4\frac{\varepsilon_i}{\beta_i}} \quad,$$

$$\bar{a}_0 = \frac{\beta_i}{\varepsilon_i}(\frac{3\gamma_i(1 - \alpha_3)}{6\beta_i + \varepsilon_i}) + \frac{\frac{1}{2}\frac{1}{\gamma_i}(1 - \alpha_3)}{\varepsilon_i} - \frac{\delta_i}{\varepsilon_i} + \frac{\frac{3}{4}\gamma_i(1 - \alpha_3)}{6\beta_i + \varepsilon_i} \quad,$$

$$\bar{a}_2 = -\frac{\frac{3}{4}\gamma_i(1 - \alpha_3)}{6\beta_i + \varepsilon_i} \quad.$$

$$T(x) = C_2 F(a_w, b_w; \frac{1}{2}; x^2) + \bar{b}_0 + \bar{b}_2 P_2(x) + \bar{b}_4(x) P_4(x) \quad, \quad 0 \leq x \leq x_i, \tag{15}$$

with

$$a_w = \frac{(\frac{1}{2} - \frac{\alpha}{2\beta_w}) + \sqrt{(\frac{1}{2} - \frac{\alpha}{2\beta_w})^2 + (\frac{\alpha - \varepsilon_w}{\beta_w})}}{2} \quad,$$

$$b_w = \frac{(\frac{1}{2} - \frac{\alpha}{2\beta_w}) - \sqrt{(\frac{1}{2} - \frac{\alpha}{2\beta_w})^2 + (\frac{\alpha - \varepsilon_w}{\beta_w})}}{2} \quad,$$

$$\gamma_0 = 1 - \alpha_1 + \frac{1}{10}\alpha_2, \quad \bar{b}_0 = \frac{\alpha(\bar{b}_2 + \bar{b}_4)}{\varepsilon_w - \alpha} + \frac{\gamma_w \gamma_0}{\varepsilon_w - \alpha} - \frac{\delta_w}{\varepsilon_w - \alpha} \quad,$$

$$\gamma_2 = \frac{1}{2}\alpha_1 - \frac{6}{7}\alpha_2 - \frac{1}{2}, \quad \bar{b}_2 = -\frac{5\alpha\bar{b}_4}{3\alpha - \varepsilon_w - 6\beta_w} - \frac{\gamma_w \gamma_2}{3\alpha - \varepsilon_w - 6\beta_w} \quad,$$

$$\gamma_4 = \frac{9}{35}\alpha_2, \quad \bar{b}_4 = -\frac{\gamma_w \gamma_4}{5\alpha - \varepsilon_w - 20\beta_w} \quad.$$

with $P_2(x)$, $P_4(x)$ the second and fourth order Legendre polynomial, respectively. Note that a_i, b_i, a_w and b_w might be complex. The three unknowns C_1, C_2 and x_i will follow from the conditions on the ice-line:

$$C_1 f_1(x_i) + (\bar{a}_0 + \bar{a}_2 x_i^2) = T_i \quad, \tag{16}$$

$$C_2 f_2(x_i) + (\bar{b}_0 + \bar{b}_2 P_2(x_i) + \bar{b}_4 P_4(x_i)) = T_i \quad, \tag{17}$$

$$\rho_w c_w h K_{hw}(C_2 f_2'(x_i) + (\bar{b}_2 P_2'(x_i) + \bar{b}_4 P_4'(x_i))) = \rho_i c_i d K_{hi}(C_1 f_1'(x_i) + 2\bar{a}_2 x_i). \tag{18}$$

Here is $f_1(x) = F(a_i, b_i; 1; 1 - x^2)$, $f_2(x) = F(a_w, b_w; \frac{1}{2}; x^2)$. In stead of solving x_i from these equations, we solve Q. In contrast to x_i, the constants

$$\gamma_w = \left(\frac{Q}{\rho_w c_w h}\right) \text{ and } \gamma_i = \left(\frac{Q}{\rho_i c_i d}\right),$$

and so Q, occur linear in the equations. It is easy to express (the known) Q as function of x_i. Define the coefficients a_0, a_2, b_0, b_2 en b_4 as $(\bar{a}_0 + \frac{\delta_i}{\varepsilon_i})/\gamma_i, \bar{a}_2/\gamma_i, (\bar{b}_0 + \frac{\delta_w}{\varepsilon_w - \alpha})/\gamma_w, \bar{b}_2/\gamma_w$

en $\bar{b}_4/\gamma_w$, respectively. Then Q follows as

$$
Q = \left[\frac{\rho_i c_i dK_{hi} f_1'(x_i) f_2(x_i)(T_i + \dfrac{\delta_i}{\varepsilon_i}) - \rho_w c_w h K_{hw} f_1(x_i) f_2'(x_i)(T_i + \dfrac{\delta_w}{\varepsilon_w - \alpha})}{A - B - C + D} \right],
$$
(19)

with

$A = K_{hi}(a_0 + a_2 x_i^2) f_1'(x_i) f_2(x_i),$

$B = 2K_{hi} a_2 x_i f_1(x_i) f_2(x_i),$

$C = K_{hw}(b_0 + b_2 P_2(x_i) + b_4 P_4(x_i)) f_1(x_i) f_2'(x_i),$

$D = K_{hw}(b_2 P_2'(x_i) + b_4 P_4'(x_i)) f_1(x_i) f_2(x_i).$

We found Q (known) as function of x_i. From this inverse nonlinear equation, we can solve x_i by some numerical technique and so C_1 and C_2. The cases $x_i = 0$ and $x_i = 1$ have to be treated separately. If $x_i = 0$, the total ocean is covered with ice. It can be shown that in that case $C_2 = 0$, together with the condition $T(0) \leq T_i$. So, for Q:

$$
a_0 Q \leq \rho_i c_i d(T_i + \frac{\delta_i}{\varepsilon_i}) \ .
$$

If $x_i = 1$, the total ocean is ice-free. Again, it can be shown that in that case $C_1 = 0$, together with the condition $T(1) \geq T_i$. So, for Q:

$$
(b_0 + b_2 + b_4)Q \geq \rho_w c_w h \left(T_i + \frac{\delta_w}{\varepsilon_w - \alpha} \right).
$$

Since we now can calculate the coefficients C_1 en C_2, the solution is known.

4. Results

The solution of the climate model has been implemented in the software package MATLAB (1992). Care has been taken to insure that for the sum-representation of the hypergeometric functions F a reasonable number of terms is satisfactory. Therefore, we have used the connection formula between hypergeometric functions with argument x and $1 - x$ (see Abramowitz and Stegun (1964), eq. 15.3.6), since for arguments close to zero the convergence properties are better. For $0 < x_i < 1$ the Q-function, eq. 19, has been evaluated. Finally, for $x_i = 0$ and $x_i = 1$ the allowed Q's are calculated. With the parameters as in Table 1, we find the following result, see Fig. 2.

According this picture, three values for x_i can be found for which the quotient of $Q(x_i)$ and the actual value of Q equals 1. At the first intersection, $x_i = 0$, there is only ice; at the second intersection there is a small zone of water and the rest of the ocean is covered with ice. The third intersection delivers a realistic approximation of the actual situation. For this realistic case, the temperature as function of latitude ϕ is given in Fig. 3.

1. Geophysics

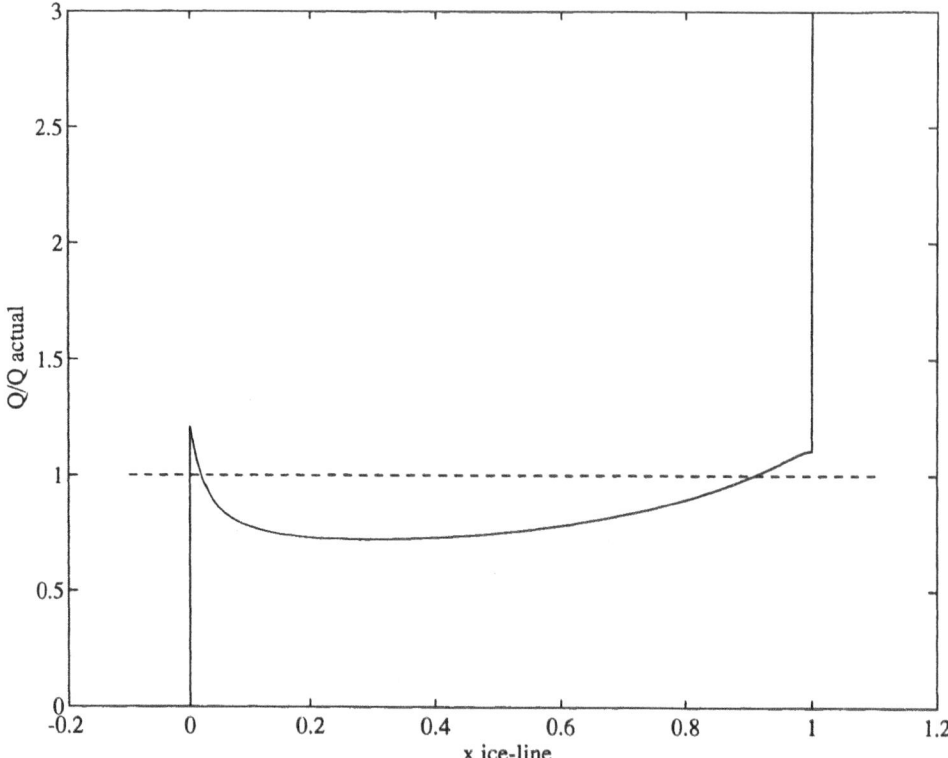

Figure 2. Q as fucntion of x_i.

The influence of the inclusion of the upwelling water has been studied and it turned out that the temperature at the equator decreases and at the pole increases, so the temperature profile flattens. As explained in North et al. (1981) this extra term results at the level of a two-node expansion in Legendre polynomials of an increase of the horizontal diffusion coefficient.

5. Stability

It depends on the parameters whether there are one or more solutions. For the values in Table 1 we found three solutions. Therefore, the question arises which solution is stable. In this section we shall perform a stability analysis. We write

$$T(x,t) = T^0(x) + \tilde{T}(x,t),$$
$$x_i(t) = x_i^0 + \tilde{x}_i(t),$$

with $T^0(x)$ and x_i^0 the stationary solutions and $\tilde{T}(x,t)$ and $\tilde{x}_i(t)$ the perturbations. It is possible to deduce partial differential equations for $\tilde{T}(x,t)$, ($0 \le x \le x_i$, and $x_i \le x \le 1$) and a set of conditions on the ice-line. Assuming that the perturbations are small, we can expand these conditions in a Taylor series. The next assumption is that $\tilde{T}$ and $\tilde{x}_i$ can be

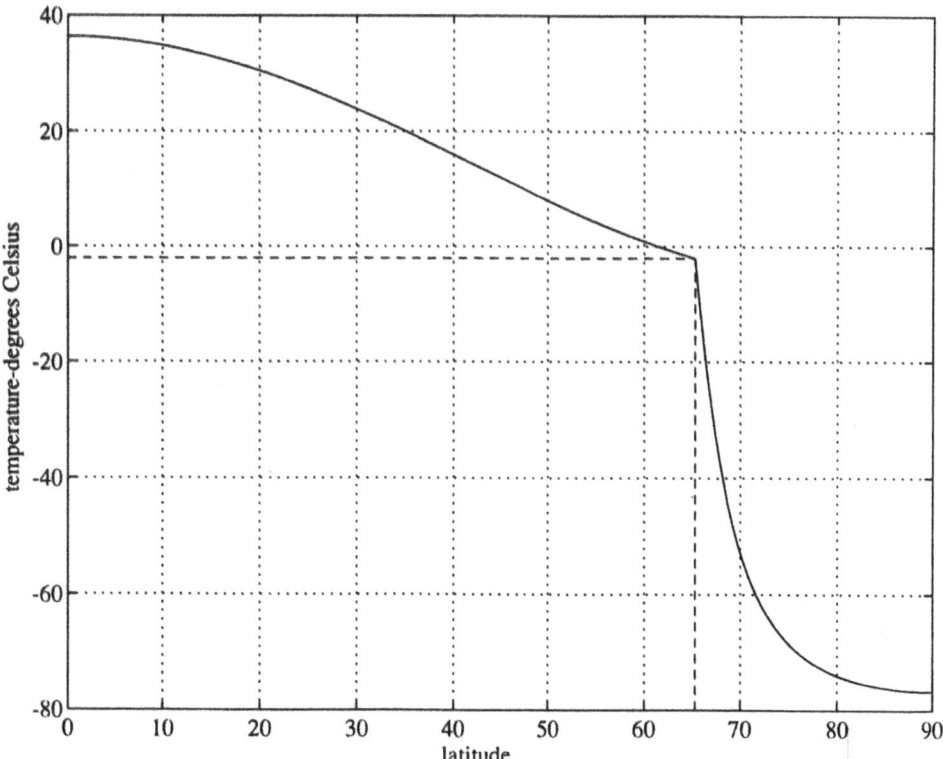

Figure 3. T as function of the latitude ϕ.

written as $\tilde{x}_i(t) = e^{\sigma t}\xi$, and $\tilde{T}(x,t) = e^{\sigma t}\hat{T}(x)$, where σ is the eigenvalue. In that case we can solve for $\hat{T}(x)$ as

$$\hat{T}_{ice}(x) = C_3 F(\hat{a}_i, \hat{b}_i; 1; 1 - x^2) \equiv C_3 f_3(x), \quad x_i \le x < 1,$$
$$\hat{T}_{water}(x) = C_4 F(\hat{a}_w, \hat{b}_w; \tfrac{1}{2}; x^2) \equiv C_4 f_4(x), \quad 0 \le x \le x_i,$$

where the coefficients are

$$\hat{a}_i = \frac{1}{4} + \frac{1}{4}\sqrt{1 - 4\frac{\varepsilon_i + \sigma}{\beta_i}}, \qquad \hat{b}_i = \frac{1}{4} - \frac{1}{4}\sqrt{1 - 4\frac{\varepsilon_i + \sigma}{\beta_i}},$$

$$\hat{a}_w = \frac{(\frac{1}{2} - \frac{\alpha}{2\beta_w}) + \sqrt{(\frac{1}{2} - \frac{\alpha}{2\beta_w})^2 + (\frac{\alpha - \varepsilon_w - \sigma}{\beta_w})}}{2}, \qquad \hat{b}_w = \frac{(\frac{1}{2} - \frac{\alpha}{2\beta_w}) - \sqrt{(\frac{1}{2} - \frac{\alpha}{2\beta_w})^2 + (\frac{\alpha - \varepsilon_w - \sigma}{\beta_w})}}{2}.$$

The interface conditions are then

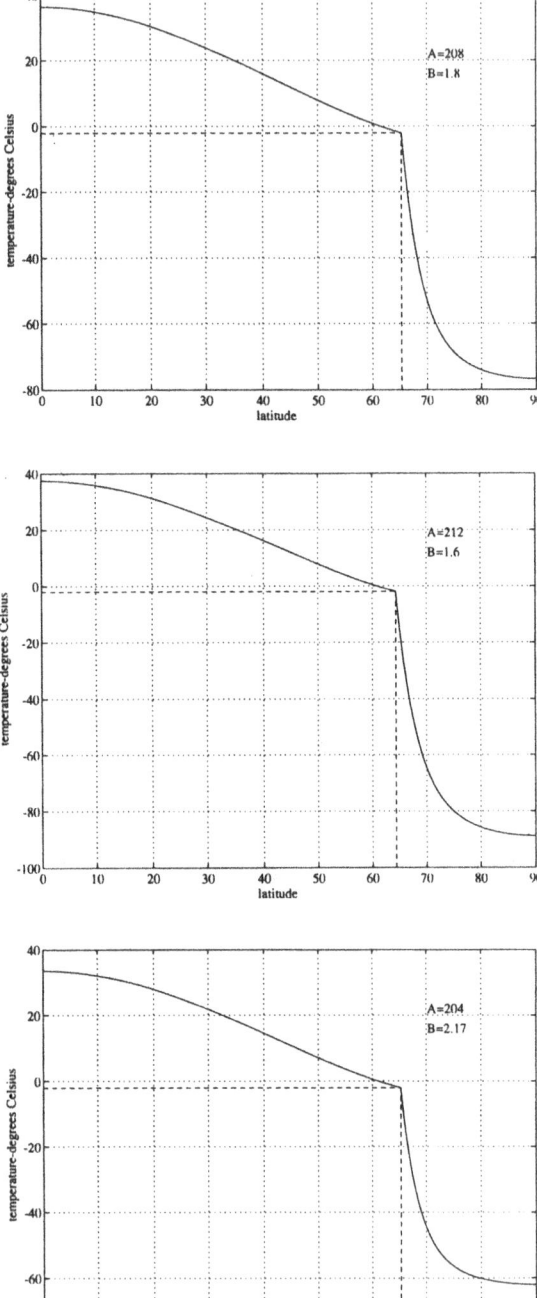

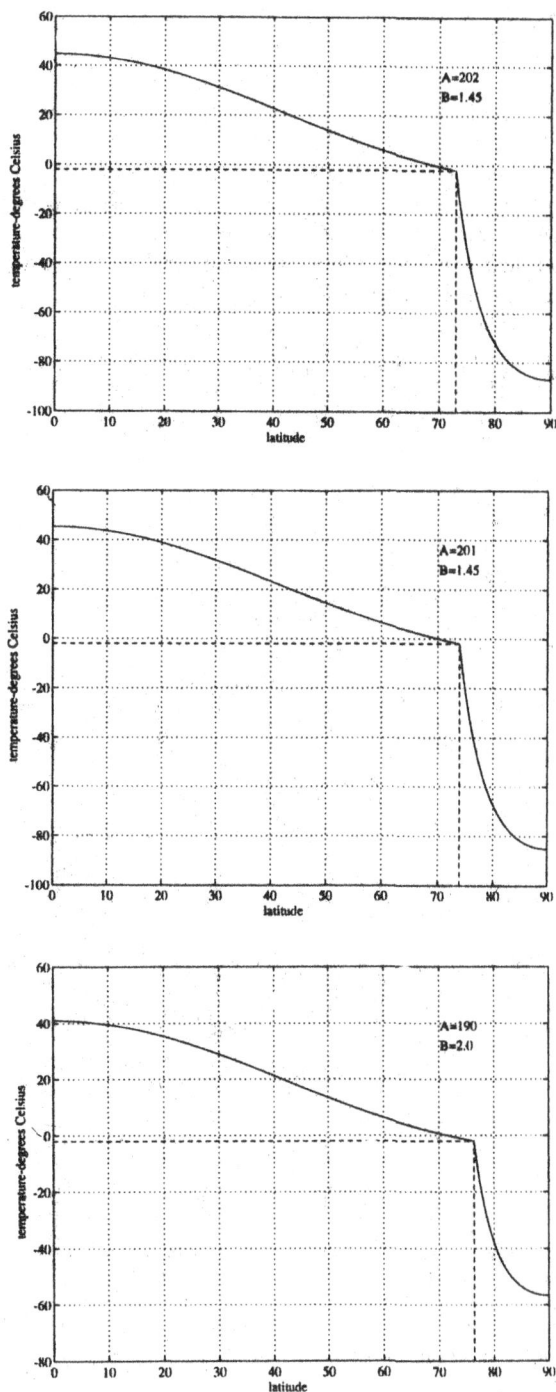

Figure 4. *T* as function of the latitude for 6 combinations of *A* and *B*.

$$\xi T^{0'}_{ice}(x^0_i) + C_3 f_3(x^0_i) = 0,$$

$$\xi T^{0'}_{water}(x^0_i) + C_4 f_4(x^0_i) = 0,$$

$$-A_w(1 - (x^0_i)^2)\{\xi T^{0''}_{water}(x^0_i) + C_4 f_4'(x^0_i)\} =$$

$$-A_i(1 - (x^0_i)^2)\{\xi T^{0''}_{ice}(x^0_i) + C_3 f_3'(x^0_i)\} + A_3\sigma\xi .$$

with $A_w = \rho_w c_w h K_{hw}$, $A_i = \rho_i c_i d K_{hi}$, $A_3 = \rho_i d R^2 L$. For these three linear homogeneous equations with three unknowns (G_3, G_4, ξ), there exists no-trivial solutions only if the determinant of this system vanishes. Using the knowledge of the stationary solutions, this condition becomes the following transcedental equation in σ

$$A_3\sigma = \left(-A_w(1 - (x^0_i)^2)T^{0'}_{water}(x^0_i)\right)\left(\frac{f_3'f_4 - f_3 f_4'}{f_3 f_4}\right)_{x=x^0_i} +$$

$$-A_w(1 - (x^0_i)^2)T^{0''}_{water}(x^0_i) + A_i(1 - (x^0_i)^2)T^{0''}_{ice}(x^0_i).$$

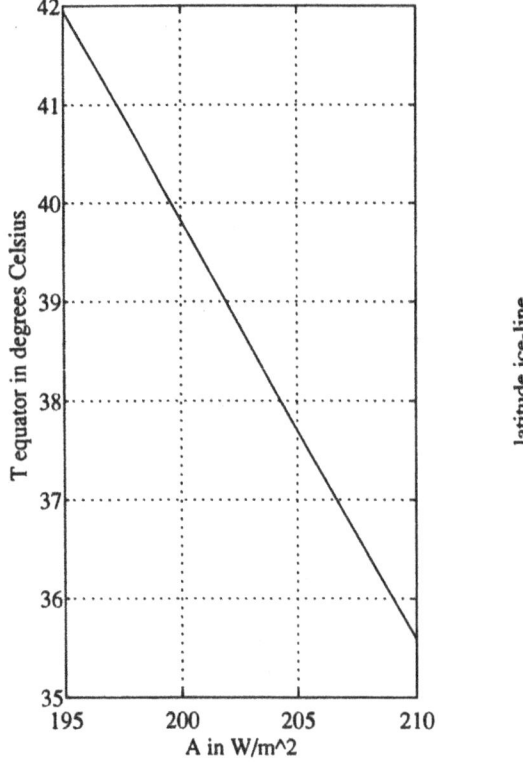 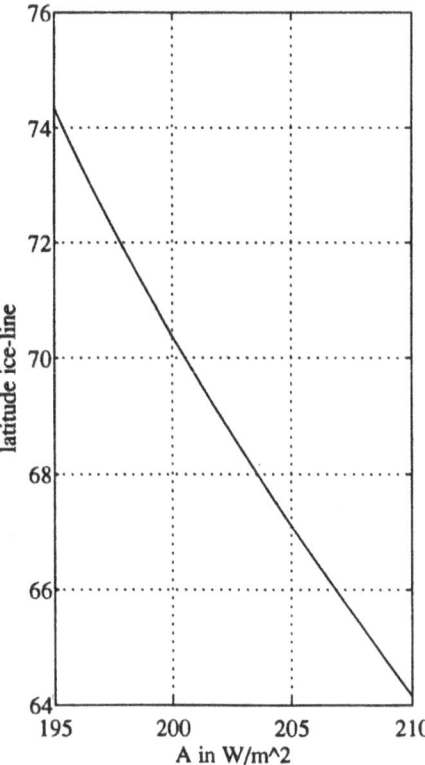

Figure 5. $T_{equator}$ and x_i as function of A.

Note that σ occurs implicitly in f_3, f_3', f_4 and f_4' in the formula above. For each value of x^0_i there exists an infinite sequence of complex numbers σ. If for all σ, $\Re(\sigma) < 0$, the solution $T^0(x)$ is stable. North et al. (1981) proved under the somewhat restrictive assumptions $u_z = 0$, $A_w = A_i$, and $L = 0$ that $T^0(x)$ is stable, provided that $Q'(x_i^0) > 0$ holds. We have implemented a numerical technique to solve the stability equation and found for the values in Table 1, $x_i^0 = 0.90851$ (so $\phi_i^0 = 65.3008^0$) and for the largest σ: $\sigma = -0.10533$, so $T^0(x)$ is stable as expected. At the other hand, the ice-line at $x_i^0 = 0.01780$ (so $\phi_i^0 =1.0198^0$) gives as largest σ: $\sigma = 0.17245$, which implies that this solution is unstable.

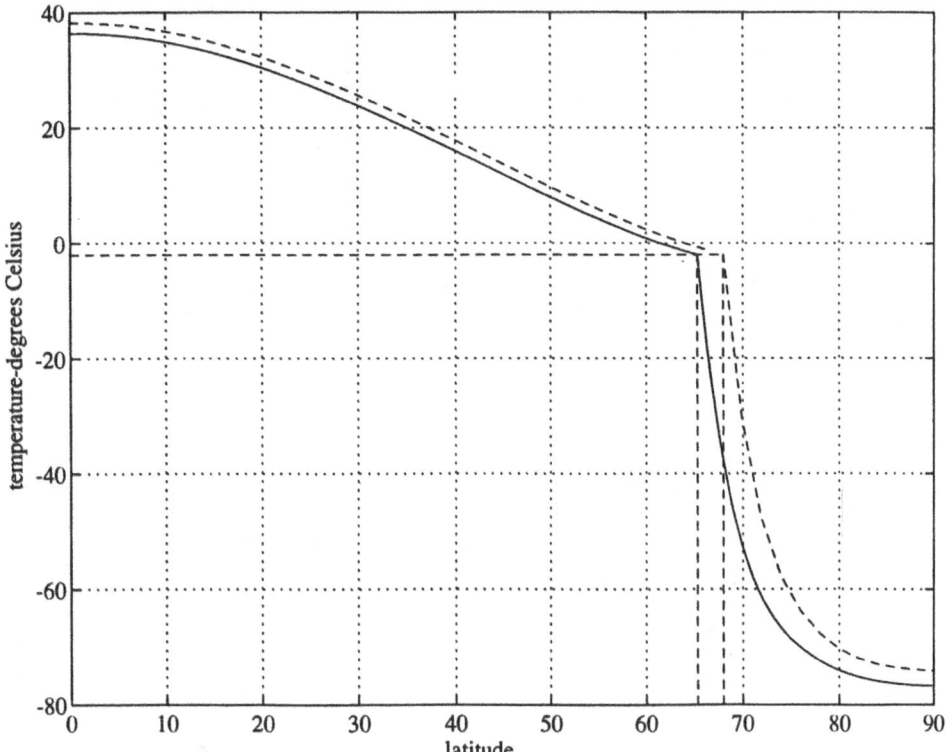

Figure 6. Actual temperature (——) and temperature for twice as large concentration CO_2 (- -).

6. Sensitivity

As a form of sensitivity analysis we exhibit some results with this model for six combinations of the essential parameters A and B, as found in the literature, see table 2.

The other parameters values can be found in Table 1. We find the following temperature curves, see Fig. 4. Since it becomes clear that for different set of parameters the results are quite different, it is interesting just to vary one parameter to see how influential that one is. Therefore we have chosen the parameter A, since that one describes the back-radiation of the earth, and is ultimately connected with an increased Greenhouse Effect. It has been found that a doubling of the concentration of CO_2 causes a reduction in A of 4.32

W/m^2 (IPCC, 1990). In Fig. 5 the temperature at the equator and the location of the ice-line x_i is given as function of A. If the concentration of CO_2 doubles, the temperature will increases globally with a value of 2 ^{0}C, and the location of the ice-line moves polewards with at least 2^0, see Fig. 6.

Table 2 Parameters from literature

A	B	found in	A	B	found in
208	1.8	Watts and Morantine (1990)	212	1.6	Cess (1976)
204	2.17	Henderson-Sellers and McGuffie (1987)	202	1.45	Budyko (1969)
201	1.45	North (1975)	190	2.0	Lin et al. (1991)

Acknowledgements

The authors thank B.J. de Haan (National Institute of Public Health and Environmental Protection, Bilthoven) and H.A. Dijkstra (Institute for Marine and Atmospheric Research, Utrecht) for a number of stimulating discussions.

References

Abramowitz, M. and I.A. Stegun (1964), *Handbook of mathematical functions*, Dover Publications, Inc, New York, U.S.A.

Budyko, M.I. (1969), 'The effect of solar radiation variations on the climate of the earth', *Tellus* **21**, 611-619.

Cess, R.D. (1976), 'Climatic change: An appraisal of atmospheric feedback mechanisms employing zonal climatology', *J. Atmos. Sci.* **33**, 1831-1843.

Henderson-Sellers, A. and K. McGuffie (1987), *A climate modelling primer*, John Wiley & Sons, Chichester, U.K.

IPCC (1990), *Climate Change: The IPCC Scientific Assessment.* Report prepared for the Intergovernmental Panel on Climate Change by Working Group 1., J.T. Houghton, G.J. Jenkins and J.J. Ephraums (eds.), Cambridge University Press, Cambridge, U.K.

Lin, R.Q., H. Kreiss, W.J. Kuang and L.Y. Leung (1991), 'A study of long-term climatic change in a simple seasonal nonlinear climate model', *Climatic Dynamics* **6**, 35-41.

MATLAB (1992), *Reference Guide*, The Math Works, Inc., Natick, U.S.A.

North, G.R. (1975), 'Analytical solution to a simple climate model with diffusive heat transport', *J. Atmos. Sci.* **32**, 1301-1307.

North, G.R., R.F. Cahalan and J.A. Coakley, Jr., (1981), 'Energy balance climate models', *Reviews of Geophysics and Space Physics* **19**, 91-121.

Rotmans, J. (1990), *IMAGE: An Integrated Model to Assess the Greenhouse Effect*, Kluwer Academic Publishers, Dordrecht, The Netherlands.

Watts, R.G. and M.E. Hayder (1984), 'A two-dimensional, seasonal, energy balance climate model with continents and ice-sheets: testing the Milankovitch theory', *Tellus* **36A**, 120-131.

Watts, R.G. and M. Morantine (1990): 'Rapid climate change and the deep ocean', *Climatic Change* **16**, 83-97.

CLIMATE MODELLING AT DIFFERENT SCALES OF SPACE

G. MARACCHI

Institute of Agrometeorology and Environmental Analysis applied to Agriculture
P.le delle Cascine, 18 - 50144 Florence, Italy.
e-mail: Maracchi@sunserver.iata.cnr.it

Summary

Climatic changes ask for new tools to understand the relationships between atmosphere, vegetation and soil. At the present new methods are made possible by applying technologies as Remote Sensing, Telecomunication, computer science and Geographical Information Systems. Applied climatology investigates the variation in space and time of meteorological parameters relevant for the biological processes in an agroecosystem. The purpose of the work is to build up a methodology to extrapolate the values of maximum and minimum temperature starting from a reference station. This was done by classifying weather types and their relationships with the topographic parameters such as elevation, slope, aspects, valley type, etc. Daily minimum and maximum temperatures are computed from the reference station. Compared with the experimental data they show a good agreement.

1. Introduction

Climatology was, up to some decades ago, mainly a field of pure scientific investigations. In the last decade there was an ongoing challenge in the direction of applications, and climatology has become a matter of operational and professional interest. This change has happened because of several factors : the interest in environmental monitoring and rehabilitation, new technologies in collecting and processing data, the evidence of possible climatic change and the development of new technologies. The increasing popularity of "applied climatology" is related to the shift from very general scale applications dealing with large regions of the earth, the macroclimate, to small scale, the topo- and microclimate. Monitoring of and spraying against pests, defining the best sites for production fruit and wine, controlling erosion and desertification, forecasting crop yields in developing countries, deciding on land utilization in industrialised and developing countries and many other useful applications in agriculture are possible fields of interest for applied climatology. Moreover, there are opportunities in other directions such as transport, pollution, building, energy and recreation.

This new role of climatology in science and its application is well attended by many international programmes and conferences within the last ten years. Examples are the European Programmes on Climatology and Natural Hazards started in 1985, on Environment started in 1991, now reinforced in the IVth Framework Program of the EEC, to be launched in 1994, the International Biosphere and Geosphere Program (IBGP), the Global Environmental Monitoring System (GEMS) of the UNEP and the WMO. The World Climate Conference, held in Geneva in 1990, acted as an intermediate summary state of the art of the work done by the Intergovernmental Panel on Climatic Change (IPCC), the Rio Conference and conventions on Climate were signed at. A similar declaration on desertification is signed in Nairobi in April 1993. This presentation aims to outline the possibilities offered by the integration of new technologies in to a well designed system so as to enlarge the practical utilization of climatology and describe the distributions in time and space of meteorological parameters, taking into account their non linear relationships. Very often these are the result of linear relationships which change in time over different meteorological scales. This assumption is at the basis of our approach.

1.1 The scale factor

Meteorology and climatology are closely related, mainly when we move from descriptive to applied climatology. We can define the latter as "the analysis of behaviour of meteorological parameters in time and space based on the study of past events which is used to forecast the probability of future events on a daily basis and at an appropriate scale of topography". To clarify the concept we can outline some practical examples:

1) In a vineyard area the agrometeorological service of a Province could forecast the probability of the occurrence of a pest, for example, Plasmopara Viticola on the basis of meteorological parameters. The model on which the advice is based considers leaf wetness and air temperature close to the leaves. On a hilly area, where vineyards are grown, leaf wetness and temperature have large spatial variability, due to land morphology, its slope, aspect, dimension of the valley, type of soil and so on. Synoptic meteorological measurements are made in standard screens, the distribution of which, even if very dense, cannot represent all situations. *Thus, the distribution of meteorological parameters over the area, is not linearly related to measurements taken in the screen.* Therefore, we need rules to convert these data in a geographically accurate distribution. Applied climatology should act as the means to produce these rules.

2) Vegetation is one of the means to control the effect of increasing CO_2, to filter the dust, to control the microclimate of urban and industrial areas. In many European countries there are not regulations to assess town and industrial planning in such a way that these problems are taken into consideration. The reasons are because the sensitivity of politicians to these problems is very low, secondly because the sensitivity of the electorate also is not yet very high with the exception of the Green movements, which are very often against the inappropriate use of new knowledge and technologies and thirdly, because, for the time being, there are very few professionals in these fields. To plan the distribution of vegetation, we need very accurate study, not only of the present situation, but also with respect to future land planning. To do this we need climate modelling methodology which considers several aspects of the system that are *often inter-related and spatially distributed non linearly.*

3) In the last decade a concern has diffused into the scientific community due to the possible effect of increasing CO_2 and other such gases as nitrogen oxide, CFCS and methane. The already mentioned programs, that strongly depend on computer simulations, are dealing with the possible effects on the earth's ecosystems of a huge climate change. In a very short time a large and even still increasing number of investigations have been financed and started to assess future scenarios. Results of modelling at different scales of space and time are at present still contradictory. However we notice how results are improving quickly due to the increased number of scientists now involved in the Global Change research community and to the increasing financial support of International Agencies such as the U.N, the E.C., and of national governments.

One problem arising from efforts to model the global circulation system is to understand the relationship between different scales of time and space in term of the driving systems and feedbacks. Increasing temperature could affect the length of growing period of crops and vegetation and therefore their evapotranspiration; the mechanisms of general circulation could be affected leading to a shift in types of climate between latitudes. For instance, in the Mediterranean area during the last five years we have experienced reduced variability of the daily weather, which is a characteristic of this climate leading to an extended period of drought in winter and long and continuos period of rainfall the autumn. The total amount and intensity of rainfall are changing. In October 1992, in Florence, we measured 550 mm of rainfall against a maximum of 323 mm in 1935 on a series of 200 years of measurements with 45 mm in a 30 minute period, a rainfall intensity at tropical proportions! Intensive exploitation of the land surface could change the albedo and consequently the energy budget of Earth. The change that we have experienced could be the consequence of the change in the energy budget of the earth in a way probably much more complicated than our models are predicting for the time being. Also in order to model the impact of climate change we need better precision in the distribution in time and space of meteorological parameters. Climatology, until recent years, dealt with average description of the climate of a region. When we apply climatological methodologies to model the impact of climate change to agriculture or to ecosystems, we take account of extreme events and the daily variability of the weather because plants and many animals are very sensitive to small changes in these parameters. Therefore we need new methodologies to simulate the distribution in space of climatic parameters.

2. The technological framework

Detailed monitoring of an area or region cannot be done via conventional measurements, such as meteorological standard screens, due to the high cost of their installation and maintainance. Therefore, there is a limit to the spatial density of a meteorological network. The development of new technologies such as remote sensing, Radar and Geographical Information Systems should help to reach a more detailed understanding of the distribution of meteorological parameters in space and time. Satellites, such as NOAA, with a pixel size of 1 km^2 measuring the surface temperature of the globe, and Landsat, can be used to study temperature patterns at this scale and their effect on vegetation using a vegetation index, NDVI. A cumulative index can also be used to assess the agroclimatic classification of an area (Fig.1). Such images can be input for a Geographical Information System. The outcome can be compared with theoretical calculation of, for instance, the distribution of solar radiation, or the morphology, elevation contours lines, width and length of valleys, etc. Also, the distribution of rainfall shows a very high degree of variability with coefficients of variation (up to 100%). A new version of compact and low cost radar could map rainfall distribution, and interpolate rainfall data from a meteorological network. Such technologies ask for new professional competence both from scientists and professionals but, first of all, from universities and technical schools to train new generations in the use of these technologies.

3. Modelling the variability of climatic parameters in space and in time

The variability in space and in time of meteorological parameters should be examined on the basis of the main categories of meteorological processes. On the basis of the following

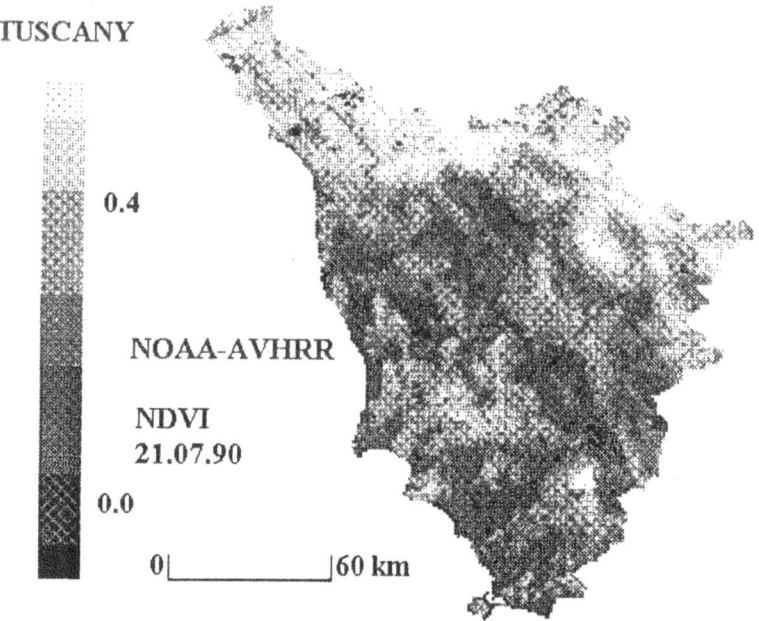

Figure 1. Normalized Difference Vegetation Index - NDVI - NOAA Satellite of Tusca-
ny - July 21th, 1990. High values NDVI are associated with moistened areas, lower values
depict dry areas.

considerations we chose maximum and minimum temperature to illustrate the methodology:
- temperature measurements are done with a higher spatial resolution than others parameters such as solar radiation, evapotranspiration or air humidity;
- temperature is the result of the type of air mass and a regional energy budget and
therefore integrates the processes of: solar radiation, turbidity of atmosphere, turbulence
due to wind, Bowen ratio, etc.,;
- temperature strongly affects the physiology of plants and animals.

Minimum temperature is related to:
- cloudiness
- air humidity
- temperature at sunset
- soil moisture
- length of the night
- type of air mass
- stability or instability
- land morphology
- land cover

Maximum temperature is related to:
- minimum temperature
- solar radiation
- turbulence
- Bowen ratio
- type of air mass
- slope, aspect and elevation of the surface

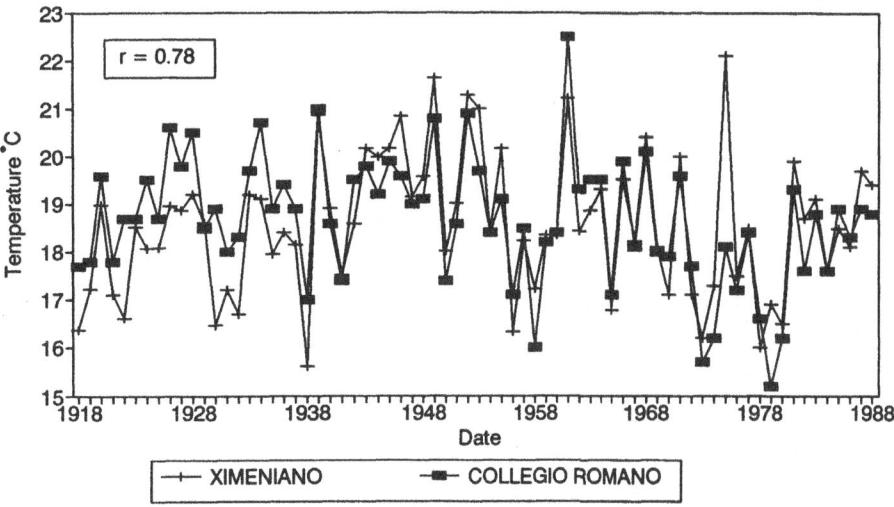

Figure 2. Monthly average maximum temperature during April for Rome (Collegio Romano) and Florence (Ximeniano) from 1918-1988.

Computation of maximum and minimum temperatures is of practical importance in view of their impact on plants in relation to drought, frost, cold damages and pest attack. The distribution of minimum and maximum temperature in space can be obtained in two ways; by interpolation between two or more points or by extrapolation from one point. Interpolation, by means of statistical methods such as the inverse of the distance, kriging, based on variogram, or spline functions. For these a network of measurements sufficiently dense is needed. This situation occurs very seldom. Extrapolation, on the other hand, needs very accurate rules to link temporal change with spatial variability and landscape features.
It is noted that interpolation and extrapolation should be considered as convergent methodologies. We dealt with the second approach.
  Change of temperature in space is not related linearly to a single parameter that describes land morphology as, for instance, elevation, and not always in the same way in time. The approach we take is based on the assumption that there are common features among points in space subjected to the same macrometeorological conditions. For instance, the relationship between the average monthly maximum temperature in April of a long time series from Rome and Florence, has a high correlation even if they are 300 km apart with a very hilly area between them and very different positions with respect to the sea (Fig. 2). On the other hand, if we compare the series of minimum temperatures of two stations 10 km apart, we can observe a low correlation despite a similar general trend (Fig.3). Since correlation is related to the value of the variation in time, we need a means to establish a correction to compute one from the other. This should be done under the following assumptions;

- variability of macroclimate in time is related to the type of weather determined by the type of air mass;
- variability of topoclimate is related to the land characteristics.
We built a matrix in which the columns represent the type of weather and the rows those land characteristics that affect the spatial distribution of temperatures. Type of weather influence minimum and maximum temperature in a different way and therefore we built a separate matrix for each parameter.

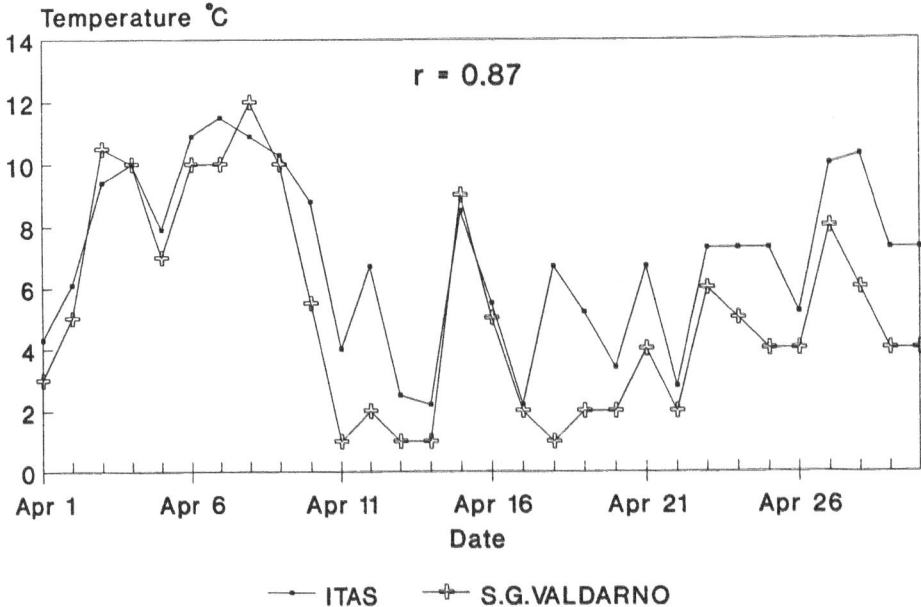

APRIL 1990
MINIMUM AIR TEMPERATURE

Figure 3. April daily minimum temperature of Florence and S. Giovanni Valdarno - 1990-91.

We applied the analysis to the province of Florence covering approximately a surface of 70 x 40 Km. We performed cluster analysis on the historical series of April maximum and minimum temperature from a reference screen at I.A.T.A (Institute of Agrometeorology and Environmental Analysis applied to Agriculture) in Florence, since in this month extreme minima can cause serious damage to vineyards, fruit trees and orchard crops. The analysis should classify different types of weather. Preliminary analysis was done on the Landsat and NOAA data in the thermal infrared (IR) channel to understand the scale of variability of ground temperature in order to choose the size of the reference pixel in our system of extrapolation. The analysis showed (Fig. 4) that at the scale of the Landsat pixel in the IR channel, (120 m x 120 m), at the time of the passage (09:30 a.m.), the effect of

GREVE

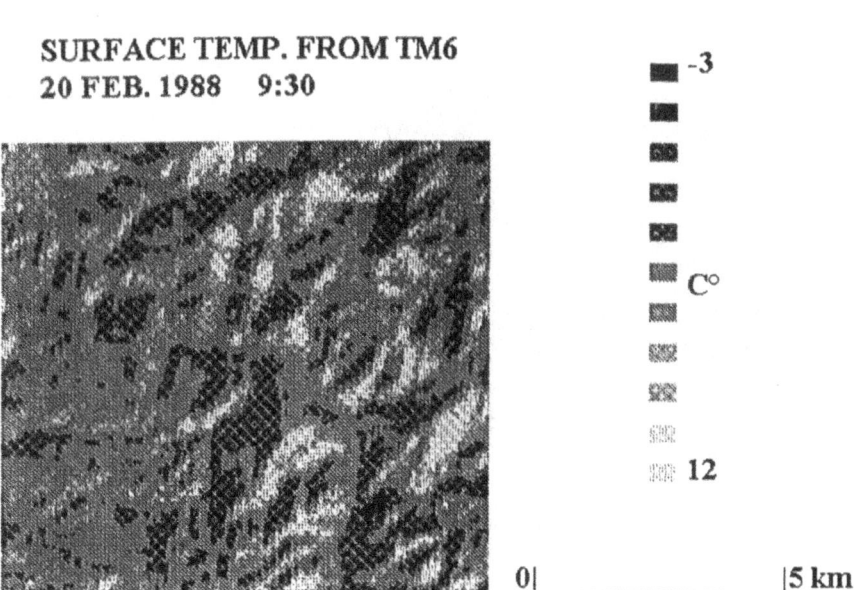

SURFACE TEMP. FROM TM6
20 FEB. 1988   9:30

-3

C°

12

0|_____|5 km

Figure 4. Landsat image in thermal IR channel for experimental area of Greve (10 km ×
10 km) - February 20th, 1988.

slope and aspect was significantly correlated with the insolation computed on a Digital
Terrain Model having the same size of pixel (Fig.5). In contrast, in the case of the NOAA
pixel, (1 km x 1 km), the distribution of temperature is much more dependent on the
general feature of the landscape: its elevation, size and shape of valleys and the distribution of hills and mountains (Fig.5, 6 and 7). Therefore we chose for this first approach a
scale of 1 km x 1 km consistent with the NOAA pixel. On this basis we built a Digital
Terrain Model for the province of Florence (Fig.8) that should be used to apply the values
of the matrix to each pixel and extrapolate the value from the standard screen of
reference.

3.1 Results

From the cluster analysis of the standard screen at I.A.T.A. we obtained six types of
weather, classified as three for the minimum and three for maximum daily temperatures
(Tab.1). We chose ten standard screens which measure maximum and minimum temperature, and are distributed in the Province area according to the main land features:
elevation, type of valley (large, narrow, very narrow), top of hill, mountains. We compared the value of minimum and maximum temperatures of the cluster with the value of the
stations in order to complete the weather matrix.

GREVE

SOLAR RADIATION
20 FEB. 1988   9:30

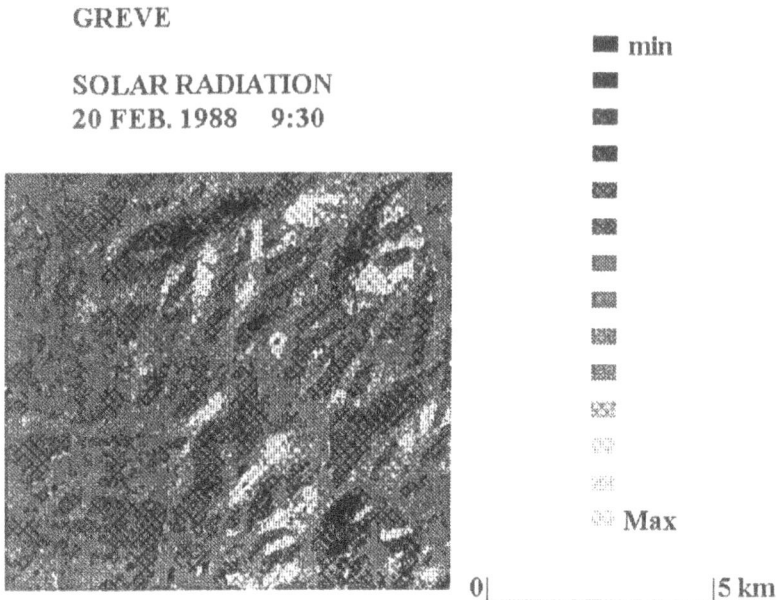

Figure 5. Solar radiation computed by a Digital Terrain Model of Greve area - with a pixel 100 × 100 m2 - February 20th, 1988.

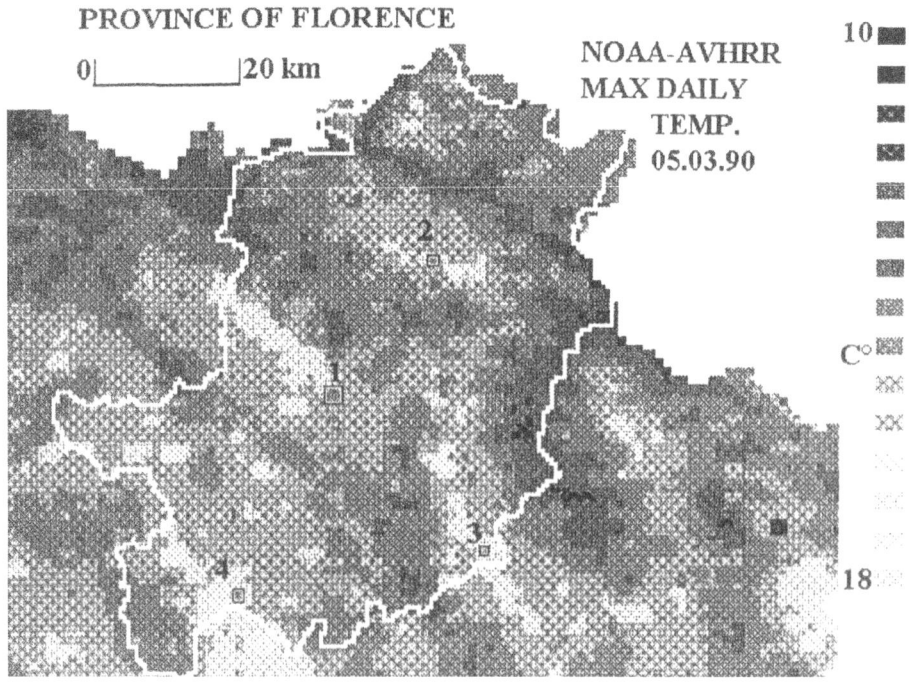

Figure 6. NOAA image in thermal IR channel with pixel 1 km × 1 km -March 5th, 1990.

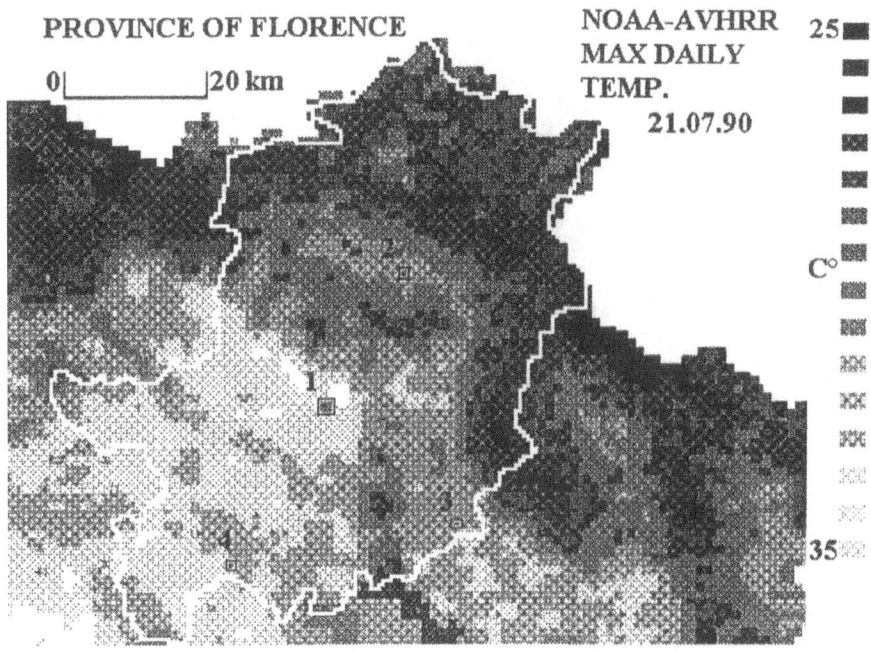

Figure 7. NOAA image in thermal IR channel with pixel 1 km × 1 km - July 21th, 1990.

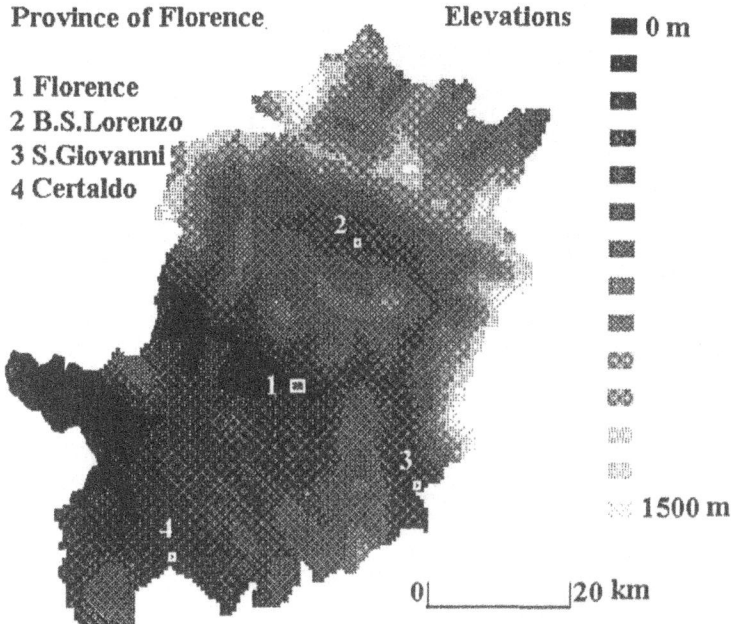

Figure 8. Digital Terrain Model of Province of Florence.

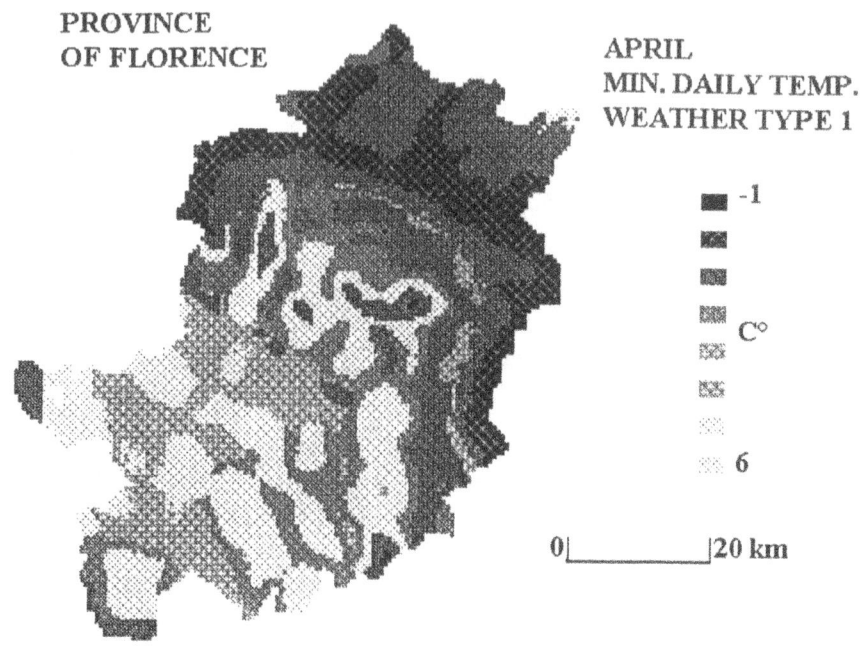

Figure 9a. Minimum daily temperature map for weather type 1 - April 26th, 1991.

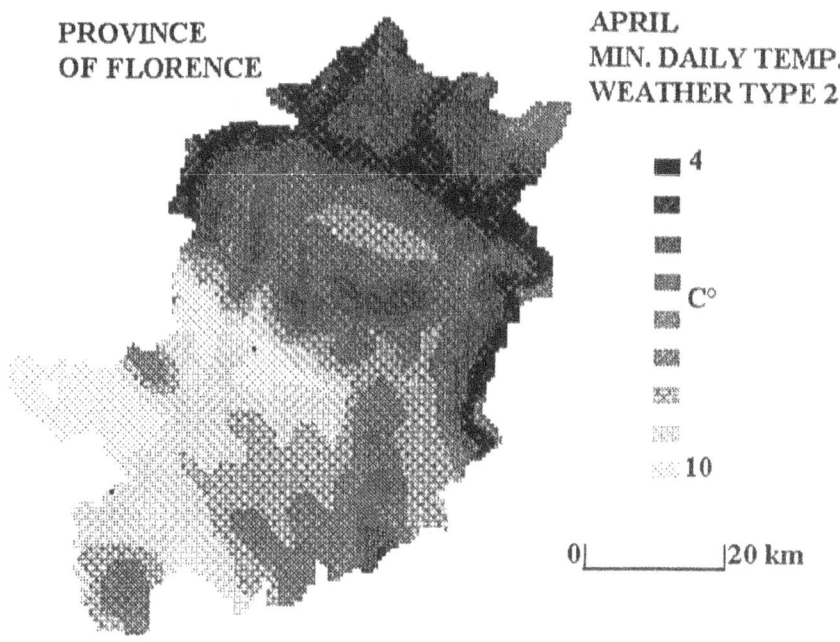

Figure 9b. Minimum daily temperature map for weather type 2 - April 27th, 1990.

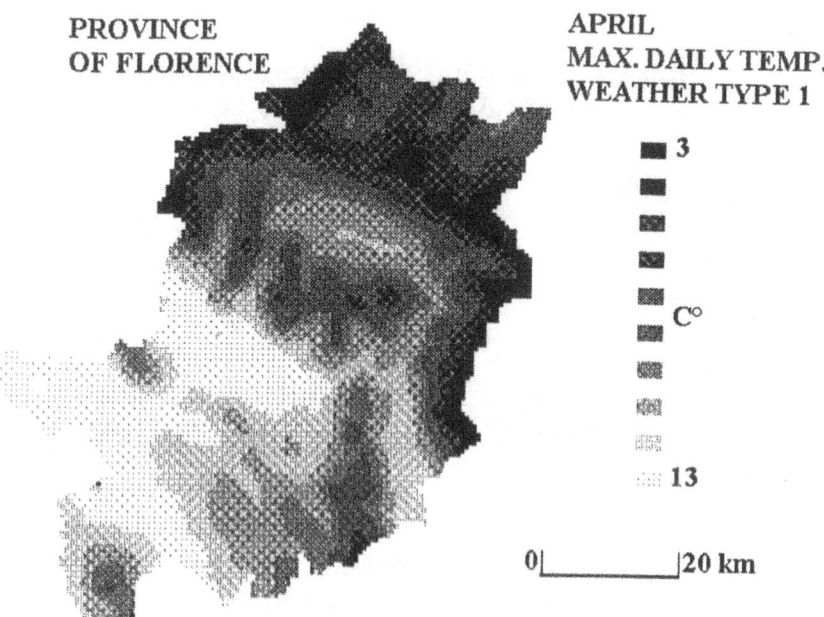

Figure 10a. Maximum daily temperature map for weather type 1 - April 16th, 1990.

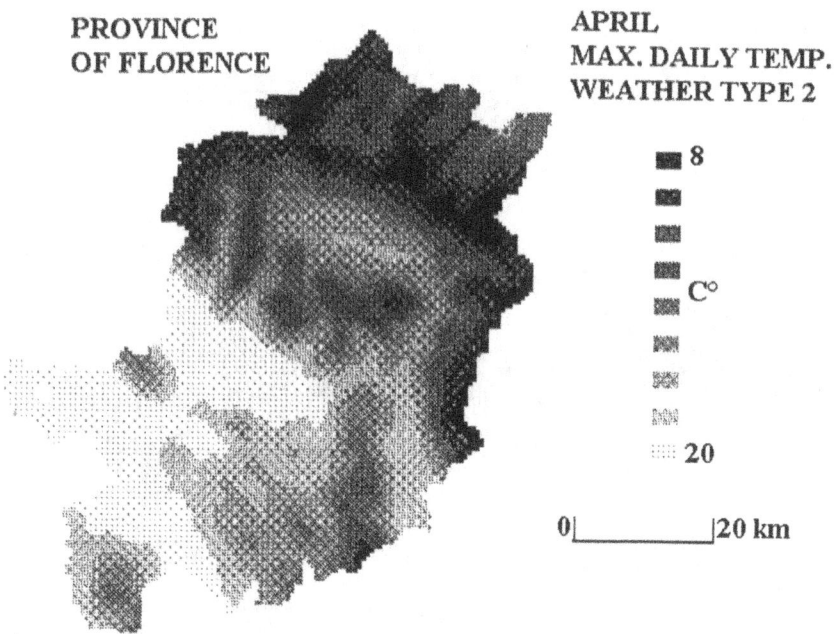

Figure 10b. Maximum daily temperature map for weather type 2 - April 21th, 1990.

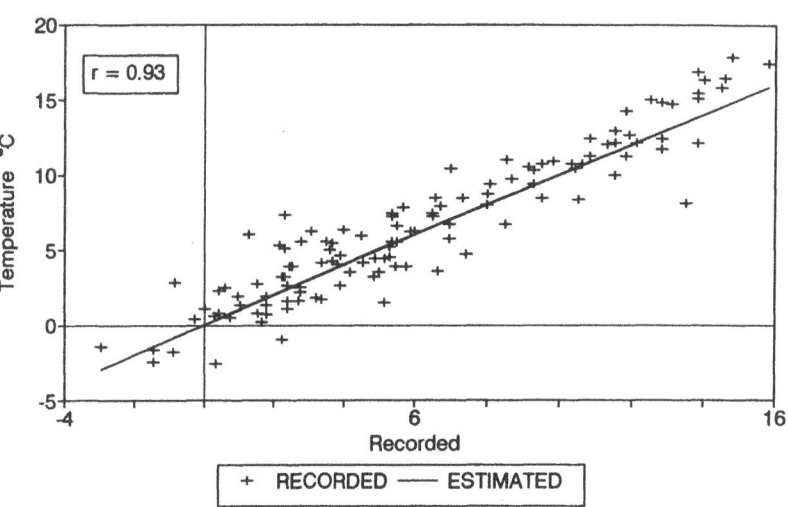

Figure 11a.   Comparison between observed and simulated April temperatures for
Vallombrosa.

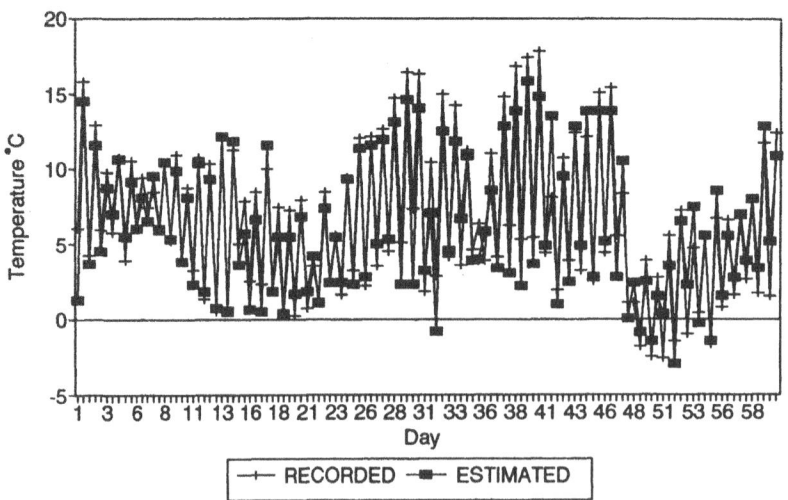

Figure 11b.   Comparison between observed and simulated April temperatures for

Vallombrosa.

We assigned each pixel of the area to one class of land and we computed the daily value of minimum and maximum for each pixel, only on the basis of the daily data of the reference screen (Fig. 5, 9 and 10). To verify the consistency of computed data with experimental data we compared the results with some stations not included in the previous group. The correlation between observed and predicted (Fig. 11), and simulation of meteorological conditions of a station at 1000 m high and 30 km from the reference screen located at 50 m, was highly consistent.

Table 1. Classification of types of weather for the April month in province of Florence, Italy for minimum and maximum temperatures

Minimum temperature						
Class	Strong	Inversion Moderate	No	Wind direction	Air humidity	Min. Temp.
1	*				V.L.	V.L.
2		*		N-NW	L.	L.
3			*	W-SW	H.	H.
Maximum temperature						
Class	Solar radiation	Wind direction	Wind speed	Air humidity	Max. Temp.	Temp. range
1	V.L.	S-SW	A.	H.	V.L.	V.L.
2	A.	W-NW	H.	A.	A.	A.
3	H.	N-NW	V.L.	V.L.	H.	H.
V.L. Are standard deviation below the mean value						
A. Are the average						
H. Are standard deviation above the mean value						

4. Conclusions

The methodology presented is based on the use of synoptic meteorological knowledge, topo- and micrometeorology together with technologies of remote sensing analysis and Geographical Information System techniques. The development of such a complex systems for analysing an area offers a new and valid approach that allows of flexible rules for the distribution in time and in space of physical parameters that can be organised in a nonlinear way. The possibility to compute accurately the conditions at each point of an area could be the basis for improving agricultural management, pollution control and also for drawing scenarios of future climatic constraints in a way more reliable than is done at present.

Acknowledgements

The author wish to thank Dr. F.Maselli, Dr. F.Pittalis and Dr. C.Palchetti for helping in image and data processing.

References

Barry, R.G. (1992), 'Mountain weather and climate'. London, Routledge.
Bindi, M., Maracchi, G., Miglietta, F. (1992), 'The effect of climatic change on biomass production in Italy: a preliminary result'. *Proc. of the 7th European Conference on Biomass for Energy and Environment, Agriculture and Industry*, Florence October 5-9 1992.
Boden, T.A., Sepanski, R.J., Stoss, F.W. (1992), 'Trends'91: a compendium of Data on Global Change - Highlights'. Carbon Dioxide Information Analysis Center, ESD Publication no.3797.
Conese, C., Maracchi, G., Maselli, F. (1991), 'Improvement in maximum likelihood classification performance on highly rugged terrain using principal component analysis'. *Int. J. Remote Sensing* (in press).
Conese, C., Maracchi, G., Maselli, F., Romani, M., Bottai, L. (1991), 'Integration of remotely sensed data into a GIS for the assessment of land suitability'. *Proc. of EARSel Workshop on 'Relationships of Remote Sensing and Geographic Information Systems'*, Hannover, Germany, September 16-18 1991.
Fantechi, R., Maracchi, G., Almeida-Teixeira, M.E.(eds.) (1991), 'Climatic change and impacts: A general introduction'. *Proc. of the European School of Climatology and Natural Hazards course*, held in Florence from 11 to 18 September 1988. Luxembourg, Commission of the European Communities.
Flohn, H.(editor) (1969), 'General Climatology, 2'. *World Survey of Climatology*, **2**, Amsterdam, Elsevier Publishing Company.
Péguy, Ch.P. (1970), 'Précis de climatologie'. Paris, Masson & Cie.
Saucier, W.J. (1989), 'Principles of Meteorological Analysis'. New York, Dover Publications.
Sneyers, R. (1975), 'Sur l'analyse statistique des séries d'observations: note technique n.143'. OMM n.415.
Yoshino Masatoshi M. (1975), 'Climate in a small area: an introduction to local meteorology'. University of Tokio Press.

SIMULATION OF EFFECTS OF CLIMATIC CHANGE ON CAULIFLOWER PRODUCTION

JØRGEN E. OLESEN[1] and KAI GREVSEN[2]

[1] Department of Agrometeorology, Research Centre Foulum,
P.O. Box 25, 8830 Tjele, Denmark
[2] Department of Vegetables, Kirstinebjergvej 6, 5792 Årslev, Denmark

e-mail: jeo@dina.foulum.min.dk

Summary

The impact of climatic changes on summer cauliflower production in northern Europe has been assessed using a dynamic crop simulation model. The sensitivity of the model to changes in temperature, global radiation and atmospheric CO_2 concentration was analyzed using historical weather data from several sites in Europe. Effects of varying the transplanting date and plant density were also studied.

Model simulations indicate that increasing atmospheric CO_2 concentration may decrease the risk of loose heads in cauliflower. Higher CO_2 concentrations may also enable a higher plant density than is currently used without detrimental effects on curd size and quality. Temperature was found to strongly affect the timing of cauliflower production, whereas the quality in terms of curd density is determined by a wider range of environmental conditions. In the model curd density is affected mainly by the balance between the source of and sink for assimilates. Plant density, atmospheric CO_2 concentration and temperature were found to be the most important variables affecting the source-sink balance.

Keywords Climate change, cauliflower, simulation model, carbon dioxide

1. Introduction

The growth and development of cauliflower plants are strongly influenced by environmental conditions such as temperature and radiation (Wurr et al., 1990). These environmental conditions also determine the quality of the produce (Wiebe and Krug, 1974). The edible part of the cauliflower plant is the inflorescence, which is harvested before it is physiologically mature. Size and quality of the curds at harvest are equally important in determining the yield in cauliflower production.

The time from transplanting to harvest can be divided into three phases (Wiebe, 1972ab; Wiebe, 1973; Wurr et al., 1981): a juvenile phase, a curd induction phase and a curd growth phase. Low temperature treatments in the juvenile phase cannot initiate curds, whereas such temperatures in the curd induction phase are effective and needed. Climatic factors affect timing and quality differently in these phases (Atherton et al., 1987; Wiebe and Krug, 1974). The interaction of the various climatic factors on crop timing and product quality of cauliflower was studied using a simulation model (Olesen et al., 1993).

The enhanced greenhouse effect may cause a global warming which will significantly affect cauliflower production. A scientific assessment of future global climate has been made by the Intergovernmental Panel on Climate Change (Houghton et al., 1990). There are large uncertainties in projecting the climate, but all current indications point at a significant temperature increase by the middle of the next century. At the same time atmospheric CO_2 concentrations are likely to increase from currently about 350 ppmv to a level of 450 - 550 ppmv by 2050 (Houghton *et al.*, 1990). Barrow (1993) found by using results from different general circulation models that the mean temperature in northern Europe by 2050 may increase by 1-2°C in the summer months and the global radiation may increase by about 2%.

The purpose of this study was to examine the effects of climatic changes on timing and yield of cauliflower crops in northern Europe. To obtain such estimates, a crop simulation model was applied to climate data from several sites in northern Europe. These data were then perturbed by changing selected climatic elements. Special emphasis was given to changes in temperature, radiation and atmospheric CO_2 concentration. In the model calculations it was assumed that the crops are otherwise grown under conditions of optimal management.

2. Material and methods

2.1 Physiological model

The model used in this study is based on the approach used in the SUCROS87 model (Spitters *et al.*, 1989). The model is briefly described below and in more detail by Olesen *et al.* (1993).

Cauliflower seedlings are assumed to have 10 initiated leaves at transplanting. A temperature sum is used to determine the duration of the juvenile phase. During the curd induction phase the rate of crop development is assumed to have an optimum temperature with symmetrically declining rates above and below the optimum. These temperature functions were determined from field data from the Netherlands (Booij, 1990) and from Denmark (Olesen *et al.*, 1993).

The model calculates the daily assimilation of the crop canopy. The effect of atmospheric CO_2 concentration on photosynthesis is modelled using the concepts of Goudriaan *et al.* (1985). The temperature dependent function for adjusting assimilation rate at light saturation was taken from Spitters *et al.* (1989). Carbohydrates resulting from the canopy assimilation in excess of maintenance and growth respiration are converted into crop structural dry matter.

The amount of dry matter available for crop growth is distributed between four plant organ types: leaves, stems, roots and curds. Before curd initiation, assimilates are only distributed among leaves, stems and roots according to a fixed partitioning scheme. After curd initiation assimilate requirements for curd growth is assumed to take priority of that of the other organs. This assimilate requirement is calculated from the maximum increase in curd volume. The curd is assumed to be shaped as the upper half of a sphere. The maximum relative growth rates of the radius of the half-sphere is assumed to be linearly related to temperature. The growth in curd volume is reduced, if there is not enough assimilates to fulfil the growth requirement.

Model parameters were estimated for several cultivars (Olesen *et al.*, 1993). In this study only the cultivar Ballade is considered. Input requirements for model calculations are: geographic latitude, daily minimum and maximum air temperature, daily global radiation, atmospheric CO_2 concentration, day of transplanting and plant density. Model output includes timing of curd initiation, curd diameter, curd volume and curd weight. A curd diameter of 13 cm is used as the harvest criterion.

2.2 Climate data

Data from four meteorological stations in Europe were used as the basis for the simulation experiments. For each station 20 to 30 years of data were available (Table 1). Daily data for air minimum temperature, maximum temperature and global radiation are used. For most stations global radiation was estimated from hours of bright sunshine using the method proposed by Rietveld (1978).

The original daily temperature and radiation data were perturbed systematically before being applied in the model. Temperature was changed by adding fixed temperature offsets, and radiation was changed by fixed percentages.

3. Results

The model was run using weather data from Rothamsted, Orleans and Pisa for 5 trans-planting dates from March to July and for applied temperature changes of -3, 0, $+3$ and $6°C$. These large temperature changes were selected to test the sensitivity of the model. Runs were also made for two different atmospheric CO_2 concentrations: 353 ppmv and 553 ppmv. Mean temperature during the period from curd initiation to harvest was calculated and then for each simulation run compared with the developmental rate of that period (Fig. 1) and with the curd density (Fig. 2).

Fig. 1 shows that the physiological model predicted an almost linear relationship between mean temperature and developmental rate over a temperature range from $8°C$ to about $20°C$. At higher temperatures there was more scatter in the data. The developmental rate was generally higher at a higher atmospheric CO_2 concentration.

Fig. 2 shows curd densities versus mean temperature from curd initiation to harvest at two CO_2 concentrations for the same data as Fig. 1. Curd density is the fresh weight of the curds divided by the volume of a curd with half-spheric shape. Curd density generally dropped with increasing temperature during the curd growth period. The drop was more pronounced at lower CO_2 concentrations.

Table 1. Meteorological data used in simulation experiments.

Station	Latitude	Longitude	Period of data
Årslev (DK)	55°18'N	10°17'E	1961-1990
Rothamsted (UK)	51°51'N	0°22'W	1951-1980
Orleans (F)	47°59'N	1°45'E	1951-1980
Pisa (I)	43°40'N	11°20'E	1958-1977

The response of the cauliflower model to changes in plant density, CO_2 concentration, temperature, global radiation and date of transplanting was examined using weather data from Årslev in Denmark. The model parameters examined were days from transplanting to harvest (curd diameter of 13 cm), curd density and leaf area. Selected results of the these analyses are presented here. The model was run for the individual years in the baseline dataset, but only the mean results are presented here.

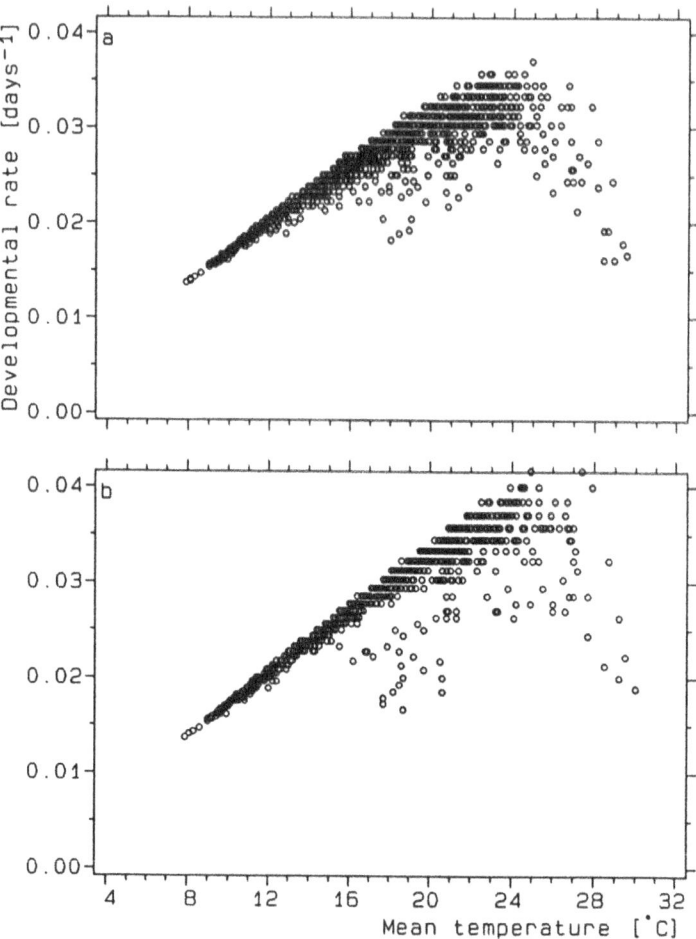

Figure 1. Simulated rate of progress from curd initiation to curd diameter of 13 cm versus mean temperature during the same period at two different atmospheric CO_2 concentrations: 353 ppmv (a) and 553 ppmv (b).

Fig. 3 shows the simulated interaction between temperature change and CO_2 concentration in terms of curd density and leaf area at harvest. Curd density increased slightly with increasing CO_2 concentration. The increase was larger at higher temperatures, where curd densities at current CO_2 concentrations were lower. There was a large increase in leaf area with increasing CO_2 concentration. The increase was slightly larger at lower temperatures.

The interaction between temperature change and radiation change on curd density and leaf area at harvest is shown in Fig. 4. Both curd density and leaf area at harvest increased with increasing radiation. The levels were different at different temperatures, but there were no apparent interactions between temperature and radiation change.

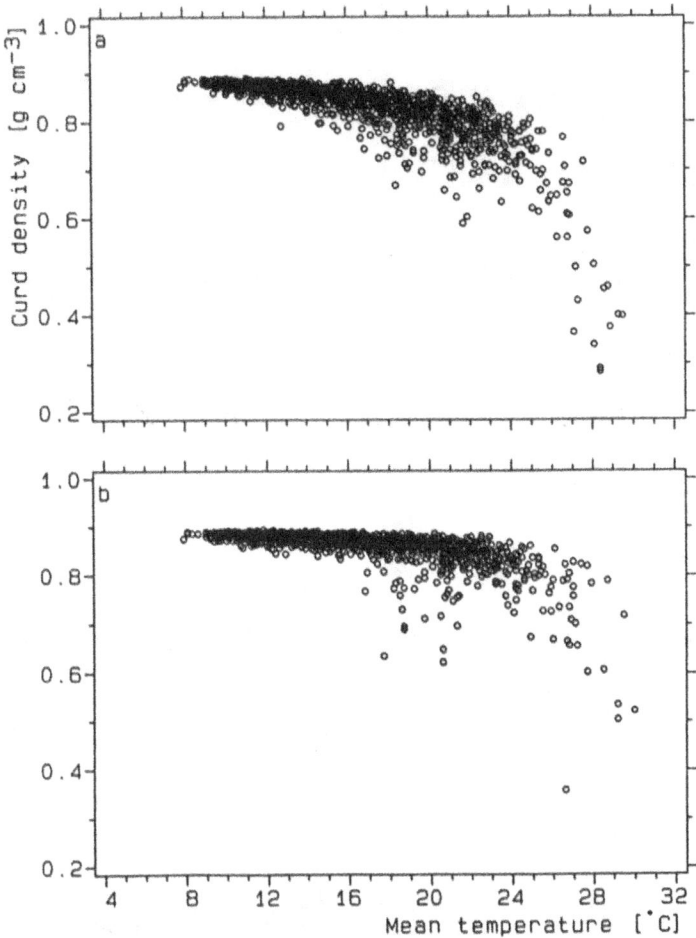

Figure 2. Simulated curd density at harvest versus mean temperature during the same period at two different atmospheric CO_2 concentrations: 353 ppmv (a) and 553 ppmv (bottom graph).

The effect of increasing atmospheric CO_2 concentration on curd density and leaf area at harvest is shown in Fig. 5 for three transplanting dates. The largest increase in curd density with increasing CO_2 concentration was obtained for the July transplanting, which at CO_2 concentrations above 500 ppmv achieved the same curd density as the June transplantings. Leaf area increased by about the same rate at all transplanting dates, with a small tendency for lower rates in the midsummer transplantings.

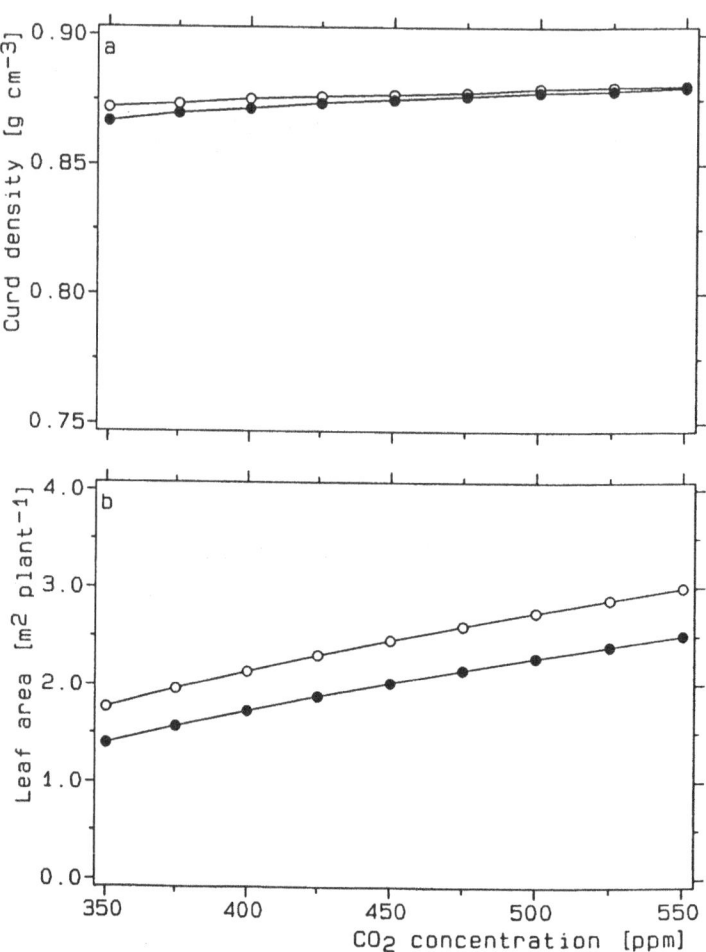

Figure 3. Simulated mean curd density (a) and leaf area per plant (b) at harvest for varying atmospheric CO_2 concentrations for two temperature changes: -1 °C (open circles) and +1 °C (filled circles). Calculated for transplanting on 1st May at Årslev with a plant density of 3 plants m⁻².

The effect of plant density on days from transplanting to harvest, curd density and leaf area at harvest is shown in Fig. 6. The results are presented for two CO_2 concentrations: 350 and 550 ppmv. The number of days from transplanting to harvest increased with increasing plant density. The increase is lower at increasing CO_2 concentration. There was an almost linear decline in curd density with increasing plant density. The decline rate was smaller at increasing CO_2 concentration. The leaf area per plant at harvest also declined with increasing plant density. This was a nonlinear relationship with higher levels at higher CO_2 concentrations.

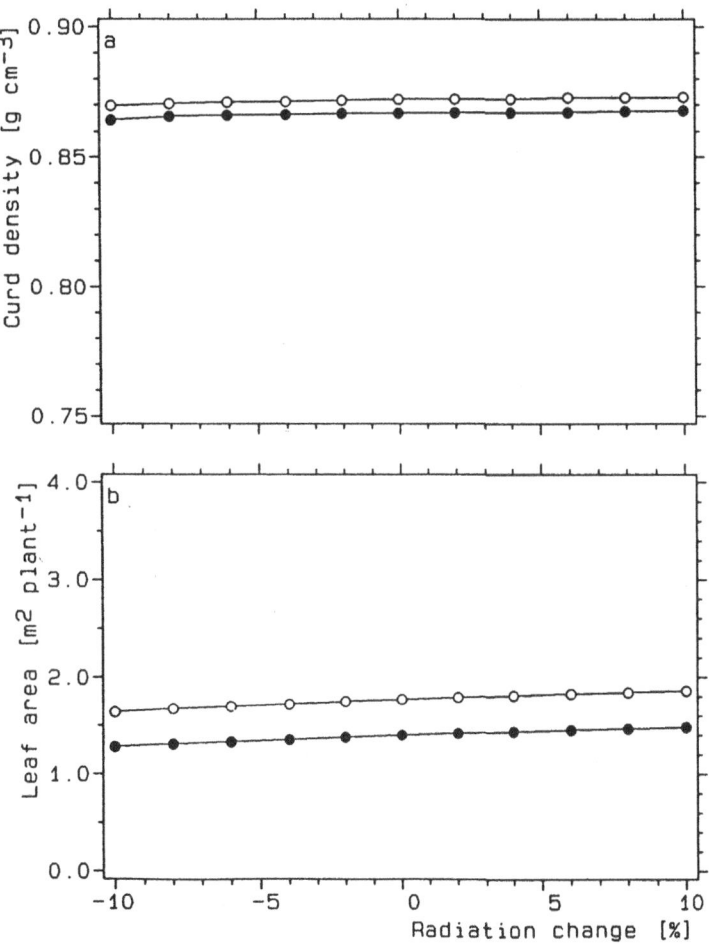

Figure 4. Simulated mean curd density (a) and leaf area per plant (b) at harvest for varying changes in global radiation for two temperature changes: -1 °C (open circles) and +1 °C (filled circles). Calculated for transplanting on 1st May at Årslev with a plant density of 3 plants m^{-2} and for a CO_2 concentration of 350 ppmv.

4. Discussion

Temperature strongly affects the timing of cauliflower production as shown in the simulations for the period from curd initiation to harvest in Fig. 1. There appears to be an optimum temperature of about 23°C at 353 ppmv and of about 25°C at 553 ppmv, below and above which the duration from curd initiation to harvest increases. This is largely a consequence of the assumptions about the effect of temperature on photosynthesis in the model. Photosynthesis was assumed to decline above 25°C, as for other temperate C_3-species (Versteeg and van Keulen, 1986). With declining rate of photosynthesis at high temperatures there is a larger risk of lack of assimilates for curd filling, which will prolong the curd growth phase (Fig. 1) and increase the risk of low curd density (Fig. 2).

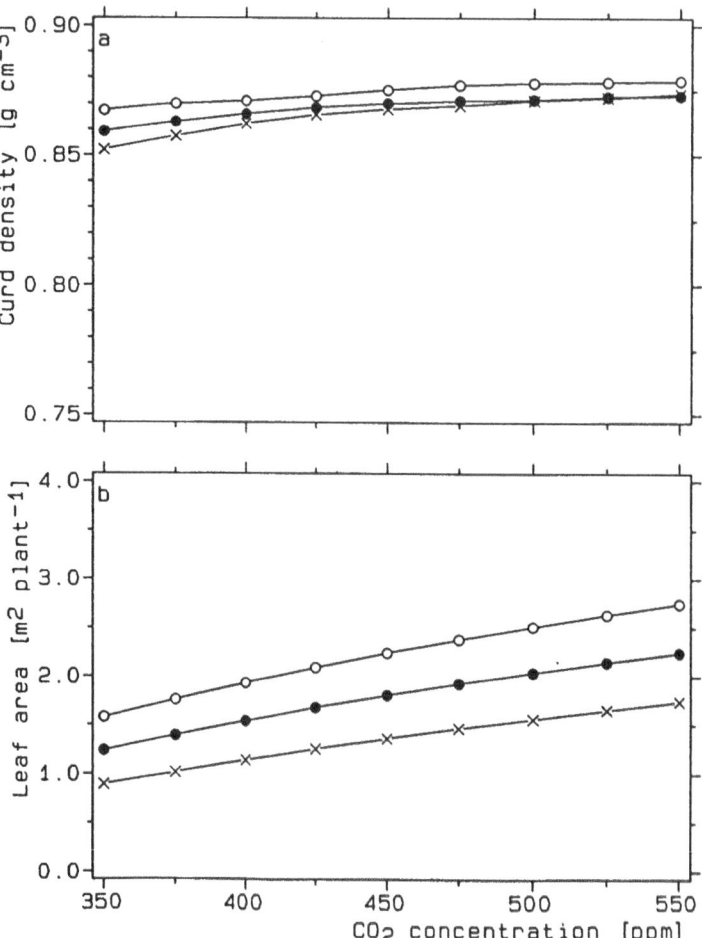

Figure 5. Simulated mean curd density (a) and leaf area per plant (b) at harvest for varying atmospheric CO_2 concentrations for three transplanting dates: 1st May (open circles), 1st June (filled circles) and 1st July (crosses). Calculated for original climate data from Årslev with a plant density of 3 plants m^{-2}.

Low curd density indicates loose heads in cauliflower, which is an important quality criterion. Fig. 2 indicates a dramatically increasing risk of getting low curd density for temperatures above about 20°C. This appears to occur at both current and increased CO_2 concentrations. Figs. 3 and 4 show only a slight effect of radiation and CO_2 concentration on curd density, whereas the response of curd density to changes in plant density is considerably higher.

These results reflect the balance between source and sink of assimilates. Global radiation, CO_2 concentration and temperature determine the available assimilate source of the crop. Temperature is assumed to be the main determinant of potential increase in curd

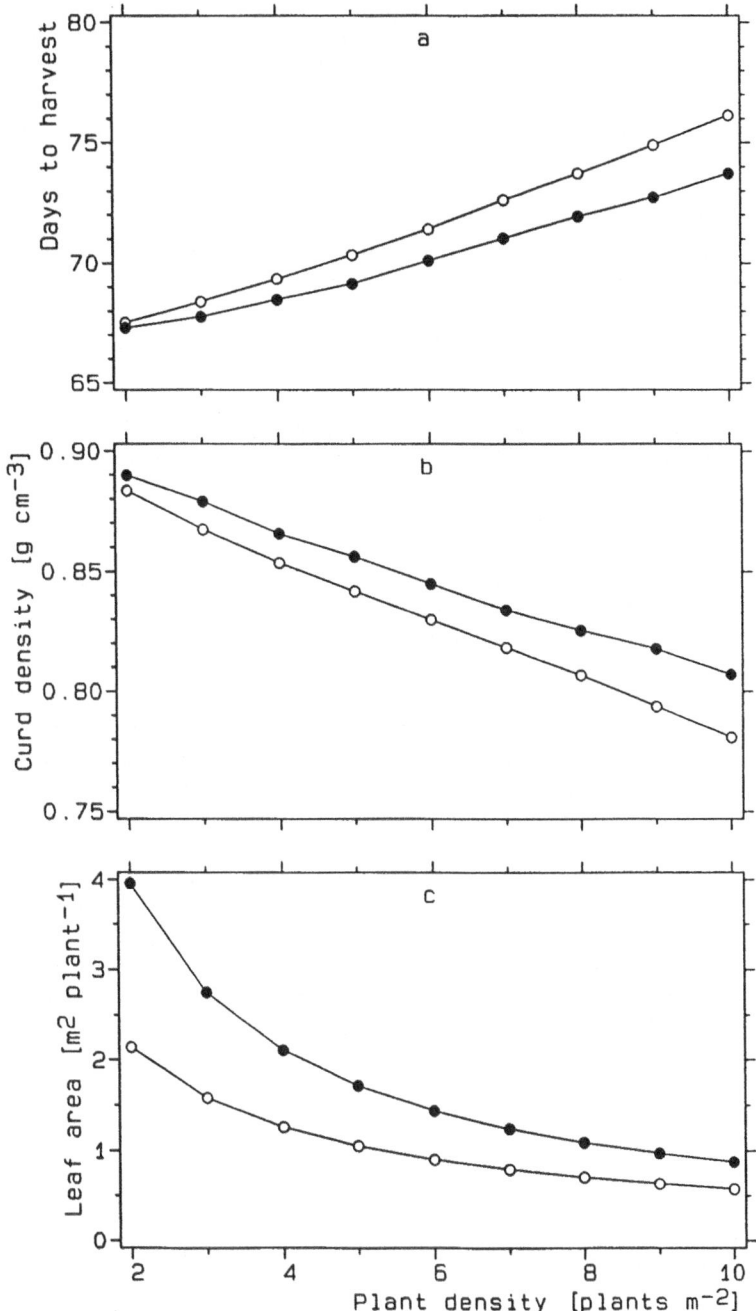

Figure 6. Simulated mean days from transplanting to harvest (a), curd density (b) and leaf area per plant (c) at harvest for varying plant densities for two different atmospheric CO_2 concentrations: 350 ppmv (open circles) and 550 ppmv (filled circles). Calculated for transplanting on 1st May at Årslev using original climate data.

volume, which determines the sink capacity per plant. The sink capacity of the entire crop is linearly related to the plant density, which is probably the reason for the almost linear relationship between curd density and plant density in Fig. 6.

The simulation results show a large increase in leaf area at harvest at increasing atmospheric CO_2 concentration. This is a result of the increased net photosynthesis at higher CO_2 concentrations. The model uses a constant specific leaf area to calculate the increase in crop leaf area. The assumption about a constant specific leaf area is probably erroneous. Morison and Gifford (1984b) thus found increases in specific leaf area of 10% to 44% for a doubling of the CO_2 concentration in several temperate broadleaf C_3-species. The leaf area was increased by 40% to 75% in the same species for a doubling of the CO_2 concentration (Morison and Gifford, 1984a). Cauliflower was not included in these experiments. It does, however, seem likely that increased CO_2 concentration will increase specific leaf area in cauliflower, but the relative effect on total leaf area will probably be larger than on specific leaf area.

Figs. 3 and 4 show that the leaf area at harvest is higher at lower temperatures. This is not an effect of temperature on photosynthesis, but an effect of a longer period from transplanting to harvest at lower temperatures, and thus a longer period for leaf growth.

Salter (1961) studied the effect of increasing plant density of cauliflower and found a decreasing leaf area per plant, similar to the simulated results in Fig. 6c. The experimental results by Salter (1961) indicated an increasing crop leaf area index [m^2 leaf per m^2 soil] when increasing the plant density from 1.3 to 10.6 plants per m^2. This was also found in the model simulations. The model thus describes the consequences on leaf area of changing plant density found in experiments.

Current cauliflower plant density in Denmark usually varies between 2.5 and 3.5 plants per m^2. In this plant density range there is only a small response of the growth duration to increased plant density (Fig. 6). At increasing plant density less assimilate is available for each curd, and the curds will become loose at a smaller size. This results in lighter heads at harvest in practice, where curds will be harvested just before they become loose. The harvest criterion in the simulations, however, was a curd diameter of 13 cm. Here, there is an almost linear relationship between plant density and curd density.

Fig. 6 shows that increasing CO_2 concentration caused both crop duration and curd density to respond less rapidly to increasing plant density. Therefore it is possible at increasing atmospheric CO_2 concentrations to increase the plant density without severely affecting timing and quality. If the criteria is to keep the same mean crop duration and curd density then an increase in atmospheric CO_2-concentration from 350 to 550 ppmv may facilitate an increase in plant density from the current 2.5-3.5 to 3.5-4.5 plants per m^2.

5. Acknowledgement

This study was funded by the Commission of the European Communities under contract number EPOC-CT90-0031 (TSTS).

References

Atherton, J.G., D.J. Hand and C.A. Williams (1987), 'Curd initiation in the cauliflower (Brassica oleracea var. Botrytis L.)', In: *Manipulation of flowering*, Butterworths,

London, pp. 133-145.

Barrow, E. (1993), 'Future scenarios of climate change for Europe', In: *The effect of climate change on agricultural and horticultural potential in Europe*, Environmental Change Unit, University of Oxford, pp. 11-39.

Booij, R. (1990), '*Development of cauliflower and its consequences for cultivation*', PhD thesis, Agricultural University Wageningen, Wageningen.

Goudriaan, J., H.H. van Laar, H. van Keulen and W. Louwerse (1985), Photosynthesis, CO_2 and plant production. In: *Wheat growth and modelling*, NATO ASI Ser., Series A: Life sciences. Vol 86. Plenum Press, New York, pp. 107-122.

Houghton, J.T., G.J. Jenkins and J.J. Ephraums (1990), '*Climate Change: The IPPC Scientific Assessment*', Cambridge University Press, Cambridge.

Morison, J.I.L. and R.M. Gifford (1984a), 'Plant growth and water use with limited water supply in high CO_2 concentrations. I. Leaf area, water use and transpiration', *Aust. J. Plant Physiol.*, **11**, 361-374.

Morison, J.I.L. and R.M. Gifford (1984b), 'Plant growth and water use with limited water supply in high CO_2 concentrations. II. Plant dry weight, partitioning and water use efficiency', *Aust. J. Plant Physiol.*, **11**, 375-384.

Olesen, J.E., E. Friis and K. Grevsen (1993), 'Simulated effects of climatic change on vegetable crop production in Europe', In: *The effect of climate change on agricultural and horticultural potential in Europe*, Environmental Change Unit, University of Oxford, pp. 177-200.

Rietveld, M.R. (1978), 'A new method for estimating the regression coefficients in the formula relating solar radiation to sunshine', *Agric. Meteorol.*, **19**, 243-252.

Salter, P.J. (1961), 'The irrigation of early summer cauliflower in relation to stage of growth, plant spacing and nitrogen level', *J. hort. Sci.*, **36**, 241-253.

Spitters, C.J.T., H. van Keulen and D.W.G. Kraalingen (1989), 'A simple and universal crop growth simulator: SUCROS87', In: *Simulation and systems management in crop protection*, Pudoc, pp. 147-181.

Versteeg, M.N. and H. van Keulen (1986), 'Potential crop production prediction by some simple calculation methods, as compared with computer simulations', *Agric. Syst.*, **19**, 249-272.

Wiebe, H.-J. (1972a), 'Wirkung von Temperatur und Licht auf Wachstum und Entwicklung von Blumenkohl. I. Dauer der Jugendphase für die Vernalisation', *Gartenbauwissenschaft*, **37**, 165-178.

Wiebe, H.-J. (1972b), 'Wirkung von Temperatur und Licht auf Wachstum und Entwicklung von Blumenkohl. II. Optimale Vernalisationstemperatur und Vernalisationsdauer', *Gartenbauwissenschaft*, **37**, 293-303.

Wiebe, H.-J. (1973), 'Wirkung von Temperatur und Licht auf Wachstum und Entwicklung von Blumenkohl. IV. Kopfbildungsphase', *Gartenbauwissenschaft*, **38**, 263-280.

Wiebe, H.-J. and H. Krug (1974), 'Wirkung der Temperatur auf Qualität und Erntedauer von Blumenkohl', *Gemüse*, **10**, 34-37.

Wurr, D.C.E., J.M. Akehurst and T.H. Thomas (1981), 'A hypothesis to explain the relationship between low-temperature treatment, gibberillin activity, curd initiation and maturity in cauliflower', *Sci. Hort.*, **15**, 321-330.

Wurr, D.C.E., J.R. Fellows R.W.P. and Hiron (1990), 'The influence of field environmental conditions on the growth and development of four cauliflower cultivars', *J. hort. Sci.*, **65**, 565-572.

VALIDATION OF LARGE SCALE PROCESS-ORIENTED MODELS FOR MANAGING NATURAL RESOURCE POPULATIONS: A CASE STUDY

R.A. FLEMING[1]
&
C.A. SHOEMAKER[2]

[1] Forest Pest Management Institute, Forestry Canada,
Box 490, Sault Ste. Marie, Ontario, CANADA P6A 5M7;
[2] School of Civil and Environmental Engineering,
Cornell University, Ithaca, New York 14853 USA

Abstract

The validation of models used in population management can be complicated by the number of component parts and by the large temporal and spatial scales often necessary. This is especially true for models developed for the analysis of management policy in forest-pest situations. In the case study considered here, a large-scale spruce budworm-forest simulation model (Jones, 1979) was tested by comparing its output with data collected annually by the Maine Forest Service survey at 1000 sites from 1975 to 1980.

In practice, model validation usually involves a comparison of observations, independent of those used to construct the model, with overall model output. This 'typical' validation was performed. In addition, separate tests were conducted on the model's major components. These components represent the forest protection policy, the budworm-forest dynamics, and pest control efficacy.

The model's output was not always consistent with the Maine survey data. In some situations, compensating inaccuracies in different model components allowed the overall model output to reasonably represent the overall behavior of the system. There were also problems in scaling the findings of studies of nonlinear population dynamics from small experimental plots to the much larger spatial scales used in the models. The results of this validation imply that some modification of the optimal management strategies suggested by the models is appropriate.

This paper concerns the validation of large-scale, process-oriented models in general. The spruce budworm-forest model is discussed as an illustrative case study.

1. Introduction

Testing models is an essential aspect of the scientific method. Often it leads to improvement of the understanding and theories on which the models are based. Testing management-oriented models has an additional role to play. It indicates the reliability of policies recommended by these models and thus helps ensure that the often considerable costs of putting recommended policies into operational practice are warranted.

Where possible, data independent of that used to form the theory or model is used for testing, and there is increasing recognition that field monitoring programs offer a valuable source of such data (e.g., Likens, 1989). In addition, it has been recommended that, because of the possibility of unconscious bias of the modeler, such testing is best performed by other individuals (Salt, 1983). At the more technical level, developing procedures for testing and evaluating simulation models is a very active area of research (Guckenheimer, 1989; Wallach and Goffinet, 1989) and there is no universally accepted methodology (e.g., Clark and Holling, 1979; Lewandowski, 1982; Reynolds and Deaton, 1982; Peters 1991).

This paper focuses on process-oriented models as opposed to purely empirical, descriptive statistical models. Specifically, it uses Jones' (1979) spruce budworm-forest simulation model as a representative case study. Process-oriented models are built with a causal architecture whereby the constituent processes define the mathematical relationships between the variables of the system. For example, key processes in the budworm-forest system include growth, reproduction, dispersal, and mortality. Because such processes are fundamental to virtually any ecological system, understanding and knowledge acquired from other ecological systems about these processes can be advantageously applied to the budworm-forest system (and *vice-versa*). Thus, a process-oriented model is built by referring to the existing inventory of the discipline's process modules, and combining and parameterizing them to suit the particular situation at hand (Holling, 1978). In this sense, the process modules provide an *a priori* structure for interpreting and organizing data.

This paper describes a process-oriented approach to validation. This approach requires familiarity with the influence of the structure of each model on its output, and familiarity with the structure of the data used to test the model. This familiarity allows one to identify relationships in the data for testing the predictive reliability of the models and some of their component submodels. For instance, in validating Jones' (1979) spruce budworm-forest model (below), we organized the output to reflect the models' representations of the forest protection policy, budworm-forest dynamics, and pest control efficacy.

In validating ecological models in general, this process-oriented approach has three principal advantages over the typical approach of comparing overall model behavior to data. First, it allows one to identify occasions where errors in component processes may compensate for each other so that model output mimics the data. The danger is that such errors may not always be mutually compensating when using the models to make predictions under different circumstances (Alcamo and Bartnicki, 1987). Second, the relatively specific identification of suspect model components often accelerates their correction, and hence, the improvement of the whole model (Caswell, 1976). Third, survey data are used to identify suspect model components, but not to make final judgements about these components. Such judgements occur only after probing into the mechanics of the processes corresponding to these suspect components. This reduces the reliance of the validation on data (e.g., survey data) collected under non-experimental procedures.

1.1 Jones' (1979) spruce budworm-forest model

The spruce budworm, *Choristoneura fumiferana* Clem. (Lepidoptera: Tortricidae) is the most damaging defoliator in North America's boreal forests. The economic impact of budworm outbreaks has prompted a variety of innovative modeling efforts to improve both our understanding (e.g., Antonovsky *et al.*, 1990; Ludwig *et al.*, 1978; Royama, 1984) and management (e.g., Baskerville and Kleinschmidt, 1981; Gage and Sawyer, 1979; Seymour *et al.*, 1985) of the budworm-forest ecosystem. The most comprehensive models were developed by Jones (1979) and Stedinger (1984) as 'laboratory worlds' for exploring the consequences of ecological hypotheses and alternative management strategies.

Jones used a rectangular grid of blocks to represent space in his large-scale model. In each block, the model, which consists of 50 difference equations, operates as a dynamic life table. (A simplified approximation of this model due to Ludwig et al. (1978) is given in the appendix). Jones calculates budworm population densities as functions of forest conditions, and as functions of budworm immigration, age specific survival rates, and previous densities. A forest protection policy submodel determines whether the budworm density and forest conditions justify spraying a model block next spring. Decisions on whether to spray a block, or not, eventually affect surrounding blocks in the model because the models allow migration of spruce budworm moths to occur from one block to another.

Growth-rate curves (Fig. 1) conveniently describe the qualitative behavior of Jones' model in a single block of forest. The shape of the curves reflect hypotheses regarding the relative importance of the various ecological processes. For instance, the 'pit' in the curves at low budworm densities represents theories about the impact of avian predation. The height of the curves depend on the maturity of the forest (Holling et al., 1979).

To follow the model's dynamics, consider the curve for the intermediate forest in Fig. 1. This curve shows an unstable budworm equilibrium at U (since slightly larger or smaller densities immediately result in continued increase or decrease, respectively) and stable equilibria at L and H. Now suppose budworm densities are initially at their low equilibrium, L. While they remain low, the recruitment curve rises slowly in response to forest aging in the model because budworm have higher survival and reproductivity in older forests. Eventually the low and middle equilibria, L and U, come together, coalesce, and then disappear as the bottom of the 'pit' rises above the equilibrium line: $R = N(t+1)/N(t) = 1$. The now mature forest presents a very good environment for the budworm so its population increases quickly, approaching equilibrium V. During an uncontrolled budworm outbreak, forests suffer considerable defoliation and many of the older trees are killed. Consequently, the budworm population can become resource-limited and the forest age structure can change to one dominated by immature trees (Blais, 1985). Jones' (1979) model represents this sudden deterioration of forest conditions by a drop of the curve from the mature forest position to the position of an immature forest where the growth-rate < 1.0 (Fig. 1). Consequently, the modeled budworm population collapses to very low densities. After the collapse, the regenerated forest slowly matures and the curve slowly rises to the position representing an intermediate forest. (To start another outbreak now, moth immigration must increase the local population from the low density equilibrium, L, to the unstable equilibrium, U). In the absence of an immigration triggered outbreak, the model represents further aging of the forest as a slow rise in the growth-rate curve to the mature forest position in Fig. 1. Thus, model budworm populations respond quickly to forest conditions (as represented by the position of the growth-rate curve). Forest aging is represented by a slow rise in this curve from the immature to the mature position (Fig. 1); the destruction of the mature forest and its replacement by an immature one are represented by a fast drop in the curve (Fig. 1).

2. Methods

Each year the Maine Forest Service (MFS) estimates the budworm population density and the percent defoliation (of needles produced that spring) at roughly 1000 sampling sites in Maine's 28,000 km^2 spruce-fir forest (Trial and Thurston, 1980). To test the models on a scale relevant to management decision-making, and because the same sites were seldom surveyed in consecutive years, we adopted the grid of 6.4 km x 8.0 km "blocks" of forest used by the MFS in mapping their surveys. Blocks were excluded from the analysis in years that they contained both sprayed and unsprayed sampling sites.

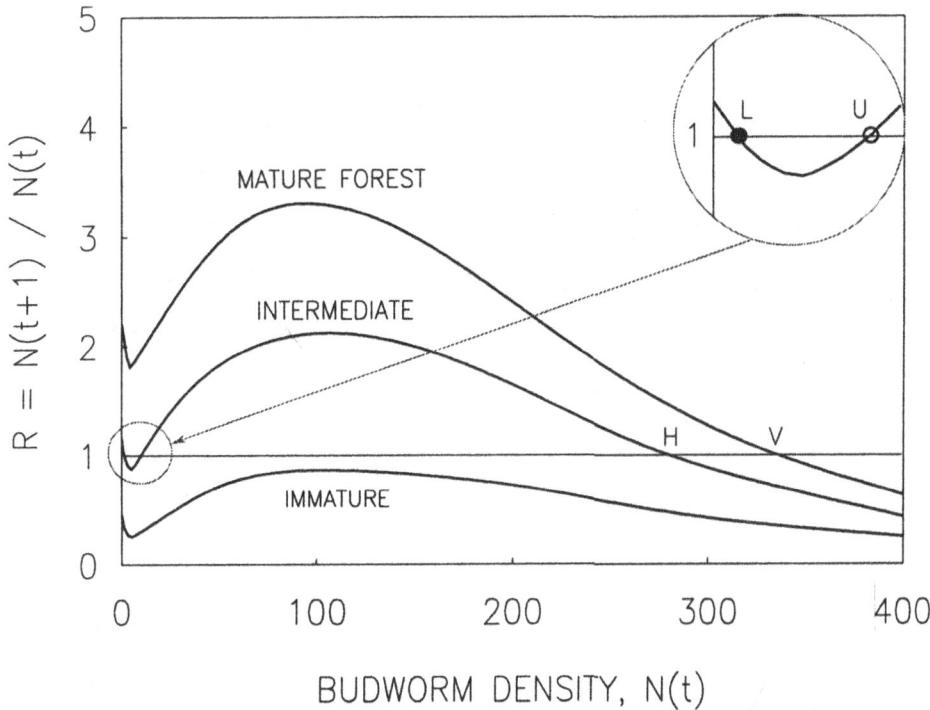

Figure 1. Curves illustrating the per-capita population growth-rate of spruce budworm at various densities (in larvae/m2 of branch area) for three different forest conditions. Equilibria occur at those densities where the current growth-rate crosses the horizontal line [where $N(t+1) = N(t)$]. (After Holling *et al.*, 1978).

In the simulations, distortions caused by transient dynamics were limited by choosing apparently reasonable initial values, and ignoring the first few simulated years of each run to allow such dynamics to damp out. In the next 100 simulated years of each run, the budworm egg-mass density and percent defoliation were recorded for each of the 80 modeled blocks of forest, producing 8000 block-years of results. Subsequent analysis determined the relationships between simulated block conditions in consecutive years. These relationships, intrinsic to the model output, were compared with corresponding relationships determined from the survey data.

3. Discussion and Results

In validating their model, the original developers have shown that, in some respects, the model as a whole has described budworm-forest dynamics in eastern North America reasonably well (Holling *et al.*, 1978; Baskerville, 1976). By contrast, we began by evaluating the reliability of the model's main components, i.e., the forest protection policy, the budworm-forest dynamics, and the pest control efficacy. Later the scope of the evaluation was broadened to deal with the model in its entirety.

3.1 Forest protection policy

Originally the forest-protection policy used in Maine was approximated by simulating insecticide applications in all modeled blocks which otherwise were likely to suffer excessive tree mortality next year (Stedinger, 1984). Such blocks are identified when their lack of foliage and budworm densities fall into a previously defined hazard zone. This policy is deterministic in the sense that all such blocks are sprayed, and only such blocks are sprayed. To test it, Fleming and Shoemaker (1992) compared the observed spray frequency with the one calculated from the model output for corresponding forest conditions (in terms of budworm density and defoliation). The model overestimated spray frequencies (by about 50%) for forest areas with poor conditions (i.e., high defoliation percentages and high budworm densities), and tended to underestimate spray frequencies (by about 10%) for forest areas in good condition.

These discrepancies may be due to overlooking the stochastic elements of the MFS forest protection policy. Environmental, financial, and political considerations all influence whether a forest area with particular infestation and defoliation conditions will be sprayed (Trial and Thurston, 1980). Hence, after a number of possible compromises proved unsatisfactory, we developed a purely stochastic policy which empirically mimics the spray rates found in the MFS survey data under the constraint that the rates do not decrease as defoliation and budworm densities increase (Table 3 in Fleming and Shoemaker, 1992). Nonetheless, this policy is an incomplete description because it excludes some factors (e.g., cumulative stand damage) which influence the planning of MFS spray operations. However, since this purely stochastic policy was the best representation of the MFS protection policy available, it was used in all simulations described below except where stated otherwise. (Sensitivity analysis suggested that reasonable alternatives to this policy had little effect on long-term model output).

3.2 Spruce budworm-forest dynamics

To examine budworm-forest dynamics in isolation from the effects of pest control operations, we concentrated on blocks of forest that were not sprayed during the years under consideration. Fig. 2 demonstrates the approach. For each year, model blocks and survey sites which were not sprayed, were classified [according to the MFS classfication system (Trial and Devine, 1982)] in terms of their budworm density, N(t). Among all the blocks and sites in each density class, we calculated the mean and standard error (SE) of the budworm densities in the following year, $t+1$. The figure shows these mean budworm densities, $N(t+1)$, and their intervals of ± 2 SE plotted against the mean of each density class in year t. The figure suggests that the model underestimates budworm population growth rates at low densities (and that it may also overestimate growth rates at high densities).

After ruling out uncertainties in sampling and moth dispersal as possible explanations for the model's underestimation of low density growth rates, spatial heterogeneity in budworm densities was considered. For instance, if the budworm density in a block of forest of intermediate age corresponds to equilibrium b in Figure 1 (i.e., $N(t) = N_b$), then the model predicts $N(t+1) = N(t) = N_b$. Now suppose this same average block density, N_b, results from a contiguous area comprising most of the block which has a density of $N_b/10$ (where $N_b/10 < N_a$), and the small remaining area of the block having a density of N_b+10 larvae/m^2. Then, if the curve for intermediate forests is separately applied to both parts of the block, the modeled populations increase throughout the block (Fleming, 1991).

This scenario is well within the range of spatial heterogeneity often observed in survey blocks (Fleming *et al.*, 1984), and the survey blocks (51 km^2) are smaller than the blocks used in Jones' (176 km^2) model. Therefore, the nonlinear dynamics of population growth in the model (Fig. 1), allow increasing local populations to swamp the effects of declining populations occupying most of a block's area. Thus, modeled budworm populations could increase more quickly from low density than they presently do if the model permitted spatial heterogeneity within blocks. By contrast, localities with declining populations would have relatively little influence in blocks dominated by sites with increasing populations. Hence, the assumption of spatial homogeneity within model blocks provides a plausible explanation for unrealistically slow increases of modeled budworm populations from low densities. Ultimately, the plausibility of this explanation depends on the allowance for spatial heterogeneity in the thousand-fold increase of spatial scale needed to translate Morris' (1963) experimental results (on which Jones based his model) from the Green River plots (.08-.10 km^2) to the 176 km^2 blocks used in the model.

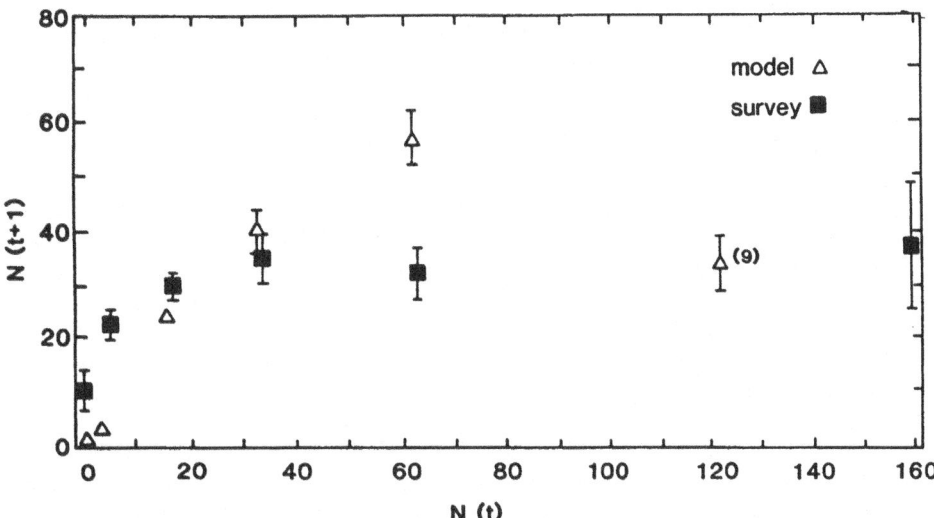

Figure 2. Relationships between budworm densities in successive years, $N(t)$ and $N(t+1)$, for blocks sprayed neither in year t nor $t+1$. Vertical bars show the extent of the intervals of $\pm\ 2\ SE$ about each mean. Sample sizes exceed 40 forest blocks for the survey data (boxes) and 125 model blocks (triangles) unless denoted otherwise in parentheses.

3.3 Pest control efficacy

Plots analogous to Fig. 2 were produced to show the relationship between forest conditions in successive years for blocks that were not sprayed the first year (t) but were sprayed the next ($t+1$). These plots suggested that the model tends to overestimate the efficacy of operational spraying at all but the largest budworm densities. The pest contol efficacy submodel uses the 80-90% rates of budworm mortality typical on balsam fir in small scale insecticide trials (Baskerville, 1976) but there are at least two reasons why these rates may overestimate operational efficacies. First, although balsam fir is the dominant tree species in Maine's forest protection area, this species comprises only about 50% by volume (60% by stems) of the spruce-fir type (Mott, 1980). Spruce is also important because a large part of the budworm popul-ation attacks spruce (Webb and Irving, 1983) and effective spray application on spruce is harder to achieve. For example, Trial and Devine (1982) report that controlled tests in eight different areas produced a median spray mortality rate of just 45% on red spruce (*Picea rubeus* Sarg.).

Second, it's inherently much easier to maintain good timing and good application procedures over the small plots used in insecticide trials than over the expansive areas of operational spraying. Unfavorable weather, logistical constraints, and asynchrony in shoot and larval development all cause problems. Ideally insecticides are applied when the budworm is in its vulnerable fourth larval instar and the balsam fir shoots have expanded sufficiently to provide an adequate spray target. Often, these events don't occur simultaneously, and even when they do, shoot development on balsam fir is usually faster than on spruce. Mistiming applications can be serious. For example, where carbaryl applications on balsam fir were delayed in 1979, 89% defoliation occurred and 2.8 larvae survived per 45 cm branch tip compared to 33% defoliation and 0.9 larvae per branch tip with proper timing (Trial and Thurston, 1980). In short, it's extremely difficult to achieve 80-90% larval mortality with operational spraying of large multispecies forests; actual operational rates in Maine probably range between 35 and 75%.

3.4 Overall model behavior

Jones was motivated to build his model by the desire to predict the characteristic behaviors of the budworm-forest ecosystem as a whole when different management policies are followed. Hence, studying model performance for this task requires output reflecting overall model behavior. For instance, spray frequencies depend on the forest protection policy, the budworm-forest dynamics, and spray efficacy. In the MFS data, an average ($\pm$ 2 SE) of 0.226 ($\pm$ 0.92) survey sites were sprayed each year. This contrasts with the model's annual rate of spraying of 0.027 ($\pm$ 0.006). A related aspect of the model's overall behavior is the frequency with which different ranges of budworm population density and defoliation occurred. Generally, the model predicted the occurrence of very low budworm density/very low defoliation conditions almost 67% more often, and the occurrence of very high defoliation almost 37% less often, than they occurred in the survey data (Fleming and Shoemaker, 1992).

This tendency of the models to underestimate both the rate of spraying and the frequency of high density/high defoliation conditions is consistent with the suggestions made above that spray efficacy is overestimated and low density budworm population growth rates underestimated in the model. Either miscalculation would unrealistically prolong the time spent by modeled budworm populations at low densities and the time taken to return to their large pre-spray densities. Consequently, the model's forest becomes quite mature before the next outbreak starts [and hence (Fig. 1), the high density population growth rates often exceed those in the survey data (Fig. 2)]. Such prolonged intervals for modeled budworm populations to return to outbreak levels, in turn, slow the rate of forest damage and delay the time when the budworm-forest system has deteriorated enough to warrant spraying again. This probably explains both the unrealistically infrequent occurrence of high budworm density/high defoliation conditions and the unrealistically long time between spray applications in the model.

Without the earlier detailed examination of the main components of the model, differences between the overall behavior of the model and the survey data could not be explained. When field data and model output disagree, the model is not necessarily at fault. The Maine Forest Service surveys were not designed for testing the model and are not perfectly suited to it. The surveys were conducted on a large scale only from 1975-1982 while the models were studied using runs of 100 years. Comparing this survey data and model output, therefore, implicitly assumes that during this period Maine's budwormforest-management system was in a state corresponding to 100 year averages of model conditions. But because Maine's forest was going through an outbreak during this period the surveys show the larger frequencies of occurrence of heavy defoliation and the smaller frequencies of low budworm densities and low defoliation associated with an outbreak, and not the ups and downs within 100 year averages. Similarly, the mean spray frequencies for the survey data apply to situations of frequent spraying. Hence, the fact that the surveys were conducted during an outbreak provides a possible explanation for some of the discrepancies between overall model output and the survey data. Thus, comparing the overall behavior of the model with the survey data is inconclusive in the absence of a more detailed examination of the model's component parts.

4. Conclusions

The ultimate goal of validation is to outline limits to, and ways to improve upon, the predictive reliability of the tested model. For management-oriented models, validation is also important in terms of a model's principal purpose as a decision support tool for policy develop-ment. Indeed, validation of Jones' (1979) spruce budworm-forest model produced modifications of the recommended policies (Fleming and Shoemaker, 1992). Three results of the validation of Jones' model, in particular, have broader implications for modeling the dynamics of renewable resource systems. First, the importance and difficulty of observing (population) dynamic processes at low densities was reinforced. Second, it was shown that spatial heterogeneity and changes of scale can interact to alter the nonlinearities of (population) dynamic processes in such a way that prediction becomes virtually impossible by standard techniques. Third, the validation illustrated a number of pitfalls that can occur in directly translating results from small scale trials and field experiments to the large scales used in models for operational management.

This interaction of spatial heterogeneity and changes of scale in nonlinear models is an important issue. In ecology, the difficulties in translating nonlinear population dynamic processes across spatial scales have recently become more widely appreciated (e.g., Turner et al., 1989; Wiens, 1989). A structure of hierarchically nested models (O'Neill et al., 1989) operating at different spatial (and temporal) scales may prove useful for making such translations. For instance, applying the nested grid approach, Odman and Russell (1991) developed a multiscale model capable both of nesting a fine-grid mesh within the course-grid mesh of a larger scale and of transferring information in both directions.

To summarize, in validating process-oriented nonlinear models, a process-oriented approach generally has three advantages over the typical one of only comparing overall model behavior with independent observations. First, this approach permits relatively specific identification of suspect model components, and thus it can accelerate the process of model improvement. Second, it allows the identification of some situations where errors in component processes might compensate for each other so that the overall model behavior closely mimics observation. A danger with such models is that the errors may not always compensate for each other when using the models to make predictions in different situations. Third, with this approach one need not rely solely on survey data or other data which may be somewhat suspect for validating a model because of the non-experimental nature of their collection. Rather, in this approach, survey data is used only to indicate suspect model components (which otherwise could have been very difficult to identify in such large and complex models). The mechanics of the processes corresponding to those suspect components are then investigated to try to resolve differences between the survey data and the model output.

References

Alcamo, J. and J. Bartnicki (1987), 'A framework for error analysis of a long-range transport model with emphasis on parameter uncertainty', *Atmospheric Environment*, **21(10)**, 2121-2131.

Antonovsky, M.Y., R.A. Fleming, Y.A. Kuznetsov and W.C. Clark (1990), 'Forest-pest interaction dynamics: the simplest mathematical models', *Theoretical Population Biology*, **37**, 343-367.

Baskerville, G.L. (1976), 'Report of the task-force for evaluation of budworm control alternatives', New Brunswick Department of Natural Resources, Fredericton, Canada

Baskerville, G. and S. Kleinschmidt (1981), 'A dynamic model of growth in defoliated fir stands', *Canadian Journal of Forest Research*, **11**, 206-214.

Blais, J.R. (1985), 'The ecology of the eastern spruce budworm: a review and discussion', In: C.J. Sanders, R.W. Stark, E.J. Mullins, and J. Murphy, editors. Recent Advances in Spruce Budworms Research, Canadian Forestry Service, Ottawa, Canada, pp. 49-59.

Casti, J. (1982), 'Catastrophes, control and the inevitability of spruce budworm outbreaks', *Ecol. Model.*, **14**, 293-300.

Caswell, H. (1976), 'The validation problem', In: Systems Analysis and Simulation in Ecology, vol. 4, B.C. Patten (ed.), Academic Press, New York, pp. 313-325.

Clark, W.C., and C.S. Holling (1979), 'Process models, equilibrium structures, and population dynamics: on the formulation and testing of realistic theory in ecology', *Fortschr. Zool.*, **25**, 29-52.

Fleming, R.A. (1991), 'Scale effects in developing models for integrated control: lessons from a case study of the eastern spruce budworm', *Mededlingen Faculteit van de Landbouwwetenschappen Rijksuniversiteit Gent*, **56(2a)**, 287-294.

Fleming, R.A. and C.A. Shoemaker (1992), 'Evaluating models for spruce budworm-forest management: comparing output with regional field data', *Ecological Applications*, **2(4)**, 460-477.

Fleming, R.A., C.A. Shoemaker and J.R. Stedinger (1984), 'An assessment of the impact of large scale spraying operations on the regional dynamics of spruce budworm (Lepidoptera: Tortricidae) populations', *Canadian Entomologist*, **116**, 633-644.

Gage, S.H. and A.J. Sawyer (1979), 'A simulation model for eastern spruce budworm populations in a balsam fir stand', In: W.G. Vogt and M.H. Mickle, editors. Modeling and Simulation. Volume 10. *Proceedings of the Tenth Annual Pittsburg Conference, Instrument Society of America*, Pittsburg, PA, USA, pp. 1103-1113.

Guckenheimer, J. (1989), 'Obstacles to modelling large dynamical systems', In: *Proc. Mathematical Approaches to Problems in Resource Management and Epidemiology, Lecture Notes in Biomathematics, vol. 10*, C. Castillo-Chavez, S.A. Levin, and C.A. Shoemaker (eds.), Springer-Verlag, Berlin, pp. 319-327.

Holling, C.S. (editor), (1978), *'Adaptive Environmental Assessment and Management'*, John Wiley & Sons, Toronto, Canada

Holling, C.S., D.D. Jones and W.C. Clark (1979), 'Ecological policy design: A case study of forest and pest management', In: G.A. Norton and C.S. Holling, editors. Pest Management. Pergamon Press, Oxford, U.K., pp. 13-90.

Jones, D.D. (1979), 'The budworm site model', In: G.A. Norton and C.S. Holling, editors. Pest Management. Pergamon Press, Oxford, U.K, pp. 9-155.

Lewandowski, A. (1982), 'Issues in model validation', *Angewandte Systemanalyse*, **3**(1), 2-11.

Likens, G.E. (editor), (1989), *'Long-term Studies in Ecology'*, Springer-Verlag, New York, N.Y.

Ludwig, D., D.D. Jones and C.S. Holling (1978), 'Qualitative analysis of insect outbreak systems: the spruce budworm and the forest', *Journal of Animal Ecology*, **47**, 315-332.

Morris, R.F. (editor), (1963), 'The dynamics of epidemic spruce budworm populations', *Memoirs of the Entomological Society of Canada*, **31**, 1-332.

Mott, D.G. (1980), 'Spruce budworm protection management in Maine', *Maine Forest Review*, **13**, 26-33.

Odman, M.T. and A.G. Russell (1991), 'A multiscale finite element pollutant transport scheme for urban and regional modeling', *Atmospheric Environment*, **25A**, 2385-2394.

O'Neill, R.V., A.R. Johnson and A.W. King (1989), 'A hierarchical framework for the analysis of scale', *Landscape Ecology*, **3**, 193-205.

Peters, R.H. (1991), *'A Critique for Ecology'*, Cambridge University Press, Cambridge, U.K.

Reynolds, M.R., Jr. and M.L. Deaton (1982), 'Comparisons of some tests for validation of stochastic simulation models', *Communications in Statistics - Simulation and Computation*, **11**(6), 769-799.

Royama, T. (1984), 'Population dynamics of the spruce budworm, (Choristoneura fumiferana)', *Ecological Monographs*, **54**, 429-462.

Salt, G.W. (1983), 'Roles: their limits and responsibilities in ecological and evolutionary research', *American Naturalist*, **122**, 697-705.

Seymour, R.S., D.G. Mott, S.M. Kleinschmidt, P.H. Triandafillou and R. Keane (1985), 'Green Woods Model - a forecasting tool for planting timber harvesting and protection of spruce-fir forests attacked by the spruce budworm: a dynamic simulation of the Maine spruce-fir forests attacked by the spruce budworm', *United States Department of Agriculture Forest Service General Technical Report NE-91*.

Stedinger, J.R. (1984), 'A spruce budworm-forest model and its implications for suppression programs', *Forest Science*, **30**, 597-615.

Trial, H., Jr. and A.S. Thurston (1980), 'Spruce budworm in Maine: 1979', *Maine Forest Service Entomology Division Technical Report 14*.

Trial, H., Jr. and M.E. Devine (1982), 'Spruce budworm in Maine: 1981', *Maine Forest Service Entomology Division Technical Report 18*.

Turner, M.G., V.H. Dale and R.H. Gardiner (1989), 'Predicting across scales: Theory development and testing', *Landscape Ecology*, **3**, 245-252.
Wallach, D. and B. Goffinet (1989), 'Mean squared error of prediction as a criterion for evaluating and comparing system models', *Ecol. Model.*, **44**, 299-306.
Webb, F.E. and H.J. Irving (1983), 'My fir lady', *Forestry Chronicle*, **59**, 118-122.
Wiens, J.A. (1989), 'Spatial scaling in ecology', *Functional Ecology*, **3**, 385-397.

Appendix

Ludwig *et al.* (1978) used three simultaneous differential equations to approximate Jones' (1979) site model. They give the annual rate of change in the number of budworm larvae per acre (1 acre = 4046.8 m^2) as

$$dB/dt = 1.52 \, B[1 - B(1 + E^2)/(335 \, E^2 S)] - 43200 \, B^2/[B^2 + (1.11 \, S)^2]. \tag{1}$$

This rate depends on forest conditions which are indicated by the size, S, and health or energy, E, of the trees. Specifically, S represents the total branch surface area per acre and

$$dS/dt = 0.095 \, S[1 - S/(25440 \, E)]. \tag{2}$$

E represents the relative 'energy reserve' of the trees and describes the general condition of the foliage and vigor of the trees. The annual rate of change of E depends on both B and S:

$$dE/dt = 0.92 \, E(1 - E) - 0.0015 \, E^2 B/[S(1 + E^2)]. \tag{3}$$

Ludwig *et al.* (1978) explain the rationale underlying the derivation of these equations and provide a qualitative analysis. Casti (1982) shows that for any realizable variation in the parameter values, this system of equations produces boom-and-bust outbreak cycles. He then uses this system to examine the feasibility of developing a control strategy which could suppress outbreaks indefinitely.

UNCERTAINTY OF PREDICTIONS IN SUPERVISED PEST CONTROL IN WINTER WHEAT: ITS PRICE AND CAUSES

WALTER A.H. ROSSING[1], RICHARD A. DAAMEN[2],
ELIGIUS M.T. HENDRIX[3] & MICHIEL J.W. JANSEN[4,5]

[1] Wageningen Agricultural University, Dept. Theoretical Production Ecology, P.O. Box 430, 6700 AK Wageningen, The Netherlands; [2] Research Institute for Plant Protection; [3] WAU, Dept. Mathematics; [4] DLO-Agricultural Mathematics Group; [5] DLO-Centre for Agrobiological Research.

e-mail: rossing@rcl.wau.nl

Summary

In supervised control, the economically optimal timing of pesticide application is equivalent with the level of pest attack where projected costs of immediate control just equal projected costs of no control. This level is called the damage threshold. Uncertainty about the costs of different strategies of chemical control of aphids (especially *Sitobion avenae*) and brown rust (*Puccinia recondita*) is calculated with a deterministic model. Sources of uncertainty, which comprise estimates of initial state and parameters, future weather, and white noise, are modelled as random inputs. Consequences of uncertainty for damage thresholds are analyzed. The relative importance of various sources of uncertainty for prediction uncertainty is calculated using a novel procedure.

Stochastic damage thresholds are lower than those calculated using average values for sources of uncertainty. Thus, uncertainty causes earlier chemical control of pests and higher input of pesticides. Due to the strongly skewed frequency distribution of costs of no control, the probability of positive return on pesticide expenditure at the stochastic damage thresholds is only 30%. White noise in the relative growth rates of both aphids and brown rust is found to be the most important source of uncertainty. More accurate estimation of parameters and initial estimates in the current model results in marginal reduction of prediction uncertainty only. Reduction of prediction uncertainty and concomitant reduction of recommended pesticide use requires reduction of the uncertainty associated with no chemical control by adopting a different approach to prediction of the population dynamics of aphids and brown rust.

Keywords Uncertainty analysis, Monte Carlo methods, non-linear systems, crop protection

1. Introduction

An important objective of pesticide application is insurance against major crop losses which occur with low probability (Norton & Mumford, 1983; Tait, 1987; Pannell, 1991). In many pathosystems pesticides are very effective in decreasing densities of growth reducing organisms. Although crop loss may occur even at low densities, the extent of loss as well as the variation in loss is usually smaller than at high pest densities. A second, potentially conflicting objective of pesticide application is maximization of (expected) return on expenditure (Norton & Mumford, 1983; Rossing *et al.*, 1993a). Extreme emphasis on insurance occurs when spraying is carried out at regular time intervals without reference to the presence of or damage by the pest. Examples of such prophylactic control strategies are found in control of late blight (*Phytophthora infestans*) in potato and fire blight (*Botrytis* spp.) in various flowerbulb species in the Netherlands.

Supervised control represents a concept of pest management in which maximization of returns on expenditure is emphasized. The level of pest attack at which the projected costs of chemical control just equal the projected costs of no control is called the damage threshold and represents the optimal time of pesticide application. Costs of decision alternatives are usually predicted with mathematical models. Recommended action is passed to farmers in decision support systems.

In this paper the importance of uncertainty for supervised control of aphids (especially *Sitobion avenae*) and brown rust (*Puccinia recondita*) in winter wheat in the Netherlands is investigated. Two questions are addressed. Firstly, to what extent do the damage thresholds for aphids and brown rust change when uncertainty about various components in the mathematical models is taken into account, or, what is the price of uncertainty. Secondly, which sources of uncertainty contribute most to uncertainty about costs associated with a control decision and how can uncertainty about the costs be reduced most effectively.

2. Research approach

2.1 Decision model

A deterministic simulation model is used to predict costs of spray strategies at given initial temperature sum and initial levels of pest attack in a winter wheat field in the Netherlands. A spray strategy consists of a time series of decisions on chemical control of aphids, brown rust or both with fixed time interval of one week. Costs of a strategy comprise the monetary equivalent of yield loss due to pest attack plus the costs of eventual chemical control.

The model consists of submodels describing crop development, population dynamics and damage per unit of pest density. Relations in the model are based on empirical data collected during several years and experiments. Crop development is calculated as a function of temperature sum above a developmental threshold of 6 °C, accumulated from crop development stage pseudo-stem elongation (DC 30, Zadoks *et al.*, 1974).

The submodel on population dynamics is initialized with field observations on aphids and brown rust incidences, i.e. percentage of sample units containing the pest. In the submodel incidence is transformed into density, which is assumed to increase exponentially with time. The relative growth rate of the aphid population is a function of crop development stage. For brown rust the relative growth rate is constant. Aphid-specific pesticides reduce the population to 85% of its pre-spray density and arrest population increase for 12 days after application. In contrast, brown rust specific pesticides do not affect the population present and arrest population increase during 18 days. Damage per pest-unit decreases with crop development stage for aphids and is constant throughout the season for brown rust. A maximum level of damage is imposed for both pests. A schematic representation of the model is shown in Fig. 1. Details are given in Rossing *et al.* (1993b).

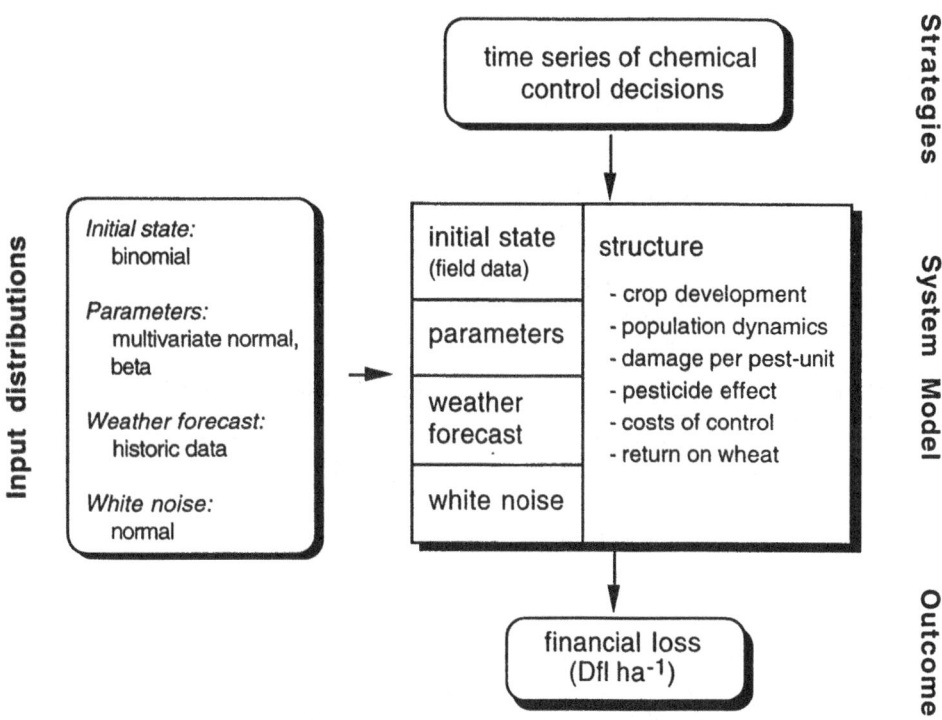

Figure 1.   Schematic outline of the decision model.

Uncertainty is modelled as random inputs into the model. Four categories of uncertainty are distinguished (Table 1). Parameters were estimated using field data and regression, the variance-covariance matrix providing a measure of uncertainty. Residual variance was ascribed to measurement effects and was disregarded for prediction. In some data sets the measurement variance could be quantified. In those cases the surplus residual variance was ascribed to natural variability and was included in the model as mutually independent, identically distributed normal variates. This source of uncertainty will be called white noise. Uncertainty about initial incidences of aphids and brown rust was modelled as binomial distributions with parameters depending on the incidence estimates and the sample sizes. Uncertainty about future average daily temperature was described by 36 years of daily minimum and maximum temperature measured in Wageningen between 1954 and 1990.

The categories parameters and estimates of the initial state represent sources of *controllable uncertainty*, since uncertainty may be reduced by collecting additional data. Uncertainty about future average daily temperature and white noise represents *uncontrollable uncertainty*, at least for a given structure of the model.

Table 1.   Sources of uncertainty in the decision model.

Category	Component	Distribution
Estimates of initial state	· incidence · temperature sum	binomial _a
Parameters	various	(multivariate) normal, beta
Gaussian white noise	· relative growth rate · incidence-density transformation · temperature sum - crop development stage transformation	normal normal normal
Future average daily temperature	-	historic data (36 years)

[a] Temperature sum is assumed to be known with negligible uncertainty.

2.2 Partitioning of model output uncertainty

The contribution of various sources of uncertainty in the model to uncertainty about costs of a spray strategy is calculated using the procedure described by Jansen *et al.* (1993). Uncertainty about model output is characterized by its variance. Sources of uncertainty are combined in Q groups which are mutually independent. In successive Monte Carlo runs new realizations of independent groups of variates are drawn by simple random sampling from the appropriate distributions, processing one group per run. After Q runs the values of all groups have been changed once, and the first cycle is completed. In total M cycles are made. Differences in model output between consecutive runs are caused by the uncertainty about one group of variates, while differences between runs i and $i+Q-1$ are due to uncertainty about all groups except one. This procedure which was termed *winding stairs sampling*, allows estimation of the full variance of mode output using the independent model outputs Q runs apart. The contribution of a group of sources of uncertainty is estimated as either its top variance, the reduction in total variance resulting from removal of uncertainty about the group, or its bottom variance, the variance remaining when uncertainty about all other groups is removed. Calculation of top variance is useful for groups of variates containing controllable uncertainty, *i.c.* parameters and estimates of the initial state. The top variance represents the maximum improvement of prediction accuracy possible for the given model structure. For sources of uncontrollable uncertainty the bottom variance is a more useful measure of uncertainty. It represents the minimum model accuracy that has to be accepted.

3. Results and discussion

3.1 The price of uncertainty

The damage thresholds for aphids and brown rust in Fig. 2 which are calculated using the

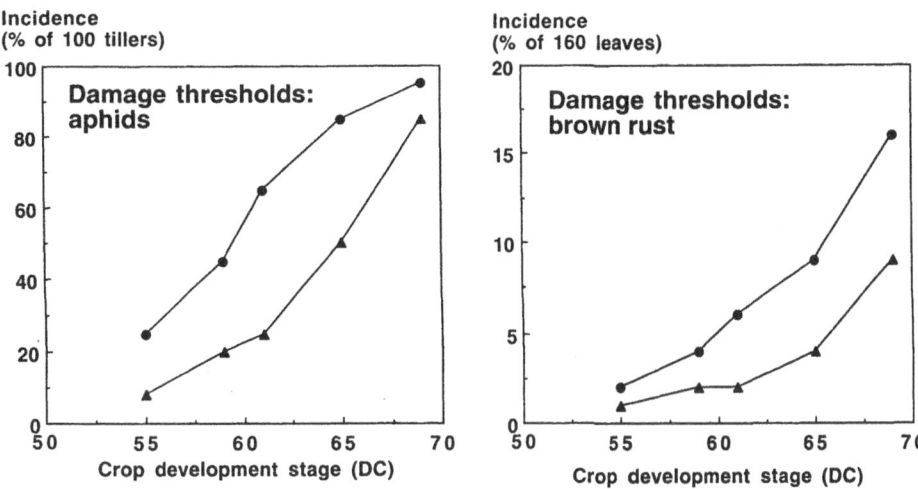

Figure 2. Damage thresholds for aphids (A) and brown rust (B) according the deterministic version of the decision model (— • —) and the stochastic version, run with $M = 500$ (— —).

decision model with random inputs are referred to as stochastic damage thresholds. Also shown are deterministic damage thresholds calculated with
average values of inputs. The stochastic damage thresholds are consistently lower than the deterministic thresholds, reflecting the convexity of the decision model.

The size of the difference between deterministic and stochastic damage thresholds is a measure for the price that has to be paid for uncertainty. Due to uncertainty pesticides are applied at lower pest incidences, leading to on average higher expenditure on pesticide input and a larger burden for the environment than would be economical with perfect information.

In Fig. 3 frequency distributions of costs associated with no chemical control at any time (NC) and immediate chemical control (IC) are shown for a single stochastic damage threshold. Potential costs associated with NC range between almost 0 Dfl ha^{-1} and 1200 Dfl ha^{-1}, with a 90%-percentile of 796 Dfl ha^{-1} (Fig. 3a). The distribution is strongly skewed to the right. In contrast, the 90%-percentile for IC is 214 Dfl ha^{-1} (Fig. 3b). The lower cost threshold of 185 Dfl ha^{-1} is due to the fixed costs of a control operation.

By definition the expected value of the difference in costs between NC and IC (Fig. 4) equals zero, since the initial state represents a stochastic damage threshold. Counter-intuitively, the probability that costs of IC are less than costs of NC is 30%, rather than the intuitive 50%. In other words, although on average immediate chemical control is an economically rational decision at initial incidences equal to or higher than the stochastic damage threshold, the majority of pesticide applications at stochastic damage thresholds are ineffective. The low probability of economic success is caused by the strong skewness of the distribution of costs of no control. This result holds for all stochastic damage thresholds, with little variation in the probability of positive return on expenditure

(Rossing *et al.*, 1993a) and warrants analysis of the causes of uncertainty about costs of NC.

3.2 Causes of uncertainty about costs of no control

In Table 2 the contribution of the various categories of uncertainty to uncertainty about predicted costs of no control of aphids and brown rust is shown for one stochastic damage threshold. Results at other damage thresholds are comparable (Rossing *et al.*, 1993c). The analysis was carried out in several steps, in each step disaggregating the sources of uncertainty. In the first step only controllable and uncontrollable sources of uncertainty are distinguished. Both for aphids and brown rust uncontrollable uncertainty was found to contribute most to total prediction

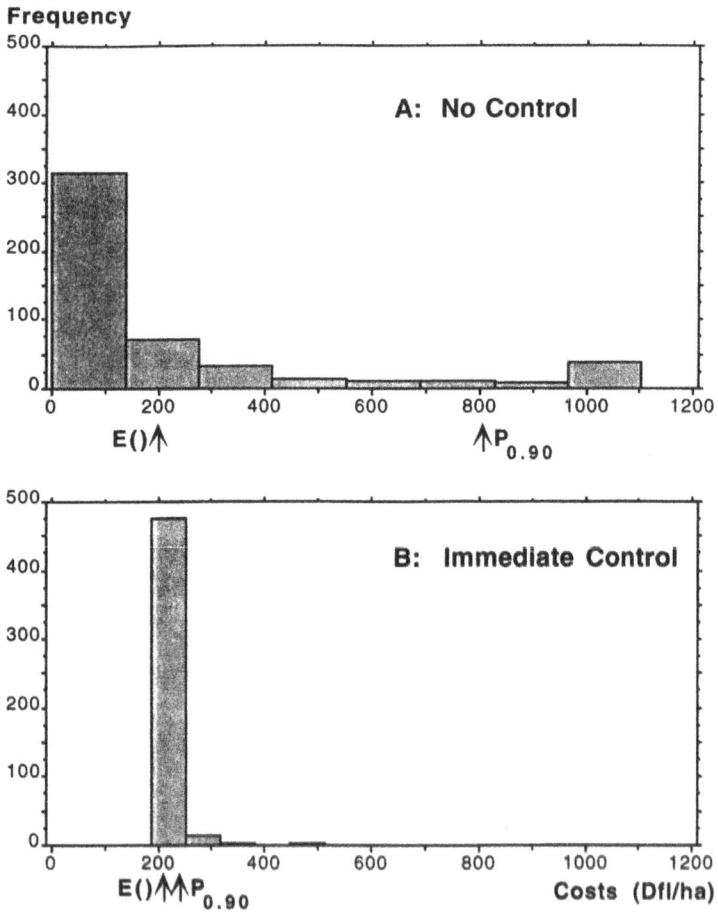

Figure 3. Frequency distributions of costs of no chemical control (A) and immediate chemical control of aphids and brown rust jointly (B) in 500 Monte Carlo runs. Initial state of the system: temperature sum 200 days, equivalent with DC±4 (se), aphid incidence 5% of 100 tillers, brown rust incidence 2% of 160 leaves. Arrows indicate the 0.90-quantile ($P_{0.90}$) or the expected value (E()) of costs.

uncertainty. Further analysis shows that among the sources of uncontrollable uncertainty white noise is far more important than predicted temperature. White noise in the relative growth rates of the pests represents the major source of

Table 2. Expected contribution to model output variance of various sources of uncertainty, as percentage of the variance of the full model, aphids and brown rust separately. Initial states: 25% aphids, 8% brown rust, crop 225 °day. No chemical control at any time.

Source of uncertainty	Contribution (%)	
	aphids	brown rust
Parameter & initial incidence estimate	26[t]	12[t]
White noise in relative growth rate	57[t]	68[t]
White noise in crop development	2[b]	6[b]
White noise in initial density	13[b]	22[b]
Future temperature	12[b]	30[b]

[t] estimate based on top variance
[b] estimate based on bottom variance

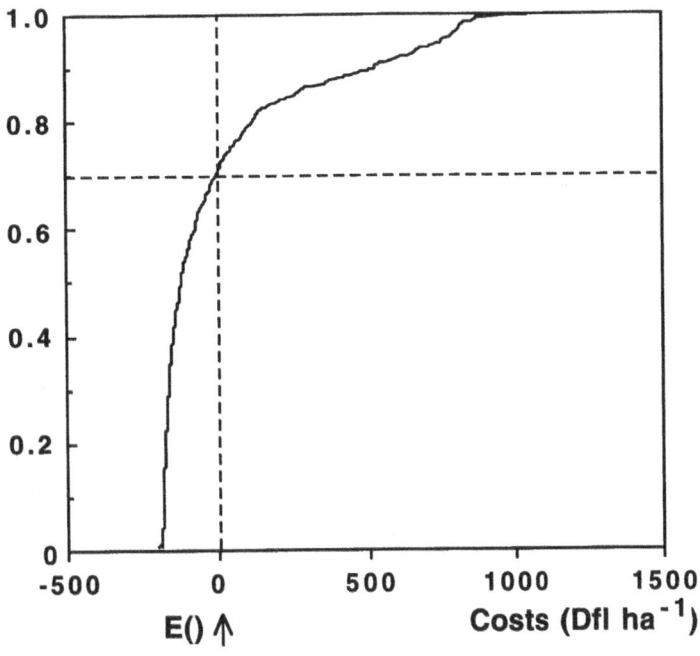

Figure 4. Cumulative relative frequency distribution of the difference of costs of no control and immediate chemical control (NC minus IC) in 500 Monte Carlo runs. For initial state of the system, see caption to Figure 3. The arrow indicates the expected value (E()) of the difference (NC minus IC).

uncertainty (Table 2). Thus, reduction of uncertainty about predicted costs of NC calls for alternative submodels describing pest population dynamics, rather than better determination of parameters and initial conditions.

3.3 Conclusions

This paper focussed on identification and quantification of sources of uncertainty and evaluation of the consequences for prediction uncertainty in pest control. The results of the analysis show that disregarding uncertainty will lead to wrong recommendations to farmers. The possibilities for improving the prediction within the constraints of the current structure of the model are nearly exhausted. Further improvement will require new concepts to be included into the model, especially concerning prediction of population dynamics. Major improvements may be expected when field-specific factors affecting relative growth rates will be taken into account, such as age distribution of the brown rust population and mortality by predators and parasites in the aphid population.

For the purpose of decision support the uncertainty in the model predictions has to be accepted. Rather than ignoring uncertainty or arbitrarily adjusting deterministic results as is done currently in many decision support systems, measures should be designed to assess the degree to which the objectives of pest control, return on expenditure and insurance, are satisfied by various decision strategies. We have proposed such framework for supporting pest control decisions under uncertainty elsewhere (Rossing et al., 1993a).

References

Jansen, M.J.W., W.A.H. Rossing and R.A. Daamen (1993), 'Monte Carlo estimation of uncertainty contributions from several independent multivariate sources', In: *Proceedings Congress Predictability and nonlinear modelling in natural sciences and economics,* Wageningen. 5-7 April 1993.

Norton, G.A. and J.D. Mumford (1983), 'Decision making in pest control', *Adv. Appl. Biol.,* **8,** 87-119.

Pannell, D.J. (1991), 'Pests and pesticides, risk and risk aversion', *Agric.Econ.,* **5,** 361-383.

Rossing, W.A.H., R.A. Daamen and E.M.T. Hendrix (1993a), 'A framework to support decisions on chemical pest control, applied to aphids and brown rust in winter wheat', *Crop Prot.,* in press.

Rossing, W.A.H., R.A. Daamen and M.J.W. Jansen (1993b), 'Uncertainty analysis applied to supervised control of aphids and brown rust in winter wheat. 1. Quantification of uncertainty in cost-benefit calculations', *Agric.Systems,* in press.

Rossing, W.A.H., R.A. Daamen and M.J.W. Jansen (1993c), 'Uncertainty analysis applied to supervised control of aphids and brown rust in winter wheat. 2. Relative importance of different components of uncertainty', *Agric.Systems,* in press.

Tait, E.J. (1987), 'Rationality in pesticide use and the role of forecasting', In: K.J. Brent and R.K. Atkin (Eds.), *Rational pesticide use,* Cambridge University Press, pp.225-238.

Zadoks, J.C., T.T. Chang and C.F. Konzak (1974), 'A decimal code for the growth stages of cereals', *Eucarpia Bull.,* **7,** 42-52.

THE IMPLICATIONS AND IMPORTANCE OF NON-LINEAR RESPONSES IN MODELLING THE GROWTH AND DEVELOPMENT OF WHEAT

MIKHAIL A. SEMENOV and JOHN R. PORTER

Dept. of Agricultural Sciences, University of Bristol,
AFRC Institute of Arable Crops Research,
Long Ashton Research Station, Bristol BS18 9AF, UK.

e-mail: semenovm@afrc.ac.uk

Abstract

Crop simulation models are used widely to predict crop growth and development in studies of the impact of climatic change. In seeking to couple meteorological information to crop-climate models it must be remembered that many interactions between crops and weather are non-linear. Non-linearity of response means it is necessary to preserve the variability of weather sequences in order to estimate the effect of climate on agricultural production and to assess agricultural risk. To date, only changes in average weather parameters derived from General Circulation Models (GCMs) and then applied to historical data have been used to construct climatic change scenarios and in only a few studies were changes in climatic variability incorporated. Accordingly, a computer system, AFRCWHEAT 3S, was designed to couple the simulation crop model for wheat, AFRCWHEAT2, with a stochastic weather generator based on the series approach. The system can perform real-time simulations of crop growth and assess crop productivity and its associated risk before harvest using recorded meteorological data from a current season supplemented by stochastically generated meteorological data. The considerable flexibility used to construct climatic scenarios, on the basis of the weather generator, makes AFRCWHEAT 3S a useful tool in studies of the impact of climatic change on wheat crops. Sensitivity analyses to changes in the variability of temperature and precipitation, as compared with changes in their mean values, were made for location in the UK for winter wheat. Results indicated that changes in climatic variability can have a more profound effect on yield and its associated risk than did changes in mean values.

Key-words crop model, weather generator, non-linearity, climatic variability

1. Introduction

Plant growth and development are connected to the environment via a combination of linear and non-linear responses (Campbell & Norman, 1989). Broadly, rates of ontogenetic processes respond linearly to temperature and photoperiod (Porter & Delécolle, 1988) whereas photosynthetic rate increases initially linearly with respect to incident photosynthetically active radiation (PAR), for a given level of CO_2, but then reaches a plateau, in C_3 species, at about 150 Wm^{-2} PAR (Marshall, 1978). Even in developmental processes where the general response is linear there are conditions for which the rate process may decline. An example is the relationship between temperature and leaf extension rate in millet (Monteith, 1987) where extension rate increased linearly with temperature between 12°C and about 30°C whereupon the rate declined but also at a linear rate. In seeking to couple meteorological information to crop-climate models it must be remembered that crop models reflect this mixture of linear and non-linear responses and, in the broadest sense, transpose a distribution of weather sequences into a distribution of total dry matter and, in the case of crop plants, a harvestable yield. Non-linearity of response means that it is necessary to preserve the variability of weather sequences in order to estimate the effects of climate on agricultural production and to assess agricultural risk, especially in studies of the impact of climatic change. Extreme weather events, such as drought or prolonged hot or cold spells, can be severe for crops and their frequency of occurrence is better correlated with changes in the variability of climate than with changes in the mean (Katz & Brown, 1992) . To date only changes in mean values have been applied to historical weather data in order to construct climatic change scenarios (Santer *et al.*, 1990; Giorgi & Mearns, 1991; Kenny *et al.*, 1993) but data more detailed than the average change in daily values are needed to build a picture of likely yield distributions. For example, changes in the variability of temperature can influence greatly dry matter production since both high and low temperatures decrease the rate of dry matter production and, at the extreme, can cause production to cease (Long *et al.*, 1988). Water deficits just prior to flowering can lead to pollen sterility and a decrease in grain set (Saini & Aspinall, 1982). Thus, a change in the variation of daily temperature level can have as important a consequence for yield as a change in average temperature. Finally there is the non-linear relationship between amount of precipitation and the water-use efficiency of plants. Average amounts of precipitation are used relatively more efficiently by plants than larger amounts because of the rapid movement of excess water to deep and unavailable layers in the soil (van Keulen & Wolf, 1986). In turn, this can lead to excess nitrate leaching to ground water especially if the rate of nitrification by soil microbes is stimulated by warmer temperatures (Porter & Miglietta, 1992).

2. AFRCWHEAT 3S: a stochastic simulation system for wheat

AFRCWHEAT 3S is a system in which a simulation growth model for wheat, AFRCWHEAT2 (Weir *et al.*, 1984; Porter, 1993), and a stochastic weather generator based on a series approach (Racsko *et al.*, 1991) are integrated. The system can perform real-time simulations of crop growth and assess crop productivity and its associated risk before harvest using recorded meteorological data from a current season supplemented by stochastically generated meteorological data. The considerable flexibility used to construct climatic scenarios, on the basis of the weather generator, makes AFRCWHEAT 3S a useful tool in studies of the impact of climatic change on wheat crops. Details of both components follow below. In current climatic projections from General Circulation Model (GCMs) distributions of weather variables are kept unchanged and this approach will likely produce errors when assessing agricultural risk. By using stochastic weather generators, instead of baseline climatic data, we can incorporate variability into climatic projections. Stochastic weather generators, based on historical weather data from a site, provide daily sequences of the main weather parameters that are statistically identical to, but unique from, the empirical data from which they were derived. Thus, they can be used as weather inputs into crop simulation models. For climatic change studies, the advantage is the possibility to change the distribution of the weather variables, for example reducing or extending their variance, or even the type of distribution itself and examine effects.

AFRCWHEAT 3S is implemented as a user-friendly computer package on an IBM-compatible PC.

The AFRCWHEAT2 wheat crop model. AFRCWHEAT2 is a complex simulation model of the growth and development of a wheat crop that describes its phenological development, dry matter production and partitioning in response to the environment using a daily timestep (Weir *et al.*, 1984; Porter, 1993). AFRCWHEAT2 includes subroutines that describe crop and soil evapotranspiration, the movement of water and nitrogen in the soil profile (Addiscott *et al.*, 1986; Addiscott & Whitmore, 1987), their uptake and effects on growth. The timing of phenological stages follows calculation of a succession of phases whose thermal duration is modified by the crop's response to daylength and vernalization. Phenology sets the time frame for other developmental processes such as leaf production and tillering (Porter, 1984). Leaf production rate is calculated as a function of temperature, modified by the rate of change of daylength at emergence. The upper limit to leaf expansion is set by temperature and the availability of assimilate, water and nitrogen determines whether or not the maximum is reached. In AFRCWHEAT2 the production of dry matter is simulated via the interception of photosynthetically active radiation (PAR) by the canopy to fix atmospheric CO_2. The photosynthesis routine describes the response of carbon fixation to both CO_2 concentration and PAR absorbed by the leaf canopy. The partitioning of dry matter between leaves, stems, roots, ears and grain is determined by partitioning coefficients whose relative values vary with phenological stage. As the crop approaches grain filling, some dry matter is diverted from stems and leaves to a labile pool that is potentially available to the grain. During grain-filling all new dry matter goes to the grain, and the labile pool can also contribute to grain mass at a temperature determined rate.

A stochastic weather generator based on the series approach. We used a version of a stochastic weather generator (Racsko *et al.*, 1991) to which technical modifications were made in order to match the output of the generator to the meteorological input data required by AFRCWHEAT2. The weather generator is based on distributions of the length of continuous sequences, or series, of dry or wet days. The distribution of other weather parameters, such as temperature and sunshine-hours are based on the current status of the wet or dry series. The amount, as opposed to the timing, of precipitation is approximated as a mixed exponential distribution. We generalized the distributions for dry and wet series to make them applicable for a wide range of locations in Europe. Daily temperature and sunshine-hours are described as stochastic processes with daily mean and standard deviation dependent on the state of the wet and dry series with Fourier-based interpolation of the distributional parameters. Other stochastic weather models, based on Markovian chains (*e.g.* Richardson, 1981) cannot approximate closely some of the parameters particularly important for plant growth and development such as long dry or wet series. In a Markovian chain the probability of occurrence of long dry or wet series decreases exponentially with the length of the series. Observed dry series in, typically, southern and central locations in Europe cannot be represented adequately by a single exponential distribution but can be approximated by a mixed exponential distribution. Even when the probability of occurrence of a long dry series is not high the consequences for agricultural risk assessment can be significant.

3. Comparison of actual and simulated weather

Generated weather sequences for the period of 30 years have been tested against observed

data in 1950-1980 at IACR-Rothamsted. The mean and standard deviations of actual and
simulated weather for daily average temperature (Fig.1) and maximum duration of dry or
wet series during the year (Fig. 2) were compared; thus covering both average and
extreme weather configurations.

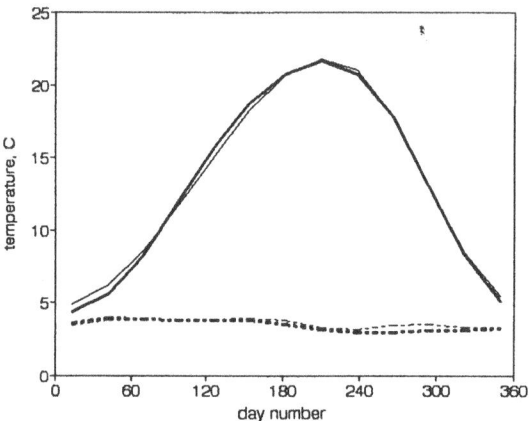

Figure 1. Fourier fitting of average maximum temperature (solid lines) and standard
deviation (dashed lines) for dry series at IACR-Rothamsted. Thick lines, actual data; thin
lines, stochastically generated data.

Another way is to use 30 years weather sequences as inputs to a simulation model and
compare the distributions of output variables such as grain yield (Fig. 3) or day of anthesis
(Fig. 4) for actual and simulated weather. Note that these two methods of comparison of
actual and simulated weather may not always give the same results for the same data set.
    The Kolmogorov-Smirnov two-sample test was used to assess the similarity between
distributions of grain yield from actual and stochastically simulated weather by calculating
the maximum distance between the cumulative distribution functions which, if sufficiently
large, allows the hypothesis that the distributions are the same to be rejected. The
difference between the two functions was 0.2 with a significance level of about 0.59 and
thus we conclude that the distributions came from the same population.   Also we
compared simulated grain yield for actual weather with the yield for mean weather
conditions (Fig. 3 and Table 1). We used the weather generator to simulate "mean"
weather, but replaced the calculated amount of precipitation, temperature and sunshine-
hours by their average values. The sequence of dry and wet series (the timing of the days
with rain) were considered as stochastic. The Kolmogorov-Smirnov test showed the
distributions of simulated grain yields for the actual and mean weather to be significantly
different. Thus, we cannot legitimately substitute stochastic weather sequences by their
mean values in attempting to assess risks to production.

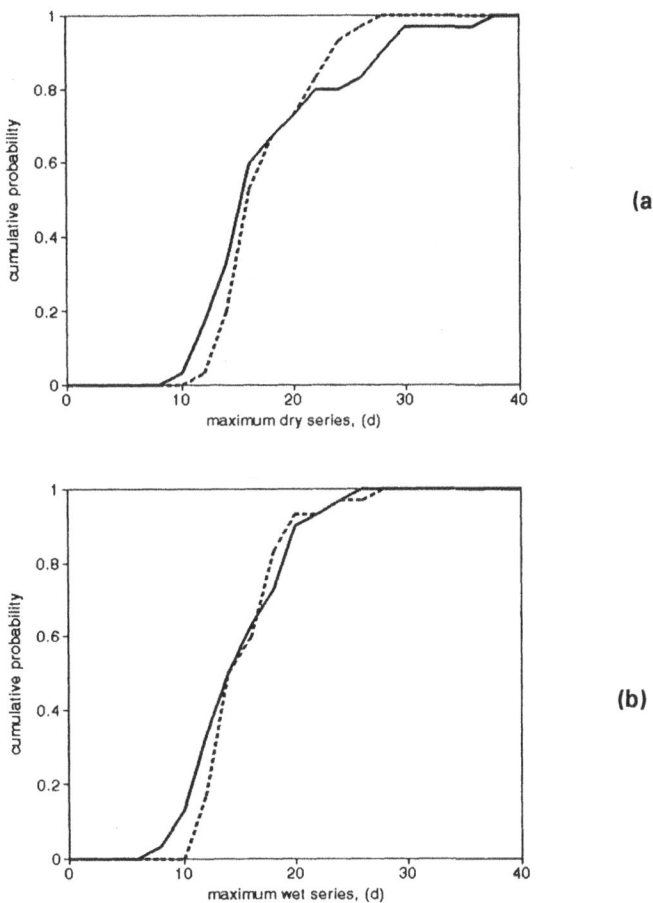

Figure 2. Cumulative distribution functions for (a) maximum dry and (b) wet series during the year for actual (solid line) and simulated (dashed line) weather data at IACR-Rothamsted.

Table 1. Simulated average yield and standard deviation for actual weather, simulated stochastic weather and "mean" weather conditions at IACR-Rothamsted, for *c.v.* Avalon.

Grain yield	Actual weather	Simulated weather	"Mean" weather
Mean, t/ha	7.15	7.52	8.97
sd	1.2	0.95	0.81

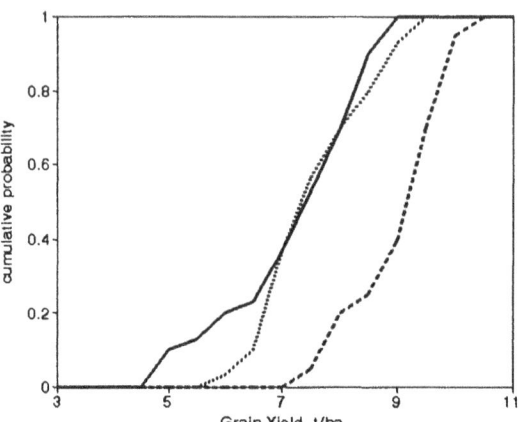

Figure 3. Cumulative distribution functions of simulated grain yield for actual weather (solid line), for weather simulated by stochastic weather generator (dotted line) and for mean weather conditions (dashed line) at IACR-Rothamsted.

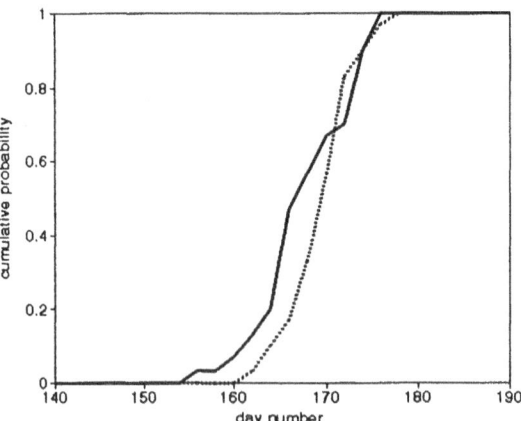

Figure 4. Cumulative distribution functions of simulated day of anthesis for actual weather (solid line), for weather simulated by a stochastic weather generator (dotted line) at IACR-Rothamsted.

4. Real-time simulation of wheat crops

Progress in the ability to follow the real-time growth and development of crops over large areas of land is desirable and is being made via the use of remote sensing (Delécolle *et al.*, 1992). In order to obtain a forward estimate of crop performance it is necessary to use process-based models of crop functioning which take account of climatic, pedological and other information. For crop models to be used, with or without conjunction to remote-sensing, to predict regional yields suitable representations of future weather are necessary. Although these cannot be obtained in an absolute sense, a range of likely prolongations can be developed using stochastic weather models. As demonstrated above using mean weather data is flawed because, clearly, the actual weather could deviate considerably from the mean values and, as stated above, crop growth is rarely a linear function of environmental variables.

AFRCWHEAT 3S has been used for real-time simulation when weather data are only available for the part of the vegetation and have to be supplemented by stochastically generated data for the rest of the season. This process preserves the temporal variability of the weather record in simulating probable values of weather variables for the near future which can be coupled to models of crop production. We analyzed how the yield distribution might change throughout the vegetation period when actual weather was supple mented by 30 generated series and how the precision of the prediction can be increased.

Table 2. Predicted average grain yield and its standard deviation (sd) using stochastically generated weather for different periods of crop development and for two years at IACR-Rothamsted using *c.v.* Avalon.

Weather generated from	1961-62	1984-85
sowing day mean yield (t/ha) sd	309 7.5 0.95	309 7.5 0.95
double ridge day mean yield (t/ha) sd	113 7.0 0.89	104 7.5 0.71
anthesis day mean yield (t/ha) sd	175 5.75 0.59	172 7.95 0.58
actual weather yield (t/ha)	5.59	7.93

Real-time simulations were made for two years, with a low grain yield 5.6 t/ha in 1961-62 and with a hight grain yield 7.9 t/ha in 1984-85 at IACR-Rothamsted, for *c.v.* Avalon with a sowing date of November 5 in both years. Three periods for the simulation of weather were considered; the whole vegetation period, the period after the double ridge stage of the main shoot apex and that after anthesis. Summarised results are presented in Table 2

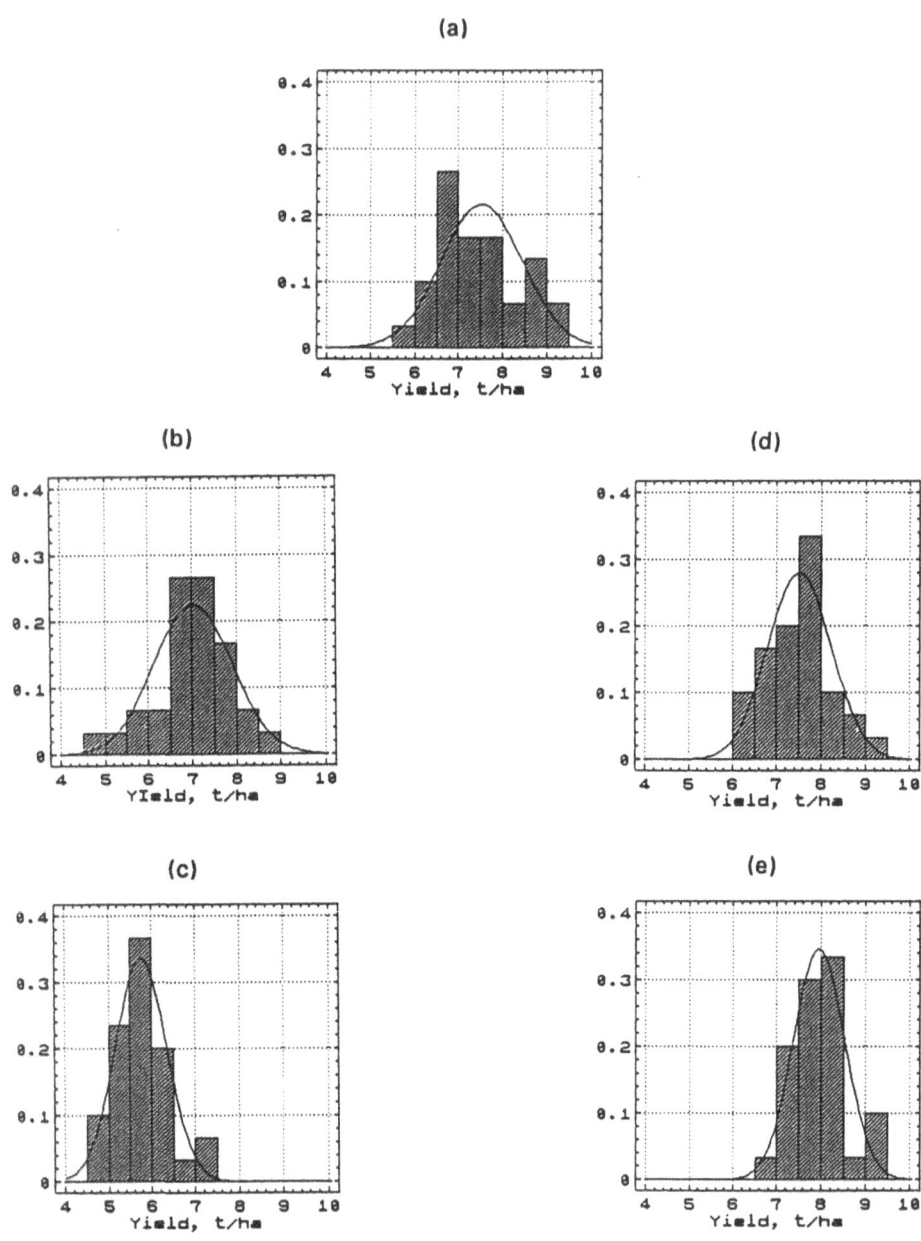

Figure 5. Simulated grain yield distributions for *c.v.* Avalon at IACR-Rothamsted. (a) weather generated for the whole vegetation, (b) 1961-62, weather generated after double ridge, (c) 1961-62, weather generated after anthesis, (d) 1984-85, weather generated after double ridge, (e) 1984-85, weather generated after anthesis.

with simulated yield distributions in Fig. 5. For 1961-62 the average predicted grain yield decreased from 7.5 t/ha at the beginning of the vegetation using completely generated weather to 5.6 t/ha for recorded weather from the whole season. The standard deviation and CV of grain yield were decreased from the beginning of vegetation until anthesis by 38% and 19%, respectively. For 1984-85 the average yield of 7.5 t/ha did not change until double ridges and had increased by 0.4 t/ha by anthesis to 7.9 t/ha. Standard deviation and CV decreased monotonically from the sowing date by 39% and 43%, respectively.

5. Effect of change in climatic variability on growth and development of wheat

AFRCWHEAT 3S has the necessary flexibility to examine the effect of changes in climatic variability on crop production. Because of the large uncertainty in how the variability of a greenhouse climate might change (Rind et al., 1989; Mearns et. al., 1990) we performed an analysis of the sensitivity of the model to changes in temperature and precipitation. The aim was to analyze the effect of changes in climatic variability on wheat growth and development as simulated by AFRCWHEAT 3S as compared with the effect on a wheat crop of changes in mean climate. The analysis was again made for IACR-Rothamsted, UK using Avalon winter wheat.

The importance of looking at the effect of climatic variability on crop growth and development arose from studies of the effects of climatic change and has been realised by different research groups (Mearns et al., 1992; Semenov & Porter, 1992; Semenov et al., 1993). To date only changes in mean weather conditions have been used to construct climatic change scenarios. Katz & Brown (1992) found that extreme climatological events are relatively more dependent on changes in climatic variability than on changes in mean values, especially for hot spells and droughts. Mearns et al. (1992) investigated how changes in climatic variability could affect wheat production and performed sensitivity analyses using the CERES-Wheat crop simulation model and historical climatic data which they perturbed in order to increase the inter-annual variance of the climatic variables.

Our studies are different in that we applied a stochastic weather generator to simulate and alter characteristics of weather sequences instead of using historical weather data alone. This approach provides a more consistent way of changing weather parameters including changes in climatic variance or their distribution and allows us to link local weather patterns with GCMs. It also provides the means to extend observed weather data over time for the assessment of risk and to simulate weather data for unobserved locations by spatial interpolation of stochastic weather model parameters (Hutchinson, 1991). The idea to use stochastic weather generators in climatic change studies was arrived at independently (Semenov & Porter, 1992; Wilks, 1992). Wilks (1992) derived altered parameters for a stochastic weather model based on a Markovian chain (Richardson, 1991) according to predicted changes in climate. In AFRCWHEAT 3S we used a weather generator based on the serial approach (Racsko et al., 1991) which simulates important weather events such as drought for crop growth and development. The possibility to alter directly parameters of the weather generator and to generate weather sequences with specific properties gives considerable freedom in the construction of the climatic scenarios.

Sensitivity to temperature. To compare the relative effects on crop growth and development of a change in mean temperature and its variance the following sensitivity analyses were performed; (i) an increase in daily average temperature of 2°C and 4°C without a change in its variance and (ii) a halving or a doubling of the daily variance of temperature without a change in mean value. We examined the predicted effects of such temperature

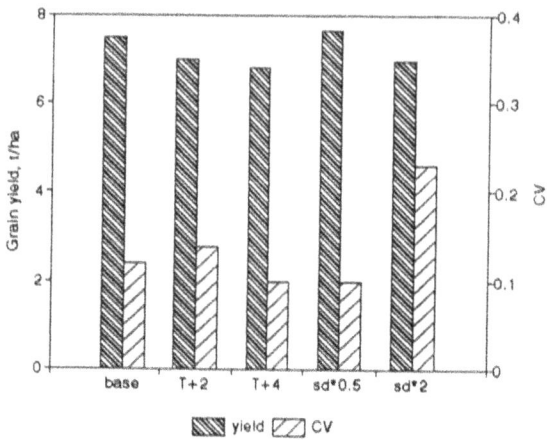

Figure 6. Temperature sensitivity analyses. Average grain yield and its CV for *c.v.* Avalon at IACR-Rothamsted for different sensitivity scenarios. T+2 and T+4, increase in mean daily temperature by 2°C and 4°C; sd*0.5 and sd*2, change in the standard deviation of temperature by x0.5 or x2.

changes on grain yield and on the timing of the developmental stages double ridges, anthesis and maturity. Changes in the variability of temperature had almost the same effect on grain yield as the above changes in mean values. Doubling temperature variability caused the same decrease in yield as an increase in mean temperature of 4°C (Fig. 6) with an associated relative increase in the CV of grain yield of 100%, a value much higher than for any scenario with only a change in mean temperature. A change in mean temperature had a predicted larger effect on the rate of development than did a change in its variance and shifted the distribution towards earlier day numbers. Changes in temperature variability altered the cumulative probability distribution of the daynumber of a stage either increasing or decreasing its variance accordingly (Fig. 7). Increasing temperature variability by a factor of two brought forward the average day number of double ridges but delayed anthesis and maturity (Table 3).

This can be explained by the fact that until double ridges both higher and lower temperatures accelerate developmental rate since low temperatures reduce the period required for vernalization whereas higher temperatures speed up the underlying rate of development. As the variance of temperature increases then these effects will become more powerful.

Sensitivity to precipitation. AFRCWHEAT 3S predicted that changes in precipitation would not have a very pronounced effect on crop growth and precipitation is not thought to be a major limiting factor for crop growth in the UK (Weir, 1988). The time course of daily precipitation can be approximated usually by exponential or gamma distributions within which there are no parameters which describe uniquely mean values or their variance and so it is not possible to construct scenarios where only the mean value or its variability alone is changed. In the weather generator the amount, and not the frequency of

Table 3. Average daynumber of stage (DR - double ridge, AN-anthesis and MA-maturity) and its standard deviation with different sensitivity analyses (Baseline, simulated baseline weather; T+2 and T+4, increase of mean daily temperature of 2°C and 4°C whilst preserving temperature variability; 0.5*sd and 2*sd, halving or a doubling of the variability of temperature whilst keeping mean values unchanged. The site was IACR-Rothamsted for *c.v.* Avalon with a sowing day date of November 5.

Stage	baseline	T+2	T+4	0.5*sd	2*sd
DR mean	105	78	58	105	102
sd	5.8	6.1	5.4	3.4	10.3
AN mean	168	152	138	168	171
sd	3.5	3.2	2.3	1.9	6.8
MA mean	224	200	183	224	228
sd	6.6	5.1	4.0	3.6	11.0

daily precipitation, is calculated via a mixture of three distributions; for amounts of precipitation less than 0.3 mm/d and less or more than 20 mm/d. In our analyses we kept unchanged the frequency of precipitation events in dry and wet series and altered only the amount of daily precipitation. Four scenarios were considered. In the first two we reduced the amount of daily precipitation by 20% and 40% and, for the other, we kept the average precipitation unchanged but altered the ratio of the probability of occurrence of the small, medium and large precipitation groups *i.e.* their variability. The probability of medium precipitation was reduced to 50% and 10% of values estimated by stochastic weather generator. The cumulative distribution functions of grain yield (Fig. 8) show that changes in precipitation distribution are not very significant when compared with changes in mean values. For other regions of the Europe where precipitation could be considered as more of a limiting factor, the effect would likely be more pronounced.

6. Discussion

Generally, the importance of non-linearity in governing the response of ecological and biological systems to their environments is being recognised increasingly. Examples occur at every scale of biological organisation from the chaotic dynamics of plant and animal populations to the feed-back inhibition within metabolic pathways. As a consequence it is clear that representations of the variability of the driving variables that act as inputs to such systems are required. For crop simulation models these requirements can be met by the incorporation of stochastic weather generators into them. AFRCWHEAT 3S is an example of a complex crop-climate simulation system which can be applied to real-time simulations and to analyze the effect of climatic variability on crops.

As a predictive scheme the next stage should be to test the system for actual crops growing in the field for which meteorological data are recorded and then passed to the computer. It would be helpful if this procedure could be applied at a number of sites simultaneously. Another useful development of the stochastic weather generator component of the system would be to use it to generalise from the present site-based approach to regional analyses using the weather generator to interpolate missing data and incorporate

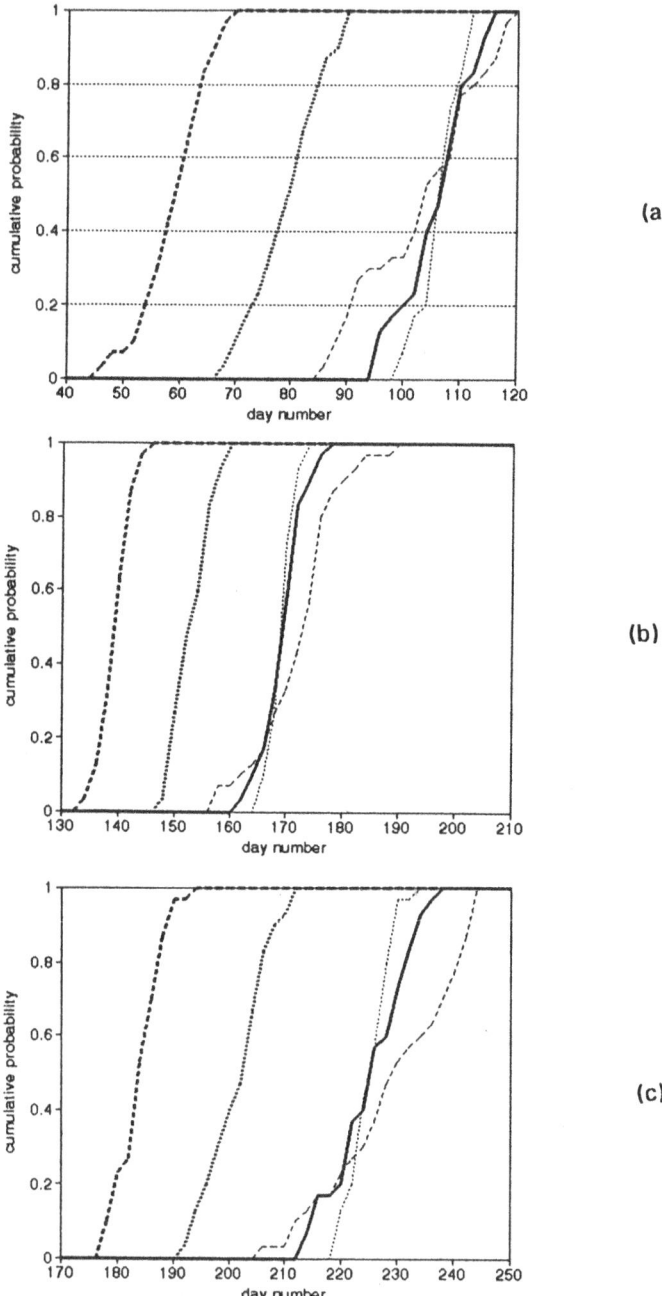

Figure 7. The effect of temperature change on wheat development. Cumulative distribution functions for (a) double ridge, (b) anthesis and (c) maturity. Heavy solid line, base climate; heavy dotted and dashed lines, increase in mean temperature of 2°C and 4°C; light dotted and dashed lines, increase in the standard deviation of temperature by x0.5 and x2.

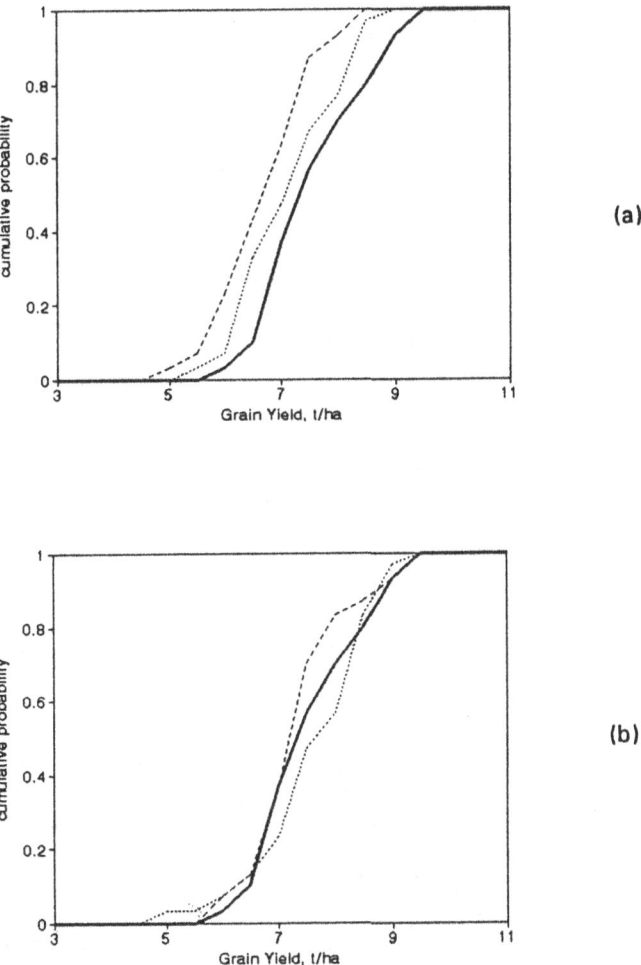

Figure 8. The predicted effect of precipitation changes on wheat grain yields as cumulative distribution functions. Solid line, baseline climate; (a) dotted and dashed lines for the scenarios reducing amount of precipitation by 0.8 and 0.6, respectively; (b) dotted and dashed line for scenarios reducing the probability of medium (0.3mm/d to 20mm/d) precipitation by factors of 0.5 and 0.1 whilst keeping average precipitation unchanged.

topographical features which bear on micro-climate. Whatever the context, be it assessment of the effects of climatic change on crop performance or the further development of real-time simulations of crop growth and development, the necessity of recognising the importance of variability and its interaction with the non-linear aspects of crop growth will remain at the core of such work.

Acknowledgements

M.A. Semenov was supported by a grant under the EPOCH programme of the Commission of the European Communities (Contract number: EPOCH-CT90-0031 TSTS) during this work. Also, we thank Dr. G.H.J. de Koning, reviewer of the paper, for his very helpful comments.

References

Addiscott, T.M., Heys, P.J. and Whitmore, A.P. (1986), 'Application of simple leaching models in heterogeneous soils', *Geoderma*, **38**, 185-194.

Addiscott, T.M. and Whitmore, A.P. (1987), 'Computer simulation of changes in soil mineral nitrogen and crop nitrogen during autumn, winter and spring', *Journal of agricultural Science, Cambridge*, **109**, 141-157.

Campbell, G.S. and Norman, J.M. (1989), 'The description and measurement of plant canopy structure', In: G. Russell, B. Marshall and P.G. Jarvis (Eds) *Plant Canopies: their Growth, Form and Function*. Cambridge: Cambridge University Press, pp. 1-19.

Delécolle R., Maas S.J., Guerif M. and Baret F. (1992). 'Remote-sensing and crop production models -present trends', *Journal of Photogrammetry and Remote Sensing*, **47**, 145-161.

Giorgi, F. and Mearns, L.O. (1991), 'Approaches to the simulation of regional climate change: a review', *Reviews of Geophysics*, **29**, 191-216.

Hutchinson, M.F. (1991), 'Climatic analyses in data sparse regions', In: R.C. Muchow and J.A. Bellamy (Eds) *Climatic risk in crop production: models and management for the semiarid tropics and subtropics*. Wallingford: CAB International, pp. 55-73.

Katz, R.W. and Brown, B.G. (1992), 'Extreme events in a changing climate: variability is more important than averages', *Climatic Change*, **21**, 289-302.

Kenny, G.J., Harrison, P.A. and Parry, M.L. (Eds) (1993), *The Effect of Climate Change on the Agricultural and Horticultural Potential in Europe*. ECU, Oxford University.

Keulen, H. van and Wolf, J. (1986), *Modelling of Agricultural Production: Weather, Soils and Crops*. Wageningen: Pudoc.

Long, S.P. and Woodward, F.I. (Eds) (1988), *Plants and Temperature*. Cambridge: The Company of Biologists Ltd.

May R.M. (1975). Deterministics models with chaotics dynamics. *Nature*, **256**, 165-166.

Marshall, B. (1978), *Leaf and ear photosynthesis of winter wheat crops*. PhD thesis, University of Nottingham.

Mearns, L.O., Schneider, S.H., Thompson, S.L. and McDaniel, L.R. (1990), Analysis of climate variability in general-circulation models - comparison with observation and changes in variability in $2xCO_2$ experiments. *Journal of Geophysical Research*, **95**, 20469-20490.

Mearns, L.O., Rosenzweig, C. and Goldberg, R. (1992), 'Effects of changes in interannual climatic variability on CERES-Wheat yields: sensitivity and $2xCO_2$ general circulation model scenarios', *Agricultural and Forest Meteorology*, **62**, 159-189.

Monteith, J.L. (1987), *Microclimatology in Tropical Agriculture*. Final Report Part 1. Introduction, Methods and Principles. London: Overseas Development Administration.

Porter, J.R. (1984), 'A model of canopy development in winter wheat', *Journal of agricultural Science, Cambridge*, **102**, 383-392.

Porter, J.R. (1993), 'AFRCWHEAT2: A model of the growth and development of wheat incorporating responses to water and nitrogen', *European Journal of Agronomy*. In press.

Porter, J.R. and Delécolle, R. (1988), 'The interaction between temperature and other environmental factors in controlling the development of plants', In: Long S.P. and Woodward F.I. (Eds) *Plants and Temperature*. Cambridge: The Company of Biologists Ltd, pp. 133-156.

Porter, J.R. and Miglietta, F. (1992), 'Modelling the effects of CO_2-induced climatic change on cereal crops', In: Y.P. Abrol (Ed) *Impacts of Global Climatic Changes on Photosynthesis and Plant Productivity*. New Delhi: Oxford and IBH Publishing Co., Ltd.

Racsko, P., Szeidl, L. and Semenov, M.A. (1991), 'A serial approach to local stochastic weather models', *Ecological Modelling*, **57**, 27-41.

Richardson, C.W. (1981), 'Stochastic simulation of daily precipitation, temperature and solar radiation', *Water Resources Research*, **17**, 182-190.

Rind, D., Goldberg, R. and Ruedy, R. (1989), 'Change in climate variability in the 21st century', *Climatic Change*, **14**, 5-37.

Saini, H.S. and Aspinall, D. (1982), 'Effect of water deficit on sporogenesis in wheat (*Triticum aestivum* L.)', *Annals of Botany*, **48**, 623-33.

Santer, B.D., Wigley, T.M.L., Schlesinger, M.E. and Mitchell, J.F.B. (1990), *Developing Climate Scenarios from Equilibrium GCM Results*. Max-Planck-Institut for Meteorologie, Report No. 47.

Semenov, M.A. and Porter, J.R. (1992), 'Stochastic weather generators and crop models: climate change impact assessment', *Effects of Global Change on Wheat Ecosystem*. *GCTE Focus 3 meeting*. Saskatoon, Canada.

Semenov, M.A., Porter, J.R. and Delécolle, R. (1993), 'Simulation of the effects of climate change on growth and development of wheat in the UK and France', In: G.J. Kenny, P.A. Harrison and M.L. Parry (Eds) *The Effect of Climate Change on the Agricultural and Horticultural Potential in Europe*. ECU, Oxford University.

Tardieu F. and Davies W.J. (1992) Stomatal response to abscisic acid is a function of current plant water status. *Plant Physiol.*, **98**, 540-545.

Weir, A.H. (1988), 'Estimating losses in the yield of winter wheat as a result of drought, in England and Wales', *Soil Use and Management*, **4**, 33-40.

Weir, A.H., Bragg, P.L., Porter, J.R. and Rayner, J.H. (1984), 'A winter wheat model without water or nutrient limitations', *Journal of Agricultural Science, Cambridge*, 102: 371-383.

Wilks, D.S. (1992), 'Adapting stochastic weather generation algorithms for climate changes studies', *Climatic Change*, **22**, 67-84.

GROWTH CURVE ANALYSIS OF SEDENTARY PLANT PARASITIC NEMATODES IN RELATION TO PLANT RESITANCE AND TOLERANCE

R.J.F. VAN HAREN[1], E.M.L. HENDRIKX[1] AND H.J. ATKINSON[2]

[1]Research Institute for Plant Protection, IPO-DLO,
P.O. Box 9060, 6700 GW Wageningen, The Netherlands

[2]Centre for Plant Biochemistry and Biotechnology,
University of Leeds, Leeds, UK, LS2 9JT

Abstract

Growth curve analysis of sedentary plant parasitic nematodes on different hosts and at different population densities is used to assess plant suitability including their resistance and tolerance. The estimated parameters of host suitability can be used in pest management programs for economic important species such as potato cyst nematodes.

1. Introduction

Sedentary plant parasitic nematodes are major pests of many world crops. For instance the need to limit economic loss from the two potato cyst nematode (PCN) species *Globodera rostochiensis* and *Globodera pallida* (Heteroderidae) impose restrictions on agricultural practices in The Netherlands and many other countries. In common with several other genera, cyst nematodes induce modified host plant cells from which the animals feed. Such feeding sites are essential for development of females which cause most of the plant damage. Large densities of the parasites present at planting cause subsequent loss of crop yield or even death of the host (Dropkin, 1989). Restrictions on the use of nematicides to control PCN are being enforced in many countries such as the Netherlands. As a consequence changes in pest management must be made to accommodate a greater dependence on integration of rotation and resistant cultivars. This requires improvements in the scientific basis for management and this is the motive underpinning the development of the approach in this work.

The development of a nematode on a plant depends upon a number of both abiotic and biotic factors. The influence of the host plant in the absence of resistance have been examined for *Meloidogyne javanica* by Bird (1972). He established that different host species support dissimilar reproductive performances by *M.javanica*. There is an influence of host nutritional status and also a density dependent relationship between the growth rate and ultimate nematode biomass (Bird, 1970). Such effects probably occur for cyst nematodes but that of host species is potentially more discernible for *Heterodera schachtii* (beet cyst nematode) with a wide host range than for PCN with its comparatively narrow host range.

Host differences influence the growth and reproduction of *G. pallida* because a range of cultivars exists that vary from highly susceptible to partially resistant. Resistance is usually assayed experimentally, using test cultivars in canisters or pots and the plants are challenged by an inoculum of different nematode populations or densities (Foot, 1977;

Kort et al, 1977; Dropkin, 1989; Arntzen & van Eeuwijk, 1992). The ratio between the density of eggs present before and after parasite reproduction provides a measure of host resistance. This approach does not discriminate among effects of host finding, penetration and subsequent parasite development. Also, it does not determine directly the reproductive performance of those individuals that establish on the plant. A more direct approach can be pertinent when resistance is expressed against the established feeding nematode. This is the principal of this work. We use a standard protocol (Robinson et al 1987, Atkinson 1992a/b) to control unwanted sources of variation before studying growth rates of sedentary nematodes.

A thorough study of growth rates requires a basis for analysis. We decided to examine the potential of a Dynamic Energy Budget (DEB) model (Kooijman, 1993) for the analysis of growth of sedentary plant parasitic nematodes. The model considers an individual as an input-output system with size, stored energy and reproductive output as state variables. Originally, the model was developed for *Daphnia magna* Straus (Kooijman, 1986a; Evers & Kooijman, 1989) and successfully applied to *Lymnaea stagnalis* (Zonneveld & Kooijman, 1989), *Mytilus edulis* (van Haren & Kooijman,1993) and micro-organisms (Kooijman et al.,1991). It permits the description of embryo development (Kooijman, 1986c, Zonneveld & Kooijman, 1993), growth (Kooijman, 1988) and body size scaling relations (Kooijman, 1986b, 1988).

The DEB model is first described and then applied to growth curves for both *Meloidogyne* on different hosts and *G. pallida* on different cultivars of potato. Our intention is to investigate the relationship between body size and reproductive output. So, if resource acquisition is coupled to growth then there are consequences at the population level that may be usefully modelled for assessing host suitability and resistance.

2. DEB model for individuals

We will restrict the present discussion to the feeding stages of the nematodes dividing them into two groups of non-reproductive juveniles and reproductive adults. Males are uncommon for most *Meloidogyne* species and they are very small in size in comparison to females in both this genus and ohter Heteroderidae. Their development has little pathological effect on plants compared with that of females. Therefore the DEB model in this work considers only the development of females.

It is assumed that growth of the structural volume and the swelling of the posterior part of the nematode body, caused by reproductive structures, is isomorphic. Due to different growth characteristics, the combined growth of structural and reproductive volume is not isomorphic. The chemical composition of the structural biomass and of stored materials are each taken to be constant, and not necessarily identical. Homeostasis is assumed for structural biomass as well as stored materials. Since the composition of stored materials will differ from that of structural biomass, and the storage density can fluctuate, homeostasis is not assumed for the combination of structural biomass and stored materials. The symbols that are used frequently in the equations are listed in Table 1.

Three state variables, volume, V (length^3), storage, E and reproductive investment R (number eggs) are distinguished. Food uptake is assumed to follow a type II Holling functional response and is taken proportional to surface area (of the stylet and/or gut), so the ingestion rate is

$$I = \{I_m\}fV^{2/3} \text{ with } f = X/(K + X), \tag{1}$$

Table 1. Variables, primary and compound parameters

symbol	dimension	interpretation
variables		
t	time	time
X	weight.length^{-3}	food density
f	–	scaled functional response: $X/K + X$
V	length^3	body volume
E	energy	energy storage
e	energy.length^{-3}	scaled energy storage density: $E/[E_m]V$
R_c	number	cumulated number of eggs
primary parameters		
V_b	length^3	volume at birth
V_j	length^3	volume at start productive stage
K	weight.length^{-3}	saturation constant
$\{I_m\}$	weight.length^{-2}.time^{-1}	maximum surface area-specific ingestion rate
$\{A_m\}$	energy.length^{-2}.time^{-1}	maximum surface area-specific assimilation rate
$[E_m]$	energy.length^{-3}	maximum storage density
$[M]$	energy.length^{-3}.time^{-1}	volume-specific maintenance costs per unit of time
$[G]$	energy.length^{-3}	volume-specific costs for growth
κ	–	fraction of utilized energy spent on maintenance plus growth
q	–	mother-egg energy conversion overhead
s	–	nutrient conducting efficiency
δ	weight.ind.$^{-1}$	resource extinction for nematodes
compound parameters		
v	length.time^{-1}	energy conductance: $\{A_m\}/[E_m]$
m	time^{-1}	maintenance rate constant: $[M]/[G]$
g	–	energy investment ratio: $[G]/\kappa[E_m]$

where $e = [E]/[E_m]$, where $[E_m]$ is the maximum storage density and $v = \{A_m\}/[E_m]$ is by definition the energy conductance (length.time^{-1}). The rate at which energy is utilized from the storage must obey the conservation law of energy:

where X is the food density, K the saturation constant and $\{I_m\}$ the maximum surface area-specific ingestion rate. The food-energy conversion is taken to be constant, $\{A_m\}/\{I_m\}$, so the assimilation energy, *i.e.* the total energy input, equals $A = \{A_m\}fV^{2/3}$, where $\{A_m\}$ is the maximum surface area-specific assimilation rate. The incoming energy adds to the reserves. The reserves follow a first order process when expressed as density, $[E] = E/V$, *i.e.* energy reserve per body volume. From the assumption of homeostasis for energy reserves it follows that the energy reserves in equilibrium are independent of the length of the nematode *i.e.*

$$\frac{de}{dt} = vV^{-1/3}(f - e) , \qquad (2)$$

$$C = A - \frac{d}{dt}([E])(V) = A - V\frac{d}{dt}[E] - [E]\frac{d}{dt}V$$
$$= e[E_m] \left(vV^{2/3} - \frac{dV}{dt}\right) . \qquad (3)$$

A fixed fraction κ of the utilized energy is spent on growth plus maintenance. The latter quantity is taken to be proportional to volume, $[M]V$. So $\kappa C = [M]V + [G]dV/dt$, where $[G]$ is the volume-specific costs for growth. Substitution gives

$$\frac{dV}{dt} = \frac{V^{2/3}ev - V gm}{e + g} , \qquad (4)$$

where the dimensionless investment ratio, $g = [G]/\kappa[E_m]$, and the maintenance rate coefficient, $m = [M]/[G]$ are compound parameters. If the food density is constant long enough, (2) states that e tends to f and remains constant as well. This turns (4) into the well known von Bertalanffy growth equation, having the solution

$$V(t) = (V_\infty^{1/3} - (V_\infty^{1/3} - V_0^{1/3})\exp\{-\gamma t\})^3 , \qquad (5)$$

where $V_\infty^{1/3} = f\kappa\{A_m\}/[M]$ is the ultimate volume$^{1/3}$ and $\gamma = (3/m + 3V_\infty^{1/3}/v)^{-1}$, the von Bertalanffy growth rate. Thus, the maximum volume$^{1/3}$ is:

$$V_m^{1/3} = \kappa\{A_m\}/[M] = v/gm ,$$

which can only be reached at prolonged exposure to abundant food.
    Back substitution of (4) into the storage utilization rate (3) gives

$$C = \frac{eg[E_M]}{e + g} (vV^{2/3} + mV) . \qquad (6)$$

So, the storage utilisation rate depends only on the volume of the organism and the energy reserves and not directly on food density. The energy allocation to reproduction equals

$$(1 - \kappa)C - \frac{1 - \kappa}{\kappa}[M]V_j \ .$$

This is a continuous energy investment. The costs for the production of an egg can be written as E_0/q, where the dimensionless factor $1/q$ between 0 and 1 relates the overhead involved in the conversion from the reserve energy from the mother initial energy available for the embryo, see Zonneveld & Kooijman (1993) and Kooijman (1993) for discussion. Substitution of equation 6 leads to a mean reproduction rate of:

$$\frac{dR}{dt} = \frac{q}{e_0 V_m} (1 - \kappa) \left(\frac{ge}{g + e}(\nu V^{2/3} + mV) - gmV_j \right). \tag{7}$$

When growth ceases, the cumulated energy drain to reproduction in animals that continue to allocate energy to reproduction under these circumstances, becomes

$$\frac{dR}{dt} = \frac{q}{e_0 V_m} gm \ (eV_m^{1/3}V^{2/3} - \kappa V - (1 - \kappa)V_j) \ . \tag{8}$$

The energy feeding the drain to reproduction accumulates during nematode development, but it is assumed to be metabolically unavailable for other purposes so

$$R_c = \int_{t_j}^{t_1} R'dt \ ,$$

where t_j denotes the time of maturation and t_1 the time of death of the female. So this energy pool differs from that in the storage pool in its availability. In PCN, eggs are retained within the body wall of the female, which protects eggs after death of the female.

3. Individual growth and reproduction, plant resistance

The feeding site, which functions as a nutrient conducting complex, is induced by the nematode upon a specific action which degrades cell walls or enlarges the initial feeding cell (Jones, 1981). Bird (1972) has measured feeding sites on different host plants and the body sizes of the nematodes grown on it. Large feeding sites result in large nematodes and small sites in small ones. The contact area of the feeding site with the vascular tissues of the plant host determines the conducting capacity of the feeding site itself. The cumulative amount of nutrients for the parasite is therefore determined by the maximum size the feeding site can reach. Host status is therefore assumed to be equal to the 'mean' scaled functional response f, which is determined by the nutrient density in the feeding site. This is preferred to other estimates such as assimilation efficiency $\{A_m\}/\{I_m\}$ or energy conductance v because the latter parameters are linked to intrinsic properties of nematodes.

Measures of size during development are the body width at the mid point of the oesophageal procorpus (oesophageal region width), length, maximum mid-body width, surface area and the volume of the nematode. The oesophageal region width is taken as a measure of structural volume, V, while the mid-body width provides a measure of the accumulated reproductive output, R_c.

Growth of G. pallida on potato cultivars of different host status Multa (partial resistant) and Maritta (susceptible) is shown in Figure 1. Bertalanffy growth equations, equation 5, are fitted to measures of the oesophageal region width during development. The fit is satisfactory for both of the cultivars, implying a more or less constant food density throughout development. PCN developed on the resistant cultivar Multa are about a factor of 2.7 smaller than individuals developed on the susceptible cultivar Maritta. In addition PCN that developed on Multa failed to reproduce. This may result from insufficient energy uptake to maintain a female reproductive system.

Different host species can also have effects on growth as is shown for volume growth, $V_i(t)$, of M. hapla in Figure 2. The fitted equation is:

$$V_i(t) = (f_i V_m - (f_i V_m - V_b) \exp\{-t/(3/m + 3f_i V_m/v)\})^3$$

with f_i the scaled functional response for the i-th cultivar. M. hapla fail to reproduce on plants when the functional response f_i is about 2.3 - 2.7 less than for susceptible hosts. This suggests that either maximum oesophageal region width or scaled functional response provides a measure of host ability to support fecund females.

Ingested energy is according to the DEB theory partitioned to three main processes: growth, maintenance and reproductive growth. The reproductive output is determined when the growth curve is known. Figure 3 shows the parallel growth of the oesophageal region width and mid-body width. Equations 4, 7 and 8 are integrated using a Douglas-Adams fifth order predictor corrector (Burden & Faires, 1990) with the assumption that the food density remains constant during development. This assumption is realistic because growth is of the von Bertalanffy type, see figures 1 and 2. The parallel action of changes in width of the oesophageal and mid body region with time confirms the supposition of a direct coupling between structural volume growth and reproductive output.

4. Competition for space in the root system, plant tolerance

The feeding site induced by a nematode occupies a root volume associated with the vascular cylinder in the root. It is assumed that the nematode ingests a fraction s of the available transported nutrients, X, in the vascular cylinder. The amount of available nutrients in the root system decreases progressively with increasing numbers of nematodes, as is shown for certain essential amino acids (Grundler and Betka, 1991). The carrying capacity of the plant for nematodes depends on the initial amount of available nutrients X and on the nutrient conducting efficiency s of the feeding site. The amount of nutrients available after the n-th feeding site is characterised by X_n which is:

$$X_n = (1 - s)^n X_0 = X_0 \exp\{-\delta n\} \text{ with } \delta = -\ln(1 - s) \tag{9}$$

This function describes a density dependent individual resource partitioning which affects the scaled functional response f_n. Decreasing functional responses affects the egg producti

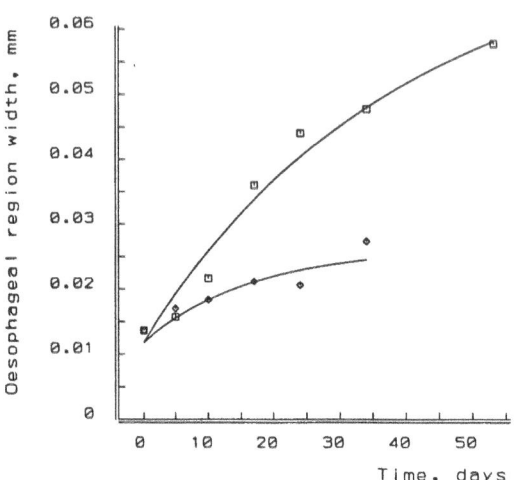

Figure 1. Oesophageal region width of *G. pallida* grown on 2 potato cultivars Multa (resistant) ◇ and Maritta (susceptible) ▫ at 20°C. Fitted curves are $L(t) = L_\infty - (L_\infty - L_0)$ $\exp\{-\gamma t\}$. Initial oesophageal region width, L_0, is a in common estimated parameter: 0.012 (SD = 6.1 10^{-4}) mm. The other estimated parameters are for the susceptible interaction: L_∞ = 0.074 (SD = 3.3 10^{-3}) mm and γ is 0.026 (SD = 2.6 10^{-3}) d^{-1}. For the resistant interaction: L_∞ = 0.027 (SD = 4.1 10^{-3}) mm and γ is 0.059 (SD = 0.030) d^{-1}.

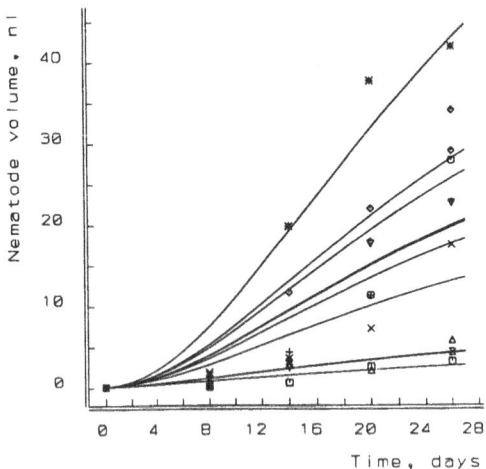

Figure 2. Volume growth of *M. hapla* on the host species varying in host suitability for the nematode. Fitted curves are cubed Bertalanffy growth equations, equation 5 with scaled food density as unique parameter per curve. No egg production was measured at values of f_i: 0.20, 0.22 and 0.23. While egg production was measured following parasitism of al eight hosts providing f_i values between 0.34 and 0.53. Data are from F.A.Al-Yayha, Centre Plant Biochem. & Biotechn. Univ. Leeds.

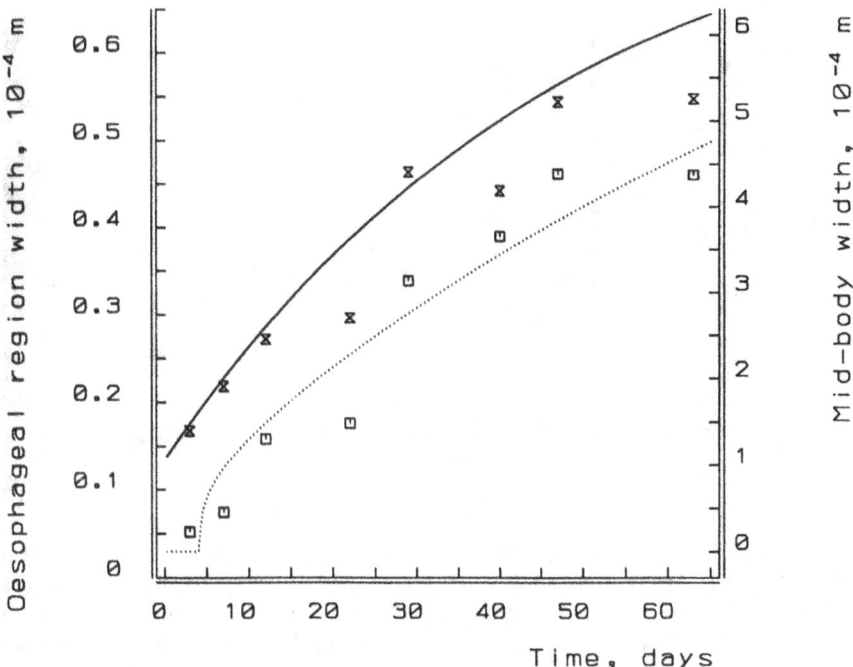

Figure 3. Growth of oesophageal region width (–) and mid-body width (···) of *G. pallida* on potato cultivar Bintje at 20°C. The equations are 4, 7 and 8 for oesophageal region width and mid-body width respectively. The estimated parameters are $m = 0.070$ (SD = 0.034) d⁻¹, $g = 6.2$ (SD = 1.7) and $\kappa = 0.09$ (SD = 0.13) with the energy conductance v is fixed on $1.3 \ 10^{-4}$ m.d⁻¹ and $f = e$ is fixed on 0.8.

on of PCN as is shown in Figure 4.

Plant yield itself is affected by large nematode burdens of the root system. The increasing damage of PCN on potato cultivar Bintje is shown in Figure 5. It is assumed that the damage to the plant is proportional to the amount of nutrients withdrawn by the feeding sites, so equation 9 can be used.

Plant tolerance to nematodes is defined as the capacity of a host to withstand a certain burden of nematodes (Trudgill, 1991). The factor δ in equation 9 can be considered to indicate tolerance since it defines a 'resource extinction' for nematodes.

The interaction between sedentary plant parasitic nematodes and its host can be characterised by two parameters: the scaled functional response, f, which can be regarded as a measure for of host suitability and δ which can be regarded as measure of tolerance of the plant to a specific nematode population. Figure 6 shows the classification in respect of resistance and tolerance of a few potato cultivars and their properties for PCN in the δ - f plane.

The simulated population dynamical consequences of different plant properties (δ - f) and intraspecific competition of the nematodes is shown in Figure 7. The left figure shows the effect on the egg production per female (cyst) while the right figure shows the same relation but now expressed as nematode egg production per plant. The maximum population

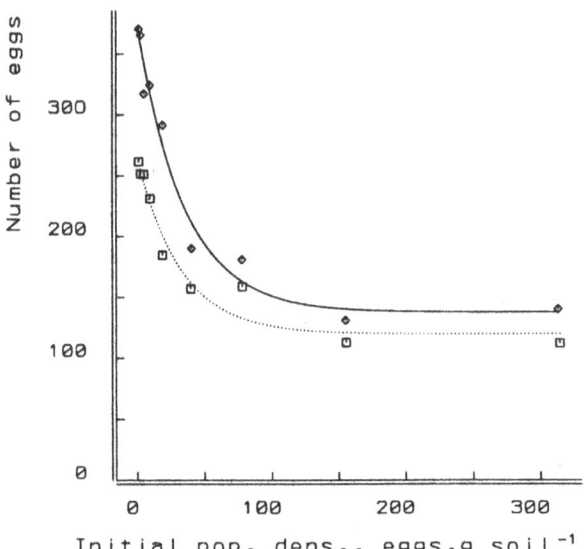

Figure 4. Number of eggs per cyst (female) as function of increasing population densities on different potato cultivars. The fitted curve is $Q = Q_\infty + Q_0 \exp\{-\delta N\}$. With Q the egg content per cyst and N the initial population densities. The estimates are respectively for potato cultivars irene and darwina: Q_∞ = 235 and 149 (SD = 11 and 9.7) eggs/female; Q_0 = 138 and 128 (SD = 8.6 and 7.6) eggs/female; δ = 0.029 and 0.039 (SD = 0.0040 and 0.0068) g soil/egg. Data are from Seinhorst (1984).

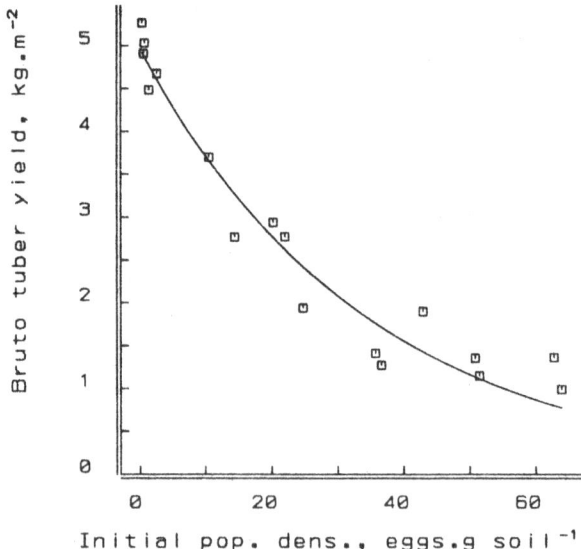

Figure 5. Yield of potato cultivar bintje as function of initial population density. The fitted curve is $Y = Y_0 \exp\{-\delta N\}$ with Y the tuber yield per area. The estimates are respectively: Y_0 = 5.0 (SD = 0.14) kg/m² and δ = 0.029 (SD = 0.0020) g soil/egg. Data are from L. Molendijk, Governmental Crop Research Station, PAGV, Lelystad, The Netherlands.

reproduction rates are at intermediate initial population densities when the negative effects of nematodes on their resource outgrow the initial population density.

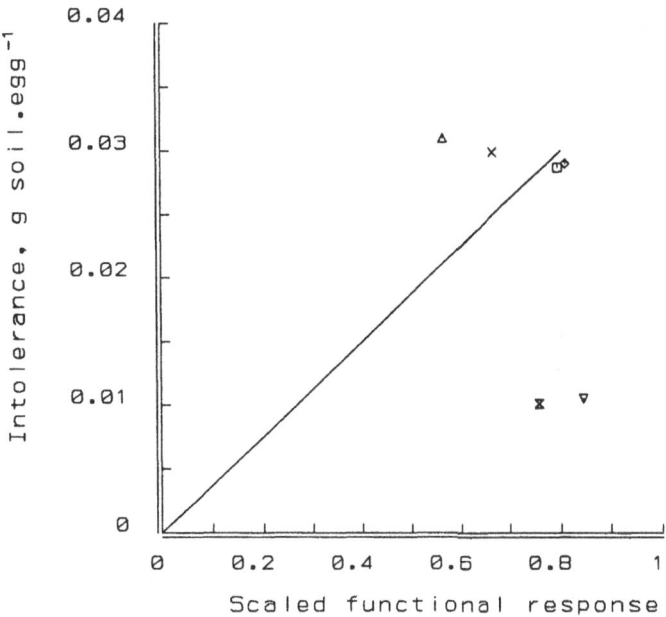

Figure 6. Scaled functional response f_i, defined as resistance and tolerance δ_i of different potato cultivars (1) susceptible: ◻, irene; ◊, bintje (2) partial resistant: ×, multa; Δ: darwina (3) tolerant: ∇, ehud and Ⅹ, astarte. The estimates are from figures 4 and 5. The x - y plane interception is used in figure 7 as y-axis.

5. Discussion

This work establishes the potential of DEB for analyzing the individual growth curves of plant parasitic nematodes. It suggests that the energy available to the nematode from the feeding site it induces in the plant, determines its growth rate and fecundity.

Apparently the animals are unable to rectify an inadequately forming feeding site by further stimulation of the plant. Consequently the growth rate of the nematode and its body size are compromised by biotic factors that influence the size of their feeding cells with the direct consequence of either a fall in fecundity or even a prevention of egg laying (Figure 1).

The DEB model provides a comprehensive approach that accommodates differences in host status of plants and both tolerance and resistance of different cultivars to one species of nematodes. With further development the model could have a range of uses. It could be of value within pest management schemes for Meloidogyne spp. to optimise cropping regimes when a number of crops of dissimilar host status are used in combination. It can also be used in a similar manner for a cyst nematode such as *H. schachtii* which may be offered different hosts such as oil seed rape and sugar beet within a rotational scheme. It has important potential for optimising the utilisation of partially resistant cultivars of crops such as potato showing at least some resistance post-establishment of PCN. A future research program could use the DEB model to provide an improved basis for ranking partially resistant cultivars or for determining the relative reproduction of different nematode populations (*i.e.* virulence) on one partially resistant cultivar.

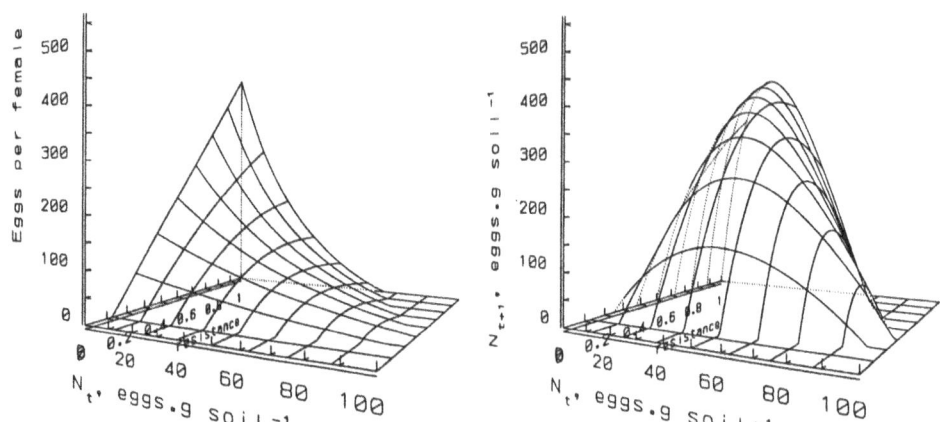

Figure 7. Left: egg content per female as function of initial population density, N_t, and of combined resistance and tolerance for the *i*-th host. The equation is: $f_i Q_0 \exp\{-\delta_i N_t\} - P$. In which P is proportional to the immature volume V_j. Right: final population density N_{t+1} as function of initial population density, N_t, and of combined resistance and tolerance. The equation is: $\phi N_i f_i Q_0 \exp\{-\delta_i N_t\} - P$ in which ϕ is the fraction invading nematodes in the root. The properties of the plant (*Y*-axis) are a combined function of resistance and tolerance corresponding with the straight line in figure 6.

References

Arntzen, F.K. & van Eeuwijk, F.A., (1992), 'Variation in resistance level of potato genotypes and virulence level of potato cyst nematode populations', *Euphytica*, **62**, 135-143.

Arntzen, F.K., Vinke J.H. & Hoogendoorn, J., (1993), 'Inheritance, level and origin of resistance to Globodera pallida in the potato cultivar Multa derived from Solanum tuburosum spp. andigena CPC 1673', *Fundamental and Applied Nematology*, **16**, 155-163.

Atkinson H.J., Gurr S.J., McPherson M.J., Macgregor, A.N. & Bowles, D.J., (1992a), 'Molecular events at nematode-induced feeding sites', *Netherlands Journal Plant Pathology*, **98 Supplement 2**, 175-181.

Atkinson H.J., (1992b), 'Nematodes', in S.J. Gurr, M.J. McPherson & D.J. Bowles (eds.), *Molecular Plant Pathology, A practical approach*, Volume **I**, IRL Press, Oxford.

Bird, A.F., (1970), 'The effect of nitrogen deficiency on the growth of Meloidogyne javanica at different population levels', *Nematologica*, **16**, 13-21.

Bird, A.F., (1972), 'Quantitative studies on the growth of syncytia induced in plants by root knot nematodes', *International Journal of Parasitology*, **2**, 157-170.

Burden, R.L. & Faires, J.D., (1989), 'Numerical Analysis', 4th edition, PWS-Kent Publ.-Company, Boston, pp.729.

Dropkin, V.H., (1989), 'Introduction to Plant Nematology', John Wiley, New York, pp.304.

Evers, E.G. & S.A.L.M. Kooijman, (1989), 'Feeding, Digestion and oxygen consumption in Daphnia magna A study in energy budgets', *Netherlands Journal Zoology*, **39**, 56-78.

Foot, M.A., (1977), 'Laboratory rearing of potato cyst nematodes; a method suitable for pathotyping and biological studies', *New Zealand Journal of Zoology*, **4**, 183-186.

Grundler, F., Betka, M. & Wyss, U., (1991), 'Influence of changes in the nurse cell system (syncytium) on sex determination and development of the cyst nematode Heterodera schachtii. I. Total amounts of proteins and amino acids', *Phytopathology*, **81**, 70-74.

Haren, R.J.F. van & Kooijman S.A.L.M. , (1993), 'Application of a Dynamic Energy Budget Model to Mytilus Edulis', *Netherlands Journal of Sea Research*, **31(2)**, 119-133.

Jones, M.G.K., (1981), 'Host cell responses to endoparasitic nematode attack: structure and function of giant cells and syncytia', *Annals of Applied Biology*, **97**, 353-372.

Kooijman, S.A.L.M., (1986a), 'Population dynamics on the basis of energy budgets', In: Metz, J.A.J. and Diekmann O. The dynamics of physiologically structured populations, *Lecture Notes in Biomathematics*, Springer Verlag, Berlin: 266-297.

Kooijman, S.A.L.M., (1986b), 'Energy budgets can explain body size relations, *Journal of Theoretical Biology*, **121**, 269-282.

Kooijman, S.A.L.M., (1986c), 'What the hen can tell about her eggs: egg development on the basis of energy budgets', *Journal of Mathematical Biology*, **23**, 163-185.

Kooijman, S.A.L.M., (1988), 'The von Bertalanffy growth rate as a function of physiological parameters: a comparative analysis', In: Hallam, T,G, Gross, L.J. and Levin, S.A. *Mathematical Ecology*. World Scientific, Singapore: 3-45.

Kooijman, S.A.L.M., (1993), 'Dynamic Energy Budgets in biological systems; Theory and applications in Ecotoxicology', Cambridge University Press.

Kooijman, S.A.L.M. & R.J.F. van Haren, (1990), 'Animal Energy Budgets affect the kinetics of xenobiotics', *Chemosphere*, **21**, 681-693.

Kooijman, S.A.L.M., E.B. Muller & A.H. Stouthamer, (1991), 'Microbial dynamics on the basis of individual budgets', *Antonie van Leeuwenhoek*, **60**, 159-174.

Kort, J., Ross, H., Rumpenhorst, H.J. & Stone, A.R., (1977), 'An international scheme for identifying and classifying pathotypes of potato cyst-nematodes Globodera pallida and G. rostochiensis', *Nematologica*, **23**, 333-339.

Reversat, (1987), 'Increase of the chemical oxygen demand during the growth in Heterodera sacchari', *Revue de Nematology*, **10**, 115-117.

Robinson, M.P. and Atkinson, H.J. and Perry, R.N., (1987), 'The influence of soil moisture and storage time on the motility, infectivity and lipid utilization of second stage juveniles of the potato cyst nematodes Globodera rostochiensis and G.pallida', *Revue de Nematology*, **10**, 343-348.

Trudgill, D., (1991), 'Resistance and tolerance of plant parasitic nematodes in plants', *Annual Review of Phytopathology*, **29**, 167-192.

Zonneveld, C & S.A.L.M. Kooijman, (1989), 'Application of a dynamic energy budget model to Lymnaea stagnalis (L.)', *Functional Ecology*, **3**, 269-278.

Zonneveld, C & S.A.L.M. Kooijman, (1993), 'Comparative kinetics of embryonic development', *Bulletin of Mathematical Biology*, **55**, 609-635.

USING CHAOS TO UNDERSTAND BIOLOGICAL DYNAMICS

BRUCE E. KENDALL[1], WILLIAM M. SCHAFFER[1], LARS F. OLSEN[2],
CHARLES W. TIDD[1] and BODIL L. JORGENSEN[2]

[1]Department of Ecology and Evolutionary Biology
University of Arizona, Tucson, AZ 85721, USA

[2]Institute of Biochemistry, Odense University
Campusvej 55, DK-5230 Odense M, DENMARK

Abstract

The application of nonlinear dynamics has begun to move beyond the problem of demonstrating the existence of nonlinearities (and chaos) in biological data. We are starting to see exciting cases where considerations of nonlinear dynamics can explain observed patterns, and give insight into the forces structuring biological phenomena. We present some examples of these, drawn from the analysis of measles epidemics and cardiac pathologies. We show that a phenomenon of many chaotic systems called transient periodicity can explain apparently qualitative shifts in the observed dynamics. Such shifts require no change of or perturbation to the system, but can be intrinsic features of purely deterministic dynamics. We show how the techniques of nonlinear forecasting can be used as analytical tools, both for quantifying the complexity of the time series in a biologically meaningful way and for determining how well a particular model accounts for the dynamics. We also post a warning for practitioners of "conventional" nonlinear time series analysis (such as dimension calculations): many biological time series are nonuniform, and this can seriously mislead the various algorithms in common use.

1. Introduction

One of the main projects in nonlinear science in the 80's was to demonstrate that chaos really exists in the world, and is not just a peculiarity of mathematical models. Much of this work focused on the analysis of time series, using techniques that purport to distinguish chaos from noise. The most commonly used methods were estimates of correlation dimension (Grassberger and Procaccia 1983) and largest Lyapunov exponent (Wolf *et al.* 1985), and more recently, characteristics of prediction accuracy (Sugihara and May 1990). Numerous studies turned up positive results (Schaffer 1984, Babloyantz *et al.* 1985, Markus *et al.* 1985, Schaffer and Kot 1985, Babloyantz and Destexhe 1986, 1988, Dvorak and Siska 1986, Mayer-Kress *et al.* 1988, Olsen *et al.* 1988, Zbilut *et al.* 1988, Frank *et al.* 1990, Olsen and Schaffer 1990, Gallez and Babloyantz 1991, Pijn *et al.* 1991, Sammon and Bruce 1991, Yip *et al.* 1991, Donaldson 1992, Gantert *et al.* 1992, Pritchard and Duke 1992). However, these techniques can be misled by certain types of stochastic data (Osborne and Provenzale 1989, Ellner 1991, Stone 1992) and so have been viewed with some skepticism. Furthermore, many biological time series are nonuniform, and as we demonstrate below, these techniques are not very reliable when applied to such data. The clearest evidence for chaos in biology is found where bifurcations can be induced as a parameter is varied (Aihara *et al.* 1985, Hayashi *et al.* 1985, Chialvo *et al.*

1990, Geest *et al.* 1992, 1993), but this type of study is limited to carefully controlled laboratory experiments. In spite of these difficulties, the project has largely been successful: a growing number of biologists are accepting the notion that chaos can often be found in the real world. However, many of them then say "So what! How can 'chaos theory' help me understand my system?"

This is a question we must address. If nonlinear dynamics is as important a structuring force in the world as we believe it to be, then biology would be greatly served if biologists turned to its tools as readily as they now turn to those of statistics. Like statistics, understanding nonlinear dynamics sufficiently well to use it sensibly requires a considerable investment of time and effort. If we are to inspire biologists to make this investment, then those of us who have already been drawn to the field must provide convincing examples of ways in which a consideration of nonlinear dynamics can enhance our understanding of the biology.

The bulk of this paper is devoted to examples of ways in which we feel these goals can be accomplished, but first we have some general comments. First of all, we need to be prepared to proceed even if we don't have unequivocal evidence that the system inquestion is chaotic. That "proof" is often just too hard to come by, and is not necessary for what follows. Rather, we *assume* that there are nonlinear dynamics underlying the process, and see if that assumption allows us to come up with testable predictions or enhanced understanding of the biology. There are a number of reasons why we might consider a system to be a good candidate for this sort of analysis. We might have a good *a priori* reason: for example, we would expect that a large part of the electrical activity in the brain should have a deterministic origin. Thus nonlinear dynamics, which assumes an underlying determinism to complex dynamics, may yield insights not attainable from stochastic theory. Alternatively, the data themselves may be suggestive of nonlinear dynamics: in the records of measles incidences, as we show below, the fluctuations are so great that random effects of tremendous magnitude would have to be added to a linear process to get the observed oscillations.

Another issue that needs to be addressed is dimensionality. The well worked-out part of the theory of nonlinear dynamics pertains to low dimensional systems. High dimensional systems are being tackled, but there are few theoretical generalizations yet available. This is a problem, for there are probably more high dimensional than low dimensional systems in biology, whether due to noise or to the inherent complexity of the system. Although this will limit the the conclusions we can draw in such situations, we expect that a low dimensional nonlinear approach may often be more rewarding than a linear or stochastic one. It also challenges us to find analytic techniques that work independent of dimension.

One technique that may satisfy this criterion is nonlinear forecasting (Farmer and Sidorowich 1987, 1988, Casdagli 1989, Sugihara and May 1990). Other articles in this volume discuss the details of how it works, as well as its efficacy as a predictive tool, and so we will not go into these topics here. Quite aside from its use in predicting the future evolution of dynamical systems, however, nonlinear forecasting can be a valuable diagnostic of the complexity of the time series, and it is in this role that it will appear in the examples we discuss here.

A mathematical object we will visit several times is the semi-attractor (Kantz and Grassberger 1985). This can best be thought of as a chaotic saddle: its core is a strange invariant set, just as in a chaotic attractor, and it has something like a basin of attraction from which trajectories come close to the semi-attractor. However, there is also a mechanism by which all trajectories except those actually on the invariant set eventually leave the neighborhood of the semi-attractor. In many cases the trajectory then moves towards a

coexisting attractor, often a simple one such as an equilibrium or limit cycle. This can
have important consequences for the observed dynamics, for if there is a perturbation
sufficiently strong to put the trajectory in the basin of attraction of the semi-attractor, the
result will be a long chaotic transient before the system returns to its attracting state.
Semi-attractors can also be found *within* a chaotic attractor, often in a "semi-periodic"
form (Kendall *et al.* 1993, Schaffer *et al.* 1993). It is well known that chaotic attractors
are formed around a "skeleton" of nonstable periodic orbits; the associated semi-attractors
can be viewed as the "flesh" of the attractor. Semi-attractors of both types are common in
nonlinear systems, and their presence has a variety of implications as we will see below.

2. Transient periodicity in biological dynamics

Time series of measles incidences in large cities reveal the presence of an epidemic nearly
every winter. In New York City from 1945 to 1963 these outbreaks exhibit a clear
biennial pattern (figure 1). This could very easily be viewed as a periodic orbit in the
presence of modest amounts of noise. However, the biennial dynamics are not characteristic of the entire time series.

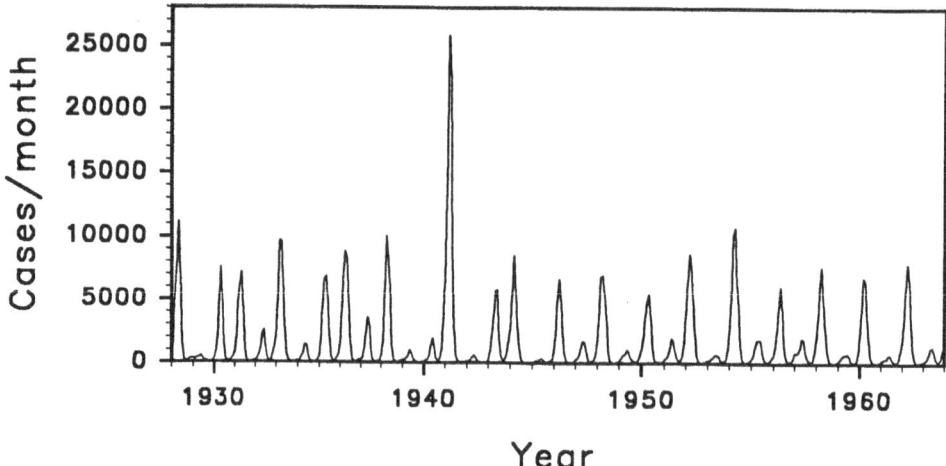

Figure 1. Number of measles cases reported per month, New York City (data from Yorke
and London 1973).

One can perhaps imagine a five-year cycle running between 1928 and 1938, but the
dynamics separating the two apparently periodic episodes seem completely irregular. In
other cities we can see shorter episodes of the biennial pattern Copenhagen 1947-1960,
Detroit 1925-1932 and 1951-1960 but the remainder of the outbreaks in these cities are
irregular (figure 2). Still other cities, such as Milwaukee (figure 3), show none of the
biennial episodes, and besides the fact of an outbreak nearly every winter, there is no
obvious pattern.

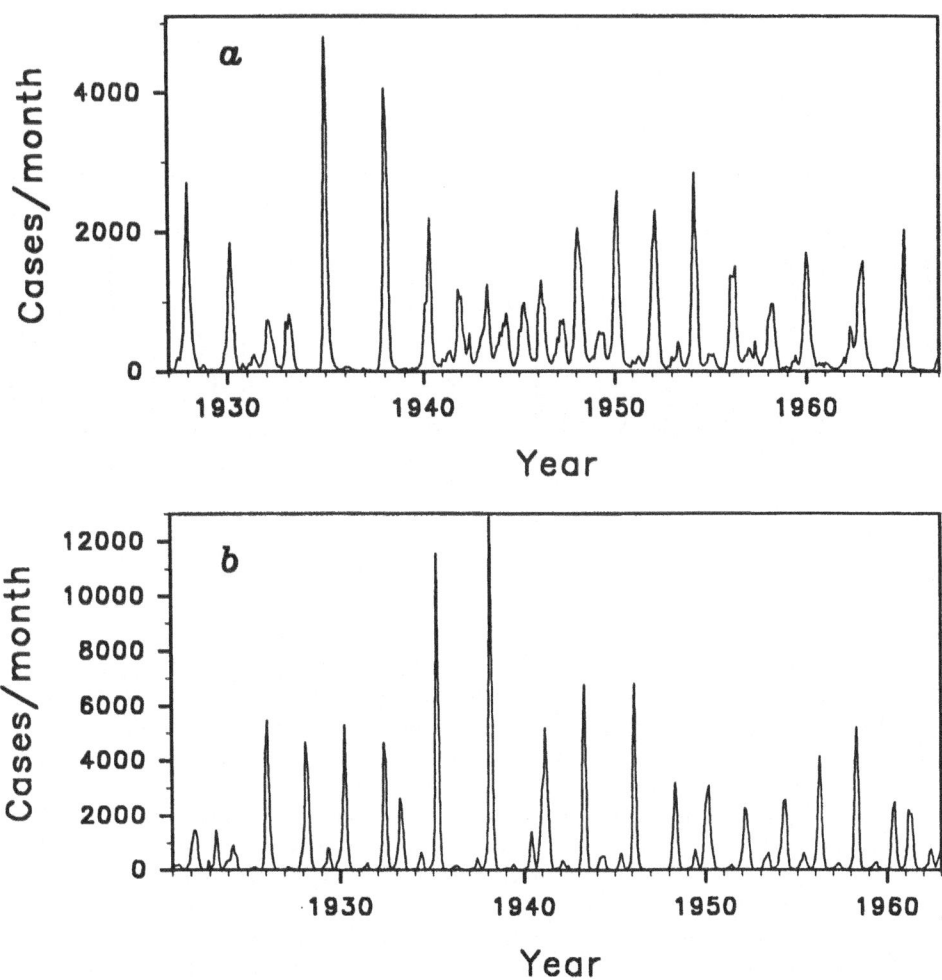

Figure 2. Number of measles cases reporter per month. (a) Copenhagen. (b) Detroit.

There has been a rather long-running (and still ongoing) debate over whether or not the measles epidemics are the result of a chaotic process (Schaffer and Kot 1985, Schwartz 1989, Dietz and Schenzle 1990, Olsen and Schaffer 1990, Sugihara and May 1990, Ellner 1991, Rand and Wilson 1991, Stollenwerk and Drepper 1992). The claims are largely based on the sorts of time series analyses described in the introduction, and thus are easilycriticized. Rather than go through all the arguments pro and con, we point out that quite large perturbations would be required for the observed data to arise from a linear periodic process, and ask whether nonlinear dynamics can provide a simpler explanation for the observed dynamics. In particular we would like to explain the curious alternation between episodes of biennial dynamics and the more irregular outbreaks. Sometimes

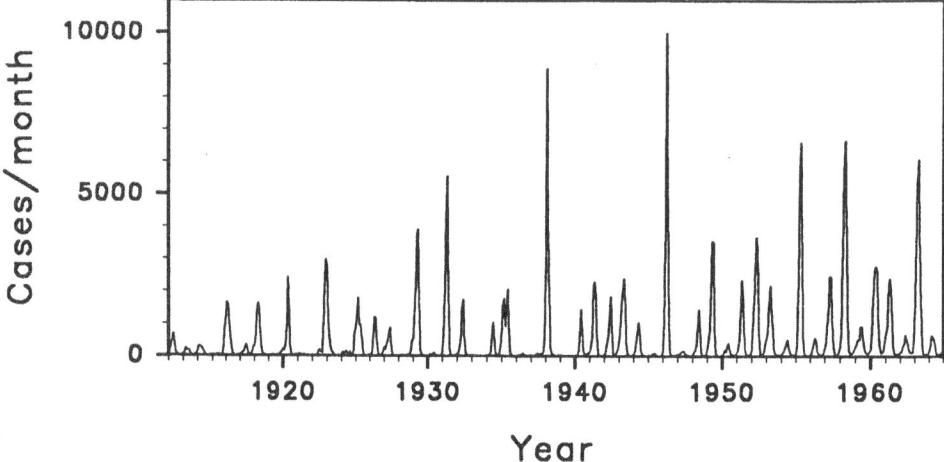

Figure 3. Number of measles cases reported per month, Milwaukee.

qualitative changes in dynamics can be attributed to shifts in some underlying parameter. Presented with only the New York City data, we might invoke such a fundamental change sometime during the early 1940's, either in the pathogen itself or in the human behavior that facilitates its spread; but the presence of biennial dynamics in other cities at other times (and in particular in Detroit where it arises twice) suggests that it might be fruitful to search for an intrinsic mechanism for the switching.

Nonlinear dynamics can in fact provide an explanation for the switching, through a dynamical mechanism we have called "transient periodicity" (Kendall *et al.* 1993, Schaffer *et al.* 1993). To demonstrate this mechanism, we would need to show bifurcation diagrams and phase space reconstructions for various parameter values. Unfortunately, we can't perform the necessary sorts of controlled experiments to produce these. The best we can do is demonstrate how this switching occurs in a model of measles epidemics, which we believe has dynamics similar to those observed in the data.

This model was first introduced by Dietz (1976). Like many epidemiological models, it considers not the pathogen itself but the host population (humans, in this case). When an individual is born he or she is susceptible to the disease. Upon contact with an infective individual, the susceptible becomes exposed. After a latency period while the pathogen proliferates in the host, the exposed individual becomes infective, capable of infecting other susceptibles. After the disease runs its course, the individual recovers, and in the case of measles, acquires a permanent immunity to reinfection. Since measles is almost never fatal for children, we only have mortality from other causes, which is assumed to occur at the same rate for all classes. This is called the SEIR model, after the four classes of the host population. A further modification involves the contact rate b. Since most exposure occurs in schools, and all the cities studied have long summer vacations, the contact rate has substantial seasonal variation (London and Yorke 1973). The exact form of the periodic function doesn't seem to matter much (F. R. Drepper, pers. comm., Kot *et al.* 1988), so for convenience we use a sinusoid. This model can be expressed as a set of coupled differential equations:

$$\frac{dS}{dt} = m(N - S) - b(t)SI$$

$$\frac{dE}{dt} = b(t)SI - (m + a)E$$

$$\frac{dI}{dt} = aE - (m + g)I$$

$$\frac{dR}{dt} = gI - mR$$

$$b(t) = \beta_0(1 + \beta_1 \cos 2\pi t).$$

We use the magnitude of the fluctuation in contact rate, β_1, for a bifurcation parameter. The other parameter values we used were $m = 0.02 \ yr^{-2}$; $a = 35.84 \ yr^{-1}$; $g = 100 \ yr^{-1}$; and $\beta_0 = 1800 \ yr^{-1}$.

For parameters in the range appropriate for measles we can get dynamics quite similar to those seen in the data (figure 4). A longer time series would show many more episodes of nearly biennial outbreaks interspersed with the irregular dynamics; the switching between the two regimes is spontaneous. How can we understand this phenomenon? To look at a bifurcation diagram, we extract a map from the flow by taking the time series of successive maxima in the time series of infectives. As β_1 increases, we see an equilibrium period-doubling to a period four orbit, and then a sudden widening to a chaotic attractor (figure 5).

What causes the sudden widening? Well, it turns out that there is a chaotic semi-attractor coexisting with the period-doubling sequence (Rand and Wilson 1991). This has two consequences. The first is that if the parameter is in the range of asymptotically periodic behavior, then a sufficiently large perturbation will give rise to a long chaotic transient. Thus we would have noise-induced transitions from periodic to irregular dynamics, followed by a deterministic decay back towards the periodic motion. The second consequence is that the destabilization of the periodic solution occurs when the attractor collides with the basin of attraction of the chaotic semi-attractor, resulting in the widening seen at $\beta_1 \approx 0.271$. In a sense, we can say that the periodic solution has also become a semi-attractor, and the new large attractor is simply the union of this new semi-attractor with the previously existing one. The "periodic" solution had already bifurcated into mild chaos before the collision: it was actually semiperiodic (Lorenz 1980), where the trajectory moves among four disjunct regions in a particular order, but the motion within each piece is chaotic. We can see this in a next-amplitude plot of the maxima (figure 6a). After the collision (which is formally known as a crisis [Grebogi et al. 1983]), we can see a concentration of points in the neighborhood of the semi-periodic semi-attractor (figure 6b). These points correspond to the episodes of nearly biennial dynamics seen in the time series. If we start a trajectory near the "period-4" semi-attractor, it exhibits approximately periodic dynamics for a time, but then "escapes" to the chaotic semi-attractor, where it exhibits large-amplitude irregular oscillations until it is drawn back to the semi-periodic semi-attractor and the whole process repeats (for the

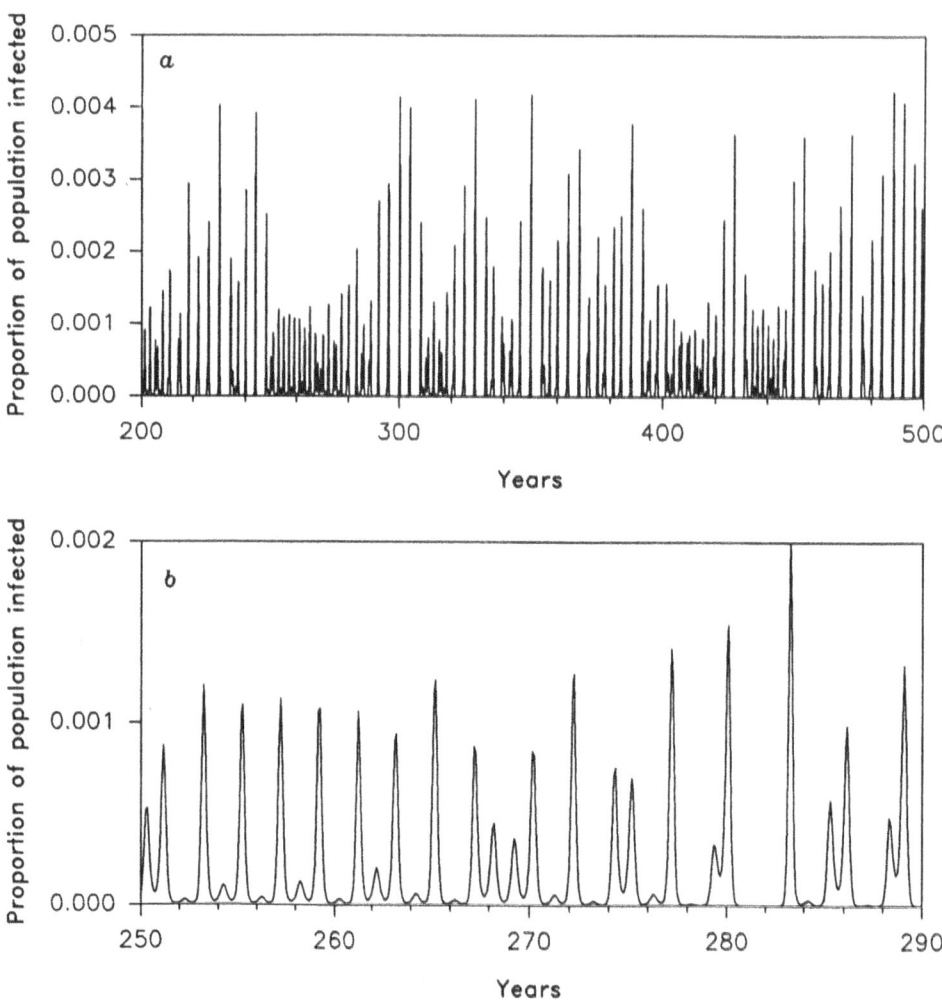

Figure 4. Time series of infectives generated by the SEIR model. (a) 300 years of simulated data. (b) Magnification of a "biennial" stretch of the time series.

mathematical details of this process, see Kendall *et al.* 1993). Thus we have spontaneous switching back and forth between the two types of dynamics. We coined the name "transient periodicity" because of the striking appearances of the transitory episodes of near-periodic motion in an otherwise aperiodic time series, although of course the same sort of switching could occur between two chaotic semi-attractors.

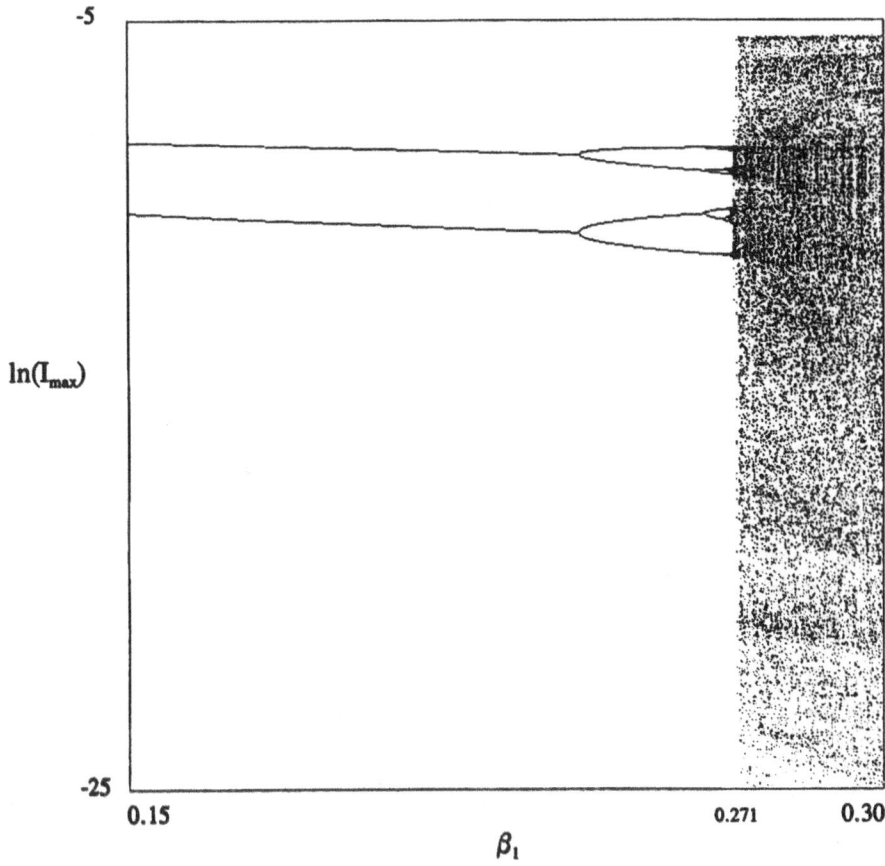

Figure 5. Bifurcation diagram for the SEIR model. The control parameter, β_1, is the magnitude of the seasonal variation in contact rate.

It is important to know whether this phenomenon is sensitive to the details of the model; if it is, then its applicability to the data is in doubt. Engbert and Drepper (1993) have introduced a modification to the SEIR model that incorporates a small immigration of infective individuals into the population of interest. This is a biologically realistic addition (New York City does not exist in isolation!) and necessary for comparisons of the deterministic equations with Monte Carlo implementations of the model, which require immigration to prevent extinction. This seemingly slight alteration to the model produces qualitatively different dynamics, including the appearance of a period three cycle and a narrowing of the chaotic attractor (Engbert and Drepper 1993). Nevertheless, preliminary analysis indicates that the chaotic attractor in this model (as well as the chaotic transient found for higher values of β_1) contains episodes of "biennial" motion, which can be explained through a similar mechanism of transient periodicity.

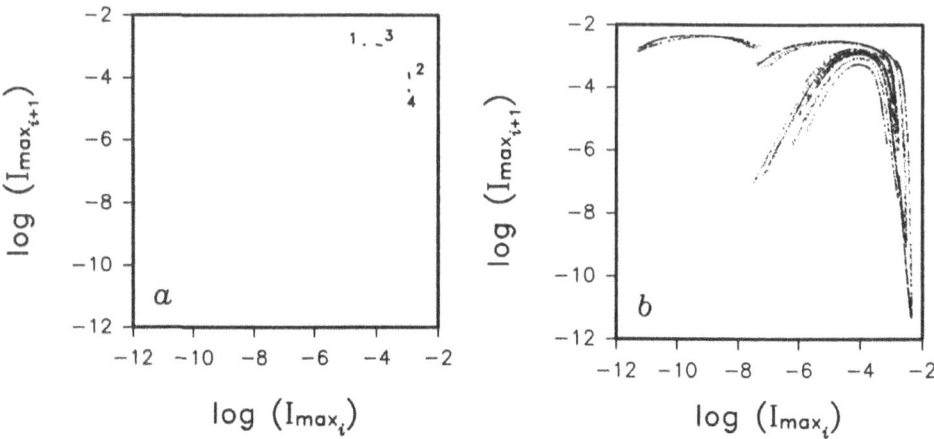

Figure 6. Next maximum map extracted from the time series of infectives generated by the SEIR model (a) $\beta_1 = 0.27$; the numbers indicate the order in which the trajetory visits the four pieces. (b) $\beta_1 = 0.28$.

Is this what is going on in the measles epidemics? To be prudent, we should say that we can't be sure, but it certainly provides a *simpler* explanation of the data than invoking major perturbations about which we have no knowledge. It is probably impossible to distinguish between deterministic transient periodicity and the noise-induced switching found for smaller values of β_1, but in a real sense the difference doesn't matter: in either case the system contains coexisting structures which govern both the "periodic" and the "aperiodic" dynamics.

3. Nonlinear forecasting and model identification

A major criticism of the SEIR model is that it is too simple (Dietz and Schenzle 1990). For example, since the transmission of measles mostly takes place in schools, preschool children should have a much lower contact rate with infectives than school children. Similarly, since most children contract measles in their pre-teen years, the few individuals who make it to secondary school without having had the disease should also have a low contact rate. Schenzle has implemented these ideas into an age-structured model (RAS) (Schenzle 1984, Bolker 1992, Bolker and Grenfell 1992). Certainly, this is a more "realistic" model of the process of measles infections, and there are certain situations where it would give better answers. For example, it does a better job of predicting the mean age of infection. However, the added details make analysis of the dynamics more difficult, so unless they substantially improve the characterization of the dynamics, we would like to use the simpler SEIR model for such analyses.

One way to compare the "dynamical accuracy" of the two models is to see whether one does a better job of forecasting the data than the other (Tidd *et al.* 1993). If the age structure, at least in the form modelled by Schenzle, is an important component of the dynamics, then this should show up in enhanced forecasting ability compared to the simple model. The forecasting algorithm is similar to one introduced by Farmer and Sidorowich (1987), using 0th order local maps. In brief, both the predictor time series and the time series to be forecast are embedded using the method of lags (Takens 1981); for each point to be forecast from, we examine the points in the predictor set that occupy a similar portion of the phase space. These points are followed forward in time until the desired prediction has elapsed, and the average of the result is used as the forecast for the original point. Details of the process can be found in Tidd *et al.* (1993).

To make this comparison we used Monte Carlo implementations of both models. This allows us to control for population size (the original ODE's are expressed in terms of fractions of the total population, but because of the extremely low minimum proportion of infectives, the importance of demographic stochasticity depends on total population size) and add stochastic effects in a non-arbitrary way. We use the squared coefficient of correlation between predicted and observed values as a measure of forecast accuracy (figure 7). The age-structured model actually makes consistently *poorer* predictions than

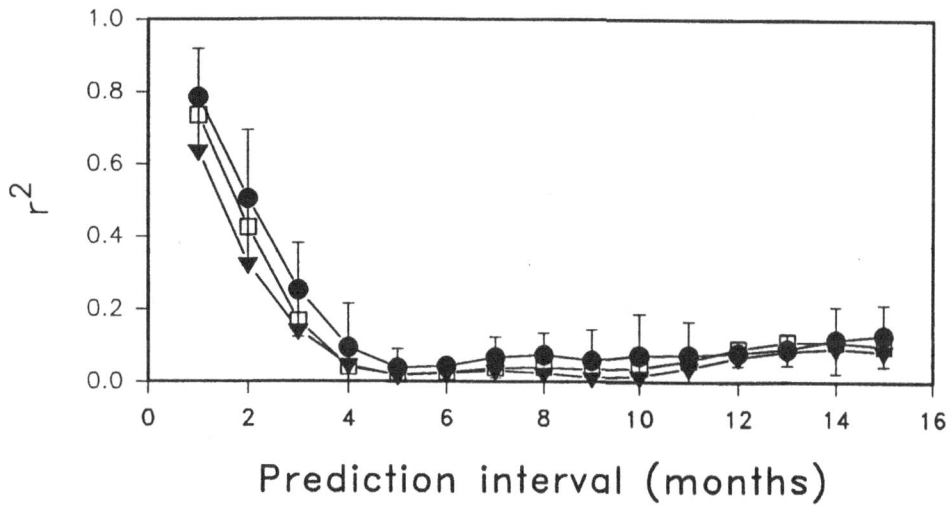

Prediction interval (months)

Figure 7. Forecasting measles data: r^2 the correlation between predicted and observed values, *vs.* prediction interval. Each point is the average of values obtained independently for Baltimore, Detroit, Copenhagen and Milwaukee. Circles: first half of the data predicting the second half; squares: SEIR model predicting the data; triangles: age structure d model predicting the data.

the simple SEIR model. The difference is small enough that we should probably not ascribe too much significance to it, but we can certainly say that the addition of age structure is not an improvement. This suggests either that age structure is not a significant factor in measles dynamics or that this particular implementation of the model is incorrect.

4. The problems of nonuniformity

There is something rather disturbing about figure 7: the predictions are terrible! The time scale is measured in months, so that after only five months, there is essentially no correlation between predicted and actual values. This is not just a problem with the models: the same problem arises when using the data itself to predict its future evolution. If nonlinear forecasting can't do any better than this, it is not much use. However, if we go back to the map extracted from the maxima of the time series we see something quite different (figure 8). Again the prediction accuracy falls off rapidly with prediction interval,

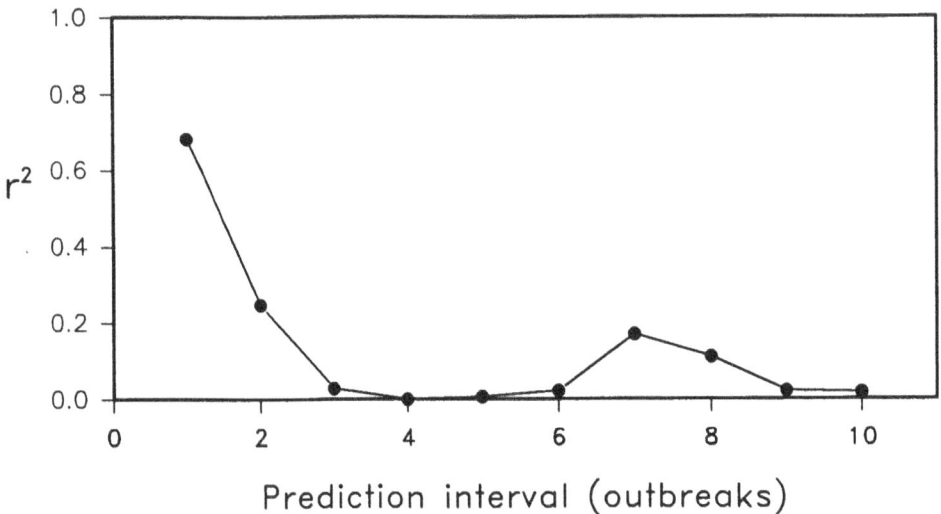

Figure 8. Forecasting measles data using only the yearly maxima.

but now the interval is measured in map units, which average 15 months, with a minimum of 6 months and a maximum of 29 months. Whereas prediction of the monthly dataset gives an r^2 of only about 0.1 over 6 months, the map's prediction of the next outbreak (which is usually a year or more in the future) has an r^2 of 0.7. In fact the map, with far fewer points, does nearly as good a job of predicting the magnitude of next year's outbreak as the full data set does of predicting the number of incidences in the next month!

How can we explain this anomaly? In the original time series, the majority of the points are near zero, while the distribution of points far from zero is quite sparse. This results in a nonuniform distribution points when the time series is embedded in a phase space. At best, we would expect all the points near the origin to contain little useful information about the dynamics, but it turns out that they actually confuse the forecasting algorithm. This is not merely a problem of too few points. While the Takens embedding theorem (Takens 1981) says that in principle we can recover the true dynamics from the time series if we have enough points, this is not necessarily true for all algorithms used for measuring those dynamics, of which nonlinear forecasting is one.

Geest *et al.* (1993) have found a similar problem in their analysis of the peroxi-dase-oxidase reaction. This is a biochemical enzyme reaction, which, when fed reactants at a constant rate, can oscillate chaotically (figure 9). Here the nonuniformity arises because each oscillation begins with a nearly identical exponential rise. This results in a three-dimensional phase portrait with a high concentration of points along the diagonal (figure 10).

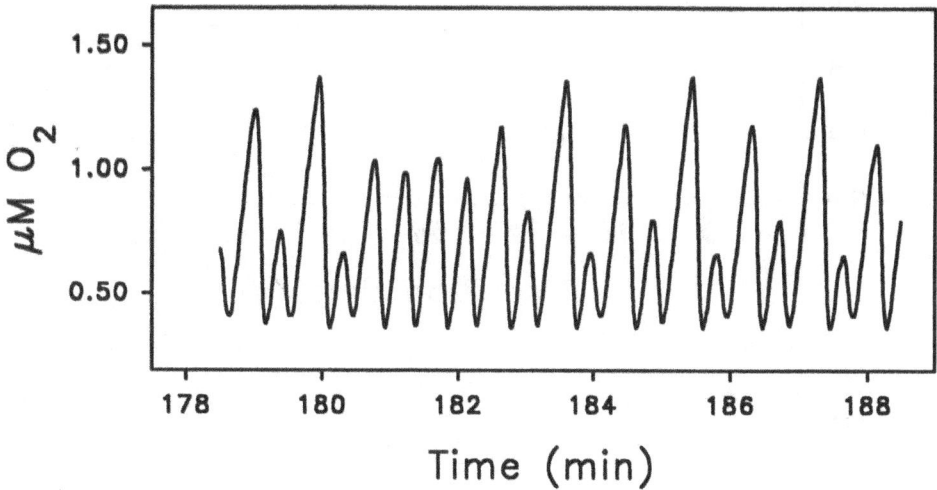

Figure 9. Time series of oxygen concentration in the peroxidase-oxidase reaction.

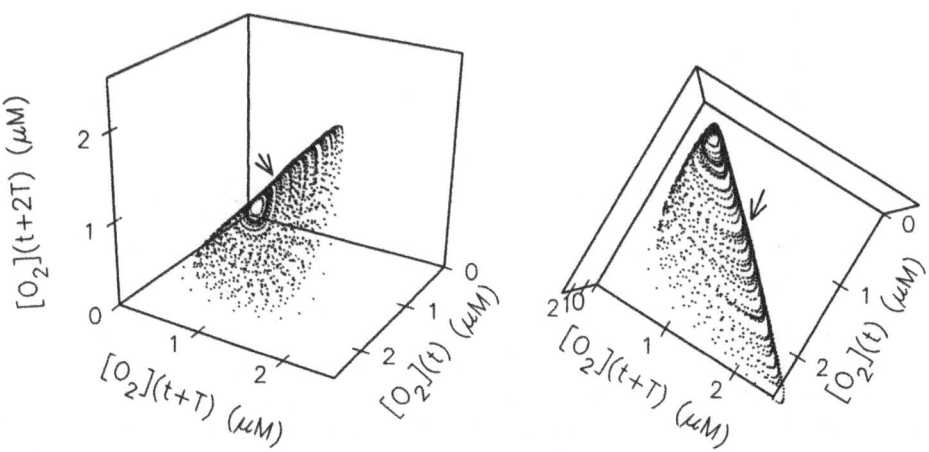

Figure 10. Three-dimensional reconstruction of the peroxidase-oxidase attractor, obtained by lagging the data in figure 9.

Once again forecasts are substantially better using the maxima of the time series than using the time series itself (figure 11).

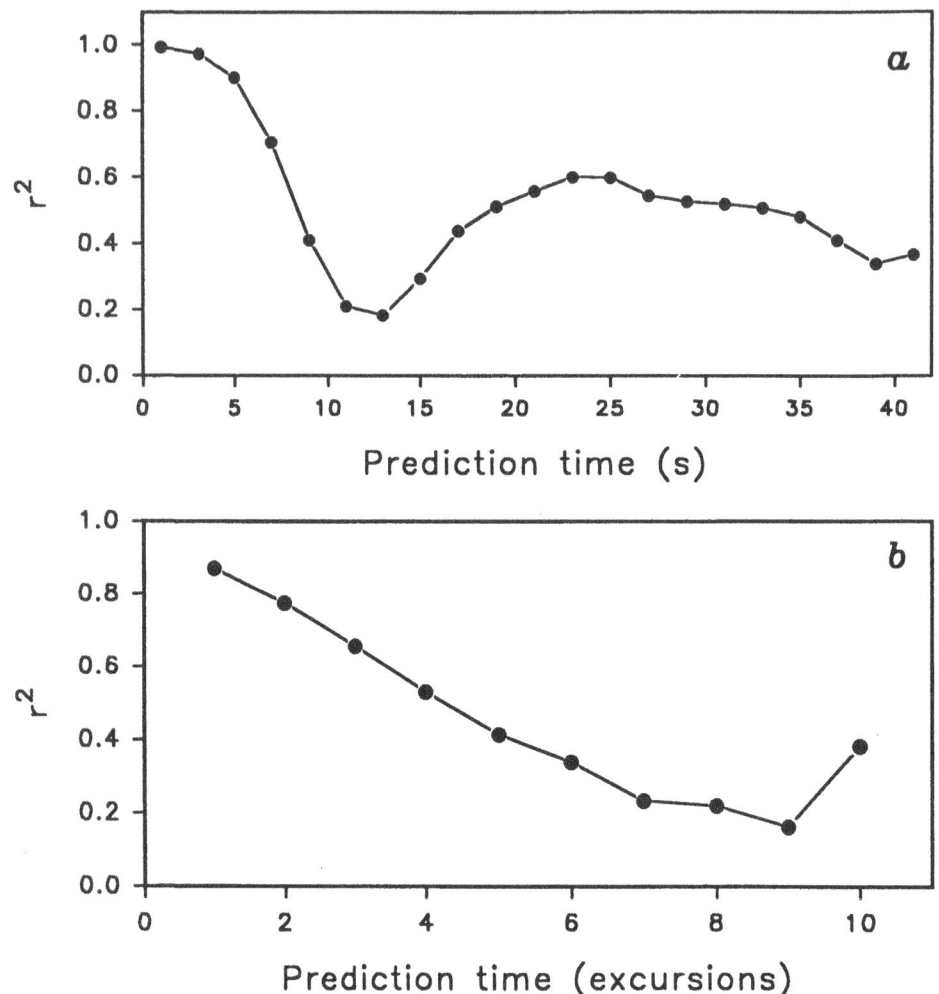

Figure 11. Forecasting the peroxidase-oxidase reaction. (a) Using all points in the time series. (b) Using only the maxima.

It is not only the nonlinear forecasting algorithm that is fooled by nonuniform data. Geest *et al.* (1993) looked at two other algorithms commonly used in nonlinear dynamics, the Grassberger-Procaccia (1983) algorithm for estimating correlation dimension and the Wolf *et al.* (1985) algorithm for estimating the largest Lyapunov exponent, and compared the output found from the time series with that found from the maxima. Ideally, a flow and the map associated with it should have the same Lyapunov exponent, and the dimension of the flow should be 1 greater than that of the map. In contrast, it was found that, relative to the flows, the maps produced dimensions that were too high and Lyapunov exponents that were too small (table 1).

Table 1. Dimensions and Lyapunov exponents from the peroxidase-oxidase reaction (λ_1 is expressed in bits/excursion). These results are from the data reported in the last three entries of table 1 in Geest *et al.* (1993).

Experiment	Flow		Map	
	D_2	λ_1	D_2	$\lambda12$
1	1.8	1.34	1.70	0.36
2	1.75	1.11	1.61	0.48
3	1.5	1.29	1.79	0.42

With this sort of discrepancy, we have to ask which set of numbers is more accurate. Geest *et al.* (1993) have also analyzed some models of the peroxidase-oxidase reaction that produce dynamics which are qualitatively similar to those of the data. Here they can calculate the spectrum of Lyapunov exponents directly from the equations of motion, and compare those to the numbers obtained by treating the output of the model as an experimental time series. They confirmed that the estimates obtained from the maps were much more accurate, even when the number of points was small.

These results are vitally important for nonlinear analyses in biology, for many, if not most, biological time series are nonuniform. Thus attempts to estimate dynamical invariants or forecast from these time series need to be alert to these sorts of problems, and it should perhaps become standard practice to extract some sort of map for the analysis. This is clearly a case where more (points) is not better. A next-maximum map is not the only type that need be used; a standard Poincaré section or a map of the time between peaks can also be useful, depending on the nature of the data and the phenomena of interest.

5. Predicting the onset of disease

One very exciting application of nonlinear science in biology that has been making rapid recent advances is predicting the onset or recurrence of pathologies of dynamic structures in the body, such as the heart. There has been some debate, essentially of a philosophical nature, over whether chaos represents a healthy state or a pathological one (Goldberger et al. 1984, 1986, Kaplan and Cohen 1990). There is almost certainly no general answer in that debate (Pool 1989); but once again, we can proceed by skipping the step of trying to "prove" that certain conditions are or are not chaotic, and see what we can learn by taking a nonlinear perspective.

An example of this approach is the work of Jørgensen *et al.* (1993). They have been studying the predictability of the heart rate, measured as R-R intervals, using techniques from nonlinear forecasting. Their goal is not prediction *per se*: knowing an individual's heartrate 20 seconds or so into the future is not very interesting. Rather they are studying the rate at which prediction accuracy falls off with forecasting interval. The falloff in prediction accuracy with time can be due either to the presence of perturbations of the

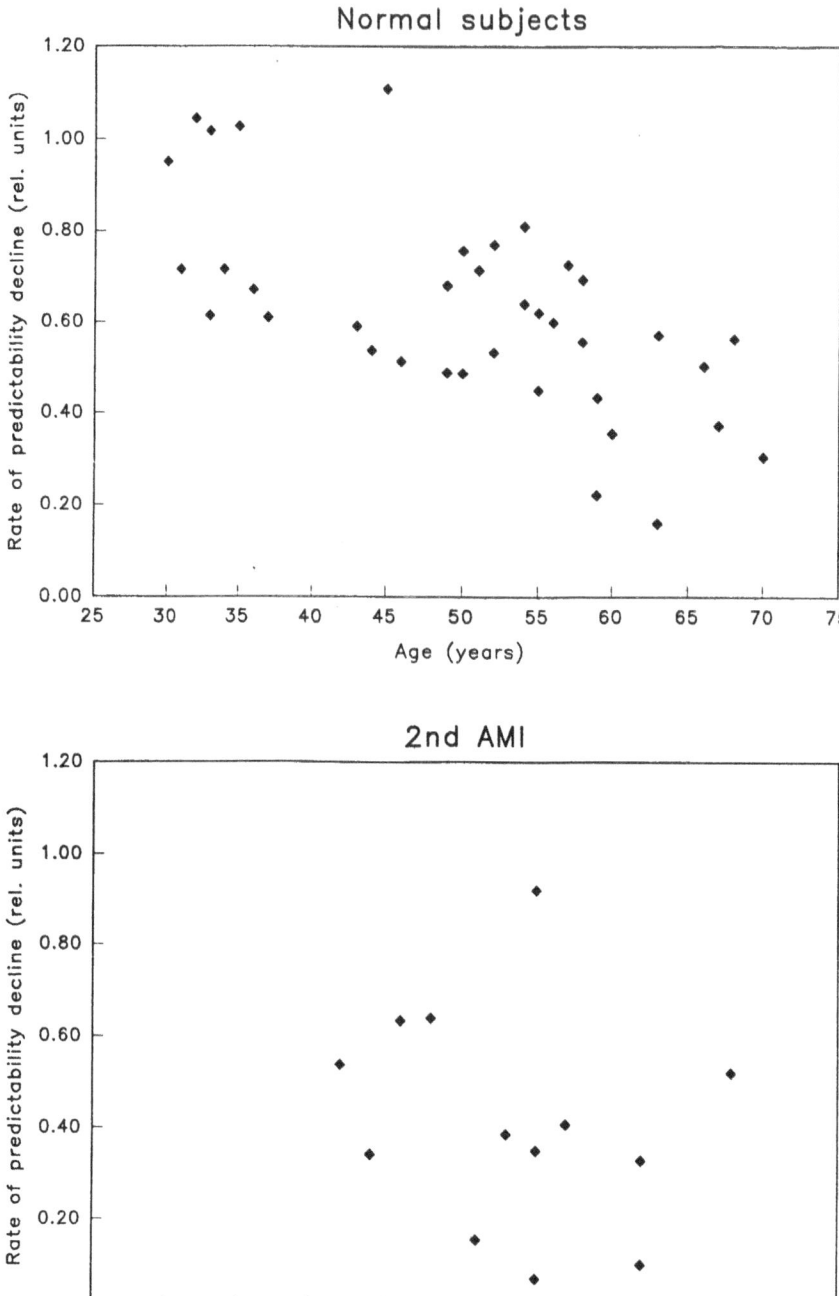

Figure 12. Rate of predictability decline in RR-intervals *vs.* age in healthy subjects (reproduced from Jørgensen *et al.* 1993).

system or to a deterministic uncertainty associated with positive Lyapunov exponents. Jørgensen *et al.* (1993) have tried to eliminate the former by studying their subjects while asleep, so to a large degree, the rate at which forecasting accuracy falls off is a measure of the dynamical complexity of the process controlling the heart rate.

They have obtained a number of interesting results from this technique. The first is that in healthy subjects, the rate with which forecasting accuracy is lost declines with age (figure 12); in other words, older people have more predictable heart rates than younger people. When they studied patients who had just survived a myocardial infarction, they discovered that, while in older patients the predictability was indistinguishable from healthy subjects, younger patients had a significantly higher predictablility, on average, than their healthy peers.

The results that have the most potential clinical value concern patients who had a second heart attack sometime during the following three years. If we look at those patients who had a second myocardial infarction or those who suffered sudden cardiac death (figure 13), we see that some of these individuals had a rate of predictability loss much lower than the range of normal subjects of the same age. This result is based on heart rate recordings

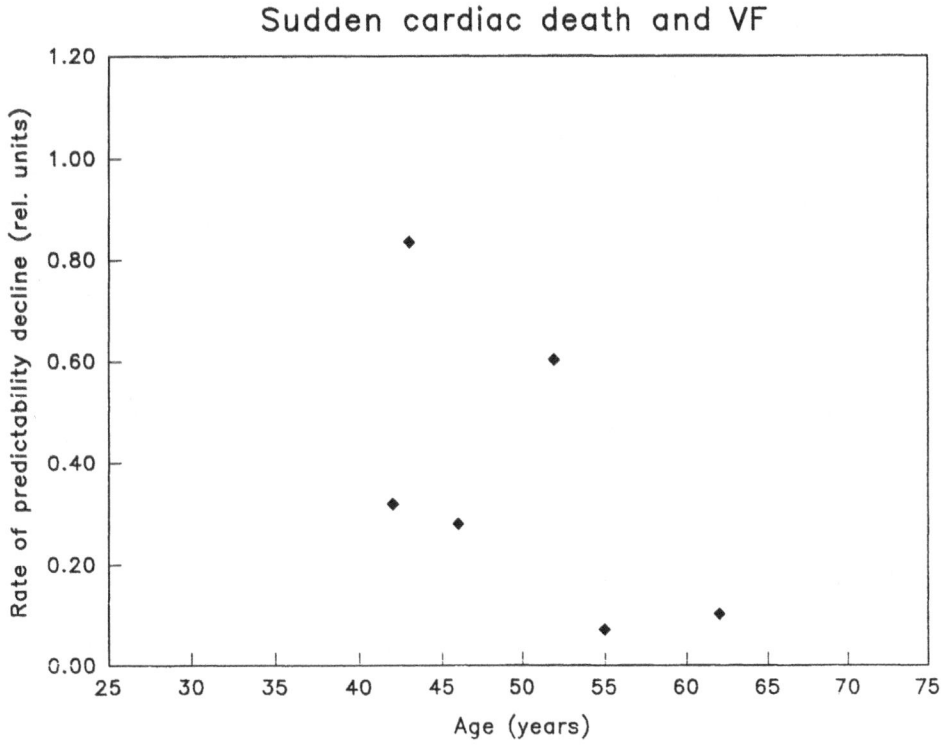

Figure 13. Rate of predictability in RR-intervals *vs.* age in moyocardial infraction patients who suffered (a) a second infarction or (b) sudden cardiac death within three years after their hospitalization (reproduced from Jørgensen *et al.* 1993).

taken only ten days after the initial myocardial infarction. Currently Jørgensen and her colleagues are looking at follow-up recordings to see if they can find a sequence through time of a decrease in heart rate predictability (for recovering patients) or a constant or even increasing predictability for patients due to suffer another attack. If they can find a consistent difference this will become an important tool for monitoring heart attack patients.

6. Conclusions

While we have focused on a few specific examples in this paper, this is by no means an exhaustive review of the exciting work currently being done. Furthermore, the examples discussed here leave open wide scope for further study. For example, we don't expect that measles epidemics are the only example of transient periodicity in biology. Certainly any time we see apparently spontaneous switching between periodic and aperiodic dynamics, we should entertain transient periodicity as a possible explanation; we are currently exploring the idea that it may be a useful way of looking at certain types of epileptic seizures (Schaffer *et al.* 1993). With regards to nonlinear forecasting, we are not yet sure what sorts of statistical tests are best suited to measuring the significance of observed differences in forecasting ability, but we feel that this will turn out to be a generally useful technique for measuring the fit of a model to the data.

There are quite a number of diseases where the pathology is accompanied by a qualitative change in the dynamics of some variable. While there are also physiological changesassociated with the disease, the time scale is often quite different: as the physiological abnormalities accumulate, the dynamics remain "nearly normal" until the sudden transition to abnormal dynamics. Understanding the dynamics in these cases may help guide our understaning of the proximal trigger for the destabilization of the nearly normal dynamics; and nonlinear analysis can help us distinguish "nearly normal" (and hence at risk) individuals from "normal" (healthy) ones. It is bad enough that the common nonlinear analysis programs yield quantitatively wrong results when applied to nonuniform data, but with such data they are even unreliable as qualitative measures for comparative purposes. These algorithms have a number of settings that have to be fiddled with to get the best estimates, and the results obtained from nonuniform data are extremely sensitive to these settings. Thus it is unlikely that the numbers obtained from two time series are biased in a similar way.

Although extracting a map from the data can reduce these difficulties, it does not eliminate them entirely. There is still often a nonuniform distribution of of points in the map, arising from transient periodicity. This does not usually give rise to the problems described in the previous paragraph, but it raises the question of whether an observed time series is "typical" of the underlying dynamics. If the time series is short and contains a longer than average episode of transient periodicity, for example, then estimates of dynamical invariants will be biased. This is a problem that can only be resolved with longer time series or multiple time series of the same process.

The presence of transient periodicity is a substantial advantage when it comes to forecasting, however. It is easy for forecasting algorithms to pick out the periodic component of the motion, and during a near-periodic episode, the error is restricted to the small-amplitude chaotic modulations. It is more difficult to predict when a particular episode of transient periodicity will end, for the ending is the expression of a chaotic process and hence sensitive to initial conditions. However, the distribution of lengths of episodes is best fit by a negative exponential (Kendall *et al.* 1993), indicating that from a statistical point of view, the probability of an episode ending is constant per unit time. If we have collected enough data to observe many episodes of transient periodicity, then we could make statements like "the system will remain nearly periodic for time t with probability p." Transient periodicity represents "islands of predictability in a sea of chaos," and may provide one answer to the challenge left by Dr. H. Tennekes at the conclusion of this conference, to find a means to forecast the accuracy of our forecasts.

Acknowledgements

We thank B. Bolker for providing us with output from his Monte Carlo implementation of the age-structure model. Some of the work reviewed in this paper was supported by a grant from the National Institutes of Health to WMS and from the Danish Research Council to LFO.

References

Grassberger, P. and I.~Procaccia (1983), 'Measuring the strangeness of strange attractors', *Physica D*, **9**, 189-208.

Wolf, A., J.B. Swift, H.L. Swinney and J.A. Vastano (1985), 'Determining Lyapunov exponents from a time series', *Physica D*, **16**, 285-317.

Sugihara, G. and R.M. May (1990), 'Non-linear forecasting as a way of distinguishing chaos from measurement error in time series', *Nature*, **344**, 734-741.

Schaffer, W.M. (1984), 'Stretching and folding in lynx fur returns: evidence for a strange attractor in nature?' *Am. Nat.*, **124**, 798-820.

Babloyantz, A., J.M. Salazar and C. Nicolis (1985), 'Evidence of chaotic dynamics of brain activity during the sleep cycle', *Phys. Lett. A*, **111**, 152-156.

Markus, M., D. Kuschmitz and B. Hess (1985), 'Properties of strange attractors in yeast glycolysis', *Biophys. Chem.*, **22**, 95-105.

Schaffer, W.M. and M. Kot (1985), 'Nearly one dimensional dynamics in an epidemic', *J. Theor. Biol.*, **112**, 403-427.

Babloyantz, A. and A. Destexhe (1986), 'Low-dimensional chaos in an instance of epilepsy', *Proc. Natl. Acad. Sciences USA*, **83**, 3513-3517.

Babloyantz, A. and A. Destexhe (1988), 'Is the normal heart a periodic oscillator?' *Biol. Cybern.*, **58**, 152-156.

Dvorak, I. and J. Siska (1986), 'On some problems encountered in the estimation of the correlation dimension of the EEG', *Phys. Lett. A*, **118**, 63-66.

Mayer-Kress, G., F.E. Yates, L. Benton, M. Keidel, W. Tirsch, S.J. Poppi and K.~Geist (1988), 'Dimensional analysis of nonlinear oscillations in brain, heart, and muscle', *Math. Biosciences*, **90**, 155-182.

Olsen, L.F., G.L. Truty and W.M. Schaffer (1988), 'Oscillations and chaos in epidemics: A nonlinear dynamic study of six childhood diseases in Copenhagen, Denmark', *Theor. Pop. Biol.*, **33**, 344-370.

Zbilut, J.P., G. Mayer-Kress and K. Geist (1988), 'Dimensional analysis of heart rate variability in heart transplant patients', *Math. Biosciences*, **90**, 49-70.

Frank, G.W., T. Lookman, M.A.H. Nerenberg, C. Essex, J. Lemieux and W. Blume (1990), 'Chaotic time series analysis of epileptic seizures', *Physica D*, **46**, 427-438.

Olsen, L.F. and W.M. Schaffer (1990), 'Chaos {\em vs.} noisy periodicity: Alternative hypotheses for childhood epidemics', *Science*, **249**, 499-504.

Gallez, D. and A. Babloyantz (1991), 'Predictability of human EEG: a dynamical approach', *Biol. Cybern.*, **64**, 381-392.

Pijn, J.P., J. van Neerven, A. Noest and F.H. Lopes da Silva (1991), 'Chaos or noise in EEG signals: dependence on state and brain site', *Electroencephalog. Clin. Neurophysiol.*, **79**, 371-381.

Sammon, M.P. and E.N. Bruce (1991), 'Vagal afferent activity increases dynamical dimension of respiration in rats', *J. Appl. Physiol.*, **70**, 1748-1762.

Yip, K.P., N.H.H. Rathlou and D.J. Marsh, (1991), 'Chaos in blood flow control in genetic and renovascular hypertensive rats', *Am. J. Physiol.*, **261**, F400-F408.

Donaldson, G.C. (1992), 'The chaotic behavior of resting human respiration', *Respiration Physiol.*, **88**, 313-321.

Gantert, C., J. Honerkamp and J. Timmer (1992), 'Analyzing the dynamics of hand tremor time series', *Biol. Cybern.*, **66**, 479-484.

Pritchard, W.S. and D.W. Duke (1992), 'Dimensional analysis of no-task human EEG using the Grassberger-Procaccia method', *Psychophysiol.*, **29**, 182-192.

Oborne, A.R. and A. Provenzale (1989), 'Finite correlation dimension for stochastic systems with power law spectra', *Physica D*, **35**, 357-381.

Ellner, S. (1991), 'Detecting low-dimensional chaos in population dynamics data; a critical review', In: *Chaos and Insect Ecology*, Va. Expt. Station Information Ser. 91-3, Blacksburg, Va., pp. 63-90.

Stone, L. (1992), 'Coloured noise or low-dimensional chaos?' *Proc. R. Soc. Lond. B*, **250**, 77-81.

Aihara, K., G. Matsumoto and M. Ichikawa (1985), 'An alternating periodic-chaotic sequence observed in neural oscillators', *Phys. Lett. A*, **111**, 251-255.

Hayashi, H., S. Ishizuka and K. Hirakawa (1985), 'Chaotic response of the pacemaker neuron', *J. Phys. Soc. Jpn.*, **54**, 2337-2346.

Chialvo, D.R., R.F. Gilmour and J. Jalife (1990), 'Low dimensional chaos in cardiac tissue', *Nature*, **343**, 653-657.

Geest, T., C.G. Steinmetz, R. Larter and L.F. Olsen (1992), 'Period-doubling bifurcations and chaos in an enzyme reaction', *J. Phys. Chem.*, **96**, 5678-5680.

Geest, T., L.F. Olsen, C.G. Steinmetz, R. Larter and W.M. Schaffer (1993), 'Nonlinear analyses of periodic and chaotic time series from the peroxidase-oxidase reaction', *J. Phys. Chem.*, in press.

Farmer, J.D. and J.J. Sidorowich (1987), 'Predicting chaotic time series', *Phys. Rev. Lett.*, **59**, 845-848.

Farmer, J.D. and J.J. Sidorowich (1988), 'Predicting chaotic dynamics', In: *Dynamic Patterns in Complex Systems*, World Scientific, Singapore, pp. 265-292.

Casdagli, M. (1989), 'Nonlinear prediction of chaotic time series', *Physica D*, **35**, 335-356.

Kantz, H. and P. Grassberger (1985), 'Repellers, semi-attractors, and long-lived chaotic transients', *Physica D*, **17**, 75-86.

Kendall, B.E., W.M. Schaffer and C.W. Tidd (1993), 'Transient periodicity in chaos', *Phys. Lett. A*, **177**, 13-20.

Schaffer, W.M., B.E. Kendall, C.W. Tidd and L.F. Olsen (1993), 'Transient periodicity and episodic predictability in biological dynamics', *IMA J. Math. Appl. Med. Biol.*, in press.

Yorke, J.A. and W.P. London (1973), 'Recurrent outbreaks of measles, chickenpox and mumps. II. Systematic differences in contact rates and stochastic effects', *Am. J. Epidemiol.*, **98**, 469-482.

Schwartz, I.B. (1989), 'Nonlinear dynamics of seasonally driven epidemic models', In: *Biomedical Modelling*, J.C. Baltzer AG, pp. 201-204.

Dietz, K. and D. Schenzle (1990), 'Discussion of the paper by Bartlett', *J. R. Statist. Soc. A*, **153**, 338.

Rand, D.A. and H. Wilson (1991), 'Chaotic stochasticity', *Proc. R. Soc. Lond. B*, **246**, 179-184.

Stollenwerk, N. and F.R. Drepper (1992), 'Evidence for deterministic chaos in empirical population fluctuations', preprint.

Dietz, K. (1976), 'The incidence of infectious diseases under the influence of seasonal fluctuations', *Lect. Notes Biomath.*, **11**, 1-15.

London, W.P. and J.A. Yorke (1973), 'Recurrent outbreaks of measles, chickenpox and mumps. I. Seasonal variation in contact rates', *Am. J. Epidemiol.*, **98**, 453-468.

Kot, M., D.J. Graser, G.L. Truty, W.M. Schaffer and L.F. Olsen (1988), 'Changing criteria for imposing order', *Ecol. Model.*, **43**, 75-110.

Lorenz, E.N. (1980), 'Noisy periodicity and reverse bifurcation', *Ann. N. Y. Acad. Sciences*, **357**, 282-291.

Grebogi, C., E. Ott and J.A. Yorke (1983), 'Crises, sudden changes in chaotic attractors and transient chaos', *Physica D*, **7**, 181-200.

Engbert, R. and F.R. Drepper (1993), 'Qualitative analysis of unpredictability: a case study from childhood epidemics', this volume.

Schenzle, D. (1984), 'An age-structured model of pre- and post-vaccination measles transmission', *IMA J. Math. Appl. Med. Biol.* **1**, 169-191.

Bolker, B. (1992), 'Chaos and complexity in measles models: a comparative numerical study', preprint.

Bolker, B. and B. Grenfell (1992), 'Chaos and biological complexity in measles dynamics', preprint.

Tidd, C.W., L.F. Olsen and W.M. Schaffer (1993), 'The case for chaos in childhood epidemics: II. Predicting historical epidemics from mathematical models', *Proc. R. Soc. Lond. B*, in press.

Takens, F. (1981), 'Detecting strange attractors in turbulence', In: *Dynamical Systems and Turbulence*, Springer-Verlag, Berlin, pp. 366-381.

Goldberger, A.L., L.F. Findley, M.R. Blackburn and A.J. Mandell (1984), 'Nonlinear dynamics in heart failure: implications of long-wavelength cardiopulminary oscillations', *Am. Heart J.*, **107**, 612-615.

Goldberger, A.L., V. Bhargava, B.J. West and A.J. Mandell (1986), 'Some observations on the question: Is ventricular fibrillation chaos?' *Physica D*, **19**, 282-289.

Kaplan, D.T. and R.J. Cohen (1990), 'Is fibrillation chaos?' *Circ. Res.*, **67**, 886-892.

Pool, R. (1989), 'Is it healthy to be chaotic?' *Science*, **243**, 604-607.

Jørgensen, B.L., A. Junker, H. Mickley, M. Møller, E. Christiansen and L.F. Olsen (1993), 'Nonlinear forecasting of RR-intervals of human electrocardiograms', In: Future Directions of Nonlinear Dynamics in Physical and Biological Systems}, Plenum Press, in press.

QUALITATIVE ANALYSIS OF UNPREDICTABILITY: A CASE STUDY FROM CHILDHOOD EPIDEMICS

RALF ENGBERT AND FRIEDHELM R. DREPPER

Arbeitsgruppe Modellierung für Umweltforschung und Lebenswissenschaften,
Forschungszentrum Jülich, D-5170, F.R.G.

e-mail: F.Drepper@kfa-juelich.de

Summary

The unpredictability of the recurrent outbreaks of childhood epidemics has been a matter of scientific dispute. At first sight the deterministic SEIR model, which divides the host population into four classes (Susceptible, Exposed, Infectious, Recovered), fails to explain the unpredictability, because realistic parameter values lead to periodic attractors. We show that these periodic attractors coexist with chaotic transients. The detailed geometrical analysis of this phenomenon suggests that chaotic transients have been underestimated in their importance for the dynamics on observable time scales. A second problem of the SEIR model is the high extinction probability of epidemics in finite populations. We argue that the immigration of infectives from outside is an essential parameter in this context. Immigration and the process of infection itself are sources of demographic fluctuations, which undergo subtle interaction with chaotic transients. The stochastic simulation of the SEIR model shows that the chaotic transients are permanently revisited. The demographic noise integrates chaotic transients and intermittent periodic episodes.

Keywords   Epidemic models, nonlinear dynamics, transient chaos, demographic stochasticity, Monte-Carlo simulation

1. Introduction

The explanation of the unpredictability of recurrent outbreaks of childhood epidemics - in particular of measles infections - in large population centres has been a subject of extensive research. Bartlett (1957) presented a stochastic (probabilistic) translation of the time independent SEIR model, which was the standard model for recurrent childhood epidemics at that time. In this model the recurrence of large epidemics is obtained as a sequence of *stochastic fade outs* of the epidemic and successive immigrations of the virus. The resulting time intervals between large epidemics were fully random, the expectation value being strongly dependent on the community size. London and Yorke (1973) showed that the *seasonal variation* of the contact rate is important to understand the basic two year period of the measles incidence pattern apparent in the power spectrum. This led to the idea that the measles dynamics is a stochastic process in the basin of attraction of a stable limit cycle with a two year period as described e.g. by Dietz (1976) and Anderson *et al.* (1984). Schaffer (1985) claimed that part of the irregularity of the empirical incidence pattern in New York City can be explained deterministically, and opened the view to the new paradigm of that time that the measles dynamics takes place in the basin of a (globally stable) *chaotic attractor*. The present study will combine elements of all three pictures into a consistent geometrically plausible model. The idea is to introduce the time dependence of the contact parameter into Bartlett's spatially aggregated model and to support the qualitative analysis of the rich spectrum of behaviour of this model by a geometric analysis of the corresponding deterministic nonlinear dynamical system. As already noticed by Tél (1990) and Rand and Wilson (1991) it is essential to include also *metastable chaotic* transients into the qualitative analysis.

In addition to helping us understand the nature of the unpredictability underlying measles epidemics the methods of analysis being used here may prove to be helpful in explaining also the unpredictability of other fluctuating populations in epidemiology and ecology.

2. The SEIR model with immigration

Most theoretical studies of the recurrent outbreaks of childhood epidemics start with the well known SEIR model (Anderson and May, 1991), which is based on the assumption of a homogeneous and uniformly mixing population. Individuals are born as susceptibles S and become exposed E by contact with infectious ones I. The average time between the point of infection and the *infectious period* ($1/g$) is the *latency period* $1/a$. After the period of being infectious the individuals are immune or recovered (R); in the case of childhood epidemics the immunity is lifelong. The birth and death rates are taken to be equal ($1/m$); in particular there is no disease induced mortality. The population size $N=S+E+I+R$ is assumed to be constant and normalised to one, so that the variables S, E, I, and R represent the corresponding fraction of the total population. The dynamics are usually formulated as a set of three ordinary differential equations (ODEs)

$$\begin{aligned}
\dot{S} &= m - (m + \lambda)S \\
\dot{E} &= \lambda S - (m + a)E \\
\dot{I} &= aE - (m + g)I \ .
\end{aligned} \tag{1a}$$

The nonlinearity of the SEIR model arises from a state dependence of the *force of infection* $\lambda = b(t)I$. The contact parameter

$$b(t) = b_0(1 + b_1\cos(2\pi t)) \tag{1b}$$

is a periodic function of time with a one year period. It represents the average number of effective contacts that one infectious individual would have per time unit in a completely susceptible population. In this study we shall use the numerical values $m=0.02$ yr^{-1}, $a=35.84$ yr^{-1}, $g=100$ yr^{-1} and $b_0=1800$ yr^{-1} (Schaffer et al., 1988). Parameter $b_1= 0$ to 0.3 is used as a *control parameter* for the qualitative analysis. Its exact numerical value is difficult to estimate; furthermore it turns out that a relatively small variation of b_1 can lead to drastic changes in the qualitative dynamics. Figure 1 shows the corresponding *bifurcation diagram*.

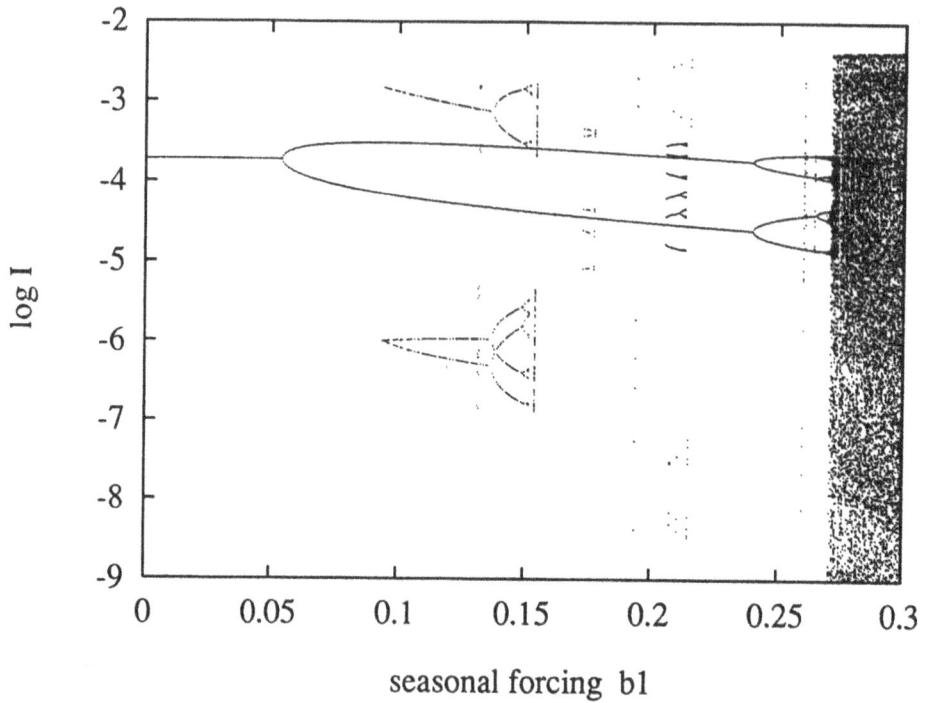

Figure 1. Bifurcation diagram of the SEIR model (1). For each parameter value b_1 we choose several random starting values, iterate 1000 years to remove transient behaviour, and plot the values of I at January 1st of the following years. The period one branch, arising from the endemic equilibrium ($b_1=0$), undergoes a period doubling cascade to chaos. There are several higher periodic orbits, which are created by saddle-node bifurcations, undergo its own period doubling routes to chaos, and disappear through sudden changes, called crises (see below).

There are several problems with the standard formulation of the SEIR model. A major inadequacy of the dynamics in the chaotic region (beyond $b_1 \approx 0.27$) is the fact that the fraction of infectives drops down to below $I \sim 10^{-6}$, so that even in the biggest population centres the *deterministic* description of the epidemics breaks down, in particular it cannot describe probabilities for local extinction of the epidemics. The same phenomenon can also be observed on the period three branch. One way to solve these shortcomings is proposed in Schwartz (1992) by *"relaxing the assumption of uniformity"*. This model introduces a coupling between a core city and its suburbs. The immigration of the virus from the outside avoids extremely low population densities of infectives.

A further step is to take into account the microscopic stochasticity due to integer population size. There have been several attempts to introduce *demographic* stochasticity via Monte-Carlo techniques (Bartlett, 1957, Olsen *et al.*, 1988, Bolker and Grenfell, 1993). For an analysis of the stochastic long term dynamics all three approaches allowed for a small influx of infectives to prevent extinction, which would be an *absorbing state* in Monte-Carlo simulations without immigration. As we shall see, this modification is also a very important one for the deterministic calculations. In fact the Monte-Carlo simulations with influx of infectives cannot converge to the deterministic simulations without immigration.

Let us assume that the epidemiological dynamics in the surroundings of the core city are desynchronised, so that the spatial average is close to the endemic equilibrium. Then there will be an interaction between individuals of the city with those from outside. This leads to a modification of the force of infection

$$\lambda = b(t) \cdot (I + wI_0) \, , \tag{2}$$

where w is the coupling constant, describing the mobility of the surrounding population in units of the central population, and I_0 is the infectious fraction of the population at endemic equilibrium. We can interpret this as an immigration process of the virus. The fact that w is assumed to be constant is an approximation similar to other simplifying assumptions necessary to derive the SEIR model.

The new parameter w is important in the same sense as b_1. It is difficult to estimate as well as essential, because it produces sudden changes in the qualitative dynamics under small variation. To demonstrate this, figure 2 shows a so-called *isoperiodic diagram* (Gallas, 1992). It summarizes a whole family of bifurcation diagrams similar to figure 1. Figure 3 gives a typical example for a small but finite value of w. Comparing figures 1 and 3 we see that the importance of the period 3 has increased in the realistic range of parameter values for b_1 (0.25 to 0.3). Figure 2 shows that even a value of w=0.0002 (684 immigrating infectives per year in a population centre of 10^7 inhabitants) is sufficient to stabilize a coexisting period 3 based attractor. In the following we will get further evidence that it is essential for a discussion of the dynamics to include the effect of immigration.

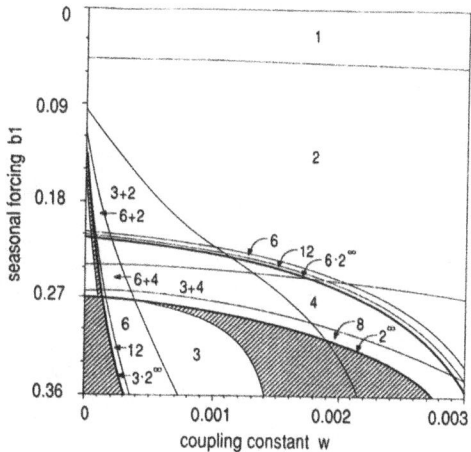

Figure 2.   Isoperiodic diagram of the modified SEIR model. For a combination of parameter values (w,b_1) we choose a random starting value, remove transient behaviour (i.e. iterating 1000 years) and calculate the periodicity of the attractor (represented by the numbers). The "+"-sign symbolises the coexistence of attractors of different periodicity. For simplicity all periodicities greater than 16 are not distinguished from chaos. The shaded region symbolises parameter values for chaotic attractors. The figure is a schematic representation. For clarity thin bands of higher periodicity and periodic windows inside the chaotic region have been removed.

3. Population biology

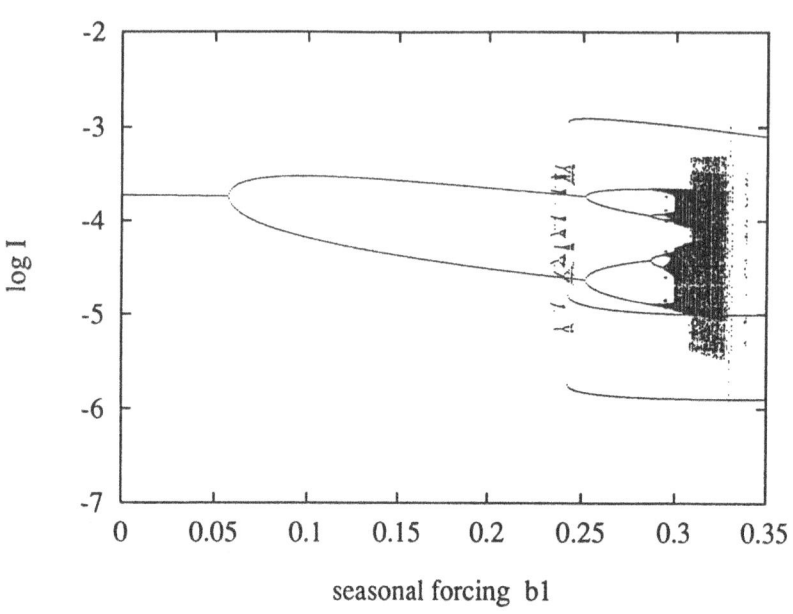

Figure 3. Bifurcation diagram of the modified SEIR model (2) for fixed w=0.0013. The period three branch is globally stable for large values of b_1.

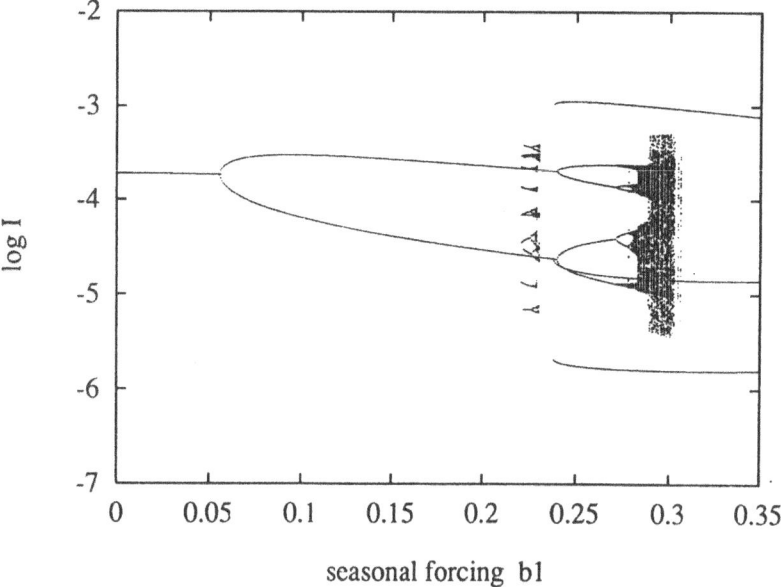

Figure 4. Bifurcation diagram of the reduced system (4) for fixed w=0.0013. Notice that the qualitative features compare well with the bifurcation diagram in figure 3 even for large values of b_1.

3. Geometric analysis of transient chaos

In this section we show that the bifurcations of the SEIR model can be completely understood from a two dimensional map. The phenomenon of coexistence of different attractors is ubiquitous in the SEIR model, as seen in the isoperiodic diagram (figure 2). We now analyse the geometric structure of a Poincaré map of (1,2).

For $b_1=0$ there are two steady states of the modified model (1,2): $(S,E,I)=(1,0,0)$, belonging to the extinction of epidemics, and to the *endemic equilibrium* (S_0,E_0,I_0),

$$S_0 = \frac{(m + a)(m + g)}{ab_0(1 + w)} \equiv 1/Q, \quad E_0 = \frac{m + g}{a}I_0, \quad I_0 = \frac{m(Q - 1)}{b_0(1 + w)}, \tag{3}$$

where Q is the *basic reproductive rate* ($Q \approx 18$ for measles).

Compared to more realistic but also more complicated modifications (Bolker and Grenfell 1993, Schenzle 1984), the advantage of the simple SEIR model (1,2) is twofold: Firstly, it captures the qualitative dynamics of the recurrent outbreaks of childhood epidemics and, secondly, it can be analysed in detail from the geometrical point of view of dynamical systems theory: All its long term dynamics can be analysed using a two dimensional map. We construct this map in to steps.

Firstly we notice that the SEIR model (1a) with seasonal forcing (1b) can be written down as a four dimensional autonomous ODE system. Following Schwartz and Smith (1983) we restrict the system to a *centre manifold* (e.g. Guckenheimer and Holmes, 1983). Introducing the small parameter $\varepsilon = b_0 (1+w) I_0$ (the force of infection at endemic equilibrium) the dynamics can be approximated by the three dimensional autonomous system (with the exceptions $\bar{b}_1 = b_1/\varepsilon^2$ and $\bar{w} = w/\varepsilon^2$ we use all other parameters in the same notation as Schwartz and Smith (1983))

$$\dot{\bar{x}} = -v\bar{y} + \varepsilon\bar{x}\,[2r - \xi_1\bar{y}\,]$$

$$\dot{\bar{y}} = v\bar{x}(1 + \bar{y}) - \varepsilon\xi_2\bar{x}^2(1 + \bar{y}) + \varepsilon v^2\bar{b}_1(1 + \bar{y})\cos(2\pi\theta) - \varepsilon v^2\bar{w}\bar{y} \tag{4}$$

$$\dot{\theta} = 1$$

for sufficiently small $\varepsilon < 1$ and b_1, $w \sim \varepsilon^2$. The relations to the physically interpretable variables are

$$S(t) = S_0(1 + \frac{\varepsilon}{v}\bar{x}(t) + O(\varepsilon^3))$$

$$E(t) = E_0(1 + \bar{y}(t) + O(\varepsilon)) \tag{5}$$

$$I(t) = I_0(1 + \bar{y}(t) + O(\varepsilon)) \,.$$

This means that after a very short transition period the relation

$$\frac{E(t)}{I(t)} \approx \frac{E_0}{I_0} \approx \frac{g}{a} \tag{6}$$

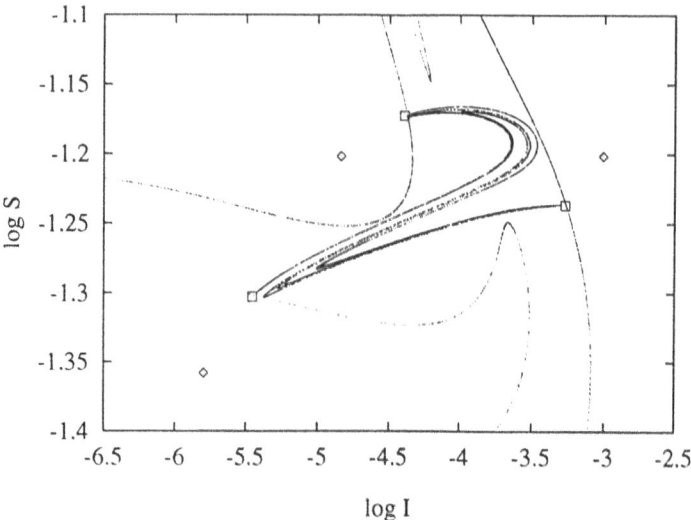

Figure 5. The phase space of the reduced SEIR model (4) for *w*=0.0013 and *b₁*=0.295. A chaotic attractor (centre) coexists with a period three attractor (◊). The stable manifold of the unstable period three fixed points (□) is the *basin boundary* between the two *basins of attraction*. It can be computed by backward iteration of a large number ($\sim 10^3$) of points with initial conditions close to the unstable period three fixed points.

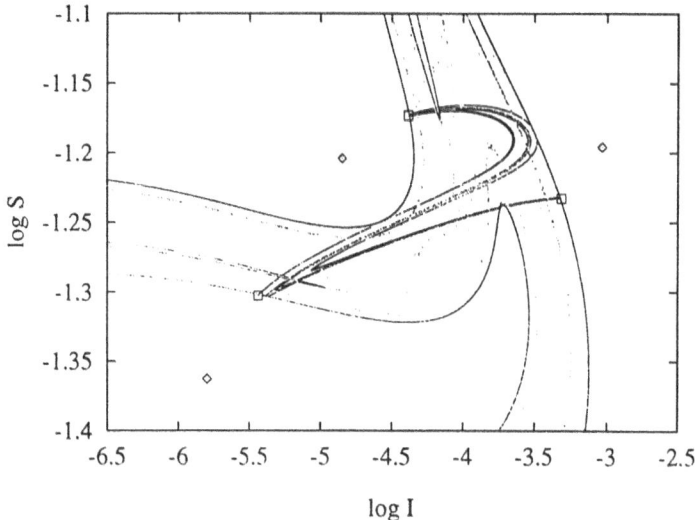

Figure 6. The phase space of the reduced SEIR model for *w*=0.0013 and *b₁*=0.31. The chaotic semi-attractor can be approximated in different ways (e.g. Tél 1990). On an intermediate time scale the trajectories, which are close to the chaotic semi-attractor, bounce around in chaotic way, indistinguishable from those on a chaotic attractor. But in the long term limit the only stable attractor is the period three orbit. Near these parameter values the basin boundary is fractal, symbolised by the scattered dots.

holds (to order ε). Without loss of generality we choose S and I as the dynamical variables. A comparison between figure 3 and 4 demonstrates the validity of the approximation (4).

For the second step we know from (4) that the SEIR model is *effectively* a three dimensional flow, from which we construct a two dimensional Poincaré map. Since the system is periodically forced, we can choose a *stroboscopic* map

$$P: \begin{pmatrix} S(t = n) \\ I(t = n) \end{pmatrix} \rightarrow \begin{pmatrix} S(t = n + 1) \\ I(t = n + 1) \end{pmatrix} \tag{7}$$

relating the values of S and I at January 1st of year n to those of year $n+1$. All long-term phenomena of the SEIR model can be read off from this map.

The effective dimensionality of the SEIR model could also be investigated by purely graphical methods. In this case one would observe that the three dimensional Poincaré map of the full system (1a) and (1b) forms a plane in the phase space because of relation (6). The explicit use of the restricted system (4) has the additional advantage that it enables us to calculate the basin boundaries in a very efficient way (see figure 5). Figure 5 and 6 give an overview of the geometry of the phase space (S_n, I_n) of the Poincaré map belonging to system (4). Below a critical threshold $b_1^{bc} \approx 0.3$ trajectories starting from randomly chosen initial values approach either the chaotic or the period three attractor (figure 5). Above this value a *boundary crisis* occurs (figure 6), where the basin boundary hits the chaotic attractor and destroys it. The remnant of the chaotic attractor is a so-called *chaotic semi-attractor*. Since the fingers of the period 3 basin are relatively thin, the escape from this semi-attractor is relatively slow. When their lifetime is long, the metastable chaotic transients in the vicinity of the semi-attractor are practically indistinguishable from orbits on a chaotic attractor. More generally, a crisis is a sudden change of a chaotic attractor, created by *"a collision between a chaotic attractor and a coexisting unstable fixed point or periodic orbit"* (Grebogi et al. 1983).

The lower state of the period 3 orbit comes close to one individual even for a town of one million people (figures 5 and 6). This means that demographic noise becomes important for the dynamics. As we will see explicitly in the next section, even for very large population centres the demographic noise destabilises the period three orbit and brings the trajectories back to the basin of attraction of the chaotic orbits. To understand the rich spectrum of behaviour of such systems one has to include the discussion of metastable transients into the geometrical analysis.

4. Microscopic description of unpredictability

Immigration and the process of infection itself are sources of demographic stochasticity. The influence of noise on chaotic transients in epidemics has been studied before by Rand and Wilson (1991). Using a *stochastic differential equation* type description (e.g. van Kampen 1981) they found a stabilisation of chaotic transients, however the mechanism for this stabilisation remains unclear in their analysis.

Following ideas of Bartlett (1957), Olsen et al. (1988) and Bolker and Grenfell (1993) the deterministic SEIR equations can be interpreted as an equation of motion for the expectation values of a Markovian stochastic process. For the computation of the realisations we use the *minimal process method*, which has been proposed independently by

Gillespie (1976) and Feistel (1977). Using a four dimensional state space $n=(s,e,i,r)$, where the components are the integer numbers of individuals in the epidemiological classes, the stochastic version of the SEIR model can be written down as a *master* equation, describing the time evolution of probability densities. The simplifying idea is to simulate single stochastic trajectories instead of computing probability densities. Assuming that the process is in the state n at time t with certainty, there is an exponentially distributed *waiting time* for transitions to all adjoining states. The simulation consists of two main iteration steps. First one has to choose a random time step according to the waiting time distribution, and secondly the adjoining state, to which the system moves, has to be determined randomly according to the relative transition probabilities.

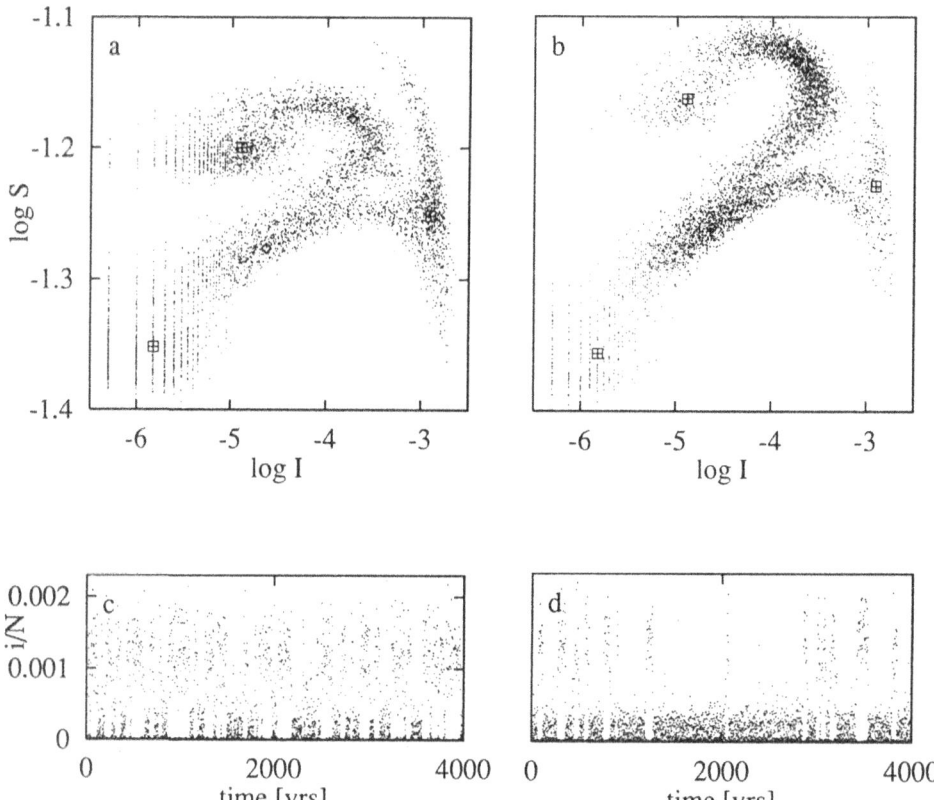

Figure 7. (a) Poincaré map (values at January 1st) of a stochastic trajectory (5000 yrs) of the modified SEIR model for $b_1=0.25$ and $w=0.0013$ and population size $N=2\cdot10^6$. At the lower left part the integer population size is visible (the lines from left to right correspond to 1, 2, 3, ... infective individuals). (b) As figure 7a, but for $N=4\cdot10^6$. (c,d) Same data as above, plotted as time series of the relative number of infectives. The hight levels of infection occur exclusively during period 3 episodes. The clustering of dots on the bottom indicates chaotic transients.

In figure 7 we compare the stochastic ("microscopic") with the deterministic ("macroscopic") dynamics. As opposed to figures 5 and 6 we use the more realistic parameter value b_1=0.25. According to figure 3 this corresponds to the region of coexistence of a period 3 and a period 4 attractor. Because of the proximity to the two boundary crises (b_1 ≈ 0.24) there are also coexisting chaotic semi-attractors. The geometric analysis of this situation will be published elsewhere (Engbert 1993). The demographic noise changes the qualitative features of the dynamics to a situation similar to figure 6 (obtained with parameter b_1=0.31).

The comparison between figure 7a,c and 7b,d shows a strong influence of the population size on the dynamics. Two different effects can be distinguished. As expected the larger population size (lower demographic noise level) brings the stochastic process closer to the deterministic dynamics. The second effect can be seen when comparing figures 7c and 7d. It shows that the decrease of the demographic noise increases the *average lifetime* of the chaotic transients, whereas the lifetime of the periodic orbit is hardly effected. Obviously the threshold population size, where the lifetime of the periodic orbit starts to diverge, is above $4 \cdot 10^6$. Even for very large population centres we get a situation of noise induced intermittency, where the dynamics switch randomly between chaotic transients and intermittent periodic episodes. In contrast to the view expressed by Rand and Wilson (1991) the stabilisation of the chaotic transients is not achieved by a prolongation of the lifetime of the chaotic transients but by a back scattering from the destabilised period 3 orbit. The decrease of the lifetime of the chaotic transients with increasing noise level is called a *noise induced crisis*. The stabilisation of the chaotic transients by destabilisation of coexisting periodic attractors may prove to be of more general importance for the understanding of population fluctuations in epidemiology and ecology.

5. Discussion

The seasonally forced SEIR model can be seen as a parent to a whole family of models. The explanatory value of the basic model derives partially from the fact that the introduction of more biological realism - e.g. the school year and the age structure - further increases the agreement between empirical data and the model (Schenzle 1984, Bolker and Grenfell 1993). The present paper does not aim at optimal biological realism, but at a better qualitative understanding of the complex interplay between the instability of the macroscopic nonlinear dynamics and the microscopic (demographic) fluctuations. The rich spectrum of behaviour of the basic model has been underestimated. Figures 4b and 4c of Bolker and Grenfell (1993) can be interpreted as an example of a noise induced crises in their realistic age-structured (RAS) model of Schenzle (1984). Their figure 4b shows the incidence pattern of a population centre of 50 million people and their figure 4c shows the more complex pattern (including a change from a period three to a period two based chaos) for a town of 1 million people. Comparing this with figures 6d and 6c we can say that the basic SEIR model (1,2) is also able to give a qualitative understanding of this threshold behaviour. Transitions between a relatively short regime of period three and longer period two based chaotic regimes can also be seen in the empirical time series of the New York City measles incidence and the ones of Copenhagen and Detroit (Kendall *et al.*, this volume). Figure 8 shows 50 years of a more disaggregated time series corresponding to the figures 7a,c.

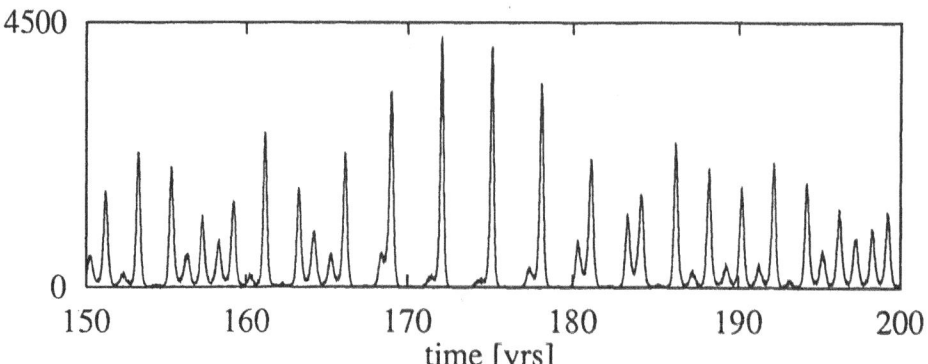

Figure 8. Time series of the number of infectives obtained from a stochastic simulation with parameter values as in figures 7a,c. A transition to (from) period three dynamics happens in year 167 (181) on the arbitrarily chosen time scale.

In summary, we can say that the deterministic SEIR model is able to give a qualitative understanding of the complex incidence pattern of measles in large cities (a) if the immigration of the virus is taken into account and (b) when the qualitative discussion is extended to the coexistence of different attractors and to metastable transients. Even for very large population centres the attractors of higher period are destabilised by demographic noise. This way the metastable transients are permanently revisited. The average lifetimes of the chaotic transients and of the intermittent periodic episodes can only be determined via stochastic simulations.

References

Anderson, R.M. and R.M. May (1991), 'Infectious diseases of humans: dynamics and control', Oxford Univ. Press.

Anderson, R.M., B.T. Grenfell and R.M. May (1984), 'Oscillatory fluctuations in the incidence of infectious disease and the impact of vaccination', *J. Hyg. Camb.*, **93**.

Bartlett, M.S. (1957), 'Measles periodicity and community size', *J. R. Statist. Soc.*, **A 120**, 48-70.

Bolker, B.M. and B.T. Grenfell (1993), 'Chaos and biological complexity in measles dynamics', *Proc. R. Soc. Lond.*, **B 251**, 75-81.

Dietz, K. (1976), 'The incidence of infectious diseases under the influence of seasonal fluctuations', *Lect. Notes Biomath.*, **11**, 1-15.

Engbert, R. (1993), diploma thesis, RWTH Aachen, in preparation

Feistel, R. (1977), 'Betrachtung der Realisierung stochastischer Prozesse aus automaten-theoretischer Sicht', *Wiss. Z. WPU Rostock,* **26**, 663-670.

Gallas, J.A.C. (1992), 'The role of codimension in dynamical systems', *Int. J. Mod. Phys.,* C (preprint).

Gillespie, D.T. (1976), 'A General Method for Numerically Simulating the Stochastic Time Evolution of Coupled Chemical Reactions', *J. Comp. Phys.*, **22**, 403-434.

Grebogi, C., E. Ott and J.A. Yorke (1983), 'Crises, sudden changes in chaotic attractors, and transient chaos', *Physica*, **7D**, 181-200.

Guckenheimer, J. and P.J. Holmes (1983), *'Nonlinear Oscillations, Dynamical Systems, and Bifurcations of Vector Fields'*, Springer Verlag, New York.

Kendall, B.E., W.M. Schaffer, L.F. Olsen, C.W. Tidd and B.L. Jørgensen (1993), 'Using Chaos to Understand Biological Dynamics', this volume.

London, W.P. and J.A. Yorke (1973), 'Recurrent outbreaks of measles, chickenpox and mumps, I. Seasonal variation in contact rates', *Am. J. Epidem.*, **98**, 453-468.

Olsen, L.F., G.L. Truty and W.M. Schaffer (1988), 'Oscillations and Chaos in Epidemics: A Nonlinear Dynamic Study of Six Childhood Diseases in Copenhagen, Denmark', *Theor. Popul. Biol.*, **33**, 344-370.

Rand, D.A. and H.B. Wilson (1991), 'Chaotic Stochasticity: a ubiquitous source of unpredictability in epidemics', *Proc. R. Soc. Lond.*, **B 246**, 179-184.

Schaffer, W.M. (1985), 'Order and chaos in ecological systems', *Ecology*, **66**, 93-106.

Schaffer, W.M., L.F. Olsen, G.L. Truty, S.L. Fulmer and D.J. Graser (1988), 'Periodic and chaotic dynamics in childhood infections', . in: Markus, M., S. Müller and G. Nicolis (eds.), *From Chemical To Biological Organization'*, Springer Verlag, Berlin.

Schenzle, D. (1984), 'An Age-Structured Model of Pre- and Post-Vaccination Measles Transmission', *IMA J. Math. appl. Med. Biol.*, **1**, 169-191.

Schwartz, I.B. and H.L. Smith (1983) *Infinite subharmonic bifurcation in an SEIR epidemic model'*, *J. Math. Biol.*, **18**, 233-253.

Schwartz, I.B. (1992), 'Small amplitude, long period outbreaks in seasonally driven epidemics', *J. Math. Biol.*, **30**, 473-491.

van Kampen, N.G. (1981), *'Stochastic Processes in Physics and Chemistry'*, North Holland, Amsterdam.

Tél, T. (1990), 'Transient Chaos', in: Hao Bai-lin (ed.), *'Directions in chaos'*, World Scientific.

CONTROL AND PREDICTION IN SEASONALLY DRIVEN POPULATION MODELS

IRA B. SCHWARTZ AND IOANA TRIANDAF

Special Project in Nonlinear Science U.S. Naval Research Laboratory
Code 6700.3 Plasma Physics Division Washington, D.C. 20375-5000 U.S.A.

email: schwartz@nls4.nrl.navy.mil and triandaf@ppdpi2.nrl.navy.mil

Summary

Recent studies have shown both theoretically and from data analysis that the dynamics of certain childhood diseases in large populations can exhibit a wide range of behavior that is periodic as well as chaotic. Mathematical models (such as the SEIR model with seasonal forcing) have been able to predict the onset of chaotic epidemics using parameters of childhood diseases such as measles, mumps, and chickenpox. A new method is presented to control certain unstable outbreaks by using inexpensive vaccine strategies. These controlled outbreaks are small in amplitude, periodic and therefore, quite predictable. New techniques for general dynamical systems will be presented and applied to this epidemiological problem in which the epidemic can be controlled and directed into a small amplitude outbreak which is regular. The incidence of the epidemic may be reduced by tracking this controlled outbreak as a function of certain vaccine strategic parameters.

Keywords: Chaos, vaccine, measles, epidemiology, control, tracking

1. Introduction

It is now well documented that childhood diseases which incur permanent immunity oscillate periodically, as well as exhibit chaotic behavior as reported in Aron (1990), Fine *et al.* (1980), London *et al.* (1973), Olsen *et al.* (1990), Schaeffer *et al.* (1985), Schwartz (1989 and 1992). Examples of such diseases are chickenpox, measles, and mumps. Another example of a disease which exhibits strong oscillations but does not fit exactly into the framework of the other childhood diseases is rubella, which exhibits outbreaks with periods as long as seven years. One common feature in each of the above-mentioned diseases is that they have a maximum peak in their respective power spectrum of one year, which is reflected in the fact that peak-to-peak outbreaks have local maxima separated by one year. This annual behavior in childhood disease has been linked to the opening and closing of schools in various cities [Fine (1980), London *et al.* (1973), Yorke *et al.* (1979)].

Given that the childhood diseases all have an annual seasonal component, spectral analysis reveals that their outbreaks can exhibit a widely varying range of longer interepidemic periods accompanied by large changes in the peak number of cases. In measles, periods on the order of 2-3 years have been observed. Mumps exhibits periods from 3-4 years, while rubella has been observed to have interepidemic outbreaks of 5-7 years. Chickenpox is the one disease which seems to exhibit a unique annual peak in its power spectrum corresponding to interepidemic outbreaks of 1 year [Olsen *et al.* (1990)].

When the diseases are examined with respect to their incidence data as a time series, further analysis may be done to see what kind of recurrent behavior the populations produce. For example, measles in New York City in the pre-vaccine years has been observed as a biennial cycle, with two year interepidemic peaks separated by what appears to be noise [Hethcote 1983]. In Baltimore, where the population is much smaller than that of New York City, measles appears to have peak cases separated by either 2 or 3 years, where the large peak-to-peak years appear as a random sequence. So depending upon the population size and structure, measles can exhibit periodic or aperiodic behavior, both of which are deterministic [Kaplan *et al.* (1992), Sugihara *et al.* (1990)]. Further analysis, using embedding techniques by Olsen *et al.* (1990) and Schaeffer and Kot (1985), reveals that measles can exhibit chaotic outbreaks based on a fundamental frequency. Their analysis suggests that the noise appearing between peaks in the outbreaks may be dominated by a deterministic process. Sugihara and May (1990) have further analyzed the data by computing decorrelation times for diseases such as measles and chickenpox by employing embedding techniques as well, and find that the data appear to be generated by a deterministic process. That is, the data have correlation times longer than that of randomly generated data. Other diseases such as mumps, also have characteristics of chaotic behavior, whereas chickenpox appears to be periodic with additive noise.

One of the goals of trying to understand the dynamics of such diseases is to develop new improved vaccine control mechanisms which have the capability of reducing the incidence of disease. In the past, theoretical models have been used to explain the dynamics of diseases. These models have also been used to exploit methods of vaccine implementation. One drawback in using a model when testing these schemes, is that the dynamics observed in nature may be qualitatively different from that which the model produces. Here we present a known method of control which has been previously implemented for maps. We use this method on a flow to control certain diseases, where the control is applied to the time series directly, without using the equations of the flow. The region in which control is successful is extended by a new method, which we call tracking. This technique works for deterministic as well as stochastic systems, and can be implemented for time series in which no model is needed. A model is not even required in the initial design stages. (See Schwartz and Triandaf (1992) for details.) The vaccine may be implemented discretely in periods of one year or shorter, or may be implemented continuously. The changes in the parameter governing vaccine control are very small, thereby potentially reducing overall cost in vaccine administration.

By using small changes in vaccine, it is also possible to improve the predictability at the peaks. This is very important in disease control when a large vaccine program is introduced. In previous work by Aron (1990), it was shown that if the vaccine pulse is given at various phases of the cycle, depending on the amplitude of the outbreak, the larger unstable periodic orbits may be excited. This in turn could lead to an enormous outbreak before the disease relaxes to a suitable small number of cases. Therefore, by using small amplitude corrections to stabilize a small periodic outbreak ensures one knows when to phase in a new large vaccine schedule with optimized results.

2. The SEIR model and the problem of vaccine control

We follow the assumptions and notation of Schwartz and Smith (1983). The reader should consult that paper for further details. Suppose the population is divided into four groups, each a function of time, t: Susceptibles $S(t)$, Exposed $E(t)$, Infectives $I(t)$, Recovered $R(t)$. The SEIR model describing the rate of change of each of the classes is:

$$S' = \mu(1 - f(t)) - \beta(t)IS - \mu S$$
$$E' = \beta(t)IS - (\mu + \alpha)E$$
$$I' = \alpha E - (\mu + \gamma)I \qquad\qquad\qquad (2.1)$$
$$R' = \gamma I - \mu R + \mu f(t).$$

In Eq. (2.1) we have made the assumptions that the population is uniformly mixing and of constant size. In particular, we assume the population is normalized to unity. The birth and death rates, μ, are equal, and $1\backslash\alpha(1/\gamma)$ is mean latent (infectious) period. The contact rate, β, is defined as the average number of effective contacts with other individuals per infective per unit time. Seasonality is incorporated by allowing β to fluctuate periodically as a function of time; i.e., we assume $\beta(t) = \beta_0(1+\beta_1\cos 2\pi t)$, where $0 < \beta_1 < 1$. This is the mechanism which simulates the annual behavior of the outbreaks which appear so regularly in the data as well as in the spectral analysis.

The function $f(t)$ is that fraction of the new susceptibles which are removed by vaccine. When $f=0$, the model predicts the dynamics of the disease as a function of time. When there is no seasonal component, f may be adjusted so that the reproductive rate of infection drops below unity. This guarantees the incidence rate will asymptote to zero. Other recent techniques show that by pulsing a vaccine once, the epidemic can be brought to small, but growing, oscillatory behavior [Agur *et al.* (1992a, 1992b)].

In contrast to this technique, our technique makes use of very small discrete changes in f to actively control a small unstable period one orbit. This has several advantages, the most obvious being that the level of incidence is much smaller than without the vaccine. Furthermore, the change of the vaccine level is very small, and is applied discretely. Secondly, if the birth rate is stochastic, the tracking method retains control without a large increase in vaccine administration. Finally, the method may be applied to the time series of cases directly, without any use of a model to set the levels of vaccine.

3. The Method of Control

In this section we describe the method of control of an unstable periodic orbit of a flow, simulated by a time series. The method applies to orbits of flows of any dimension, having one unstable direction; i.e., one Floquet multiplier lies outside the unit circle. (Although the flow used in this example has unstable periodic orbits with one unstable direction only, the method of control can be generalized to orbits which have more than one unstable direction [Auerbach (1992), Romeiras (1992)].) In this method of control one stabilizes a periodic orbit of a flow through only small time-dependent perturbations of an accessible system parameter. In the case of vaccine control, the parameter is f.

The method is derived from it's counterpart for maps which was introduced in Ott et al. (1990). A similar method for flows was described in Nitsche and Dressler (1992). To fix ideas we will present the control method applied for stabilizing a period one orbit of the flow (2.1) by adjusting the external parameter f; i.e., by monitoring small fluctuations in the number of susceptibles receiving the vaccine. The algorithm performed well also when the control parameter was the contact rate, β_1. However, in practice this is a parameter that cannot be controlled. In addition we will assume that like in a real situation, noise is present so that the equations now read:

$$S' = \mu(1 - f(t)) - \beta(t)IS - \mu S + \varepsilon(t)$$
$$E' = \beta(t)IS - (\mu + \alpha)E + \varepsilon(t) \qquad (3.1)$$
$$I' = \alpha E - (\mu + \gamma)I + \varepsilon(t)$$

where ε denotes the noise term. Since

$$R(t) = 1 - S(t) - E(t) - I(t),$$

the system is reduced to a system of three equations and we need to consider only three variables.

In what follows let us denote $\phi = (S,E,I)$ and let us rewrite (3.1) in compact form as follows:

$$\phi' = F(t,\phi).$$

Starting from the flow (3.1) one first generates a map by taking a Poincaré section of the flow. In our case, since we are interested in stabilizing a period one orbit of the flow, the map defined by:

$$x_{n+1} = T(x_n, f_0) = \phi(1,0;x_n, f_0) \qquad (3.2)$$

will define a Poincaré section of the flow close to the period one orbit. This map acts as follows. Each iterate, x_n, is taken as initial condition, and the flow is calculated with this initial condition one period later. This determines the value of the next iterate x_{n+1}. The period one orbit we want to stabilize represents a fixed point of this map at $f = f_0$, $x_F(f_0) = T(x_F(f_0), f_0)$. The Jacobian of the map T at the fixed point is

$$A = \frac{\partial \phi}{\partial x_n} (1,0;x_n, f_0)\big|_{x_n = x_F},$$

which can be calculated by solving the system of first variation of (3.1) about the periodic solution. Introducing the new variable, $\xi(f) = x(f)-x_F(f_0)$ shifts this fixed point to the origin when $f = f_0$. In a neighborhood of the fixed point, the map can then be approximated by the linear map:

$$\xi_{n+1} = T(\xi_n, f) \approx \xi_F(f) + A(\xi_n - \xi_F(f)) . \qquad (3.3)$$

Here it is understood that f is close to f_0. The period one orbit we want to stabilize is a saddle, where the Jacobian has two eigenvalues λ_{s_1} and λ_{s_2} inside the unit circle and one eigenvalue λ_u with $|\lambda_u| > 1$. Iterating without control, the fixed point would approach the saddle along the stable eigendirections e_{s_1}, e_{s_2}, but will drift away along the unstable one e_u.

Control is implemented by adjusting the parameter f at each iterate so that the next iterate of the map will fall on the stable manifold of the saddle. We will derive below the way in which f must be chosen following Ott *et al.* (1990). When $f_0 = 0$, no vaccine is present and the attractor is chaotic. A picture of the projection of this attractor in the *S-I* plane is shown in Fig 1a, along with a picture of the chaotic time series for the infectives in Fig 1b. For this time series, we introduce a small level of vaccine, say $f_0 = 0.01$, and use the control derived below to stabilize a period one orbit. This orbit is close to the period one orbit when $f_0 = 0$.

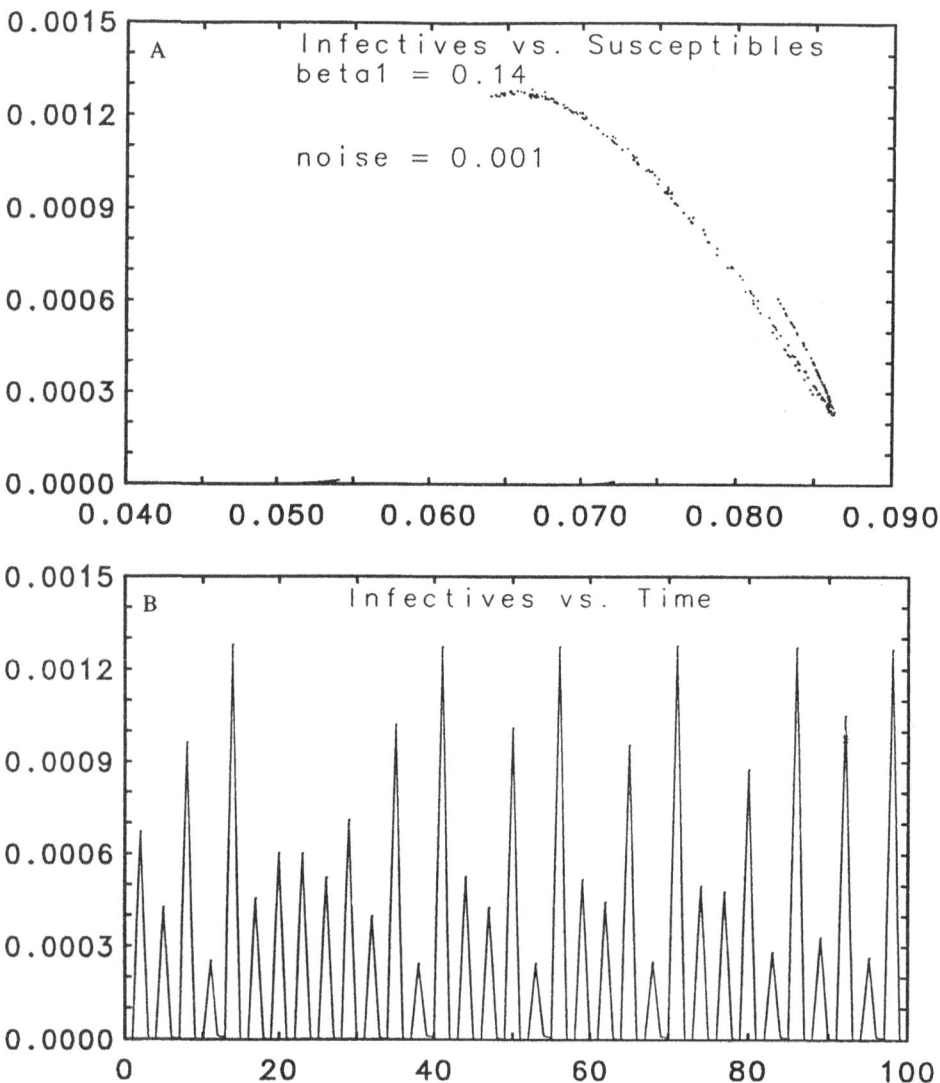

Figure 1. The projection of the chaotic attractor in *IS* plane, $f_0 = 0.0$, $\beta_1 = 0.14$ noise = 0.001 (a) and the corresponding chaotic time series for I (b).

When initializing this method on a computer, one first localizes the period one unstable orbit inside the chaotic attractor using a Newton-type scheme to solve the equation

$$x = T(x, f_0) \tag{3.4}$$

for a fixed point corresponding to a period one orbit. We then approximate the derivative of the fixed point with respect to the parameter as follows:

$$g \equiv \left. \frac{\partial \xi_F(f)}{\partial f} \right|_{f = f_0} \cong \frac{1}{\bar{f}} \xi_F(\bar{f}),$$

where $\xi_F(\bar{f}) = x_F(\bar{f}) - x_F(f_0)$ and can be found by solving:

$$x_F(\bar{f}) = T(\bar{f}) = \phi(1,0;x_F(\bar{f}),\bar{f}). \tag{3.5}$$

Here $\bar{f}$ is very close to f_0, and is used to approximate g. Equation (3.2) becomes:

$$\xi_{n+1} \cong f_n g + A \cdot (\xi_n - f_n g) \tag{3.6}$$

where f_n is given below. The flow we consider is a three dimensional flow, so that the matrix A will be of the following form:

$$A = \lambda_u e_u f_u + \lambda_{s_1} e_{s_1} f_{s_1} + \lambda_{s_2} e_{s_2} f_{s_2} ,$$

where f_{s_1}, f_{s_2} and f_u are contravariant basis vectors defined by:

$$f_{s_1} \cdot e_{s_1} = f_{s_2} \cdot e_{s_2} = f_u \cdot e_u = 1, \quad f_{s_1} \cdot e_u = f_{s_2} \cdot u_u = f_u \cdot e_{s_1} = f_u \cdot e_{s_2} = 0. \tag{3.7}$$

To choose f_n we require:

$$f_u \cdot \xi_{n+1} = 0,$$

which ensures that ξ_{n+1} lies in the plane defined by e_{s_1} and e_{s_2}. Solving in the above for f_n and taking into account (3.6) we get:

$$f_n - f_0 \equiv \frac{\lambda_u \xi_n \cdot f_u}{(\lambda_u - 1)g \cdot f_u} \equiv C \cdot \xi_n .$$ (3.8)

We assume $g \cdot f_u \neq 0$, and that the range in which f_n is allowed to vary is given by:

$$|f_n - f_0| < f_* ,$$ (3.9)

where f_* is small and represents the parameter range in which the linear approximation 3.5 holds. For f_n outside this range we set $f_n = f_0$. After choosing f_n according to equations (3.8) and (3.9) we introduce these values in the flow and reiterate the map T. We remark

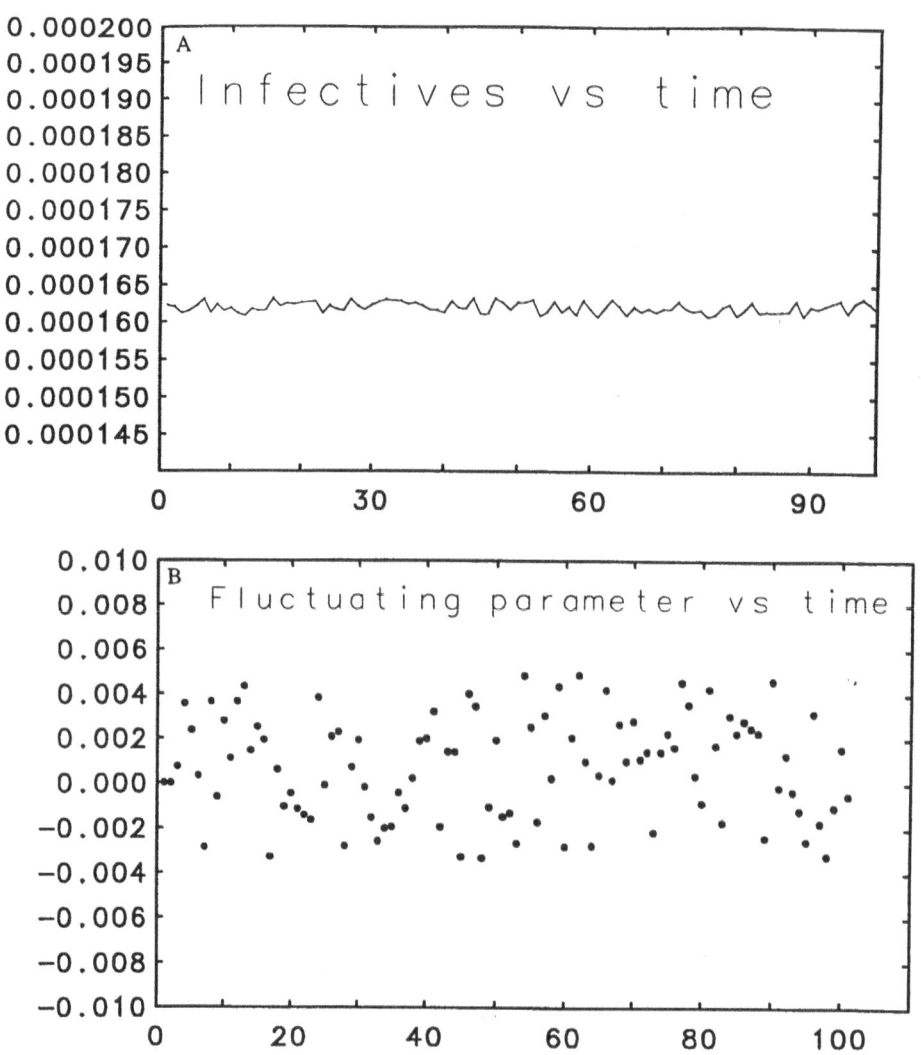

Figure 2. The controlled fraction of infectives vs n, $f_0 = 0.01$, $\beta_1 = 0.14$ noise = 0.001 (a) and f vs time (b).

that the equations of the flow were used only initially to localize the unstable orbit; i.e., for solving (3.4) and (3.5) and for finding the matrix A. These quantities are then kept constant and used for determining f_n as we iterate. The control itself is done by fluctuating the parameter according to formula (3.8) and then adjusting this parameter in the real time series without using the explicit equations of the flow.

In real situations where the data is given only by a time series, both the reference point and the gain C used in control can be computed without a model. From a practical point of view, this is important because the model may not be accurate enough to predict the relevant control parameters. Although we use a model here to generate a time series, the method may be applied directly to data to get all of the essential information from a nonlinear dynamics point of view to implement the control. That means in (3.8) the value of f_n is determined from real data and not by using the equations of the flow.

In Fig 2a the controlled fixed point is shown as a function of the iterate. This represents one of the infinitely many periodic orbits close to the period one orbit embedded in the chaotic time series shown in Fig 1b. Comparing these two figures we see that by controlling with very small levels of vaccine, we reduced by almost an order of magnitude the high peaks appearing in the chaotic time series. In Fig 2b the corresponding fluctuations in the vaccine parameter are shown as a function of the iterate. We chose to fluctuate about a vaccine parameter value of $f_0 = 0.01$. This numerical example shows that once the number of infectives reached a desired value, this can be maintained by small smart changes in the fraction of the population receiving the vaccine. We would like to achieve this stability as a larger and larger fraction of the population is vaccinated.

In order to maintain the control as f_0 is increased, we introduce another algorithm which will be presented in the next section.

4. The Tracking Algorithm

In this section we introduce our tracking algorithm which allows us to trace an orbit of a flow as a function of a parameter. Our method greatly extends the region over which control is possible, and is our main contribution in this paper. This is done only by monitoring an external system parameter in the real time series. We will describe the method for Eq. (3.1) and track the orbit we controlled in section 2 as a function of the parameter f. So far the control worked only for a fixed value of f. The control region will be extended to hold when the parameter f is increased continuously.

As in the previous section we first form the map T by taking a Poincaré section of the flow. The fixed point of this map will be tracked as a function of the parameter f by using a prediction-correction technique. This technique was introduced in Schwartz and Triandaf (1992) for tracking orbits of maps. We will briefly review this technique below and examplify it for tracking the period one orbit which we controlled in section 2.

The method consists in a prediction step followed by a correction step. As the parameter f_0, is increased, let us say to $f_0+\delta f$ the new position of the orbit is predicted. This can be done in various ways which can be found in Schwartz and Triandaf (1992). In our case, since the variables $S, E,$ and I vary slowly with respect to f a very simple prediction was enough. The predicted value of the orbit, at the new parameter value $f_0+\delta f$ is taken to be the value of the orbit previously determined, i.e. $x_F(f_0)$. The prediction alone will not allow tracking very far. In addition a correction must be made. The correction makes use of the following observation: as we control about the correct value of the orbit, the mean value in the corresponding fluctuations in the parameter will be close to zero. This can be seen also in Fig. 2b. If we control for several iterates when an error is present in determining the orbit, then the mean of the corresponding fluctuations in the parameter will change proportionally to the error between the real fixed point and the control point. Therefore, when we predict an orbit the error can be corrected by minimizing the mean in the fluctuations of the parameter. The predicted point is slightly modified until we bring the mean of the fluctuations in the parameter f as close to zero as possible. This will be taken as the value of the orbit at $f_0+\delta f$. Another prediction step then follows and the procedure is repeated. In Fig. 3 the orbit which we controlled in section 2 is tracked as the parameter f is increased continuously.

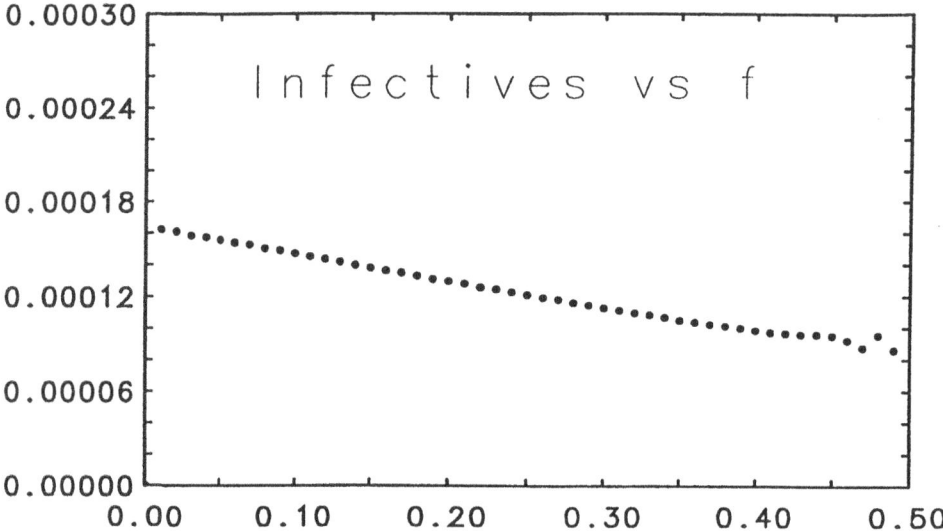

Figure 3. The fraction of infectives vs f using deterministic tracking, $\beta_1 = 0.14$ noise = 0.001.

In implementing the method, we iterated only 5 times around the predicted value, before correcting it. So in practice a relatively small number of guesses in the fraction of the population to receive the vaccine, will bring us back to the desired stabilized state.

5. The Stochastic Tracking

In Triandaf and Schwartz (1993) an algorithm was introduced, similar to the tracking algorithm where an orbit of a map was tracked as a function of a stochastic parameter. This algorithm was referred to as random walk control or stochastic tracking and is used to cancel drift, which is unavoidable to some level in real time series. In this section we present the extension of this algorithm to flows and apply it to tracking the same period one orbit of the SEIR flow as in the previous section. This time f will vary stochastically.

Having localized a desired orbit of a flow corresponding to a certain parameter value, let us say f_0, we would like to keep the flow on this state as f is changed to some value nearby value $f_0+\delta f$ where δf is random. In our code δf is not chosen deterministically, but is picked by a random number generating function. Only the maximum amplitude of this step is fixed.

In order to achieve this we will first form the map T. The random walk control can then be applied to this map. To fix ideas suppose that we want to stabilize the same period one orbit x_F as in the previous sections. Once an orbit of the map T is localized and controlled, f is randomly varied. We need to determine the orbit at the new parameter value $f_0 + \delta f$. In order to do that we first consider an approximation of the orbit at this parameter value.

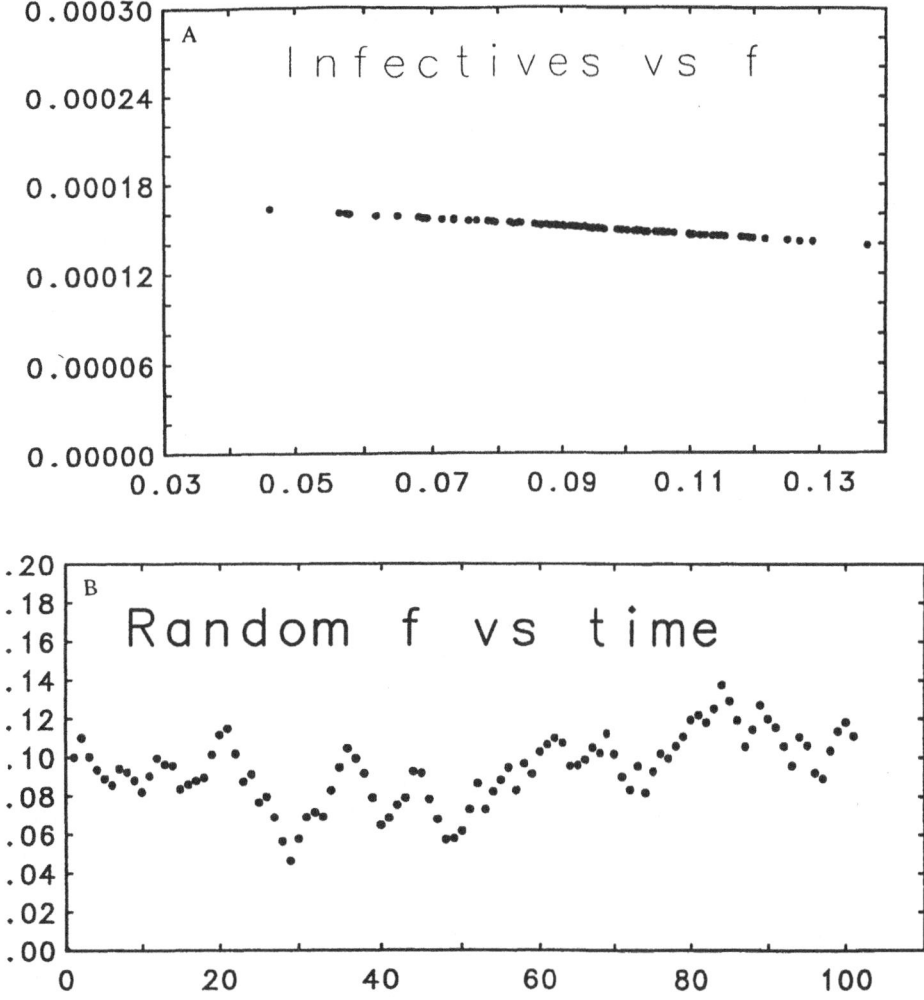

Figure 4. The fraction of infectives vs f, using stochastic tracking $f_0 = 0.1$, $\beta_1 = 0.0.14$and noise $= 0.001$ (a) and the random parameter f vs the number of random steps (b).

This corresponds to the prediction step in the tracking algorithm. As an approximation of the orbit at $f_0 + \delta f$, for the SEIR flow it was enough to take the previously controlled orbit, $x_F(f_0)$. We then control taking this approximate value as reference point. As in the tracking algorithm, the mean in the fluctuations in the parameter, reveals how far away our approximation is from the orbit. We correct as in the tracking algorithm. Another random change in the parameter then follows and the procedure is repeated. A picture of the orbit obtained by this algorithm can be seen in Fig 4a. In Fig 4b the corresponding random steps are shown as a function of the iterate, i.e. in the order they ocurred. In this example we controlled only 10 times before correcting.

6. Conclusions

We have introduced a new efficient numerical algorithm for stabilizing and tracking orbits of a flow. One first forms a suitable map from the flow, the orbit of the flow we are interested in tracking corresponding to a fixed point of the map. The algorithm stabilizes the fixed point of the map over a wide parameter range. In the method, we combined the original OGY algorithm with the tracking technique introduced in Schwartz and Triandaf (1992). The novelty of this method consists in that we were not only able to control, but also track an orbit over a wide parameter range. Moreover this can be done with the real time series; i.e., without using the equations of the flow.

This method was demonstrated on the SEIR epidemic model giving performance that is realistic for practical applications. Controlling an orbit of this flow amounts in practice to avoiding high-peak bursts in the number of infectives and maintaining this as a higher percentage of the population receives the vaccine. For the flow we considered, using the original OGY method of control gave the expected performance, even in the presence of noise so we did not need to use the modification of this algorithm introduced in Nitsche *et al.* (1992).

Acknowledgment

Dr. I. Triandaf gratefully acknowledges the support of the Office of Naval Technology for conducting this research.

References

Z. Agur, L. Cojocaru, G. Mazor, R.M. Anderson, and Y.L. Danon (1992a) , 'Pulse mass measles vaccination across age cohorts', preprint, The Weizmann Institute of Science.

Z. Agur, G. Mazor, Y.L. Danon,, R.M. Anderson, L. Cojocaru, and R.M. May (1992b), 'Measles immunization strategies for an epidemiogically heterogeneous population: the Israeli case study', preprint, The Weizmann Institute of Science.

Aron, J.L.(1990), 'Multiple attractors in the response to a vaccination program', *J. Theor. Pop. Biology*, **38**, 58-67.

Auerbach, D., Grebogi, C., Ott, E., and Yorke, J.A. (1992), 'Controlling chaos in high dimensional systems', *Phys. Rev. Letts.* **69**, 3479-3482.

Fine, P. and Clarkson, J.(1980), 'Measles in England and Wales. I. An analysis of factors underlying seasonal patterns'. *Int. J. Epidemiology* **11**, 5-14.

Hethcote, H. W.(1983), 'Measles and rubella in the United States', Am. J. Epidemiology **117**, 2-13.

Hethcote, H.W.(1991), 'Periodicity in epidemiological models', Preprint .

Kaplan, D. T. and Glass, L. (1992), 'Direct test for determinism in a time series', PRL **68**, 427-430.

London, W.P., Yorke, J.A.(1973), 'Recurrent outbreaks of measles, chickenpox, and mumps. I. Seasonal variation in contact rates'. *Am. J. Epidemiology* **98**, 453-468.

Nitsche, G., and U. Dressler (1992), 'Controlling Chaotic Dynamical Systems Using Time Delay Coordinates', *Phys. Rev. Lett.* **68**, 1.

Olsen, L.F., Schaffer, W.M., Truty, G.L., and Fulmer, S.L. (1990), 'The dynamics of childhood epidemics', preprint.

Ott, E.C., and C. Grebogi and J.A. Yorke (1990), 'Controlling Chaos' *Phys. Rev. Lett.* **64**, 1196.

Romeiras, F. J., Grebogi, C., Ott E., and Dayawansa, W.P. (1992), 'Controlling chaotic dynamical systems', *Physica D* **58**, 165-172.

Schaeffer, W.M., Kot, M.(1985), 'Nearly one dimensional dynamics in an epidemic', *J. Theor. Biol.* **112**, 403-427.

Schwartz, I.B., Smith, H.L. (1983), 'Infinite subharmonic bifurcations in an SEIR model', *J. Math. Biol.* **18**, 233-253.

Schwartz, I.B.(1985), 'Multiple recurrent outbreaks and predictability in seasonally forced nonlinear epidemic models', *J. Math. Biol.* **21**, 347-361.

Schwartz, I.B. and Triandaf, I. (1992), 'Tracking unstable orbits in experiments: A new continuation method', *Phys. Rev. A* **46**, 7439 .

Schwartz, I.B.(1989), 'Nonlinear dynamics of seasonally driven epidemic models', in: Biomedical Modelling and Simulation, 201-204, J. Eisenfeld and D.S. Levine eds., New York: Scientific Publishing Co..

Schwartz, I.B.(1992), 'Small amplitude, long period outbreaks in seasonally driven epidemics', *J. Math. Biol.,* **30**, 473-492.

Sugihara, G., May, R.M.(1990), 'Nonlinear forecasting as a way of distinguishing chaos measurement error in time series', *Nature* **344**, 734-741.

Triandaf, I. and Schwartz, I.B.(1993), 'Stochastic tracking in nonlinear dynamical systems', Phys. Rev. A. in press.

Yorke, J.A., Nathanson, N., Pianigiani, G. (1979), 'Seasonality and the requirements for perpetuation and eradication of viruses in populations'. *Am. J. Epidemiology* **109**, 103-123.

SIMPLE THEORETICAL MODELS AND POPULATION PREDICTIONS

ALAN A. BERRYMAN[1] and MIKAEL MUNSTER-SWENDSEN[2]

[1] Dept. Entomology, Washington State University,
Pullman, WA 99164, USA.
[2] Inst. Population Biol., University of Copenhagen, 15,
Universitetsparken, DK-2100 Copenhagen, Denmark;

e-mail: berryman@wsuvm1.csc.wsu.edu

Summary

Using census data of spruce needleminer populations and their natural enemies from one stand of Norway spruce trees to estimate the parameters of a complex simulation model and two simple theoretical models, we predict spruce needleminer abundance in three other stands and compare the predictions with census data. The simple theoretical models were as good as the complex model in forecasting needleminer population numbers one year ahead. The reason may be that simple theoretical models capture the dominant structure of the dynamic system in an unbiased way.

1. Introduction

It has been claimed that simple theoretical models are not only cheap and easy to fit to commonly available data (e.g., census data), but that they are also more precise than complex simulation models. For example, Berryman (1978, 1991) found that a discrete logistic equation with time-lag described the cyclic dynamics of Douglas-fir tussock moth populations in Western North America and correctly forecast that an outbreaking population would not reach economic damage levels.

Ludwig and Walters (1985) showed that a simple Ricker equation, which is also a discrete version of the classical logistic model, gave better estimates of optimal fishing effort than a more complex age-structured model, even when the data were generated by the complex model itself. Albrecht (1992) demonstrated that small simple (aggregated) models provided better prognoses and deeper insights into complex ecological and economic systems than complicated simulation models.

Finally, Munster-Swendsen and Berryman (unpublished) compared the predictions of a complex, biologically explicit, simulation model with those of simple theoretical models. In this paper we briefly describe these results and then attempt to explain why simple models are not only more "beautiful" but often "better" and certainly "cheaper" than high-resolution simulation models.

2. Complex simulation model

Munster-Swendsen (1985, 1991) reported the results of a 20-year study of the spruce needleminer (Epinotia tedella) infesting a Norway spruce (Picea abies) stand in Denmark. Nineteen years of annual census data from this stand (stand A) were used by Munster-Swendsen to construct a detailed simulation model of needle miner population dynamics. This model describes all the biological interactions (parasitism, diseases and predation) as a sequence of mortalities in each generation. These calculations make use of the so-called pseudo-interference submodel (Munster-Swendsen 1985) which describes mortality (k) as a function of the density (P) of the killing agent (adult parasitoids or infectious stage of diseases); i.e.,

$$\log k = a + b(\log P). \tag{1}$$

The simulation model included submodels for parasitism (two prime parasitoids and one hyperparasitoid species), a fungal disease and a protozoan disease that affects fertility (Munster-Swendsen 1991). Weather acts as a driving variable on the model parameters.

3. Simple theoretical models

Berryman (1991, 1992a) has argued that simple models developed upon sound theoretical foundations offer the most promise for analyzing population fluctuations and predicting changes in population abundance from year to year. Here we use a one-species discrete-time, nonlinear, time-delayed, generalization of the Ricker or "logistic" equation

$$Nt = N_{t-1} \exp\{A[1 - (N_{t-d}/K)^q], \tag{2}$$

where N_{t-d} is population biomass density in generation (or year) t-d, d is the time-delay in the negative feedback (density dependent) response, A is the maximum instantaneous per-capita rate of increase of the needleminer, K is the equilibrium density, and q is a coefficient of nonlinear density dependence (see Cook 1965, Nelder 1961, Thomas et al. 1980, Berryman 1991, 1992a). This equation is easily extended to two species or even food webs (Berryman 1992b). The two-species predator/prey equations are

$$N_t = N_{t-1} \exp\{A_n - B_nN_{t-1} - C_nP_{t-1}/N_{t-1}\}, \tag{3a}$$

$$P_t = P_{t-1} \exp\{A_p - B_pP_{t-1} - C_pP_{t-1}/N_{t-1}\}, \tag{3b}$$

where N_t is the density of prey, P_t the density of predators, B_i ($i=n,p$) the coefficients of intraspecific interaction, and C_i the coefficients of interspecific interaction. These models were fit to the data from Stand A by calculating the per-capita instantaneous rate of increase, $\log_e(Nt/Nt\text{-}1)$, and regressing it against the correctly lagged population density (N_{t-2} in the one-species model), or the density of the prey or predator populations (N_{t-1} or P_{t-1}) and the proportion parasitized (P_{t-1}/N_{t-1}) in the two-species model (using the POPSYS microcomputer system; Berryman and Millstein 1989, and Berryman 1990). The fitted parameter values were, for the one species model $A = 2.27$, $K = 790.65$, $q = 0.34$, $d = 2$, $r^2 = 0.65$ and for the two species model $A_n = 3.25$, $B_n = 0.000075$, $C_n = 6.985$, $r^2 = 0.88$ for the spruce needleminer equation, and $A_p = 3.6$, $B_p = 0.000016$, $C_p = 7.938$, $r^2 = 0.9$ for the parasitoids.

4. Predicting needleminer abundance

During his studies on spruce needleminer populations in Denmark, Munster-Swendsen also collected data from several other spruce stands. Of these, three had reasonably long series of observations; stand D = 13 years, stand E = 11 years, and stand H = 9 years. We used these independent data sets to test the predictions of the complex simulation model and the two simple theoretical models. Predictions of larval densities in a particular year, N_t, were calculated from actual larval densities in the two previous years, N_{t-1} and N_{t-2}, in the one-species model, from larval density and number parasitized in the previous year, N_{t-1} and P_{t-1}, in the two-species model, and from the densities of needleminers and two parasitoid species and one hyperparasitoid in the simulation model. Because disease incidence was not measured in the test stands, and because weather could not be known a priori one year ahead, these variables were set at their average values in the simulation model.

A total of 27 predictions were made by each model (11 years in Stand D, 9 years in Stand E, and 7 years in Stand H). All models performed quite well at forecasting the qualitative dynamics of spruce needleminer populations in all three stands. As a measure of qualitative precision, we counted the number of times the models correctly forecast the direction of population change (increase, decrease or no change). The complex simulation model was correct 86% of the time and the two simple models 84% of the time. We used the absolute difference between predicted and observed densities to compare quantitative predictions of the models. The simple 2-species model performed best in 12 cases (out of 27), the simulation model was better in 8 cases, and the simple 1-species model was best in 7 cases. As would be expected, all models performed poorly when the observed data exhibited extreme fluctuations, often due to abnormal weather conditions.

5. Discussion

Our results generally support the contention that simple models built upon sound theoretical foundations are at least as good as complex, biologically explicit simulation models in forecasting the direction and magnitude of changes in biological populations. Why is this so? The first reason is that the simple models used in this exercise are not arbitrary empirical equations, but are based on the theory of population dynamics, a theory which has been derived by mathematical and logical reasoning (Lotka 1925, Berryman and Stenseth 1984, Royama 1992). The advantage of theoretical models over arbitrary equations is that they capture the fundamental processes underlying the data and thereby set the bounds of model behavior so that biologically ridiculous forecasts are discouraged. Thus, although purely empirical models may predict quite well within the domain of the data, extrapolation outside of this domain is often risky or nonsensical. The second reason is that complex simulation models usually contain many linked and interlinked functions, all of which have to be estimated independently. Any errors in estimation can be magnified and exacerbated as the signals pass through these multiple computations. On the contrary, the functions of simple theoretical models represent the aggregate of many real processes or, more probably, only those functions that dominate the behavior of the real system. If the latter is true, and there is some reason to believe it is (Berryman 1993), then the advantage of simple models may be that they capture the dominant structure in an unbiased way, because structure is determined by fitting to data, while the structure of complex simulation models is determined by the experiences (and biases) of the model builder. Finally, our results suggest that theoretical models fit to commonly available survey or census data can provide surprisingly accurate predictions of population changes one year or generation into the future, and thus can be useful tools in the hands of resource or pest managers. Although the simplest 1-species model was generally the least accurate at predicting changes in needleminer abundance, it may be the most useful to managers because the increase in precision obtained by adding a second predictor variable in the 2-species model, or three more predictors in the simulation model, hardly seems to justify the additional cost of measuring these variables.

References

Albrecht, K.-F. (1992), 'Problems of modelling and forecasting on the basis of pheno-menological investigations', *Ecol. Modell.* **63**, 45-69.

Berryman, A.A. (1978), 'Population cycles of the Douglas-fir tussock moth (Lepidoptera: Lymantriidae): the time-delay hypothesis', *Canad. Entomol.* **110**, 219-227.

Berryman, A.A. (1990), 'POPSYS Population analysis system: Series 2, Two-species analysis', *Ecological Systems Analysis*, Pullman, WA.

Berryman, A.A. (1991), 'Population theory: an essential ingredient in pest prediction, management, and policy making', *Amer. Entomol.* **37**, 138-142.

Berryman, A.A. (1992a), 'On choosing models for describing and analyzing ecological time series', *Ecology* **73**, 694-698.

Berryman, A. A. (1992b), 'The origins and evolution of predator-prey theory', *Ecology* **73**, 1530-1535.

Berryman, A.A. (1993), 'Food web connectance and feedback dominance, or does everything really depend on everything else?', Oikos, in press.

Berryman, A.A. and J.A. Millstein (1989), 'POPSYS Population analysis system: Series 1, One-species analysis', *Ecological Systems Analysis*, Pullman, WA.

Berryman, A.A. and N.C. Stenseth (1984) 'Behavioral catastrophes in biological systems', *Behav. Sci.* **29**, 127-137.

Cook, L.M. (1965), 'Oscillations in the simple logistic growth model', *Nature* **207**, 316.

Lotka, A.J. (1925), 'Elements of physical biology', Williams and Wilkins, Baltimore.

Ludwig, D. and C.J. Walters (1985), 'Are age-structured models appropriate for catch-effort data?' *Canad. J. Fish. Aquat. Sci.* **42**, 1066-1072.

Munster-Swendsen, M. (1985), 'A simulation study of primary-, clepto- and hyperparasitism in Epinotia tedella (Lepidoptera, Tortricidae)', *J. Anim. Ecol*, **54**, 683-695.

Munster-Swendsen, M. (1991), 'The effect of sublethal neogregarine infections in the spruce needleminer, Epinotia tedella (Lepidoptera: Tortricidae)', *Ecol. Entomol.* **16**, 211-219.

Nelder, J.A. (1961), 'The fitting of a generalization of the logistic curve', *Biometrics* **17**, 89-110.

Royama, T. (1992), 'Analytical population dyunamics', Chapman and Hall, London.

Thomas, W.R., M.J. Pomerantz and M.P. Gilpin (1980), 'Chaos, asymmetric growth and group selection for dynamical stability', *Ecology* **61**, 1312-1320.

INDIVIDUAL BASED POPULATION MODELLING

S.A.L.M. KOOIJMAN

Department of Theoretical Biology, Free University
de Boelelaan 1087, 1081 HV Amsterdam, the Netherlands
email: bas@vu.bio.nl

Summary

Realistic models for the dynamics of populations of animals or bacteria should minimally account for uptake and use of resources by individuals. In field situations, it is usually necessary to implement also more advanced behaviour, such as interactions between individuals and spatial and temporal inhomogeneities. The dynamics of many heterotrofic systems can be understood by focusing on energy fluxes only, because mass fluxes tend to be closely coupled to them. Realistic and relatively simple descriptions of energy uptake and usage by individuals appeared to be possible for this purpose. Surface area related uptake, volume related maintenance and storage dynamics are the main key elements. These non-specific descriptions distinguish three energy-defined life stages of an animal (embryo, juvenile and adult) and allow the derivation of body size scaling relations of parameter values. Consequent application of the first law of thermodynamics at both the individual and the population level proves to restrict oscillations considerably in comparison with for instance Lotka--Voterra-based population dynamics. The dynamics of populations of energy-structured individuals can to some extent be simplified to a description of the energy uptake and use by the population in terms of that by individuals. These new objects, populations, can be linked into food chains and food webs to explore potential dynamics of ecosystems. Realistic descriptions of a three-step microbial food chain have been obtained. Body size scaling relations can be used to reduce the number of parameters of the system. The specification of ecosystem dynamics then reduces to that of particle size distributions. In this way it proved to be possible to explain for instance, why food chains cannot have many links.

1. Introduction

Theories and models for the dynamics of unstructured populations usually have an attractive simplicity, but the implicit assumption that individuals are identical restricts their value in understanding the dynamics of real world populations. This has stimulated research in structured populations, where individuals differ in one or more characteristics, DeAngelis and Gros, 1992, Ebenman and Persson, 1988, Lomnicki, 1988, Metz and Diekmann, 1986). As soon as individuals are allowed to differ in feeding rates, survival probabilities and reproduction rates, as a first step to improve realism, several problems show up. The first problem is to obtain a realistic and simple description of the performance of individuals. The second one is to evaluate the consequences for population dynamics of such individuals, and the third one is to judge to what extent this complexity of model structure did indeed improve the realism of the model.

A realistic description of the performance of individuals in terms of input/output devices has of course a much wider bearing than to population dynamics alone. Done in a mechanistic way, it provides an interface between physiology and ecology and can give a firm basis to the comparison of species in an evolutionary context. Requirements for adequate descriptions are conflicting; realism and physiological studies ask for a lot of detail, *i.e.* many state variables and parameters, while ecological and evolutionary studies ask for simplicity. Energetics proved to be a fruitful approach to reduce the physical/chemical complexity of organisms in a way that still allows useful descriptions of feeding, reproduction and survival.

This contribution discusses our attempts at the Vrije Universiteit to build a theory in the physical tradition for the energetics of individuals, and our experiences with population models on the basis of these individuals. The theory has been successfully applied to micro-organisms Hanegraaf, 1993, Kooijman *et al.* 1991; Muller, 1993; Stouthamer and Kooijman, 1993), oligochaetes (Ratsak *et al.*, 1992; Ratsak, 1993), daphnids (Evers and Kooijman, 1989; Kooijman, 1986b), pond snails (Visser *et al.*, 1993; Zonneveld and Kooijman, 1989; Zonneveld *et al.*, 1993; Zonneveld, 1992), mussels (Haren and Kooijman, 1993), birds (Kooijman, 1992b; Zonneveld and Kooijman, 1993a), embryonic development of variety of organisms (Kooijman, 1986c; Zonneveld and Kooijman, 1993b), growth of a wide variety of organisms (Kooijman, 1988b) and body size scaling relationships (Kooijman, 1986a. The evaluation of implications of the theory for population dynamics (Kooijman 1986b; Kooijman, 1992a; Metz *et al.*, 1984; Kooijman *et al.*, 1989; Kooi and Kooijman, 1992; Kooi and Kooijman, 1993a; Kooi and Kooijman, 1993b) and ecotoxicology (Kooijman, 1985b; Kooijman, 1985a; Kooijman, 1988a; Kooijman, 1991; Kooijman and Haren, 1990; Kooijman and Metz, 1983; Kooijman *et al.*, 1987; Bedaux and Kooijman, 1993b; Bedaux and Kooijman, 1993a; Hallam *et al.*, 1989) are still in a developmental phase. The results and developments of the theory have been brought together and expanded in (Kooijman, 1993), including numerous tests for realism on the basis of experimental results at all levels of biological organisation, ranging from the molecular to the population level.

2. Individual budgets

Most work concentrated so far on chemoheterotrophs, which extract free energy from organic matter. These organisms show a close coupling between energy and mass fluxes. It can be shown that the specification of energy fluxes completely specifies mass fluxes at the same time, via the conservation law for mass. The general approach is illustrated in diagram 1, which stresses the dynamic character of energy budgets, that is how and why the budget changes during the life cycle.

The main energy fluxes through an individual are indicated in diagram 2 in a bit more detail. Although such diagrams seem to tell little, in combination with the list of assumptions, *i.e.* axioms, it completely specifies the mathematical structure of the energetics for individuals. It is quite a bit of work, however, to show exactly how, and even more work to evaluate the implications. The list of assumptions is in fact the end result of all the work that has been done on the Dynamic Energy Budget DEB theory by a varying group of people during the last 15 years; it represents the biological insight obtained so far. The core of the work has been to test the realism of the assumptions and to show how a wide variety of biological phenomena can be understood quantitatively in terms of combinations of the assumptions that are listed in table 1. The present contribution presents some basic points to give a general idea.

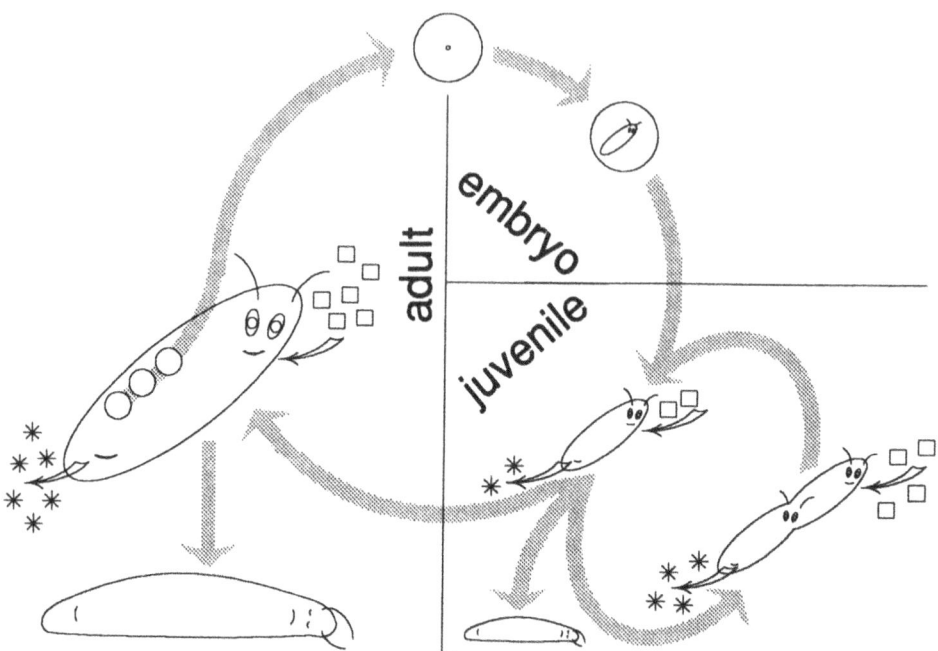

Figure 1. Dynamic Energy Budget theory aims to quantify the energetics of heterotrophs as it changes during life history. The key processes are feeding, digestion, storage, maintenance, growth, development, reproduction and aging. The theory amounts to a set of simple rules, summarized in table 1, and wealth of consequences for physiological organization and population dynamics. Although some of the far reaching consequences turn out to be rather complex, the theory is simple, with only one parameter per key process. Intra- and inter-specific body size scaling relationships are in the core of the theory and include dividing organisms, such as microbes, by conceiving them as juveniles.

2.1 Surface area/volume relationships

Since maintenance requirements are proportional to volume and uptake is proportional to surface area, surface area/volume relationships are of importance. The DEB theory has been set up for 3D-isomorphs with volume as the basic state variable. 3D-isomorphs are organisms that do not change in shape during growth. The change in shape during growth of some organisms is too large to be neglected. The parameter that stands for the surface area-specific uptake rate of such organisms has to be multiplied, therefore, with the so-called shape correction function; It stands for the ratio of the real surface area of the organism (as far as relevant for uptake processes) and the isomorphic one as a function of volume. This calls for an arbitrarily chosen reference volume at which the shape of the

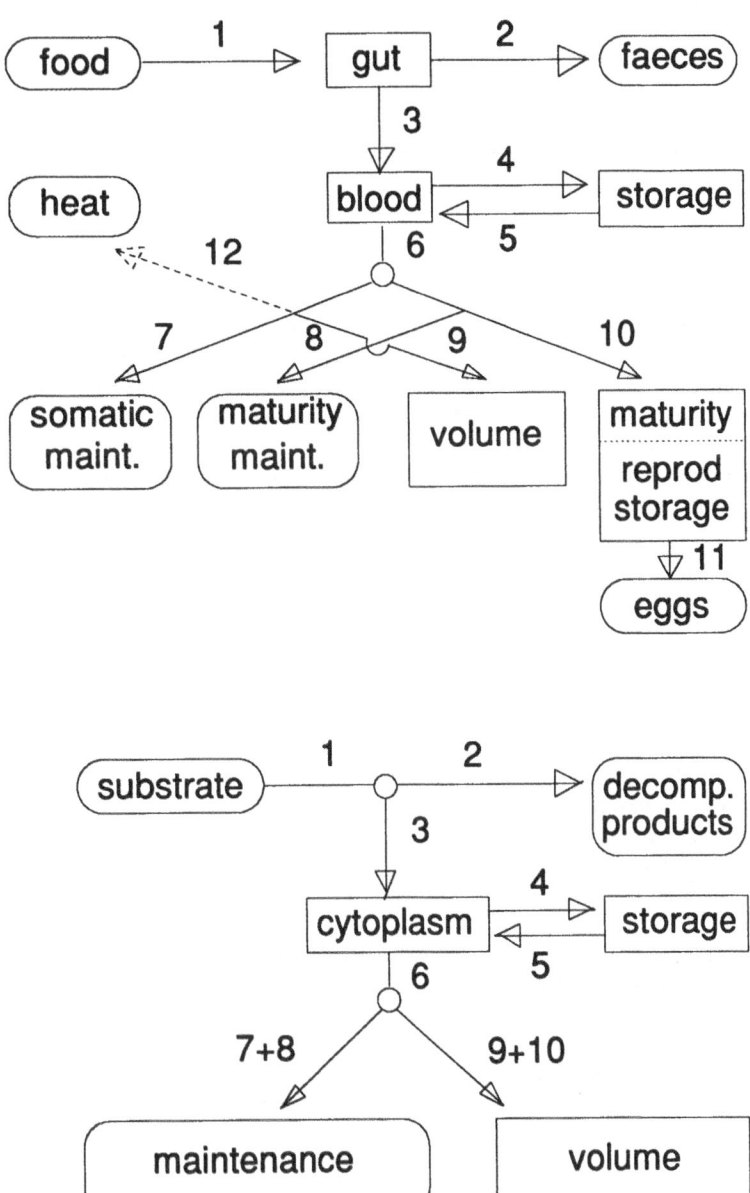

Figure 2.   Energy fluxes through a heterotroph: 1 ingestion (uptake), 2 defecation, 3 assimilation, 4 demobilization, 5 mobilization, 6 utilization, 7 maintenance, 8 maturation, maintenance 9 growth investment, 10 maturation, 11 reproduction, 12 heating (endotherms only). The rounded boxes indicate sources or sinks. Rates 3, 7, 8, 9 and 10 contribute to a bit to heating, this is not indicated to simplify the scheme.

Table 1. The assumptions that lead to the DEB model as formulated for multicellular animals and modified for unicellulars.

---

1 Body volume, stored energy density and accumulated damage are the state variables.

2 Three life stages can exist: embryos, which do not feed, juveniles, which do not reproduce, and adults. The transition between stages depends on the cumulated energy invested in maturation.

3 The feeding rate is proportional to surface area and depends hyperbolically on food density.

4 Food is converted to energy at a fixed efficiency and added to the reserves.

5 The dynamics of energy density in reserve is a first order process, while maximum density is independent of the volume of the individual and homeostasis is observed.

6 A fixed fraction of energy, utilized from the reserves, is spent on somatic maintenance plus growth, the rest on maturity maintenance plus maturation or reproduction.

7 Somatic and maturity maintenance are proportional to body volume, but maturity maintenance does not increase after a given cumulated investment in maturation.
   7a Heating costs for endotherms are proportional to surface area.

8 Costs for growth are proportional to volume increase.

9 The energy reserve density of the hatchling equals that of the mother at egg formation, the embryo beginning at an infinitesimally small size.
   9a Foetuses develop in the same way as embryos in eggs, but at a rate unrestricted by energy reserves.
   9b Unicellulars divide a fixed time after initiation of DNA duplication, which occurs upon exceeding a certain volume.

10 Under starvation conditions, individuals always give priority to somatic maintenance and follow one of two possible strategies:
   10a They do not change the reserve dynamics (so continue to reproduce).
   10b They cease energy investment in reproduction and maturity maintenance (thus changing reserves dynamics).
   10c Most unicellulars and some animals shrink during starvation, but do not gain energy from this.

11 Aging related damage accumulates in proportion to the concentration of damage inducing compounds, which accumulate in proportion to the volume-specific metabolic rate. For unicellulars damage is lethal, therefore it does not accumulate.

12 Apart from 'accidents', the hazard rate is proportional to accumulated damage, but death occurs if somatic maintenance costs can no longer be paid.

(abstract) 3D-isomorph is taken equal to the real shape. For 2D-isomorphs, *i.e.* organisms that do not grow along one spatial axis and that grow isomorphically along the other two axes, the shape correction function works out to be $(V\,V_d)^{-1/6}$, where V denotes volume and V_d the reference volume. For 1D-isomorphs, *i.e.* organisms that only grow along one spatial axis, the correction function is $(V/\,V_d)^{1/3}$, and for 0D-isomorphs, *i.e.* (abstract) organisms that do not grow in surface area, but only in volume, the correction function is $(V/\,V_d)^{-2/3}$. A geometrical representation of a 0D-isomorph is something as a sheet of paper, where we place a second sheet on top of it. The volume has increased by a factor 2, but the surface area did not change. Biofilms on flat hard surfaces approach 0D-isomorphs, because the surface area across which substrate is taken up does not change as the film grows in thickness. Rods, *i.e.* bacteria that resemble a short cylinder that grows in length only, represent a static mixture of a 0D- and a 1D-isomorph. The 0D-component relates to the caps, the 1D-component to the growing cylinder. The mixture is static because the weight coefficients do not depend on volume; they depend only on the morphology of the caps and the aspect ratio of the cell. Colonies of bacteria that grow on agar plates represent dynamic mixtures of 0D- and 1D-isomorphs. The 0D-component is the center of the colony that does not change its surface area and only grows in height, the 1D-component is the periphery where surface area is proportional to volume. During growth, the colony transfers 1D-material to 0D-material and the 1D-material decreases relative to the 0D-material.

Most changing shapes can be conceived as mixtures of isomorphs of various dimensions, but 1D-isomorphs are of special interest. This is because surface area is for them proportional to volume, thus uptake to maintenance, which implies that they grow exponentially (at a rate that might be a function of time). This makes that the partition of biomass over individuals is irrelevant for population growth, one huge individual will grow in the same way as a population of many small ones with the same total volume. The structured population reduces to the unstructured case in its dynamics. This reduction to unstructured populations does not apply in detail to mixtures between 0D- and 1D-isomorphs, such as rods, but they can made arbitrary close to 1D-isomorphs by decreasing the 0D-component, so by decreasing the aspect ratio of the cells. This suggests a perturbation technique for the analysis of structured populations in terms of unstructured ones, but such a technique still has to be worked out.

2.2 Reserves and mass fluxes

Reserves are essential as a state variable for several reasons, one of which being the combination of a continuous maintenance requirement and a discrete arrival process of food particles. Since the level of reserves depend on the feeding level, new interpretations of measurements, such as wet and dry weights, are necessary in terms of contributions of the structural body mass and reserves. The idea of homeostasis, *i.e.* constant chemical composition, is essential to relate the abstract state variables structural body volume and energy reserves to measurements. Since the chemical composition of the structural component and the reserves can differ, particular shifts in body composition are possible. The definition of reserves is in their dynamic behaviour: those compounds that increase in density with increasing feeding level. The two-way classification of compounds into structural body mass and reserves can in principle be extended to more classes on the basis of the residence times of the compounds in the body, but this simple two-way classification is already rich enough to cover a wide variety of experimental results. The change in body composition as a function of feeding level can be used to decompose the total body into a structural and a reserve component in a unique way. Reserve compounds differ from structural body mass in the way compounds are temporary. Reserves are subjected to continuous supply and use, while structural body mass is subjected to maintenance, which includes a continuous turnover of proteins.

Mass fluxes, such as respiration, *i.e.* the use of oxygen or the production of carbon dioxide, must in steady state situations be proportional to energy fluxes. The conservation law for mass defines all proportionality constants. For multicellulars that do not feed or digest, respiration rate is about proportional to the use of energy from the reserves. It can be shown to be a weighted sum of a surface area and a volume, with simple relations between the weight coefficients and energy fluxes to maintenance (plus heating in case of endotherms), growth, development and reproduction. This explains why respiration is found to be approximately proportional to weight$^{0.75}$, which is known as the 'Kleibers law', and why intra-species comparisons work out a bit different from inter-species comparisons.

Reserves hold the key to many other intra- and inter-species body size scaling relationships pertinent to energetics and life history. A lot of literature describes such relationships empirically, but the DEB theory allows a mechanistic derivation. The core of the argument for inter-species scaling relationships is that ultimate size can be expressed as a simple function of energy parameters and that all parameters can be classified as either intensive or extensive. Intensive parameters relate to densities of compounds, and are therefore independent of body size. Extensive parameters relate to physical design and are proportional to (volumetric) length. The extensive parameters are the saturation constant for food uptake, maximum surface area-specific ingestion and assimilation rate, the maximum volume-specific reserve capacity and the volumetric lengths at birth and at puberty, *i.e.* the transition from embryo to juvenile and from juvenile to adult respectively. The idea is to write a quantity of interest, such as the maximum reproduction rate, as a function of parameters, multiply the extensive parameters by volumetric length and the body size scaling relation results.

The effect of the introduction of reserves in population dynamics is an increased tendency to oscillate. This is because predators can depress declining prey populations to extra low levels because they can survive and reproduce on the expense of reserves for a while. Predators also follow rapidly growing prey populations with some delay, because they have to fill their reserves first. This allows the prey population to reach extra high densities. Energy and mass balances are usually not observed explicitly in most population dynamical models in the literature. Introduction of such balances restricts oscillations.

2.3 Aging

The dynamics of the aging process can well be described as a result of free radicals, which couples the process to respiration and so to reserve dynamics and energetics. Although this mechanism has been proposed some time ago Harman, 1962; Harman, 1981; Tice and Setlow, 1985; Finch, 1990), the quantitative aspect is new. The hazard rate has just one pure aging parameter, the aging acceleration, one energy parameter, the maintenance rate coefficient and an integral of body volume over age. At the cellular level cells are either affected by aging, which prohibits further growth and leads to death, or they are unaffected. Multicellular organisms can accumulate the effects of aging and can dilute them by growth. This explains why rapidly growing individuals, such as embryos, are hardly affected by aging. It proved to be possible to estimate the maintenance rate coefficient from survival data of cohorts of individuals, where other causes of death have been excluded. (The maintenance rate coefficient is a compound energy parameter standing for the ratio of the maintenance costs and the costs for growth.) Difference in survival of species that show sex dimorphism, such as daphnids where the male is substantially smaller than the female, can be traced back to differences in energetics only. The effect of feeding levels on survival and body size scaling relations for life span is well captured with the assumptions 11 and 12.

Dynamic Energy Budgets only depend on structural body volume and energy reserves as state variables in the present formulation of the theory, which leads to quite realistic descriptions for many aspects of energetics. More detail can be obtained be letting energy parameters depend on accumulated damage to model the senile state. Such a coupling would introduce additional parameters, of course, and has for this reason not been worked out yet.

The population dynamical significance of aging is in its consequence that no individual has eternal life. Limited life spans are essential to obtain characteristics of populations that are independent of the inoculum. Most biologists will think probably that this is trivial, from the point of view of an applied mathematician, this property is essential. In most field situations, aging has a minor role in the death rate.

3. Population budgets and entropy

To reduce the problem to model population dynamics, as a first step it is assumed that individuals of a population only interact by feeding on the same resource in a chemostat environment. These assumptions might not be realistic for field situations, but if we fail to understand population dynamics in such simple situations, we will never understand more realistic dynamics. Populations can be conceived as super individuals that feed and produce as indicated in figure 3. The feeding and production rates of the population depend on the distribution of the individuals the state space spanned up by volume, reserves and cumulated damage.

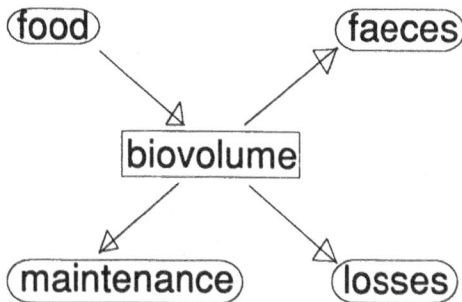

Figure 3. The population, quantified as the sum of the volumes of the individuals, converts food into faeces, while extracting energy. Part of this energy becomes lost in maintenance processes and part of it is deposited in losses, i.e. the cumulated harvest. The harvesting effort determines the allocation rules and sets the population size and so its impact on resources.

Recently we succeeded to work out the dynamic mass and energy balances for populations of 1D-isomorphs Hanegraaf, 1993), which means that the kinetics of energy uptake and use has been placed into a thermodynamic framework. This cross linking is of importance, because the thermodynamical aspects of growth has been approached so far via the route of non-equilibrium thermodynamics. Such approaches combine poorly with mechanistic arguments. Battley, who has done a great deal of work on the thermodynamical aspects of microbial growth, even states that the concept of 'maintenance' might follow from kinetic arguments, but that it is incompatible with thermodynamic arguments, (Battley, 1987). The thermodynamic framework allows a separation of overhead costs for growth, for instance, from energy that has been deposited in living biomass. This invites for the definition of efficiency measures to compare species in an evolutionary context. Many thermodynamic efficiency measures have already been defined, but none make the link with kinetic theories, based on mechanisms.

The DEB theory also provides a method to determine the entropy of living systems. Existing ideas on the value of entropy differ widely; the entropy of bacteria is even higher than that of its energy substrate (succinate), according to Battley (Battley, 1993), while Ling (Ling, 1984) suggests that cells must have extremely low entropy values, because of their complex structure. The method to determine the entropy of living systems is on the basis of Gibbs' relationship, via enthalpy (which can be measured directly) and free energy. The free energy of biomass is derived via dissipating heat and knowledge of the free energies of all simple compounds. The DEB theory ties dissipating heat to the overhead costs in all energy fluxes (uptake, maintenance and growth in case of micro-organisms), which allows to distinguish reserves from structural biomass by making use of shifts in these fluxes as a function of the population growth rate. We expect that the entropy of reserves is high and of structural biomass low, but numerical estimates are not available yet.

3.1 Steady states

If the chemostat is in steady state, the frequency distribution of the individuals over the state space can be obtained analytically, and the behaviour of statistics can be studied as function of the control variables of the chemostat: food density in the influx and the throughput rate. An important statistic is the conversion efficiency of food (substrate) into biomass: the yield coefficient. This dimensionless statistic can be expressed on the basis of structural biovolume, see figure 4, C-moles or free energy, each with its own merits. Maintenance and reserves causes that this conversion coefficient is not constant. The yield coefficient is taken to be constant in the standard models of Lotka--Volterra and Monod. Only a few elements of individual budgets seem to determine yield coefficients; the change in shape of dividing organisms, for instance, is not very important.

Yield coefficients can be of direct practical interest in relation with sewage treatment for instance, which is aiming at the combination of a high degree of mineralization of organic substrate and a low yield coefficient, or in relation with bioproduction processes, which is aiming at high yield coefficients. The significance of yield coefficients for fundamental science is in the reduction of complexity at the population level. This is essential for further steps towards the dynamics of food webs. An intriguing aspect of yield coefficients is that is the value for populations is mainly determined by processes feeding and harvesting, while for individuals, physiological processes are in control. This is obvious from a thought experiment where food is supplied to a closed population where no individual dies. Such a population will grow to a carrying capacity where all incoming food is used for maintenance, and all growth and reproduction is ceased. So food is supplied, but no biomass comes out, which gives a yield coefficient of zero. For increasing death and/or harvesting rate, the yield coefficient will increase till some maximum value, since the population actually converts food to dead or harvested individuals. Physiological processes define the maximum yield coefficient and the maximum harvesting rate, but ecological processes define the current value.

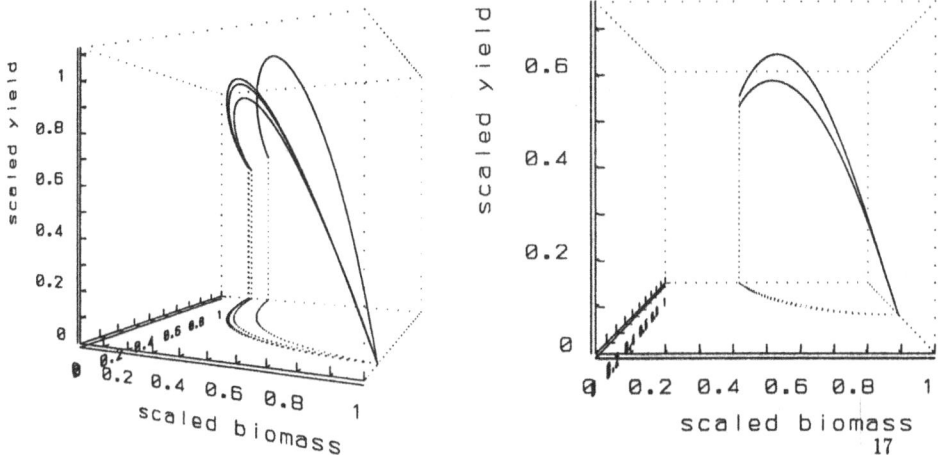

Figure 4. Left figure: The scaled yield factor as a function of the scaled total biovolume of the population and the scaled functional response for populations subjected to a constant food input and different harvesting efforts at steady state. The four curves relate from low to high maxima to dividing filaments, rods and isomorphs, and reproducing isomorphs (right curve). Right figure: the yield factor for reproducing isomorphs subjected to random and age-specific harvesting.

We observed that individuals in computer simulation studies tend to synchronize, *i.e.* after a few generations, all individuals have the same age, size, energy reserves etc. It seems that a combination of two elements of the behaviour of individuals are responsible: the growth rate decreases with body volume and propagation starts at a fixed body volume. The synchronization pulse becomes effective as soon as the population approaches its carrying capacity and the reproduction rate becomes low. If reproduction occurs in clutches, as in daphnids, the individuals synchronize rapidly. Budding yeasts, which produce one bud at a time, the synchronization is somewhat slower. 1D-isomorphs do not synchronize because the growth rate does not decrease with body volume, while it can be expected that 2D- and 0D-isomorphs synchronize better than 3D-isomorphs. This has still to be worked out further.

3.2 Transient states

When substrate or food densities change slowly or very rapidly in comparison with changes in the frequency distribution of individuals over the state space, the dynamics of structured populations can still be modelled by a set of coupled ordinary differential equations for food density and total biovolume with a yield coefficient that varies in time. For intermediate rates of change of food densities, the dynamics of structured populations can become quite complex. Two routes exist to evaluate the dynamics by computer simulations.

The first route is via a partial differential equations for the change in frequency of individuals over the state space, coupled to ordinary differential equations for food density and propagation (Metz and Diekman, 1986). De Roos has designed an efficient algorithm to integrate the system (Roos, 1988), the computing time being roughly proportional to the number of cells one has to define to discretize the state space for individuals. A basic problem of this approach is in the approximation of numbers of individuals by continuous variables, which becomes a problem as soon as the reproduction rate per individual decreases to very low levels. The behaviour of a population where a large number of individuals together produce few offspring is quite different from a population where each individual by itself fails to acquire enough resources to produce a single young.

The second route is to evaluate the changes of each individual. The flexibility of this approach comes with an computing effort that is roughly proportional with the number of individuals. For a low number of individuals stochastic effects via survival becomes important. Small differences in parameter settings for each individual proved to have substantial effects on population trajectories. Such a step towards realism is important because the mentioned synchronization between the individuals comes with single generation oscillations that are too strong to be realistic. Differences between individuals can break the synchronization to some extend.

For 1D-isomorphs, where the structured populations reduce to unstructured ones, analytical results can be obtained. It can be shown that fed-batch cultures of 1D-isomorphs grow logistically if the saturation constant is high and the reserve capacity is low or vice versa. This suggests that the way back, from observations of population behaviour to propositions about individual energetics is difficult or even impossible. The main problem in the study of structured population dynamics is in the reduction of relevant parameters.

4. Food chains

We started to study food chains of {\sc deb}-structured populations. Bob Kooi has been able to show that chains of length four can have quite complex behaviour (Kooi and Kooijman, 1993a), but no chaotic behaviour showed up so far. Chaotic behaviour has been found for food chains of length 3 with logistic growth of the lowest level, (Hastings and Powell, 1991) and with an oscillating supply of an inert lowest level, (Kot et al., 1992). The most important reason is in the dynamics of the lowest level. In all models that respect mass and energy balances, the lowest level is inert, i.e. the 'individuals' do not propagate. This is because the propagation rate must depend on food uptake because of the mass balance equation but by definition, no food exists for the lowest level. Since almost all models for food chains have propagation in the lowest level, this means that these models do not observe mass balances explicitly. Chaotic behaviour can easily arise in food chains for length four or longer. This leads to another important observation, namely that

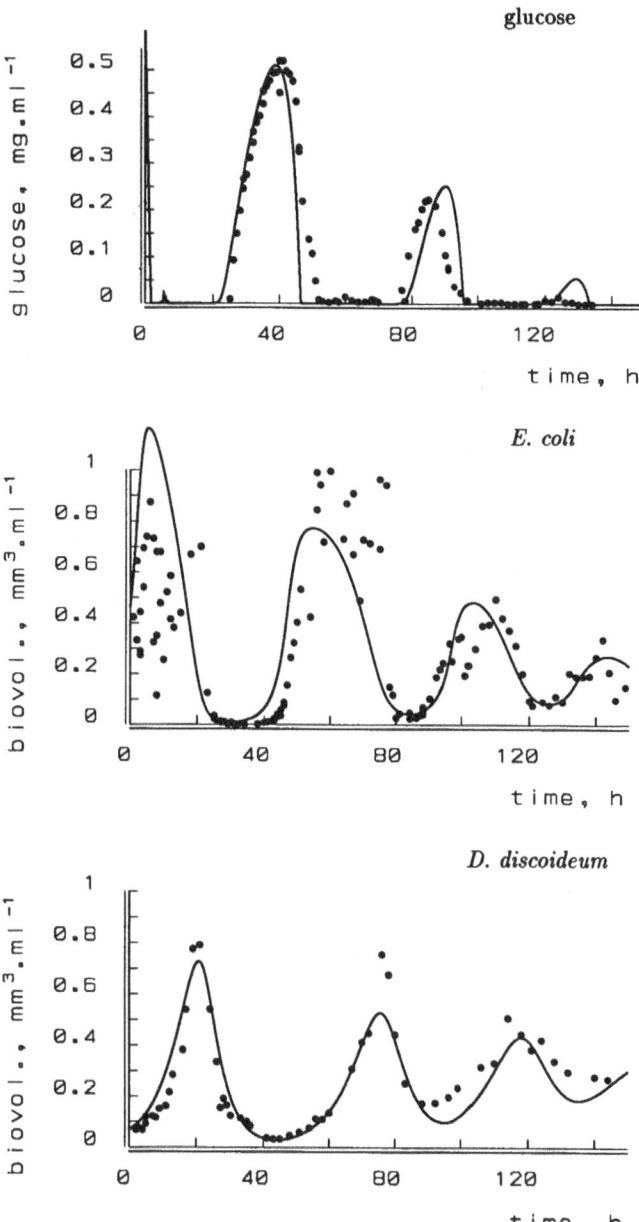

Figure 5. A chemostat with a three-step food chain of glucose, the bacterium *Escherichia coli* and the cellular slime mold *Dictyostelium discoideum* at 25°C, throughput rate 0.064 h^{-1} and a glucose concentration of 1 mg,ml^{-1} in the feed. Data from Dent *et al.* Dent *et al.* 1976).

it proved to be very difficult to obtain parameter combinations where such long chains can exist in homogeneous environments. We used several strategies to get rid of the large amount of parameters that is involved. One of them is to tie the parameters according to body size scaling relations and work with chosen maximum body sizes for each step in the chain. We are still working on the mathematics of how the parameters of individuals determine the maximum chain length, if it really exists theoretically, but this is far from simple, even for 1D-isomorphs. Simulation studies indicate that chains of an even number of steps behave quite different from chains of an odd number of steps. This is because an abundant predator depresses its prey, so that the prey of the prey is abundant again. Such a geometry complicates the analyses.

The first preliminary results with practical applications suggest that changes in shape during growth hardly contribute to the dynamics of populations, and so of food chains, if the organisms divide in two parts; the range of body volumes of a factor 2 is too small to note effects of changes in shape. This means that size-structure in micro-organisms is not very important and their population dynamics can be understood within the context of unstructured population dynamics. Figure 5 gives a practical example for the glucose-bacterium-myxamoeba food chain in a chemostat. The significance of this particular example is that it has been studied extensively (Bazin *et al.*, 1974; Bazin and Saunders, 1978; Bazin and Saunders, 1979; Dent *et al.*, 1976; Saunders, 1980), with the main conclusion that Lotka-Volterra type of models cannot not be fitted to this data-set and that application of catastrophy theory strongly suggests an interaction between the myxamoeb-es, such that the feeding rate per myxamoeba is inversely proportional to their density. Bob Kooi has been able to fit the DEB model (for 1D-isomorphs) to this data-set with remarkable success. Even the cyclic changes in mean cell sizes of the bacteria and the myxamoebes fit well, due to the mechanism described in assumption 9b of table 1. The details are given in \cite{KoKo93a}. Our conclusion is that both maintenance and reserves are necessary to obtain a good fit, and that there is no need to assume interactions or other species-specific behaviour.

5. Conclusion

In conclusion I think, that we make good progress in improving the physiological basis of population dynamics, and that the near future will bring more valuable results for populations in simple environments, such as in reactors. We are still far from useful theory for populations in more complex environments, or where more subtle interactions occur. The meaningful reduction of the number of parameters and the generality of such theory are the main problems.

References

Battley, E.H. (1987), 'Energetics of microbiol growth'. J. Wiley & Sons, Inc.

Battley, E.H. (1993), 'Calculation of entropy change accompanying growth of Escherichia coli K-12 on succinic acid'. *Biotechnol. Bioeng*, **41**, 422-428.

Bazin, M.J., V. Rapa, and P.T. Saunders (1974), 'The integration of theory and experiment in the study of predator-prey dynamics'. In M. Usher and M. Williamson (Eds.), *Ecological stability*. London: Chapman & Hall.

Bazin, M.J. and P.T. Saunders (1978), 'Determination of critical variables in a microbial predator-prey system by catastrophe theory'. *Nature (Lond.)*, **275**, 52-54.

Bazin, M.J. and P.T. Saunders (1979), 'An application of catastrophe theory to study a switch in Dictyostelium discoideum'. In R. Thomas)Ed.), *Kinetic Logic- a Boolean Approach to the Analysis of Complex Regulatory Systems*. Berlin: Springer-Verlag.

Bedaux, J.J.M. and S.A.L.M. Kooijman (1993a), 'Hazard-based analysis of bioassays'. *J. Environ. Stat.* to appear.

Bedaux, J.J.M. and S.A.L.M. Kooijman (1993b), 'Stochasticity in deterministic models'. In C. Rao, G. Patil, and N. Ross (Eds.), *Handbook of Statistics 12*: *Environmental Statistics*, volume 12. North Holland.

DeAngaelis, D.L. and L.J. Gross, Eds. (1992), 'Individual-based models and approaches in ecology'. Chapman & Hall.

Dent, V.E., M.J. Bazin, and P.T. Saunders (1976), 'Behaviour of Dictyostelium discoideum amoebae and Escherichia coli grown together in chemostat culture'. *Arch. Microbiol.*, **109**, 187-194.

Ebenman, B. and L. Persson (1988), 'Size-structured populations'. *Ecology and evolution*. Springer-Verlag.

Evers, E.G. and S.A.L.M. Kooijman (1989), 'Feeding and oxygen consumption in Daphnia magna; A study in energy budgets'. *Neth. J. Zool.*, **39**, 56-78.

Finch, C.E. (1990), 'Longevity, senescence, and the genome'. Unviersity of Chicago Press.

Hallam, T.G., R.R. Lassiter and S.A.L.M. Kooijman (1989), 'Effects of toxicants on aquatic populations'. In S. Leven, T. Hallam, and L. Gross (Eds.), *Mathematical Ecology* 352-382. Springer-Verlag.

Hanegraaf, P.P.F. (1993), 'Coupling of mass and energy yields in micro-organisms. Application of the Dynamic Energy Budget model and an experimental approach'. PhD. thesis, Vrije Universiteit, Amsterdam. in prep.

Haren, R.J.F. van and S.A.L.M. Kooijman (1993), 'Application of a dynamic energy budget model to Mytilus edulis'. *Neth. J. Sea Res.*, **32**. to appear.

Harman, D. (1962), 'Role of free redicals in mutation, cancer, aging and maintenance of life'. *Radiat. Res.*, **16**, 752-763.

Harman, D. (1981), 'The aging process'. *Proc. Nat. Acad. Sci. U.S.A.* **78**, 7124-7128.

Hastings, A. and T. Powel (1991), 'Chaos in a three-species food chain'. *Ecology*. **72**, 896-903.

Kooi, B.W. and S.A.L.M. Kooijman (1992), 'Existence and stability of microbial prey-predator systems'. submitted.

Kooi, B.W. and S.A.L.M. Kooijman (1993a), 'Many limiting behaviours in microbial food-chains'. In O. Arino, M. Kimmel, and D. Axelrod (Eds.), *Proceedings of the 3rd conference on mathematical population dynamics.*, Biological Systems: Wuerz. to appear.

Kooi, B.W. and S.A.L.M. Kooijman (1993b), 'A quantitative explanation for the singular behaviour of myxamoebae'. submitted.

Kooijman, S.A.L.M. (1985a), 'Toxiciteit op populatie niveau'. *Vakbl. Biol.*, **23**, 163-185.

Kooijman, S.A.L.M. (1985b), 'Toxicity at populations level'. In J. Cairns (Ed.) *Multispecies toxicity testing* 143-164. Pergamon Press.

Kooijman, S.A.L.M. (1986a), 'Energy budgets can explain body size relations'. *J. Theor. Biol.*, **121**, 269-282.

Kooijman, S.A.L.M. (1986b), 'Population dynamcis on the basis of budgets'. In J. Metz and O. Diekmann (Eds.), *The dynamics of physiologically structured populations*, Springer Lecture Notes in Biomathematics, 266-297. Springer-Verlag.

Kooijman, S.A.L.M. (1986c), 'What the hen can tell about her egg; egg development on the basis of budgets'. *Bull. Math. Biol.*, **23** 163-185.

Kooijman, S.A.L.M. (1988a), 'Strategies in ecotoxicological research'.   *Environ. Asp. Appl. Biol.*, **17**(1), 11-17.

Kooijman, S.A.L.M. (1988b), 'The von Bertalanffy growth rate as a function of physiological parameters: A comparative analysis'. In T. Hallam, L. Gross, and S. Levin (Eds.), *Mathematical ecology* 3-45. Singapore: World Scientific.

Kooijman, S.A.L.M. (1991), 'Effects of feeding conditions on toxicity for the purpose of extrapolation'. *Comp. Biochem. Physiol.*, **100**C(1/2), 305-310.

Kooijman, S.A.L.M. (1992a), 'Biomass conversion at population level'. In D. DeAngelis and L. Gross (Eds.), *Individual based models; an approach to populations and communities* 338-358. Chapman & Hall.

Kooijman, S.A.L.M. (1992b), 'Effects of temperature on birds'. In *Birds Numbers 1992; 12th internat. conf. of IBCC and EOAC*. Noordwijkerhout. to appear.

Kooijman, S.A.L.M. (1993), 'Dynamic Energy Budgets in Biological Systems; Theory and Applications in Ecotoxicology'. Cambridge Unviersity Press.

Kooijman, S.A.L.M. and R.J.F. van Haren (1990), 'Animal energy budgets affect the kinetics of xenobiotics'. *Chemosphere*, **21**, 681-693.

Kooijman, S.A.L.M., N. van der Hoeven, and D.C. van der Werf (1989), 'Population consequences of a physiological model for individuals'. *Funct. Ecol.*, **3**, 325-336.

Kooijman, S.A.L.M. and J.A.J. Metz (1983), 'On the dynamics of chemically stressed populations; the deduction of population consequences from effects on individuals'. *Ecotox. Environ. Saf.*, **8**, 254-274.

Kooijman, S.A.L.M., E.B. Muller, and A.H. Stouthamer (1991), Microbial dynamics on the basis of individual budgets. *Antonie van Leeuwenhoek*, **60**, 159-174.

Kot, M., G.S. Sayler, and T.W. Schultz (1992), 'Complex dynamics in a model microbial system'. *Bull. Math. Biol.*, **54**, 619-648.

Ling, G.N. (1984), 'In search of the physical basis of life'. Plenum Press.

Lomnicki, A. (1988), 'Population ecology of individuals'. Princeton University Press.

Metz, J.A.J. and O. Diekmann (1986), 'The dynamics of physiologically structured populations', **68**, of *Lecture Notes in Biomathematics*. Springer-Verlag.

Metz, J.A.J., O. Diekmann, S.AL.M. Kooijman, and H.J.A.M. Heijmans (1984), 'Continuum population dynamics, with applications to Daphnia magna'. *Nieuw Arch. Wisk.*, **4**, 82-109.

Muller, E.B. (1993), 'Minimisation of sludge production at the treatment of domestic waste water by membrane retention of activated sludge'. PhD thesis, Vrije Universiteit, Amsterdam. in prep.

Ratsak, C.H. (1993), 'Reduction of activated sludge by protozoa and matazoa'. PhD thesis, Vrije Universiteit, Amsterdam. in prep.

Ratsak, C., S.A.L.M. Kooijman, and B.W. Kooi (1992), 'Modelling the growth of an oligochaete on activated sludge'. *Water Res.*, **27**, 739-747.

Roos, A. de (1988), 'Numerical methods for structured population models; The escalator boxcar train'. *Num. Meth. Part. Diff. Eq.*, **4**, 173-195.

Saunders, P.T. (1980), 'An introduction to catastrophe theory'. Cambridge University Press.

Stouthamer, A.H. and S.A.L.M. Kooijman (1993), 'Why it pays for bacteria to delete disused DNA and to maintain megaplasmids'. *Anthonie van Leeuwenhoek*, **32**, 39-43.

Tice, R.R. and R.B. Setlow (1985), 'DNA repair and replication in aging organisms and cells'. In C. Finch and E. Schneider (Eds.), *Handbook of the biology of aging*. 173-224. New York: Van Nostrand.

Visser, J.A.G.M. de, A. ter Maat, and C. Zonneveld (1993), 'Energy budgets and reproductive allocation in the simultaneous hermaphrodite Lymnaea stagnalis (L.): a trade-off between male and female function'. subm.

Zonneveld, C. (1992), 'Animal energy budgets: a dynamic approach'. PhD thesis, Vrije Universiteit, Amsterdam.

Zonneveld, C. and S.A.L.M. Kooijman (1989), 'The application of a dynamic energy budget model to Lymnaea stagnalis', *Funct. Ecol.*, **3**, 269-278.

Zonneveld, C. and S.A.L.M. Kooijman (1993a), 'Body temperature affects the shape of avian growth curves'. submitted.

Zonneveld, C. and S.A.L.M. Kooijman (1993b), 'Comparative kinetics of embryo development'. *Bull. Math. Biol.*, **55**, 609-635.

Zonneveld, C., A. ter Maat, and J.A.G.M. de Visser (1993), 'Food intake, growth and reproduction as affected by day length and food availability in the pond snail, Lymnaea stagnalis'. submitted.

ECOLOGICAL SYSTEMS ARE NOT DYNAMIC SYSTEMS: SOME CONSEQUENCES OF INDIVIDUAL VARIABILITY

VOLKER GRIMM[1] and JANUSZ UCHMAŃSKI[2]

[1] Centre for Environmental Research Leipzig-Halle (UFZ)
Section Ecosystem Analysis (ÖSA)
Permoserstr.15, O-7050 Leipzig, F.R.G.

[2] Institute of Ecology
Polish Academy of Sciences
05092 Łomianki, Poland

Summary

Ecological systems are not dynamic systems (they are not "state variable models"), because they consist of individuals, which are different. An individual-based model of single population of competing individuals is presented to explain this. The model illustrates the meaning of "density dependence" on the level of individuals. Two time scales on which density dependence is operating can be distinguished: "between generation" and "within-generation". For the latter one, individual variability is an important factor.

Keywords Population dynamics, individual-based model, competition, individual variability

1. Introduction

The aim of this paper is to formulate a thesis which may be of fundamental importance to general methods of mathematical modelling of ecological systems. It can be phrased in the following way: ecological systems are not dynamic systems. We discuss and illustrate this problem by means of different descriptions of single populations. Generally, in contemporary theoretical ecology, there are two ways of describing the dynamics of single populations. One is the classical approach applied, for instance, to the logistic equation. The other approach is used for the construction of the so-called individual-based models (DeAngelis and Gross 1992).

2. Classical models

Classical theoretical ecology assumes that ecological systems are dynamic systems with population size N as a state variable (referred to by DeAngelis /1992/ as "state variable models"). In biological language it means that individuals are identical or it is sufficient to describe population dynamics by means of average properties of its individuals, properties of individuals are constant in time and individuals are most often described as identical molecules. The basic example of the model produced under the above assumptions are the logistic equations in their differential form.

Owing to this abstraction it became possible to investigate the meaning of such concepts as "regulation". Classical models describe ecological systems as systems which have equilibrium states. The ecological system as described by classical theoretical ecology remains in the equilibrium or aspires towards it.

In order to reconcile theory with reality, many different aspects of real populations have been taken into consideration in population models, e.g. time delays, spatial extent of population, heterogeneity of the environment, demographic stochasticity, etc. They lead to a variety of dynamic behaviours of model populations: from an asymptotic increase of density and converging oscillations, through oscillations, deterministic chaos and stochastic fluctuation, and as far as extinction of the population.

But one aspect has thus far received only little attention from theorists: individual variability in even-aged populations. Although considerable progress in the modelling of age- and size-structured populations was made during the last decade (Metz and Diekmann, 1986), the source of individual variability in most of these models is difference in age. Mathematical ecology continues to be a consequence of the averaging of individual variability.

3. Individual-based models

In individual-based models the description of the dynamics of the ecological system is based on the description of individuals and is a by-product of the latter. Individuals are elements from which the population is built, but they are not identical (even in the same age) and each of them is characterized by a number of properties, including consumption, assimilation, respiration, reproduction, growth, weight, etc. Which of these properties are important in each particular case depends on the problem to be solved. There are also resources in individual-based models. Most often they are modelled explicitly.

Building individual-based models of ecological systems has some general background. Individuals are only single biological entities which can be investigated in experiments, while population is not more than a group of individuals considered together with interactions between them. However, there is no point in constructing individual-based models if they produce the same results and lead to the same concepts that the classical models based on average properties of individuals yield. There is no abstract general individual and therefore there can be no generally valid individual-based model. But some suggestions can be derived from examples of individual-based models.

4. An example of individual-based model

4.1 Description of the model

Individuals are characterized by their weight w. The growth of single isolated individual can be described by the balance-equation (Zaika 1975, 1985, Zaika and Makarova 1971, Sibly and Calow 1986, Reiss 1989):

$$\frac{dw}{dt} = a_1 w^{b_1} - a_2 w^{b_2} , \tag{1}$$

where a_1, a_2, b_1, b_2 are constant. The first power term describes the gain in body mass due to assimilation of resources, while the second one describes loss due to respiration. The solution of eq. (1) is a sigmoid growth curve characterized by a final weight. From the analysis of energetic budgets of isolated individuals, which grew under different resource conditions V, it is known (Sushchenya 1975, Sushchenya, Khmeleva 1967) that a_2, b_1 and b_2 are constant, but only the a_1 coefficient varies with V:

$$a_1 = a_1(V) \ . \tag{2}$$

A function introduced by Ivlev (1961) can be used here:

$$a_1 = a_{1,max}(1 - e^{-sV}) \ . \tag{3}$$

The parameter s determines, how fast assimilation reaches its maximum value when resource level V increases.

For N competing individuals we have to specify the way the individuals interact with each other. We restrict our analysis to pure exploitation competition. This assumption excludes most terrestrial animals where interferences among individuals play an important role. Competition in our model is global whereas interference competition mostly is local. Hence, the following set of equations portrays a hypothetical population of N aquatic organisms: individuals grow according to eq. (1) and interact with each other by exploitation of common resources:

$$\frac{dw_1}{dt} = a_{1,1}w_1^{b_1} - a_2w_1^{b_2} \ , \tag{4a}$$

$$\dots$$

$$\frac{dw_N}{dt} = a_{1,N}w_N^{b_1} - a_2w_N^{b_2} \ , \tag{4b}$$

$$\frac{dV}{dt} = g - u \sum_{i=1}^{N} a_{1,i}w_i^{b_1} \ . \tag{4c}$$

Resources are replenished by linear inflow g and are reduced by the consumption of all individuals. Parameter u is the inverse of the coefficient of assimilation efficiency.

Individual variability is incorporated into the model via the coefficient $a_{1,i}$. The following assumptions are basis for calculation of $a_{1,i}$: (1) The initial weight w_0 of individuals is distributed symmetrically; (2) At optimum resource conditions all individuals assimilate with nearly the same rate, i.e. a_1 is the same for all individuals. The resulting distribution of final weights is also symmetric, its variability is small; (3) At deteriorating resource conditions the assimilation rate of individual with an initial advantage in weight is less reduced than the assimilation of smaller individual. This kind of "asymmetric competition" leads to variability in a_1 which in turn leads to positively skewed distribution of final weights (Uchmański 1985, 1987, Uchmański, Dgebuadze 1990).

We used a graphical model to calculate a_1. The value of s in eq. (3) and in turn a_1 is specified only for the smallest and highest possible initial weight w_0:

$$s = \begin{cases} s_{\min} & for \quad w_0 = w_{0,\min} \; , \\[2mm] s_{\max} & for \quad w_0 = w_{0,\max} \; . \end{cases} \qquad (5)$$

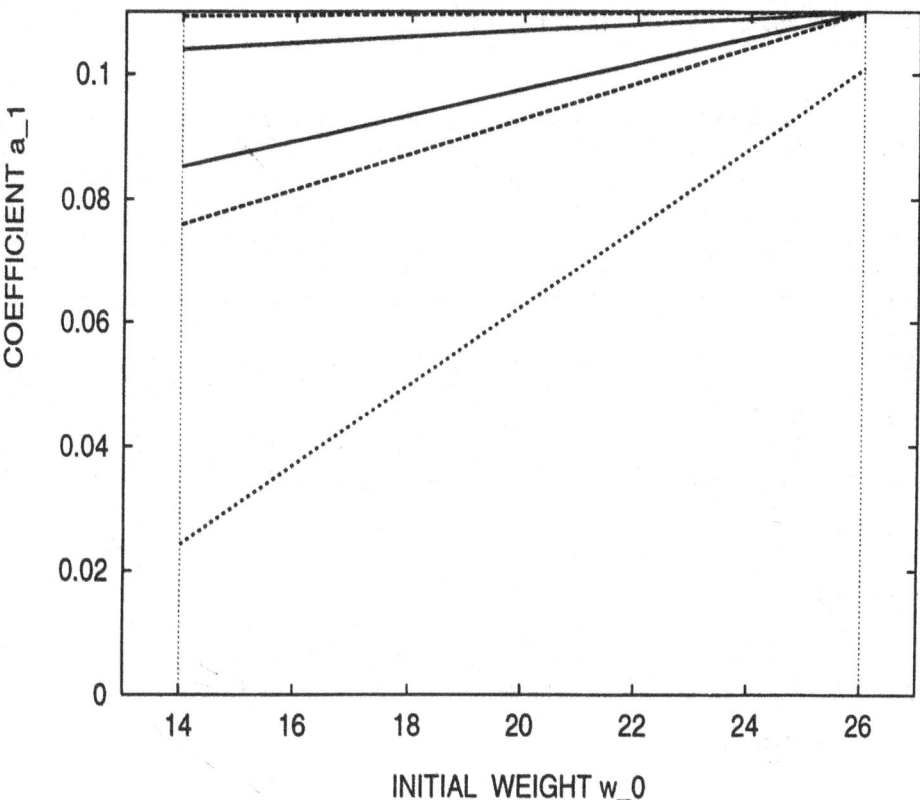

Figure 1. Graphical model for the calculations of the coefficient $a_1(w_0,V)$. The values for $a_1(w_{0,\min},V)$ and $a_1(w_{0,\max},V)$ are specified according to eq. (3). For all other initial weights, a_1 is determined by linear interpolation. The different lines correspond to different resource conditions. Solid lines - highest (upper line) and lowest (lower line) resource level V in the dynamics presented in Fig. 3; dashed lines - corresponding to Fig. 5. The resource conditions are (from top to bottom): 2.04×10^6, 1.94×10^6, 0.99×10^6, 0.47×10^6, 0.1×10^6. The lower, dotted line belongs to extremely bad resource conditions.

For initial weights between $w_{0,\min}$ and $w_{0,\max}$, a_1 is determined by linear interpolation (Fig. 1). From premise (3) (see above) it follows that $s_{\min} < s_{\max}$. Thus, at each moment during growth of each individual, a_i is determined by the individuals initial weights and by the actual amount V of resources:

$$a_1 = a_1(w_0, V).\tag{6}$$

Ranking in a_i according to the distribution of initial weights means that ranking is unchanging during growth, which of course is only an approximation to conditions in real populations. A more realistic description of ranking would include changes in absolute ranking due to stochastic effects and changes in relative ranking due to the positive feedback between a_i and the results of assimilation, i.e. gain in body mass.

At the end of the growth period individuals produce an integer number z_i of offspring which is proportional to the difference between final wight w_{end} and some threshold weight $\hat{w}$ (Łomnicki 1978, Kooijman and Metz 1984):

$$z_i = Round(c(w_{i,end} - \hat{w})),\tag{7}$$

where function "Round" rounds a real number to the nearest integer. Individuals with final weight less then $\hat{w}$ die without leaving progeny. $\hat{w}$ was defined as a fraction of, for example, 65% of the maximal possible final weight (see eq. (1)).

As a result of consumption resource conditions may become so deficient that individuals lose weight because the "balance" between assimilation and respiration is negative. We introduced mortality due to starvation into the model by applying the following rule: if an individual loses more then $m\%$ (for instance 20%) of its previous maximum weight, it will die.

The simulation procedure was as follows. At the beginning, all parameters and initial values for the number of individuals N_0 and resources V_0 were provided. The initial weights of the individuals were taken from a normal distribution with mean $w_{0,mean}$ and variance σ^2. Then in each generation, individuals grew according to eqs. (4) and (6) and to the procedure described above. The number of offspring produced at the end of the growth period was determined and the simulation cycle was started anew. The simulation was stopped when there were no more resources or individuals.

We chose parameter values and initial values for N_0 and V_0 which quarantee nearly optimal resource conditions during the whole growth period in the first generation. The following parameter set was used: $a_{1,max} = 0.11$, $a_2 = 0.03$, $b_1 = 0.7$, $b_2 = 0.9$, $w_{min} = 14$, $w_{max} = 26$, $w_{mean} = 20$, $\sigma = 2$, $u = 1$, $s_{min} = 1.5 \times 10^{-6}$, $s_{max} = 1.5 \times 10^{-5}$, $c = 0.02$, $\hat{w} = 0.65 \times 662$, $g = 150$, $m = 0.8$. Initial values: $N_0 = 30$, $V_0 = 1.5 \times 10^6$.

4.2 Results of the model

A detailed analysis of the model would go beyond the scope of this contribution. We will present only some results which demonstrate the potential relevance of individual variability with respect to regulation of the population.

Figures 2a and b show the growth of individuals for good and for insufficient resource conditions, respectively. In Figure 2a, at the beginning all individuals grow with nearly maximal assimilation rate. As long as individuals are small, consumption and inflow of resources are approximately balanced. But with an increase in body weight also consumption increases and, in turn, resources decrease. This leads, according to eq. (3) and Fig. 1, to an increase in individual variability with respect to a_i. As a result the growth curves fan out. But in Fig. 2a, the effect is still minimal. Variability in final weight is small and all individuals are able to produce two or more offspring.

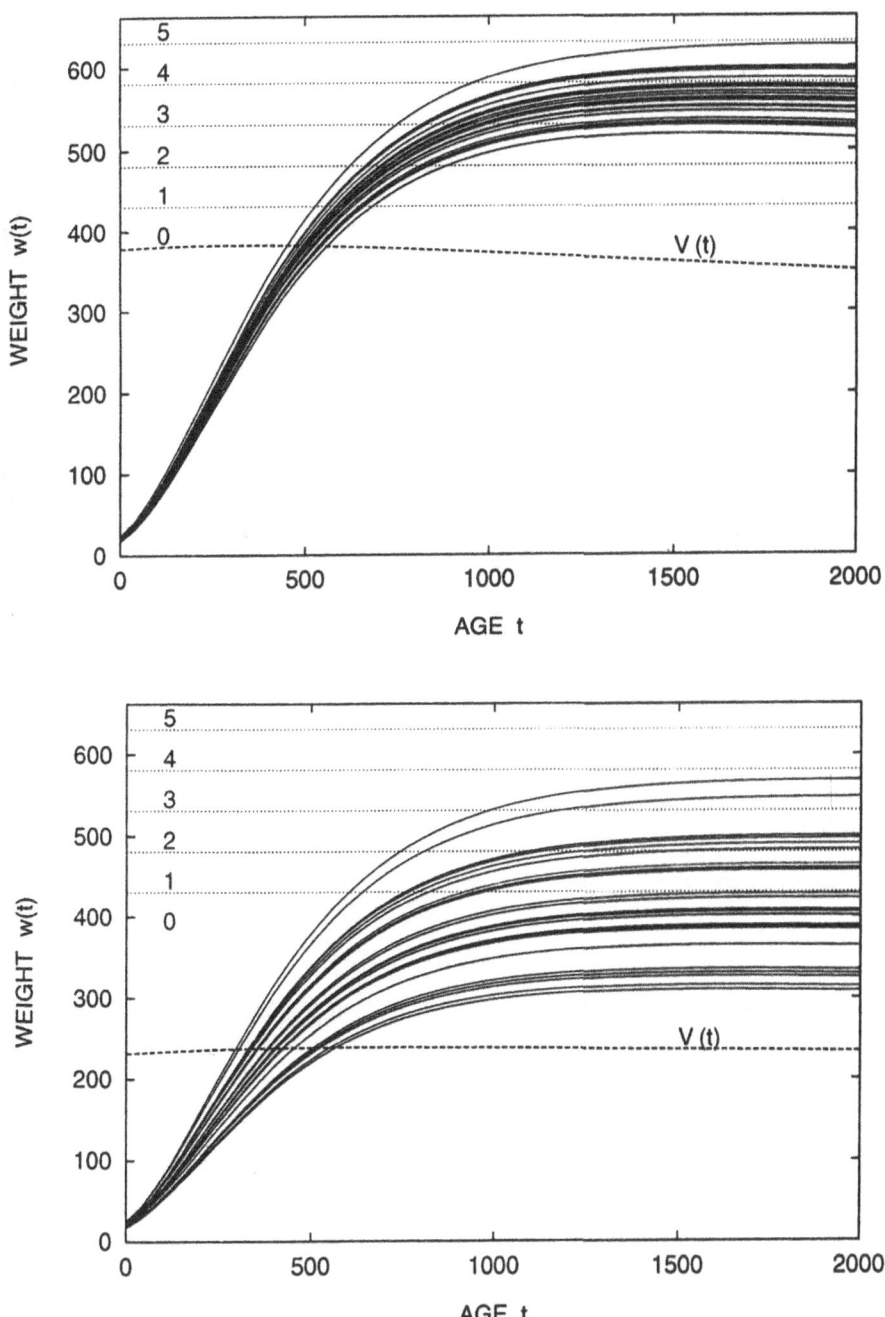

Figure 2. Individual growth curves (solid lines) and resource dynamics (dashed line, x2.0x10⁴) within one generation of Fig. 3. The parameter and initial values are given in the text. The dotted lines parallel to the *t*-axis and the labels mark the number of offspring produced by surviving individuals with a certain final weight at the end of the generation: a - 31th generation, b - 14th generation (see Fig. 3).

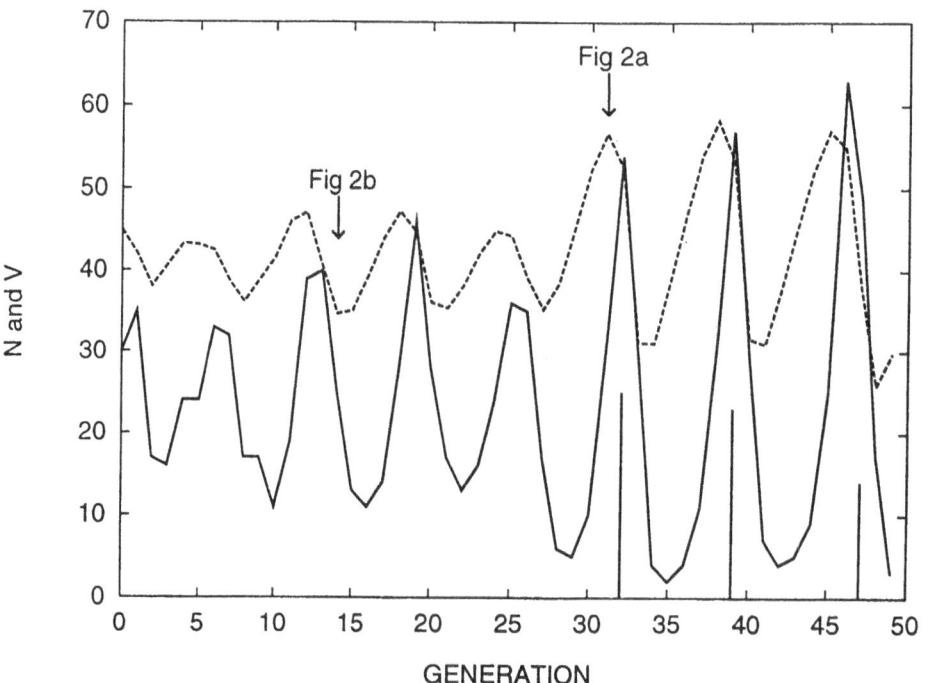

Figure 3. Dynamics of the number N of surviving individuals in each generation (solid line), and resources V at the beginning of each generation (dashed line, x3.0x10$_5$). The impulses indicate the number of individuals that died due to starvation in the generation. In this particular run of the model, after 50 generations the population is extinct.
demonstrated mechanism of density dependence is very effective in ensuring the persistence of the population. In 49 out of 50 simulation runs, the population survived for at least 300 generations.

In contrast in Fig.2b resource condition are deficient from the beginning. Variability in a_1 is high and therefore also variability in final weights. The bulk of individuals reaches a final weight which is not sufficient for reproduction, but there are still individuals which reproduce. This is first of two regulatory mechanisms based on individual variability we want to demonstrate. Under bad resource condition the "strong" individuals, i.e., in our case, individuals with an initial advantage in weight, grow at the expense of smaller ones. This may ensure the survival of the population although resources for an average individual would be insufficient for its reproduction. A typical resulting population dynamics is shown in Fig. 3. The number of surviving (but not necessarily reproducing) individuals, and the amount of resources oscillate. After 50 generations, the population is extinct. Since the initial weight distribution is taken as a random sample a normal distribution, the resulting population dynamics varies from simulation to simulation. The mean time to extinction in hundred runs of the model was 48 generations (standard deviation 30 generations).

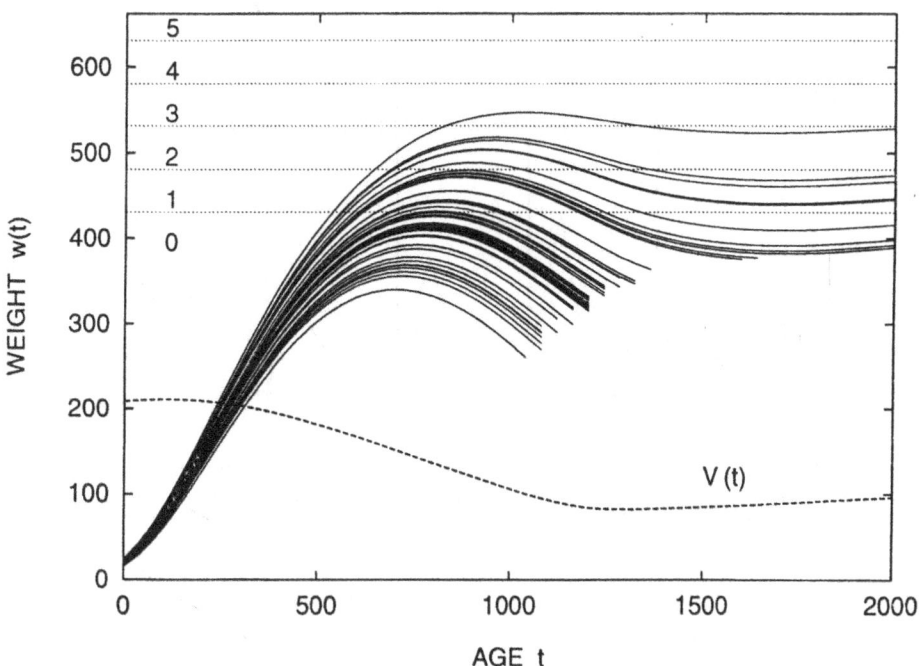

Figure 4. Individual growth curves (solid lines) and resource dynamics (dashed lines, x2.0x10⁴) within the 52th generation of Fig.5. The parameter and initial values as for Fig. 3, except s_{min} = 2.5x10⁻⁶, s_{max}=2.5x10⁻⁵ and g = 300. Only a randomly taken sample of one third of all growth curves in this generation (156 individuals at the beginning of the generation) is presented.

In Fig. 4, the same parameter set was used as in Figs. 2 and 3, except s_{min} = 2.5x10⁻⁶, s_{max} = 2.5x10⁻⁵ and g = 300. Because the product sV (see eq.(3)) determines a_i, s defines which amount of resources V are "insufficient" or "optimal". Therefore, an increase in s_{min}

and s_{max} means that individuals grow faster. To meet the resulting higher demands of resources, also a higher value for g has been chosen.

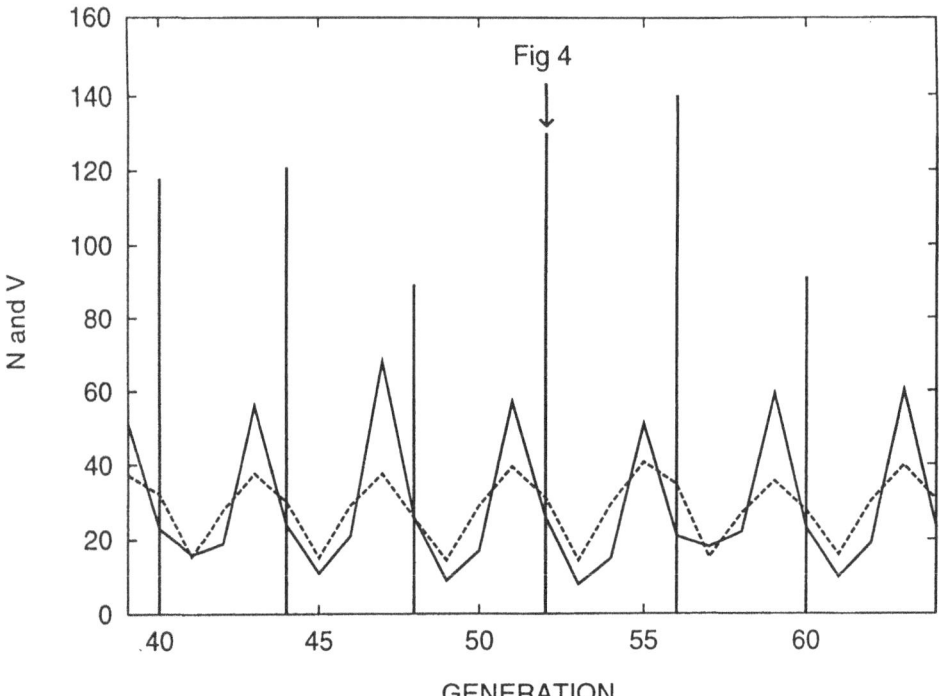

Figure 5. Population and resource dynamics for the parameters and initial values of Fig.4. See Fig. 3 for an explanation.

The main difference between Fig. 2a and Fig. 4 is that in Fig. 4 there is an "overcrowding", i.e. there are 156 individuals at the beginning, instead of 30 in Fig. 2a. After the initial phase, consumption reduces the amount of resources to such an extent, that for small individuals a_i becomes too small to compensate for the loss in body weight due to respiration. Individuals which loose more then 20% of their previous maximum weight will die. This process, which in plant ecology is called "self-thinning", causes successive improvement of resource conditions for individuals with higher weight until consumption and the inflow of resources are balanced again. In fact, among individuals, which are able to survive the "self-thinning" phase there are individuals, which will reproduce at the end of generation, but also individuals, which are not able to reproduce. In general, the described process is the second regulatory mechanism based on individual variability which we wanted to demonstrate.

The resulting population dynamics still shows oscillations (Fig. 5), but is much more regular than the dynamics in Fig. 3, where no "self-thinning" mortality occurred during growth except in the last generation before extinction. The impulses in Fig. 5 indicated the number of individuals that died during growth. The effect of high mortality is to prevent the oscillations in resources from becoming more and more violently. The

5. Discussion and conclusions

Individual-based models do produce different result than classical models. In most cases, also for parameters values different from those used in this contribution, the individual-based model presented above showed more or less regular fluctuations of individualnumber with intermediate down to small amplitudes. The dynamics of such systems has no equilibrium state or limit cycle. It exhibits rather "imperfect regulation" (Uchmański and Grimm, in press). All the attributes needed for negative feedback between individual number and reproduction are present: reproduction depends on the amount of resources, while utilisation of resources is influenced by the number of individuals. But it is not sufficient to produce "regulation of population" in the classical sense. The two assumptions, namely that an individual can produce zero, one, two... offspring and that each individual produces different number of offspring (and additionally some random effects), preclude such a precise relationship between individual number and reproduction rate, which produces asymptotic stability of population.

The new results produced by individual-based models influence our understanding of basic ecological concepts. We discussed earlier (Uchmański and Grimm, in press) some problems concerning "regulation" and "stability" of ecological systems described by individual-based models. We would like to add some comments to ongoing in ecological literature, heated discussion about "density dependence". The chance to find a new argument in this discussion, confining oneself to the population level, is minimal. In this paper, we present one possible way out of this dilemma: turning attention to the level of individuals. We don't argue for blind reductionism, but want to demonstrate that averaging individual variability may cancel some important mechanisms of "density dependence".

At the level of individuals, "density dependence" is not an abstract, phenomenological metaphor, but the result of individual, intraspecific competition for resources. In the context of our model, density affects individuals via resources. The amount of resources available for one individual at a given time instant depends on two factors: firstly, on the absolute amount of resources, and secondly, on the number of individuals. The absolute amount of resources is determined by the history of the population, i.e. by the consumption of resources in the preceding generations. Hence, via the absolute amount of resources density affects individuals on a "between-generation" time scale. Most models of theoretical population ecology describe "density dependence" on this time scale. But the number of individuals affects - at least in our model with global exploitation competition - individuals immediately while they are growing. Our model suggests that with respect to this "within-generation density dependence" (Cappuccino 1992) individual variability should be of crucial importance.

"Between-generations density dependence" includes a time lag between density and the effect of density. Under certain circumstances, time lags in regulation cause oscillations or even chaotic fluctuations of population size. But although ecologists are aware of the theoretical chaos-generating potential of density dependence, they do not find chaos in real populations. One possible mechanism which may prevent chaos is "within-generation" density dependence based on individual variability.

If individual variability really is an essential property of natural populations, this would question the suitability of mathematical modelling with only one variable, N, i.e. density. The complexity and the number of processes which describe the individual and its relationship with other individuals and environment can not be in principle included in a simple dynamical system. Hence, ecological systems are not dynamic system, because they consist of individuals, which are different. We know no answer to the question: is it possible to approximate the individual-based model by some simpler dynamic system? This problem is of minor importance from the ecological point of view, but it may have some value for the mathematical side of the modelling process. We suggest that time series analysis of ecological data should be performed very carefully because it often implicitly assumes identification of some dynamic system. "State-variable"-modelling will continue in playing an important part in ecology, but in many cases it may be necessary to choose an individual-based approach.

Models of classical mathematical ecology can be divided into two classes. Models which are far removed from individual-based models belong to the first category. The logistic equation is the best example here. The second category consists of models, such as the models of birth and death processes, which remain at the border between the classical approach and the individual-based model. They can serve as a null approximation of individual-based models and as a transition stage between the classical and non-classical approaches in ecology.

The model presented in this contribution is far from being all-out. The basic idea is that individual variability changes in a predictable way with the amount of available resources and with density. At insufficient resource conditions small initial differences between individuals may reinforce during growth due to a positive feedback mechanism. Within plants this kind of asymmetric competition is well documented, but its relevance for animal populations has been questioned recently by Latto (1992). The only way to decide this question is to replace our phenomenological submodel by a particular mechanistic description of competition between individuals. Only a detailed analysis of real populations could serve as a point of departure for such an mechanistic description. Our work suggests, as does Latto's paper, that empirical analysis of this kind would be worthwhile.

Acknowledgements

J. Uchmański was financially supported by Umweltforschungszentrum Leipzig-Halle during working on this subject.

References

Cappuccino, N. (1992), 'The nature of population stability in Eurosta *solidaginis*, an outbreeding herbivore of goldenrod', *Ecology*, **73**, 1792-1801.

DeAngelis, D.L. (1992), 'Mathematics: a bookkeeping tool or means of deeper understanding of ecological systems?' *Varhandlungen der Gesellschaft für Ökologie*, **21**, 9-13.

DeAngelis, D.L. and L.J. Gross [Eds] (1992), 'Individual-based models and approaches in ecology. Populations, communities and ecosystems', Chapman and Hall, New York,

London.

Grimm, V., E. Schmidt and C. Wissel (1992), 'On the application of stability concepts in ecology', *Ecological Modelling*, **63**, 143-61.

Ivlev, V.S. (1961), 'Experimental ecology of the feeding of fishes', Yale University Press, New Haven.

Kooijman, S.A.L.M., J.A.J. Metz (1984), 'On the dynamics of chemically stressed populations: the deduction of population consequences from effects on individuals', *Ecotoxicology and Environmental Safety*, **8**, 254-274.

Latto, J. (1992), 'The differentiation of animal body weight', *Functional Ecology*, **6**,386-395.

Łomnicki, A. (1978), 'Individual differences between animals and the natural regulation of their numbers', *Journal of Animal Ecology*, **47**, 461-475.

Metz, J.A.J. and O. Diekmann [Eds] (1986), 'The dynamics of physiologically structured populations', Springer-Verlag, Berlin.

Reiss, M.J. (1989), 'The allometry of growth and reproduction', Cambridge University Press, Cambridge.

Sibly, R.M. and P. Calow (1986), 'Physiological ecology of animals', Blackwell, Oxford.

Sushchenya, L.M. (1975), 'Kolicestvennyje zakonomernosti pitanija rakoob raznych', Nauka i Technika, Minsk.

Sushchenya, L. M. and N.N. Khmeleva (1967), 'Potreblenie piscy kak funkcja vesa tela u rakoobraznych', *Dokl. Akad. Nauk. SSSR*, **76**, 1428-1431.

Uchmański, J. (1985), 'Differentiation and frequency distributions of body weights in plants and animals', *Philosophical Transactions of Royal Society London, Ser. B*, **310**, 1-75.

Uchmański, J. (1987), 'Resource partitioning among unequal competitors', *Ekologia polska*, **35**, 71-87.

Uchmański, J. and J. Dgebuadze (1990), 'Factors affecting skewness of weight distributions in even-aged populations: a numerical example', *Polish Ecological Studies*, **16**, 297-311.

Uchmański, J and V. Grimm (in press), 'The concept of "regulation" revised: a model study of the significance of individual variability', In: *Proceedings of the 6th European Ecological Congress, Marseille*.

Zaika, V.E. (1975), 'Balansovoye uravnenye rosta', In: *Kolichestvenniye aspekty rosta organizmov*, Nauka, Moscow, pp. 25-33.

Zaika, V.E. (1985), 'Balansovaya teoria rosta zhivotnykh', Naukova Dumka, Kiev.

Zaika, V.E. and N.P. Makarova (1971), ,Biologichesky smysl parametrov vkhody ashchykh v uravnenye rosta Bertalanfi', *Dokl. Akad. Nauk SSSR*, **199**, 242-244.

SPATIO-TEMPORAL ORGANIZATION MEDIATED BY A HIERARCHY IN TIME SCALES IN ENSEMBLES OF PREDATOR-PREY PAIRS

CLAUDIA PAHL-WOSTL

Swiss Federal Institute for Environmental Science and Technology (EAWAG)
and Swiss Federal Institute of Technology, Zürich (ETH)
CH-8600 Dübendorf, Switzerland
e-mail: pahl@eawag.ch

Summary

Spatio-temporal organization refers to functional changes in a system's interaction network mediated by system dynamics. Complex, even chaotic, dynamics are claimed to be characteristic for an ecological community as a whole. In systems of this type the emphasis should be shifted from predicting individual trajectories to investigating organizational properties at the level of the system as a whole such as the relationship between dynamic diversity, spatio-temporal organization and system function. Such effects are illustrated with a model where a hierarchy of different time scales is introduced into an ensemble of predator-prey pairs (PPP) by distributing the latter along the body weight axis. The PPPs share a common pool of a limiting resource. Model versions comprising a single PPP are characterized by a time-invariant steady state. As soon as further PPPs are added the system becomes unstable exhibiting first periodic and then chaotic oscillations. In spite of the chaotic and unpredictable dynamics of the single PPPs, a number of system properties were found to be independent of the initial conditions chosen. The efficiency of resource utlization increases with an increasing number of PPPs due to the associated increase of the temporal organization of the network as a whole. The effects of spatio-temporal organization on system function are further illustrated by results from model versions where a dimension of space was introduced by assuming that the species diffuse along one spatial dimension.

1. Introduction

Hutchinson (1959) was one of the first to point out the inability of a view based on equilibrium assumptions to explain the coexistence of the myriad of different species encountered in ecosystems. Meanwhile investigations of ecological processes over a range of spatial and temporal scales constitute an integral part of ecological research. Such investigations have revealed the importance of spatio-temporal variability for ecosystem function and species' coexistence (e.g., Chesson and Huntly 1988; Steele 1990).Variability may be imposed onto the system exogenously e.g. by seasonal variations in the environment. In the present paper the emphasis is put on variability that is generated endogenously within the system by interactions of the component species. The relevance of complex dynamics in ecological communities is still a matter of debate (review in Pimm 1991). Complex dynamics in general and chaotic dynamics in particular have often been accused to be deterimental to natural populations. Meanwhile, there is increasing theoretical evidence that systems with chaotic dynamics have even a higher potential for adapting to changing environmental conditions and for resisting to perturbations (Pacala *et al.* 1990; Wilson, 1992; Allen, 1993; Hastings, 1993).

It has been conjectured that complex dynamics are characteristic for species communities allowing for temporal organization and an efficient exploitation of limiting resources. (Pahl-Wostl 1990, 1993a,b). The effects of such behaviour were studied with multispecies models of the general type represented in Fig. 1a. An ensemble of compartments shares a common resource pool. The resource is recycled within the system that is open exchanging energy and matter with an external environment. The network in Fig. 1a represents a redundant system where n compartments perform the same function. However, once the dynamics of the interactions are resolved along the dimension of time, the network may appear as in Fig. 1b where the activity periods of the single compartments are confined to different time intervals. The original redundancy in the time averaged representation is resolved.

(a)

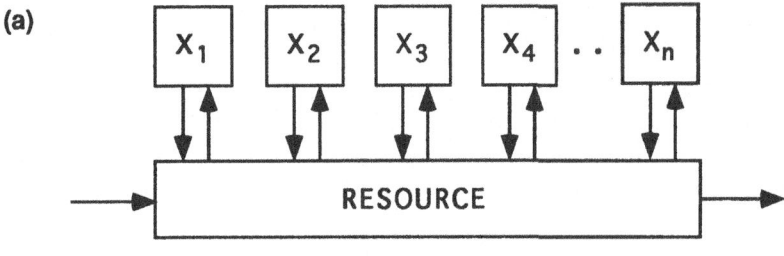

(b)

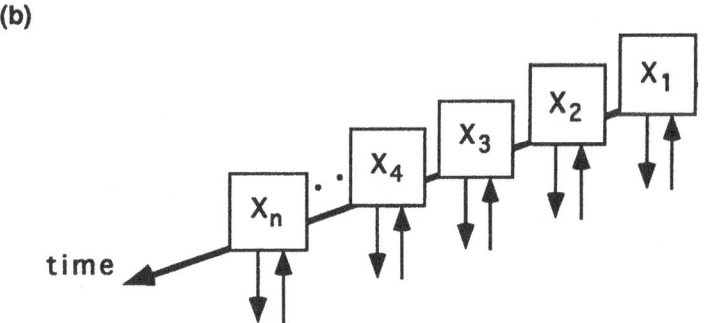

Figure 1. (a) Network representation of a general type of models where a group of functionally redundant compartments shares a common pool of a limiting resource. (b) Possible pattern of activity when the time averaged representation in (a) is resolved along the dimension of time. This reduction in functional redundancy in the temporally resolved relative to the time averaged network is referred to as temporal organization.

Such behaviour is referred to here as temporal organization of a network. It may derive from both periodic and chaotic oscillations. A measure to quantify the temporal organization, I_t, of flow networks was introduced by (Pahl-Wostl 1990, 1992):

$$I_t = \sum_{j=0}^{n} \sum_{i=1}^{n+2} \sum_{k=1}^{r} \frac{T_{jik}}{T} \log \left(\frac{T_{jik}^2 T}{T_{ji.} T_{j.k} T_{.ik}} \right), \tag{1}$$

where T_{jik} denotes the flow from compartment j to compartment i during the kth time interval in a system with n internal compartments. The indices 0, $n+1$, and $n+2$ refer to exchanges with the environment. The dimension of time is discretized into r time intervals. A point denotes always a summation over the corresponding index. The three points in the index were omitted in case of the total system throughput, T, corresponding to the sum of all flows in the system. Correspondingly the measure for the spatio-temporal organization, Its, yields:

$$I_{tS} = \sum_{j=0}^{n} \sum_{i=1}^{n+2} \sum_{k=1}^{r} \sum_{l=1}^{s} \frac{T_{jikl}}{T} \log \frac{T_{jikl}^2 T}{T_{ji..} T_{j.kl} T_{.ikl}}, \tag{2}$$

where T_{jikl} denotes the flow from compartment j to compartment i during the kth time interval in the lth spatial grid cell. The dimension of space is discretized into s spatial cells. The temporal (spatio-temporal) organization is high if the temporal (spatio-temporal) patterns of the network flows result in a decrease of the functional redundancy in the network averaged over time and space. The derivation of Eqs. (1,2) and more explanations about their meaning are given in (Pahl-Wostl 1990, 1992).

The characteristics of overall system dynamics are discussed now in more detail using a specific type of model where each compartment represents a predator-prey pair (PPP). The model is considered to be of a more general nature investigating the dynamics of an ensemble of functionally identical units operating on different time scales. In particular the model was developed with reference to ensembles of phytoplankton-herbivore pairs sharing a pool of a limiting nutrient. In pelagic systems energy flow is directed mainly along a gradient of increasing body weight. Taking into account the influence of body weight on rates of processes in and interactions among organisms, body weight may be used as central variable to develop generic energy transfer models (e.g. Platt and Denman, 1978, Pahl-Wostl 1993a).

2. Model description

Allometric rules describing the scaling of organismal properties with body weight were derived from observations that physiological rates, r, vary as a power function of the body weight, w:

$$r = aw^{-\varepsilon}, \tag{3}$$

where r has dimensions $[T^{-1}]$ for mass-specific rates, and T may be any unit of time, w is body mass $[M]$, ε is the allometric exponent and is dimensionless, a is the rate coefficient and has dimensions $[T^{-1}M^{-\varepsilon}]$. The exponential constant ε proved to be remarkably constant at a value of around 0.25 to 0.3 for a large range of organisms of different weight and taxa (e.g., Peters 1983).

Table I. Symbols used throughout the text, model parameters and numerical values (nondimensional). Rates in multiples of the basic growth rate g_0.

I_t			measure of temporal organization as defined in Eq. (1)
I_{ts}			measure of spatio-temporal organization as defined in Eq. (2)
n			total number of predator-prey pairs in a simulation run
PPP			predator-prey pair
$< >$			time average
i			index of a PPP with the prey, B_i, belonging to weight class i and the predator,
Pi,			belonging to weight class $i + q$.
D			diffusion coefficient in the spatial model - ranges from 0 to 10^{-4}.
r	=	0.412	rate of respiration.
b	=	0.010	rate of loss from the nutrient pool
K	=	2.550	half saturation of predator's functional response
I_n	=	0.200	external input of nutrient
l_{ext}	=	0.050	quadratic loss term of predator
ε	=	0.250	allometric exponent
α_k	=	$2^{-\varepsilon k}$	allometric factor for species in the kth weight class
q	=	10	predator-prey weight class difference
γ	=	$2^{-\varepsilon q}$	allometric factor for a predator relative to its prey (= 0.177)

In the model presented here a basic parameter set is chosen and the species specific parameters are derived from the basic set by allometric rules. A PPP is then characterized by its location along the body weight axis. The body weight, w_i, of the prey B_i and the body weight, w_{i+q}, of its predator P_i are assigned according to:

$$w_k = 2^k w_0 . \tag{4}$$

Hence, the species are representative of weight classes equally spaced on a logarithmic scale. The weight ratio between neighbouring classes, w_{k+1}/w_k, which was chosen here to be equal to 2, determines the base of the logarithm. The difference in weight classes between a predator and its prey, q, was assumed to be constant and to be equal to 10. A difference of ten in weight class corresponds to a weight ratio of about 1000, and an allometric factor $2^{-\varepsilon q}$ equal to 0.177. In what follows this allometric factor, which describes the difference in time scale between a predator and its prey, will be referred to as γ. The respiration rate, r_k, and the growth rate, g_k, of a species in weight class k are derived from the corresponding rates of a species in weight class 0 by combining Eqs. (3) and (4):

$$g_k = \alpha_k g_0 \quad \text{and} \quad r_k = \alpha_k r_0 \tag{5}$$

where w_k/w_0 was replaced by $\alpha_k = 2^{-\varepsilon k}$, the allometric factor of the weight class k. Incorporating "type II" saturating functional responses for both prey and predator species the model takes the form:

$$\frac{dB_i}{dt} = \alpha_i \{ f(N)B_i - rB_i - \gamma h(B_i)P_i \} \ , \tag{6}$$

$$\frac{dP_i}{dt} = \alpha_{i+q} \{ h(B_i)P_i - (r + 1_{ext}P_i)P_i \} \ , \tag{7}$$

where $i = 0, 1, ..., q-1$ and

$$f(N) = \frac{N}{N+1} \ ,$$

$$h(B_i) = \frac{B_i}{B_i + K} \ ,$$

$$\frac{dN}{dt} = In + r \sum_{i=0}^{q-1} \alpha_i (B_i + \gamma P_i) - bN - \sum_{i=0}^{q-1} \alpha_i f(N)B_i \ . \tag{8}$$

Dimensionless variables were chosen by expressing all concentrations in multiples of the half saturation constant of the prey's growth response, and by expressing time in multiples of g_o^{-1}, the basic turnover time of a species in weight class 0. In represents a time-invariant external input into the nutrient pool. The behaviour of a similar version of this model that comprised an additional inactive nutrient pool was discussed in detail by (Pahl-Wostl 1993a). The qualitative system dynamics, the most important characteristics of which will be summarized, are not changed by omitting the inactive nutrient pool. In addition, to illustrate the effects of introducing a spatial dimension results from a one-dimensional spatial version of the simplified model will be presented.

3. Results of model simulations

Model simulations were performed for an increasing number of PPPs, referred to as n. Each new pair was shifted one weight class up relative to the previous one. The initially stable system ($n=1$), exhibits periodic oscillations ($n=2,3$), to become finally chaotic ($n>3$). Fig. 2 shows the temporal variation of the nutrient pool N observed in model simulations for (a) $n=3$, and (b) $n=5$. One notes the breakdown of the regular pattern with increasing n, what is also supported by the autocorrelation functions in Fig. 2(c). Further, there exists a lower threshold for N corresponding to the concentration N_{cr} where in the absence of predation the growth of a prey species equals its losses - from Eq. (6):

$$N_{cr} = \frac{r}{1-r} = 0.7. \tag{9}$$

Both mean and variability of the nutrient concentration decrease with increasing n.

    The chaotic behaviour may be explained by PPPs with different dynamics being coupled nonlinearly through the nutrient pool. Chaos arises because the characteristic frequencies of the subsystems are incommensurate (the period of one oscillation is not some multiple

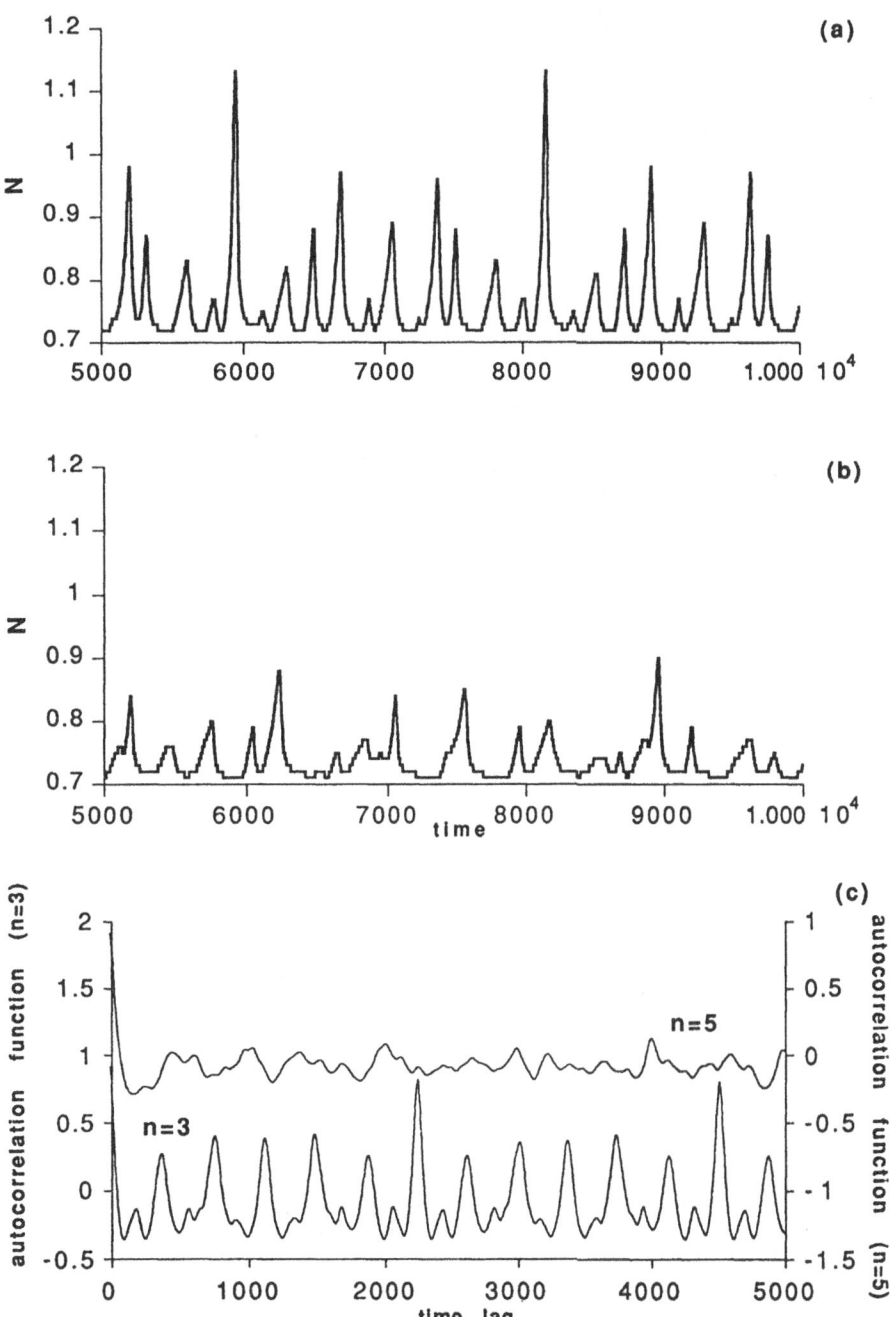

Figure 2. Temporal variation of the concentration of the nutrient obtained in simulations with the spatially homogenous model for (a) three (*n*=3) and (b) five (*n*=5) pairs. (c) Autocorrelation functions of (a) and (b) revealing the breakdown of the periodic oscillations with an increasing number of PPPs.

of other frequencies). Such a statement is supported by the absence of chaos in a model
where the pairs are identical (Pahl-Wostl 1993a). Further support for the chaotic nature of
the dynamics was provided by the presence of sensitivity to initial conditions. Neighbouring trajectories were observed to diverge exponentially until the distance fluctuated
around the average distance of two points on the attractor. In the case of chaotic dynamics
it is not possible to predict the system's trajectory in the phase space spanned by the
different pairs. However, the state of the system converges asymptotically to an attractor
and the variability of overall system properties (e.g. nutrient concentration in the pool) was
observed to decrease.

Such a behaviour can be explained by the fact that the envelope of the sum of the
species activities is smoothed with an increasing number of pairs corresponding to an
increasing diversity in time scales. Whereas a wide range of fluctuations in biomass were
observed, the total metabolic activity obtained by weighing each prey's biomass by its
allometric factor was observed to remain close to an upper threshold determined by the
critical nutrient concentration N_{cr}.

The regularities observed and the increasing efficiency in the exploitation of the resource
are attributed to the temporal organization of the system effected by the time-sharing of
the limiting nutrient. The temporal organization as quantified by ΔI_t is represented in Fig.
3 as a function of the number of pairs. ΔI_t refers to the increase in the measure of
temporal organization as defined in Eq. (1) relative to the organization in the time
averaged network. The increase in temporal organization corresponding to a decrease in
functional redundancy with increasing n is associated with a decrease in the average
concentration of the nutrient in the pool, $<N>$, corresponding to an increase in the
efficiency of nutrient utilization.

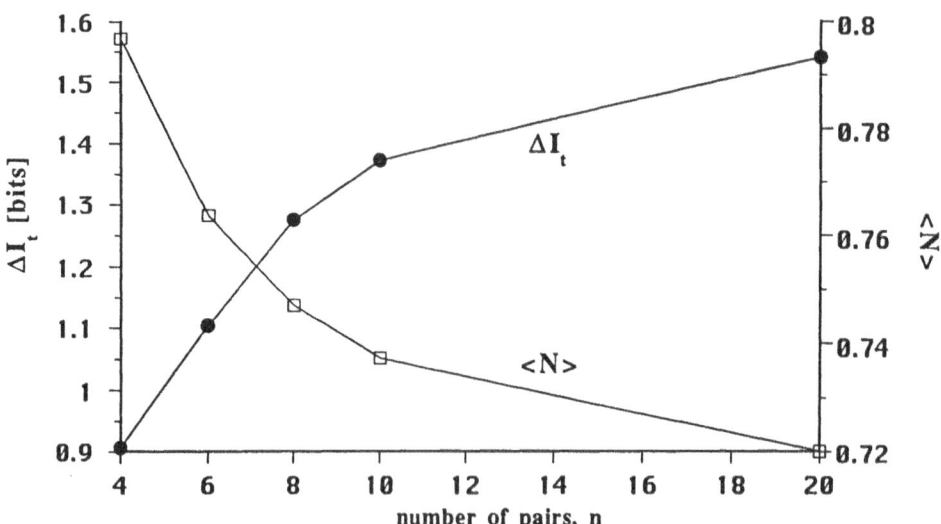

Figure 3. The increase in temporal organization, ΔI_t, and the average concentration of
nutrient in the pool, $<N>$, as a function of the number of PPPs.

Further support of the importance of spatio-temporal organization for system function is provided by first results from a 1-dimensional spatial model.

4. One-dimensional spatial model

After introducing a single spatial dimension the corresponding equations are obtained by adding diffusion terms to Eqs. (6) to (8):

$$\frac{\partial B_i}{\partial t} = (6) + D\frac{\partial^2 B_i}{\partial x^2} \ , \tag{6a}$$

$$\frac{\partial P_i}{\partial t} = (7) + D\frac{\partial^2 P_i}{\partial x^2} \ , \tag{7a}$$

$$\frac{\partial N}{\partial t} = (8) + D\frac{\partial^2 N}{\partial x^2} \ . \tag{8a}$$

The spatial dimension was scaled by setting the length scale of interest equal to 1. The reflecting boundary conditions chosen yield at $x=0$ and $x=1$:

$$\frac{\partial B_i}{\partial x} = \frac{\partial P_i}{\partial x} = \frac{\partial N}{\partial x} = 0 \quad \text{for all } t \text{ and for } i = 1,...,n. \tag{10}$$

In the first simplifying approach presented here the same diffusion coefficient was applied for all species as well as for the nutrient in the pool. The equations were solved numerically using 100 spatial grid sites. The initial values of all state variables were assigned at random.

Fig. 4 represents the temporal variations of the spatial averages of the biomass summed over all prey $<B_{tot}>_s$ and over all predator species, $<P_{tot}>_s$, respectively, obtained in a model with 2 pairs for decreasing D. For $D = 10^{-5}$, the periodic temporal oscillations are spatially synchronized as becomes obvious in Fig 4a. No spatial pattern is observed because the exchange rate between adjacent sites (equal to $D*10^4$) is in the same order of magnitude as the growth rates of the species that range from 1 to 0.1. The other extreme is given in Fig. 4e for $D = 0$ corresponding to a superposition of 100 independent grid sites. The spatial synchronization obtained for $D = 10^{-5}$ breaks down when D is decreased to 10^{-6} (Fig 4b). In this case the spatial exchange rates are an order of magnitude slower than the growth rates. Hence, spatial inhomogeneities are maintained. Diffusion mediates chaotic behaviour in time at a single spatial location as demonstrated by the autocorrelation functions represented in Fig. 5. The dashed curve, obtained for $D = 0$, is characteristic for periodic temporal oscillations whereas the full curve, obtained for $D = 10^{-6}2$, reveals the chaotic nature of the temporal oscillations.

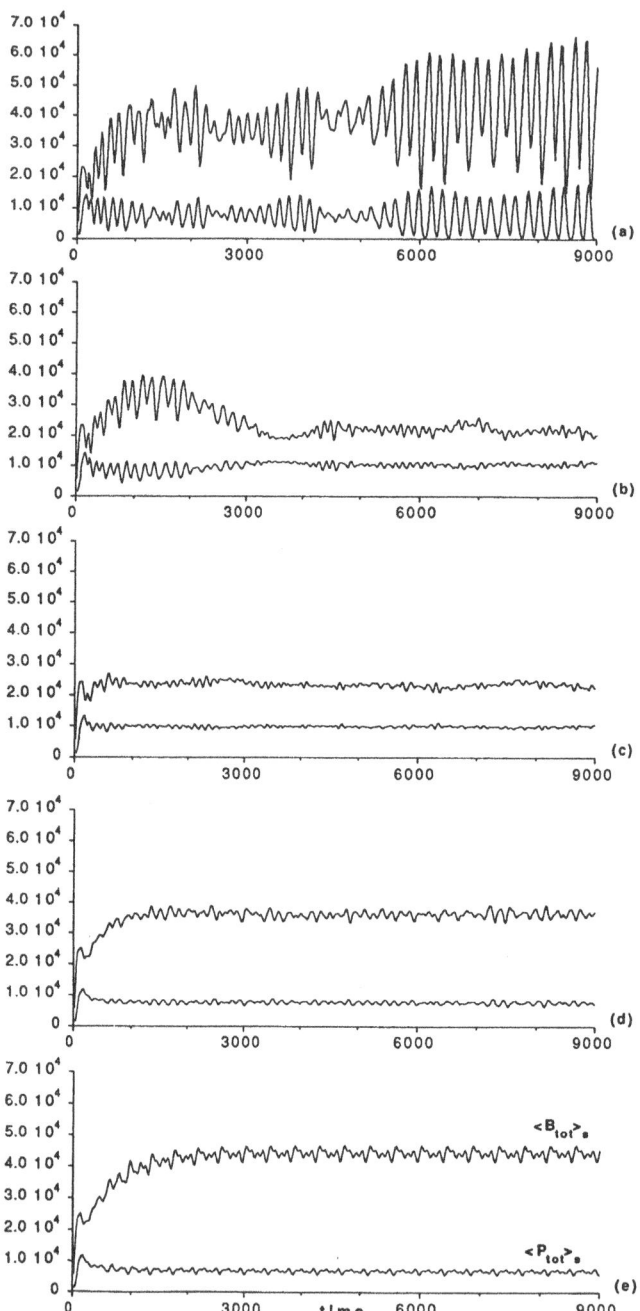

Figure 4. Time course of the spatially averaged biomass of all prey and predator species, respectively, obtained with a one-dimensional model with two PPPs. The diffusion coefficient, *D*, was varied: *D* = (a) 10^{-5}; (b) 10^{-6}; (c) 10^{-7}; (d) 10^{-9};(e) 0.

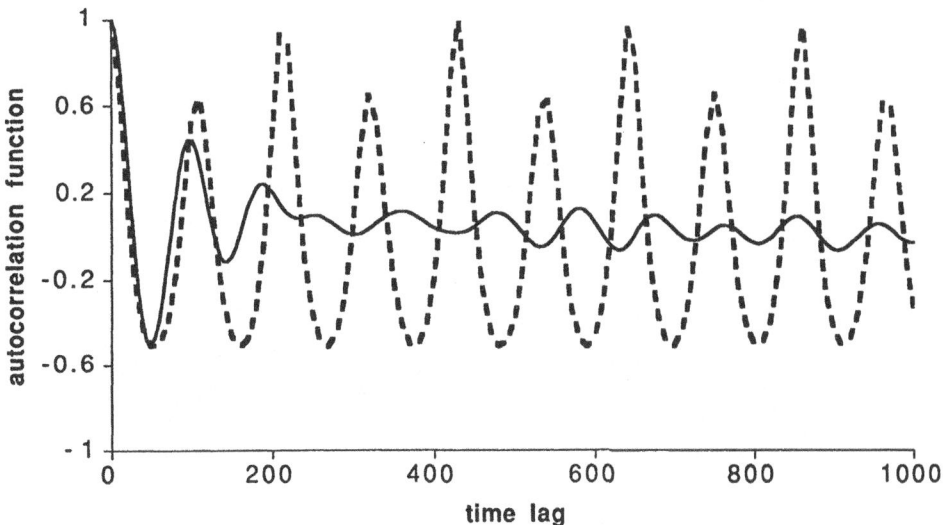

Figure 5. Autocorrelation function from time series of the nutrient, N, at a single spatial location for: $D = 0$ - dashed curve, $D = 10^{-6}$ full curve.

The fact that low diffusion values mediate temporal chaos at a fixed spatial location in an otherwise periodic system was also observed and studied in detail by Pasucal (1993) for a single predator-prey model.

One observes a regular behaviour of the spatio-temporal averages of the statevariables as a function of D. Fig. 6a shows a drop in the total biomass of all prey species ($<B_{tot}>$), a significant increase in the nutrient pool and a moderate increase in total predator biomass ($<P_{tot}>$) at the threshold where the spatial synchronization was observed to break down. Efficient spatio-temporal organization is impeded as can be noted by the drop in the quantitative measure of a network's spatio-temporal organization, I_{ts}, defined in Eq. (2). For a further decrease in D the various parameters recover until for $D=0$, corresponding to an ensemble of independent cells, they yield again the values of the spatially synchronized situation.

One finds a direct correlation of the measure of spatio-temporal organization, I_{ts}, with functional attributes as shown in Fig. 6b. The increase in the efficiency of nutrient utilization with increasing Its is reflected in the decrease of the concentration in the pool and an increase in total system activity, T. The same correlation was observed in a system with 5 PPPs as represented in Fig. 6c. Travelling waves extending over the whole spatial dimension were observed for $D = 10^{-5}$. Decreasing D resulted in a progressive breakdown of the spatial correlations.

The results obtained indicate that the potential to engage into effective spatio-temporal organization depends on the relationship of the temporal and spatial scales of a system. Here might be a clue to understand functional patterns in space and time and their mutual relationship. This is quite intriguing an idea even when one cannot derive any general conclusions from the preliminary results presented here. The characteristics of the relationship between space and time remain to be investigated and conditions for effective spatio-temporal organization remain to be derived.

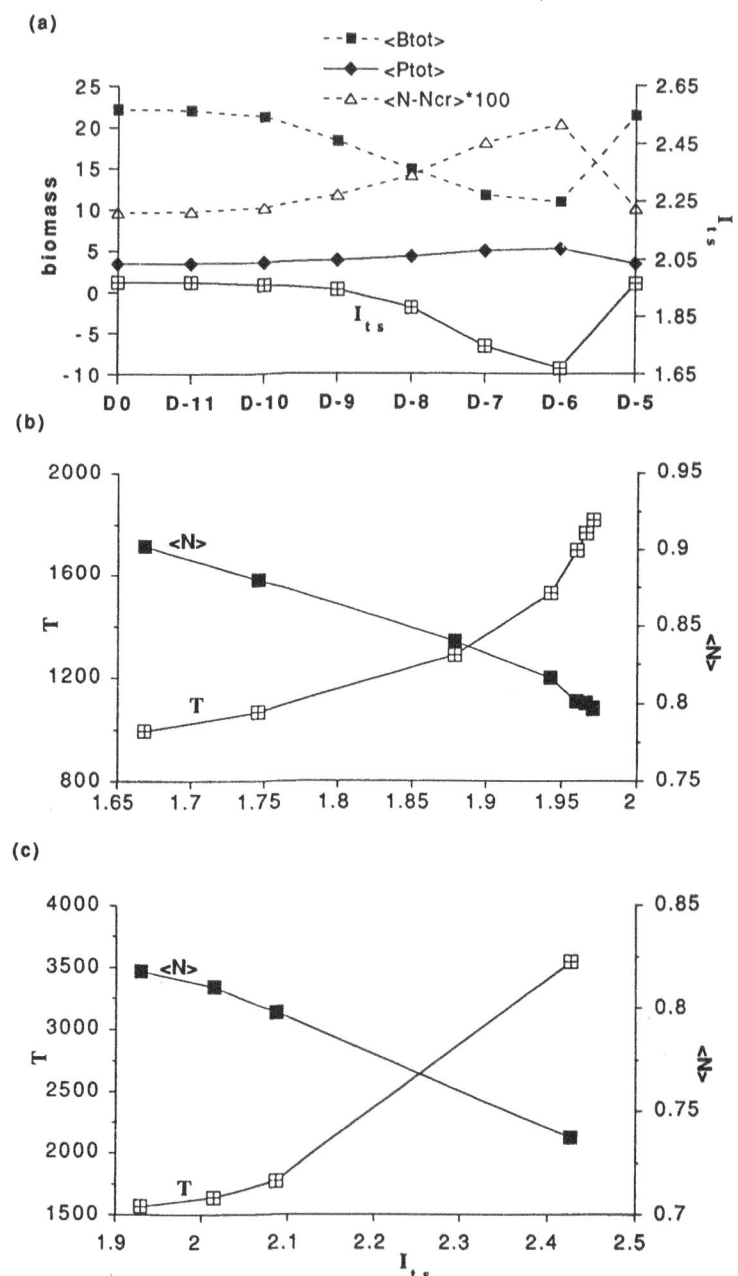

Figure 6. (a) Spatio-temporal organization, I_{ts}, spatio-temporal averages of: $\langle N{-}N_{cr}\rangle$, the biomass of all prey, $\langle B_{tot}\rangle$, and all predator, $\langle P_{tot}\rangle$, species, obtained in simulations with $n{=}2$ for different values of D. (b), (c) Nutrient concentration, $\langle N\rangle$, and total system throughput, T, as a function of I_{ts} obtained in model simulations for (b) $n{=}2$ and (c)$n{=}5$.

5. Discussion and conclusions

In addition to the results presented above, the concept of spatio-temporal organization in feedback systems was illustrated previously by a set of models with different sources for the spatio-temporal variability (Pahl-Wostl 1990, 1992, 1993a,b,c). The following properties are common to all the systems studied:

- An ensemble of compartments X_i shares a common resource R.

$$- \quad \frac{dX_i}{dR} < 0 \quad \text{and} \quad \frac{dR}{dX_i} > 0 \quad \Rightarrow \quad \frac{dX_i}{dX_i} < 0 \; ,$$

hence

$$\frac{dX_i}{dt} > 0 \quad \Rightarrow \quad \text{negative feedback of } X_i \text{ on } X_j ,$$

$$\frac{dX_i}{dt} < 0 \quad \Rightarrow \quad \text{positive feedback of } X_i \text{ on } X_j .$$

The total activity in a system, A, is limited by resource availability and cannot exceed a threshold A_{max}. An increase in the number of dynamically distinct interacting units results in a minimization of the time averaged difference $A_{max} - A$. The interacting units are identified as positive feedback cycles coupled by the common resource pool. One may better talk about positive feedback spirales in time (and/or space), because the temporal (spatial) pattern is essential for positive feedback to occur. Regardless of the source of variability, it was observed for all models that an increase in temporal organization mediated by an increase in dynamic diversity leads to an increase in the efficiency of resource utilization and a decrease in the variability of global system properties such as primary production or the resource concentration in the pool. This may be identified as an increase in autonomy reflected in the decrease of the ratio of the external input of the resource to the amount of the resource stored in the total living biomass of the system.

Kaneko and Ikegami (1992) concluded from results of model simulation that an evolutionary system with many species maintains its stability at a weak high-dimensional chaotic state, rather than in a fixed point or strong chaos. They called this state homeochaos. Despite the differences in the basic model assumptions - Kaneko and Ikegami used host-parasite models where the network was generated and changed through evolutionary processes - some parallels can be seen to the decrease in the variability of system function with an increasing number of pairs observed here. Chaos may indeed gain a new significance when viewed from a larger network perspective instead of focussing on single species.

Based on the results obtained one may suggest a general principle for ecosystem organization. By processes of temporal and spatial self-organization, an ecosystem's development is directed towards a state of higher autonomy as expressed by the independence of external influence. Such a development derives from the increase in the temporal and spatial organization of the system, thereby leading to a decrease in the variability of global system variables and an increase in the variability of local system variables. The terms "global" and "local" refer here to phase space as well as geometric space. It is the very increase in the local variability by processes of invasion, adaptation and self-organization that enables the efficient exploitation of the resources and that results in the homeostasis of the global system properties. A system's development must be viewed from a dualistic perspective in that processes at the species level permanently change macroscopic properties of the system, which then impose new constraints on the species themselves.

The ability to organize in time hinges on the presence of a diversity of functionally similar but dynamically distinct units. More emphasis must be given to the diversity with respect to dynamic characteristics both in model development and in the application of diversity measures. The information theoretical method introduces should be especially useful for characterizing complex networks. Phase space representations based on state variables only are not sufficient to reflect the essentials of a system's organization (Pahl-Wostl 1993c).

References

Allen, J.C., Schaffer, W.M. and Rosko, D. (1993), 'Chaos and extinction in ecological populations', *Nature*, **364**, 229-232.

Chesson, P. and Huntly, N. (1988), 'Community consequences of life-history traits in a variable environment', *Ann. Zool. Fennici*, **25**, 5-16.

EbenhÜh, W. (1992), 'Temporal organization in a multi-species model', *Theor. Popul. Biol.*, **42**, 152-171.

Hastings, A. and Powell, T. (1991), 'Chaos in a three-species food chain', *Ecology*, **72**, 896-903.

Hastings, A. (1993), 'Complex interactions between dispersal and dynamics: lessons from coupled logistic equations', *Ecology*, **74**, 1362-1372.

Hutchinson, G.E. (1959), 'Homage of Santa Rosalia or why are there so many kinds of animals', *Amer. Nat.* **93**, 145-160.

Kaneko, K. and Ikegami, T. (1992), 'Homeochaos: dynamic stability of a symbiotic network with population dynamics and evolving mutation rates', *Physica D*, **56**, 406-429.

Pacala, S., Hassell, M. and May, R. (1990), 'Host parasitoid associations in patchy environments' *Nature*, **344**, 150-153.

Peters, R. (1983), 'The implications of body size', Cambridge University Press, Cambridge, England.

Pahl-Wostl, C. (1990), 'Temporal organization: a new perspective on the ecological network', *Oikos*, **58**, 293-305.

Pahl-Wostl, C. (1992), 'Information theoretical analysis of functional temporal and spatial organization in flow networks', *Mathl. Comp. Mod.*, **16**, 35-52.

Pahl-Wostl, C. (1993a), 'The influence of a hierarchy in time scales on the dynamics of, and the coexistence within, ensembles of predator-prey pairs', *Theor. Popul. Biol.*, **43(2)**, 184-216.

Pahl-Wostl, C. (1993b), 'Food webs and ecological networks across spatial and temporal scales', *Oikos*, **66**, 415-432.

Pahl-Wostl, C. (1993c), The description of dynamic systems from the perspective on a network of interactions. *Int. J. Systems Sci.*, **24**, 1301-1316.

Pascual, M. (1993), 'Diffusion-induced chaos in a spatial predator-prey system', *Proc. R. Soc. Lond. B*, **251**, 1-7.

Pimm, S. (1991), 'The balance of nature?', University of Chicago Press, Chicago, 433p.

Platt, T. and Denman, K. (1978), 'The structure of pelagic marine ecosystems', *Rapp. P.-V. Reun., Cons. Int. Explor. Mer.*, **173**, 60-65.

Steele, J. (1990), 'The ocean landscape', *Landscape Ecology* , **3**, 185-192.

Wilson, D.S. (1992), 'Complex interactions in metacommunities, with implications for biodiversity and higher levels of selection', *Ecology*, **73**, 1984-2000.

CONTINENTAL EXPANSION OF PLANT DISEASE: A SURVEY OF SOME RECENT RESULTS

F. VAN DEN BOSCH[1], J.C. ZADOKS[2] AND J.A.J. METZ[3]

[1] Department of Mathematics, Wageningen Agricultural University
Dreijenlaan 4, 6703 HA Wageningen, The Netherlands.

[2] Department of Phytopathology, Wageningen Agricultural University
P.O. Box 8025, 6700 EE Wageningen, The Netherlands.

[3] Institute of Evolutionary and Environmental Sciences, Leiden University
Kaiserstraat 63, 2311 GP Leiden, The Netherlands.

Abstract

In this paper we show how the continental expansion of fungal plant diseases can be modeled. The model takes the form of a set of two integral equations. The equations describe a systematic book-keeping of the dynamics of the number of foci in host fields. Using results on related models for the spatial expansion of epidemics, the velocity of expansion can be calculated. We present explicit formulae for the expansion velocity , both within one growing season and for successive growing seasons. The method thus developed is applied to the invasion of *Phytophthora infestans*. This example serves as a first verification of the model. The example also shows what type of data are needed to get insight in the development of quarantine pests.

Keywords fungal plant disease, spatial spread, integral equation, *Phytophthora infestans*

1. Introduction

Plant diseases newly introduced to an area often have devastating consequences. No invasion of a plant disease is so well know as the potato blight (*Phytophthora infestans* (Mont.) de Bary) epidemic. This 'new' disease was first discovered in 1843 near the great ports of the east coast of the United states. It was probably brought in by a ship from South-America. From 1843 on an epidemic wave overran eastern United States with a velocity of approximately 295 km/year (Stevens, 1933). In 1845 it reached the geographical limits within which *P. infestans* is still a serious problem today. Due to this disease severe losses, up to 40%, were reported. Agricultural practice in the U.S. was sufficiently divers (wheat and maize were also grown). Therefore the epidemic only had moderate social consequences (Schumann, 1991).

In Europe the situation was different. At the start of the 19th century a large part of the human population depended almost completely on potato for their nutrition. Grain, also grown in small quantities, were needed to pay the rent to the landlords (Schumann, 1991). *P. infestans* was found first in 1845 near the west coast in Belgium, probably brought in by a ship from America. Within one season the epidemic spread, with a velocity of approximately 12.9 km/day, all over western Europe. Losses up to 60% were reported. Losses continued to be severe during the following years (Bourke, 1964). Much is known about the social impact of this epidemic in Ireland. Many farmers who could not pay their rents were evicted from their land and homes. In those days eviction was most likely followed by death from starvation. From the eight million people in Ireland, approximately one million died due to starvation and mal nutrition related diseases. An other 1.5 million managed to emigrate, mostly to the United States and Canada (Schumann, 1991).

The spectacular *P. infestans* epidemic is just one out of many well documented examples of invasions of plant diseases on a continental scale. Expensive quarantine measures are taken all over the world to prevent such unwanted invasions. Quarantine, containment and eradication programs are developed for each disease on an ad-hoc basis. Knowledge of the processes regulating the expansion of plant disease on a continental scale is virtually absent. Estimation of the relative danger posed by a quarantine pest and the design of emergency measures for eradication and containment would profit from models of pandemics with at least some predictive value (McGregor, 1978, Heesterbeek and Zadoks, 1987, Zadoks, 1988). In this paper we survey some recent results concerning the expansion velocity of air-borne fungal plant diseases on a continental scale. Using a mathematical model we show how the expansion velocity depends on the underlying processes. Recent reviews of the general theory of spatial epidemic spread can be found in van den Bosch *et al.* (1990), Mollison (1991) and Metz and van den Bosch (in prep).

2. Expansion during one growing season

2.1 The model

When modelling the dynamics of a (pathogen) population one usually takes the individual as the basic unit. Plant diseases often develop so called foci or hot-spots. Foci originate from a single infected plant. From this centre of infection the disease builds up a, radially expanding, patch of high disease intensity, called a focus. We will use this focus as our generalised individual. The model is a systematic book keeping of the number of foci in fields distributed over the continent.

Spores staying inside the canopy layer contribute to the development of the focus. Only spores leaving the canopy layer can cause new foci. The number of new foci started per unit of time in a field at position x at time t is denoted by $b(t,x)$. $b(t,x)$ equals the number of spores deposited in the field per unit of time multiplied by the probability that such a spore will actually start a new focus (ψ). The number of spores deposited in the field is the sum of the spores originating from the field itself that are 're-deposited' in the field (S_1) and the number of spores originating from other fields (S_2).

$$b(t,x) = \psi(S_1 + S_2) . \tag{1}$$

Note that we assumed ψ does not depend on the number of foci in the field. This assumption amounts to a linear model. For a justification of using a linear model for the present purpose see Metz and van den Bosch (in prep).

The number of spores produced per unit of time that leave the canopy layer is denoted by $A(t,x)$. From the 'above canopy layer' spores a fraction κ is re-deposited in the field. This gives us

$$S_1 = \kappa \, A(t,x) \; . \tag{2}$$

The number of spores deposited per unit of time at position x, originating from other fields is denoted by $v(t,x)$.

$$S_2 = v(t,x) \; . \tag{3}$$

This completes the equation for $b(t,x)$.

A field at position ξ at time t produces $A(t,\xi)$ spores above the canopy layer per unit of time. A fraction $1\text{-}\kappa$ of these spores is dispersed over large distance. We will denote by $D_1(x,\xi)$ the probability per unit area that a spore produced at ξ is deposited at x. We assume that this dispersal density is rotationally symmetric. Moreover we assume that the fields are homogeneously distributed over the continent. This amounts to assuming that $D_1(x,\xi)$ is translation invariant. These assumptions yield $D_1(x,\xi)=D(|x\text{-}\xi|)$. The number of spores produced at ξ and deposited at x per unit of time per unit of area is thus given by $(1\text{-}\kappa)A(t,\xi)D(|x\text{-}\xi|)$. To calculate the number of spores deposited at x originating from all possible fields, $v(t,x)$, one has to add all contributions of fields at all possible positions ξ. The fraction of the surface area of the continent covered by host fields is denoted by ϕ. The expression for $v(t,x)$ thus is:

$$v(t,x) = \phi \; (1-\kappa) \int_{R^2} A(t,\xi) \; D\big(|x-\xi|\big) \; d\xi \tag{4}$$

This completes the formula for $v(t,x)$.

The number of 'above canopy layer' spores produced per unit of time by a focus having age a, is denoted by $g(a)$. The number of 'above canopy' spores produced by foci of age a in a field at x is thus given by $b(t\text{-}a,x)g(a)$. To calculate the total 'above canopy layer' spore production of this field we have to add all contributions of foci of various ages. We can thus write

$$A(t,x) = \int_0^t b(t - a,x) \; g(a) \; da \; . \tag{5}$$

This completes the equation for $A(t,x)$.

Substituting equation (5) in (1) and (4) the model finally takes the form

$$b(t,x) = \psi v(t,x) + \psi\kappa \int_0^t b(t - a,x)g(a)da \; ,$$

$$v(t,x) = \phi(1 - \kappa)\int_{R^2}\int_0^t b(t - a,\xi)g(a)D(|x - \xi|)dad\xi \tag{6}$$

To be able to calculate the velocity of spatial expansion of the epidemic we still have to specify the functions $g(a)$ and $D(x)$. For these ingredients a number of alternative special cases can be developed. Realising that data on all processes underlying continental epidemic spread are extremely scarce, it is obvious that we have to restrict to simple, parameter sparse submodels to have any chance of being able to apply the model to real life situations. Here we will use two simple one-parameter functions.

For the above canopy spore production of a focus we assume

$$g(a) = \alpha \, a \,. \tag{7}$$

For spore dispersal we assume $D(x)$ to be described by a Bessel-density (Broadbent & Kendall, 1953, van den Bosch *et al.*, 1990). The marginal density of a Bessel-density is the double exponential density

$$D_1(\xi) = \frac{1}{2}\frac{1}{\sigma} \exp\left(-\sqrt{2}\frac{1}{\sigma}|\xi|\right). \tag{8}$$

The parameter σ is the standard deviation of the double-exponential density. This parameter will appear in the equation for the velocity of epidemic spread.

2.2 The velocity of expansion

Our model is a linear version of the epidemic model extensively studied by Diekmann (1978, 1979), Thieme (1977, 1979), Van den Bosch *et al.*(1990), Mollison (1991) and Metz and van den Bosch (in prep). We state some of the results of our model obtained by applying the ideas of Diekmann and Thieme.

The model admits travelling wave solutions. Such solutions can be visualised as disease profiles with a fixed shape in space that are translated at a constant velocity through space. For our linear model the disease profile has exponential shape. The travelling wave is given by

$$v(t,x) = N \, e^{\lambda(Ct - x)} \,, \tag{9}$$

where C is the velocity of spatial expansion of the epidemic, and λ is a parameter describing the steepness of the disease profile.

Applying the procedures developed by Diekmann (1978) and Van den Bosch *et al.* (1990) it can be show that for $\kappa \ll 1$,

$$C \approx \sqrt{2 \, \phi \, \psi \, \alpha \, \sigma^2} \,, \quad \lambda \approx 2\sqrt{\frac{\phi}{\kappa}\frac{2}{\sigma^2}} \,. \tag{10}$$

3. Expansion during successive seasons

So far we only considered the epidemic expansion within one growing season. Expressing the velocity C in the units km per day the total distance travelled by the epidemic wave in a growing season of T days is $C*T$. Next the crop is harvested and much of the disease mass is removed from the field. During the crop free period a focus has to survive on a secondary host, on a cull pile, or on plant parts left behind in the field. Only a very small fraction of the foci will survive the crop free period. This fraction is denoted by ε. From the exponential shape of the disease profile and the probability of a focus to survive, ε, it is calculated that the velocity of spatial expansion of the continental epidemic in successive years, V, is given by

$$V = \sqrt{2 \, \phi \, \psi \, \alpha \, \sigma^2} \, * \, T - \frac{1}{2\sqrt{2}} \sqrt{\frac{\kappa}{\phi}\sigma^2} \, \ln\!\left(\frac{1}{\varepsilon}\right) \tag{11}$$

4. *Phytophthora infestans* revisited

In this section we discuss the *P. infestans* epidemic once more. We show how the relevant parameters can be estimated and consider the predictability of the velocity of spatial expansion. Such an exercise has two major goals. Firstly, it serves as a first verification of the model. Secondly, it shows what type of data are needed.

The parameter most accurately and straightforward to estimate is ϕ, the fraction of the continent covered with host fields. In their agrarian history of England and Wales, Thirsh and Mingay (1989) report $18.8 \ 10^3$ km^2 of potato field around 1850. Assuming that this is characteristic for Europe we find an estimate of $\phi \approx 0.11$. For Northern United States, Bidwell and Falconer (1941) give estimates of the total potato production in 1839, and of the productivity per acre. From these data we find $\phi \approx 0.008$.

Phytophthora infestans overwinters on cull piles (van der Plank, 1963). The probability that a focus survives through winter, ε, is estimated from the total number of foci in a certain area at the end of the growing season and the number of 'primary' foci developed from the infected cull piles next growing season. Such data are unfortunately not available. In his aerial photography surveys Brenchley (1964, 1966) studied, among other things, the development of *P. infestans* epidemics in the UK. He reported, in 1963, in 450 potato fields only three initial foci at the beginning of the growing season (Brenchley, 1964). From the photographs from his work in 1964 we see that a potato field, which is probably not well kept, contains at the end of the season 20 foci. These are the only data on which we can base our 'guestimate' of ε. We find $\varepsilon \approx 0.0003$.

Brenchley (1966) also studied the development of a *P. infestans* epidemic within fields in detail. In a large area with only one initial focus he counted the number of newly established foci. Inoculum blown in from distant fields was considered to be negligible. From his figures one can count the number of daughter foci developed from a mother focus. In our notation the number of daughter foci, $N(a)$, is given by

$$N(a) = \int_0^a \kappa \psi g(\sigma)\, d\sigma = \frac{\kappa \psi \alpha}{2}\, a^2 \, , \tag{12}$$

where a is the age of the mother focus. Using this formula and the data from Brenchley we find $\kappa \psi \alpha \approx 0.002$.

To apply the formulae for the velocity of epidemic expansion we have to know κ and $\psi \alpha$ separately. In the fifties the Dutch National Agricultural Advisory Service had a yearly survey program on *P. infestans*. They give data on the number of fields infected in various areas of the Netherlands (Lint and Meyer, 1956, Anonymus, 1953). At the beginning of the growing season the number of infected fields increases exponentially in time. The exponential growth parameter, r, is approximately 0.15. It is reasonable to assume that the initial inoculum is distributed homogeneous since the epidemic is endemic for more than a century. Thus $b(t,x)$ and $v(t,x)$ in equation (6) are independent of x. Integrating over space we arrive at a non-spatial variant of the model. The rate of increase in the number of infected fields can be found from:

$$1 = \phi \frac{1-\kappa}{\kappa} \frac{\kappa \psi \alpha}{r^2 - \kappa \psi \alpha} \quad \Longleftarrow\Longrightarrow \quad \kappa = \frac{\phi[\kappa \psi \alpha]}{r^2 - [\kappa \psi \alpha](1-\phi)} \, . \tag{13}$$

We arrive at an estimate of $\kappa \approx 0.01$.

The only remaining parameter is the standard deviation of the marginal dispersal density. Data on the dispersal of spores on an inter-regional scale are not available. There have been some model studies on this problem. Aylor (1986) developed a detailed model for the spread of spores. Using his model it should in principle be possible to estimate σ. Aylor however shows that dispersal distance is very sensitive to UV damage of the spores. We do not know of data on the sensitivity of *P. infestans* spores to UV. To estimate σ we took recourse to calculating this parameter on the basis of our model and the velocity estimated for the expansion in Europe. Applying formula (10) and the parameters estimated we find $\sigma \approx 61.5$ km.

Now make the assumption that all parameters estimated for the European continent are also valid on the American continent, except for ϕ. Using formula (11) and a *P. infestans* season of 110 days (Hirsr and Stedman, 1960, Lint and Meyer, 1956) we get a predicted velocity of spatial expansion of 182 km per year. Compared to the observed velocity of 295 km per year the 'predicted' velocity deviates 40%.

5. Discussion

In this paper we have shown how the continental expansion of plant disease epidemics can be modelled. From this model we derived explicit formulae for the velocity of spatial expansion, both within one growing season and over successive growing seasons. These formulae are valid for $\kappa \ll 1$. In the example of *P. infestans* on potato $\kappa \approx 0.01$, implying that the formulae can be used.

To calculate the expansion velocity, submodels for the rate of above crop spore production of a focus and for the inter-regional dispersal of spores were chosen. For both, simple one parameter models were taken. It is possible that other choices lead to different velocities. The question thus is; how robust are our velocity formulae? Work on this point is needed. We will return to this point in the near future.

Estimating the parameters for *P. infestans* we experienced that accurate data are very scarce. We even had to make the assumption that the parameters estimated for the European continent also hold for the American continent. From a biological viewpoint this assumption is questionable, since there are major climatic differences between the two areas of invasion. Considering the need to develop a theory for quarantine pests we conclude that much experimental and field work has to be done to improve our knowledge of the processes underlying continental expansion. The example also made clear what type of data are needed. Specially the aerial photography work of Brenchley showed to be very useful. In this light it is surprising that this work has hardly had any follow-up. The more so, if one realises the recent interest in techniques as remote-sensing, aerial photography and geographical information systems. We think that use of these techniques in the studies on quarantine pests could result in a significant contribution in the next few years.

There are several studies on the spatial expansion of newly introduced species. Some studies compare model predictions of the velocity of expansion with observed velocities. For the velocity of focus expansion, where all parameters can be estimated accurately, the predicted velocities deviate between 10 and 30% from the observed ones (Van den Bosch *et al.*, 1988). For mammals, birds and insects, for which the parameters are less accurately known, predicted velocities deviate between 20 and 60% (Andow *et al.*, 1990 ,Van den Bosch *et al*, 1992). In this light a deviation of 40 % for the *P. infestans* invasion is surprisingly small. Specially if one realises the inaccuracy of the parameter estimates. This good fit might be a matter of luck only. Presently the authors are working on other examples to get more insight in the accuracy with which the expansion velocity can be predicted. We will return to this point in the near future.

More work has to be done before we can draw general conclusions. The first results however show that we have been able to capture the essential mechanisms behind continental expansion of plant diseases. The *P. infestans* example shows that it is in principle possible to estimate all relevant parameters. A first prediction of the expansion velocity showed to be surprisingly good. We thus conclude that using the approach developed in this paper, the velocity of continental spread of plant disease is open to investigations. Such work is bound to result in useful information on important quarantine pest problems.

References

Andow, D.A., P.M. Kareiva, S.A. Levin and A. Okubo. (1990), 'Spread of invading organisms', *Landscape Ecology*, **4**, 177-188.

Anonymus (1953), 'Verslag van de enquete over het optreden van de aardappelziekte in 1953', Jaarboek 1953, *Versl. en Meded. Plantenziektek. Dienst*, **124**, 34-46.

Aylor, D.E. (1986), 'A framework for examining inter-regional aerial transport of fungal spores', *Agricultural and Forest Meteorology*, **38**, 263-288.

Bidwell, P.W. and J.I. Falconer (1941), 'History of Agriculture in the Northern United States 1620-1860', Peter Smith, New York. 506 pp.

Bosch, F. van den, H.D. Frinking, J.A.J. Metz and J.C. Zadoks (1988), 'Focus expansion in plant disease: III. Two experimental examples', *Phytopathology*, **78**, 919-925.

Bosch, F. van den, J.A.J. Metz and O. Diekmann (1990), 'The velocity of spatial population expansion', *Journal of Mathematical Biology*, **28**, 529-565.

Bosch, F.van den, R. Hengeveld and J.A.J. Metz (1992), 'Analyzing the velocity of animal range expansion', *Journal of Biogeography*, **19**, 135-150.

Bourke, P.M.A.(1964), 'Emergence of potato blight, 1843-1846', *Nature*, **203**, 805-808.

Brenchley, G.H. (1964), 'Aerial photography for the study of potato blight epidemics', *World Review of Pest Control*, **3**, 68-84.

Brenchley, G.H. (1966), 'Aerial photography in agriculture', *Outlook on Agriculture*, **5**, 258-265.

Broadbent,S.R. and Kendall. D.G. (1953), 'The random walk of Trichostrongylus retotaeformis', *Biometrics*, **9**, 460-465.

Diekmann, O. (1978), 'Thresholds and travelling waves for the geographical spread of infection', *Journal of Mathematical Biology*, **6**, 109-130.

Diekmann, O. (1979), 'Run for your life', *Journal of Differential Equations*, **33**, 58-73.

Heesterbeek, J.A.P. and J.C. Zadoks (1987), 'Modelling pandemics of quarantine pests and diseases: problems and perspectives', *Crop Protection*, 21.

Hirst, J.M. and O.J. Stedman (1960), 'The epidemiology of *Phytophthora infestans*. II. The source of inoculum', *Annales Applied Biology*, **48**, 489-517.

Lint, M.M. de, and C.P. Meyers (1956), 'Resultaten van de Enquete over het optreden van de aardappelziekte in 1955', Jaarboek 1956, *Verls.en Meded. Plantenziektek. Dienst*, **128**, 116-133.

McGregor, R.C.(1978), 'People placed pathogens: The emigrant pests', In: *Plant Disease. An Advanced Treatise*, Vol.2, p.383. (ed. by J.G. Horsfall and E.B. Cowling). New York: Academic Press.

Metz, J.A.J. and F. van den Bosch (1993), 'Velocities of epidemic spread', submitted for publication.

Mollison, D. (1991), 'Dependence of epidemic and population velocities on basic parameters', *Mathematical Bioscience*, **107**, 225-287.

Plank, J.E. van der (1963), 'Plant Disease: Epidemics and control', Academic Press, New York and London. 349 pp.

Roughgarden, J. (1979), 'Theory of Population Genetics and Evolutionary Ecology: An introduction', MacMillan Publishing Co. New York. 634 pp.

Schumann, G.L. (1991), 'Plant Diseases: Their biology and social impact', The American Phytopathological Press, Minnesota, USA. 397 pp.

Stevens, N.E. (1933), 'The dark ages in plant pathology in America: 1830-1870', *Journal of the Washington academy of sciences*, **23** (9), 435-447.

Thieme, H.R. (1977), 'A model for the spread of an epidemic', *Journal of Mathematical Biology*, **4**, 337-351.

Thieme, H.R. (1979), 'Asymptotic estimates of the solutions of nonlinear integral equations and asymptotic speeds for the spread of populations', *Journal fur Reine und Angewandte Mathematiek*, **306**, 94-121.

Thirsk, J. (general editor) and G.E. Mingay (Vol. editor) (1989), 'The Agrarian History of England and Wales', Volume VI 1750-1850. Cambridge University Press.

Zadoks, J.C. (1988), 'Twenty-five years of botanical epidemiology', *Phil. Trans. R. Soc. Lond.*, B **321**, 377-387.

MODELING OF FISH BEHAVIOR

JENS G. BALCHEN

Department of Engineering Cybernetics
The Norwegian Institute of Technology
7034 Trondheim, Norway

Summary

The paper gives a survey of some fundamental principles which could form the basis for mathematical modeling of the large scale behavior (schooling) of fish in an ocean. The principle of *maximization of comfort* is presented as an acceptable fundamental hypothesis explaining behavior and the consequences of using this principle are discussed. A standard Kalman filtering structure is suggested for the estimation and prediction of fish behavior from observations made by research vessels etc.

Keywords Modeling, prediction, fish behavior, fish physiology.

1. Introduction

The subject of mathematical modeling of the behavior of animals is certainly both difficult and controversial. Therefore also very little is available in the litterature on this subject. The behavior of individual animals is governed by so many external influences and the animal itself demands so high dimensionality in a model that the modeling endeavour seems to be rather useless.

The situation becomes different however when attention is directed towards the study of groups of individuals of lower complexity such as schools of fish. If the modeling of behavior is limited to large scale motions such as migration, rather than the detailed movements of the individuals, there is a certain hope to arrive at models which could be useful. Models of fish behavior could have an application in fisheries technology in the development of fishing gear, in aquaculture for the development of large scale fish farming and for the planning of fishing operations in the utilization of pelagic species of fish.

2. General modeling principles

The establishment of mathematical models of physical, biological or social systems can be done in a number of entirely different ways.

One such way is to adopt *first principles*, that is to utilize apriori detailed knowledge of the *fundamental mechanisms* involved in the processes governing the behavior of the system. Such is the case when describing the motion of bodies subject to external forces moving through media of different kinds. The laws of physics often are known with an adequate degree of accuracy so that matematical models can be developed which will predict the future behavior with good precision. Examples of this kind of models are found in space technology, robotics, electrical engineering, chemical engineering, hydrodynamics, meteorology etc.

Another approach is to disregard any knowledge based on first principles and utilize only *observations* and construct a mathematical model with certain structural constraints e.g. linearity, with a behavior which will fit the observed data in a certain sense. A multitude of techniques are available for such *model fitting*. Typical fields where this technique has been applied, is in biology, sociology and economics. A general experience has been that the precision which can be achieved in prediction with such models usually is *not very good*.

Obviously the two points of view expressed by the two techniques above should not be contradictary. In trying to develope mathematical models of complex systems both *apriori knowledge* and *empirical knowledge* should be utilized.

The apriori knowledge can take the form of *structural knowledge* and *parametric knowledge*. The structural knowledge maps the interactions between the quantifiable phenomena describing the system behavior, such as the laws of physics governing motion of bodies, thermodynamics, hydrodynamics etc. The parametric knowledge expresses the particular numbers quantifying the strength in the relationships such as the size of a mass, heat conductivity, friction coefficients etc.

Often the *structural* apriori knowledge forms the basis of a model whereas some of the *parameters* are left free to be adapted so that the behavior of the model will fit to an *observed* behavior. This is typically the case in models of chemical engineering processes where the structural knowledge is acceptable whereas certain parameters like heat conductivity, activation energy or friction coefficients should be adjusted to fit the observed behavior. It is referred to as model identification.

When modeling complex processes employing *first principles* the modeller could present a number of alternative *hypotheses* which could be tested against observed behavior. Each of the hypotheses could be selected from an acceptable set offering different, but relevant descriptions of system behavior (Lund *et.al.* 1992). Multiple hypothesis modeling has proven useful in many technical applications such as fault detection and process control.

Model validation is the task that follows the model structural and parametric adaptation to the observed data and comes before the application to prediction of future behavior. In model validation the model is used to generated some entirely new behavior to be compared with observed behavior which has not been used before in constructing and adapting the model. In other words *the predictability* is tested before the model is actually used. A number of important issues enter into the problems of *model adaptation, model validation* and *model predictability*. As the time axis moves towards the *future*, data from the *near past* are used to adapt the model parameters (and possibly structure by hypothesis testing). From the present time the model can be asked to predict system behavior into the future with increasing prediction horizon. If the model is implemented in a computer, it can calculate future behavior quantities assuming different future external influences. Such predictions can be used for future model validation when actual data are available for comparison with the computed predictions.

3. Modeling fish behavior

Certain species of fish (herring, mackerel, etc.) exhibit distinctive schooling behaviors. The consequence of this is that the behavioral tendencies of an individual is being suppressed by the school so that the school behaves like a "big fish" with a behavior in terms of physiology which is similar to that of an individual fish (Balchen 1972, Balchen 1975). The *internal dynamics* of schools of fish has been the subject of research for many years and the theories have been verified by experiments to an extent that they can be used for further studies of the overall behavior of the school itself. This means that we may regard the school as an object in itself (Shaw 1978, Balchen 1979, Sannomiya 1984).

A *school of fish* is regarded as a collection of a large number of individuals which moves in the water masses as one body with a *center of gravity* determining its collective motion and *an average diameter* as a measure of its geometric extension. Each individual in the group at different locations may have different velocities and directions of motion because of the internal dynamics of the school, but the nearest neighbours always tend to move almost identically.

The motion of a school of fish can be defined in *an earth fixed* coordinate system in which also the *geophysical environment* around the fish moves. Some aspects of the fish behavior can be explained relative to the earth topography itself such as the bottom, coast line and rivers, whereas other aspects must be defined relative to the motion of the water masses because the fish behaves relative to temperature gradients, passively drifting plankton etc.

One very important aspect of the behavior of fish is that each species can have a *predator-prey* relationship to other species. Therefore, in such cases the modeling of interacting species must be done together as one large system. The simplest case is obtained when considering a species such as herring feeding on plankton which in turn is not exhibiting other than simple light controlled behavior. The principles of *multi-species* behavior modeling are not any different from *the single species* modeling, but the models become more complex.

For single species behavior modeling a number of general behavioral hypothesis have been suggested which are intended as bases for the overall behavior description (Weihs 1973a, 1973b, 1982, Brett 1971). In the following the principle of *maximization of comfort*, as introduced by Balchen (1976) and (1979) is adopted.

According to the principle of maximization of comfort: *the school of fish will at any time move in such a way (direction and velocity) so as to maximize its "comfort"*.

The comfort is defined as a physiological sensation of *well being* which is a function of *the physiological state* ($\underline{x}$) of the fish (the individuals are assumed identical) and of the *geophysical* and the *biological environment* ($\underline{v}$) of the fish such as the water temperature, light intensity, food concentration etc.

The physiological state of the fish is an outcome of the past development of the fish in its environment and can be described by a state space of reasonable dimension. In Balchen (1979) it is suggested that five state variables ($x_1...x_5$) should describe energy related variables such as stomach content (satiation), reversible and irreversible energy content etc. whereas three state variables ($x_6...x_8$) are allocated for endocrinologically related variables closely associated with behavior. A more comprehensive model of fish physiology is presented in Olsen and Balchen (1992).

The physiological model of a school of fish thus shall be given by

$$\frac{d}{dt}(\underline{x}) = \underline{\dot{x}} = \underline{f}(\underline{x}, \underline{v}, |\underline{\dot{r}}|, t) \, , \tag{1}$$

where

$\underline{x}$: physiological state of the fish
$\underline{v}$: geophysical and biological state of the fish environment
$|\dot{r}|$: magnitude of velocity of motion
t : time.

The environment of the fish which determines both its physiological changes and its behavior is composed of a combination of phenomena some of which are of *scalar* character (nondirectional) and some which are *vectorial* (directional). Typically temperature is a scalar quantity whereas light may be either scalar or vectorial. The environmental state is a scalar or vectorial *field* of the three-dimensional coordinate vector ($\underline{z}$) defining the position in earth fixed coordinates.

The modeling of the *motion behavior* of a school of fish regarded as a point in space and based on the principle of *maximization of comfort* requires a definition of the term *comfort*. The comfort $C(\underline{x}, \underline{y})$ is a scalar function of the physiological state ($\underline{x}$), the environmental state ($\underline{v}$) and the genetic parameters. The behavior model is based on the hypothesis that the fish, at any time, moves with a velocity ($\dot{r}$) which is a function of the gradient of the field of comfort such that

$$\underline{\dot{r}} = \underline{g}(\frac{\partial C(\cdot)}{\partial \underline{r}}) \; , \tag{2}$$

where $\underline{r}$ is three-dimensional position vector of the school of fish in the earth fixed coordinate system ($\underline{z}$). A simple approximation of the general function ($\underline{g}(\cdot)$) in (2) could be a linear transformation

$$\underline{\dot{r}} = G\frac{\partial C(\cdot)}{\partial \underline{r}} \; , \tag{3}$$

where G is a matrix of appropriate dimension in general dependent upon the space coordinate $\underline{z}$.

The comfort function $C(\cdot)$ has been studied in some detail (Balchen 1979) and it has been suggested that it under certain conditions can be decomposed in a number of specialized comfort functions which depend only on very few variables. Examples of such specialized comfort functions are shown in Figure 1-3.

In Figure 1 is shown the contribution to the comfort from the environmental temperature (v_1) and the "hormonal" state x_7. As can be seen, both high and low temperatures are associated with low comfort i.e. there is a range of temperatures preferred by the fish dependent upon the "hormonal" state x_7.

In Figure 2 the comfort associated with the physiological state variable x_1 (satiation) and the environmental state variable v_2 (concentration of food) is indicated. It shows that the attraction to food is dependent both on the level of satiation and the concentration of food.

Finally the contribution to the comfort associated with the general light intensity (v_3) is shown in Figure 3 as $C_3(\cdot)$. Here it is indicated that one of the physiological states (x_8) (sexual hormone) is contributing in the way that it shifts the preferred light intensity and modifies the sensitivity to changes in light. The relationships depicted in Figures 1-3 are interpretations of information available in the literature.

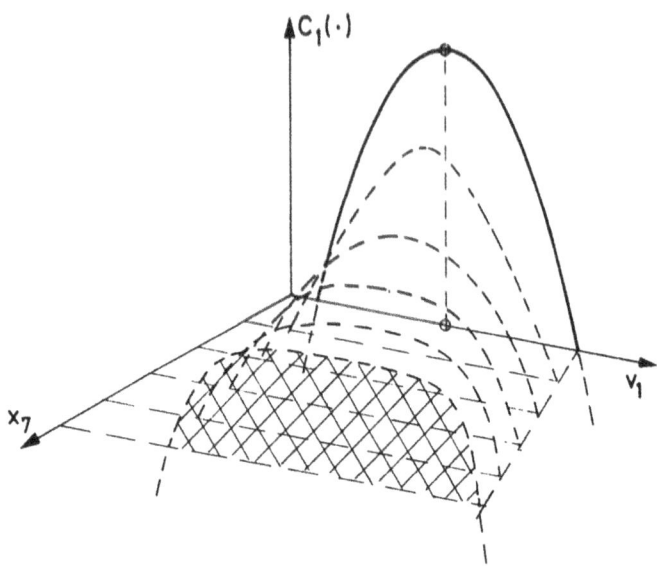

Figure 1. The comfort associated with temperature (v_1) and a physiological state ($x_7 \approx$ adrenocortical steroids).

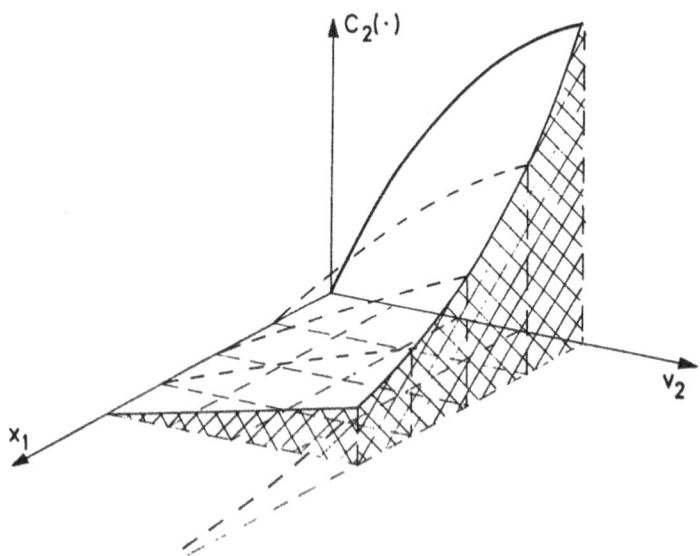

Figure 2. The comfort associated with concentration of food (v_2) and a physiological state ($x_1 =$ satiation).

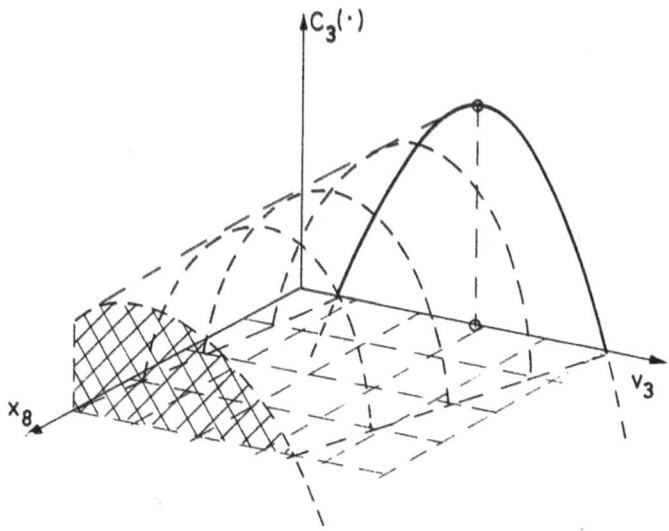

Figure 3. The comfort associated with the general light intensity (v_3) and a physiological state $(x_8 \approx$ sexual hormon).

The total comfort function $C(\cdot)$ can be regarded as a *product* of the specialized comfort functions $C_1(\cdot)$, $C_2(\cdot)$ etc. Alternatively the total comfort could be modelled as the *sum* of the specialized comfort functions (note that by taking the logarithm of a product one gets a sum).

The environmental state $(\underline{v}(\underline{z}, t))$ which influences both the physiological state of the fish and the behavior of the fish through the comfort function can be modelled by itself. In most cases there will be a negligible influence by the fish upon the environmental state.

The environmental model consists of different layers:

- A hydrodynamic model in three-dimensional coordinates describing the flow of water.
- A thermodynamic model describing the energy state of the water.
- A model describing the distribution and transportation of nutrients in the water masses.
- A model describing the growth, distribution and transportation of phytoplankton.
- A model describing the growth, distribution and transportation of zooplankton.

Such models have been under construction for many years by many people e.g. (Nihoul and Djenidi 1986, Støle-Hansen and Slagstad 1991, Slagstad and Støle-Hansen 1991, Ebenhöh 1980) and are even implemented in large scale commercial computer systems available for the real time simulation of fairly large ocean areas.

Combining the hierarchy of environmental models with the behavior model described above yields a system for the total simulation of the behavior of a single school of fish and schools of noninteracting fish. This has been illustrated for the Barents Sea capelin (*Mallotus Villosus*) in Reed and Balchen (1982).

The total state of a school of fish $(\underline{x}^t)$ can be defined as an augmented vector consisting of the physiological state $(\underline{x})$ derived from (1) and the position state $(\underline{r})$ derived from (2). Thus we have

$$\dot{\underline{x}}' = \begin{bmatrix} \dot{\underline{x}} \\ \dot{\underline{r}} \end{bmatrix} = \begin{bmatrix} \underline{f}(\underline{x},\underline{v},|\dot{\underline{r}}|,t) \\ \underline{g}(\dfrac{\partial C}{\partial \underline{r}}) \end{bmatrix} = \tilde{\underline{f}}(\underline{x}',\underline{v},t) \ . \tag{4}$$

A block diagram illustrating (5) is shown in Figure 4. For computational purposes the differential equation of (4) must be discretized:

$$\underline{x}'(k + 1) = \underline{F}(\underline{x}'(k), \ \underline{v}(k),k) \cong \underline{x}'(k) + \Delta t \ \tilde{\underline{f}}(\underline{x}'(k),\underline{v}(k),k) \ , \tag{5}$$

where Δt is the time increment, k the increment number, and

$$\underline{v}(k) = \underline{v}(\underline{x},t)|_{\substack{t = k\Delta t \\ \underline{z} = \underline{r}}} \ .$$

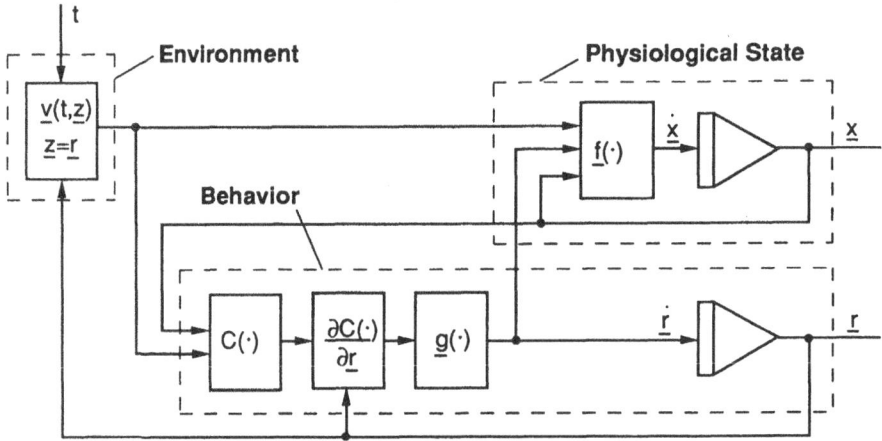

Figure 4.   Block diagram of the total system model.

4. The predictability of models for fish behavior

The above outline of a concept for modeling fish behavior contains many uncertain factors. Therefore one should expect that the predictability of the system of models would be rather low. In this context the predictability is defined as the time development of the error in the expected location and of the variance of the expected location of a school of fish.

Well established estimation theory (e.g. Sage and Melsa 1971) reveals that the expected value of the state of a system when predicting into the future will be given by the dynamic model of the system (1) whereas the covariance of the estimation error will grow mono-tonically into the future with a rate which is directly related to the dynamics (eigenvalues) of the system (1).

However, as is well known among fishermen and people living along the coasts of our fishing nations, the migration of our most important species of fish is a rather well predictable phenomenon because it repeats quite precisely during a one year geophysical cycle. Most schooling and migrating fish are guided by geophysical cycles resulting in temperature gradients, salinity gradients etc. and leading them to feeding grounds and spawning grounds in a very cyclic manner. Therefore the modellers have very much information available on observed behavior of both the physical and biological environment (the ocean) and the behavior of the fish in this environment. The fisheries authorities of our fishing nations have well-equipped research vessels observing the fish motion and measuring its population density and physiological development. Therefore an enormous amount of data is available for the updating of both the environmental models, the physiological models and the behavior models.

The consequence of the above reflections is that in the prediction of future behavior of fish in terms of motion there are two dominating components influencing the result:

- The behavior of the physical and biological environment of the fish.
- The behavior of fish relative to the physical and biological environment.

Which of these two components that contributes most to the uncertainty (error) in the prediction is not entirely clear, but it is evident that the larger a gradient of the comfort functions is, the more precisely the fish will be navigating and thus the errors due to prediction of the physical and biological environment will be dominating. This can also be concluded by arguing that in an ocean without "signals" the fish will disperse evenly.

The model of the total state of a school of fish can be used in a scheme for estimating the present and future behavior of the fish utilizing standard estimation methods, for example Kalman filtering techniques.

Many of the variables constituting x' may be measured at certain times, but with uncertainties associated with each measurement to account for sampling errors. Observations made by research vessels using echo-sounding equipment and samples of individuals in schools of fish are examples of real measurements. These measurements are modeled by

$$\underline{y}(k) = \underline{h}(x'(k)) + \underline{w}(k) \tag{6}$$

where $\underline{h}(\cdot)$ may well be a simple linear transformation matrix H, and $\underline{w}(k)$ represents the measurement uncertainty.

Standard Kalman-filtering techniques using (6) and (7) can produce an estimate of the present total state of a school of fish as shown in Figure 4. $\underline{\bar{x}}(k)$ is the *apriori* estimate of the total system state; $\underline{\hat{x}}(k)$ is the *aposteriori* estimate; $\underline{\bar{yy}}(k)$ is the estimated measurement; and $K(k)$ is the Kalman filter gain matrix. The well-established recursive algorithm for determining $K(k)$ is reviewed:

$$\underline{\bar{x}}'(k + 1) = \underline{F}(\underline{\hat{x}}'(k), \underline{v}(k), k) \ , \tag{7}$$

$$\underline{\hat{x}}'(k) = \underline{\bar{x}}'(k) + K(k)(\underline{y}(k) - \underline{\bar{y}}(k)) \ , \tag{8}$$

$$\underline{\bar{y}}(k) = \underline{h}(\underline{\bar{x}}'(k)) \ , \tag{9}$$

$$K(k) = \bar{X}(k)H^T(k)(H(k)\bar{X}(k)H^T(k) + W(k))^{-1} \ , \tag{10}$$

$$\bar{X}(k + 1) = \phi(k)\hat{X}(k)\phi^T(k) + \Omega(k)V(k)\Omega^T(k) \ , \tag{11}$$

$$\hat{X}(k) = (I - K(k)H(k))\bar{X}(k) \ , \tag{12}$$

where

$$\phi(k) = \frac{\partial \underline{F}(\cdot)}{\partial \underline{\hat{x}}'(k)}, \ H(k) = \frac{\partial \underline{h}(\cdot)}{\partial \underline{\bar{x}}'(k)}, \ \Omega(k) = \frac{\partial \underline{F}(\cdot)}{\partial \underline{v}(k)} \ , \tag{13}$$

$$\hat{X}(k) = cov(\underline{x}'(k) - \underline{\hat{x}}'(k)), \ \bar{X}(k) = cov(\underline{x}'(k) - \underline{\bar{x}}'(k)) \ ,$$

$$W(k) = cov(\underline{w}(k)), \ V(k) = cov(\underline{v}(k)) \ . \tag{14}$$

Figure 5 illustrates the structure of the Kalman filter.

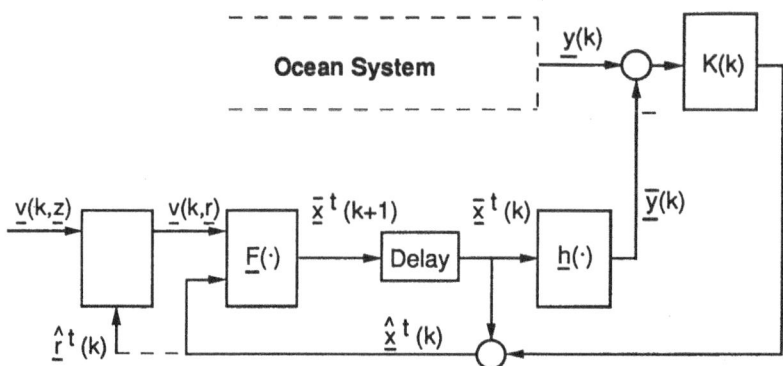

Figure 5.   Block diagram of total system state estimator.

When a prediction is desired, the model is used to continue the calculation starting with the estimated present state. Note that future values of the environmental state will drive the model of the state of the fish since it has been assumed that the environment is estimated using a scheme similar to that used for the other subsystems. If future values of the environment can not be predicted, then even a mathematical prediction of the fish behavior will be uncertain. Figure 6 shows a possible outcome of the estimation and prediction of the migration of two aggregates of pelagic fish in an ocean. The small circles indicate that the aggregates have a certain horizontal distribution. The dark circles indicate updating points where observations of the school locations are made.

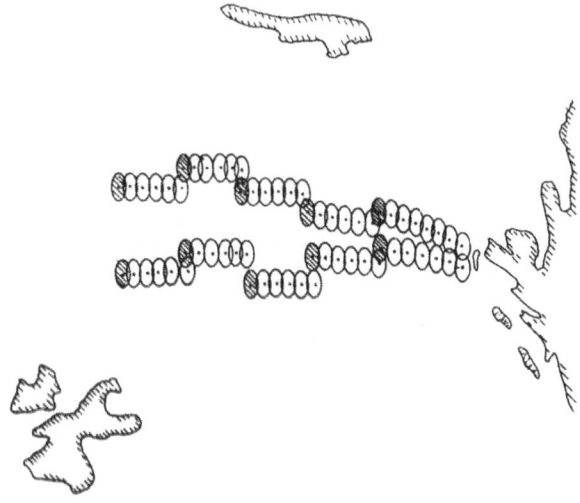

Figure 6. Estimation and prediction of the migration of two aggregates of pelagic fish in an ocean.

5. Applications of dynamic models of fish behavior

Numerous applications of dynamic models of fish behavior can be suggested such as

- In estimators to predict future behavior in order to
  - direct fishing operations
  - develope strategies for environmental development (polution control)
- In stocking operations to enhance population.
- Online applications in large scale aquaculture facilities.
- Simulation studies for scientific purposes.

It is envisaged that tomorrows fisheries research vessels cruising in the oceans to record the location and concentration of the fish populations will have online contact with a *dynamic map* (dynamic model) of the *ocean state* and the *fish population* state which is updated by means of new information from each vessel at regular intervals. This will improve the ability to forecast when fishing operations can take place and thereby reduce operational costs and increase the probability of catch.

6. Conclusions

Mathematical modeling of fish behavior is both fascinating and potentially useful in many respects. First principles understanding of why fish behave the way they do, forms the basis for establishing mathematical models that can be realized in modern high capacity computer systems and become useful in controlling fish behavior and planning the utilization of fish populations.

References

Balchen, J.G. (1972), 'Feedback Control of Schooling Fish', *Proc. IFAC 5th World Congress*, 5th Instr. Soc. Am. USA.

Balchen, J.G. (1975), 'Mathematical Modelling of Fish Behavior. Principles and Applications', *Proc. IFAC 6th World Congress*, Instr. Soc. Am. USA.

Balchen, J.G. (1976), 'Principles of Migration in Fishes', *Report   STF48 A76045*, Foundation of Scientific and Industrial Research (SINTEF), Trondheim, Norway.

Balchen, J.G. (1979), 'Modeling, Prediction and Control of Fish Behavior' in *C.T. Leondes, editor, Control and Dynamic Systems*, **15**, 99-146, Academic Press, New York, USA.

Balchen, J.G. (1981), 'Mathematical and Numerical Modeling of Physical and Biological Processes in the Barents Sea'. In Chapman, D.G.; V.C. Gallucci (eds.). *Quantitative Population Dynamics*. International Co-operative Publ. House, Burtonsville, Maryland.

Brett, J.R. (1971), 'Energetic response of salmon to temperature: A study of some thermal relations in the physiology and fresh water ecology of sockeye salmon (Oncorhyncus nerka)', *Am. Zool*, **11**, 99-113.

Ebenhöh, W. (1980), 'A Model of the Dynamics of Plankton Patchiness', *Modeling, Identification and Control (MIC)*, *1*, **2**, 69-92, Oslo, Norway.

Lund, E.J., J.G. Balchen, B.A. Foss (1992), 'Monitoring Processes with Structurally Different Operating Regimes'. Preprints IFAC Symposium: On-Line Fault Detection and Supervision in the Chemical Industries, Newark, Delaware, pp 126-131.

Nihoul, C.J., S. Djenidi (1986), 'Perspectives in Three-dimensional Modelling of the Marine System'. Eighteenth International Liege Colloquium on Ocean Hydrodynamics, Belgium.

Olsen, O.A., Balchen, J.G. (1992), 'Structured Modeling of Fish Physiology'. *Mathematical Biosciences* **112**, 81-113, Elsevier Science Publishing Company, New York, USA.

Reed, M., Balchen, J.G. (1982), 'A Multi-Dimensional Continuum Model of Fish Population Dynamics and Behavior: Application to the Barents Sea Capelin (Mallotus Villosus)', *Modeling, Identification and Control (MIC)*, *3*, **2**, 65-110, Oslo Norway.

Sage, A.P., J.L. Melsa (1971), 'Estimation Theory with Applications to Communication and Control'. McGraw-Hill Book Co., New York.

Sannomiya, S., Matuda, K. (1984), 'A Mathematical model of fish behavior in a water tank'. *IEEE Trans. Systems, Man, and Cybernetics*, **14**, 157-162.

Shaw, E. (1978), 'Schooling Fishes'. *American Scientist*, **66**, 166-175.

Slagstad, D., K. Støle-Hansen (1991), 'Dynamics of plankton growth in the Barents Sea: Model Studies', *Polar Research* **10 (1)**, pp 173-186.

Støle-Hansen, K., D. Slagstad (1991), 'Simulation of currents, ice melting and vertical mixing in the Barents Sea using a 3-D baroclinic model', *Polar Research* **10 (1)**, pp 33-44.

Weihs, D. (1973a), 'Mechanically Efficient Swimming Techniques for Fish with Negative Buoyancy', *Journal of Marine Research*, **31**, 194-209.

Weihs, D. (1973b), 'Hydromechanics of Fish Schooling', *Nature*, London, 241, 290-291.

Weihs, D. (1982), 'Bioenergetic Considerations in Fish Migration'. In: McCleave, J.D. et.al. eds. *Mechanisms of Migration in Fishes*, pp. 487-508. Plenum Press, New York, USA.

UNDERSTANDING UNCERTAIN ENVIRONMENTAL SYSTEMS

M. B. BECK

Warnell School of Forest Resources, University of Georgia
Athens, Georgia 30602-2152, USA

Summary

Developments over the past two decades in the identification of models of environmental systems are reviewed, with special reference to the quality and pollution of surface freshwaters. As in so many fields, the early 1970s were a time of great expectations: it would not be long, we believed, before the admittedly less well defined problems of environmental systems analysis would nevertheless yield to the already vast array of methods available from applied mathematics and control theory (which had been so successful in their application, for example, to the analysis of aerospace systems). Such a yielding has still to come to pass, at least for multivariable models of more than, say, five or six state variables. In the past decade, because of the seemingly insuperable difficulties of model identifiability, we have promoted the pragmatic view that what really matters is the ability to generate "robust" predictions that are maximally insensitive to a lack of identifiability. Such pragmatism, coupled with a continuing dearth of successful techniques of system identification, does not bode well. The digital computing technology on which we are able to realise our "set of concepts" (our models) continues to expand rapidly. A similar expansion, although less dramatically so, is apparent in the technology of instrumentation and remote sensing, through which our "given data" are acquired in ever greater volumes. No such expansion is evident in the capacity of the brain to juggle with disparate facts and figures until the ever more comprehensive, given data can be reconciled with the increasingly massive sets of concepts. Whither, then, is environmental system identification bound in the next decade? A modest attempt to answer this question will be made, by way of conclusion.

Keywords Kalman filter, system identification, identifiability, predictability, uncertainty, water pollution

1. Introduction

In a recent article -- on interactive computing as a teaching aid - MacFarlane (1990) has presented a three-element characterisation of knowledge. According to the American philosopher Lewis these three elements are (as reported by MacFarlane):

(i)   the given data;

(ii) a set of concepts; and

(iii) acts which interpret data in terms of concepts.

It is readily apparent that the problem of system identification (the derivation of a model whose behaviour bears the closest possible resemblance to the observed behaviour of the actual system) is covered exactly by the third of the three elements. What then are the prospects for success in these "acts which interpret data in terms of concepts", that is, for success in reconciling the candidate model with the given data? In other words, what are the prospects for progress in understanding uncertain environmental systems?

From contemporary experience we know that the scope and resolution of both the "given data" and the "set of concepts" necessary for understanding the behaviour of environmental systems are expanding at an increasing rate. We know too that this rate of expansion, if anything, is greater in respect of the set of concepts (the model) . The General Circulation Models (GCMS) of climatology and meteorology are of suitably massive proportions, with typically seven or more state variables (wind velocities, air temperature, and so on; Folland *et al.*, 1991) to be accounted for at over 10^5 spatial locations -- and doubtless soon to be still more. The computing capacity now available for realising models of the behaviour of a system offers us a truly staggering, expanding universe of possibilities. Indeed, in the popular scientific press this potential is neatly captured in headlines such as "Is It Real, Or Is It A Cray?" (Pool, 1989) or "Speculating In Precious Computronium" (Amato, 1991).

Technical support for manipulating the logical consequences of our "set of concepts" is thus assured. Technical support for acquiring the "given data" is likewise assured, although it might always be argued to be (relatively) inadequate. For example, in the area of modelling ocean circulation patterns it has been said that the rate of expansion in computing power (for realising ever more refined models) will bring about a need for a ten-fold increase every six years in data acquisition (for defining boundary and initial conditions). Given that US$2 billion per annum are reported to be required to service the current data-retrieval systems for monitoring ocean circulation, it is almost inconceivable that access to data will ever expand at a rate faster than the access to computing power.

No such assurances as these exist for development of the technical support necessary for engaging the model in a meaningful interpretation of the data. Indeed, how does such "interpretation" come about? It is a result of juggling with, and sifting through, a unique assortment of disparate facts and figures assembled by the individual, upon which some kind of order is eventually imposed. It is a subjective mental process. News of advances in computational capacity is abundant; news of advances in the technology of instrumentation and remote sensing is commonplace; news of the *increasing* capacity of the brain to juggle with disparate facts and concepts is non-existent.

Furthermore, what form of technical support would be desirable for promoting, provoking, or stimulating acts that interpret the data in terms of a set of concepts? This review of the tentative attempts to answer such questions is organised on a simple chronological basis, beginning with the hope of enlightenment in the 1970s,, passing through the clouds of uncertainty gathering during the 1980s and now, in the 1990s, looking forward to the prospect of rekindled hopes of further enlightenment.

2. Enlightenment: a better sense of the problems

Few of us -- working in the early 1970s -- could have guessed at the richness of "paradigms" now available for description and computerbased realisation of our theories about the behaviour of environmental systems. There are, for example, the following options: of classical differential calculus; of qualitative simulation (and the calculus of fuzzy logic); of cellular automata; and of pictorial simulation. All but the first of these have either been enabled or profoundly influenced by developments of just the past decade in the hardware and software of electronic computing. All are applicable, in principle, to the characterisation of problems associated with the contamination of surface water systems (Camara *et al.*, 1987, 1990; Castro *et al.*, 1993).

But such choice was not available to us two decades ago. Systems were conceived of as assemblies of mechanisms, characterised by state variables (x) subjected to input load disturbances (u) that generate various forms of output response (y). It was almost beyond question that description of their behaviour should take the form of a set of algebraic and/or differential equations. In terms of Lewis's three-element characterisation of knowledge, albeit perhaps with some licence:

(i)   the input-output data [u,y) constituted the external description of the system's behaviour and were the "given data";

(ii) the states and parameters (α) of the model, i.e., [x,α], constituted the internal description of the system's behaviour and were therefore a formal realisation of the "set of concepts"; and

(iii) calibration of the model was the "act which interpreted the data in terms of the set of concepts".

In respect of this last, it was generally assumed that all of the appropriate constituent hypotheses of the model would already have been assembled and correctly expressed; all that remained was to tune the parameters of the model, as an instrument requiring calibration for subsequent prediction. In retrospect, of course, one can ask whether the fine tuning of an already well structured instrument is the essence of progress in understanding, especially in the presence of gross uncertainty (attaching both to one's prior theories and to the observations of behaviour).

This restatement of Lewis's characterisation of knowledge presumes an important distinction between the input-output space (u,y] and the state-parameter space [x,α].

Such a distinction is important on two accounts. First, the recourse to a state-parameter space description of the system's behaviour suggests that the objective is indeed to reconcile an assembly of constituent hypotheses (the model, or the set of concepts) with a set of observations. This does not suggest that a model cast in the input-output space is devoid of, or unrelated to, a set of concepts -- as we shall see later -- but that it may not ultimately be the most useful vehicle for interpretation of the field data. The inputoutput models of time-series analysis are rather primitive vehicles for such interpretation. They might best be used as the means of preparing the data for subsequent interpretation through some other form of model (Beck, 1991). Second, for reasons of academic discipline, a state-parameter space description is by far the more popular form of expression of the models of environmental systems analysis.

Choosing thus a middle course, between the partial differential equations of what might frequently be referred to as a "physicsbased" model and the algebraic, discrete-time equations of the "black-box" models of control theory and time-series analysis, we have the following form of "conceptual" model

$$\dot{x}(t) = f\{x,u,\alpha;t\} + \xi(t) \qquad (1a)$$

$$y(t_k) = h\{x,\alpha;t_k\} + \eta(t_k) \qquad\qquad\qquad (1b)$$

in which the dot notation in $\dot{x}(t)$ denotes differentiation with respect to time t, ξ is a vector of unknown disturbances (or errors of model structure, or errors of observation of u) associated with the state vector dynamics and η is a vector of errors associated with observations of y, assumed pragmatically to have been made at discrete instants in time t_k.

2.1 Filtering theory as the conceptual framework

One can look at the problem of calibration as a matter of signal processing. Such a view is entirely sympathetic to the notion of models as vehicles ' for the interpretation of data. And the classic solution to the problem of reconstructing information about $[x,\alpha]$ from information about $[u,y]$ in the presence of uncertainty (or noise), as all control theorists will know, is the filtering theory of Kalman (Kalman, 1960; Kalman and Bucy, 1961). Quintessentially, the filter reconciles a prediction from the model with an observation of the system through a process of feedback (to the model!), in which the account taken of the mismatch between theory and observation is modulated according to the balance of uncertainties attaching to these two "elements of knowledge".

Over the years filtering theory has come to reflect something of a universal framework for exploring and formally defining -- yet not necessarily solving (as will become readily apparent) -- many of the most interesting sub-problems of uncertainty, identifiability and predictability. This is, however, a very personal, perhaps idiosyncratic, view (Beck, 1987).

Within this framework, and with reference to the model of equation (1), the problem of calibration can formally be defined as the problem of combined state-parameter estimation:

Problem #1: State-parameter estimation

Given the observations and the model structure, i.e., given $[u,y;f,h]$, and given assumptions about the various sources of uncertainty, i.e., $\{P_{xx}(t_0), P_{\alpha\alpha}(t_0), S(t_k), Q(t_k), R(t_k)\}$, determine the best estimates of $[x,\alpha]$ and the uncertainty attaching to these estimates, i.e., $\{P_{xx}(t_N). P_{\alpha\alpha}(t_N)\}$.

Here, P_{xx} is defined as the variance-covariance matrix of the state estimation errors, $P_{\alpha\alpha}$ is the variance-covariance matrix of the parameter estimation errors, S and R are respectively the variance-covariance matrices of the input and output observations, and Q is the variance-covariance matrix of the process (excluding the errors now accounted for explicitly under S). t_0 and t_N represent respectively the discrete instants of time at the beginning and end of the observation period, with $k = 0, 1, 2, ..., N$ sampling instants. In fact, to be pedantic, prior estimates of $[x,\alpha]$ at t_0 must also be assumed in the above problem definition.

The obvious flaw in this statement of calibration -- as the *first* act of reconciling the model with the data -- is that the model structure, as reflected in (f,h) (and also in the choices of the elements in $[x,\alpha]$), is not given a priori. *Before* determination of a best $[x,\alpha]$ the problem of model structure identification, defined formally as follows, must be addressed:

Problem #2: Model Structure Identification

Given the observations $[u,y]$ and given assumptions about the various sources of uncertainty, i.e., $\{P_{xx}(t_0),\ P_{\alpha\alpha}(t_0),\ S(t_k),\ Q(t_k),\ R(t_k)$ determine $[f,h;x,\alpha]$ and the accompanying $\{P_{xx}(t_N),\ P_{\alpha\alpha}(t_N)\}$.

Between these two statements (of state-parameter estimation and model structure identification), the degree of belief in the likely success of our prior theories in describing in general the behaviour of an environmental system has diminished. The burden of specifying $[f,h]$ correctly has shifted, from an almost complete reliance on prior theory and conjecture, to some engagement of the field observations in this process (which would in fact be more consistent with Lewis's three-element characterisation of knowledge). This change, and elucidation of the problem of model structure identification, did not arise from any philosophical consideration, but rather from a case study of organic waste degradation in a stretch of lowland river in eastern England (Beck and Young, 1976).

In short, the instrument of prediction may need more than just finetuning; it may need substantial re-design. The questions of interest to reconciliation of the model with the data are: which design is closest to the "truth"; how can we approach this "truth" at the fastest possible rate from some starting point; and what is a useful starting point for the prior model structure?

Exactly how one might go about answering these questions -- within the framework of filtering theory -- is summarised in Beck (1986, 1987) . The fact that one can obtain recursive estimates $[[\hat{x}(t_k),\ \hat{\alpha}(t_k)]$ across the period of the observed record from t_0 to t_N has been crucial, however, to insights about the nature of the problem of model structure identification. And this capacity to provide temporally varying parameter estimates is in turn the distinctive feature of a filtering-like algorithm, which thereby sets it apart from any other approach to a solution of this problem. But like any other approach, the filter cannot directly identify the "true" structure $[f,h]$, since this implies, inter alia, a means of estimating integer values for the numbers of differential and algebraic equations in the model. Its distinction lies in revealing unreasonable fluctuations in the recursive estimates of $\hat{\alpha}(t_k)$ that result from significant discrepancies between the structure underlying the observed field data $[f,h]$ and the structure of the candidate model, let us say $[f',h']$. In fact, we would hope these fluctuations are only superficially "unreasonable" and that behind them lies a plausible explanation.

Since individual parameters relate to constituent model hypotheses we have, in principle, a means to establish the "success" or failure of these *individual* hypotheses (as opposed to the more customary assessment of whether the model *as a whole* succeeds or fails) . Furthermore, after working on a number of case study problems (Beck and Young, 1976; Beck, 1982, 1985) it was possible to distil out a more systematic organising principle for the procedure of model structure identification, which comprised the following elementary questions of (Beck, 1986):

(i)   how to expose the *failure* (inadequacy) of the constituent hypotheses of a model structure;

(ii)   how to *infer* the form of an improved model structure from diagnosis of the failure of an inadequate structure?

Preparing tests of the model structure in order to answer these questions can again be realised within the framework of filtering theory. Respectively:

(i)   It can be assumed that the constituent parameter is unknown but invariant with time, i.e., α, with the expectation that a variable parameter estimate, i.e., $\hat{\alpha}(t_k)$, will deny that prior assumption; or

(ii)   It can be assumed that the parameter is variable, but varying in an unknown (random-walk) fashion, i.e., $\alpha(t_k)$, with the expectation of the posterior result that a more useful model of the parameter variations can be postulated through interpretation of $\hat{\alpha}(t_k)$.

Access to these tests is gained via the use of an assumed intensity of random perturbation of the parameter dynamics, by analogy with the state vector dynamics. Thus, distinguishing now between Q_{xx}, as the variance-covariance matrix of the unknown state perturbations, and $Q_{\alpha\alpha}$, as the variance-covariance matrix of a corresponding set of unknown parameter perturbations, the specification of the latter can be equated with a quantitative characterisation of the degree of confidence attaching to each constituent model hypothesis.

Moreover, setting $Q_{\alpha\alpha} = 0$ is the most dramatic way of formulating the test of (i) above; it ought to have the greatest possibility of exposing unambiguously the failure of a constituent model hypothesis. The ability of a physical engineering structure to resist deformation when placed under a test load is, by analogy, dependent upon the mechanical properties of its structural members; and these can be likened to the degree of confidence attached to each constituent hypothesis in an abstract model. The more confidently, or the more boldly the hypotheses are assumed to be stated, so the model structure is less flexible, more rigid, more brittle, and the more demonstrative should be the failure of the test structure.

Quite the opposite, however, is needed for the test of (ii) above. In this, speculation about a possibly improved specification of the model structure is the objective. The test draws its strength from the inherent flexibility of the model structure, which can be easily moulded to the patterns in the data and which can, therefore, be suggestive of ways in which to modify hypotheses.

All this, of course, is fine in principle, but not in practice.

2.2 Towards the limits

It is "fine in principle" because of the philosophical underpinnings that can be attached to what is in effect a Popperian programme of falsifying boldly stated, constituent hypotheses. And such association has uniquely been enabled as a result of using filtering theory as the conceptual framework for grappling with the problems of identification. In terms of understanding uncertain environmental systems and the evolution of knowledge MacFarlane (1990) equally so establishes a strong association between the work of Lewis and Popper:

Popper's and Lewis's approaches to knowledge are essentially the same, with different emphases. Both split concepts from interpretations, and both emphasise the distinct role of the individual mind in generating and using knowledge. Both regard the acquisition of knowledge as an iterative feedback process. Popper emphasises the objectivity of concepts, and Lewis emphasises the pragmatic role of the individual mind.

It is poor in practice because, first, too many prior assumptions - on $\{P_{xx}(t_0), P_{\alpha\alpha}(t_0), S(t_k),$ $Q(t_k), R(t_k)\}$ not to mention the prior estimatexs of $[x,\alpha]$ at t_0 -- must be made in order to implement the filter.

Second, it is equally poor in practice because of the difficulty of absorbing and interpreting the sheer volume of diagnostic information yielded from the filter (i.e., in the temporal variations of $\{P_{xx}(t_k), P_{\alpha\alpha}(t_k)\}$ and the recursive estimates $[\hat{\alpha}(t_k), \hat{\alpha}(t_k)]$ themselves). This was especially true in the more ambitious exercises in model structure identification attempted at the time, principally in respect of higher-order, state-space descriptions of the degradation of organic material and the proliferation of algal populations in the Bedford Ouse River (Beck, 1982; 1983). In anything but the smallest of models it is difficult to determine unambiguously where the constituent model hypotheses can be said to have failed. A lack of variation with time of a parameter estimate can result from two causes: the associated hypothesis is crucial and "correct"; or it is simply redundant (not identifiable), and no relevant information has been transferred from the external description of the system to this particular constituent of its internal description. One might then be able to resolve this ambiguity of interpretation through inspection of the variations with time of the diagonal, and then off-diagonal, elements of $P_{\alpha\alpha}(t_k)$ (although these possibilities were not greatly exploited at the time). But in short, the self-same sophistication of the questions that could be asked through the framework of filtering theory had become the barrier to progress in unscrambling the answers so generated. Precisely the same advantage that had led to insights into what the problems actually were, had become the cause of downfall in progress towards their practical solution.

Third, for a programme of research that had shunned the use of larger-scale, arguably "physics-based", models as vehicles for the interpretation of field data, not least because of the presumed difficulty of discriminating between key and redundant hypotheses, limits to the notion of the alternative "small being beautiful" had become apparent. From where, for example, would one pick a (posterior) hypothesis for replacing a demonstrably failed hypothesis in a prior model of insufficient content? This is no easier to answer than the alternative of identifying and casting out a redundant hypothesis from a prior model with surplus content.   Unlike the input-output models of time-series analysis, there are no systematic rules for extension or reduction of the number of terms (hypotheses) in the structure of these conceptual, state-space models.

Fourth, the power of the classical experiments of laboratory science lay presumably in promoting the possibility of "acts which interpret data in terms of concepts" by reducing the "set of concepts" under scrutiny to as small a set as possible and by maximising the scope for acquiring a large volume of the "given data". In principle, the possibility of progress in the identification of a model should be enhanced when the order of $[u,y]$ is very much greater than the order of $[x,\alpha]$ or, more succinctly, $O[u,y] \gg O[x,\alpha]$ (where order (O) increases with the number of elements in the respective vectors and, for the external description, is an increasing function of the density of temporal sampling). Achieving the condition of $O[u,y] \gg O[x,\alpha]$ can rarely be the case in the analysis of environmental systems. But where it is, as in the paucity (*not* simplicity) of the set of concepts required to describe pollutant transport and dispersion in a river, in combination with the facility of implementing repeatable dye-tracing experiments with high-frequency sampling, progress can be dramatic (Beer and Young, 1983; Young and Wallis, 1986). It may even prompt a shift in paradigm of the set of concepts (Young and Lees, 1993), though notably through the formal identification of input-output models cast resolutely in terms of the external description of the system's behaviour $[u,y]$. Such models do not directly permit reconciliation of the data with the set of concepts, in the sense intended here. They enable a translation from a "raw", $[u,y]$, to a "refined", $[\hat{u}, \hat{y}]$, external description of the system's observed behaviour, where the latter may provoke insights into the possible nature of $[x,\alpha]$ that are different from those apparent from the former (as, for example, in understanding what may lie beneath the now celebrated time-series of carbon dioxide variations in the upper atmosphere; Young *et al.*, 1991).

Such, however, is not the norm, the epitome of which is quite the opposite, i.e., $O[u,y] \ll O[x,\alpha]$. Above all, it was this that rendered the conceptual framework of filtering theory "poor in practice" or, rather more correctly, poor in *widespread* practice.

We had come to understand better what the problems were, in the process of assisting these "acts which interpret data in terms of concepts"; some principles for their more systematic resolution had emerged; yet even in the analysis of the relatively small-scale systems upon which the development of these principles had ·been based, it was extremely difficult to make them work well.

3. Uncertainty: the escape to pragmatism

By the mid-1980s the search for a method of resolving the problem of model structure identification, as posed above, had had to be put to one side. So too, though less consciously and with less speed, had the notion that there might be some "true" structure to be revealed 'through systematic reconciliation of the model with the observed patterns of behaviour.

In retrospect, our view of what was possible had been drifting: away from these more philosophical and "absolutist" ideas; towards a more pragmatic and "relativistic" position. This was not without purpose (nor without success) . After all, in the light of the above, it was natural to ask just what -- at bottom -- would be achievable? And still more so was this questioning needed, given that even for hydrological systems with copious volumes of data, it had long been found that it was not possible to recover a uniquely best set of parameter estimates allowing a match between the model and the observations (Ibbitt and O'Donnell, 1971; Johnston and Pilgrim, 1976; Sorooshian and Gupta, 1983). Given a candidate specification of the model structure, $[x,\alpha]$, the imperative was to establish what choice and form of observation $[u,y]$ would make this internal description identifiable? Or, put the other way round, what $[x,\alpha]$ could be recovered unambiguously from a given $[u,y]$? In fact, could we ever recover the "true" values of the parameters? And would it matter if we could not?

3.1 Identifiability: theoretical bounds on the possible

These questions, though seemingly theoretical, have a certain practical significance. In seeking to understand the acidification of surface waters, correct identification of the paths along which water flows from its impact with the ground and subsequent entry into a stream is crucial. These flow paths determine the soils and minerals with which the water has contact, the nature of the chemical interactions experienced and the duration of these interactions. Their significance is reflected in the values assumed by the various parameters in a conceptual state-space model. Hence there is a distinct need to know what form of observational data $[u,y]$ -- on water flows, on which natural chemical "tracers", at what sampling frequency, in what sequence, and with what degree of confidence -- will enable *in theory* the determination of a set of unambiguously estimated values for these parameters.

In other words, by a suitable rearrangement of the sets of "knowns" and "unknowns" in the previous definitions of state-parameter estimation and model structure identification, it is possible to develop a definition of a priori identifiability as follows:

Problem #3: A Priori Identifiability

Given a candidate model structure $[f,h]$ and given assumptions about $\{P_{xx}(t_0), P_{\alpha\alpha}(t_0), Q(t_k)\}$ determine which candidate set of observations $[u,y]$, with what degree of uncertainty $\{S(t_k), R(t_k)\}$, will allow the estimation of $[x,\alpha]$ with an acceptable degree of uncertainty $\{P_{xx}(t_N), P_{\alpha\alpha}(t_N)\}$.

This clearly goes beyond the notion of a priori identifiability in the deterministic sense of Bellman and Åström (1970), Pohjanpalo (1978), or Godfrey *et al.* (1982). Here "a priori" connotes simply the act of establishing what is possible *before* implementation of the progamme of observations in the field, thus giving the problem a title not strictly in conformity with usage elsewhere (as in Walter and Pronzanto, 1990).

In welcome contradistinction to what has been said earlier, Problem #3 is a problem to which the framework of filtering theory allows a successful solution (Beck *et al.*, 1990; Kleissen *et al.*, 1990). This success is bought at some expense, however, principally in that: (i) the analysis must be conducted for specific parameterisations $[\hat{\alpha}]$, and may therefore be but localised; (ii) $[\hat{\alpha}]$ must be prevented (within the filter) from varying with time, if there is to be any clarity in interpreting the results of the analysis (this is accomplished by assuming the hypothetical observations (u,y) available to the filter are identical with those generated from the nominal reference trajectory of the model's solution); and (iii) the identifiability of a parameter must be equated with expansion and contraction over time of the elements of the estimation error variance-covariance matrix $P_{\alpha\alpha}(t_k)$.

The test of Problem #3 seeks not to determine whether convergent estimates of the model's parameters can as such be obtained but is instead designed to explore the way in which uncertainty is propagated through the model and, in particular, how information (as the reciprocal of uncertainty) is transferred from the external description of the system $[u,y]$ to the component parts of its internal description $[x,\alpha]$. In contrast to the problems of model calibration and model structure identification, whose focus is on the nature of the estimates of $[x,\alpha]$, the focus in this test is on the properties of $P_{\alpha\alpha}(t_k)$. Whether, and to what extent, the elements of this matrix contract or expand will be indicative of whether the constituent model parameters are more or less identifiable. The information-transcribing mechanism, uniquely associated with the filtering algorithm, is itself revealing of how access to which observations, and which combinations of field conditions and naturally perturbing events, can enhance (or corrupt) the confidence attaching to the various parts of the model.

Yet such analysis, in the context of needing to understand the behaviour of uncertain environmental systems, is in the end only shadow-boxing. It does not facilitate acts which interpret data in terms of concepts. Rather, since it deals with hypothetical data from contemplated experiments, it merely enables bounds to be set on what is possible for such reconciliation in the best of all theoretical worlds. This is not, however, without practical implications, as, for example, in the design of better pumping tests for the identification of aquifer parameters in the characterisation of groundwater fields.

In the real world, as well as in this theoretical world, we now know that the bounds on what is possible are close by. At most just a handful of parameters α can be recovered unambiguously from a single input-output pair, monitored with a high frequency (Hornberger *et al.* 1985; Beven, 1989; Jakeman *et al.*, 1990; Kleissen *et al.*, 1990). This has the ring of relativism about it, in the sense that we may expect it to hold irrespective of the scale of resolution of the observations. Yet it may not be a linear property, in the sense that two input-output pairs will probably not permit as many as two handfuls of model parameters to be recovered. And there is evidence of a theoretical nature suggesting that the presence of nonlinearities in the candidate model will degrade still further the "degree" of identifiability of the parameters (Kleissen *et al.*, 1990).

A number -- and it is notably small -- has been placed on what is achievable.

3.2 Sparseness of the data: progress under substantial uncertainty

We might draw at least three conclusions from this experience, that:

(i)   most, if not virtually all, of the conceptual state-space models used to describe the behaviour of environmental systems are not identifiable and are therefore incapable of unambiguous reconciliation with the given data (which is not to say that they are not useful models);

(ii)   the entire concept of identifiability is, as a consequence of (i), simply not useful (or that our quantitative analysis of it is defective); or

(iii) from the pragmatic point of view of making predictions, this inability to eliminate ambiguity is not material (what matters is that the ambiguity is apparent and its implications quantifiable).

This last is an area in which progress has been made, although it might in fact be seen as something of a retreat: from seeking the ideal of acquiring a uniquely *optimal* set of parameter estimates; through a search for a cluster of merely *relatively good* candidate parameterisations of the model; to being content with just an *acceptable* set of such candidate parameterisations.

This retreat was not actually a matter of conscious pursuit. Its origins were born of other motivations. One of these was the recognition (in the late 1970s) that $O[u,y]$ may be so small as to amount to little more than merely a subjective, expert, qualitative appreci-ation, $["u","y"]$ say, of the external description of the system's behaviour. So common was (and still is) this the case -- of "rich" sets of concepts confronted with "impoverished" sets of sparse data -- that the absence until then of systematic methods enabling some kind of progress to be made under these conditions, is a curiosity (Hornberger and Spear, 1981). Having thus shifted from a quantitative to a qualitative description of observed behaviour, expectations of what is possible from the acts which interpret the field data in terms of a set of concepts must similarly so be drawn back. One can at most investigate which, among the many constituent mechanisms (hypotheses) in this rich set of concepts, are key (and which redundant) to discriminating between the model's capacity to generate what is defined as acceptable behaviour and its complement (not-the-behaviour). Crucially, it would then be to a better understanding of the nature of these key mechanisms so identified that the inevitably limited capacity for implementing field experiments should gainfully be directed (Hornberger and Spear, 1980; Spear and Hornberger, 1980).

In the face of gross uncertainty the distinction sought in this analysis, between key and redundant mechanisms, must necessarily be based on a sufficiently large sample of randomly generated candidate model parameterisations. Each parameterisation classified as acceptable constitutes an equally probable interpretation of past behaviour so that, with a twist to the original intention of the analysis, the ensemble of such acceptable parameterisations may be used for computing an ensemble of predictions (Fedra *et al.*, 1981). Any ambiguity, distortion, or uncertainty residing in the model following this justifiably passing attempt at its reconciliation with the data, is thereby apparent and quantitatively reflected in its predictions of future behaviour.

3.3 Identifiability, predictability and pragmatism

We can thus escape to the pragmatism of making predictions; and there have been many subsequent examples of precisely this (Keesman and van Straten, 1991; Klepper *et al.*, 1991; Beck and Halfon, 1991; and Beven and Binley, 1992). Again, in the case of surface water acidification, all the attraction of so doing is readily evident. For a situation of $O[u,y] \gg O[x,\alpha]$ even the best attempts at identification lead to ambiguous, if not contradictory, interpretations of past behaviour: $[x,\alpha]^1$ and $[x,\alpha]^2$. In effect, it is found that the water has contact either strictly with an upper soil horizon alone or strictly with both an upper and lower soil horizons (Beck *et al.*, 1990). Either could be used for predictive purposes.

The interesting question -- called herein a question of "predictability" (albeit not in the terms discussed by Wegman (1989), for example) -- is whether this makes any material difference.   More formally, and once more rearranging the sets of "knowns" and "unknowns" in the preceding problem definitions, we have:

Problem #4: Predictability

Given $[f,h;u]$, together with assumptions about $\{Q(t_k), S(t_k), R(t_k)\}$, and two (or more) interpretations of past behaviour crystallised through $[x,\alpha]^{1,2}$ and $\{P_{xx}(t_N), P_{\alpha\alpha}(t_N)\}^{1,2}$, determine whether $[y]$ differs significantly from $[y]^2$ in the light of $\{T(t_k)\}^1$ and $\{T(t_k)\}^2$.

Here $[y]^{1,2}$ are understood to be output responses of interest, upon which decisions may be based, and which are generated from $[x,\alpha]^{1,2}$ respectively; $\{T(t_k)\}^{1,2}$ are the respective error variance-covariance matrices associated with $[y]^{1,2}$. It should also be noted that t_k now refers to discrete instants in time over a forecasting horizon starting at t_N, for which a future pattern of input disturbances u is assumed to be known, albeit with a degree of uncertainty (as expressed by $S(t_k)$).

It will come as no surprise that this problem can be approximately solved within the context of filtering theory, simply by assuming that the next sampling instant for observing the system's behaviour is infinitely far into the future (Beck, 1983; Beck and Halfon, 1991).

4. Contemporary scene

If there is a "true" structure believed to govern the behaviour of the system -- which is essentially the assumption that motivated the developments of the 1970s (Section 2 above) -- our goal may not necessarily be to have this "truth" revealed by systematic and protracted attempts at reconciliation of the model with the data. Our outlook on the way in which we seek to understand the world around us and, perhaps more pragmatically, to utilise this understanding, has changed. Progress throughout the 1980s (as set out in Section 3) has brought to us to a position in which we recognise that each successive attempt at reconciling the model with the data will result in a distorted interpretation of what is observed; this distortion can be quantified (not least within the conceptual framework of filtering theory); its effects can be faithfully reflected in the predictions generated from the distorted model structure; and the task is then to choose between alternative courses of policy and regulatory action in a manner that is maximally insensitive to this ambiguity and distortion. In a sense too, the field has drawn back from the search for optimality in the performance of its models.

4.1 Behaviour under novel conditions: reachable futures

There is a paradox. The greater the degree of extrapolation from past conditions, so the greater must be the reliance on a model as the instrument of prediction; hence, the greater the desirability of being able to quantify the reliability of the model, yet the greater is the degree of difficulty in doing just this. What is more, the contemporary problems of environmental protection are increasingly of such a form where prediction of behaviour under quite novel conditions is called for. For example, there is a need to predict the fate of utterly novel chemicals in the environment before they are released into that environment. Or alternatively, in the case of surface water acidification, there is a need to extrapolate from small-scale catchments observed over a matter of years to behaviour over entire continental regions and over decades into the future.

We are strongly accustomed to the idea of behaviour being specified in terms of time-series of observations of the system's inputs and outputs $[u,y]$. This is not only rarely the case, as already acknowledged in the work of Hornberger, Spear and Young (Young *et al.*, 1978; Hornberger and Spear, 1980; and Spear and Hornberger, 1980), it is also very restrictive. The notion that $[u,y]$ can be replaced by more qualitative, linguistic descriptors ["u" , "y"] may come to have a profoundly liberating influence on the subject of understanding uncertain environmental systems.

The analyst has immense freedom to be creative in defining the task or purpose, of a model. Fitting the historical data as closely possible has been a traditional such purpose, although this was not an end in itself, merely a means to a better understanding of the system's behaviour. In their seminal work Young *et al.* (1978) were concerned to locate a sample of randomly generated values for the model's parameters that enabled the model outputs to satisfy certain crude constraints, i.e., ["u", "y"], on what is defined (not actually observed) to be an acceptable statement of past behaviour. Yet if behaviour can be so defined for the past, so too can it be for the future (Beck, 1987, 1991), such that the task shifts to that of locating a sample of randomly generated values for the model's parameters that enable the model outputs to match certain crude constraints on what has been defined to be radically different behaviour of the system in the future.

The questions of interest become ones of whether and how a prespecified pattern of future behaviour is, as it were, "reachable" (Beck, 1991). They can formally be defined thus:

Problem #5: Reachable Futures

Given $[f,h;"u"]$ and $["y"]$, a prespecified pattern of *future* output responses (that may be radically different from those of the past), determine from $["x", "\alpha"]$ which constituent parameters $[\alpha]^K$ are key to enabling the model to generate $["y"]$ and which $[\alpha]^R$ are redundant.

Here $["u", "y"]$ has been assumed, loosely speaking, to subsume the previous use of $[u,y]$ and $\{S(t_k),R(t_k)\}$ and likewise $["x", "\alpha"]$ subsumes $[x,\alpha]$ and $\{P_{xx}(t_k), P_{\alpha\alpha}(t_k), Q(t_k)\}$. Though this may seem a novel problem at first sight, it is not. It is merely a modest rearrangement of the more familiar control problem of finding what input, regulatory action $[u]$ (as onnosed to $[\alpha]$) will transfer $[y]$ to some desired performance level $[y^d]$. In practice, answers to such questions are of current interest in determining, for example, what aspects of a lake's biochemistry, possibly in combination with which changes in the lake's local climate, will lead to an expressly feared radical change of behaviour in the future.

But to turn matters entirely on their head, if such future responses $["y"]$ are truly radically different from those of the past, then in theory the values of $[\alpha]^k$ thus identified ought strictly not to be identifiable (or at most barely identifiable) from the observed record of past behaviour. Now this, if a sensible assertion, would make a virtue out of a lack of model identifiability! Indeed, it provides pointers for where to search -- within the low amplitude, perhaps relatively low-frequency noise of the short records of past behaviour -- for the barely identifiable seeds (constituent mechanisms) of the radically different behaviour that is feared in the future. We have been preoccupied with identifying the major, dominant modes of behaviour captured unambiguously in the "signal". Yet it may well be the minor modes of behaviour, buried within the "noise" at the fringes of our understanding, that have the capacity of becoming the dominant modes of future behaviour.

4.2 Filtering theory: a renaissance

In the end, then, there is a distinct impression of things turning back on themselves, perhaps nowhere more so than in the revival of interest in filtering theory as a means of solving the problem of model structure identification. For we also need pointers, as observed in Section 2, for where to search for a (posterior) hypothesis for replacing a demonstrably failed hypothesis in a prior model of insufficient content.

In 1979 Ljung published a paper on a modified form of extended Kalman filter (let us say LEKF for short) that would improve the estimation of parameters in a conceptual, state-space model (Ljung, 1979). The attraction and potential of this algorithm for the purposes of identifying the model's structure were immediately obvious (Beck, 1987). It held out the possibility, among other advantages, of changing significantly the composition of two of the most important practical difficulties of solving this problem, because (with reference back to Section 2):

(i)   It reduced the number of arbitrary prior assumptions that had been necessary to implement the EKF, specifically from the plethora of $\{P_{xx}(t_0),\ P_{\alpha\alpha}(t_0),\ S(t_k),\ Q(t_k),\ R(t_k)\}$ to just $\{P_{\alpha\alpha}(t_0),\ S(t_k),\ R(t_k)\}$;

(ii) It provided access to estimates not only of $[x(t_k),\ \alpha(t_k)]$ but also the elements of the gain matrix $K(t_k)$, the focal point of the feedback mechanism whereby the filter modulates the account taken of the mismatch between theory and observation according to the balance of the uncertainties attaching to the model and the data.

The balance between "prior assumptions" and "posterior performance", as gauged not just by $[\hat{x}(t_k),\ \hat{\alpha}(t_k)]$, nor merely with $\{P_{\alpha\alpha}(t_k)\}$ in addition, but now also with $K(t_k)$, has been markedly shifted towards the latter. Indeed, the very way in which the set of concepts are reconciled with the given data -- through use of the filter's gain matrix -- has been made more sensitive to the confrontation of the one "element of knowledge" with the other.

The potential of Ljung's algorithm seems to have remained largely unrealised since its publication, and undoubtedly so in the field of environmental systems analysis, at least until very recently, that is (Stigter, 1993). In fact, liberated thus from the confines of a conventional view of filtering theory, all manner of interesting questions may be opened up. For example:

(i) Does the LEKF have a useful directional property, in the sense of revealing through its "posterior performance" (rather than by "prior assumption") to which constituent hypotheses in the model the failure to match the given observations is due?

(ii) Should the gain matrix of a filter be chosen not -- as is conventionally the case -- in order to minimise the variance of the state-parameter estimation errors but rather to maximise sensitivity to the detection of a structural error or the "seeds" of a structural change?

(iii) Or indeed, should the gain matrix be specified through a neural net trained to detect structural anomalies?

These, however, are already in the realm of speculation for the distant future.

5. In conclusion

Progress can seem painfully slow at times. For many of the ideas reviewed towards the end of this paper are not new. It is really rather disquieting, for example, to have to draw attention to the gap of fourteen years between the publication of Ljung's algorithm and its subsequent successful implementation on a problem of interest to this paper. What is rather different from previous such reflections is the organisation of this review around Lewis's three-element characterisation of knowledge: (i) the given data; (ii) a set of concepts; and (iii) acts which interpret data in terms of concepts. It is this last that has been of principal interest, albeit within the context of making predictions of future behaviour, and inevitably under uncertainty.

At the heart of filtering theory -- as admittedly with any algorithm of system identification -- is a feedback mechanism of correction, adaptation and learning. Drawing a parallel with Lewis's third element of knowledge, what the filter achieves so elegantly is modulation (through its gain matrix) of the account taken of the mismatch between the data and the set of concepts according to the balance of whether the data are believed to be less uncertain than the set of concepts, or vice-versa. Moreover, in contrast to other algorithms, estimated values of the model's parameters can be generated as a function of time. These estimates are accompanied by estimates of the variance of the parameter estimation errors, which are also available as a function of time.

The conceptual framework of filtering theory facilitates insights of a general nature into the mechanics of reconciling the set of concepts with the given data. Fluctuations over time of the reconstructed model parameter estimates are indicative of the failure (or success) of the constituent hypotheses of the model. By analogy with a physical engineering structure, these fluctuations are conditioned upon the strength of the hypotheses and their ability to withstand the various loads imposed on the structure (as a consequence of the mismatch between the model and the data). Such an insight has a very strong association with Popper's view on the acquisition of knowledge. The fundamental difficulties with this, however, are that: (i) the principles of model structure identification so derived can barely be made to work successfully for even the simplest of models; (ii) much of any distortion of the model's structure may be governed by a plethora of notoriously arbitrary prior assumptions; (iii) the evidence gathered from the test is voluminous and not easily distilled into the essence of understanding how the model's structure might be improved; and (iv) the algorithms of filtering theory are being used for a purpose for which they were never intended.

Given this impasse it was easier to make progress on other fronts.

Thus first, decisions are made on the basis of predictions. All predictions are subject to uncertainty; this uncertainty derives in part from the residual uncertainty in the model; and the pattern of residual uncertainty in the model is a function of whatever distortion, or ambiguity, remains as a result of all the preceding attempts to reconcile the set of concepts with the data. What matters is whether or not the same decision should be made in the light of these distortions and ambiguities. This is a line of enquiry we can successfully pursue, using, if nothing else, the ubiquitous Monte Carlo simulation. Pragmatically, it may not matter that the distortions are incapable of rectification and the ambiguities incapable of resolution. It is crucially important that they can be quantified, however, through measures of model uncertainty, and their consequences accounted for in the propagation of prediction errors.

Second, and entirely theoretically, if there were a need to resolve the ambiguities in understanding the past, what form of field observations with what accuracy would be needed? This too is a question for which there is a means of obtaining an answer. Contraction and expansion of the uncertainty attaching to the model's parameter estimates, as uniquely illuminated through the estimated variance-covariance matrices of a filtering algorithm, determine when information in the external description of the system (the data) can be productively and counter-productively brought to bear on the constituents of its internal description (the model).

Third, the obligation of forecasting the future under conditions substantially different from those of the past will be the downfall of most, if not all, models. There is a dilemma (Beck, 1983). The "large" model -- that will result from including everything of conceivable relevance to the problem at hand -- may indeed be capable of predicting "correctly" such radically different behaviour; but we would place little confidence in this prediction. The "small" model -- that will result from any honest analysis of the past data -- may quite "incorrectly" predict behaviour in the future essentially similar to that of the past; worse still, we might place great confidence in this erroneous prediction. There is, of course, no way in which we could have fore-knowledge of either of these results. Instead, more fruitful progress can be made in identifying which constituent mechanisms in the model, and/or what degree of climate change, for instance, may be key to the reaching (or not) of some predefined radically different "target" behaviour of the given system in the future.

Last, and turning to matters of contemporary interest, progress may now be possible in two of the principal areas where the original insights of filtering theory failed nevertheless to enable the development of practical solutions to the problems of interpreting the given data in terms of a set of concepts. Certain modifications of the basic algorithms allow us to dispense with some of the arbitrary prior assumptions and to have access to a greater variety of fee ' dback, diagnostic information. It remains to be seen whether there are ways of manipulating all of this information for the purposes of accelerating a reconciliation of the model with the data.

Acknowledgements

The work reported on in this paper has been supported in part by the Visiting Scientists and Engineers Program of the United States Environmental Protection Agency (project title "The Analysis of Uncertainty in Environmental Simulation"). The paper is also a contribution to the International Task Force on Forecasting Environmental Change, which is currently receiving support from the National Water Research Institute of Environment Canada, the National Institute of Public Health and Environmental Protection (RIVM) of the Netherlands, and the International Institute for Applied Systems Analysis, Laxenburg, Austria.

The author is currently Visiting Professor in the Department of Civil Engineering at Imperial College, London.

References

Amato, I. (1991), 'Speculating in precious computroniuml', *Science*, **253**, 23 August, 856-857.

Beck, M. B. (1982), 'Identifying models of environmental systems behaviour', *Mathematical Modeling*, **3**, 467-480.

Beck, M.B. (1983), 'Uncertainty, system identification, and the prediction of water quality', in *Uncertainty and Forecasting of Water Quality* (M.B. Beck and G. van Straten, eds), Springer, Berlin, pp 3-68.

Beck, M.B. (1985), 'Lake eutrophication: identification of tributary nutrient loading and sediment resuspension dynamics', *Appl Math Comput*, **17**, 433-458.

Beck, M.B. (1986), 'The selection of structure in models of environmental systems', *The Statistician*, **35**, 151-161.

Beck, M.B. (1987), 'Water quality modeling: a review of the analysis of uncertainty', *Wat Res Res*, **23(8)**, 1393-1442.

Beck, M.B. (1991), 'Forecasting environmental change', *J Forecasting*, 10(1&2), 3-19.

Beck, M.B. and E. Halfon (1991), 'Uncertainty, identifiability and the propagation of prediction errors: a case study of Lake Ontario', *J Forecasting*, 10(1&2), 135-161.

Beck, M.B. and G. van Straten (eds) (1983), *'Uncertainty and Forecasting of Water Quality*, Springer, Berlin.

Beck, M.B. and P.C. Young (1976), 'Systematic identification of DO-BOD model structure", *Proc Am Soc Civil Eng, J Environ Eng Div*, **102(EE5)**, 902-927.

Beck, M.B., F.M. Kleissen and H.S. Wheater (1990), 'Identifying flow paths in models of surface water acidification', *Rev Geophys*, **28(2)**, 207-230.

Beer, T. and P.C. Young (1983), 'Longitudinal dispersion in natural streams" *J Environ Eng*, **109(5)**, 1049-1067.

Bellman, R. and K-J. Åström (1970), 'On structural identifiability', *Math Biosci*, **7**, 329-339.

Beven, K.J. (1989), 'Changing ideas in hydrology - The case of physically-based models', *J Hydrol*, **105**, 157-172.

Beven, K.J. and A.M. Binley (1992), 'The future of distributed models: model calibration and uncertainty prediction', *Hydrological Processes*, **6(3)**, 279-298.

Camara, A.S., F.C. Ferreira, D.P. Loucks and M.J.F. Seixas (1990), 'Multidimensional simulation applied to water resources management', *Wat Res Res*, **26(9)**, 1877-1886.

Camara, A.S., M. Pinheiro, M.P. Antunes and M.J.F. Seixas (1987), 'A new method for qualitative simulation of water resources systems. 1. Theory', *Wat Res Res*, **23(11)**, 2015-2018.

Castro, A.P., D.L. Gallagher and A.S. Câmara (1993), 'Dynamic water quality modeling using cellular automata', *Technical Report*, Virginia Polytechnic Institute, Blacksburg, Virginia.

Fedra, K., G. van Straten and M.B. Beck (1981), 'Uncertainty and arbitrariness in ecosystems modelling: a lake modelling example', *Ecol Modelling*, **13**, 87-110.

Folland, C., J. Owen, M.N. Ward and A. Colman (1991), 'Prediction of seasonal rainfall in the Sahel Region using empirical and dynamical methods', *J Forecasting*, **10(1&2)**,21-56.

Godfrey, K.R., R.P. Jones, R.F. Brown and J.P. Norton (1982), 'Factors affecting the identifiability of compartmental models', *Automatics*, **18(3)**, 285-293.

Hornberger, G.M. and R.C. Spear (1980), 'Eutrophication in Peel Inlet, I. Problem-defining behaviour and a mathematical model for the phosphorus scenario', *Wat Res*, **14**, 29-42.

Hornberger, G.M. and R.C. Spear (1981), 'An approach to the preliminary analysis of environmental systems', *J Env Management*, **12(1)**, 7-18.

Hornberger, G.M., K.J. Beven, B.J. Cosby and D.E. Sappington (1985), 'Shenandoah Watershed study: calibration of a topographybased, variable contributing area hydrological model to a small forested catchment', *Wat Res Res*, **21(12)**, 1841-1850.

Ibbitt, R.P. and T. O'Donnell (1971), 'Fitting methods for conceptual catchment models', *Proc Am Soc Civil Eng, J Hydr Div*, **97(HY9)**, 1331-1342.

Jakeman, A.J., I.G. Littlewood and P.G. Whitehead (1990), 'Computation of the instantaneous unit hydrograph and identifiable component flows with application to two small upland catchments', *J Hydrol.*, **117**, 275-300.

Johnston, P.R. and D.H. Pilgrim (1976), 'Parameter optimization for watershed models', *Wat Res Res*, **12(3)**, 477-486.

Kalman, R.E. (1960), 'A new approach to linear filtering and prediction problems', *Am Soc Mech Eng Trans, J Basic Eng*, **82**, 3545.

Kalman, R.E. and R.S. Bucy (1961), 'New results in linear filtering and prediction theory', *Am Soc Mech Eng Trans, J Basic Eng*, **83**, 95-108.

Kleissen, F.M., M.B. Beck and H.S. Wheater (1990), 'The identifiability of conceptual hydrochemical models', *Wat Res Res*, **26(12)**, 2979-2992.

Klepper, O., H. Scholten and J.P.G. van de Kamer (1991), 'Prediction uncertainty in an ecological model of the oosterschelde Estuary', *J Forecasting*, **10(1&2)**, 191-209.

Ljung, L. (1979), 'Asymptotic behaviour of the extended Kalman filter as a parameter estimator', *IEEE Trans Automat Contr*, **24**, 36-50.

MacFarlane, A.G.J. (1990), 'Interactive computing: a revolutionary medium for teaching and design', *Computing & Control Engineering Journal*, **1(4)**., 149-158.

Pohjanpalo, H. (1978), 'System identifiability based on the power series of the solution', *Math Biosci*, **41**, 21-33.

Pool, R. (1989), 'Is it real, or is it a Cray?', *Science*, **244**, 23 June, 1438-1440.

Sorooshian, S. and V.K. Gupta (1985), 'The analysis of structural identifiability: theory and application to conceptual rainfallrunoff models', *Wat Res Res*, **21(4)**, 478-495.

Spear, R.C. and G.M. Hornberger (1980), 'Eutrophication in Peel Inlet, II. Identification of critical uncertainties via generalised sensitivity analysis', *Wat Res*, **14**, 43-49.

Stigter, J.D. (1993), 'Application and further development of the extended Kalman filter in a river water pollution problem', *Graduation dissertation*, Department of Applied Mathematics, University of Twente, Enschede, The Netherlands.

van Straten, G. and K.J. Keesman (1991), 'Uncertainty propagation and speculation in projective forecasts of environmental change: a lake-eutrophication example', *J Forecasting*, **10(1&2)**, 163-190.

Walter, E. and L. Pronzanto (1990), 'Qualitative and quantitative experiment design for phenomenological models - A survey', *Automatics*, **26**, 195-213.

Wegman, E.J. (1988), 'On randomness, determinism and computability.', *J Statistical Planning and Inference*, **20**, 279-294.

Young, P.C., and M. Lees (1993), 'The active mixing volume: a new concept in modelling environmental systems', in *Statistics and the Environment* (V Barnett and R Turkman, eds) , Wiley, Chichester (to appear).

Young, P.C., and S.G. Wallis (1986), 'The aggregated dead zone (ADZ) model for dispersion in rivers', in *Water Duality Modelling in the Inland Natural Environment*, British Hydraulics Research Association, Cranfield, England, pp 421-433.

Young, P.C., G.M. Hornberger and R.C. Spear (1978), 'Modelling badly defined systems - some further thoughts', in *Proceedings SIMSIG Simulation Conference*, Australian National University, Canberra, pp 24-32.

Young, P.C., C.N. Ng, K. Lane and D. Parker (1991), 'Recursive forecasting, smoothing and seasonal adjustment of non-stationary environmental data', *J Forecasting*, **10(1&2)**, 57-89.

SYSTEM IDENTIFICATION BY APPROXIMATE REALIZATION

CHRISTIAAN HEIJ

Econometrics Institute, Erasmus University Rotterdam
P.O. Box 1738, 3000 DR Rotterdam
The Netherlands

Abstract

We discuss the use of state variables in time series modelling. Current procedures are mostly based on realization theory. First certain parameters are estimated which describe the process, e.g., the systems impulse response or autocorrelations. These parameters are then transformed into an approximate state space model. In this note we suggest an opposite procedure. First an approximate state trajectory is estimated, and in a second stage a corresponding state space model is determined. This method allows to infer several structural properties of the process from the observed data, in particular the dynamical structure (length of the involved time lags), causality (which variables are inputs and outputs), and the noise (whether it is stochastic or not).

1. Introduction

Time series modelling involves the selection of a model structure and the estimation of parameters describing the process. In the classical, statistics oriented approach these two aspects are mostly considered separately. A popular method is that proposed by Box and Jenkins (1970). First the orders of an ARIMA model are somehow determined from the data. Once the model structure is fixed, the remaining parameters are estimated. The choice of the model orders can then be subjected to statistical tests, and one can repeat the procedure if the model would not be acceptable. Another method for selecting the model structure is to use a criterion taking the model fit and model complexity jointly into account. A well-known example is the Akaike information criterion, cf. Akaike (1976).

Another approach for modelling observed time series is by means of state space models. Probably the most well-known use of such models in econometrics is the Kalman filter for maximum likelihood estimation of ARIMA models, cf. Harvey (1989). This method also requires that the model structure has been selected in a preliminary analysis. We will focus on alternative state space modelling techniques where the selection of the model structure and the estimation of the model parameters are treated as two interconnected aspects of the same modelling problem. We briefly discuss a balancing method for stochastic time series modelling as developed by Aoki (1987). This method is based on deterministic realization theory. The estimated autocorrelations are used in an approximate realization procedure to estimate the parameters of a state space model. As an alternative we describe a so-called subspace identification procedure. In this method, first approximate state variables are estimated, which are then used to identify a state space model. The advantage of this method is that one need not decide a priori whether the model deviations are of a stationary stochastic or of a non-stochastic nature. The procedure delivers so-called driving variables, and one can test whether these correspond to noisy or more systematic factors influencing the observations. This subspace identification procedure is based on exact state space modelling of time series as developed in Willems (1986). Corresponding procedures have been developed for input-output systems, see, e.g., De Moor (1988).

The attractive aspect of our procedure is that the choice of the model structure is to a large extent data determined. The methods developed by Aoki, De Moor and others start from the assumption that the data have either a stochastic or a deterministic character. Further, the causality structure is imposed as inputs (exogenous variables) and outputs (endogenous variables) have to be selected. In our method we only make the a priori restriction, like do the other methods, that the model is linear, time invariant, and with finite lags. The choice of inputs and outputs and of the residual properties can be made ex post, after the model has been determined. As we impose less prior restrictions this means that we also have less information on the formal properties of the procedure. For example, classical identification procedures assume random noise which makes it possible to analyse asymptotic statistical properties like consistency. Currently there is also a growing interest in nonlinear deterministic dynamics and the corresponding limit theory. In our opinion these assumptions are in general rather artificial. We see it as an attractive aspect that, by means of the unstructured driving variables, we leave certain aspects of the dynamics unmodelled. We regard model deviations as a consequence of necessary simplifications in modelling which need not exhibit any clear probabilistic or nonlinear deterministic regularity. In this respect we take a more pragmatic standpoint.

2. Realization Theory

We will briefly review some elements from deterministic and stochastic realization theory.

Suppose that a linear input-output system of the following type is given

$$y(t) = \sum_{k=0}^{\infty} G(k)u(t-k). \tag{1}$$

Here y denotes the outputs and u the inputs of the system. The problem in realization theory is to represent this system by means of a state space model

$$x(t+1) = Ax(t) + Bu(t), \tag{2}$$
$$y(t) = Cx(t) + Du(t). \tag{3}$$

Here A, B, C, D are matrices of appropriate dimensions. The variables in x are called state variables. The state $x(t)$ at time t summarizes the information from the past inputs which is of relevance for the future evolution of the process. Indeed, the values of the future outputs $\{y(s); s \geq t\}$ are completely determined by the current state $x(t)$ and the future inputs $\{u(s); s \geq t\}$.

The state is an artificial variable which is auxiliary in describing the dynamical relationship between inputs and outputs. As a consequence the parameters A, B, C are not identified. The model (2), (3) is a realization of (1) if and only if $G_0 = D$ and $G(k) = CA^{k-1}B$, $k \geq 1$.

It is well-known, cf. Kalman et al. (1969), that for minimal realizations, i.e., with the number of states as small as possible, the parameters are unique up to a choice of basis in the state space. A particular basis is obtained by choosing so-called balanced coordinates, cf. Moore (1981). In these coordinates the state components are somehow ordered in decreasing importance. The contribution of each state component to the dynamical evolution of the process is measured by the so-called (Hankel) singular values. This leads to straightforward model reduction procedures by neglecting the states corresponding to small singular values, cf. also Glover (1984).

Next we consider stochastic systems. Assume that there are no inputs and that the system variables y follow a stationary process. We restrict attention to the linear dynamical interrelationships between the variables as expressed in the autocovariances

$$R(k) := E\{y(t)y'(t-k)\}, \quad k \in Z. \tag{4}$$

The stochastic realization problem consists of representing such a process in the form

$$x(t+1) = Ax(t) + B\varepsilon(t), \tag{5}$$
$$y(t) = Cx(t) + D\varepsilon(t). \tag{6}$$

Here ε is a white noise process and x is called the state. We suppose that A is a stable matrix, in which case the state process has the Markov property. As is well-known, see e.g. Davis and Vinter (1985), a process y can be represented in state space form (5), (6) if and only if y is an ARMA process. Minimal realizations are again non-unique. Apart from a choice of basis in the state space, there is a choice of basis in the noise space and a choice of the covariance matrix of the state, cf. Faurre et al. (1979). A uniquely defined minimal realization can be obtained by means of canonical correlation analysis, as is exposed in Akaike (1975). This can also be seen as a choice of stochastically balanced coordinates, see Desai and Pal (1982). As before, this leads to a simple model reduction procedure by neglecting the states with small canonical correlation coefficients.

This classical realization theory is, one could say, a transformation theory. Given a representation of the system, by the input-output map (1) or the autocovariances (4), the question is to determine another system representation in terms of state space models. In the next section we describe a method in terms of arbitrary observed data.

3. Signal-oriented Realization Theory

An alternative realization approach has been developed by Willems (1986). This approach could be called signal-oriented, as it is based on an arbitrary observed time series, denoted by

$$w = \{w(t); \ t \in Z\}. \tag{7}$$

The realization problem consists of representing this time series by a state space model of the type

$$x(t+1) = Ax(t) + Bv(t), \tag{8}$$
$$w(t) = Cx(t) + Dv(t). \tag{9}$$

Here x and v are both artificial, auxiliary variables. No dynamical restrictions on v are supposed, and these are called the free, driving variables. Then the process x has again the state property. Restricting to minimal realizations, i.e., with the number of states and driving variables as small as possible, the parameters are non-unique. The freedom consists of a choice of basis in the state and driving variable spaces and of a state feedback, cf. Willems (1986).

In this signal-oriented setting, a minimal realization can be obtained in terms the spaces of past and future observations. Define the s-shift of the observation by $w_s(t) = w(t+s)$, $t \in Z$. Let W_- denote the linear space generated by the components of all past shifts $\{w_s; s \leq -1\}$, and let W_+ denote the corresponding space of future shifts $\{w_s; s \geq 0\}$. So W_+ and W_- are subspaces of the set of all univariate discrete time series. Then the state space is given by the common features of past and future, i.e.,

$$X = W_- \cap W_+. \tag{10}$$

By choosing a basis in the state space X we obtain the state trajectory x. The driving variables v and the parameters in model (8), (9), can than be obtained by determining a linear space L such that

$$(w(t), x(t), x(t+1)) \in L. \qquad \text{for all} \quad t \in Z. \tag{11}$$

This can subsequently be transformed into a model of the type (8), (9).

We remark that this set-theoretic approach to deterministic realization in some respects resembles the geometric realization theory for stochastic systems, for which we refer to Lindquist and Picci (1985).

4. Approximate State Space Modelling

In sections 2 and 3 we discussed three types of realization theory, that is,

(i)   balanced realization of deterministic input-output systems;
(ii)   stochastically balanced realization of purely stochastic systems;
(iii) set-theoretic realization of an observed time series.

For corresponding realization algorithms we refer, respectively, to Silverman (1971), Akaike (1975), and Willems (1986). These algorithms are developed for the ideal situation that the system parameters in (1) or the autocovariances in (4) are known. Further they assume that exact finite dimensional state space representations (2), (3) or (5), (6), or (8), (9) exist. In practice the information is more limited, in which case one should look for approximate realizations. Model reduction procedures can, by definition, be applied for simplifying given state space models. It depends on the structure of the corresponding algorithms whether they can also be used in less ideal situations for the approximate realization of observed data. This is indeed possible for the algorithms for deterministic input-output systems. This area has seen rapid developments in recent years, starting with the work of Glover (1984). Concerning the stochastic balancing method of Desai and Pal (1982), it seems at this moment unclear how this algorithm could be amended to make it useful for the approximate stochastic realization of observed data. Other methods based on canonical variables have been developed by, e.g., Akaike (1976).

Aoki has introduced a deterministic balancing method for modelling observed stochastic processes, see Aoki (1987). In this approach the observed data are used to estimate the covariances (4). These parameters are then employed in an approximate deterministic balancing procedure. This exploits the fact that these algorithms, for the deterministic case, are capable of dealing with data which do not exactly satisfy a state space description. A disadvantage is, however, that this method neglects the so-called positive real condition of stochastic systems. For further details we refer to Heij *et al.* (1991). There it is also shown that this method often encounters serious problems in estimation. If the method works, however, then this gives a combined estimation of the model structure and the model parameters. The model structure is determined by the number of states which are accepted in the model description. This choice is based on the systems singular values.

An alternative method for joint model selection and parameter estimation is obtained by approximate algorithms based on the signal-oriented realization theory described in section 3. In this case an approximate state trajectory can be constructed from an approximate intersection of the past and future spaces W_- and W_+. Such an approximate intersection can be based on canonical correlation analysis, cf. De Moor (1988), see also Van der Veen (1991) for an overview. Optimal approximate state trajectories are based on the magnitude of the angles between the spaces W_+ and W_- and on the amount of energy which the observations exhibit in the different directions in these spaces. Once the state has been determined, approximate restrictions of the form (11) can be based on a singular value decomposition of the data. We refer to Heij and Roorda (1991) for further details. We mention that the number of state variables is based on the canonical correlation coefficients in (10), and the number of driving variables on the singular values in (11).

5. Examples

In this section we describe two simple examples which illustrate the approximate algorithm of section 4 based on the signal-oriented realization theory of section 3. In particular we will show that this method can be applied to deterministic and stochastic systems alike. The choice of the character of the driving variables can be made a posteriori, after the model has been determined. The same holds true for the choice of inputs and outputs.

5.1 Example 1

We first consider a deterministic system, the so-called "Mexican hat". This contains two variables, w_1 and w_2 which are related by

$$w_2(t) = -\frac{d^2}{dt^2}\left\{\int_{-\infty}^{\infty}\varphi(x)w_1(t-x)dx\right\},\tag{12}$$

where $\varphi(x)=(2\pi)^{-1/2}\exp\{-1/2x^2\}$ denotes the standard normal density. So w_2 indicates the longer term changes in w_1 after the short term oscillations have been smoothed. Note that this is not a causal input-output system, as w_2 depends not only on the past but also on the future of w_1. We consider the data of figure 1, where w_1 corresponds to a change in level under noise with w_2 the corresponding response. These sampled data consist of 81 observations.

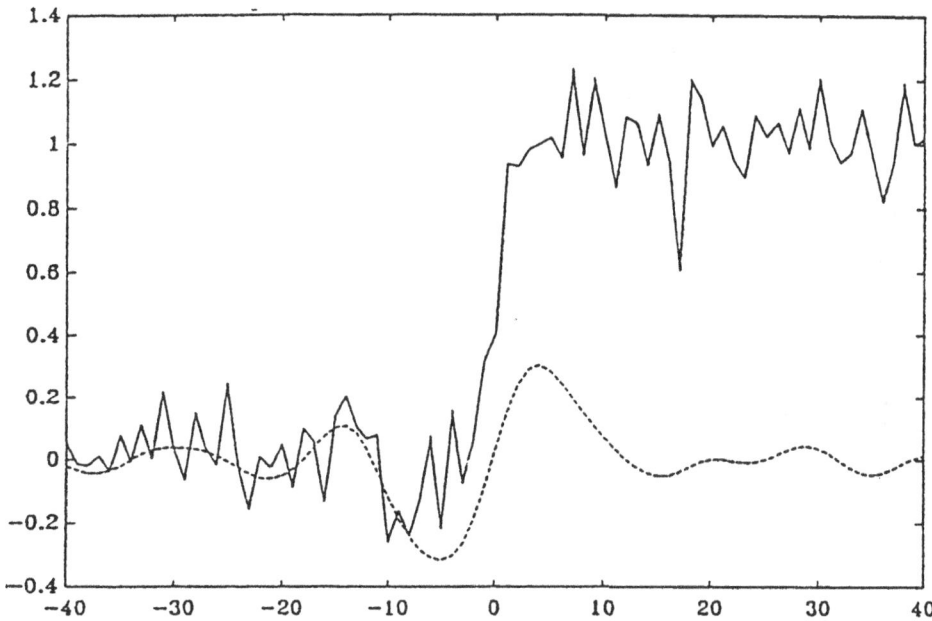

Figure 1.   Data in example 1 (w_1 ——— , w_2 ---).

In this case the canonical angles in (10) suggest a state dimension of 8. The state then consists of the first eight canonical variables. The singular values in (11) suggest nine relations. In terms of the state space model (8), (9), this leaves one component of w unrestricted, so we obtain a system with a single input and a single output.

In order to measure the quality of the estimated model, we compare it with an optimal approximation. Note that the system (12) can be written in terms of a causal and an anticausal part. For each we determined a fourth order optimal Hankel norm approximation, cf. Glover (1984), resulting in a combined eighth order approximation of the system (12). The estimated system is compared with this optimal Hankel norm approximation by considering the responses in w_2 of the systems if a pulse w_1 is applied. Figure 2 shows the simulation errors in w_2 for the two approximations of (12). These errors are of the order 10^{-4}, which is very small as compared with the systems response (12) to a pulse, which is of the order 10^{-1}.

We conclude that the algorithm works fairly well in modelling observed data in this case. The result is even comparable to optimal Hankel norm approximation. This last method, however, requires that inputs and outputs have been selected and, moreover, that the observed input is a pulse and the output the corresponding impulse response. Our method is much less restrictive, as it can be applied for arbitrary data, and it still leads to comparable results.

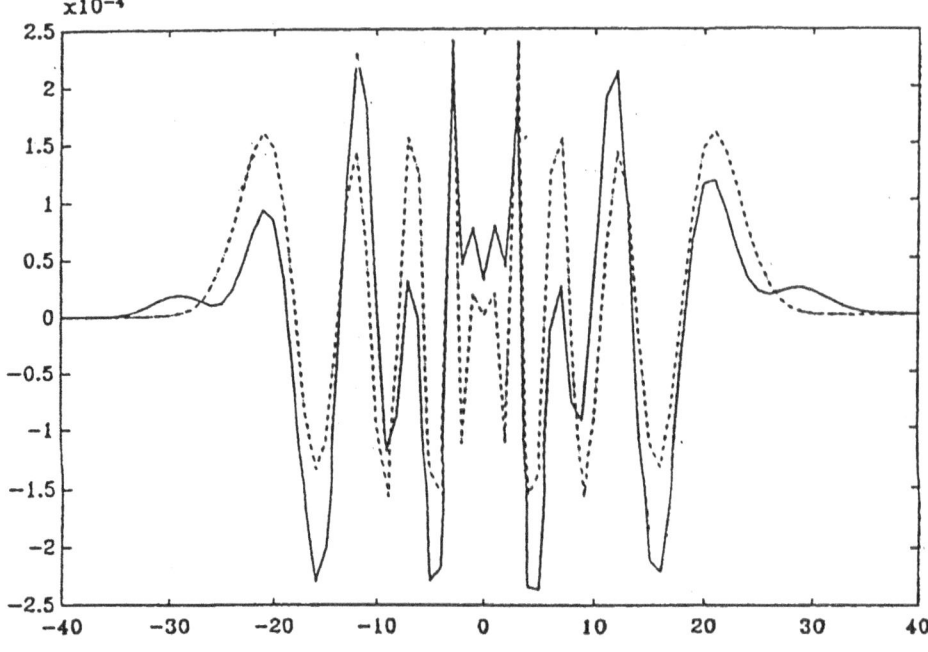

Figure 2.   Spectral factors (system ——— , model ---).

5.2 Example 2

As a second example we consider a stochastic ARMA (2,2) model, specified by

$$y(t) = -0.4y(t-1) + 0.32y(t-2) + \varepsilon(t) - 0.2\varepsilon(t-1) - 0.63\varepsilon(t-2). \tag{13}$$

The roots of the autoregressive part are -0.8 and 0.4, and those of the moving average part are 0.9 and -0.7. The data consists of a univariate time series of length 80, generated by (13) with ε an unobserved sample of a white noise process with mean zero and variance one. The canonical angles suggest a state dimension one. Choosing the state trajectory accordingly, the singular values in (11) are (0.7, 13.2, 13.9), which suggests a single relationship. This means that, apart from $x(t)$, an additional factor is needed in order to explain $(y(t), x(t+1))$ by the state space model (8), (9). If we require that this factor is orthogonal to the explanatory state trajectory, then an optimal factor can be computed by a singular value decomposition of the data. The resulting estimated model (8), (9) can be rewritten as the following ARMA (1,1) model,

$$y(t) = 0.37y(t-1) + v(t) - 0.91v(t-1). \tag{14}$$

In figure 3 we show the spectral factors of the original system (13) and of the estimated model (14), i.e., the impulse responses from ε to y and from v to y respectively. This seems to be a reasonable first-order approximation. Figure 4 gives the noise ε and the constructed variables v. This shows that the algorithm determines a very accurate estimate of the unobserved noise components. So, on the basis of the observed data, the models

(13) and (14) can not be distinguished by any criterion based on the model residuals v. This indicates that the algorithm also works fairly well for stochastic systems.

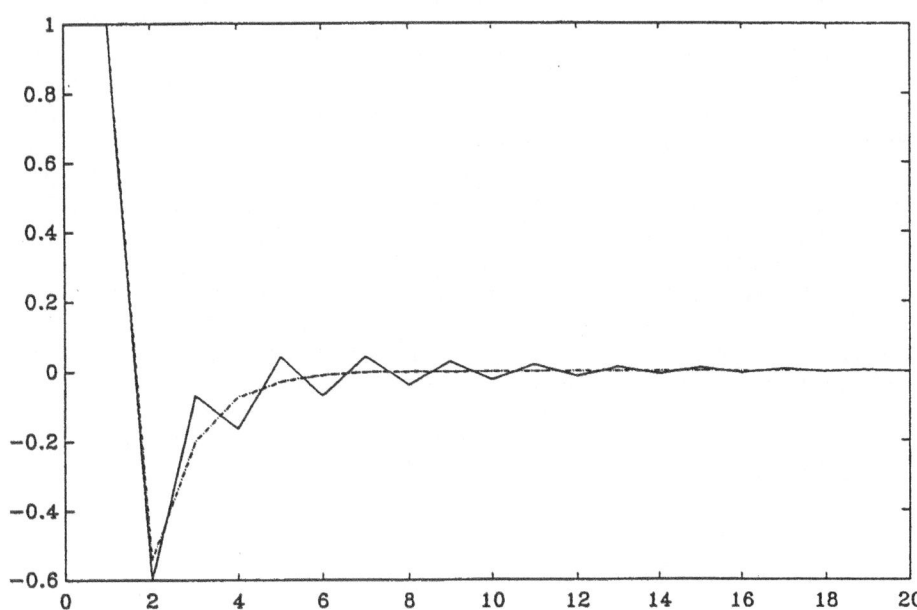

Figure 3.   Errors in impulse response (identified ——— , optimal Hankel ---).

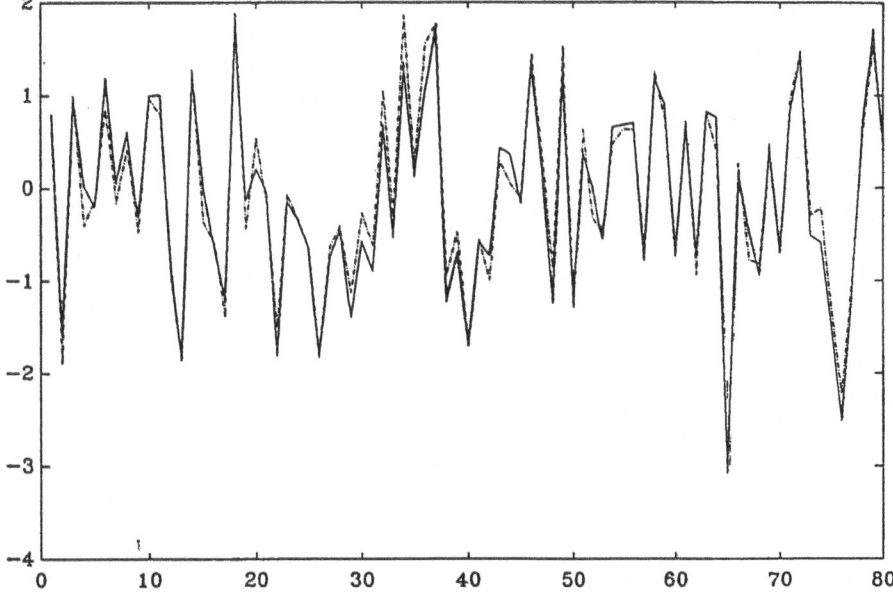

Figure 4.   Noise ε and estimated driving variables v (ε ——— , v ---).

6. Conclusions

Traditionally, econometrics and other areas involved in time series modelling seem to have focused mainly on the statistical aspects of data modelling, and systems theory on the dynamical aspects of a priori specified models. In practice a stochastic framework need not always make sense, and the dynamical structure of the model is not given but it has to be determined from the observed data. State space methods take the aspects of structure selection and parameter estimation simultaneously into account.

In this note we outlined a signal oriented approach for the construction of approximate states and their use in time series modelling. This procedure can be applied to deterministic as well as to stochastic systems. Some examples indicate rather good properties of this procedure. The method is primarily data oriented. The restriction is that only linear, time invariant models with finite lags are considered. The method is free of assumptions on causality (inputs and outputs, exogenous and endogenous variables), noise structure (deterministic or stochastic), and dynamical structure (length of the lags involved, number of states). Instead, these system properties are determined from the data.

In our opinion, the explicit construction of state trajectories is a valuable tool in modelling dynamical relationships. In this note we tried to give an idea of this method, which is a topic of ongoing research.

References

Akaike, H., (1975), 'Markovian representation of stochastic processes by canonical variables'. *SIAM Journal on Control and Optimization*, **13**, 162-173.

Akaike, H. (1976), 'Canonical correlation analysis of time series and the use of an information criterion', in *R.K. Mehra and D.G. Lainiotis (eds.), System Identification and Case Studies*, Academic Press, New York, 27-96.

Aoki, M. (1987), 'State Space Modelling of Time Series. Springer', Berlin.

Box, G.E.P., and G.M. Jenkins (1970), 'Time Series Analysis, Forecasting and Control', Holden-Day, San Francisco.

Davis, M.H.A., and R.B. Vinter (1985), 'Stochastic Modelling and Control', Chapman and Hall, London.

De Moor, B. (1988), 'Mathematical Concepts and Techniques for Modelling of Static and Dynamic Systems', Thesis, University of Leuven.

Desai, U.B., and D. Pal (1982), 'A realization approach to stochastic model reduction and balanced stochastic realization', *Proceedings 16th Annual Conference on Information Science and Systems*. New Jersey & Princeton.

Faurre, P. M. Clerget and F. Germain (1979) 'Opérateurs Rationnels Positifs', Dunod, Paris.

Glover, K. (1984), 'All optimal Hankel norm approximations of linear multivariate systems and their L^{∞} error bounds', *International Journal of Control*, **39**, 1115-1193.

Harvey, A.C. (1989), 'Forecasting, Structural Time Series Models and the Kalman Filter', Cambridge University Press, Cambridge.

Heij, C., T. Kloek and A. Lucas (1992), Positivity conditions for stochastic state space modelling of time series', *Econometric Reviews*, **11**, 379-396.

Heij, C., and B. Roorda (1991), 'A modified canonical correlation approach to approximate state space modelling', *Proceedings 30th CDC*, IEEE, Brighton, 1343-1348.

Kalman, R.E., P.L. Falb and M.A. Arbib (1969), 'Topics in Mathematical System Theory', McGraw-Hill, New York.

Lindquist, A., and G. Picci (1985), 'Realization theory for multivariable statioary gaussian processes', *SIAM Journal on Control and Optimization*, **23**, 809-857.

Moore, B.C., (1981), 'Principal component analysis in linear systems: controllability, observability, and model reduction', *IEEE Transactions on Automatic Control* **26**, 17-32.

Silverman, L.M., (1971), 'Realization of linear dynamical systems'. *IEEE Transactions on Automatic Control* **16**, 554-567.

Van der Veen, A.J. (1991), 'SVD-based estimation of low-rank system parameters', in *F. Deprettere and A.J. van der Veen (eds.), Algorithms and Parallel VLSI Architectures*, vol. A., Elsevier, Amsterdam, 203-228.

Willems, J.C. (1986), 'From time series to linear system', part I: Finite dimensional linear time invariant systems, part II: Exact modelling. *Automatica*, **22**, 561-580, 675-694.

SENSITIVITY ANALYSIS VERSUS UNCERTAINTY ANALYSIS: WHEN TO USE WHAT?

JACK P.C. KLEIJNEN

School of Management and Economics, Tilburg University
(Katholieke Universiteit Brabant), 5000 LE Tilburg, The Netherlands
e-mail: kleijnen@kub.nl

Summary

Decision makers and other users of models are interested in model validity. From their viewpoint the important model inputs should be split into two groups, namely inputs that are under the decision makers' control versus (environmental) inputs that are not controllable. Specifically, users want to ask 'what if' questions about global (not local) sensitivities: what happens if controllable inputs are changed (scenario analysis), what if model parameters and structure change? Among the techniques to answer these questions are statistical design of experiments (such as fractional factorial designs) and regression analysis. These techniques may show that some non-controllable inputs of the model are important; yet these inputs may not be known precisely. Then risk or uncertainty analysis becomes relevant. Its techniques are Monte Carlo sampling, including variance reduction techniques (such as Latin hypercube sampling), possibly combined with regression analysis. Controllable inputs can be optimized through Response Surface Methodology (RSM).

Keywords: sensitivity analysis, uncertainty analysis, risk analysis, validation, experimental design, regression, screening, Latin hypercube sampling, optimization

1. Introduction

The *analysis methods* discussed in this paper are known in the literature under such names as sensitivity, what-if, perturbation, risk, and uncertainty analyses. Definitions of these terms vary; for example, Helton, Garner, McCurley, and Rudeen (1991) use a definition of 'sensitivity analysis' that differs substantially from the one used in this report. We define the keyterms in the title of this paper as follows (we shall elaborate these definitions in the next sections). *Sensitivity analysis* or *what-if analysis* is the systematic investigation of the reaction of model outputs to *extreme* values of the model inputs and to drastic changes of the model structure. For example, what are the effects if in a queueing simulation the arrival rate doubles; what if the priority rule changes from FIFO to LIFO? So we examine *global*, not local sensitivities. In *uncertainty analysis* the model inputs are sampled from certain distributions, to quantify the consequences of the uncertainties in the model inputs, for the model outputs. Our conclusion will be that sensitivity analysis should precede uncertainty analysis. Each type of analysis may apply its own statistical techniques; for example, sensitivity analysis may use 2^{K-p} designs, whereas uncertainty analysis applies either crude Monte Carlo sampling or more sophisticated variance reduction techniques such as Latin hypercube sampling. Some techniques are applied in both analyses; for example, regression modelling.

The *issues* to be solved by sensitivity and uncertainty analyses are discussed under such headings as validation and optimization. These analyses and issues are studied in all scientific disciplines that use mathematical models. The theme of this conference, however, is limited to dynamic models in the 'natural' sciences -especially geophysics, agriculture, environmental sciences, and ecology- and economics. Unfortunately, nobody can be an expert in all these disciplines. This paper is *biased* by more than 25 years of experience with the technique of simulation, especially its statistical aspects and its application to problems in business, environmental, agricultural, military, and computer systems; see Kleijnen and Van Groenendaal (1992). Hence, models in natural sciences and economics are under-exposed in this contribution. However, the experience of the other participants in the Wageningen conference is reflected in the various contributions to this book. The reader of this contribution is assumed to have a basic knowledge of mathematical statistics, especially elementary regression analysis.

Not only is there a variety of related methods and issues, there is also much *software*. This software greatly simplifies the implementation of these methods. We shall refer to software throughout this paper. Unfortunately, easy-to-use software might also be dangerous (see §3.1).

All scientific methods are based on *assumptions*, which limit the applicability of these methods. These assumptions may be documented explicitly or they may be left implicit. Many practitioners do not know *when* to use *what* method. The goal of this paper is to explain which questions may be asked in practice, and which methods can answer these questions.

These questions are also discussed in Downing, Gardner, and Hoffman (1985, 1986), Easterling (1986), Iman and Conover (1980), and McKay (1992). This paper, however, is not a recapitulation of those publications: sensitivity and risk analyses remain *controversial* topics. For example, in §3.1 we shall claim that Latin hypercube sampling (LHS) should not be applied as a screening technique. (Controversies were also observed at an earlier workshop on 'Uncertainty analysis' organized in 1989 by the Dutch 'National Institute of Public Health and Environmental Protection', abbreviated in Dutch to 'RIVM'.)

This paper is organized as follows. In §2 we discuss model validation and what-if analysis of controllable and non-controllable inputs. This includes 1) obtaining real-world data, which may be scarce or abundant, 2) simple statistical tests for comparing simulated and real data, and 3) sensitivity analysis using statistical design of experiments with its concomitant regression analysis. This sensitivity analysis estimates which inputs are important. If these inputs are controllable, then they may be optimized. Otherwise uncertainty analysis may be applied. In §3 we address uncertainty analysis. This encompasses 1) the basics of that analysis, including applications in economics and the natural sciences, and 2) uncertainty analysis of stochastic simulation models. In §4 we give conclusions.

2. Model Validation and What-if Questions

Validation is concerned with determining whether the conceptual model is an accurate representation of the system under study. Validation is necessary whenever a model is meant to answer questions about a real-life system. Kleijnen (1993) gives a survey of the validation (and verification) of models -especially simulation models- which guides the discussion in this section. The experience of other researchers is reported in other contributions to this book.

There is no standard theory on validation, neither is there a standard 'box of tools'. The emphasis of this paper is on *statistical techniques*, which may yield reproducible, objective, quantitative data about the quality of models (other techniques -such as computer animation- are discussed in Kleijnen 1983). Data on model inputs and outputs may be available in different quantities. *Data availability* may be used to classify validation techniques.

2.1 Obtaining real-world data

To obtain a valid model, the analysts should try to measure the inputs and outputs of the real system, and the intermediate variables.

1) Sometimes it is *difficult or impossible* to obtain relevant data. For example, in simulation studies of nuclear war, it is (fortunately) impossible to get the necessary data. In natural sciences and economics there may be no data on future situations (but there may be data on past situations that may be extrapolated; also see the paragraph following 4) below).

2) Usually, however, it is possible to get *some* data. Typically the analysts have data on the existing system variant. For instance, for the existing manufacturing system the current scheduling rule is well known. And in an ecological model there may be data on the existing situation ('status quo' scenario).

3) In the military it is common to conduct *field tests* in order to obtain data on *future* variants. For example, real mine fields are created, not by the enemy but by the friendly navy; next a mine hunt is executed in this field to collect data.

4) In some applications there is an *overload* of data, namely if these data are collected electronically. For instance, point-of-sale (POS) systems at the supermarket check-outs record all transactions.

The further the analysts go back into the past, the more data they get and the more powerful the validation test will be, *unless* they go so far back that different laws governed the system. For example, many econometric models do not use data prior to 1945, because the economic infrastructure changed drastically during World War II. (Also see the discussion under 1) above.)

Moreover the data may show *observation error*, which complicates the comparison of real and simulated time series. Barlas (1989, p. 72) and Kleijnen and Alink (1992) discuss observation errors in a theoretical and a practical situation respectively. Also see the other contributions to this book.

2.2 Simple techniques for comparing model and real data

Suppose the analysts have succeeded in obtaining at least some data on the real system (if not, the sensitivity analysis of the next subsection can be used). They should then feed real-world input data into the model, in *historical* order. After running the simulation program, the analysts obtain a *time series* of simulation output and compare that time series with the historical time series for the output of the existing system. Kleijnen (1993) discusses several simple techniques, for example, the use of the t statistic to test whether the expected values of the simulated and the real time series are equal. More sophisticated methods are discussed elsewhere in this book (see Klepper and Slob).

Often simulation is meant to predict *relative responses*, not absolute responses. For example, what is the effect of adding one server to a queueing system; what if different scenarios are implemented in an environmental simulation? The analysts may then test whether simulated and real responses are positively correlated (without having the same means). Kleijnen (1993) shows how to estimate and test this correlation, using elementary regression analysis.

Sometimes simulation is meant to predict *absolute responses*. For example, in mine hunting the question may be whether the probability of detecting mines is so high that it makes sense to do a mine sweep; see Kleijnen and Alink (1992). An environmental example might be: do the costs of a certain scenario outweigh the benefits of that same scenario? The analysts may then combine the test on means with the test on correlation; see Kleijnen (1993).

Sometimes there are very many observations. Then not only the means of the simulated and the real time series and their (cross)correlation can be compared, but also their autocorrelations (with lag 1, 2, etc.). *Spectral analysis* is a sophisticated technique that estimates the autocorrelation structure of the simulated and the historical time series respectively, and compares these two structures. Unfortunately, that analysis is rather difficult (and -as stated- requires long time series).

2.3 Sensitivity or what-if analysis

Models and submodels (modules) with *unobservable* inputs and outputs can not be subjected to the tests of the preceding subsection. The analysts should then apply sensitivity analysis, in order to determine whether the model's behaviour agrees with the judgments of the experts (users and analysts). In case of observable inputs and outputs, sensitivity analysis is also useful, as this subsection will show.

The techniques for sensitivity analysis discussed in this paper, are *design of experiments and regression analysis*. These techniques and their application to a variety of problems are discussed in Kleijnen and Van Groenendaal (1992, pp. 147-186). Regression analysis and their application to simulation in radioactive waste disposal is examined in Helton, Garner, Rechard, Rudeen, and Swift (1992). Alternative techniques are reviewed in these two publications and in Downing *et al.* (1985), Helton *et al.* (1991, chapter II), and McKay (1992).

Most practitioners apply an inferior design of experiments: they change one model input at a time. Such a design gives estimated effects of input changes that have relatively high variances (low accuracies). Moreover, such a design cannot estimate interactions among inputs. Unlike 'one at a time' designs, *factorial* designs change several inputs (or factors) simultaneously. For example, a 2^K design consists of all 2^K combinations of K inputs, each input being studied at two values. So if there are three inputs ($K = 3$), then eight combinations are examined. For high values of K, however, too many combinations result and too much computer time is needed to simulate all these combinations. *Fractional factorial* designs consider only a fraction of all combinations. For instance, 2^{3-1} designs take only half ($1/2 = 2^{-1}$) of all 2^3 combinations. In general, 2^{K-p} designs consider only a 2^{-p} fraction of all 2^K combinations. So a 2^{8-2} design includes 64 of the 256 combinations of two values per input; these 64 combinations permit the estimation of interactions between inputs. There are also designs that consider more than two values per input; for example, *central composite* designs use five values (in order to estimate second order effects, to which we shall return). Tables and software help to decide *how many* input combinations to investigate and *which* combinations to simulate. Software is advertised in, for example, *OR/MS Today*.

How can the results of such experiments with models be analyzed and used for interpolation and extrapolation? Practitioners often *plot* the model output (say) y versus the input x_k, one plot for each input k with $k = 1,...,K$. (For example, if the arrival and service rates are changed in a queueing simulation, then $K = 2$.) More refined plots are conceivable, for instance, superimposed plots.

This practice can be formalized through *regression analysis* (in experimental design, this analysis is also known as Analysis of Variance -or ANOVA- with 'fixed effects'; if inputs are sampled, then ANOVA with random effects may apply; also see Kleijnen 1987, pp.285-293). We start with simple situations: single simulation response, regression model without curvature. So let y_i denote the *response* of the simulation model (for example, average waiting time per day or carbon dioxide concentration at the end of the simulated period) in combination i of the K model inputs, with $i = 1,...,n$ where n denotes the total number of simulated combinations (for example, $n = 2^{K \cdot p}$). If the model is deterministic, then each combination is run only once. Deterministic models are abundant in the natural sciences, because the classic laws of nature are deterministic and simple deterministic models are deemed valid. If the model is stochastic, then combination i is run (say) m_i times with $m_i \geq 1$ and y_i denotes the average of these m_i runs; see Kleijnen and Van Groenendaal (1992). Stochastic simulation is used in queueing analysis and in environmental and public health studies accounting for spatial and interindividual variability.

Let x_{ik} be the 'standardized' value of model *input* k in combination i, that is, if L_k denotes the lowest value of the original input (say) z_k, and H_k its highest value, then $x_{ik} = (z_{ik} - b_k)/a_k$ with $a_k = (H_k - L_k)/2$, which measures the dispersion of the original input k, and $b_k = (H_k + L_k)/2$, which quantifies the location of the original input k. See Kleijnen and Van Groenendaal (1992, pp. 177-179, 183-185), and also Downing *et al.* (1985. p.156) and Helton *et al.* (1992, chapter 6, p. 4).

Let β_k denote the main or first order *effect* of the standardized input k; that is, it measures how much the response changes as the original input changes from its lowest to its highest value, ignoring high order effects of inputs. Let $\beta_{kk'}$ designate the (two-factor) interaction between the inputs k and k'. We ignore interactions among three or more inputs, because they are hard to interpret. At this stage of our exposition we also ignore quadratic effects β_{kk}, which measure curvature: in sensitivity analysis we are interested in main effects only. However, we do not want the estimators of main effects to be biased by interactions. Quadratic effects -if present- bias the overall effect or grand mean β_0, which is of no interest in sensitivity analysis (β_0 is important when predicting the simulation response through a regression model). There are designs that require only $2K$ combinations to obtain unbiased estimators of the K main effects, in the presence of $(K(K - 1)/2)$ two-factor interactions; these designs are called 'resolution 4' designs; see Kleijnen (1987, p. 301).

Finally, let e_i represent the fitting error in combination i when approximating the simulation model by a simple regression model (without quadratic effects and without interactions among three or more inputs).

Then the input/output behaviour of the simulation model is approximated through the regression (meta)model or response surface

$$y_i = \beta_0 + \sum_{k=1}^{K} \beta_k x_{ik} + \sum_{k=1}^{K-1} \sum_{k'=k+1}^{K} \beta_{kk'} x_{ik} x_{ik'} + e_i. \tag{1}$$

Note that y and x_k may also denote the *ranks* of the original variables. This rank regression is popular in risk analysis. Saltelli and Homma (1992, pp. 79-82) give an elementary survey.

The least squares criterion applied to the simulation input/output data (y, X) gives estimated effects $\hat{\beta}$. Of course, the validity of the resulting approximation ($\hat{y} = X\hat{\beta}$) must be tested. The simplest measure is the well-known R^2 coefficient (the closer R^2 is to one, the better). *Cross-validation* is more in the spirit of validation of models in general. First it deletes simulation input x_i and the concomitant output data y_i. Next it estimates the regression parameters from the remaining data, which yields $\hat{\beta}_{(-i)}$. Then it employs the resulting estimated regression model to compute the forecast $\hat{y}_i$ for the input combination x_i. The comparison of forecasted output $\hat{y}_i$ and simulated output y_i is used to validate the approximation. This procedure can be repeated for all input combinations. The implementation of cross-validation is simple, since modern regression software provides statistics known as PRESS, DEFITS, DFBETAS, and Cook's D. See Kleijnen and Van Groenendaal (1992, pp. 156-157).

Inputs may be *qualitative*. Examples are the priority rule in a queueing simulation and the scenario in environmental simulation. Technically, this requires binary variables (x_{ik} is zero or one); see Kleijnen (1987).

A *case study* illustrating the application of experimental design and regression analysis is provided by Kleijnen, Rotmans, and Van Ham (1992). They apply these techniques to several modules of IMAGE, a deterministic simulation model developed at RIVM for the greenhouse effect of carbon dioxide (CO_2) and other gases. This approach gives estimates of the effects of the various inputs. These estimated effects should have the right *signs*: the policy analysts (not the statisticians) know that certain inputs increase the global temperature. Wrong signs indicate computer errors or conceptual errors. Indeed Kleijnen *et al.* (1992, p. 415) give examples of estimated sensitivity estimates with the wrong signs, which lead to correction of the simulation model. The remaining estimated effects show which inputs are important. One more example is the case study in Kleijnen and Alink (1992), concerning mine hunting at sea by means of sonar. The role of experimental design in the validation of simulation models is also discussed in Pacheco (1988).

Classic experimental designs, however, may require too much computer time, when the simulation study is still in its early (pilot) phase. Then *very many inputs* may be conceivably important. Bettonvil and Kleijnen (1991) present a *screening* technique based on sequential experimentation with the model. They split up the aggregated inputs as the experiment proceeds, until finally the important individual inputs are identified and their effects are estimated. They apply this technique to the RIVM model with 281 inputs. It is remarkable that the statistical technique identified some inputs that were originally thought to be unimportant by the policy analysts.

Before executing the experimental design, the analysts must determine the experimental domain or experimental frame. The design tells *how* to explore this domain, using the expertise of the statistician. Zeigler (1976, p. 30) defines the *experimental frame* as 'a limited set of circumstances under which the real system is to be observed or experimented with'. He emphasizes that 'a model may be valid in one experimental frame but invalid in another' (also see Klepper and Slob's contribution to this book). This paper (§2.1) has already mentioned that going far back into the past may yield historical data that are not representative of the current system; that is, the old system was ruled by different laws. Similarly, a simulation model is valid only if its input data remain within a certain area. For example, Bettonvil and Kleijnen's (1991) screening study shows that the greenhouse simulation is valid, only if its input values range over a relatively small area; outside that area the resulting simulation responses (CO_2) had magnitudes that were immediately declared unrealistic by the analysts. In general, it is difficult to develop valid models for completely new situations!

Mathematically the experimental frame may be defined as the hypercube formed by the k standardized inputs x_{ik} of the model. In practice, some corner points (combinations of extreme values) may be unrealistic, that is, they fall outside the experimental frame; see Janssen, Heuberger, and Sanders (1991, chapter 5) and Kleijnen (1987, pp. 318-319).

So any simulation model is valid only for a certain area of its inputs. Within that area the (valid) simulation model's input/output behaviour may vary. For example, a *first* order regression model (see equation 1 with the double summation term eliminated) is a good approximation of the input/output behaviour of a queueing simulation model, only if the traffic load is 'low'. When traffic is heavy, a second order regression model (which includes curvature) or a logarithmic transformation may apply. Some researchers fit a (meta)model to the simulation input/output data that holds over the whole experimental area. For example, Sacks, Welch, Mitchell, and Wynn (1989) apply covariance-stationary processes or Kriging to approximate deterministic simulation models. Barton (1992) surveys many new alternatives to the polynomial metamodel presented in (1). These approaches are so new that definitive evaluations cannot be given yet. Morover, they may aim at fast and accurate prediction (interpolation), not at sensitivity analysis.

The *magnitudes* of the estimated effects β show which inputs are *important*. Since the regression model is only an approximation to the simulation model, false conclusion are possible. For example, an input might have an unimportant first order effect but an important quadratic effect in (1), given a certain experimental area: non-monotonic reaction of simulation response to simulation input. Cross-validation might fail to reject the regression metamodel. We consider this example to be 'pathological'. Our approach - like any other approach with finite sample sizes- can not guarantee correct conclusions. Also see Saltelli, Andres, and Homma (1993) 's comparison of the performances of different sensitivity analysis techniques.

Mathematically all inputs are called x_k. However, general systems theory (GST) splits these inputs into (i) inputs that are under the decision makers' control and (ii) 'environmental' inputs, which (by definition) are not controllable.

The *controllable inputs* should be steered -by the decision makers- into the right direction. For example, in the greenhouse case the governments should restrict emissions of the gases concerned. There are several *optimization* techniques for simulation models. These models may have multiple responses that are nonlinear, possibly stochastic, complicated functions of their inputs. *Response Surface Methodology* (RSM) is a heuristic sequential technique that combines experimental design (especially 2^{k-p} and central composite designs), regression analysis, and steepest ascent, in order to find the model inputs that give a better (possibly the maximum) model response, in terms of a specific performance criterion; see Kleijnen and Van Groenendaal (1992, pp. 181-185). This reference also gives an RSM case study in steel tube manufacturing, namely a production planning system with 14 controllable inputs and several response types.

For the important *environmental inputs* the analysts should try to collect data on the (input) values that may occur in practice. If they succeed, then the validation techniques of the preceding subsection can be applied. If they do not succeed in getting accurate information, then the uncertainty analysis of the next section may be used.

3. Uncertainty Analysis

The analysts may be unable to collect reliable data on the values of the important environmental inputs that may occur in practice (see §2). Then *uncertainty analysis* or *risk analysis* may be applied.

3.1 The basics of uncertainty analysis

First the analysts derive a *probability function* for the input values (we use the following terminology: a random variable is characterized through its probability function or its cumulative probability function, called the distribution function; two or more random variables are characterized through their joint probability or distribution function). That distribution may be estimated from sample data, if those data are available; otherwise that distribution must be based on subjective expert opinions (also see Helton *et al.* 1992, chapter 2, p. 4). Usually these inputs are assumed to be statistically independent (so their joint distribution equals the product of their marginal distributions). Correlated inputs are discussed in Helton *et al.* (1992, chapter 3, p. 7) and Reilly, Edmonds, Gardner, and Brenkert (1987).

Next the analysts use pseudorandom numbers to sample input values from those distributions: *Monte Carlo or distribution sampling*. Variance reduction techniques (VRTs) are possible: for example, *Latin hypercube sampling* (LHS) forces the sample (of size n) to cover the whole experimental area (for example, in case of a single input, that input's domain is partitioned into, say, s subintervals and each subinterval is sampled s/n times). We recommend LHS as a VRT, not as a screening technique. For screening we recommend changing the inputs to their extreme values (specified by a fractional factorial design) and testing if at those values the outputs also change. Also see Banks (1989) and Bettonvil and Kleijnen (1991) versus Downing et al. (1985) and McKay (1992).

These *n* sampled input values are fed into the simulation model. First we consider *deterministic* simulation models. Such models are run with sampled input values; that is, *during* a simulation run all inputs are constant (no randomness), but from run to run they vary. This yields an estimated distribution of output or response values. That distribution may be characterized by its location (measured by the mean, modus, and median) and its dispersion (quantified by the standard deviation and various quantiles or percentiles, such as the 90% quantile). For a basic introduction to risk analysis see Kleijnen and Van Groenendaal (1992, pp. 75-78).

Which quantities sufficiently summarize a distribution function, depends on the users' *risk attitude*: risk neutral (then the mean is a statistic that characterizes the whole distribution sufficiently), risk aversion, or risk seeking; see Balson, Welsh, and Wilson (1992). They further distinguish between *risk assessment* (defined as risk analysis in this paper) and *risk management* (risk attitude, possible countermeasures).

Combining uncertainty analysis with *regression analysis* gives estimates of the effects of the various inputs, that is, regression analysis shows which inputs contribute most to the uncertainty in the output. Mathematically, this means that in eq. (1) the deterministic independent variables x_{ik} are replaced by random variables. Helton et al. (1991, 1992) call this combination of uncertainty and regression analysis 'sensitivity analysis'.

Risk analysis is used in *economics*. Hertz (1964) introduced this type of analysis in capital investments: what is the probability of a negative Net Present Value?. Krumm and Rolle (1992) give recent applications in the Du Pont company. Risk analysis in business applications may be implemented through add-ons (such as @RISK) that extend spread-sheet software (such as Lotus 1-2-3).

In the *natural sciences*, uncertainty analysis is also popular. For example, in the USA the Sandia National Laboratories have performed many uncertainty analyses for nuclear waste disposal; see Helton *et al.* (1991, 1992). Oak Ridge National Laboratory investigated radioactive doses absorbed by humans; see Downing *et al.* (1985). Nuclear reactor safety was investigated for the Commission of the European Communities in Ispra (Italy); see Olivi (1980) and Saltelli and Homma (1992). Uncertainty analysis is also performed extensively at RIVM; see Harbers (1993), Janssen *et al.* (1992), and several contributions to this book. Three environmental studies for the electric utility industry are presented in Balson *et al.* (1992). Uncertainty analysis in the natural sciences has been implemented through software such as LISA (see Saltelli and Homma 1992, p. 79), PRISM (see Reilly *et al.* 1987), and UNCSAM (see Janssen *et al.* 1992).

Note that risk analysis is also used in the analysis of computer security; see Engemann and Miller (1992) and FIPS (1979).

As we saw, a basic characteristic of uncertainty analysis is that information about the inputs of the simulation model is not reliable; so the analysts do not consider a single 'base value' per input variable, but a distribution of possible values. Unfortunately, the form of that distribution must be specified (by the analysts together with their clients). This specification may be 'software driven', that is, the analysts concentrate on the devel-opment of software that implements a variety of statistical distributions, but their clients are not familiar at all with the implications of these distributions; also see Easterling (1986). Bridging this gap requires intensive collaboration between model users, model builders,and software developers.

Consequently, it may be necessary to study the effects of the *specification* of the input distributions (and of other types of inputs such as scenarios). This type of sensitivity analysis may be called *robustness analysis*. Examples can be found in Helton *et al.* (1992, section 4.6); also see Janssen *et al.* (1992) and Kleijnen (1987, pp. 144-145). Faster sam-pling techniques for robustness analysis are discussed by Beckman and McKay (1987).

3.2 Uncertainty analysis of stochastic models

The type of question answered by uncertainty analysis is 'what is the chance of...?' So the model must contain some random element. Typically, however, that randomness is limited to the inputs of the model, whereas the model itself is deterministic. A simple example is: if the day and location on earth is known, then the time of sunrise is also known; otherwise that time is not exactly known.

Some models, however, are intrinsically *stochastic*; that is, without the randomness the problem disappears. For example, in queueing models the problem is the waiting times of customers or jobs and the utilization degrees of servers or machines. These stochastic models have random variables, which have certain distributions. For example, in queueing models the interarrival times may be independent drawings from a single exponential distribution with rate (say) λ (so the mean time between the arrival of two successive customers is $1/\lambda$). This rate λ is an input of the model. That model generates a stochastic time series of waiting times. The question may be 'what is the probability of customers having to wait longer than 15 minutes?'. For simple models this question can be answered analytically or numerically. For more realistic models, simulation is used. Mathematical statistics is used to determine how many customers must be simulated in order to estimate the response with prespecified accuracy; see Kleijnen and Van Groenendaal (1992, pp. 187-197).

Risk analysis has not been applied to stochastic economic and business models (all kinds of sensitivity analysis have been employed; see Kleijnen 1987, pp. 241-242, and Kleijnen and Van Groenendaal 1992, pp. 147-186). Yet Kleijnen (1983) explained how in a queueing simulation, risk analysis can be applied.

Helton *et al.* (1991, 1992) briefly discuss uncertainty analysis of stochastic models in the natural sciences (nuclear power plants, the spreading of nuclides). We also mention environmental health modelling. Note that spatial variation may be modeled through random fields, discussed at length by Cressie (1991).

4. Conclusions

This paper surveyed sensitivity and uncertainty analyses of mathematical models. It emphasized mathematical techniques, particularly statistical procedures, which yield reproducible, objective, quantitative results.

Sensitivity analysis (as defined in this paper) means that the model is subjected to 'extreme value' testing. The model is valid only within its experimental frame (defined in §2.3 as the limited set of circumstances under which the real system is to be observed or experimented with). Mathematically that frame may be defined as the hypercube formed by the K standardized inputs x_{ik} of the model. Experimental designs (such as 2^{K-p} fractional factorials) specify *which* combinations (the fractions 2^{-p} of the 2^K corner points of that hypercube) are actually observed or simulated. The n observed input combinations and their corresponding responses are analyzed through a regression (meta)model. That regression model is an approximation of the simulation model. That regression model gives quantitative measures of the importance of the simulation inputs.

From the users' viewpoint the important inputs should be split into two groups, namely inputs that are under the decision makers' control versus so-called environmental inputs, which are not controllable. Information on the important environmental variables is wanted. If the value of an important environmental variable is not well-known, then the chances of various values can be quantified through a probability function. If a sample of data is available, then this function can be estimated objectively, applying mathematical statistics. Otherwise subjective expert opinions are used.

Uncertainty or risk analysis measures the uncertainties of the model responses that result from the uncertainties in the model inputs, which are due to the limited availability of data (see §2.1). Response uncertainties are quantified through the joint statistical distribution of those responses.

Combining uncertainty analysis with regression analysis shows which non-controllable inputs contribute most to the uncertainty in the output. (Some authors call this combination 'sensitivity analysis'.)

The controllable inputs should be steered into the right direction. Response Surface Methodology (RSM) is a heuristic technique that combines experimental design, regression analysis, and steepest ascent, in order to find the model inputs that give better model responses, possibly the maximum response.

Our conclusion is that sensitivity analysis should precede uncertainty analysis. Each type of analysis may apply its own set of statistical techniques (for example, 2^{K-p} designs in sensitivity analysis; Latin hypercube sampling in uncertainty analysis); some techniques (such as regression modelling) are applied in both analyses.

Acknowledgement

Detailed comments made by two anonymous referees helped to improve this paper.

References

Balson W.E., Welsh J.L., and Wilson D.S. (1992), 'Using decision analysis and risk analysis to manage utility environmental risk'. *Interfaces*, **22(6)**, 126-139.

Banks J. (1989), 'Testing, understanding and validating complex simulation models', *Proceedings of the 1989 Winter Simulation Conference*.

Barlas Y. (1989), 'Multiple tests for validation of system dynamics type of simulation models', *European Journal of Operational Research*, **42(1)**, pp. 59-87.

Barton R. (1992), 'Metamodels for simulation input-output relations'. *1992 Winter Simulation Conference Proceedings*, Association for Computing Machinery, New York.

Beckman R.J. and McKay M.D. (1987), 'Monte Carlo estimation under different distributions using the same simulation'. *Technometrics*, **29(2)**, 153-160.

Bettonvil, B. and Kleijnen, J.P.C. (1991), 'Identifying the important factors in simulation models with many factors'. Tilburg University.

Cressie N.A.C. (1991), *Statistics for spatial data*. John Wiley & Sons, Inc.

Downing D.J., Gardner R.H., and Hoffman F.O. (1985), 'An examination of response-surface methodologies for uncertainty analysis in assessment models'. *Technometrics*, **27**, 151-163.

Downing D.J., Gardner R.H., and Hoffman F.O. (1986), 'Response to Robert G. Easterling'. *Technometrics*, **28(1)**, 92-93.

Easterling, R.G. (1986), 'Discussion of Downing, Gardner, and Hoffman (1985)'. *Technometrics*, **28(1)**, 91-92.

Engemann K.J and Miller H.E. (1992), 'Operations risk management at a major bank'. *Interfaces*, **22(6)**, 140-149.

FIPS (1979), *Guidelines for automatic data processing risk analysis*. FIPS PUB 65 (Federal Information Processing Standards Publication), Washington.

Harbers A. (1993), *Onzekerheidsanalyse op een simulatiemodel voor het broeikaseffect*. (Uncertainty analysis of a simulation model for the greenhouse effect.) Tilburg University.

Helton J.C., Garner J.W., McCurley R.D. and Rudeen D.K. (1991), *Sensitivity analysis techniques and results for performance assessment at the waste isolation pilot plant*. Sandia Report, SAND90-7103.

Helton J.C., Garner J.W., Rechard R.P., Rudeen D.K., and Swift P.N. (1992), *Preliminary comparison with 40 CFR part 191, subpart B for the waste isolation pilot plant, vol. 4: uncertainty and sensitivity analysis*. Sandia Report, SAND91-0893/4.

Hertz D.B. (1964), 'Risk analysis in capital investments'. *Harvard Business Review*, 95-106.

Iman, R.L. and Conover W.J. (1980), 'Small sample sensitivity analysis techniques for computer models, with an application to risk assessment'. (Including comments and rejoinder). *Communications in Statistics*, A9, no. **17**, 1749-1874.

Janssen, P.H.M., P.S.C. Heuberger, and R. Sanders (1992), *UNCSAM 1.0: a software package for sensitivity and uncertainty analysis manual*. National Institute of Public Health and Environmental Protection, Bilthoven, The Netherlands.

Kleijnen, J.P.C. (1983), 'Risk analysis and sensitivity analysis: antithesis or synthesis?'

Simuletter, **14(1-4)**, 64-72.

Kleijnen J.P.C.(1987), *Statistical Tools for Simulation Practitioners*. Marcel Dekker, Inc., New York.

Kleijnen J.P.C. (1993), 'Verification and validation of simulation models'. *European Journal of Operational Research* (accepted).

Kleijnen, J.P.C. and Alink, G.A. (1992), 'Validation of simulation models: Mine-hunting case study'. Tilburg University.

Kleijnen J.P.C. and W. Van Groenendaal (1992), *Simulation: a Statistical Perspective*. Wiley, Chichester.

Kleijnen, J.P.C., Rotmans, J. and Van Ham, G. (1992), 'Techniques for sensitivity analysis of simulation models: a case study of the CO_2 greenhouse effect'. *Simulation*, **58(6)**, 410-417.

Krumm, F.V. and Rolle, C.F. (1992), 'Management and application of decision and risk analysis in Du Pont'. *Interfaces*, **22(6)**, 84-93.

McKay M.D. (1992), 'Latin hypercube sampling as a tool in uncertainty analysis of computer models'. *1992 Winter Simulation Conference Proceedings*, Association for Computing Machinery, New York.

Olivi L. (1980), *Response surface methodology in risk analysis; synthesis and analysis methods for safety and reliability studies*. Edited by G. Apostolakis, S. Garibra and G. Volta, Plenum Publishing Corporation, New York.

Pacheco, N.S. (1988), 'Session III: simulation certification, verification and validation', *SDI Testing: the Road to Success; 1988 Symposium Proceedings International Test & Evaluation Association*, ITEA, Fairfax (Virginia 22033).

Reilly J.M., Edmonds J.A., Gardner R.H., and Brenkert A.L. (1987), 'Uncertainty analysis of the IEA/ORAU CO_2 emissions model'. *The Energy Journal*, **8(3)**, 1-29.

Sacks J., Welch W.J., Mitchell T.J., and Wynn H.R. (1989), 'Design and analysis of computer experiments (includes comments and rejoinder)'. *Statistical Science*, **4(4)**, 409-431.

Saltelli A., T.H. Andres, and Homma T. (1993), 'Sensitivity analysis of model output: an investigation of new techniques'. *Computational Statistics and Data Analysis*, **15**, 211-238.

Saltelli A. and Homma T. (1992), 'Sensitivity analysis for model output: performance of black box techniques on three international benchmark exercises'. *Computational Statistics and Data Analysis*, **13**, 73-94.

Zeigler, B. (1976), *Theory of Modelling and Simulation*. Wiley Interscience, New York.

MONTE CARLO ESTIMATION OF UNCERTAINTY CONTRIBUTIONS FROM SEVERAL INDEPENDENT MULTIVARIATE SOURCES

MICHIEL J.W. JANSEN[1] & WALTER A.H. ROSSING[2] & RICHARD A. DAAMEN[3]

[1] Agricultural Mathematics Group-DLO, Box 100, 6700 AC Wageningen, The Netherlands; Centre for Agrobiological Research;

[2] Wageningen Agricultural University, Department of Theoretical Production Ecology;

[3] Research Institute for Plant Protection

Summary

An efficient random sampling method is introduced to estimate the contributions of several sources of uncertainty to prediction variance of (computer) models. Prediction uncertainty is caused by uncertainty about the initial state, parameters, unknown (e.g. future) exogenous variables, noises, etcetera. Such uncertainties are modelled here as random inputs into a deterministic model, which translates input uncertainty into output uncertainty. The goal is to pinpoint the major causes of output uncertainty. The method presented is particularly suitable for cases where uncertainty is present in a large number of inputs (such as future weather conditions). The expected reduction of output variance is estimated for the case that various (groups of) inputs should become fully determined. The method can be applied if the input sources fall into stochastically independent groups. The approach is more flexible than conventional methods based on approximations of the model. An agronomic example illustrates the method. A deterministic model is used to advise farmers on control of brown rust in wheat. Empirical data were used to estimate the distributions of uncertain inputs. Analysis shows that effective improvement of the precision of the model's prediction requires alternative submodels describing pest population dynamics, rather than better determination of initial conditions and parameters.

1. Introduction

After the structure of a deterministic model has been decided upon, much has to be done before the model can be used for prediction in a given situation. In this process random phenomena are encountered. Initial values are measured more or less accurately. Parameters have to be estimated, usually in a series of experiments on submodels. During the estimation of submodel parameters, for instance by regression, more random effects turn up. Often parameters will show 'natural' variation from case to case. The general setting of a parameter could be the population-parameter: the mean over a population of application-cases. The population-parameter is estimated with some error, and the parameters of particular cases will differ from the population-parameter. During submodel regression, lack of fit may become apparent, in the sense that differences between observations and fit are larger than can be ascribed to observation-error. If one does not succeed in modifying the regression equation so that the lack of fit disappears, and if the lack of fit is non-systematic, one might incorporate submodel-errors as random effects in an extended model (the case of systematic lack of fit will not be treated in this paper). One might, for instance, account for temporal variability in subprocesses by white noise inputs. In addition, exogenous variables processed by the model may not be known when predictions are made. We will account for all such effects by assuming a deterministic model with random inputs. This enables an a priori investigation of prediction precision, based on the assumption that the structure of the model is correct.

Usually, the widely different nature of the inputs has implications for their mutual (in)dependency. The whole state of affairs will hinge on the model and its application, but we will try to give some examples. If various sets of parameters are estimated in different experiments, estimation errors of parameters from different sets are likely to be stochastically independent, whereas parameter estimation errors within a set will often be dependent. Natural variation of parameters has little to do with errors in estimating population means. Errors in the measurement of the initial state are independent of errors in parameter estimates. Random exogenous variables and noise inputs, in their turn, may often be assumed to be independent of errors in parameter estimation and in measurement of initial values.

The situation is formalized as follows. A scalar model output Y is assumed to depend on a number of stochastically independent random input vectors, say $X_1, X_2, \ldots X_n$:

$$Y = f(X_1, X_2, \ldots X_n).$$

The vectors X_i may have different lengths. The function f is deterministic, usually it is evaluated by simulation; f may represent a single output or a combination of outputs. Fixed inputs are not represented in the argument list, being of no interest in the present context.

In this paper, Y's variability will be characterized by its variance. It will be assumed that Y has finite mean and variance; this is guaranteed for instance when f is bounded. The use of the variance as measure of uncertainty has an economic rationale: if the loss caused by a prediction error is proportional to the square of that error, the expected loss is proportional to the variance.

Uncertainty analysis consists of the investigation of the output distribution, given the model and the distribution of the inputs. One may investigate the *full variance*, that is the variance of Y induced by all sources X_i collectively. Let U denote a group of one or more sources of uncertainty X_i; then, by assumption, U is independent of the complementary sources. With respect to U, two variance components are particularly interesting. Firstly, the *top marginal variance* from U, which is defined as the expected reduction of the variance of Y in case U should become fully known, whereas the other inputs remain as variable as before. Secondly, the *bottom marginal variance* from U, defined as the expected value of the variance of Y in case all inputs except U should become fully known, U remaining as variable as before. Since one does not know in advance at what value the sources will become fixed, one can only determine the distribution of the two variances mentioned; we will content ourselves with the mean of these distributions. Iman and Hora (1990) define uncertainty components in a similar way. A source will hardly ever become fully known, so the two variance components represent limits to what may be achieved: the top marginal variance is a maximal variance reduction, the bottom marginal variance is a minimal residual variance. It will appear that the top marginal variance from a source can never be greater than its bottom marginal variance.

Since the sources of uncertainty are closely related to the methods of data acquisition, knowledge of the variances mentioned can be used to assess the potential benefits of further research. One may pinpoint sources of uncertainty that are worthwhile to determine more accurately, namely those sources that one *can* determine better, and that have a large top marginal variance. On the other hand, there may be sources which cannot become fixed; the bottom marginal variance from the collective of such sources is the best one could ever expect to reach.

In section 2 an overview is given of some current types of uncertainty analyses. The anatomy of uncertainty from two independent complementary sources is discussed in section 3. In section 4 we introduce a new sampling method for uncertainty analysis. The method is applied to an example from agronomy in section 5.

2. Taxonomy of uncertainty analyses

With respect to the *mathematical techniques* used, one may discern two major types of uncertainty analyses: analyses based on an approximation of the model (most often a low-order polynomial of the inputs, sometimes a nonparametric regression function), and analyses based directly on the model. Analyses based on an approximation may be attractive because of computational efficiency. Apart from that, an approximation may be interesting for its own sake. But an approximation-based analysis depends critically on the quality of the approximation. One has to check the approximation in each new case, and will not always be satisfied. A good linear approximation, for instance, may simply not exist over the entire range of the inputs. Moreover, if the model has high-dimensional inputs, such as exogenous variables or noises, an approximation is not readily available. (See also Iman & Helton, 1988, and Morris, 1991.) In contrast, methods based directly on the model are more easily applicable over a wide range of circumstances. They may require more computation, but in many cases this extra effort is far from unsurmountable.

With respect to the *questions* addressed, one may also distinguish two major types of uncertainty analyses: analyses in which only the full uncertainty is investigated, the uncertainty when all inputs are variable; and analyses in which one also investigates the decrease of uncertainty if one or more inputs were to become known more accurately. The latter subject is sometimes called the analysis of uncertainty contributions.

The *dependency* between the groups of inputs distinguished in the analysis of uncertainty contributions brings about another division in the realm of uncertainty analyses. Unfortunately, something simple like *the* uncertainty contribution of an input does not exist, except for rare occasions. Usually the variance reduction due to the fixing of an input depends on which inputs have already been fixed. The concept of uncertainty contribution of an input becomes extra complicated when the inputs distinguished are mutually dependent (Janssen, 1993). Latin hypercube samples, for instance, are often used as data for an uncertainty analysis (McKay & Beckman & Conover, 1979). This method was originally designed for independent scalar inputs: the samples being constructed by independent association of stratified samples of scalar inputs. Occasionally one may solve dependency problems by using a bit more technique or by giving up a bit of exactness (Iman & Conover, 1982; Stein, 1987). Technically and conceptually, uncertainty analysis becomes much simpler if one aggregates the inputs into independent groups, and restricts the analysis to those groups. By doing so one does not sacrifice relevance, since the dependencies are functionally linked with the current method of data acquisition.

The uncertainty analysis proposed will be restricted to independent groups of inputs: the smallest possible independent input sets are to the analysis like atoms to a chemist. This restriction enables an analysis based directly on the model, in which one can estimate the top marginal variance and the bottom marginal variance from each source X_i and from some pooled sources. Thus, the analysis proposed calculates, without recourse to an approximation, the uncertainty contributions of stochastically independent groups of inputs while allowing for dependency within groups. As far as we know, such analyses do not yet exist.

3. Anatomy of uncertainty from independent sources

In this section we study the decomposition of the prediction variance caused by two stochastically independent vector inputs. Extension to a larger number of independent vector inputs is straightforward. Stochastic variables and vectors will be written in upper case letters, the values they assume in lower case. If an input vector, say U, becomes fully known at value u, whereas the other input vector, say V, independent of U, remains as uncertain as before, the best prediction of Y will be the mean of $f(u,V)$ (*best* in the least squares sense). The situation is illustrated in Table 1 for the case that U and V can assume a finite number of equiprobable values.

Table 1

	v_1	v_2	v_3	v_4	v_5	
u_1	y_{11}	y_{12}	y_{13}	y_{14}	y_{15}	$y_{1.}$
u_2	y_{21}	y_{22}	y_{23}	y_{24}	y_{25}	$y_{2.}$
u_3	y_{31}	y_{32}	y_{33}	y_{34}	y_{35}	$y_{3.}$
u_4	y_{41}	y_{42}	y_{43}	y_{44}	y_{45}	$y_{4.}$
	$y_{.1}$	$y_{.2}$	$y_{.3}$	$y_{.4}$	$y_{.5}$	$y_{..}$

A dot index indicates that the mean has been taken over the index; $f(u_i,v_j)$ is denoted by y_{ij}. The left column and the upper row contain the values assumed by U and V. The best predictions are conditional means of the model output Y. The bottom right element $y_{..}$ is the best prediction when neither U nor V are known. The right column and the bottom row contain the best predictions at the given value of U respectively V. The output y_{ij} can be decomposed into general mean, main effects and interactions, as usual in analysis of variance:

$$y_{ij} = y_{..} + (y_{i.}-y_{..}) + (y_{.j}-y_{..}) + [y_{ij}-y_{..}-(y_{i.}-y_{..})-(y_{.j}-y_{..})].$$

In the general case, the function $f(u,v)$ can be decomposed as follows. Let f_0 denote the best prediction when U and V are unknown:

$$f_0 = E f(U,V).$$

The best predictions when U or V have become fixed at u or v respectively, are given by $E f(u,V)$ and $E f(U,v)$ respectively. Let $f_u(u)$ and $f_v(v)$ denote the corrections to f_0 when U respectively V get fixed:

$$f_u(u) = \mathrm{E}\, f(u,V) - f_0, \qquad\qquad f_v(v) = \mathrm{E}\, f(U,v) - f_0$$

and let $f_{uv}(u,v)$ denote what is left. Then $f(u,v)$ may be decomposed:

$$f(u,v) = f_0 + f_u(u) + f_v(v) + f_{uv}(u,v),$$

which is sometimes called the *analysis of variance decomposition* of f. Accordingly f_0 is called the general mean, $f_u(u)$ and $f_v(v)$ are called main effects of u and v, while $f_{uv}(u,v)$ is called the interaction of u and v. The *full variance* of f neatly falls apart:

$$\mathrm{Var}\, f(U,V) = \mathrm{Var}\, f_u(U) + \mathrm{Var}\, f_v(V) + \mathrm{Var}\, f_{uv}(U,V).$$

If U were to become fixed at u, the best prediction would be $f_0 + f_u(u)$, leaving $f_v(V) + f_{uv}(u,V)$ as prediction error, with reduced variance $\mathrm{Var}\, f_v(V) + \mathrm{Var}\, f_{uv}(u,V)$, which is a function of u. It is not known in advance at which value U should become fixed. One might wish to calculate the distribution of this reduced variance, but we will be content with the mean of the reduced variance over U, that is $\mathrm{Var}\, f_v(V) + \mathrm{Var}\, f_{uv}(U,V)$. Accordingly, the *top marginal variance* from U, the expected variance reduction due to the fixing of U while V remains as variable as before, is given by $\mathrm{Var}\, f_u(U)$. Similarly, the *bottom marginal variance* from V, the expected variance left over when only V remains uncertain, equals $\mathrm{Var}\, f_v(V) + \mathrm{Var}\, f_{uv}(U,V)$. The top marginal variance from U is seen to be the variance of the main effect of U, whereas the bottom marginal variance from V is equal to the sum of the variances of the main effect of V and the interaction between U and V.

Let u_1 and u_2 denote two independent realizations of U, and let v denote some fixed value that can be assumed by V. Then $f(u_1,v)$ and $f(u_2,v)$ are independent realizations of $f(U,V)$ given $V=v$. Thus $d \equiv f(u_1,v) - f(u_2,v)$ has expectation 0, while its variance, i.e. its expected square, is twice the variance of $f(U,V)$ given $V=v$. So $\tfrac{1}{2}d^2$ is an unbiased estimate of this latter variance. It follows that if v is a random realization of V, $\tfrac{1}{2}d^2$ is an unbiased estimate of what we defined as the bottom marginal variance from source U. The top marginal variance from U is obtained by subtracting the bottom marginal variance from V from the full variance of Y.

4. A new sampling method

In order to estimate top marginal variances and bottom marginal variances, we need a sample with frequent occurrences of blocks with values of $f(X_1, X_2,...X_n)$ within which one argument is constant while all other arguments vary randomly; and of blocks within which one argument varies randomly while all other arguments are constant. The arguments that are constant within blocks should vary randomly between blocks.

In an ordinary Monte Carlo sample a new realization of the model output f is obtained by drawing new values for the inputs $X_1 \ldots X_n$, and calculating the output f after *all* new drawings. Such a sample contains no information about the role of the individual inputs. In the method introduced here, f is calculated after *each* drawing of a new value of an individual source. New input values are sampled in some fixed *cyclic* order. In this way, one obtains a stationary sequence y_i ($i=1,2...$) of realizations of Y. After each cycle, n steps, an independent realization is obtained. Differences between realizations within a cycle hold information about the contributions of the various inputs to the full variance. Such a sample will be called a *winding stairs* sample. Although the sample is best represented on a cylinder (the 'wall of the staircase') it may well be represented in rows and columns, as illustrated in Table 2. A row consists of the steps required to make a full turn, so that a column consists of steps connected by a vertical line on the wall of the staircase.

Table 2. A winding stairs sample for a model with three random inputs X_1, X_2 and X_3. Independent realizations of X_i are denoted by $x_{i,1}$, $x_{i,2}$, etcetera. The sample consists of eight cycles.

$f(x_{1,1}, x_{2,1}, x_{3,1})$	$f(x_{1,1}, x_{2,2}, x_{3,1})$	$f(x_{1,1}, x_{2,2}, x_{3,2})$
$f(x_{1,2}, x_{2,2}, x_{3,2})$	$f(x_{1,2}, x_{2,3}, x_{3,2})$	$f(x_{1,2}, x_{2,3}, x_{3,3})$
$f(x_{1,3}, x_{2,3}, x_{3,3})$	$f(x_{1,3}, x_{2,4}, x_{3,3})$	$f(x_{1,3}, x_{2,4}, x_{3,4})$
$f(x_{1,4}, x_{2,4}, x_{3,4})$	$f(x_{1,4}, x_{2,5}, x_{3,4})$	$f(x_{1,4}, x_{2,5}, x_{3,5})$
$f(x_{1,5}, x_{2,5}, x_{3,5})$	$f(x_{1,5}, x_{2,6}, x_{3,5})$	$f(x_{1,5}, x_{2,6}, x_{3,6})$
$f(x_{1,6}, x_{2,6}, x_{3,6})$	$f(x_{1,6}, x_{2,7}, x_{3,6})$	$f(x_{1,6}, x_{2,7}, x_{3,7})$
$f(x_{1,7}, x_{2,7}, x_{3,7})$	$f(x_{1,7}, x_{2,8}, x_{3,7})$	$f(x_{1,7}, x_{2,8}, x_{3,8})$
$f(x_{1,8}, x_{2,8}, x_{3,8})$	$f(x_{1,8}, x_{2,9}, x_{3,8})$	$f(x_{1,8}, x_{2,9}, x_{3,9})$

4.1. Variance estimation

We will show that a winding stairs sample enables unbiased estimation of the full variance, of the top marginal variance and of the bottom marginal variance, for single inputs as well as for groups of adjacent inputs. A more detailed discussion, including consideration of sample size requirements, will be given elsewhere.

The first column, the sequence y_{kn+1} ($k=0,1,2...$), consist of mutually independent realizations of Y. So the variance of this sequence estimates the *full variance*. The same applies to the other columns, leading to n (dependent) estimates which can be pooled.

The sequence $d_k \equiv y_{kn} - y_{kn+1}$ ($k=1,2,3...$) consists of differences between two values of the output with independent realizations of X_1, the other inputs having the same random value. The expectation of d_k equals 0. The expectation of $\frac{1}{2}d_k^2$ is equal to the *bottom marginal variance* from X_1; so the latter is estimated without bias by the mean of the sequence $\frac{1}{2}d_k^2$. Successive values of $\frac{1}{2}d_k^2$ are dependent. However, d_k and d_{k+m} are independent if $|m| > 1$; the dependence has a short range, from which it follows that the variance estimator is asymptotically normal. The accuracy of the variance estimator can also be estimated from the sequence $\frac{1}{2}d_k^2$. The range-1 dependence has to be taken into account in the calculations, which can be done with techniques similar to those used with first order moving-average time-series: the correlation matrix has the same form as in such series.

The *top marginal variance* from X_1 can be estimated similarly from the differences $\delta_k \equiv y_{kn+1} - y_{(k+1)n}$ ($k=0,1,2...$). Subtracting the mean of $\frac{1}{2}\delta_k^2$ from the full variance estimate produces an estimate of the top marginal variance from X_1. The same procedure can be applied to the other inputs, and to groups of adjacent inputs.

A significance test on the difference of two variance estimates is possible, since the estimates are asymptotically normally distributed, and since the accuracy of a sample difference can be estimated just as before, with the only modification that the dependence in the sequence used may now have range 2.

4.2. Efficiency

The efficiency of our sampling method is of a different type than the efficiency of, say, block designs or the latin hypercube method, where adverse effects on the accuracy are suppressed. The efficiency of our method lies mainly in the multiple use of model evaluations: one evaluation is used six times if all kinds of uncertainty contributions are estimated. The price paid is a dependency between estimates and within the sequences that provide the estimates. But these dependencies pose no serious problems. The extensibility of the sample also contributes to the efficient use of the method. One may start with a moderately small sample and append more cycles only if necessary in view of the accuracy of the results. Moreover, one may start with a small number of steps in a cycle, inserting new steps only at places where the effect of current steps is found to be large.

4.3. Related approaches

Owen (1992) shows how to estimate variances of latin hypercube sample means. A remarkable spin-off of Owen's exercise is a method to estimate top marginal variances of independent one-dimensional inputs.

Morris (1991) introduces random sampling plans for sensitivity analysis. The plans are made for independent one-dimensional inputs. Apart from these differences in application, the plans show a great similarity with our winding stairs samples: both consist of one-factor-at-a-time designs, and the efficiency of both methods lies in the multiple use of function evaluations.

5. Example

The winding stairs method is applied to a model used to advise farmers on control of brown rust in winter wheat. The model calculates the loss due to brown rust in Dutch guilders per hectare. We will very briefly sketch the model; a detailed description is given in Rossing *et al.* 1993a, 1993b and 1993c. The *incidence* of brown rust is observed as the number of leaves with rust spots in a sample of 160 leaves. By an empirical relation, the incidence is translated into a *density* (mean number of spots per leaf); a random effect is added to account for lack of fit in the density function. The density grows exponentially, the relative growth rate varying in time as a constant plus a white noise. The growth stops when the crop is harvested, that is when the crop reaches the mature stage of development. The stage of development is a function of the cumulative temperature, with a random effect to account for lack of fit. Finally, the loss is calculated as a bounded function of the integral over time of the number of brown rust spots.

The relations used in the model have been investigated experimentally. The parameters were estimated by regression: the resulting parameter error distributions are used in the uncertainty analysis. In some submodels, undeniable lack of fit showed up, which appeared to be non-systematic and could not be ascribed, for instance, to natural parameter variation (which we did not observe). The variances were estimated, and the effects were incorporated as random inputs. The temperature uncertainty is represented by 36 years of historical temperature data.

The construction of an approximation of the model was hampered by the numerous temperature data and even more by the noise inputs. These problems were aggravated by the fact that the analyses had to be performed under a broad range of circumstances, each of which required construction and quality inspection of an approximation. For these reasons, an analysis without an approximation was attractive.

In the uncertainty analysis the following six inputs were assumed independent, and were randomly sampled in the following cyclic order: (1) errors in estimating parameters; (2) error in estimation of initial state; (3) noise on relative growth rate; (4) non-systematic error in crop development function; (5) non-systematic error in the incidence-density relation; (6) a season of daily temperatures. The sampling order was chosen to enable the grouping of inputs into methodologically interesting classes. With some effort, the first two inputs might well become known more accurately. The subprocess errors and noises, however, will remain in force as long the model's structure is not modified; and the temperature is notoriously unpredictable.

The logarithm of the loss due to brown rust appears to be approximately normally distributed, although the right tail is somewhat shortened. In Table 3 we give the results of the analysis of a 250-cycle winding stairs sample of the (base 10) logarithm of the loss for one specific situation. The results of the analysis are formulated in terms of top marginal variances. This is always possible, since the top marginal variance of a source is complementary to the bottom marginal variance of the complementary sources.

Table 3. Estimated top marginal variances of various groups of uncertainty sources. The top marginal variances are expressed in percents of the estimated full variance (.544 with a relative standard error of 8%). The greatest variance in a column is marked with an asterisk if it is significantly greater than the other variances (separate tests, 5%-two-sided).

1. estimation error in parameters		11	11
2. measurement error in initial value			
3. noise on relative growth rate	80*	63*	45*
4. error on crop development function			7
5. error on density function			22
6. temperature	20	20	20

From the first column in Table 3 it may be seen that the full variance might be reduced up to some 80% in the imaginary case that all inputs could be fixed except the totally unpredictable temperature. Thus a worthwhile variance reduction is not a priori impossible. Subsequently the non-temperature sources are inspected more closely: a split is made between parameters and initial values on the one hand and subprocess errors and noise on the other hand. From the second column it is seen that one will reach a reduction of some 11% at most by very accurate experiments to determine parameters and very accurate measurement of the initial conditions. For that reason these two sources were not considered separately. A reduction of the submodel errors and noise, representing imperfections in the current model structure, is much more promising. These three sources are separated in the third column. The non-systematic error on the incidence-density relation appears to be of some importance. But the noise on the relative growth rate is seen to be the main offender: up to 45% reduction of prediction variance might possibly be obtained if this 'noise' could be deterministically accounted for. That can only be imagined, however, with a more comprehensive model.

The sum of the true relative top marginal variances in a column is at most 100%, the deficit being caused by interactions, but due to sampling error the sum of the estimates of the relative top marginal variances may be greater than 100%. In Table 3 the column sums are close to 100%: the interaction between the sources is small. This implies that the logarithm of the loss can be well approximated by a sum of the form $\Sigma \ g_i(X_i)$, in which X_i denote the sources of variation distinguished, while the functions $g_i()$ denote arbitrary, presumably non-linear, functions.

Better prediction precision leads to more adequate control of brown rust, which implies less spraying. Rossing et al. (1993a) contains an investigation of some economic consequences of the current uncertainty. It remains an open question, however, whether a more comprehensive model can be constructed which would realize the maximal uncertainty reductions calculated above.

6. Discussion

The winding stairs uncertainty analysis proposed in this paper enables assessment of uncertainty contributions from stochastically independent sources while allowing for dependency within sources. The restriction to independent sources is no sacrifice of practical relevance since the dependencies are functionally linked with the method of data acquisition. The analysis is not based on an approximation of the model. A good approximation enables a very efficient uncertainty analysis. Occasionally, however, one will not succeed in obtaining an adequate approximation. Large input variation and model non-linearities may frustrate the construction of approximations. The presence of uncertainty in a large number of inputs, such as noise signals, may pose additional problems. Except for the case when the approximation is overwhelmingly good, one will always have to deal with the question wether the approximation is adequate for the analysis. All such problems can be circumvented with Monte Carlo techniques, at the price of a larger computational effort. The winding stairs strategy is a Monte Carlo sampling technique designed to capitalize on the computations. This is achieved by multiple use of model evaluations and by the extensibility of the sample. One may start with a moderately small sample and append new cycles if required in view of the precision of the uncertainty estimates. Moreover, one may start with a small number of steps in a cycle, inserting new steps only at places where the effect of current steps is found to be large.

References

Iman, R.L. and W.J. Conover (1982), 'A distribution free approach to inducing rankcorrelations among input variables', *Communications in Statistics*, **B 11**, 311-334.

Iman, R.L. and J.C. Helton (1988), 'An investigation of uncertainty and sensitivity analysis techniques for computer models', *Risk Analysis*, **8**, 71-90.

Iman, R.L. and S.C. Hora (1990), 'A robust measure of uncertainty importance for use in fault tree system analysis', *Risk Analysis*, **10**, 401-406.

Janssen, P.H.M. (1993), 'Assessing sensitivities and uncertainties in models: a critical evaluation', In: *Proc. Conference on Predictability and Nonlinear Modelling in Natural Sciences and Economics*.

McKay, M.D., R.J. Beckman and W.J. Conover (1979), 'A comparison of three methods for selecting values of input variables in the analysis of output from a computer code', *Technometrics*, **21**, 239-245.

Morris, M.D. (1991), 'Factorial sampling plans for preliminary computational experiments', *Technometrics*, **33**, 161-174.

Owen, A.B. (1992), 'A central limit theorem for latin hypercube sampling', *J.R.Statist.Soc.*, **B 54**, 541-551.

Rossing, W.A.H., R.A. Daamen, E.M.T. Hendrix and M.J.W. Jansen (1993a), 'Prediction uncertainty in supervised pest control in winter wheat: its price and major causes', In: *Proc. Conference on Predictability and Nonlinear Modelling in Natural Sciences and Economics*.

Rossing, W.A.H., R.A. Daamen and M.J.W. Jansen (1993b), 'Uncertainty analysis applied to supervised control of aphids and brown rust in winter wheat. 1. Quantification of uncertainty in cost-benefit calculations', To appear in: *Agricultural Systems*.

Rossing, W.A.H., R.A. Daamen and M.J.W. Jansen (1993c), 'Uncertainty analysis applied to supervised control of aphids and brown rust in winter wheat. 2. Relative importance of different components of uncertainty', To appear in: *Agricultural Systems*.

Stein, M. (1987), 'Large sample properties of simulations using latin hypercube sampling', *Technometrics*, **29**, 143-151.

ASSESSING SENSITIVITIES AND UNCERTAINTIES IN MODELS: A CRITICAL EVALUATION

P.H.M. JANSSEN

National Institute of Public Health and Environmental Protection (RIVM),
P.O. Box 1, 3720 BA Bilthoven, The Netherlands
e-mail: cwmpj@rivm.nl

Summary

Sensitivity and uncertainty analysis are important ingredients of the modelling process, and contribute substantially to a reliable and efficient development, assessment and application of mathematical models. Quantifying how much the concerned model components contribute to the sensitivity and uncertainty in the model outputs is an essential issue in these analyses. An overview is given of various measures which are commonly used for assessing these contributions; their main features are discussed and critically evaluated.

Keywords: Sensitivity, uncertainty, regression, correlation.

1. Introduction

Mathematical models are useful tools in the study of many real life problems. However, our knowledge and information on these problems is typically limited, uncertain and poor. In order to enable a reliable development and application of such models, it is therefore often imperative to perform a thorough *sensitivity and uncertainty analysis* of the model to clarify the crucial aspects/factors of the model and to convey the origins and effects of model uncertainties, see Heuberger and Janssen (1993). The obtained results moreover lead to increased insight in the model and provide us with useful suggestions for further model development, calibration and data acquisition.

Sensitivity analysis is primarily concerned with the question how model outputs are affected by variations in the values of the *model components* (e.g. model parameters, initial conditions etc.), and renders useful information in situations where these components are incompletely known or subject to changes or misspecifications (e.g. in the context of model calibration, design, optimization). In uncertainty analysis the focus is somewhat different. Situations are considered where uncertainty and/or risk play a prominent role. This is reflected by assuming that the values of the model components are uncertain, e.g. due to incomplete knowledge (e.g. measurement errors) and/or natural variability. Uncertainty analysis studies how this uncertainty in the model components affects the uncertainty in the model outputs. A variety of techniques has been proposed for performing these analyses, and many applications have been reported, illustrating the relevance of these studies (Iman and Helton (1988), Janssen *et al.* (1990)).

A central issue in sensitivity and uncertainty analysis is the ranking of the importance of the individual model components by assessing their contribution to the sensitivity or uncertainty of the model output(s). Various measures for this assessment have been proposed in literature, see Iman and Helton (1988), Janssen *et al.* (1990), Saltelli and Marivoet (1986), Elston (1992), but their properties and interrelationships are still rather unclear and badly understood. This paper tries to improve on this highly undesirable situation by proposing a general theoretical framework for discussing these measures, and by presenting important results. The exposition is mainly theoretical; part of the material is illustrated by an example. Finally, some important practical implications of the obtained results are indicated.

2. Framework

The measures will be presented in the context of *deterministic* models, i.e. replicate simulations of the model with the same values for the model components yield identical model outputs. The individual model components (parameters, initial conditions) are assumed to be scalar quantities $x_i \in \mathbb{R}$ for $i = 1, ..., p$, which can take values on a continuous range. The associated model output y is also supposed to be a *scalar* quantity $y \in \mathbb{R}$, given as (deterministic) function of the model components x_i, $i = 1, ..., p$:

$$y = H(x_1, ..., x_p). \tag{1}$$

Multivariate model outputs can be treated in a similar fashion, e.g. by considering each component separately. The possible dependence of y on time or on other model quantities (e.g. inputs) is not explicitly indicated in the notation. The specific form of the function $H(.)$ depends on the model under study, and is kept deliberately general.

Uncertainty will be expressed in a *probabilistic* way, by considering the model components x_i, $i = 1, ..., p$ as random variables. The resulting model output y will consequently be also a random variable (or a random process, if time is involved). It is implicitly assumed that first and second (central) statistical moments of the quantities x_i and y exist. They are denoted by Exp(.) (expectation) and Var(.) or σ^2(variance). The standard deviation is denoted by σ. The uncertainty content is expressed in terms of the variance. The results are formulated on the population level (i.e. asymptotic situation) instead of on the finite sample level. A further notational convention is that x^T denotes the transpose of vector x; the inverse of a matrix A is denoted by A^{-1}. The (i,j)-th entry of a matrix A is denoted by $[A]_{ij}$.

3. Sensitivity measures

Sensitivity measures intend to express how much the model output y is affected by variations of the model components x_i. Central questions in this context are: (i) How is x_i varied?, (ii) Which combinations of variations are allowed?, and (iii) How to assess their effect on y?

A common approach is to consider variations (say Δx_i) of x_i around a certain *nominal* (i.e. reference) value $x_i^{(o)}$. The obtained results consequently have a *local* character, and their generality depends on the representativity of the employed nominal point. Moreover the results will depend on the *size* of the employed variation Δx_i, which is usually taken small to enable a linear approximation.

The variations Δx_i for $(i = 1, ..., p)$ can be combined individually or simultaneously, and freely or in a restricted (e.g. correlated) sense (i.e. question (ii)). Variations are typically taken as independent in *pre*-calibration sensitivity studies, which are aimed at detecting 'insensitive' parameters to be excluded from the subsequent calibration. However, in *post*-calibration studies, where one likes to know whether the obtained parameter estimates are sufficiently accurate for the model to yield reliable and accurate predictions, correlations between variations should be considered to obtain meaningful results (Elston (1992)).

Various ways can be used to express the influence of variations Δx_i in x_i on the model output y (i.e. question (iii)), ranging from a *direct* evaluation of the model for '*individual* parameter perturbations', to *approximate* approaches relying on e.g. Taylor series expansions or numerical as well as statistical approximation techniques (curve-fitting, splines, sequential linearization, regression analysis and response surface design, statistical interpolation; see e.g. Downing *et al.* (1985), Kleijnen (1992), Sacks *et al.* (1989)). These approximate approaches are usually more convenient, less limited and laborious (e.g. interactions and simultaneous variations of the components can be more easily studied) than the direct approach. They are based on an approximation of the original relationship (1) by a functional expression (so called *metamodel*) which is more easy to handle. Important issues are the form of the metamodel and the domain on which it will be a valid approximation of the original model. Popular metamodels are the straightforward first-order model:

$$y \approx \beta_0 + \sum_{i=1}^{p} \beta_i x_i \tag{2}$$

and the second-order model

$$y \approx \beta_0 + \sum_i \beta_i x_i + \sum_j \beta_{jj} x_j^2 + \sum_{k,l;k<l} \beta_{kl} x_k x_l \tag{3}$$

The coefficients β. in these metamodels are determined e.g. on basis of (approximate) first and second order derivatives of $H(.)$ (Taylor series expansion) or on basis of regression analysis, curve-fitting etc. of the model components and the associated model output, for various settings $[x_1(k), ..., x_p(k)]$, $(k = 1, ...,N)$ of the model components.

The first-order model forms an easy basis for expressing the influence of the *individual* variations Δx_i on y. The second-order model offers in addition the possibility to consider *interactions* (i.e. non-additive combined effects of the separate variations) between the components x_i. If these interactions are important on the domain where the variations take place, then it is obvious that the *individual* effect of Δx_i on the model output y can not be neatly separated from the effects of the settings x_j and/or the variations Δx_j of the other components, $j \neq i$. These effects will typically occur in a combined fashion. The best which often can be done in this situation, is to express instead these *combined* effects, e.g. in terms of the coefficients β_{kl} in the metamodel (3). Similar remarks hold for the quadratic (β_{jj}) and higher order terms. If metamodel (2) or (3) do not render suitable approximations of the relationship (1), alternative metamodels have to be considered (e.g. consisting of other non-linearities or of transformations of the x_i's and y's), or one must even resort to the direct approach. All this will, however, further complicate (or even preclude) a straightforward and unequivocal assessment of the sensitivity of y to *individual* variations Δx_i.

As a consequence most sensitivity measures employed in practice are therefore based on the linear metamodel (2). Needless to say, this restricts their validity to situations where this linear approximation is appropriate, which often implies that the considered variations should be small. In this case the assessment of sensitivity will be unambiguous and straightforward:

Sensitivity measures for free variations

Three common schemes are typically used for varying the x_i independently:

* If an *absolute* variation $\Delta x_i = f$ is applied, then the associated variation $[\Delta y]_i$ of the output y is approximately:

$$[\Delta y]_i \approx ASC_i \cdot f = ASC_i \cdot \Delta x_i \qquad (4)$$

where ASC_i is the *absolute sensitivity coefficient* which is equal to β_i in eqn. (2).

* If the variation is considered *relative* to its *nominal* value $x_i^{(0)}$, i.e. $\Delta x_i = f \cdot x_i^{(0)}$, then the corresponding variation of the model output y can be expressed relatively with respect to its nominal value $y^{(0)}$ $(=H(x_1^{(0)}, ..., x_p^{(0)}))$ [Note: $y^{(0)}$ and $x_i^{(0)}$ are assumed to be non zero]:

$$\frac{[\Delta y]_i}{y^{(0)}} \approx RSC_i^{(nom)} \cdot f = RSC_i^{(nom)} \cdot \frac{[\Delta x_i]}{x_i^{(0)}}, \qquad (5)$$

where $RSC_i^{(nom)}$ denotes the *relative sensitivity coefficient* given by (use eqn. (4)):

$$RSC_i^{(nom)} := \frac{ASC_i \cdot x_i^{(0)}}{y^{(0)}}. \qquad (6)$$

This quantity is also used in the multiplicative sensitivity analysis of Majkowski *et al.* (1981).

* In situations where *uncertainty* in the model components is of concern, the employed variation is often taken as a fraction of the standard deviation in the model components, i.e. $\Delta x_i = f \cdot \sigma_{x_i}$. The associated variation of y, say $[\Delta y]_i$, is related correspondingly to its standard deviation σ_y, and results in

$$\frac{[\Delta y]_i}{\sigma_y} \approx RSC_i^{(sdev)} \cdot f = RSC_i^{(sdev)} \cdot \frac{[\Delta x_i]}{\sigma_{x_i}}, \qquad (7)$$

where the *relative sensitivity coefficient* $RSC_i^{(sdev)}$ is defined by (use eqn. (4)):

$$RSC_i^{(sdev)} := \frac{ASC_i \cdot \sigma_{x_i}}{\sigma_y}. \qquad (8)$$

Sensitivity measures for restricted (correlated) variations

In situations where variations do not vary freely, alternative measures should be employed to yield meaningful results. E.g. in post-calibration sensitivity studies, variations are considered under the constraint that they refer to equally probable values, e.g. in terms of the likelihood characterizing the confidence region of the estimated parameters. We employ elliptical confidence regions of the form:

$$(\Delta x)^T V^{-1} (\Delta x) \leq f^2 , \tag{9}$$

where V denotes the estimated covariance matrix of the parameters. Although the elliptic confidence regions apply, strictly speaking, only for linear models with normal errors they can often be used as adequate approximations for non-linear models, especially if a transformation has been performed to reduce the non-linear effects. In line with Elston (1992), three possible measures for sensitivity can be proposed (see figure 1):

Single Constrained Parameter Variation (SCPV): This measure expresses how much the output y changes ($[\Delta y]_i \approx \text{SCPV}_i \cdot f$), when varying only a *single* component x_i under the constraint (9), whilst keeping the other components constant. Consequently Δx_i will maximally be $f/\sqrt{[V^{-1}]_{ii}}$. Inserting this in eqn. (4), it is obvious that the associated sensitivity coefficient SCPV_i is equal to:

$$\text{SCPV}_i = \frac{\text{ASC}_i}{\sqrt{[V^{-1}]_{ii}}} \tag{10}$$

* *Maximal Constrained Parameter Variation (MCPV)*: This measure expresses the change ($[\Delta y]_i \approx \text{MCPV}_i \cdot f$) of the model output y due to *maximal* variation of the component x_i within the restrictions (9), *allowing for simultaneous variations* of all other components x_j, $j \neq i$. The MCPV_i can be expressed in terms of ASC and V:

$$\text{MCPV}_i = \frac{\sum_{j=1}^{p} [V]_{ij} \cdot \text{ASC}_j}{\sqrt{[V]_{ii}}} \tag{11}$$

This follows from the general fact that the maximum of $[\gamma^T u]$ under the constraint $[u^T A u \leq \Delta^2]$ is attained for $u = \Delta A^{-1} \gamma / \sqrt{\gamma^T A^{-1} \gamma}$, and will hence be equal to $\Delta \cdot \sqrt{\gamma^T A^{-1} \gamma}$. Here u and γ denote n-dimensional real vectors and A is a positive definite symmetric $n \times n$-matrix. This result can be proved easily by applying Lagrange multipliers. See also Janssen (1993).

* *Maximal Constrained Output Variation (MCOV)*: Allowing for a *completely free* variation of *all* x_i within the limitations imposed by (9), the MCOV expresses how much y will vary *maximally* ($[\Delta y]_{max} \approx \text{MCOV} \cdot f$). MCOV is equal to

$$\text{MCOV} = \sqrt{\sum_{i=1}^{p} \sum_{j=1}^{p} [V]_{ij} \cdot \text{ASC}_i \cdot \text{ASC}_j} \tag{12}$$

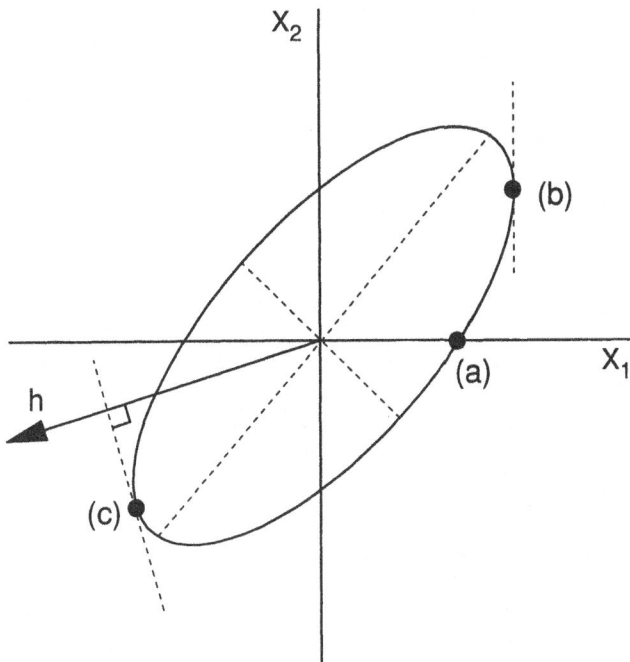

Figure 1. Points in the parameter space which correspond to the sensitivity measures for correlated variations (see eqn. (10)-(12)): **(a)**: SCPV; **(b)**: MCPV; **(c)**: MCOV. The ellipsoid represents the constrained variations of eqn. (9); the vector h characterizes the linear metamodel of eqn. (2) (i.e. $y = h^T \cdot x$).

4. Uncertainty measures

Uncertainty measures intend to express how much the uncertainty in each component x_i contributes to the uncertainty in the model output y. A multitude of measures has been proposed for this purpose. In the sequel we briefly review the most common ones, classifying them according to the techniques which have been used to derive them. Their relationships will be indicated in section 5.

4.1 Measures based on regression analysis

The key idea behind this approach is to apply a simple regression metamodel to approximate the relationship (1). A common choice is to use a linear regression model (compare eqn. (2)):

$$y = \beta_0 + \sum_{i=1}^{p} \beta_i x_i + \hat{\varepsilon} = :\hat{y} + \hat{\varepsilon} \tag{13}$$

Considering x_i as random variables, β_0, β_1 ,...., β_p denote the *ordinary regression coefficients (ORC)*, obtained by minimizing the mean square prediction error:

$Exp_x[y - \beta_0 - \Sigma_{i=1}^{p} \beta_i x_i]^2$.

In eqn. (13) $\hat{y}$ denotes the best linear predictor and $\hat{\varepsilon}$ is the associated prediction error. The goodness of the linear approximation (13) can be assessed by the R^2 of regression:

$$R^2: = \frac{\sigma_{\hat{y}}^2}{\sigma_y^2} = 1 - \frac{\sigma_{\hat{\varepsilon}}^2}{\sigma_y^2} \tag{14}$$

R^2 is a number between 0 and 1, expressing the fraction of the variance explained by the linear approximation; $R^2 \approx 1$ indicates that a linear approximation is adequate.

Using standard properties of the regression it can be easily established that (Janssen *et al.* (1990), Janssen (1993)):

$$R^2 = \sum_{i=1}^{p} \sum_{j=1}^{p} \beta_i^{(s)} \beta_j^{(s)} \rho_{x_i x_j} , \tag{15}$$

where $\rho_{x_i x_j}$ is the correlation between x_i and x_j, and

$$\beta_i^{(s)} = SRC_i := \beta_i \frac{\sigma_{x_i}}{\sigma_y} \qquad (i = 1,...,p) \tag{16}$$

denotes the so called *standardized regression coefficient (SRC)*, i.e. the regression coefficient which results from applying linear regression to the standardized quantities y/σ_y and x_i/σ_{x_i}.

Expression (15) illustrates that SRC_i uniquely characterizes the (linear) *uncertainty contribution* of x_i to y, *if x_i is uncorrelated (!!)* with the other components x_j (i.e. $\rho_{x_i x_j} = 0$, $i \neq j$). Moreover it clarifies that correlations between components *preclude* an unequivocal determination of the uncertainty contribution of an *individual* component x_i: the contribution of x_i can not be neatly separated from the ones of the correlated components x_j, $j \neq i$. In the literature various ways have been proposed to 'cope' with this problem:

1.   *Discarding x_i from the (linear) regression*, and measuring the associated loss in explained variance (Dale *et al.* (1988)): Let $\hat{y}_{/i-}$ denote the associated linear predictor, when predicting y on basis of x_1, ..., x_{i-1}, x_{i+1}, ..., x_p, and let $\sigma_{\hat{y}_{i-}}^2$ be its variance. Then the relative loss in explained variance, due to discarding x_i from the linear regression, is expressed as (*RPSS= Relative Partial Sum of Squares*):

$$RPSS_i: = \frac{\sigma_{\hat{y}}^2 - \sigma_{\hat{y}_{i-}}^2}{\sigma_y^2} . \tag{17}$$

2.   *Considering only x_i in the (linear) regression*, and determining its contribution to the variance: This leads to a linear regression model in x_i:

$$y = \beta_0^{(i)} + \beta_1^{(i)} x_i + \varepsilon^{(i)} = :\hat{y}_{(i)} + \hat{\varepsilon}_{(i)} \tag{18}$$

where $\beta_1^{(i)} = \rho_{yx_i} \cdot \sigma_y / \sigma_{x_i}$ with ρ_{yx_i} denoting the correlation between y and x_i. The (relative) linear contribution of x_i to the variance is now expressed as

$$\frac{\sigma_{\hat{y}_{(i)}}^2}{\sigma_y^2} = \frac{\left[\beta_1^{(i)}\right]^2 \sigma_{x_i}^2}{\sigma_y^2} = [\rho_{yx_i}]^2 . \tag{19}$$

3.   *Accounting for the associated correlated components:* An alternative way (Janssen *et al.* (1990)) of determining the uncertainty contribution of x_i is to express how much (i.e. $\Delta\sigma_y$) the 'uncertainty' σ_y in the output y will be affected by an incremental change of the 'uncertainty' in x_i, i.e. from σ_{x_i} into $\sigma_{x_i}(1 + f)$, thereby accounting for the induced change in the correlated components x_j $(j \neq i)$:

$$\frac{\Delta\sigma_y}{\sigma_y} \approx PUC_i f = PUC_i \frac{\Delta\sigma_{x_i}}{\sigma_{x_i}} . \tag{20}$$

The resulting quantity PUC_i (*Partial Uncertainty Contribution*) is a combination of regression and correlation quantities (Janssen *et al.* (1990), Janssen (1993)):

$$PUC_i: = \sum_{j=1}^{p} \beta_j^{(s)} \rho_{yx_j} \rho_{x_i x_j}^2 . \tag{21}$$

4.   *Orthonormalizing the components:* The correlated components are first transformed into uncorrelated components $\tilde{x}_i$ with unit variance. The uncertainty contribution of these 'orthonormal' components   is $\tilde{x}_i$ assessed on basis of linear regression between y and $\tilde{x}_k$; finally, by back-transformation of these results, the contribution of the *original* components x_i is determined. It can easily be shown, see Janssen (1993), that the thus obtained uncertainty measure is equivalent to the correlation coefficient ρ_{yx_i}.

5.   *Step-wise regression*: By entering variables in a sequential fashion into the regression, studying the associated improvement of the regression fit R^2, and regarding the relative size of the SRC's in the final model, one achieves an importance ranking of the components. This step-wise regression (Iman *et al.* (1981)) can be performed in various ways. The computed uncertainty contributions depend on the already selected model components and hence depend on the applied step-wise regression scheme.

4.2 Measures based on correlation analysis

The key idea behind this approach is to measure the association between the component x_i and the model output y. The most simple form is to use the *linear correlation coefficient* ρ_{yx_i} (*LCC_i*). Since the LCC_i also incorporates the influence of the other correlated components, Iman and Helton (1988) proposed the use of the *partial correlation coefficient (PCC)*, which represents that part of the association between y and x_i which is not due to correlations with the remaining x_j, $j \neq i$. The PCC is obtained by correcting first y and x_i for all linear correlation effects of the remaining x_j, $(j \neq i)$, and then evaluating the correlation coefficient for the corrected quantities $\tilde{y}_i$, $\tilde{x}_i$, i.e. PCC$_i := \rho_{\tilde{y}_i \tilde{x}_i}$.

Use of this measure however hampers a fair comparison between the various components, since different model outputs (i.e. the corrected $\tilde{y}_i$) are considered for the different components x_i. Therefore an alternative measure was proposed in Janssen *et al.* (1990) by correcting only the component x_i, and correlating the corrected quantity $\tilde{x}_i$ to the *original* model output y, i.e. SPC$_i := \rho_{y\tilde{x}_i}$. This quantity is termed the *semi-partial correlation coefficient (SPC)*, and is closely related to the RPSS (see the subsequent theorem 1).

4.3 Measures based on conditioning

Assuming that the value of the component x_i is known, the corresponding reduction of the uncertainty was expressed in Iman and Hora (1990) as:

$$\Delta_{x_i} := \text{Var}(y) - \text{Var}(y|x_i) , \tag{22}$$

where $\text{Var}(y \mid x_i)$, is the variance of y conditional on the specific value of x_i. The *expected reduction of variance (ERV)* associated to ascertaining the true value of x_i over its domain/distribution, is then defined as

$$\text{ERV}_i := Exp_{x_i}\{\text{Var}(y) - \text{Var}(y|x_i)\} = \text{Var}_{x_i}\{Exp(y|x_i)\} \tag{23}$$

i.e. the variance of the conditional expectation. This last equality is a well known fact in probability theory (Rao (1973)). Jansen *et al.* (1993) have independently discovered this ERV measure as an adequate tool to assess the uncertainty contribution for independent multivariate components.

The associated quantity *(RERV=Relative Expected Reduction of Variance)*:

$$\text{RERV}_i := \frac{\text{ERV}_i}{\sigma_y^2} = \frac{\text{Var}_{x_i}\{Exp(y|x_i)\}}{\sigma_y^2} \tag{24}$$

can be considered as the relative uncertainty contribution of the component x_i. Unlike the previous methods, the RERV applies also to *non-linear* models. Its computation can however be very laborious or complicated. This can partly be alleviated by applying an efficient Monte Carlo technique (winding stairs sampling for independent components; see Jansen *et al.* (1993)) or more sophisticated regression ideas (Iman and Hora (1990)).

4.4 Alternative measures

The regression and correlation measures in section 4.1 and 4.2 are only valid when non-linearities are minor (reflected by a large R^2). When encountering important *non-linearities*, one should resort to other methods, e.g. using the RERV (section 4.3), considering more *sophisticated (non linear) regression* models, or applying suitable *transformations* (e.g. rank-transformations or continuous non-linear transformations) on y and x_i before performing regression and correlation analysis (see Janssen *et al.* (1990), Iman and Helton (1988), Janssen (1993)). When applying the two last mentioned alternatives, it will however often be difficult (or even impossible) to relate the finally obtained results to straightforward statements on the uncertainty contribution of the *original* components x_i.

5. Properties of the measures

The next theorems (cf. Janssen *et al.* (1990), Janssen (1993)) present various important results on the relations between the presented measures:

Theorem 1 (a) The various measures defined in section 4 are related as:

$$LCC_i = \sum_{j=1}^{p} SRC_j \cdot \rho_{x_i x_j} , \tag{25}$$

$$SPC_i = \frac{SRC_i}{\sqrt{[C_x^{-1}]_{ii}}} , \tag{26}$$

$$RPSS_i = (SPC_i)^2 , \tag{27}$$

$$PCC_i = \frac{SPC_i}{\sqrt{(SPC_i)^2 + (1 - R^2)}} , \tag{28}$$

where C_x denotes the correlation matrix of x_i's. C_x^{-1} denotes its inverse.

(b) If regression model (13) is used as the linear metamodel on which the sensitivity measures in section 3 are based, and if the matrix V in (9) is equal to the covariance matrix of the model components x_i, then the measures in section 3 and 4 are related as:

$$RSC_i^{(sdev)} = SRC_i; \; SCPV_i = SPC_i \cdot \sigma_y; \; MCPV_i = LCC_i \cdot \sigma_y; \; MCOV = R \cdot \sigma_y . \tag{29}$$

∎

Theorem 2 Suppose that the model output y and the model components $x_1, ..., x_p$ are related as:

$$y = \beta_0 + \sum_{j=1}^{p} \beta_j x_j + r , \tag{30}$$

where $\beta_0, ..., \beta_p$ are constants, and r is a random variable which is independent from x_j, $j = 1, ..., p$. Suppose additonally that the mutual relationship between the x_j, x_k, for $j,k = 1, ..., p$, is in essence linear, i.e.

$$x_j = \gamma_{j,k} x_k + \varepsilon_{j,k} , \quad j \neq k \tag{31}$$

for certain $\gamma_{j,k} \in \mathbf{R}$, with $\varepsilon_{j,k}$ being independent from x_k. Then

$$RERV_i = (\rho_{yx_i})^2 \ . \tag{32}$$

∎

These theorems establish important results between the various measures and render moreover a useful sensitivity interpretation of the uncertainty measures SRC, LCC and SPC (see theorem 1 (b)). They also form the foundations for a critical evaluation of the presented uncertainty measures:

As was pointed out in section 4.1, the choice of an adequate uncertainty measure is straightforward in case of *uncorrelated* components: the SRC appears to be a good candidate in case that $R^2 \approx 1$. Since theorem 1 shows that the SRC, the LCC and the SPC are equal for uncorrelated components, the LCC and SPC will also be adequate measures in these circumstances.

Difficulties occur when *correlations* exist between the various components, since the individual uncertainty contributions of the components cannot be neatly separated and the SRC is no longer a reliable indicator. The other measures which have instead been proposed, show also deficiencies (see the example below):

1.   The RPSS is an easy interpretable measure, in terms of the loss of explained variance when discarding the specific component x_i. Therefore a *large* $RPSS_i$ indicates that the component x_i is important (discarding x_i from the regression results in a considerable deterioration of the 'fit'). *However*, a *small* $RPSS_i$ does *not* justify the conclusion that the contribution of x_i is insignificant: e.g. discarding an 'important' component x_i (in terms of the SRC) from the regression can be nearly fully compensated by other (possibly less important) components if they are highly correlated with x_i. One should be aware of these effects when using and interpreting the RPSS. In order to arrive at more definite conclusions, additional information on the correlation structure and the size of the SRC's should be taken into account (see Janssen (1993)). Due to property (27) similar statements hold for the SPC.

2.   Due to the properties (27) and (28), the PCC behaves similar to the RPSS and the SPC in the *ordinal* sense (if $R^2 < 1$), and therefore leads to the same importance ranking of the components. Its numerical values however do not always offer a clear discrimination between important and unimportant components, e.g. all PCC will be near to 1 or -1 if $R^2 \approx 1$ (due to (28)); see also the subsequent example.

3.   The correlation based measure LCC shows even a more pessimistic picture: correlated components can reduce or intensify the role of a specific component x_i; hence the size of the LCC needs not be indicative for the relevance of x_i.

4.   Due to the apparent similarity between LCC and (R)ERV for the specific situation sketched in theorem 2, the (R)ERV has similar drawbacks as the LCC in case of correlated components.

5.   Neither does the PUC render a unique and straightforward indication for the uncertainty contribution: due to the fact that the influence of correlated components is taken into account, it can occur that 'weak' components (i.e. having a small SRC) will score high on the PUC-measuring scale if they are strongly correlated with 'strong' components (i.e. having a large SRC). An additional drawback is that the PUC is derived under conditions which will not always be fulfilled in practice: only 'small' changes in the uncertainty are considered and it is assumed that the specified correlation structure remains fixed during these changes. Moreover the PUC does not adequately take into account the *mutual* effects which the other correlated x_j, x_k $(j,k \neq i)$ have on each other. See Janssen *et al.* (1990) and Janssen (1993) for further details.

Example 1 This example serves to illustrate some of the deficiencies mentioned in the foregoing discussion. The model output y is given as a simple linear function $y = \gamma_1 x_1 + \gamma_2 x_2 + \gamma_3 x_3$ of the components x_1, x_2, x_3, which all have zero mean and variance 1. Moreover it is assumed that x_1 is correlated with x_2 with correlation coefficient ρ_{12}, and that x_3 is uncorrelated to x_1, x_2. For the choices of γ_1, γ_2, γ_3, ρ_{12} three cases are distinguished, as presented in table 1. Notice that case A and B are characterized by a high (negative) correlation between x_1 and x_2. Since the coefficients γ_1 and γ_2 are equal in case A, the influence of x_1 and x_2 will therefore nearly cancel out.

Table 1. Settings for uncertainty analysis in example 1.

	Case A	Case B	Case C
ρ_{12}	-0.99	-0.99	0.50
γ_1	10	10	10
γ_2	10	1	20
γ_3	2	2	1

The results on the uncertainty, which can be computed explicitly, are displayed in table 2, where the contributions of the various components is evaluated in terms of the LCC, the SPC and the PUC. The PCC's have been omitted from the table, since they will be -1, or +1 in this example (correcting y for all linear correlation effects of x_j, $j \neq i$, renders $\tilde{y}_i = \gamma_i \cdot \tilde{x}_i$ hence $\rho_{\tilde{y}_i \tilde{x}_i}$ is +1 or -1).

The first two rows of the table display the uncertainty contribution of the combined correlated components x_1 and x_2, and of the individual uncorrelated component x_3. Studying these results, various problems can be distinguished:

4. Systems sciences

Table 2. Results of uncertainty analysis in example 1.

	Case A	Case B	Case C
$Unc(x_1 \& x_2)$	2	81.2	300
$Unc(x_3$	4	4	1
SPC_1	0.576	0.1528	0.4992
SPC_2	0.576	0.0153	-0.9984
SPC_3	0.8165	0.2167	0.0576
LCC_1	0.0408	0.9761	0.0000
LCC_2	0.0408	-0.9642	-0.8646
LCC_3	0.8165	0.2167	0.0576
PUC_1	0.3300	0.9551	0.2492
PUC_2	0.3300	0.9320	0.9967
PUC_3	0.6667	0.0469	0.0033

*   Case A illustrates that the SRC's do not necessarily yield reliable expressions for the uncertainty content in case of correlations: Since the SRC_i's are in this example proportional to the γ_i's (due to $\sigma_{x_1} = \sigma_{x_2} = \sigma_{x_3} = 1$) the SRC_1 and SRC_2 are substantially larger than SRC_3. This however does not imply that x_1 and x_2 contribute far more to the uncertainty of the output y than x_3. Compare the first two rows in table 2 which show that the joint uncertainty contribution of x_1 and x_2 is even exceeded by the contribution of the uncorrelated component x_3! This is due to the fact that x_1 and x_2 nearly cancel each other.

Neither do the LCC's offer clear indications: the extremely low values of the LCC of x_1 and x_2, as compared to the LCC of x_3, would suggest that the uncertainty contribution of x_1 and x_2 is strongly outperformed by the one of x_3. The first two rows of table 2 indicate, however, that this is far less dramatic than suggested by the LCC. Observe that the SPC's and the PUC's, on the other hand, render more reliable information on the actual contributions.

*   Case B illustrates, however, that also the SPC's and the PUC's have their shortcomings: a small value of the SPC does not justify the conclusion that the associated component is relatively unimportant. E.g. although the sizes of the SPC for x_1, x_2 are lesser than the one of x_3, the uncertainty contribution of the first two components exceeds the latter considerably. This deficiency is due to the potential compensatory effects of the correlated model components (see item 1 in the above discussion).

These correlation effects can likewise cause problems for the PUC: Notice that the PUC of x_2 is in the same range as the one of x_1, although x_2 is far less important than x_1 when judged by the coefficients γ_i in the regression model. The PUC thus outweighs weak components which are strongly correlated with strong components. When applying the PUC these (artifical) effects have to be interpreted carefully. See e.g. the practical applications reported in Traas and Aldenberg (1993), Heuberger and Janssen (1993), Kros *et al.* (1993).

Notice that the LCC in case B has similar deficiencies as the PUC.

*   Case C finally illustrates once again (compare case A), that small values of the LCC are not conclusive: the LCC of x_1 is even zero, although this component clearly plays a role.

                                                                   ∎

6. Final discussion

A large number of sensitivity and uncertainty measures (see table 3) has been presented, and their main features and interrelationships have been discussed.

Table 3. Glossary of presented sensitivity and uncertainty measures.

Acronym	Meaning	Definition
ASC	Absolute Sensitivity Coefficient	see (4)
$RSC^{(nom)}$	Relative Sensitivity Coefficient	see (5)
$RSC^{(sdev)}$	Relative Sensitivity Coefficient	see (7)
SCPV	Single Constrained Parameter Variation	see (10)
MCPV	Maximal Constrained Parameter Variation	see (11)
MCOV	Maximal Constrained Output Variation	see (12)
SRC	Standardized Regression Coefficient	see (16)
RPSS	Relative Partial Sum of Squares	see (17)
PUC	Partial Uncertainty Coefficient	see (21)
LCC	Linear Correlation Coefficient	ρ_{yx_i} (see section 4.2)
PCC	Partial Correlation Coefficient	$\rho_{\hat{y}_i \hat{x}_i}$ (see section 4.2)
SPC	Semi-Partial Correlation Coefficient	$\rho_{y\hat{x}_i}$ (see section 4.2)
RERV	Relative Expected Reduction of Variance	see (24)

This discussion illustrates that:

1.   Most *sensitivity* analyses have a local character, pertaining to some nominal point around which variations are considered. If a linear metamodel provides a suitable approximation, the assessment of the sensitivity contribution is straightforward and unambiguous, for unrestricted as well as for restricted (correlated) variations. The various measures presented in section 3 are suitable tools in this context. However, in situations where a linear approximation is inadequate the situation is less favorable: due to the presence of interactions and higher order effects the influence of *individual* variations can no longer be easily and clearly assessed. Combined or compounded effects should be considered instead.

2.   Measuring the contribution of model components in *uncertainty analysis* is only
      straightforward in situations where the mutual correlations between model compo-
      nents are *minor*:

 (i)   If the model is suitably approximated by a regression based *linear metamodel*, the
       associated SRC's provide adequate tools for measuring the uncertainty contribution.
       The alternative measures LCC, SPC (RPSS) and PUC are closely related to the
       SRC and can thus be used equally well.
 (ii) If *non-linearities* are substantial, the RERV appears to be an adequate and intuiti-
       vely appealing measure.

3.   In situations where *substantial correlations* occur, the uncertainty contributions of
      the *separate* components can not be disentangled properly. All proposed measures
      contain some artificial and biased elements in the way in which they (nonetheless)
      try to express the contribution of an individual component:

 (i)   In situations where the *linear* regression metamodel is an adequate approximation
       of the original model, the easily interpretable RPSS (or equivalently the SPC)
       seems the most appropriate candidate amongst the many alternative measures.
       Large sizes of the RPSS (or SPC) indicate important 'sources'. However, low
       values of the RPSS are less conclusive, and do not necessarily indicate insignificant
       'sources'. In these situations the RPSS should be used in judicious combination
       with additional information on e.g. the correlation structure and the SRC's, to
       prevent erroneous conclusions.
 (ii) In situations where *non-linearities* are substantial the problem of assessing the
       uncertainty contributions is even more complex and nearly untouched.

Due to the intrinsic impossibility to separate the uncertainty of the *individual* components in situations with substantial correlations, one should instead consider *groups* of
strongly correlated components together, and try to express their joint uncertainty
contribution. Issues like 'How (and in which order) to group them and how to assess
and compute their joint contribution?' are central in this context. Also the problem of
'How to interpret and use this information?' is relevant and deserves further study. This
concerns e.g. important items like 'Which groups, and which components within these
groups, should be determined more accurately (e.g. by additional measurement, research,
or calibration), to diminish the uncertainty in the model outputs?'
The expected reduction of variance associated to 'ascertaining' the i-th group G_i can, in
complete analogy with the derivation of the ERV-measure in section 4.3, be proposed as
an adequate uncertainty measure in this context. This quantity appears to be equal to the
variance of the conditional expectation $Var_{G_i}[Exp(y|G_i)]$.

The presented material focussed mainly on theoretical aspects, in order to obtain a more
thorough and deeper understanding of the properties and relationship of the various
measures. Although this provided useful guidelines for which measures to use in practice,
other crucial practical issues, like 'How to compute the measures, and how to decide on
the importance of correlations and non-linearities etc.?' have not been adressed. Therefore
we like to conclude with a brief discussion of some practical items, indicating moreover
important issues which require additional research in the future:

-a-   How to compute these sensitivity and uncertainty measures?
Iman and Helton (1988), Janssen *et al.* (1990) illustrate that the Monte Carlo
approach is suitable to perform the computations for sensitivity or uncertainty
analysis of the model. By generating N samples from specified distributions of the
model components x_1, ..., x_p, and subsequently simulating the model output y for
these settings, N combinations $[y(k), x_1(k), ..., x_p(k)]$, $(k = 1, ..., N)$ are obtained
which can be subjected to regression and/or correlation analysis. Most measures are
based on linear regression and correlation analysis, and can thus be easily derived
in this way. The software package UNCSAM offers adequate tools to perform
these computations for a large variety of models (Heuberger and Janssen (1993)).
Computation of the RERV-measure can however be rather cumbersome or laborious; one should resort to non-linear regression (Iman and Hora (1990)) or use
special sampling techniques (Jansen *et al.* (1993)).

-b-   How to decide when correlations are substantial?
Correlations can severely affect the outcomes of uncertainty analysis, and will
typically preclude an unequivocal determination of the uncertainty contribution of
separate components. If the linear regression model of eqn. (13) is an adequate
approximation to the model (i.e. $R^2 \approx 1$), then $\sigma_y^2 \approx \Sigma_{i,j} \beta_i \sigma_{x_i} \cdot \beta_j \sigma_{x_j} \cdot \rho_{x_i x_j}$. This expression provides a straight-forward way to assess the influence of the correlations on
the uncertainty in the model output, and to study its major causes. However, when
the linear approximation (13) is inadequate, it becomes more difficult to perform
an appropriate assessment. Application of more advanced regression models or
transformations can sometimes be helpful; otherwise one has to resort to performing various analyses with or without correlations between the components, and to
studying their differences.

-c-   How to determine whether a linear approximation is adequate?
In uncertainty analyses the R^2 (see (14)) expresses the fraction of the output
variance which is explained by the linear model; therefore the nearness of R^2 to 1
provides a suitable measure for the adequacy of the linear approximation. However,
this interpretation is strictly spoken only valid in the asymptotic case (i.e. on the
population level); for the finite sample situations encountered in practice, other
measures are more appropriate e.g. the adjusted R^2, Mallows' C_p statistic, PRESS-
statistic etc. (see Seber (1977)).

In sensitivity analyses, these statistical quantities are often not completely appropriate, since they heavily rely on probabilistic arguments. It is recommended to use
instead more numerically based quantities to judge the adequacy of the linear
approximation (2), e.g. by trying to find (tight) upper bounds for the deviation
between the model output and the linear approximation over the domain of
approximation. However, since this can be a difficult and laborious task, one
nevertheless often resorts to using the above mentioned goodness-of-fit measures
(R^2, adjusted-R^2 etc.).

-d-   How to proceed if the proposed measures do not work?
In situations where e.g. non-linearities and/or correlations are substantial, the
proposed measures often will fail. Ad hoc solutions can be tried, consisting e.g. of
employing advanced non-linear regression models or appropriate transformations.

If these solutions also fail, then, for the time being, no suitable results can be obtained. Future research should focus on these situations.

-e-   How to treat multivariate outputs?

The proposed measures can be employed for each model output component separately. Since this will typically yield a bulk of (diverse) information, especially if model outputs are time series, there is an obvious need to summarize this information adequately. The way in which this should be done will certainly depend on the aspects which are of main interest for the application at hand. An additional point of concern is to group model components (e.g. parameters) which have similar effects on the various outputs. This is particularly relevant in calibration applications in order to prevent identifiability problems. A related question is how such a group should be treated in the actual calibration. These items require extra study in the future.

-f-   How to develop alternative expressions for describing the uncertainty content?

The presented measures for uncertainty analysis use the *variance* to express the uncertainty content. This can lead to instable estimates in situations where fat-tailed distributions occur (i.e. the values can vary largely from the considered sampled set of values). In Iman and Hora (1990) an alternative measure is presented, based on sample *quantile* estimates which are more robust. These estimates however have the drawback that they can be strongly dependent on the considered nominal point where the separate x_i are evaluated. Also this issue deserves more future attention.

References

Dale, V.H., H.I. Jager, R.H. Gardner and A.E. Rosen (1988), 'Using sensitivity and uncertainty analysis to improve predictions of broad-scale forest development', *Ecological Modelling*, **42**, 165-178.

Downing, D.J., R.H. Gardner and F.O. Hoffman (1985), 'An examination of Response-Surface methodologies for uncertainty analysis in assessment models', *Technometrics*, **27**, 151-163.

Elston, D.A. (1992), 'Sensitivity analysis in the presence of correlated parameter estimates', *Ecological Modelling*, **64**, 11-22.

Heuberger, P.S.C. and P.H.M. Janssen (1993), 'UNCSAM: a software tool for sensitivity and uncertainty analysis of mathematical models', *Conference on Predictability and Nonlinear Modelling in Natural Sciences and Economics*, Wageningen, The Netherlands. 5-7 April.

Iman, R.L. and J.H. Helton (1988), 'An investigation of uncertainty and sensitivity analysis techniques for computer models', *Risk Analysis*, **8**, 71-90.

Iman, R.L., J.H. Helton and J.W. Campbell (1981), 'An approach to sensitivity analysis of computer models. Part 2: Ranking of input variables, response surface validation, distribution effect and technique synopsis', *Journal of Quality Technology*, **13**, 232-240.

Iman, R.L. and S.C. Hora (1990), 'A robust measure of uncertainty importance for use in fault tree system analysis', *Risk Analysis*, **10**, 401-406.

Jansen, M.J.W., W.A.H. Rossing and R.A. Daamen (1993), 'Monte Carlo estimation of uncertainty contributions from several independent multivariate sources', *Conference on Predictability and Nonlinear Modelling in Natural Sciences and Economics*, Wageningen, The Netherlands. 5-7 April.

Janssen, P.H.M. (1993), 'Assessing sensitivities and uncertainties in models: theoretical results and practical examples', RIVM report (in preparation), Bilthoven, The Netherlands.

Janssen, P.H.M., W. Slob and J. Rotmans (1990), 'Sensitivity and Uncertainty Analysis: an Inventory of Ideas, Methods and Techniques' (in Dutch), *RIVM-report nr. 958805001*, Bilthoven, The Netherlands.

Kleijnen, J.P.C. (1992), 'Sensitivity analysis of simulation experiments: regression analysis and statistical design', *Mathematics and Computers in Simulation*, **34**, 297-315.

Kros, J., W. de Vries, P.H.M. Janssen and C.I. Bak (1993), 'The uncertainty in forecasting trends of forest soil acidification', *Water, Air, and Soil Pollution*, **66**, 29-58.

Majkowski, J., J.M. Ridgeway and D.R. Miller (1981), 'Multiplicative sensitivity analysis and its role in development of simulation models', *Ecological Modelling*, **12**, 191-208.

Rao, C.R. (1973), 'Linear Statistical Inference and Its Applications', Second edition, John Wiley & Sons, New York.

Sacks, J., S.B. Schiller and W.J. Welch (1989), 'Designs for computer experiments', *Technometrics*, **31**, 41-47.

Saltelli, A., and J. Marivoet (1986), 'Performances of nonparametric statistics in sensitivity analysis and parameter ranking', *Report EUR 10851 EN*, Commission of the European Communities.

Seber G.A.F. (1977), 'Linear Regression Analysis', John Wiley & Sons, Inc, New York.

Traas, Th.P. and T. Aldenberg (1993), 'Combination of uncertainty analysis and risk assessment: application to a bioaccumulation model', *Conference on Predictability and Nonlinear Modelling in Natural Sciences and Economics*, Wageningen, The Netherlands. 5-7 April.

UNCSAM: A SOFTWARE TOOL FOR SENSITIVITY AND UNCERTAINTY ANALYSIS OF MATHEMATICAL MODELS

P.S.C. HEUBERGER and P.H.M. JANSSEN

National Institute of Public Health and Environmental
Protection (RIVM),
P.O. Box 1, 3720 BA Bilthoven, The Netherlands
e-mail: cwmheub@rivm.nl

Summary

The paper addresses the important role of sensitivity and uncertainty analysis in the mathematical modeling process and discusses guidelines to perform these analyses. The main features are presented of the software package UNCSAM, which applies efficient Monte Carlo sampling in combination with regression and correlation analysis to perform sensitivity and uncertainty analyses on a large variety of simulation models. The use of UNCSAM is illustrated by an environmental application study.

Keywords Sensitivity, uncertainty, Latin Hypercube sampling.

1. Introduction

The use of mathematical models is widespread in many sciences. They are generally considered as useful tools in the study of many real life problems. Since, however, in general our knowledge and information on these problems is limited, uncertain and poor it is often necessary to perform a thorough *model analysis* to gain insight into the reliability of the model(s). Two major aspects of a model analysis are

* *sensitivity analysis*: the study of the influence of *variations* in model parameters, initial conditions etc. on model outputs. Such an analysis is especially important in situations where model parameters etc. are incompletely known or subject to changes or misspecifications.

* *uncertainty analysis*: the study of the *uncertain* aspects of a model, and of their influence on the (uncertainty of the) model outputs. Especially in situations where uncertainty and/or risk play a prominent role, uncertainty analysis is indispensable.

These analyses serve to clarify the crucial aspects of the model, and to convey the origins and effects of model uncertainties. The results will moreover render increased insight into the model, and can provide useful suggestions for further model development, calibration and data acquisition.

During the last decades a variety of techniques has been proposed for these analyses; for an overview see Iman and Helton (1985,1988), Janssen *et al.*(1990). The majority of techniques aims at deterministic (simulation) models; comparative evaluation studies show that in particular the efficient Monte Carlo based method which uses the recently developed Latin Hypercube sampling technique (McKay *et al.* (1979), Iman and Conover (1980)) is an appropriate candidate to analyze these models.

Based on these observations it was decided at the RIVM to develop a comprehensive and flexible software package UNCSAM (UNCertainty analysis by Monte Carlo SAMpling techniques) to perform sensitivity and uncertainty analyses on a large variety of mathematical models, on basis of the Monte Carlo oriented approach. Before describing the main features of UNCSAM, we first present background material on sensitivity and uncertainty analysis. Finally we illustrate the use of UNCSAM by an environmental application study, and end up with some conclusions. UNCSAM is commercially available.

2. Sensitivity and uncertainty analysis

2.1 Strategy

The process of performing a sensitivity- and/or an uncertainty analysis typically starts with formulating why such an analysis is needed, which information should be obtained, and in which form it should be reported. Subsequently, the various potential sensitivity or uncertainty 'sources' in the model have to be specified and quantified; this concerns the (a) model structure; (b) model inputs ('external forcings'); (c) boundary and/or initial conditions; (d) model parameters. For sensitivity analysis nominal values of these 'sources' are quantified together with appropriate variations around these values.

In an uncertainty analysis usually probability distributions and mutual correlations are specified, characterizing the uncertainty in the parameters, model inputs, boundary and initial conditions. Variations and/or uncertainties in model structure are difficult to quantify; in practice one therefore usually resorts to a comparative study of feasible alternative model structures.

The next step is to evaluate the effects of the above mentioned variations and uncertainties, on the model outputs. In addition to quantifying the 'overall' effect of the combined 'sources' (e.g. by presenting the range of the model outputs, their mean, variance, percentiles, distribution functions, histograms), it will also be of interest to determine the influence of the individual 'sources', and to rank their importance accordingly. Various sensitivity and uncertainty measures can be employed to quantify these individual contributions; see Janssen (1993) for an overview. The obtained results can be reported numerically (e.g. by presenting tables), as well as graphically.

The results of the sensitivity or uncertainty computations depend on the choices and specifications of the various 'sources'. Since these choices can be rather subjective, it is important to check finally the robustness of the obtained results with respect to these choices.

2.2 Monte Carlo based methods

Monte Carlo based methods for sensitivity and uncertainty analysis are simple and straightforward, and can be applied to a wide variety of models. They rely on the assumption that variations and uncertainties in the 'sources' can be suitably described by specifying probability distributions and mutual correlations. Sampling is performed from these distributions, resulting in a set of values for the various 'sources' (Monte Carlo sampling). In order to limit the computational load in the subsequent computations, it is recommended to use an efficient sampling technique, e.g. the recently developed Latin Hypercube Sampling technique (see McKay *et al.* (1979), Iman and Conover (1980), Stein (1987), Owen (1992)).

The sampled values are subsequently used to simulate the model outputs (Monte Carlo simulation), saving these results for further analysis. This analysis consists of computing and showing basic statistical information, and of evaluating the sensitivity and uncertainty measures of the various 'sources'.

2.3 Sensitivity and uncertainty contributions

Various statistics can be employed to quantify the sensitivity and uncertainty contribution of the 'sources' to the model outputs. See Janssen (1993) for an overview and a critical evaluation. For the purpose of this study two easy computable uncertainty measures will be presented which are based on linear regression analysis.

Suppose that N samples have been drawn of the p sources $x_1,...,x_p$, resulting in N combinations $x_1(k),...,x_p(k)$ for $k = 1,...,N$. For each of these settings the model output has been simulated, resulting in N associated values $y(k)$ for $k = 1,...,N$. By performing straightforward least-squares linear regression on these values, the regression equation:

$$y(k) = \beta_0 + \sum_{i=1}^{p} \beta_i x_i(k) + \hat{e}(k) = : \hat{y}(k) + \hat{e}(k) \quad k = 1,...,N \tag{1}$$

is obtained (*metamodel*). The quantities $\beta_0, \beta_1,...,\beta_p$ denote the ordinary regression coefficients, and $\hat{e}(k)$ denotes the regression residual. The associated R^2 of regression is equal to:

$$R^2 : = \frac{S_{\hat{y}}^2}{S_y^2} = 1 - \frac{S_{\hat{e}}^2}{S_y^2} \tag{2}$$

where S^2 denotes the sample variance of the associated quantity. R^2 is a number between 0 and 1, expressing the validity of the regression model $\hat{y}(k)$ to approximate the original model output $y(k)$; $R^2 \approx 1$ indicates a good approximation.

On the basis of equation (1) the uncertainty in y (as expressed by its variance S_y^2) is readily expressed in terms of $\beta_i S_{x_i}$:

$$S_y^2 = \sum_{i=1}^{p} \sum_{j=1}^{p} (\beta_i S_{x_i})(\beta_j S_{x_j}) r_{x_i x_j} + S_{\hat{e}}^2 \tag{3}$$

Here $r_{x_i x_j}$ denotes the (sample) correlation coefficient between the uncertainty 'sources' x_i and x_j. If x_i is uncorrelated with the other 'sources', it is obvious that the quantity $\beta_i S_{x_i}$ measures the linear uncertainty contribution of the 'source' x_i. The related standardized quantity

$$\beta_i^{(s)} = \beta_i \frac{S_{x_i}}{S_y} \quad (i = 1,...,p) \tag{4}$$

is called the standardized regression coefficient (SRC). It expresses the linear relationship between the standardized 'source' x_i/S_{x_i} and the standardized output y/S_y. From (1) it is obvious that

$$\frac{y(k)}{S_y} = \frac{\beta_0}{S_y} + \sum_{i=1}^{p} \beta_i^{(s)} \frac{x_i(k)}{S_{x_i}} + \frac{\hat{e}(k)}{S_y}.$$

Notice from (3) that the SRC also measures the fraction of the uncertainty in y which is contributed by x_i (if $r_{x_i x_j} = 0$ for $i \neq j$). The SRC's will therefore only give a valid impression of the uncertainty contribution if the 'sources' show no substantial correlation, and if the regression model is a fair approximation of the original model output y (i.e. $R^2 \approx 1$).

Expression (3) shows moreover, that correlations between 'sources' make an unequivocal determination of the uncertainty contribution of an individual 'source' impossible, since it can not be neatly separated from the contributions of the correlated 'sources'. Various alternative measures have been proposed to cope with this problem (see Janssen (1993)). One of these measures, the so called partial uncertainty contribution (PUC), tries to express the relative change $(\Delta S_y/Sy)$ in the uncertainty of the model response y, due to a relative change in the uncertainty $(\Delta S_{x_i}/S_{x_i})$ of the individual 'source' x_i, accounting for the induced change of the correlated 'sources' x_j ($j \neq i$):

$$\frac{\Delta S_y}{S_y} \approx PUC_i \frac{\Delta S_{x_i}}{S_{x_i}} \tag{5}$$

The PUC can be expressed as a combination of regression and correlation quantities (Janssen et al. (1990)):

$$PUC_i := \sum_{k=1}^{p} \beta_k^{(s)} r_{x_k y} (r_{x_i x_k})^2 \tag{6}$$

Here $r_{x_k y}$ denotes the (sample) correlation between the 'source' x_k and the model output y. In practice the square root RTU (RooT of Uncertainty):

$$RTU_i := \sqrt{|PUC_i|} \tag{7}$$

is often used as a measure for the uncertainty contribution, since it is closely related to the SRC for uncorrelated 'sources' (observe that the $RTU_i = |SRC_i|$ if x_i is uncorrelated with x_k, $k \neq i$).

Being based on the regression model (1), these measures only measure the linear uncertainty contribution. Their usefulness therefore strongly hinges on the condition that the non-linearities in the relationship between the 'sources' x_i and the model output y are minor (reflected by $R^2 \approx 1$). When encountering important non-linearities one should therefore resort to other methods, e.g. considering more sophisticated (non linear) regression models, or applying suitable transformations on the data (i.e. y and x_i's); see Janssen (1993).

3. The software package UNCSAM

UNCSAM is a flexible software package to perform sensitivity and uncertainty analyses of mathematical models, based on Monte Carlo techniques in combination with regression and correlation analysis. The main features of UNCSAM are as follows:

Application Area: UNCSAM can be used for a broad spectrum of (simulation) models, largely independent of their form and implementation.

Basic components: UNCSAM,
Version 1.1, comprises a collection of programs, developed for the various activities needed in sensitivity and uncertainty analysis:

* Sampling:
  Sampling can be done from a variety of continuous probability distributions: (log)uniform, (log)normal, (log)triangular, exponential, logistic, Weibull, Beta, histogram. Use can be made of simple random sampling and of Latin Hypercube sampling. User-specified parameter correlations are taken into account; a correction for spurious correlations can be applied (Iman and Conover (1982)).

* Basic statistical analysis:
  Information on mean, variance, median, skewness, kurtosis, percentiles, correlations and covariances can be computed, and shown in tabular or graphical form. Moreover empirical distribution functions and histograms can be determined.

* Confidence bounds for estimated quantities:
  The aforementioned statistical quantities are computed as estimates on basis of a finite number of samples. Information on the accuracy of these statistics can be obtained by computing their confidence bounds. UNCSAM offers the opportunity to compute the confidence bounds for the empirical cumulative distribution function.

* Determination of sensitivity and uncertainty contributions:
  A variety of measures has been proposed in literature for evaluating the sensitivity and uncertainty contributions of the individual 'sources' (Janssen (1993)). Most measures are based on regression and correlation analysis of the sampled model 'sources' (e.g. parameters) and the associated simulated model outputs, or their (rank)-transformed counterparts. UNCSAM offers the possibility to explicitly compute 20 different sensitivity and uncertainty measures. The manual (Janssen *et al.*) (1992)) presents guidelines for using and interpreting these measures.

* General:
  UNCSAM is developed in a model independent way, irrespective of how the user's model is implemented. This offers the user the flexibility to use the package for many application studies. An interface between UNCSAM and the model will however be required, in order to pass data files (of appropriate format) from the package to the model and vice versa. UNCSAM provides interface programs which are widely applicable. Moreover, additional programs are available for file and data manipulation, graphics etc.

Operational mode: UNCSAM can be used in batch mode as well as in interactive mode. Software is available to generate instruction files for batch mode execution. A simple menu-driven option is available when working in interactive mode.

Data Input: The various programs which constitute UNCSAM require input files containing the necessary information. In most cases these files have been generated before by other programs of UNCSAM. In a few situations the input-files have to be constructed explicitly by the user (e.g. by using a wordprocessor or text-editor).

Data Output: The final results of the analyses are stored in ASCII-files in the form of tables, or in the form of plot instructions for graphical presentation. The tabular information can be displayed on screen, or printed on a hardcopy device. The graphical information can be made visible by the graphical software at hand.

Hardware Requirements: The software is written in Ansi standard Fortran 77, embedded in an Ansi-C environment. Being written in a portable code, it can thus be used on a variety of devices, such as PC-XT/AT/286/386/486/OS2, VAX, SUN workstations etc. It is advised to use UNCSAM on machines which are equiped with a mathematical co-processor. The PC version of UNCSAM will occupy about 3Mbyte of memory.

Documentation: A comprehensive users manual is available (Janssen *et al.* 1992). A simple case study is included in the manual as an illustration.

Availability: An executable version of UNCSAM (version 1.1) along with documentation is available from the RIVM.

In offering the above mentioned features, a large number of tasks in sensitivity and uncertainty analysis can be performed in a standardized, automatized and time-efficient way when applying UNCSAM. This alleviates the user from much unnecessary burden, and offers him/her the opportunity to focus on the important activities of problem formulation, data collection and result interpretation, activities which typically require much insight and expert knowledge, and which therefore cannot be (fully) automatized.

4. Application study

Presently, UNCSAM has been used in various environmental application studies (e.g. sensitivity or uncertainty analysis of acidification models, atmospheric process model, global warming model, pesticide leaching model; risk assessment studies in ecological settings; see e.g. Johansson and Janssen (1993), Krol (1993), Tiktak *et al.* (1993), Traas and Aldenberg (1993)). Moreover it serves as an aid when performing sensitivity analyses for calibration studies. The package is also applied in the field of public health (uncertainties in predicting the spread of infectious diseases, e.g. AIDS).

In order to illustrate the use of UNCSAM, we present some results of a recent uncertainty analysis in the context of the acidification problem; see Kros *et al.* (1993). This application study concerns the soil acidification model RESAM (REgional Soil Acidification Model), developed at the Winand Staring Centre for Integrated Land, Soil and Water Research. Primary aim of the model is to make predictions of long-term environmental effects of acid deposition on forest soils on a regional scale. RESAM is a dynamic, one-dimensional, multi-layer, process-oriented model, based on the conceptual relationship between forest element cycling and soil acidification. It simulates the major biogeochemical processes occurring in the forest canopy, litter layer and mineral soil horizons. The model consists of a set of mass balance equations, equilibrium equations and rate-limited equations (mostly first order reactions); the model input includes atmospheric deposition and hydrological data; see de Vries and Kros (1989) for more complete information. We will now discuss the various steps which constitute the uncertainty analysis:

1. Problem formulation: Since the results of RESAM are used to analyze the acidification problem and to evaluate the benefit of potential abatement strategies (policy applications), it is imperative that the uncertainty of the model predictions is analyzed, particularly due to the lack of sufficient data to calibrate the long-term model. Major aim of the uncertainty analysis is therefore the quantification of the uncertainty in the model responses to a given deposition scenario, due to uncertainty and spatial variability in the data. The secondary aim is to determine which additional data most improve the reliability of predictions. The analysis is restricted to one specific forest soil ecosystem: a leptic podzol with Douglas fir, subject to a reducing deposition scenario. The soil profile consists of four distinguished horizons (layers): A0 (litter layer, 4cm), A1 (15 cm), Bh (25 cm) and C (20cm). The investigated model responses are pH, Al/Ca ratio and NH_4/K ratio in the root zone. These variables are generally used as indicators for forest soil acidification and for potential forest damage. All variables are yearly water-flux averaged quantities, considered on a regional scale (the Netherlands is divided in 20 typical receptor areas), during the time period 1987-2010.

2. Inventory of uncertainty 'sources': In fact one tries to capture two different forms of uncertainty: one form is due to our lack of knowledge, and results in a specification of a distribution reflecting our 'degree of belief'. The other form is due to the natural variability which is exhibited by many processes, and is reflected in a distribution describing the 'probability of occurrence'. In the present study the uncertainty in the model structure is not considered; this will be subject of a future study where the impact of various process formulations is investigated. Due to the prohibitively large number of potential uncertainty 'sources' ($\geq$ 200), a considerable reduction was attempted by neglecting those 'sources' which were a priori regarded as insignificant for the investigated model responses. The number of considered uncertainty 'sources' was further reduced to about 80, by applying various additional assumptions (e.g. steady state nutrient cycling; no feedback mechanisms between reducing depositions of N and S, and their contents in needles; constant hydrology). These 'sources' concerned (see Kros *et al.* (1993) for complete information):

1.   Input terms: Both wet and dry deposition of SO_2, NO_2, and NH_3, and wet deposition of the base cations (Ca, Mg, K, Na) and chloride are considered as uncertain. The applied deposition scenario for SO_2, NO_2 and NH_3 is based on intended emission reductions in the Netherlands in the time periods 1987-2000 and 2000-2010. The deposition of the base cations and chloride are considered as constant over this time range. The uncertainty in all input scenarios is restricted to their initial values (1987). Once having sampled this initial value, the remaining time traject of the scenario is uniquely determined. Additional uncertain influences in this traject, due to e.g. political and technological factors, are hence not considered.

2.   Initial conditions: The amounts and element contents in needles, roots, stems and root distribution for the Douglas fir tree species are considered as uncertain. This comprises 33 'sources'.

3.   Model parameters: Parameters which characterize the relevant soil- and vegetation processes are considered as uncertain. This comprises 34 sources.

3. Quantification of uncertainty 'sources': The required information on the uncertainty was obtained from expertise judgment, extensive field research and literature data. Characterization of these uncertainties was an elaborate task; (log)normal as well as uniform distributions were applied. Moreover the most obvious correlations between the 'sources' were specified, resulting in 9 mutual correlations between deposition quantities, turnover parameters of roots and needles, and selectivity constants (soil processes); see Kros *et al.* (1993) for details.

4. Evaluation of the effects on the model outputs: Since one typical simulation run with RESAM took approximately 30 minutes on a VAX-mainframe computer, it was decided to apply the efficient Latin Hypercube Sampling (LHS) technique. A choice of $N > 4/3 \cdot p$ samples, where p is the number of parameters, usually gives satisfactory results for LHS (see Iman and Helton (1985)). Hence 150 samples were determined by UNCSAM, sampling with the LHS technique from the given distribution and correlation specifications. The associated model simulations were subsequently performed. Attention was especially focused on the uncertainty over the time range 1987-2010 for the pH, the Al/Ca-ratio and the NH_4/K-ratio in the top of the root zone (layer 1; A1, 15cm), and the bottom of the root zone (layer 3; C, 20cm). This uncertainty is presented by:

* showing tables with the mean, standard deviation, and coefficient of variation at the beginning (1987), halfway (2000), and at the end (2010) of the simulation period.

* showing plots with the trajectories of the mean, median (50 percentile), 2.5 and 97.5 percentile during the simulation period. Moreover the model response is given for a so called reference run, obtained by simulating the model with the parameter values set at their mean value. Motivation for this last information is to check whether it is possible to suffice with only one single (reference) simulation run to obtain a realistic impression of the mean behavior in extensive regional applications.

* showing the histograms of the quantities in 1987, 2000 and 2010.

In addition to this 'overall' uncertainty, the contribution of the individual 'sources' to the uncertainty is evaluated by computing appropriate uncertainty measures and displaying them in tables and in plots.

We only discuss the results for the pH, focusing mainly on layer 3; see Kros *et al.* (1993) for the other situations. Considering the results in Table 1 it is obvious that the absolute uncertainty (standard deviation) of the pH in the subsoil is slightly higher than in the topsoil, whereas the opposite is true for the relative uncertainty (coefficient of variation). Both the absolute and relative uncertainty remain fairly constant in both layers during the simulation period. The mean value of the pH in both layers increases during the simulation period, due to the reducing deposition scenario.

Table 1. Mean, standard deviation and coefficient of variation of the pH in layer 1 and 3 in 1987, 2000 and 2010.

	Layer 1			Layer 3		
	1987	2000	2010	1987	2000	2010
Mean	2.96	3.10	3.22	4.10	4.21	4.33
Standard deviation	0.106	0.111	0.126	0.131	0.123	0.138
Coef. of variation	0.036	0.036	0.039	0.032	0.031	0.032

This behavior also shows up in the trajectories of the mean and percentiles in both layers during the simulation period; see Fig. 1 for the results of layer 3. This figure also shows that the reference run and the mean correspond reasonably well. Notice also a slight difference between the median and the mean, which indicates that the pH distribution is skewed to the left. This is confirmed by the histograms in Fig. 2.

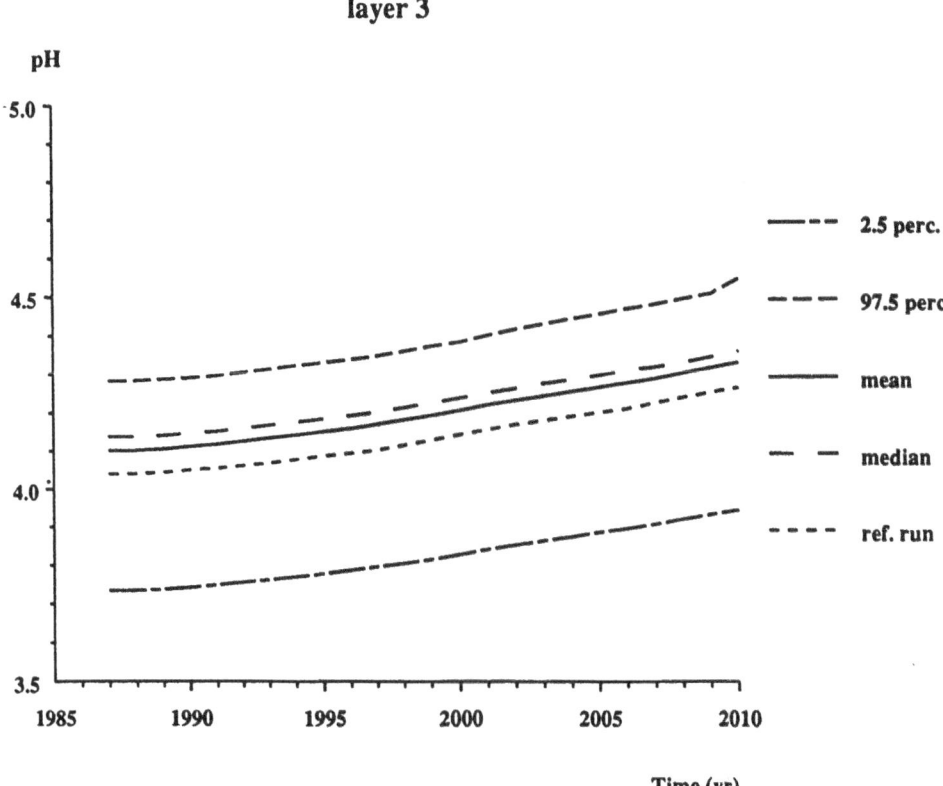

Figure 1. Temporal evolution of the mean, median, 2.5 and 97.5 percentile, and the reference run of the pH in layer 3.

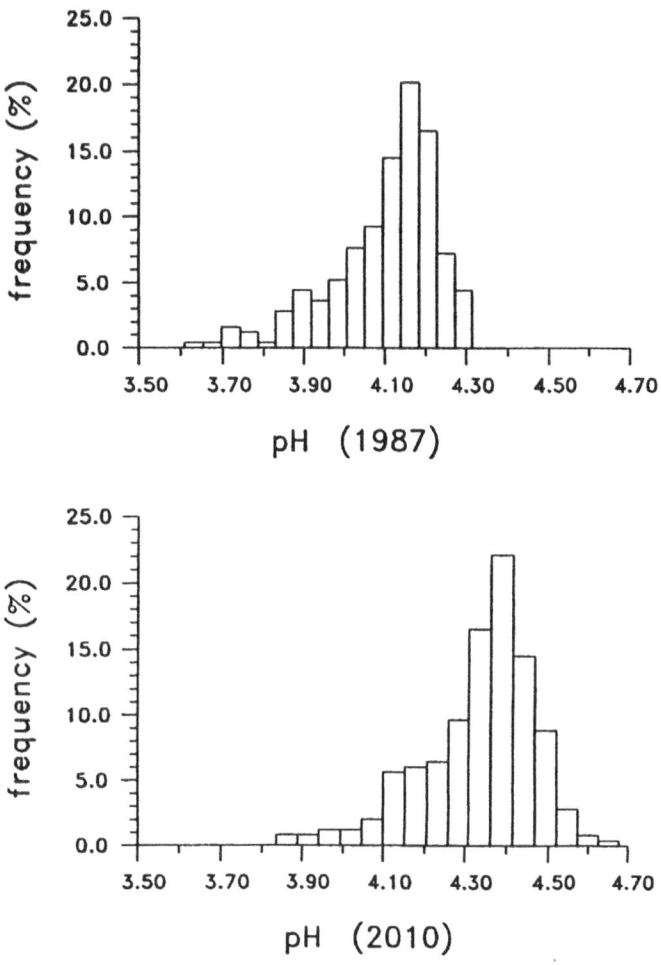

Figure 2. Histograms of the pH in layer 3 at the beginning and at the end of the simulation period.

Before studying the uncertainty contribution of the individual 'sources', one should first check whether the calculated linear regression model, which is used for evaluating these contributions, gives an adequate fit to the simulated model responses. The results on the R^2 of the linear regression in Fig. 3 show that this is indeed the case (a R^2 of 0.9 indicates that the linear regression model accounts for 90% of the uncertainty (variance) of the model output (see eqn. (2)), which is quite reasonable).

Since the RTU and SRC measure render useful information on the uncertainty contribution, and on the direct (linear) relation between 'sources' and outputs, a table is presented showing the values of the RTU and SRC at the beginning, middle and the end of the simulation period. Moreover the trajectory of the RTU for the three most important sources is plotted to obtain a more complete picture over the full time range.

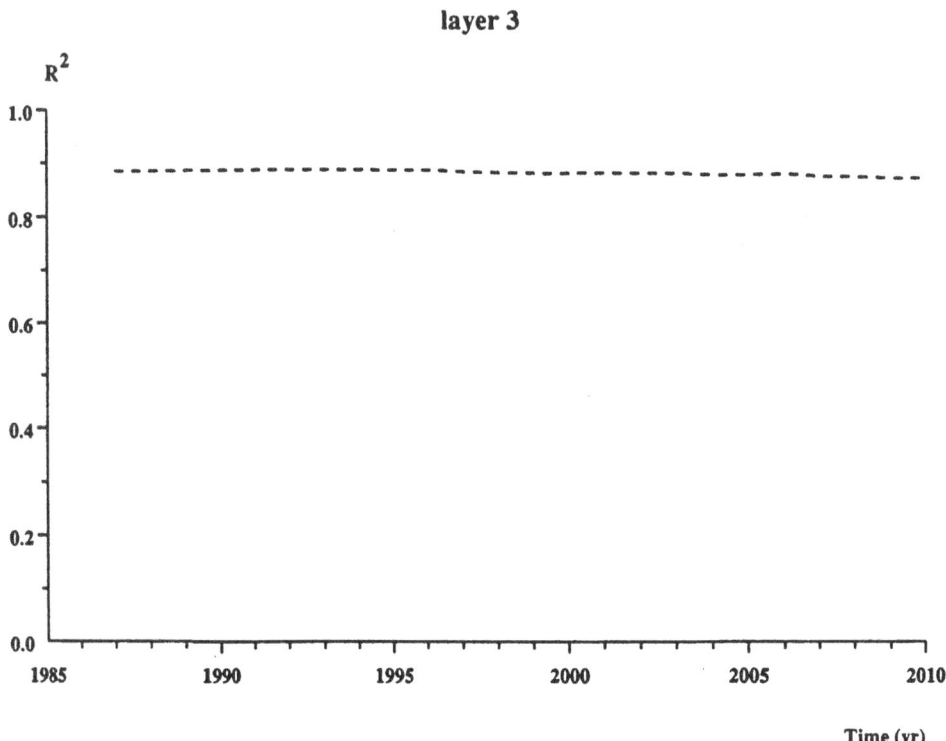

Figure 3. Temporal evolution of the R^2 of regression for the pH in layer 3.

According to the results in Table 2 and Fig. 4, the uncertainty of the pH in layer 3 is mainly determined by the aluminum dissolution dynamics, characterized by the equilibrium constant of aluminum hydroxide (KAl_{ox}) in layer 3 (saturation). This quantity determines the H^+-buffering by aluminum hydroxides. Its uncertainty contribution decreases during the simulation period, due to the decreasing deposition.

Next to the parameter KAl_{ox}, the uncertainty in pH, as measured by the RTU, is mainly caused by the dry deposition flux of NH_3 ($FNH_{3,dd}$) and SO_2 ($FSO_{2,dd}$), which are the major contributors to the external acid load. Since the RTU, unlike the SRC, explicitly accounts for the effects of mutual correlations between the 'sources', the values of the RTU can differ considerably from those of the SRC for 'sources' which are mutually correlated. This is clearly illustrated by the above mentioned fluxes $FNH_{3,dd}$ and $FSO_{2,dd}$: the SRC for $FSO_{2,dd}$ is considerably smaller than the SRC for $FNH_{3,dd}$, indicating that $FSO_{2,dd}$ contributes less directly to the pH than $FNH_{3,dd}$. But, due to its high correlation (0.89) with the important $FNH_{3,dd}$ 'source', the $FSO_{2,dd}$ scores rather high on the RTU-scale.

Table 2. Uncertainty contribution and rankings for the uncertainty 'sources' for the pH in layer 3 in 1987, 2000 and 2010.

	1987		2000		2010	
	RTU	SRC	RTU	SRC	RTU	SRC
KAl_{ox}	0.90 (1)	0.91 (1)	0.89 (1)	0.89 (1)	0.80 (1)	0.81 (1)
$FNH_{3,dd}$	0.16 (2)	-0.17 (2)	0.16 (2)	-0.19 (3)	0.21 (3)	-0.18 (4)
$FSO_{2,dd}$	0.15 (3)	0.05 (17)	0.15 (3)	0.08 (8)	0.21 (4)	0.01 (67)
$FNO_{2,dd}$	0.12 (4)	-0.09 (5)	0.12 (4)	-0.10 (6)	0.18 (5)	-0.12 (8)
$KK_{ex,1}$	0.10 (5)	-0.14 (3)	0.10 (6)	-0.14 (4)	0.09 (10)	-0.12 (7)
kNa_v	0.09 (6)	0.09 (6)	0.10 (7)	0.10 (7)	0.15 (7)	0.15 (5)
$KNH_{4,ex,3}$	0.07 (7)	-0.05 (14)	0.08 (9)	-0.02 (36)	0.08 (11)	-0.01 (66)
$k_{ni,3}$	0.07 (8)	-0.06 (12)	0.04 (22)	-0.03 (22)	0.05 (30)	-0.04 (31)
$KNH_{4,ex,1}$	0.06 (9)	0.05 (18)	0.06 (12)	0.05 (15)	0.06 (19)	0.06 (12)
FCl_{dw}	0.06 (10)	-0.11 (4)	0.11 (5)	-0.20 (2)	0.23 (2)	-0.36 (2)

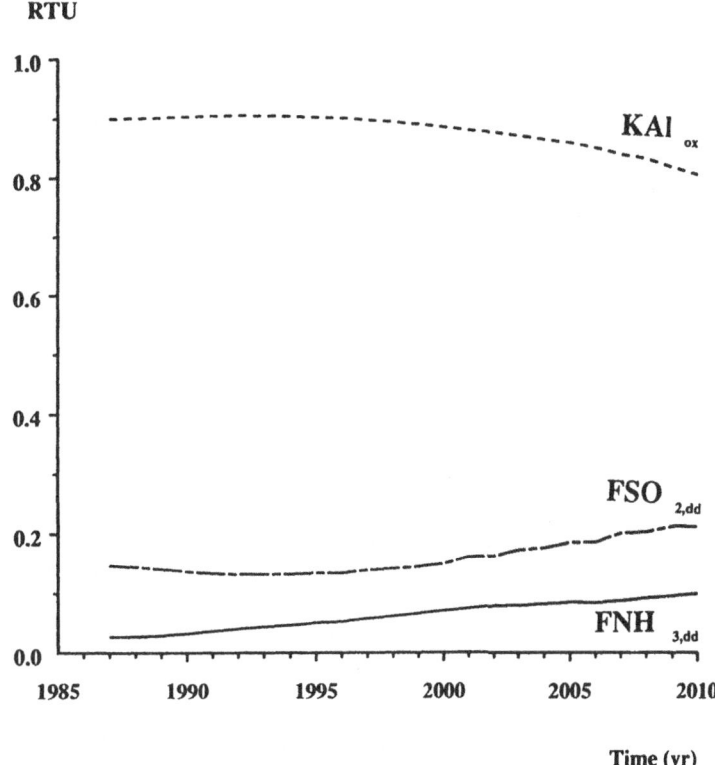

Figure 4. Temporal evolution of the RTU measure of model parameters for the pH in layer 3.

Table 2 illustrates moreover that the wet chloride deposition FCl_{dw} contributes substantially to the uncertainty of the pH at the end of the simulation period. This remarkable result is an artefact, caused by the way in which the wet deposition of Cl and Na are related in this study. Instead of establishing a functional relationship between FCl_{dw} and FNa_{dw} (on basis of the concentrations of sea salt), a correlation (of 0.8) was imposed between these fluxes. As a consequence the Cl depostion needs not be fully compensated by Na and the other base cations. In this situation the non-compensated part of Cl is directly compensated by H, thus creating the (artificially) high contribution of FCl_{dw}.

A more elaborate treatment and discussion of all these uncertainty aspects for other model outputs and layers can be found in Kros *et al.* (1993). The results render a complete picture of the uncertainty and of its main contributors. They show that the uncertainty is most influenced by the input-depositions of SO_x, NO_x, and NH_x ('external acid load') and the parameters and variables (contents in needles; amounts in roots i.e. initial conditions) which determine the nitrogen and aluminium dynamics ('internal acid load'). These contributions will depend on the considered model outputs, soil compartment and time. The obtained information can be used as a basis for further model development and data collection. E.g. processes characterized by relatively certain parameters can possibly be aggregated, and important parameters whose uncertainty is due to incomplete knowledge are primary candidates for calibration.

5. Performing a robustness study: Various ways can be followed to conduct a robustness-study to investigate the influence of different specifications of distributions and correlations: e.g. by performing a completely novel uncertainty analysis with UNCSAM on the basis of the new specifications. This requires additional, computationally expensive, model simulations of RESAM. Less costly alternatives are: (a) the application of the linear metamodels, if adequate, which have already been obtained during the old uncertainty analysis; (b) performing a smart reweighing of the old simulation results (cf. Beckman and McKay (1987)). Due to time-restrictions imposed on the uncertainty analysis of RESAM, it was however not possible to perform a thorough robustness study.

5. Conclusions

Sensitivity and uncertainty analysis should be an essential part of the model building activities, especially in the context of ill-defined, uncertain, or information poor application areas. Performing these analyses in a structured way will not only convey the crucial aspects of a model and the origins and effects of model uncertainties, but will also lead to increased insight in the model and its reliability/validity, and provide useful suggestions for well-structured further research (model development, calibration, data acquisition).

The software package UNCSAM, which is based on Monte Carlo techniques in combination with regression and correlation analysis, is a useful tool for actually performing these analyses on a large variety of simulation models, largely independent of the way in which the models are implemented. Sampling can be performed from a large number of distributions, and many ways are offered, graphically as well as tabulary, to obtain a clear and complete picture of the uncertainties and their main contributors.

Acknowledgement

We thank J. Kros and W. de Vries of the Winand Staring Centre for Integrated Land, Soil and Water Research for their permission to present part of the results on the uncertainty analysis of RESAM, and for providing the illustrations in the text.

References

Beckman, R.J. and M.D. McKay (1987), 'Monte Carlo estimation under different distributions using the same simulation', *Technometrics*, **29**, 153-160.

Iman, R.L. and W.J. Conover (1980), 'Small sample sensitivity analysis techniques for computer models, with an application to risk assessment', *Communications in Statistics*, **A9**, 1749-1842.

Iman, R.L. and W.J. Conover (1982), 'A distribution free approach to inducing rank correlations among input variables', *Communications in Statistics*, **B11**, 311-334.

Iman, R.L. and J.H. Helton (1985), 'A comparison of uncertainty and sensitivity analysis techniques for computer models', *Report NUREG/CR-3904*, U.S. Nuclear Regulatory Commission, Washington.

Iman, R.L. and J.H. Helton (1988), 'An investigation of uncertainty and sensitivity analysis techniques for computer models', *Risk Analysis*, **8**, 71-90.

Janssen, P.H.M. (1993), 'Assessing sensitivities and uncertainties in models: a critical evaluation', *Conference on Predictability and Nonlinear Modelling in Natural Sciences and Economics*, Wageningen, The Netherlands. 5-7 April.

Janssen, P.H.M., P.S.C. Heuberger and R. Sanders (1992), 'UNCSAM 1.1: a Software Package for Sensitivity and Uncertainty Analysis: Manual', *RIVM-report nr. 959101004*, Bilthoven, The Netherlands.

Janssen, P.H.M., W. Slob and J. Rotmans (1990), 'Sensitivity and Uncertainty Analysis: an Inventory of Ideas, Methods and Techniques' (in Dutch), *RIVM-report nr. 958805-001*, Bilthoven, The Netherlands.

Johansson, M. and P.H.M. Janssen (1993), 'Uncertainty analysis on critical loads for forest soils', *Conference on Predictability and Nonlinear Modelling in Natural Sciences and Economics*, Wageningen, The Netherlands. 5-7 April.

Krol, M.S. (1993), 'Uncertainty analysis for the computation of green-house gas concentrations in IMAGE', *Conference on Predictability and Nonlinear Modelling in Natural Sciences and Economics*, Wageningen, The Netherlands. 5-7 April.

McKay, M.D., R.J. Beckman and W.J. Conover (1979), 'A comparison of three methods for selecting values of input variables in the analysis of output from a computer code', **21**, 239-245.

Kros, J., W. De Vries, P.H.M. Janssen and C.I. Bak (1983), 'The uncertainty in forecasting trends of forest soil acidification', *Water, Air and Soil Pollution*, **66**, 29-58.

Owen, A.B. (1992), 'A central limit theorem for Latin Hypercube sampling', *J.R. Statist. Soc. B*, **54**, 541-551.

Stein, M. (1987), 'Large sample properties of simulations using Latin Hypercube Sampling', *Technometrics*, **29**, 143-151.

Tiktak, A., F.A. Swartjes, R. Sanders and P.H.M. Janssen (1993), 'Sensitivity analysis of a model for pesticide leaching and accumulation', *Conference on Predictability and Nonlinear Modelling in Natural Sciences and Economics*, Wageningen, The Netherlands. 5-7 April.

Traas, Th.P. and T. Aldenberg (1993), 'Combination of uncertainty analysis and risk assessment: application to a bioaccumulation model', *Conference on Predictability and Nonlinear Modelling in Natural Sciences and Economics*, Wageningen, The Netherlands. 5-7 April.

De Vries, W. and J. Kros (1989), 'The long-term impact of acid deposition on the Aluminum chemistry of an acid forest soil', in *Regional Acidification Models*, ed. Kämäri, J., D.F. Brakke, A. Jenkins, S.A. Norton and R.F. Wright, pp. 113-128, Springer-Verlag, Berlin Heidelberg.

SET-MEMBERSHIP IDENTIFICATION OF NON-LINEAR CONCEPTUAL MODELS

KAREL J. KEESMAN

Dept. Agric. Engng. and Physics, Wageningen Agricultural University
Bomenweg 4, 6703 HD Wageningen, The Netherlands
e-mail: karel.keesman@user.aenf.wau.nl

Summary

Identification of conceptual models nonlinear in the parameters from bounded-error data is considered. The assumption that errors are point-wise bounded implies that a set of parameter vectors is found instead of an 'optimal' parameter estimate. For our class of models, the Monte Carlo Set-Membership algorithm is appropriate to approximate the exact solution set by a number of feasible realizations. In addition to the feasible parameter set, representing the parametric uncertainty, information about the modelling uncertainty is also provided. In order to obtain realistic predictions both uncertainty sources must be quantified from the available data and evaluated over the prediction horizon. Three 'real-world' examples will illustrate the features of this set-membership approach to system identification and prediction.

Keywords Set-membership identification, nonlinear systems, Monte Carlo methods

1. Introduction

For nontechnical systems, like environmental and economical systems, mathematical models have mainly been developed for simulation and (forecasting) prediction. For the development of these models two approaches can be distinguished. The first one, usually indicated as *(theoretical) modelling*, represents a speculative way of model building on the basis of "laws of nature" and additional relationships. The second route is called *system identification*, and comprises two steps: (a) selection of an appropriate model structure from limited prior system knowledge, intended use of model and experimental data, and (b) estimation of the unknown model parameters. In Fig. 1 various components in the identification process are presented.

Since our prior knowledge as well as the data are uncertain, the model as a result of system identification must reflect the uncertainties in order to be realistic. The most common way to deal with uncertainties is to assume that it is of random nature, which can be described in terms of statistical properties as mean, variance, probability density functions, and uncorrelatedness. If, however, such detailed characterization of the uncertainty is not possible due to rather short data records, or if the uncertainty has non-random components as a result of model simplifications or systematic measurement errors, a statistical approach will give unreliable results. Under these circumstances, a deterministic error characterization in terms of lower and upper bounds only will be a good alternative. Overviews of (nonlinear) time-domain estimation methods within the so-called set-membership or bounded-error context, which was initiated by Schweppe (1968) and Witsenhausen (1968), can be found in Norton (1987), Walter *et al.* (1991) and Milanese and Vicino (1991).

In this paper we will focus on the identification of nonlinear conceptual models, represented as *output error* model structures, for prediction from point-wise bounded error data. Due

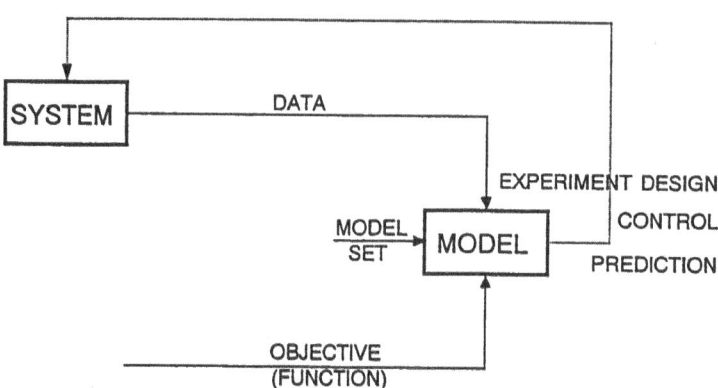

Figure 1. System identification scheme.

to its iterative solution of the parameter estimation problem, the Monte Carlo Set-Membership (MCSM) algorithm (Keesman, 1990; see also Spear and Hornberger, 1980; Fedra *et al.*, 1981) is applicable to this class of models, as will be demonstrated by the application to three 'real-world' examples. These examples will clearly illustrate the features of a set-membership approach to system identification and prediction.

2. Basic concepts and notation

A general nonlinear output error model representation of a finite-dimensional continuous-discrete time system is,

$$\frac{dx(t,p)}{dt} = f[x(t,p),u(t),t;p] , \tag{1a}$$

$$x(t_0,p) = x_0 , \tag{1b}$$

$$y(t_k) = g[x(t_k,p)u(t_k),t_k;p] + e(t_k) \quad k = 1,...,N , \tag{1c}$$

where x is a state and u an input vector. In what follows, expressions with respect to the uncertainty for both the output data vector $y \in \mathbf{R}^s$ and the parameter vector $p \in \mathbf{R}^n$ will be given. Each element of the output error vector, $e_j(t_k)$ for $j=1,...,s$, contains errors from both the measurement and the modelling process. The bounded-error approach implies the assumption that, for symmetric bounds,

$$|e_j(t_k)| \quad \leq \quad \varepsilon_{jk} \qquad \qquad \text{for } j=1,...,s \text{ and } k=1,...,N . \tag{2}$$

It should be noted that symmetry is usually assumed due to lack of more detailed information. On the basis of this error interval model we define the *behaviour set* as,

$$\Omega_y = \cup_{k=1}^N \{\bar{y} \in \mathbb{R}^s \colon \bar{y} = y(t_k) - e(t_k); \; e(t_k) \in \Omega_e(t_k) \},\tag{3}$$

where $\Omega_e(t_k)$, the *error set*, contains all error vectors which fulfil the condition of Eqn. (2). In addition to this behaviour set we define the *prior parameter set* in terms of lower and upper bounds on the individual parameters,

$$\Omega_p \quad = \quad \{p \in \mathbb{R}^n \colon p^m \le p \le p^M\},\tag{4}$$

which contains *a priori* parametric information from, for instance, preceding estimation procedures or literature. The ultimate aim is then to identify the set of parameter vectors $\tilde{\Omega}_p$, which is consistent with model (Eqn. 1a-c), behaviour set (Eqn. 3) and prior parameter set (Eqn. 4). This set with feasible parameter vectors, which explicitly represents the parametric uncertainty,

$$\tilde{\Omega}_p = \{p \in \mathbb{R}^n \colon y(t_k) - g[x(t_k,p),u(t_k),t_k;p] \in \Omega_e(t_k); \; p \in \Omega_p, \; k = 1,...,N\}\tag{5}$$

is called the *posterior* or *feasible parameter set*. The MCSM algorithm can handle this nonlinear parameter estimation problem. The key idea behind this algorithm is that randomly selected parameter vectors which result in a model response consistent with the behaviour set belong to the feasible set. In order to relate this feasible parameter set to an output set in the measurement space an additional definition is required. The so-called *feasible model output set* is defined as,

$$\tilde{\Omega}_{\bar{y}} = \cup_{k=1}^N \{\hat{y} \in \mathbb{R}^s \colon \hat{y}(t_k|p) = g[x(t_k,p),u(t_k),t_k;p]; \; p \in \tilde{\Omega}_p\},\tag{6}$$

where $\tilde{\Omega}_{\bar{y}} \subseteq \Omega_y$.

3. Examples

3.1 Respiration rate of activated sludge

In the first example, the identification of parameters in an activated sludge process model from bounded data is demonstrated. The actual respiration rate (r_{act}) of activated sludge can be measured with an on-line respiration meter (Spanjers and Klapwijk, 1987). The measured data is represented in Fig 2.

The model that describes the process in the respiration meter is:

$$\frac{dS_s(t)}{dt} = -\left[\frac{Q(t)}{V} + \frac{\hat{\mu}_{max}X_B(t)}{K_s + S_s(t)}\right]S_s(t) + \frac{Q(t)}{V}\eta S_{s,in}(t),\tag{7a}$$

$$r_{act}(t) = \frac{\hat{\mu}_{max}X_B(t)}{K_s + S_s(t)}S_s(t) + bX_B(t), \tag{7b}$$

$$y(t_k) = r_{act}(t_k) + e(t_k), \tag{7c}$$

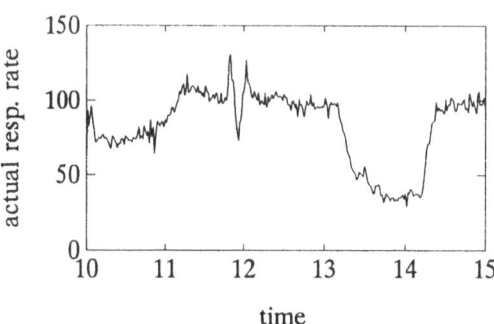

Figure 2. Measured actual respiration rate.

where S_s is the substrate concentration (g O_2 m^{-3}). The model inputs are: substrate concentration of the influent $S_{s,in}$ (g COD m^{-3}), biomass concentration X_B (g MLSS m^{-3}) and flow Q (m^3 s^{-1}). The volume of the respiration meter is denoted by V (m^3). Our objective is then to identify the model parameters: maximum specific growth rate of the bacteria ($\hat{\mu}_{max}$), Monod constant for substrate (K_s), specific decay rate (b) and yield factor (η).

From Fig. 3 it can be seen that individual values for both K_s and b cannot properly be estimated from the respiration rate data with error bounds equal to 25 g O_2 m^{-3} s^{-1}, since their feasible realizations can be found on the complete, and rather large, range of predefined values. On the contrary, we notice that only certain combinations of η, K_s and b give feasible realizations. Similar conclusions in terms of parameter correlations have been obtained from the results of extended Kalman filtering, an identification method within a stochastic context. Notice also that η is dominant in controlling the model behaviour, which could also have been seen from a comparison between $S_{s,in}$ and the measured respiration rate. It should also be noted that the feasible parameter set in Fig. 3 is directly related to the error bound. For instance, for much smaller error bounds, say 10 g O_2 m^{-3} s^{-1}, not a single feasible parameter vector will be found.

Hence, in general, there is an error bound for which $\tilde{\Omega}_p$ reduces to a singleton; a special case which will be commented on in the third example. The choice of an appropriate error bound for the identification of $\tilde{\Omega}_p$ will be dealt with in the second example after having presented a measure of the modelling error.

Apart from inspections in the parameter space, we also examine the relation between measurements and the feasible model output set, which is a subset of the behaviour set (Fig. 4). If all measurements are contained in the feasible output set associated with reasonably specified error bounds, we could say that the measurements, within certain bounds, have been generated by the model. Hence, the model is a valid representation of the system.

Measurements that are not contained in the set indicate then outliers or, if this holds for a series of consecutive measurements, it shows that most likely the model is invalid.

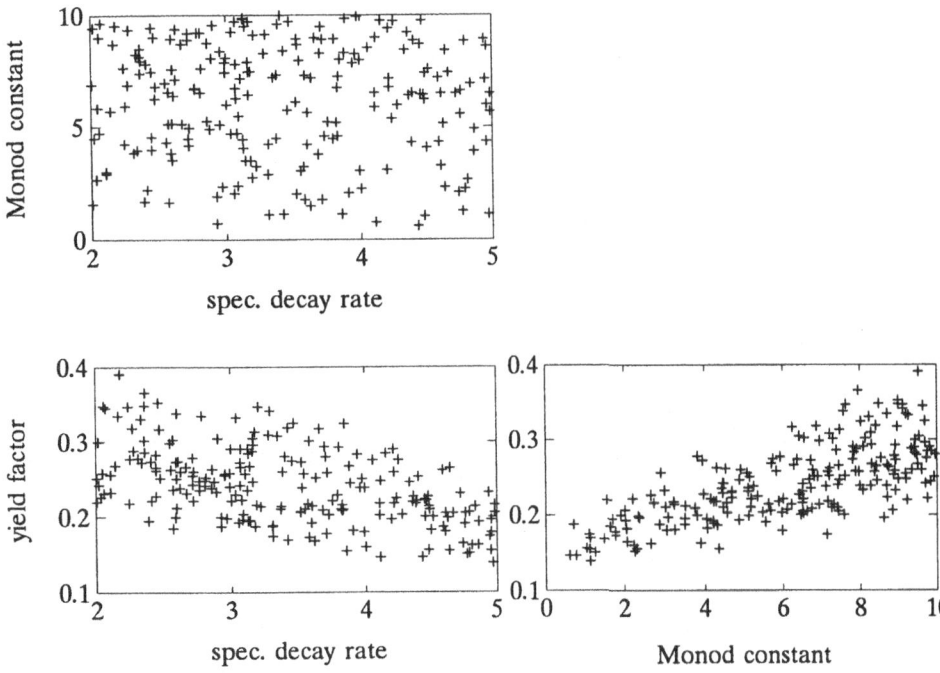

Figure 3. Set-membership parameter estimation results.

In our example, it appears that almost all measurements are contained in the feasible model output set, except those measurements which characterise the strong decrease of r_{act} at about 13.00 hrs and the strong increase at 14.00 hrs. Careful inspection shows that from 13.00 hrs the model output is delayed with respect to the measured data. It has been found that this delay was caused by an incorrect calculation of the input sequence $S_{s,in}$ from different inflows. In the next example we will focus on this kind of "misfit" due to modelling errors in more detail.

3.2 Dissolved oxygen in lake

In the second example, we will deal with the identification of a model describing the dissolved oxygen (DO) concentration in a lake (see also Keesman and van Straten, 1987; 1989). And, in particular, we will demonstrate how to deal with modelling error.

It should first be noted that the parametric uncertainty in terms of $\tilde{\Omega}_p$ of the preceding example includes, at least partially, a modelling error because of our output-error model formulation. The non-compensated modelling error, w_j for $j=1,...,s$, can be estimated from the set of estimated errors, the residual set. In an earlier paper (Keesman and van Straten, 1989) we have derived the following expression for the modelling error:

$$w_j \in \tilde{\Omega}_w(j) = \{ w_j \in \mathrm{R}: |w_j| \le w_j^M \} , \tag{8}$$

where the bounds on the set are found from,

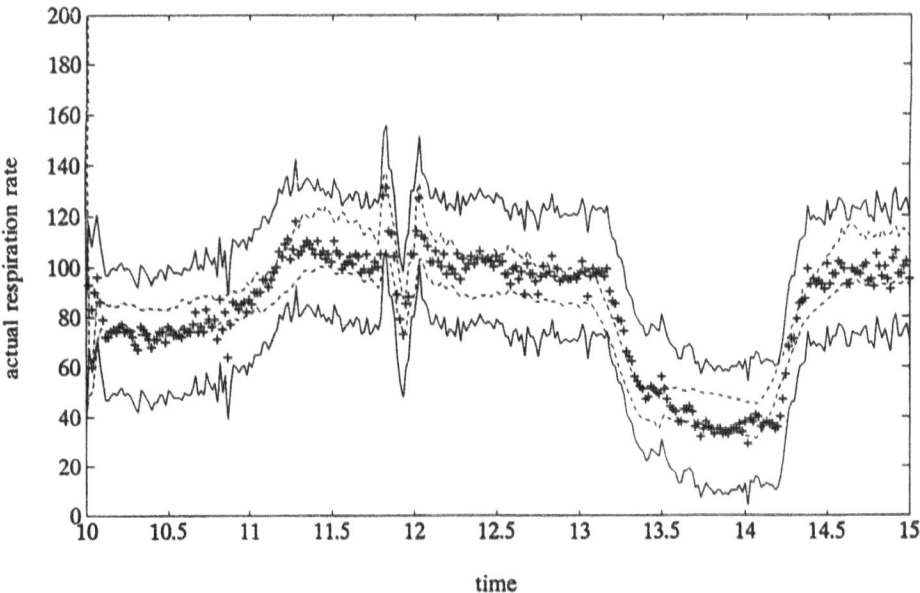

Figure 4. Measurements (+), behaviour set (−) and feasible model output set (--).

$$w_j^M = \max_k \hat{w}_j(k) , \tag{9}$$

$$\hat{w}_j(k) = \min_{i=1,...,M} |y(t_k) - g[x(t_k,p_i),u(t_k),t_k;p_i]| , \tag{10}$$

where $M = \text{card } \tilde{\Omega}_p$. In order to take into account this modelling error in, for instance, model predictions, it can simply be added to the feasible model output trajectories because of the output-error model formulation. Hence, for model predictions including modelling errors Eqn. (6) is extended to,

$$\tilde{\Omega}_{\hat{y}}^* = \cup_{k=1}^{N+P} \{\hat{y} \in \mathbf{R}^s: \hat{y}(t_k|p) = g[x(t_k,p),u(t_k),t_k;p] + w; \; p \in \tilde{\Omega}_p \; and \; w \in \tilde{\Omega}_w \} , \tag{11}$$

where P is the prediction horizon.

Clearly, w_j^M is an upper bound, and thus a rather conservative estimate, of the modelling error. In order to reduce the effect of high frequent measurement errors on this estimate, it has been suggested to prefilter the output data. Notice that $\tilde{\Omega}_w(j)$ as well as $\tilde{\Omega}_p$ (see Eqn. 5) strongly depend on the predefined error bounds ε_{jk}. Let us illustrate this for a single output system with constant error bound ε. Then, ε is a design parameter by which we can weigh the trade-off between parametric and non-compensated (nonparametric) modelling uncertainty. The modelling error bound is maximal when $\tilde{\Omega}_p$ reduces to a singleton and it is equal to zero when all measurements, which are not detected as an outlier, are contained in $\tilde{\Omega}_{\hat{y}}$. In case the outputs are linearly related to the parameters, we can easily derive a lower bound on ε for which holds that w^M is nihil. Moreover, it can be proved that under weak conditions

$$\tilde{\Omega}_{\hat{y}}^{*}(k|w^{M}=0) \supseteq \tilde{\Omega}_{\hat{y}}^{*}(k|0<w^{M}<w^{M+}) \supseteq \tilde{\Omega}_{\hat{y}}^{*}(k|w^{M}=w^{M+}) \qquad (12)$$

where w^{M+} is the maximum modelling error bound.

As an alternative to the set-membership description of the modelling error given in (8)-(10), which implies a constant upper error bound, we may reduce the conservatism in the estimate by taking into account the time structure of $\hat{w}_j(k)$. In other words, the non-compensated modelling error will be treated as a finite time series. Let us assume that $\hat{w}_j$ can be represented by the linear, time-invariant black-box model structure,

$$A(q)\hat{w}_j(k) = \frac{q^{-k}B(q)}{F(q)}u_j(k) + \frac{C(q)}{D(q)}v_j(k), \qquad (13)$$

where A, B, ..., F are polynomials in q^{-1}, the backward-shift operator, and either A or F must be equal to one. The first term at the right-hand side computes the undisturbed or noise-free output; the last term, the so-called noise model, describes the effect of disturbances on the process, where $\{v_j\}$ is an independent driving input sequence with zero mean. The input sequence $\{u_j\}$ can be selected from observed system inputs, preferably those which are highly correlated with $\{\hat{w}_j\}$, or it can just be a suitable predefined function. The problem then is how to identify the best model structure with associated parameters. In Ljung (1987) several approaches which solve this problem can be found. In a forthcoming paper (Keesman, 1993) we will further elaborate on this.

Let us now consider the DO concentration (c) in a lake which is modelled as,

$$dc(t)/dt \quad = \quad K_r[C_s(t) - c(t)] + \alpha I(t) - R, \qquad (14a)$$

$$y(t_k) \quad = \quad c(t_k) \quad + e(t_k), \qquad (14b)$$

where $C_s(t)$ is the saturated DO concentration and $I(t)$ the radiation. The parameters K_r, α and R represent the reaeration coefficient, the photosynthetic production rate and the oxygen consumption rate, respectively. After identifying these parameters for $\varepsilon_k=2$ g DO m^{-3}, we find a feasible model output set with residual set and non-compensated modelling error as is represented in Fig. 5. In order to amplify the low frequency components in $\{\hat{w}\}$ for our prediction purpose, an AR process has been selected to describe the modelling error,

$$\hat{w}(k) = \frac{1}{1 - 0.62q^{-1} - 0.16q^{-2} + 0.08q^{-3} + 0.01q^{-4} - 0.34q^{-5}} v(k). \qquad (15)$$

In summary, we can say that the dissolved oxygen concentration has been modelled by a conceptual ('grey box') and a black box model part (Eqns. 14-15), which can be put together when making model predictions.

3.3 Lake eutrophication

The last example deals with the identification and prediction of shallow lake eutrophication (see also Keesman and van Straten, 1991; van Straten and Keesman, 1991). In addition to the preceding steps made within a set-membership identification procedure, in this section we will introduce some fuzzy set-theoretic concepts for computational efficiency.

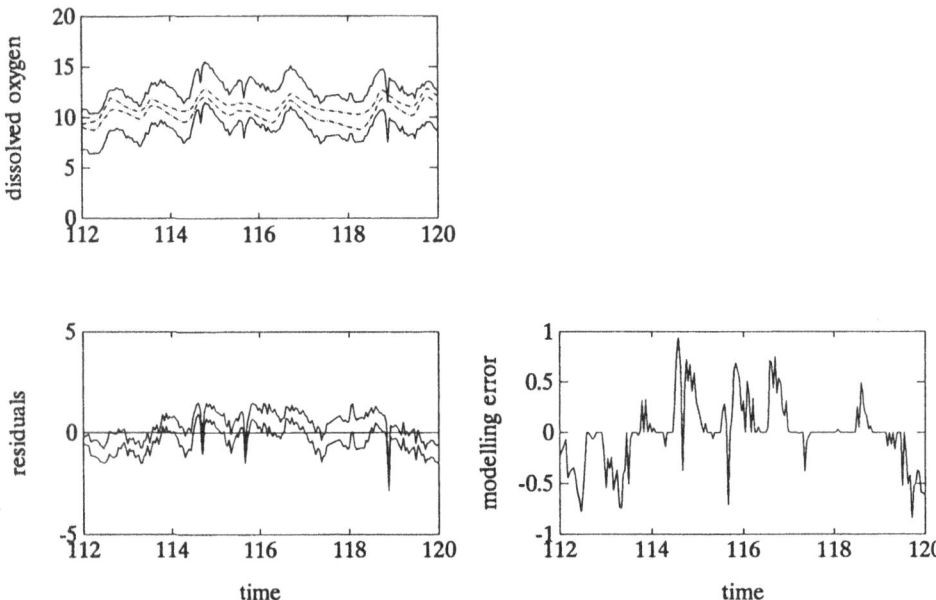

Figure 5. Feasible model output set, residual set and modelling error for DO-model.

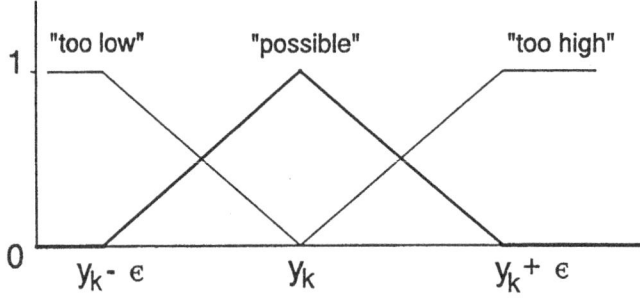

Figure 6. Triangular membership function for Y_k.

So far, it has been assumed that any realization belongs to either a feasible or, its complement, a nonfeasible set. However, a more general approach is obtained by introducing the concept of a fuzzy set (Zadeh, 1965). A fuzzy set X is represented by a membership function u_X, such that ($u_X: X \rightarrow [0,1]$). The values $u_X(x)$ give the grade of membership of an element x in X. If it is assumed that the noise-free measurement has a diminishing possibility of occurrence if it is further away from the actual measurement, the membership function $u_Y(y_k)$ related to y_k is triangular (Fig. 6) instead of rectangular as originally defined in Eqn. (2) and (3). In addition to this, a criterion function has to be defined in order to determine the grade of membership of a realization in the parameter space. On the basis of fuzzy set-theoretic considerations a min-operator membership criterion function,

$$M_m(p) = \min_{1 \leq k \leq N}\{u_Y(y_k; p)\} \tag{16}$$

is a good choice.

This concept will now be illuminated to the identification of a eutrophication model, which is expressed in terms of state equations for the algae-P fraction, the detritus fraction and the ortho-P fraction,

$$dA(t)/dt = \text{growth} - \text{death} - \text{outflow}$$

$$dD(t)/dt = \text{death} - \text{mineralization} - \text{sedimentation} + \text{inflow} - \text{outflow} \tag{17a}$$

$$dP(t)/dt = -\text{growth} + \text{mineralization} + \text{exchange sediment} + \text{inflow} - \text{outflow}$$

and a measurement equation, which relates the observations to the states A, D and P,

$$\begin{bmatrix} \text{Chl-a} \\ \text{ortho-P} \\ \text{total P} \end{bmatrix}_{t_k} = \begin{bmatrix} 1.05 & 0 & 0 \\ 0 & 0 & 1 \\ 1 & 1 & 1 \end{bmatrix}\begin{bmatrix} A(t) \\ D(t) \\ P(t) \end{bmatrix}_{t_k} \tag{17b}$$

The feasible model output set related to chlorophyll-a (Chl-a) for different error bounds is presented in Fig. 7.

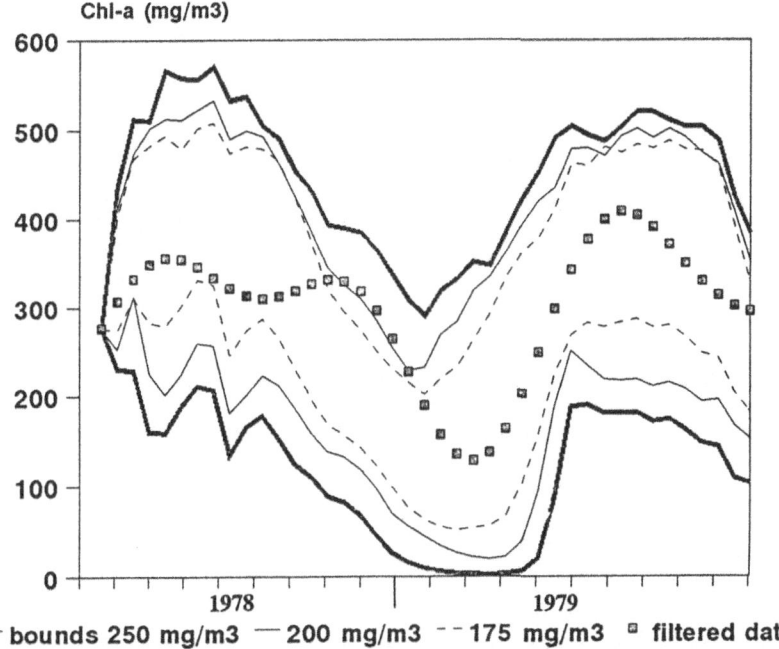

Figure 7. Feasible output sets Chl-a for 250, 200 and 175 mg P m^{-3} error bounds.

The subsets for 200 and 175 mg P m^{-3} error bounds can be easily obtained from the set for 250 mg P m^{-3} . It should be noticed then that for 250 mg P m^{-3} error bounds $M_m(p)>0.0$ for all $p \in \tilde{\Omega}_p$. For feasible model responses associated with the 200 mg P m^{-3} error bounds, and using a triangular membership function (see Fig. 6), it holds that $M_m(p) \geq 1\text{-}200/250$. Hence, selection of an arbitrary value $M_m > 0.0$ leads to a subset of $\tilde{\Omega}_p$ (M_m-level set) and thus to smaller model output set. This implies that it is not necessary to repeat all calculations if error bounds are squeezed from large values selected initially, to avoid the problem of empty parameter sets, back to more "realistic" values.

In line with this we can squeeze the error bounds to a limit where only one parameter vector remains, if the problem allows a unique solution. Notice, then, that due to the sampling procedure this parameter vector is a suboptimal estimate of the max-min estimate p^* obtained by maximizing $M_m(p)$,

$$p^* = \arg\ \max\ \overline{M_m(p)}, \tag{18}$$

where the overbar denotes an overall membership function value for the multiple output case. The optimal estimate can then be found from a formal estimation procedure using the suboptimal estimate, as an initial guess, and the parameter interactions resulting from the MCSM-algorithm. This max-min estimate can be viewed as the centre of $\tilde{\Omega}_p$. This max-min estimation, in addition to the parameter set estimation is very valuable. First, it reveals erroneous assumptions with respect to the predefined parameter ranges (Eqn. 4) if an optimal dominant parameter value is on one of the bounds. Secondly, the associated residual sequence reveals "hard" information about the discrepancy between model and data. If, for instance, the largest residual is larger than the assumed measurement error bound, we must conclude that the model is not able to describe the observed system behaviour in all respect. And, thirdly, the residual sequence can be correlated with observed system inputs in order to improve the model.

Finally, we will discuss some aspects of set-membership prediction. If the main goal of our identification procedure is prediction, it is of the greatest importance to quantify the uncertainty originating from the modelling and measurement process as good as possible. On the basis of these uncertainties, the prediction uncertainty propagation can then be evaluated. The contribution of the parametric uncertainty to the prediction uncertainty is obtained by evaluating all $p \in \tilde{\Omega}_p$, for $k=N+1,...,N+P$ where P is the prediction horizon. Moreover, the modelling error, expressed as an output term, can be added to the model responses in the prediction space (see Eqn. 11). Notice that this approach implies that processes dominating the model behaviour during the identification stage are directly extrapolated via the identified set $\tilde{\Omega}_p$.

In Fig. 8a the predicted effects of flushing on the total P concentration are shown. In addition to these extrapolation results, predictions on the basis of a modification in the parameter that represents the background P concentration in the sediment are presented, too (see Fig. 8b). These and other subjects related to the prediction of environmental systems are also discussed by Beck (1991)

4. Conclusions

The presented examples have shown the possibilities of a set-membership approach to identification. The identification, even for complex systems with highly nonlinear interactions, can be performed in a rather simple and robust way using the Monte Carlo Set-Membership algorithm. Obviously, instead of the Latin hypercube sampling scheme utilized in the MCSM algorithm other random search algorithms, based on more effective sampling schemes, could be applied as well (see for instance Brooks, 1958; Bekey and Masri, 1983; Klepper and Hendrix, 1993).

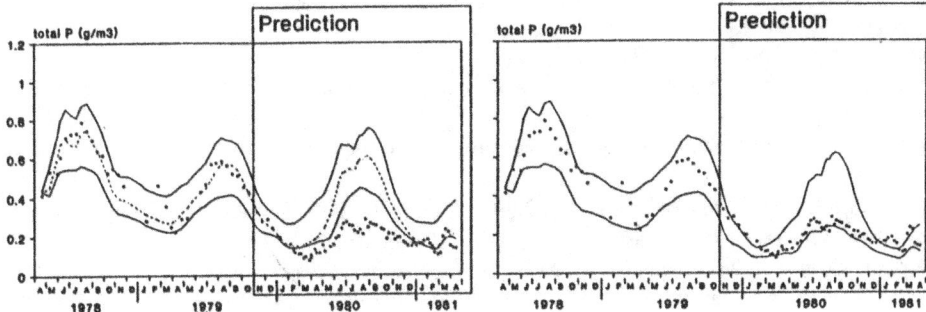

Figure 8. Bounded model predictions.

References

Schweppe, F.C. (1968), 'Recursive state estimation: unknown but bounded error and system inputs', *IEEE Trans. Autom. Control*, **13**, 22-28.

Witsenhausen, H.S. (1968), 'Sets of possible states of linear systems given perturbed observations', *IEEE Trans. Autom. Control*, **13**, 556-558.

Norton, J.P. (1987), 'Identification and application of bounded-parameter models', *Automatica*, **23**, 497-507.

Walter, E. and H. Piet-Lahanier (1990), 'Estimation of parameter bounds from bounded-error data: a survey', *Math. Comp. Simul.*, **32**(5&6), 449-468.

Milanese, M. and A. Vicino (1991), 'Optimal estimation theory for dynamic systems with set membership uncertainty: an overview', *Automatica*, **27**(6), 997-1009.

Keesman, K.J. (1990), 'Membership-set estimation using random-scanning and principal component analysis', *Math. Comp. Simul.*, **32**(5&6), 535-543.

Spear, R.C. and G.M. Hornberger (1980), 'Eutrophication in Peel Inlet, II, Identification of critical uncertainties via generalized sensitivity analysis',

Fedra, K., G. van Straten and M.B. Beck (1981), 'Uncertainty and arbitrariness in ecosystems modelling: a lake modelling example', *Ecol. Model.*, **13**, 87-110.

Spanjers, H. and A. Klapwijk (1987), 'On-line meter for respiration rate and short-term biochemical oxygen demand in the control of the activated sludge process', In: *Proc. Workshop on Instrumentation, Control and Automation of Water and Waste Water Treatment and Transport Systems*, Kyoto, Japan, pp. 67-77.

Keesman K.J. and G. van Straten (1987), 'Modified set-theoretic identification of ill-defined water quality system from poor data', In: *Proc. IAWPRC Symp. on Systems Analysis in Water Quality Management*, Pergamon Press, New York, pp. 297-308.

Keesman K.J. and G. van Straten (1989), 'Identification and prediction propagation of uncertainty in models with bounded noise,' *Int.J.Control*, **49**(6), 2259.

Ljung, L. (1987), '*System Identification: Theory for the User*', Prentice Hall, New Jersey.

Keesman, K.J. (1993), 'Model parameter and modelling error identification in nonlinear conceptual models from bounded-error data', To appear in *Proc. Int. Congress on Modelling and Simulation*, Perth, Australia, 6-10 December.

Keesman K.J. and G. van Straten (1990), 'Set-membership approach to identification and prediction of lake eutrophication', *Water Resour. Res.*, **26**(11), 2643-2652.

Van Straten, G. and K.J. Keesman (1991), 'Uncertainty propagation and speculation in projective forecasts of environmental change: a lake-eutrophication example', *J. Forecasting*, **10**(1&2), 163-190.

Zadeh, L.A. (1965), 'Fuzzy sets', *Inf. Control*, **8**, 338-353.

Beck, M.B. (1991), 'Special Issue: Forecasting Environmental Change', *J. Forecasting*, **10**(1&2), pp. 230.

Brooks, S.H. (1958), 'A discussion of random methods for seeking maxima', *Oper. Res.*, **6**, 244-251.

Bekey, G.A. and S.F. Masri (1983), 'Random search techniques for optimization of nonlinear systems with many parameters', *Math. Comp. Simul.*, **25**, 210-213.

Klepper, O. and E.M.T. Hendrix (1993), 'A method for robust calibration of ecological models under different types of uncertainty', *First Workshop on Forecasting Environmental Change*, IIASA Laxenburg, Austria, 22-24 February.

PARAMETER SENSITIVITY AND THE QUALITY OF MODEL PREDICTIONS

HANS J. POETHKE, DETLEF OERTEL & ALFRED SEITZ

Institute of Zoology, University of Mainz,
Saarstrasse 21, D-55099 Mainz, Germany
e-mail: achim@uzomzc.biologie.uni-mainz.de

Summary

Using SIM-PEL, a comprehensive model for the pelagic compartment of lake ecosystems, we analyse synergistic toxicant effects in lake ecosystems. We show, that - even for a rather simple model - model predictions may be strongly dependent on the time horizon of the prediction and on the quality of input parameters. For longer time spans, small errors in parameter estimation may lead to qualitatively wrong prediction of toxicant effects. Monte Carlo simulations allow to take errors in parameter estimation into account, but they need rather good estimates of parameter variance.

1. Introduction

Data of ecological processes are always produced with a certain error (Jorgensen, 1979; see also Rose,1985). In the modelling process these errors will be propagated to the model parameters. Particularly in highly integrated models of ecosystems, errors in model parameters will be much greater then those of ecosystem-data themselves, since model parameters will integrate over several different elementary processes, and often model parameters have to compensate for shortcomings of too simplified model constructs. Implications of these errors on the quality of model predictions are unravelled by an adequate sensitivity analysis.

In the following, we will demonstrate for a model of the pelagic compartment of standing water bodies some approaches to the analysis of parameter sensitivity. In chapter 3.1 we will derive the coefficients of sensitivity for two model parameters, which are of particular importance in ecotoxicological studies. Nonlinearities in of parameter sensitivity are shown. In chapter 3.2 we explore the dependence of these nonlinearities on the time horizon of simulation studies, and in chapter 3.3 we present our results in a two dimensional parameter space. A short discussion of sensitivity analysis in Monte Carlo simulation models follows in chapter 3.4.

2. The model

For our numerical experiments, we used SIM-PEL (Fig.1), a model for the pelagic compartment of still water bodies (Oertel, 1992). SIM-PEL has been constructed for the analysis of toxicant effects in deep Eifelmaar-lakes. It is a simplified descendant of models like CLEAN (Park *et al.*, 1974; Scavia *et al,*. 1974) and SWACOM (Bartell *et al.*, 1988). Remineralisation of nutrients is modelled according to Rose (1985). To enable the phenomenon of luxury consumption to be modelled, we represented nutrient uptake dependent on internal and external nutrient concentrations (for a comparison of nutrient uptake models, see Morrison, 1987).

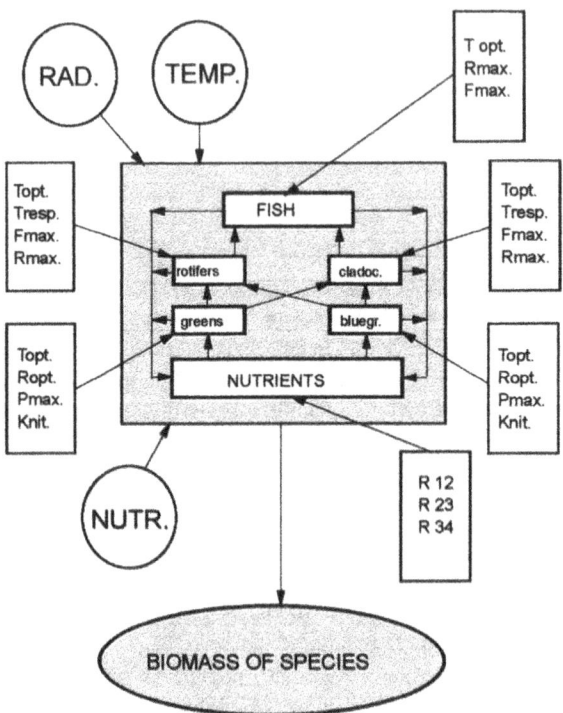

Figure 1. Schematic representation of energy- and information fluxes in the model SIM-PEL.

The model comprises of 16 state variables: Three for the different fractions of phosphorus, five for the fractions of nitrogen, two for the biomasses of two modelled filter feeders and six for the biomasses and the phosphorus- and nitrogen-concentrations in the two modelled species of algae. The influence of fish on the food web is modelled as a time dependent feeding pressure on filter feeders. Additional forcing functions are temperature, radiation and nutrient input into the system.

3. Modelling indirect toxicant effects

3.1 The coefficient of sensitivity S

Using SIM-PEL, we analysed the influence of the respiration rate of zooplankton (r_{max}) and the photosynthetic rate of algae (p_{max}) on the maximum annual phytoplankton biomass. In ecotoxicological experiments, an increase of r_{max} may be the direct effect of an insecticide, whereas a decrease of p_{max} may directly result from a herbicide. Thus, a modification of these two parameters may be used to simulate the influence of an insecticide or herbicide on the ecosystem.

Obviously, SIM-PEL is far to simple to take into account all processes that might be relevant for the propagation of toxicant effects in the pelagic compartment, and each pesticide has direct effects that can only be accurately described by a simultaneous modification of several parameters (Bartell *et al.*, 1988). Consequently, our simulation results can only give a first qualitative estimate about indirect effects that might occur if a toxicant is released into a lake, and they may help us to understand the limitations of predictive modelling in this field.

Taking the yearly maximum of phytoplankton biomass as endpoint of our study and the parameters r_{max} and p_{max} as input to our model, the coefficient of sensitivity

S = (rel. change of endpoint)/(rel. change of parameter)

gives an estimate of the influence of these two input parameters on predicted phytoplankton biomass. On the one hand, this allows to estimate the influence of errors in the determination of input parameters on the quality of model predictions. On the other hand, the sensitivity coefficients may be used to make qualitative predictions of the indirect (synergistic) effects of a toxicant on the pelagic compartment of lake ecosystems.

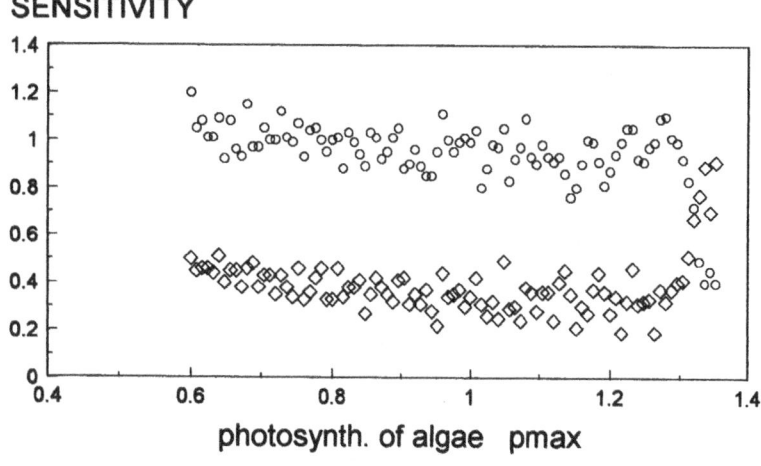

Figure 2. Dependence of the sensitivity coefficients S_r (circles) and S_p (squares) on the maximum photosynthesis of algae p_{max} (r_{max} was kept fixed at r_{max}=1.0) for year 1 (result of 100 simulation runs).

If we vary r_{max} and p_{max} by 1% and simulate the behaviour of the lake for one year, we get S_r= 0.99 for the respiration rate of zooplankton and S_p = 0.34 for the photosynthesis. Obviously our model is rather insensitive with respect to both parameters, and, as can be seen in Fig.2 for a modification of rmax, sensitivity is relatively independent of the parameter values. Consequently our sensitivity function is nearly linear with respect to the parameter variation, and we may conclude, that a herbicide, which reduces the photosynthetic activity of algae by 10% would consequently reduce the peak phytoplankton density in the first year after the introduction of a toxicant by 3.4% only, and an insecticide augmenting the respiration rate of the two zooplankton groups by 10% would cause a 9.9% increase in peak phytoplankton density.

3.2 The dependence of parameter sensitivity on the time span simulated

The relatively low level of sensitivity only holds for the first simulated year. If the simulation is continued, S_r and S_p show strong fluctuations in the course of time (Fig.3), and in the long run S_r even becomes negative. I.e., for longer time spans, an insecticide which causes an increase in rmax would cause a decrease of peak phytoplankton density. This is the opposite of what has been the result for the first year and what we would have assumed as plausible.

On the other hand, this result only holds for $r_{\mathrm{max}} = 1.0$. Even small modifications of this (or other) parameters may result in qualitatively different predictions (Fig.4). Based on this type of analysis, a general prediction of the toxicant effect is not possible, unless the parameter values for the undisturbed system are known with unlimited precision. But this is evidently not possible for any ecosystem, no matter how much effort is put into the acquisition of data (Seitz & Ratte, 1991).

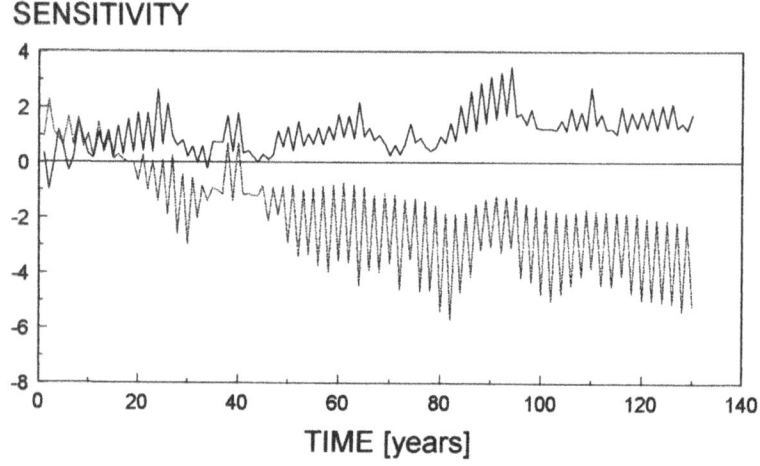

Figure 3. Dependence of the sensitivity coefficients S_r (dotted line) and S_p (full line), for r_{max}=1.0 and p_{max}=1.0, on the time passed since the model parameter was modified (result of a simulation run covering 130 years).

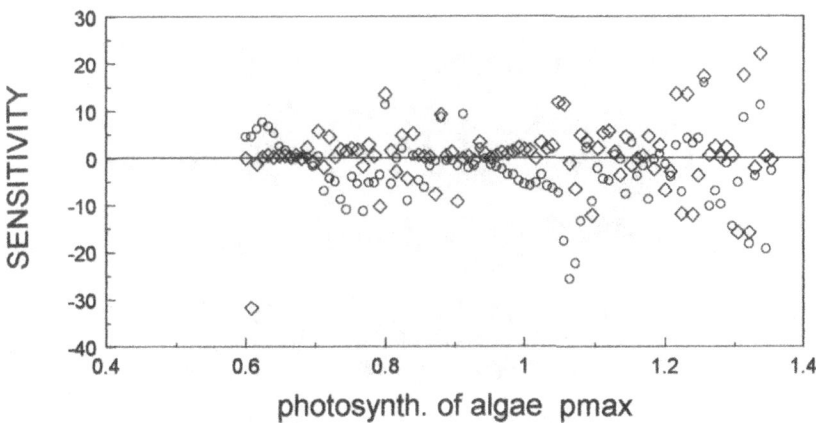

Figure 4. Dependence of the sensitivity coefficients S_r (circles) and S_p (squares) on the maximum photosynthesis of algae p_{max} (r_{max} was kept fixed at $r_{max}=1.0$) for year 130 (result of 100 simulation runs).

3.3 Exploring the parameter space

Some insight into the reasons for the strong fluctuations of S is gained, if we present the simulation results in a two dimensional "parameter-space" (Fig.5). In the first simulated year (Fig.5a), the structure of the resulting phytoplankton density surface is very smooth. Peak densities increase towards higher values of r_{max} as well as p_{max}. For this time horizon, we predict an increase of peak phytoplankton density, if either rmax or pmax are increased. Since sensitivity coefficients are smaller than 1.0 this increase will not be very strong.

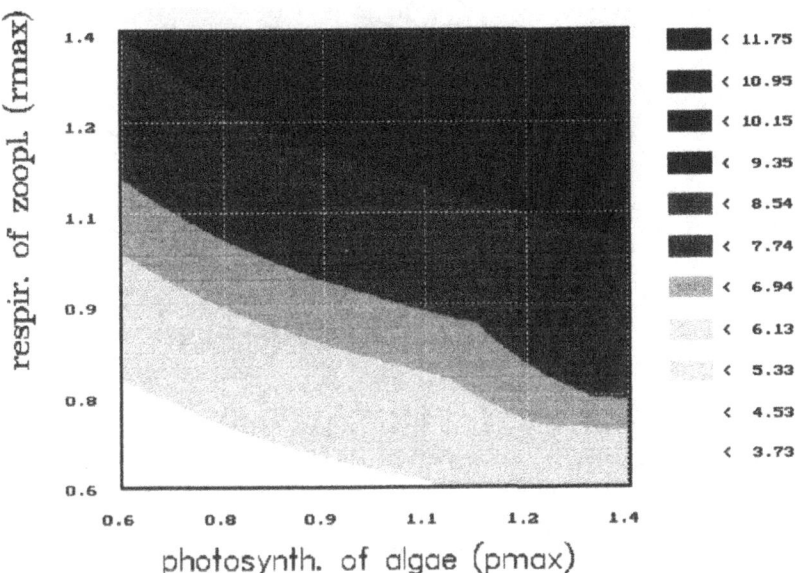

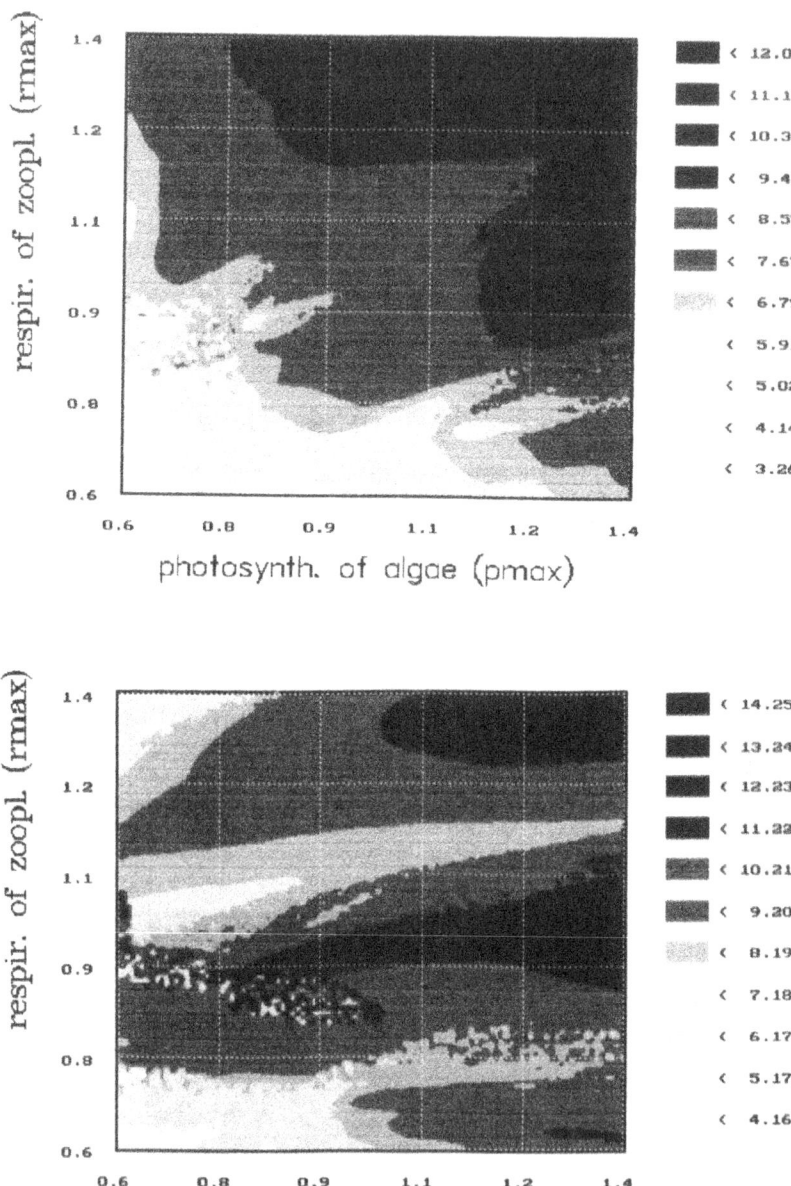

Figure 5. Dependence of the predicted yearly peak phytoplankton biomass on the two model parameters r_{max} and p_{max}. Results of 10,000 simulation runs (100 values for rmax and 100 values for p_{max}). The different figures give the biomasses in the first (A), the 10th (B) and the 130th (C) year of the simulation. Biomass values [mg/l] are given as different shades of grey ranging from white (low biomass) to black (high biomass). Standard value for both parameters is 1.0.

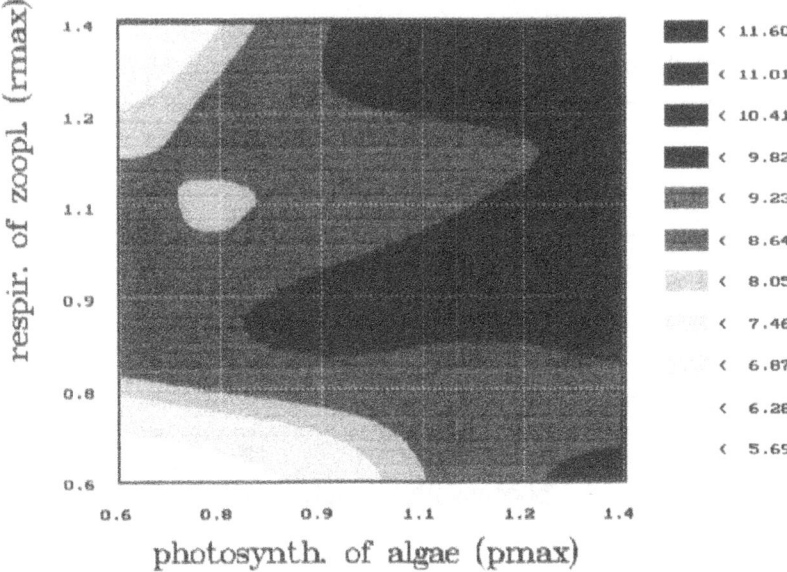

Figure 6. As Fig.5C. but now each point gives the mean of peak phytoplankton biomass from 400 Monte Carlo-simulations (r_{max} and p_{max} are normally distributed around the indicated value with 10% standard deviation).

In the following years, this smooth surface gets more and more structured (Figs.6b and 6c). Now predictions get more and more dependent on the initial parameter values. Dotted regions indicate deterministic chaos, which is a well-known phenomenon in periodically driven systems (Peitgen et al., 1992). This explains the enormous fluctuations of S_r and S_p in Figs 4. and 5.. In particular in these regions, deterministic predictions of toxicant effects are meaningless, since even extremely small modifications of model parameters may result in large variation of model predictions. In this region of the parameter space sensitivity coefficients are one order of magnitude bigger than in the first simulated year (see Figs. 2. and 4.).

3.4 Monte Carlo (MC)-simulation

The true parameters of the modelled ecosystem are never known with much precision (Rose, 1985; Poethke *et al.*, 1991, 1993; Ratte *et al.*, 1993; Seitz & Ratte, 1991). MC-simulation and risk analysis (O'Neill *et al.*, 1982) offer techniques to tackle this problem. Since MC-simulation averages out over a certain range of the parameter space, the structure of the resulting biomass surface (Fig.6) is rather smooth. This is reflected in the sensitivity coefficients, too (Fig.7). They are one order of magnitude smaller than in the deterministic case (Fig.4). Evidently, this reduction of sensitivity is critically dependent on the chosen standard deviation and predictions may even quantitatively be influenced by the choice of parameter variation for MC-simulations (Poethke, 1993).

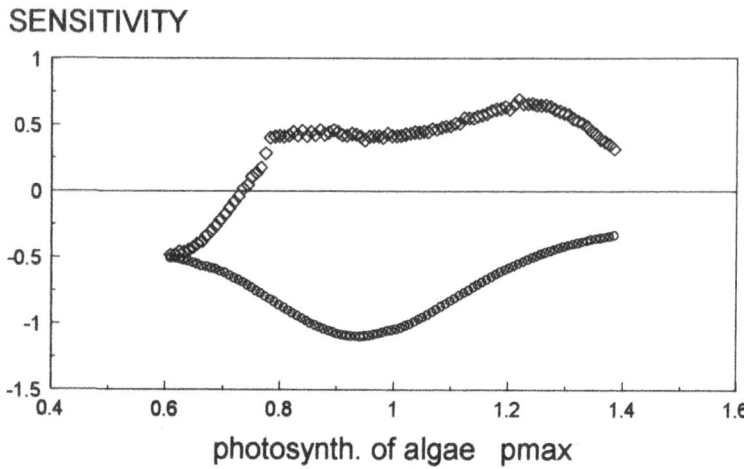

Figure 7. Dependence of the sensitivity coefficients S_r (circles) and S_p (squares) on the maximum photosynthesis of algae p_{max} (r_{max} was kept fixed at $r_{max}=1.0$) for year 130 (result of 100X400 MC-simulation runs)

4. Conclusion

We have shown, that model predictions of indirect effects of toxicants on lake ecosystems may be strongly dependent on the time horizon of the prediction, as well as on the exact parameters used to describe the undisturbed system. Some indirect toxicant effects develop with characteristic times of tenth of years. Models which intend to be valid for such time scales need to take into account species succession and even evolutionary processes like changes in the genetic composition of the species in the system (Seitz & Ratte, 1991). But, even if we do not take such processes into account, predictions are very sensitive to an exact description of the initial state of our system. Usually we will not be able to give a sufficiently precise definition of the initial state to make deterministic predictions of the long term dynamic behaviour of the system. MC-simulation allows to take into account the fuzzy definition of initial conditions, but parameter variance may as well have a strong influence on model predictions.

References

Bartell, S.M., R.H. Gardner and R.V. O'Neill (1988), 'An integrated fates and effects model for estimation of risk in aquatic systems', In Adams, W.J., Chapman, G.A., and Landis, W.G. (eds.): Aquatic Toxicology and Hazard Assessment. American Society for Testing and Materials Special Technical Publication, **971**, 261-274.

Jorgensen, S.E. (1979), 'Handbook of environmental data and ecological parameters',

International Society of Ecological Modelling.

Morrison, K.A., N. ThΘrien and B. Marcos (1987), 'Comparison of six models for nutrient limitations on phytoplankton growth', *Can. J. Fish. Aquat. Sci.*, **44**, 1278-1288.

Oertel, D. (1992), 'Beiträge zu Methoden, Möglichkeiten und Grenzen der Simulation von Schadstoffauswirkungen in aquatischen Ökosystemem', Ph.D. Thesis. University of Mainz.

O'Neill, R.V., R.H. Gardner, L.W. Barnthouse, G.W. Suter, S.G. Hildebrand and C.W. Gehrs (1982), 'Ecosystem risk analysis: a new methodology', *Environmental Toxicology and Chemistry,* **1**, 167-177.

Peitgen, H.O., H. Jürgens and D. Saupe (1992), 'Chaos and fractals - New frontiers of science', Springer Verlag, Berlin.

Poethke, H.J., D. Oertel and A. Seitz (1991), 'Risk assessment of toxicants to pelagic food webs: a simulation study', In: Moeller, D.P.F., Richter, O. (eds.) Analyse dynamischer Systeme in Medizin, Biologie und ſkologie. Springer Verlag, Berlin. 192-199.

Poethke, H.J., D. Oertel and A. Seitz (1993), 'Models for hazard assessment: problems and perspectives. In Hill, I.R. et al. (eds.). *EWOFFT: Proceedings of the European Workshop of Freshwater Field Tests.* Lewis Publ. Inc, Michigan. in press.

Ratte, H.T., H.J. Poethke, U. Dülmer and U. Hommen (1993), 'Modelling of aquatic fielt test for hazard assessment', In Hill, I.R. et al. (eds.). *EWOFFT: Proceedings of the European Workshop of Freshwater Field Tests.* Lewis Publ. Inc, Michigan. in press.

Rose, K.A. (1985), 'Evaluation of nutrient-phytoplankton-zooplankton models and simulation of ecological effects of toxicants using laboratory microcosm ecosystems', Ph.D. Thesis. University of Washington.

Seitz, A. and H.T. Ratte (1991), 'Aquatic ecotoxicology: On the problems of extrapolation from laboratory experiments with individuals and populations to community effects in the field', *Comp. Biochem. Physiol.*, **100C** (1/2), 301-304.

TOWARDS A METRICS FOR SIMULATION MODEL VALIDATION

HUUB SCHOLTEN[1], MARCEL W.M. VAN DER TOL[2]

[1] Wageningen Agricultural University, Department of Computer Science
Dreijenplein 2, 6703 HB Wageningen, The Netherlands
e-mail hscholten@rcl.wau.nl

[2] Rijkswaterstaat, Tidal Waters Division, P.O. Box 20907,
2500 EX The Hague, The Netherlands

Abstract

A large group of nonlinear dynamic simulation models can be seen as intermediates between hard (physical) and soft (management science) models, because they are based on insufficient or not generally accepted theories and hypotheses. This type of models (ecological, environmental, and economic) is characterized by highly uncertain outcomes, due to an uncertain, unidentifiable model structure, not well known model parameters and uncertain model inputs. Most validation techniques offer merely a terminology and a procedural validation approach without any metrics. Let S be a part of reality, which satisfies the constraints of a relevant experimental frame (specification of time, location, experimental conditions and relevant state variables). S can only be known by making observations of the real system. Any simulation model of the real system S has to be based on the available theoretical and other *a priori* knowledge. Each source of uncertainty will influence the model outcomes. Let O be the set of observations and M the set of model results, both within the same experimental frame, and both including uncertainty ranges, then validation tests for the fit between O and S. In the terminology of Popper most models of this class are invalid (no perfect match of O and S) and have to be rejected. This paper suggests to test for the usefulness of a model in terms of model adequacy (which part of the system can be adequately simulated) and model reliability (which part of the model outcome matches system behavior). The test on model usefulness instead of model validity provides a metrics which helps to determine the scope of the model and increases its acceptability. The method is illustrated with examples.

Key-words   simulation, model adequacy, model reliability, validation metrics

1. Introduction

Environmental, economic and (agro)ecologic nonlinear dynamic simulation models are often intermediates between hard (physical) and soft (management science) models. Physical models are usually based on sufficient and generally accepted theories and hypotheses. Management science models include highly unpredictable humans as part of the system. The present study centers upon the class of models which are built on an insufficient theoretical framework of (partly) controversial hypotheses and which are characterized by highly uncertain model outcomes.

 In general model behavior is influenced by model structure (including spatial aspects), parameter values, and model inputs (initial conditions, forcing functions and boundary conditions). Each of these factors contribute to model result uncertainty (O'Neill and Gardner, 1979, Walters, 1986). Simulation model validation has to deal with this problem by comparing uncertain model results (model behavior) with uncertain observations of the real system (system behavior). A standard definition of validation is given by Schlesinger *et al.* (1979):

> *Substantiation that a computerized model within its domain of applicability possesses a satisfactory range of accuracy consistent with the intended application of the model.*

Many validation techniques have been suggested (Hermann, 1967, Wigan, 1972, Sargent, 1982, 1984, 1989a, 1989b, Lewandowski, 1982, Young, 1983, Reckhow, 1989, Summers *et al.*, 1993), but most authors offer merely a terminology instead of a methodology.

Most intermediate hard simulation models can be easily invalidated or falsified, in the terminology of Popper (1959). Mankin *et al.* (1975) suggested that this does not mean that such invalidated models could not be useful to some extent. Its usefulness depends on the kind of answers which are expected of the model and thus of its objective. If some conditions are fulfilled, the model behavior can be judged as good or preferably useful for some of the state variables, whereas a model can be considered not useful under different circumstances or for other variables. Models which are not only intended to summarize research results, but also to predict future behavior (management tool), have to be validated: which part of future model behavior will be predicted correctly (Mankin *et al.*, 1975)? Two aspects of validation, proposed by Mankin *et al.* (1975) and recently used by Scholten and Van der Tol (1993), are discussed here. First, one has to determine which part of the system behavior can be adequately simulated with the model (model adequacy). Secondly, which part of the model outcome matches system behavior (model reliability).

2. Methods

2.1 Concepts

Let S be the hyperspace of relevant observable entities of the real system, with time and space as further dimensions. Our knowledge of S is collected in two ways: by making observations and by running the model. O is the a subspace of S of all observed data by lab experiment, direct measurement or even estimated with some other model, as is frequently done. M is a subspace of S of all model results, i.e. model outcomes for a series of output variables associated with the model objective. Model results M are rather uncertain, because M depends on *a priori* model parameters, which are not well known, the choice of model structure and uncertain model inputs like forcing functions (O'Neill and Gardner, 1979, Walters, 1986). The incomplete knowledge of the badly known parameters is collected by measurements, experiments and from literature. If a set of values for each of these badly known parameters is called a parameter vector, then there are many parameter vectors (the hyperspace P) in agree with the (limited) *a priori* knowledge. Each of these parameter vectors will give different model results, which together span up the hyperspace M. Any statement on the real system (S) relies on observations of the reality (O) or on the *a priori* knowledge and general theories built into the model (M). O largely matches the real system S (except errors in O). M, on the other hand coincides at least partly with S. In Fig. 1a model results (M) intersect with the real system (S) and with observations (O). Note that some model results coincide with the unobserved real system.

a

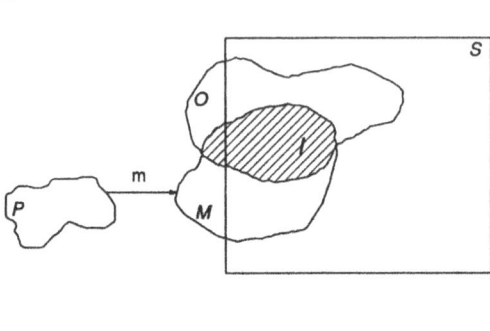

b

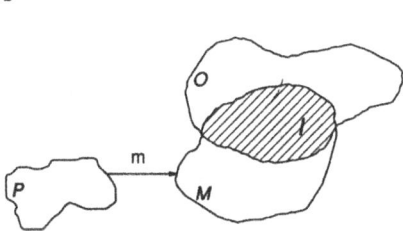

Figure 1. Two-dimensional projected hyperspaces of the real system (S), observed data (O), model results (M), and, *a posteriori* known, parameter uncertainty (P), with known S (a) and unknown S (b). The shaded area is the overlap between O and M, which has to be maximized.

Model calibration aims to find those parameter vectors or the part of the parameter hyperspace which maximizes the intersection of the model hyperspace (M), the real system hyperspace (S) and the observation hyperspace (O). Because S is unknown, calibration will be restricted to maximizing the intersection of M and O (the shaded area I in Fig. 1b).

With this concept, model adequacy (which part of the observed system behavior can be simulated with the model) is defined as the volume of the hyperspace I compared to the volume of the hyperspace O and model reliability (which part of the model simulations is also observed) as the volume of the hyperspace I compared to the hyperspace M. Mankin *et al.* (1975) proposed these definitions, but did not develop a method how to calculate adequacy and reliability in a multi-variate (many states) case, using a discrete set of observations with errors and a set of discrete model outcomes with parameter uncertainty. This question will be addressed here.

The ultimate goal of calibration, related to the inverse problem theory (Tarantola, 1987, Klepper and Rouse, 1991) is to find those parameter vectors P', which realize perfectly adequate and perfectly reliable model results (an estimate for the *a posteriori* parameter set). Following Mankin (1975) a model is called valid if $M - I = \varnothing$ and $I \neq \varnothing$ (the model corresponds to system behavior under all conditions), but the model will be useful if

$M - I \neq \emptyset$ and $I \neq \emptyset$ (it represents some of the ecosystem behavior). Probably all ecosystem models fall into the latter useful, but not valid category of models.

2.2 Calculating model uncertainties

In the case of a nonlinear model, methods for estimating model uncertainty fall into two major groups: first- or second-order variance propagation and Monte Carlo methods (Beck and Young, 1976, Carver, 1980, Draper and Smith, 1981, Birta, 1984, Beck, 1987, Tarantola, 1987, Summers *et al.*, 1993). For first-order variance propagation the model has to be locally linearized, which has several limitations, as the final results depend strongly on initial guesses and local minima in the distance between model results and data. This approach will often lead to unrealistic results (Klepper *et al.*, 1991). Most methods in this category have to be specially programmed for each model and the labor costs will be often prohibitive (Summers *et al.*, 1993).

Monte Carlo ('brute force') methods also have drawbacks, as these use ample computation time. More sophisticated Monte Carlo methods will complete calibration within a reasonable time frame. The calibration method used for this study is quite satisfactory (Klepper, 1989, Scholten *et al.* 1990a, Klepper *et al.* 1991). This method, called 'Controlled Random Search' (CRS) was originally proposed by Price (1979) as a constrained polynomial function-minimization and it solves complex optimization problems with many skewed or disjoint distributed parameters and with multi-extremal object functions. Further on it is related to optimization algorithms as 'Simulated Annealing' (Metropolis, 1953, Kirkpatrick *et al.*, 1983, Press *et al.*, 1986, Tarantola, 1987, Davis, 1987) and 'Genetic Algorithms' (Holland, 1975, Davis, 1987, Davis and Coombs, 1989, Goldberg, 1989).

Calibration with CRS starts with a random or Latin Hypercube sampled set of parameter vectors with no significant covariances between parameters (McKay *et al.*, 1979, McKay, 1988, Iman and Helton, 1988). The result of the first (sampling) stage is a set of parameter vectors, each with a calculated Lack of Fit between the model outcome of this vector and the observed data. This penalty function (Lack of Fit) is used in subsequent calibration runs to update the set of parameter vectors. After applying CRS the parameters are strongly correlated and the covariances between the parameters are used for a Monte Carlo uncertainty analysis with a limited number of model runs. For all variables the results of each run are sorted at each point in time and stored as quantiles for graphical representation or further analysis.

2.3 Calculating model adequacies and reliabilities

Our knowledge of O and M is restricted to the sets O and M, which are assumed to represent both hyperspaces. If m is the model with the chosen modelstructure and it includes the underlying theoretical aspects, then: $M = m(P')$, with P' the *a posteriori* estimated set of parameter vectors selected during calibration with a penalty function. The set O contains samples at discrete points in time for the relevant state variables and other system entities of interest. Often there exist no proper estimate of the errors in O. Instead of errors for each variable in O, 'desired accuracies' can be defined. The ensemble of observations in time and accuracies will characterize O sufficiently. The set M includes model generated uncertainties for the states at discrete times (the quantiles, see section

2.2).

  In order to decide to what extent the models is useful, the multivariate problem will be
divided in a series of univariate problems. Linear interpolation provides trajectories for
each state and other relevant system entities. Using these trajectories the univariate
adequacy and reliability are calculated. The set of all adequacies and reliabilities summarizes the usefulness of the model.

Scholten & Van der Tol, fig.2

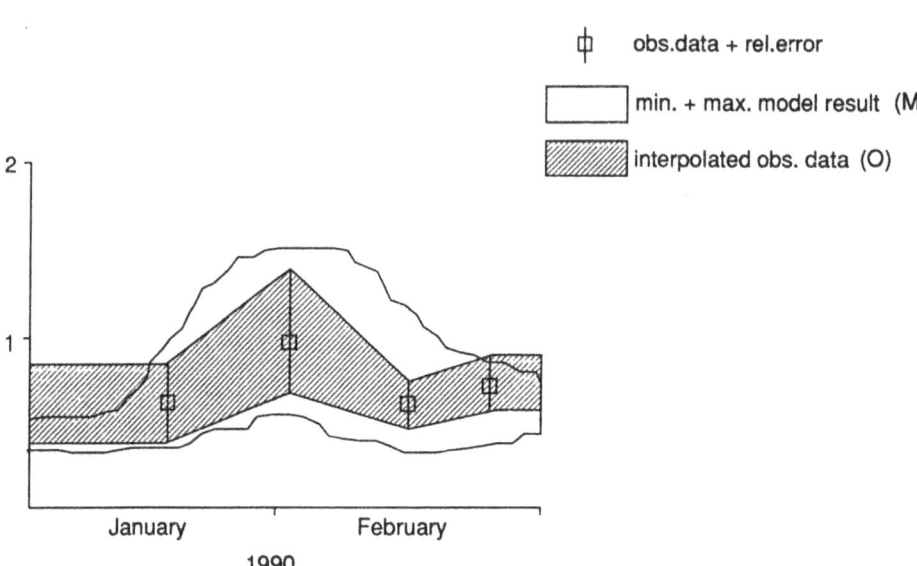

Figure 2. Model generated uncertainty ranges (empty), observed data and error bounds
for observed data (hatched), calculated from observed data with a relative error and linear
interpolation between dates with observed data. Adequacy is calculated as the area of the
intersection (I_{area}) divided by the area of the error bands of the observations (O_{area}), and the
reliability as the area of the intersection (I_{area}) divided by the area of the model generated
uncertainty bands (M_{area}).

  In this study we choose a geometric approach to calculate the univariate adequacies and
reliabilities. First we have to estimate the intersections of the hyperspaces M, O and I with
the 2-dimensional variable-time plane. The area enclosed by the (linear interpolated) 1st
and 100th percentiles (minimum and maximum) is used as a univariate estimate for M and
is defined as M_{area}. O is estimated as the area (O_{area}) enclosed by the observation error
bands, which are calculated from the observation errors by linear interpolation. The
intersection area of M_{area} and O_{area} is I_{area} and it is used to estimate I.

2.4 Applying validation metrics

The method is applied on two different models. DEMO (DEmonstration MOdel) is not a model of a real system, but it is intended as a test model of the simulation software package SENECA (Scholten *et al.*, 1990a). Because the observed data are in fact generated by the model, using one perfectly known parameter vector, this model is suited to test calibration, uncertainty analysis and model validation. The second model, SMOES (Simulation Model of the Oosterschelde EcoSystem), is a complex ecosystem model reported in detail in Klepper (1989), Klepper *et al.* (1993), and Scholten and Van der Tol (1993). This model also is used to evaluate ecosystem properties (Scholten *et al.*, 1990b, Herman and Scholten, 1990, Van der Tol and Scholten, 1992).

2.5 How to interpret the results

In general an uncertainty analysis of a calibrated model will produce four types of results in terms of model adequacy and model reliability (Fig. 3).

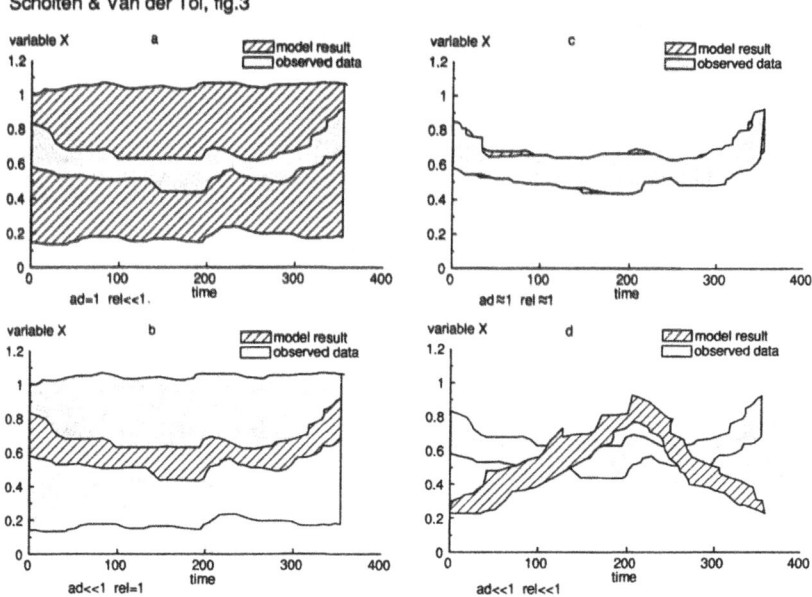

Figure 3. Model validation with high adequacy and low reliability (a), low adequacy and high reliability (b), high adequacy and high reliability (c), and low adequacy and low reliability (d).

Modelers, who want to play safe, will present simulation model results with wide uncertainty bands (Fig. 3a). This conservative approach has a low resolution: different future scenario's will produce almost equal (largely overlapping) predictions. A substantial fraction of the model results will not be found in the observed system. Decision makers using the model will normally not accept this kind of predictions, as no conclusions can be drawn on the effects of the scenario's.

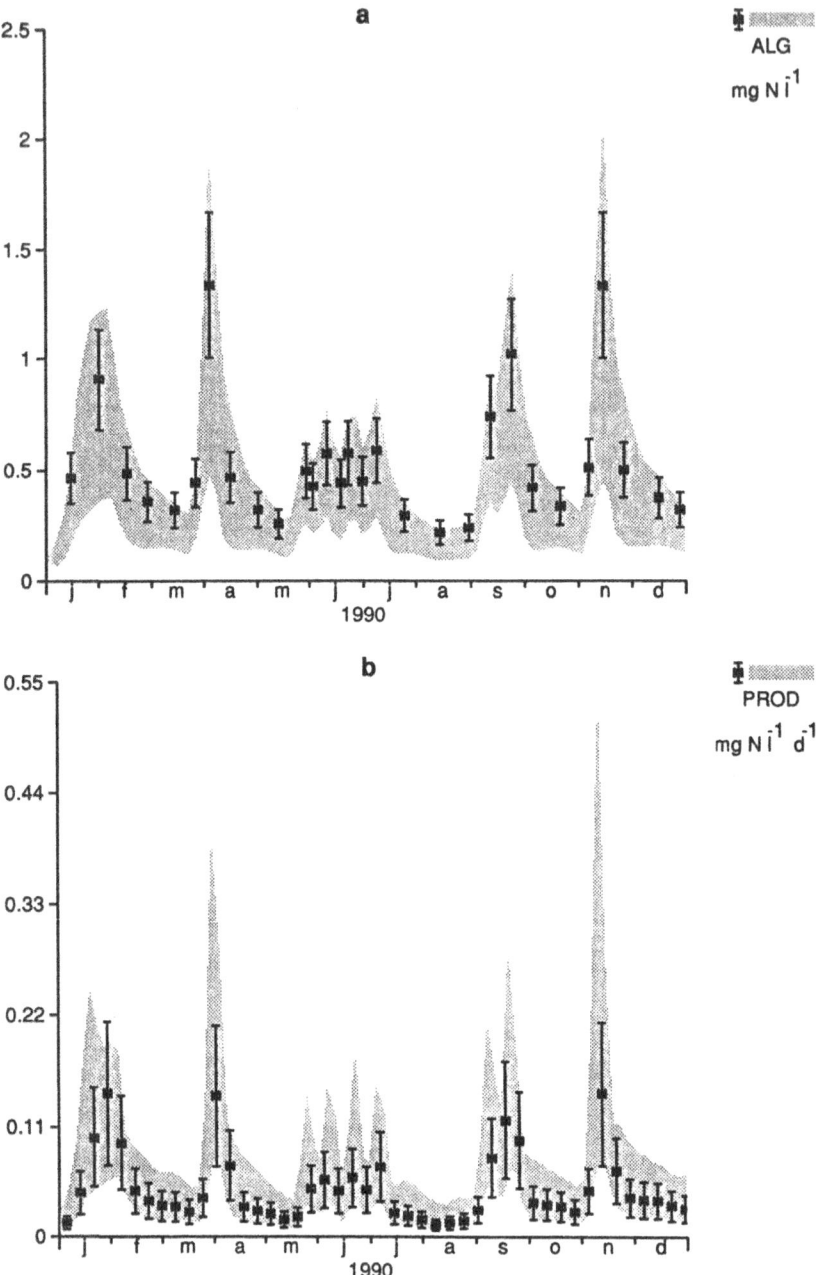

Figure 4. Validation results of DEMO after a preliminary calibration and uncertainty analysis for the variables ALG (biomass of algae expressed as nitrogen and a relative error of 0.25) with an adequacy of 0.84 and a reliability of 0.51 (a), and PROD (primary production of algae and a relative error of 0.5) with an adequacy of 0.95 and a reliability of 0.61 (b).

Narrow ranges in the model results compared to wide ranges in the observations are common practice in ecological, (eco)toxicological and environmental models (Fig. 3b). Often model results are presented as single output trajectories. Such results suggest a very high accuracy of the prediction, which is not supported by observations. Any trajectory through the observed data will give reliable results. Research should be focussed on the reduction of the uncertainty in the observations, but model uncertainty should be assessed too. Using this model to investigate the effects of management scenario's will lead to erroneous conclusions, as almost every scenario will have a significantly different effect. Nevertheless, most decision makers will appreciate this kind of predictions, because these allow firm, although incorrect, conclusions.

A model, which describes adequately and reliably what is known by observing the real system should be the goal of every simulation modelling exercise, both from the modeler's and from the decision maker's point of view (Fig. 3c).

Any modeler or model user is familiar with model results which are totally wrong (Fig. 3d). These lack adequacy and reliability. Apparently the knowledge of the system incorporated in the model is insufficient for an accurate prediction of this variable, but this does not mean that the entire model is useless. Other model variables may describe system behavior more adequately and more reliably.

3. Results and discussion

Applying the method presented in this paper to the DEMO-model showed a high adequacy for the uncalibrated model and a low reliability, as was expected (not shown). The uncertainty bands of the calibrated model fitted the system observations sufficiently, which means that the model is an adequate and rather reliable description of the system (Fig. 4).

Calculating the adequacy and reliability of several variables of SMOES, showed that the calibrated model is invalid in the Popperian sense, but useful to some extent for many of its variables, with adequacies and reliabilities ranging between 0 and 1. As an example the primary production of algae before (Fig. 5a) and after calibration (Fig. 5b) are shown. Of five important variables adequacies and reliabilities are calculated before (Fig. 6a) and after calibration (Fig. 6b). Calibration is responsible for a reduction in adequacy of all five variables, but it increases reliability. Future calibration exercises with SMOES should concentrate on higher model reliabilities, but not at the expense of lower model adequacies. Some other validation results of SMOES are discussed in detail in Scholten and Van der Tol (1993).

The method presented here to quantify simulation model adequacy and reliability in the validation process, is a first step to an objective validation, opposite to a completely subjective 'face validation' (Hermann, 1967, Sargent, 1984b). The latter method is often used to estimate model quality, even for complex models with many variables predicting system behavior. Obviously, face validation will lead to confusing and woolly conclusions on model quality. The results of the method presented here, are in agree with domain expert judgements and correspond with intuitive interpretation of simulation modelling results.

Although the method is an attempt to objective validation, it contains subjective, arbitrary elements. These elements have to be attributed to arbitrary aspects of CRS (the number of stored parameter vectors, weights used in the penalty function, a proper stopcriterion, and the number of parameter vectors used to calculate the parameter vector centroid, see Price, 1977, Scholten and Van der Tol, 1993) and to insufficient knowledge

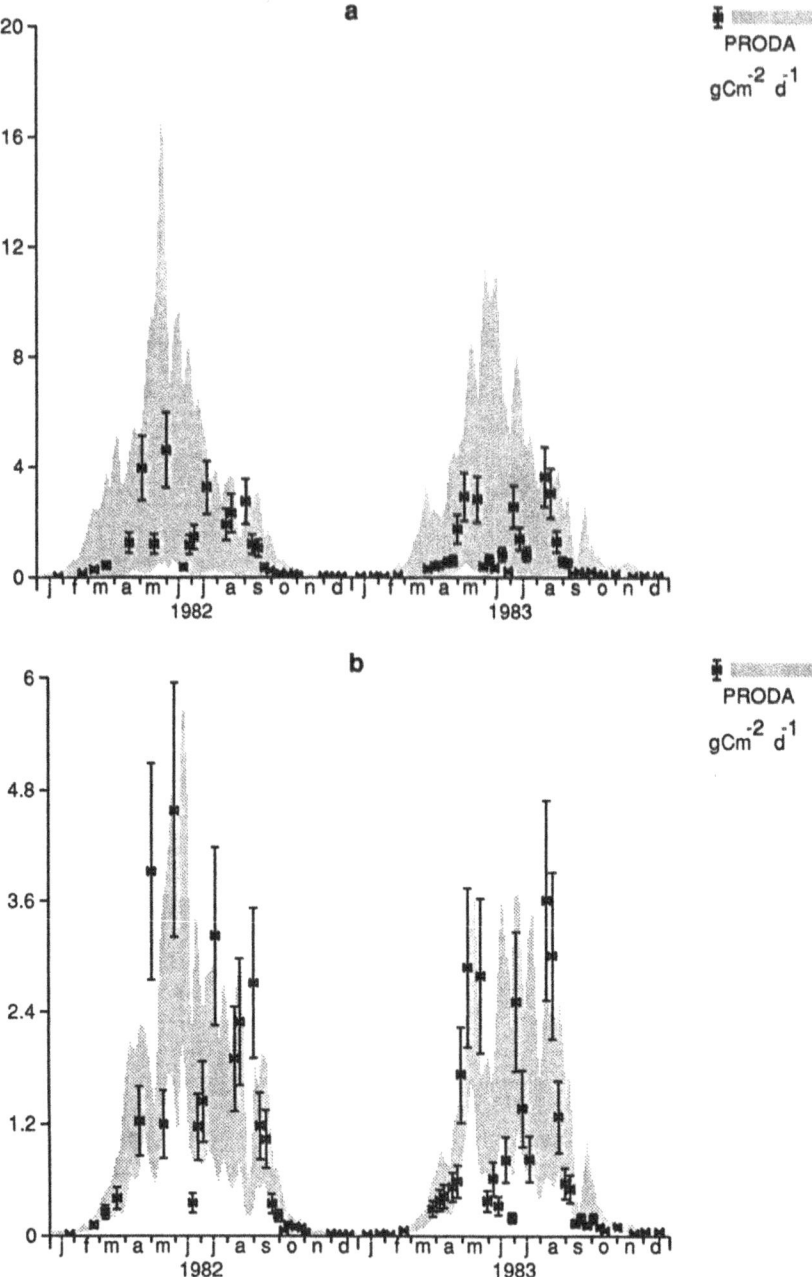

Figure 5. Uncertainty analysis results for the variable PRODA (primary production of two groups of algae) before the building of the storm-surge barrier in the seaward spatial compartment of the Oosterschelde (SW Netherlands). Uncertainty bands calculated by the simulation model SMOES (shaded area) before (a) and after calibration (b). See Fig. 6 for corresponding adequacy and reliability values.

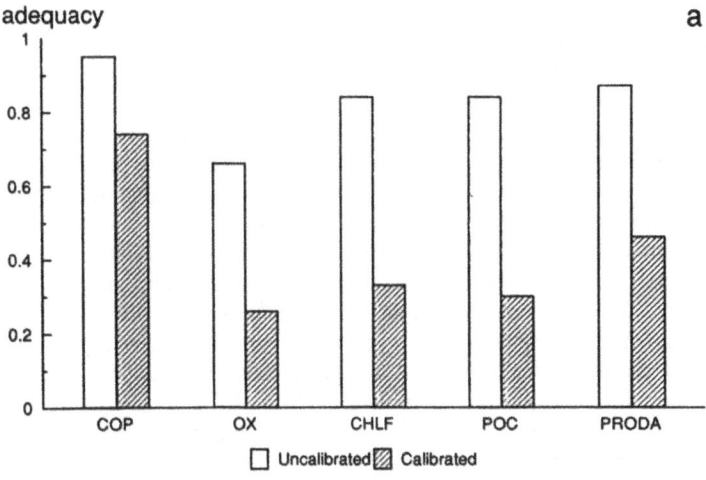

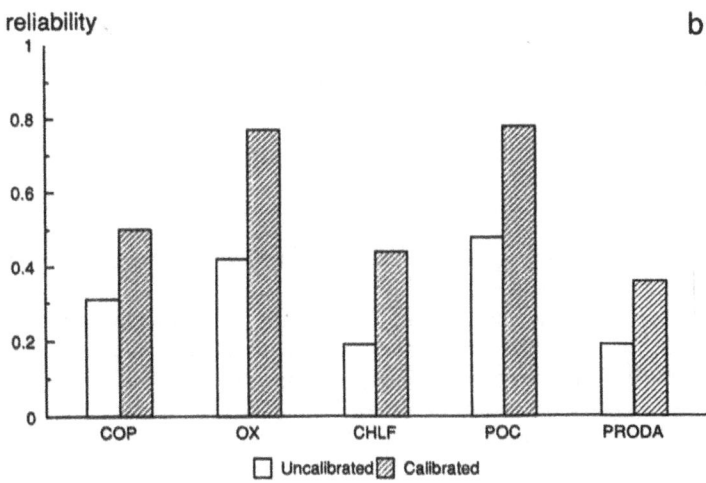

Figure 6. Validation results for copepod biomass (COP, r.e. 0.5), oxygen concentration (OX, r.e. 0.1), chlorophyll concentration (CHLF, r.e. 0.5), particulate organic carbon (POC, r.e. 0.5), and primary production of the two groups of planktonic algae (PRODA, r.e. 0.3) in the seaward spatial compartment of the Oosterschelde before the building of the storm-surge barrier, expressed in terms of adequacy (a) and reliability (b).

on errors in the observations. The latter aspect forced us to use desired accuracies instead of error distributions of each observation. Despite its arbitrary aspects, the method developed and applied in this study makes validation at least explicit.

One of the main problems of the intermediate hard models, for which the present method has been developed, is the contradiction between comprehensive, but unidentifiable models and simpler models which can be identified, as is discussed in Beck (1987). Comprehensive models of the first category are characterized by overparameterization and contain a surplus of information, which is not supported by observations. Excluding all redundant hypotheses and the associated parameters is not always advisable, as in these models the parameters often have a physical, chemical or biological meaning. The unidentifiability will prevent the discovery of a single best parameter vector, which would allow a different approach of the validation problem than suggested here. The surplus of knowledge incorporated will, on the other hand, sometimes be an advantage, when the model has to predict different future behavior. Knowledge, which is redundant in the description of observed past system behavior, will perhaps predict future behavior better than a simpler (identified) model. The comprehensive model includes perhaps some of the processes and mechanisms relevant to predict the future. If this will be true, can only be answered, when we know the system behavior in the different future.

Future model analysis studies would benefit from improvements in the method presented here. Adequacy and reliability can be used during calibration instead of more classic penalty or object functions. The geometric approach we used here, can easily be replaced by a method, which evaluates adequacy and reliability for each single observation. The rather arbitrary linear interpolation (with the associated risk of making large errors) can be avoided in this way. This is especially important for variables with a highly dynamical behavior and for variables, which have not been measured with a sufficient sampling frequency (Jørgensen, 1986).

Summarizing it can be stated that an evaluation fo model usefulness is a promising alternative for model validation. The adequacy and reliability of model states fit intuitively in the vocabulary of both modelers and domain experts. These two simulation model quality parameters allow a quantitative discussion of model results.

References

Beck M.B. (1987), 'Water quality modeling: a review of the analysis of uncertainty', *Water Resources Research*, **23**, 1393-1442.

Beck M.B. and P.C. Young (1976), 'Systematic identification of DO-BOD model structure', *J. Env. Eng. Div., Proc. Am. Ass. Civil Eng.*, **102, no. EE5**, 909-927.

Birta L.G. (1984), Optimization in simulation studies, In: Oren T.I., B.P. Zeigler and M.S. Elzas (Eds.), *Simulation and model-based methodologies: an integrative view*, 10 in the series: NATO ASI Series F: Computer and System Science, Springer-Verlag, Berlin, Heidelberg, 451-473.

Carver M.B. (1980), 'Parameter optimization in the continuous simulation packages FORSIM and MACKSIM', *Mathematics in Computers and Simulation*, **22**, 298-318.

Davis L., Ed. (1987), '*Genetic algorithms and simulated annealing*', Pitman, London.

Davis L. and S. Coombs (1989), 'Optimizing network link sizes with genetic algorithms', In: Elzas M.S., T.I. Oren and B.P. Zeigler (Eds.), *Modelling and simulation methodology*, 4 in the series: Modelling and simulation, North-Holland, Amsterdam, 317-331.

Draper N.R. and H. Smith (1981), '*Applied regression analysis*', 2nd edition, in the series: Wiley series in probability and mathematical statistics, Wiley, New York.

Goldberg D.E. (1989), '*Genetic algorithms in search, optimization, and machine learning*', Addison-Wesley Publishing Company Inc., Reading, Massachusetts.

Herman P.M.J. and H. Scholten (1990), 'Can suspension-feeders stabilise estuarine ecosystems?', In: Barnes M. and R.N. Gibson (Eds.), *Trophic relations in the marine environment*, in the series: Proc. 24th Europ. Mar. Biol. Symp, Aberdeen University Press, Aberdeen, 104-116.

Hermann C.F. (1967), 'Validation problems in games and simulation with special reference to models of international politics', *Behavior Science*, **12**, 216-231.

Holland J.H. (1975), '*Adaptation in natural and artificial systems*', The University of

Michigan Press, Ann Arbor .

Iman R.L. and J.C. Helton (1988), 'An investigation of uncertainty and sensitivity analysis techniques for computer models', *Risk Analysis*, **8**, 71-90.

Jørgensen S.E. (1986), '*Fundamentals in ecological modelling*', 9 in the series: Developments in environmental modelling, Jorgensen S.E. (Ed.), Elsevier, Amsterdam.

Kirkpatrick S., C.D. Gelatt and M.P. Vecchi (1983), 'Optimization by simulated annealing', *Science*, **220**, 671-680.

Klepper O. (1989), '*A model of carbon flows in relation to macrobenthic food supply in the Oosterschelde estuary*', DGW-LUW, Wageningen, thesis.

Klepper O. and D.I. Rouse (1991), 'A procedure to reduce parameter uncertainty for complex models by comparison with real system output illustrated on a potato growth model', *Agricultural Systems*, **36**, 375-395.

Klepper O., H. Scholten and J.P.G. Van de Kamer (1991), 'Prediction uncertainty in an ecological model of the Oosterschelde estuary, S.W. Netherlands', *Journal of Forecasting*, **10**, 191-209.

Klepper O., M.W.M. Van der Tol, H. Scholten and P.M.J. Herman (1993), 'SMOES: A simulation model for the Oosterschelde EcoSystem. Part I: description and uncertainty analysis', *Hydrobiologia*.

Lewandowski A. (1982), 'Issues in model validation', *Angewandte Systemanalyse*, **3**, 2-11.

Mankin J.B., R.V. O'Neill, H.H. Shugart and B.W. Rust (1975), The importance of validation in ecosystem analysis, In: Innis G.S. (Ed.), *New directions in the analysis of ecological systems, Part 1*, vol.5, No. 1 in the series: Simulation Councils Proc. Ser, Simulation Councils Inc., Lajolle, California, USA, 63-71.

McKay M.D. (1988), Sensitivity and uncertainty analysis using a statistical sample of input values, In: Ronen Y. (Ed.), *Uncertainty analysis*, CRC Press, Inc., Boca Raton, Florida, 145-186.

McKay M.D., W.J. Conover and R.J. Beckman (1979), 'A comparison of three methods for selecting values of input variables in the analysis of output from a computer code', *Technometrics*, **21**, 239-245.

Metropolis N., A.W. Rosenbluth, M.N. Rosenbluth, A.H. Teller and E. Teller (1953), 'Equation of state calculations by fast computing machines', *The Journal of Chemical Physics*, **21**, 1087-1092.

O'Neill R.V. and R.H. Gardner (1979), Sources of uncertainty in ecological models, In: Zeigler B.P., M.S. Elzas, G.J. Klir and T.I. Oren (Eds.), *Methodology in systems modelling and simulation*, North-Holland Publ. Co., Amsterdam, 447-463.

Popper K.R. (1959), '*The logic of scientific discovery*', 2nd edition, Unwin Hyman Ltd, London.

Press W.H., B.P. Flannery, S.A. Teukolsky and W.T. Vetterling (1986), '*Numerical recipes: the art of scientific computing*', Cambridge University Press, Cambridge.

Price W.L. (1977), 'A controlled random search procedure for global optimisation', *The Computer Journal*, **20**, 367-370.

Reckhow K.H. (1989), Validation of simulation models: philosophy and statistical methods of confirmation, In: Singh M.G. (Ed.), *Systems and control encyclopedia: theory, technology, applications*, vol.6, Pergamon Press, Oxford, etc., 5011-5015.

Sargent R.G. (1982), Verification and validation of simulation models, In: Cellier F.E. (Ed.), *Progress in modelling and simulation*, Academic Press, London, etc, 159-169.

Sargent R.G. (1984a), Simulation model validation, In: Oren et al T.I. (Ed.), *Simulation and model-based methodologies: an integrative view*, Springer Verlag, Berlin Heidelberg, 537-555 (NATO ASI Series F10).

Sargent R.G. (1984b), A tutorial on verification and validation of simulation models, In: Sheppard S., U. Pooch and D. Pegden (Eds.), *Proceedings of the 1984 winter simulation conference*,,, 115-121.

Sargent R.G. (1989a), Validation of simulation models: general approach, In: Singh M.G. (Ed.), *Systems and control encyclopedia: theory, technology, applications*, vol.6, Pergamon Press, Oxford, etc., 5008-5011.

Sargent R.G. (1989b), Validation of simulation models: statistical approach, In: Singh M.G. (Ed.), *Systems and control encyclopedia: theory, technology, applications*, vol.6, Pergamon Press, Oxford, etc., 5015-5019.

Schlesinger S., R.E. Crosbie, R.E. Gagne, G.S. Innis, C.S. Lalwani, J. Loch, R.J. Sylvester, R.D. Wright, N. Kheir and D. Bartos (1979), 'Terminology for model credibility', *Simulation*, **32**, 103-104.

Scholten H., B.J. De Hoop and P.M.J. Herman (1990a), '*SENECA 1.2: a Simulation Environment for ECological Application (Manual)*', DIHO, Yerseke, Ecolmod report EM-4, ISBN 90-9003978-3.

Scholten H., O. Klepper, P.H. Nienhuis and M. Knoester (1990b), 'Oosterschelde estuary (S.W. Netherlands): a self-sustaining ecosystem?', *Hydrobiologia*, **195**, 201-215 (Suppl. North Sea-estuaries interactions).

Scholten H. and M.W.M. Van der Tol (1993), 'SMOES: a Simulation Model for the Oosterschelde EcoSystem. Part II: calibration and validation', *Hydrobiologia*, in press.

Summers J.K., H.T. Wilson and J. Kou (1993). 'A method for quantifying the prediction uncertainties associated with water quality models', Ecological Modelling, **65**, 161-176.

Tarantola A. (1987), '*Inverse problem theory. Methods for data fitting and model parameter estimation*', Elsevier, Amsterdam.

Van der Tol M.W.M. and H. Scholten (1992), 'Response of the Eastern Scheldt ecosystem to a changing environment: functional or adaptive?', *Netherlands Journal of Sea Research*, **30**, 175-190.

Walters C. (1986), '*Adaptive management of renewable resources*', in the series: Biological resource management, MacMillan Publ. Co., New York.

Wigan M.R. (1972), 'The fitting, calibration and validation of simulation models', *Simulation*, **18**, 188-192.

Young P. (1983), The validity and credibility of models for badly defined systems, In: Beck M.B. and G. Van Straten (Eds.), *Uncertainty and forecasting of water quality models*, Springer-Verlag, Berlin, Heidelberg, New York, 69-98.

USE OF A FOURIER DECOMPOSITION TECHNIQUE IN AQUATIC ECOSYSTEMS MODELLING

I. MASLIEV

International Institute for Applied Systems Analysis
A-2361 Laxenburg, Austria

e-mail: masliev@iiasa.ac.at

Summary

A quasilinear system of ordinary first-order differential equations of the type frequently used in ecosystems modelling (including mathematical models of aquatic ecosystems) is considered. It is assumed that the system is subject to periodical changes in coefficients and/or right-hand sides (due to diurnal or seasonal character of the described ecological processes). The periodic component of the state variables caused by these disturbances is considered to be small enough to allow usage of first-order Taylor formulae. Under these assumptions a decomposition of the system dynamics into "the slow motion" component and first-order Fourier harmonics is performed. The resulting set of equations can be solved with large time steps, still preserving information on the periodic as well as the smooth average components of dynamical behavoir of the initial system. The performance of the method is evaluated using an algae growth equation, the only growth limiting factor being that of light availability. The results acquired suggest the proposed method is useful both for adjusting the average motion component and for evaluation of the diurnal dynamics of algae. Further uses of the method are discussed and proposed.

Keywords Numerical methods, nonlinear systems, mathematical modelling, aquatic ecosystems

1. Introduction

Mathematical modelling of aquatic systems (especially nutrient-phytoplankton models) is widely used in water resources planning, eutrophication analysis and policy making (Somlyódy and Van Straten, 1986). Recently, numerous advanced methodologies have become available in the field, allowing uncertainty analysis with Monte-Carlo simulations, dynamic programming and other techniques (M.B.Beck, 1987). Application of such methods requires repeated simulations of the system in question. Calibration of model parameters on the basis of measured data is likewise based on repeated simulations of the aquatic system. If diurnal changes of the component values are considerable, then the time step used in simulations must be small enough to reproduce within-day behavoir of state variables (for example, to account for the diurnal changes in solar radiation, which is important for phytoplankton growth). In pursuing increased time steps for simulation, the generally used approach is to integrate the right-hand sides of the governing equations over a daily time interval and then solve for the seasonal dynamic (Thomann and Mueller, 1987). However, the governing equations of the phytoplankton-nutrient system are nonlinear, and simple averaging of the right-hand sides can lead to poor results. But the diurnal changes of system state variables values are often small with respect to the daily averages, and this can be used in formulating equations for Fourier components of the system state variables. Exact reproduction of system dynamics can be achieved only if the series of Fourier components are infinite. In practice, however, it is often enough to retain zero and first order terms only. Closed systems of equations for Fourier components can then be constructed, using the assumption that diurnal variations of state variables are small relative to their daily average values. Zero-order components could then be used in the analysis as the daily averaged values of the corresponding state variables, and first-order components give an apprehension of the periodic dynamics of the system, including its magnitude and phase. This approach is similar to a well-known method from analytical mechanics dealing with a system disturbed by a periodical external force. Originally proposed by Bogolyubov and Krylov (1937), this method was later widely used in particle and plasma physics and in many other areas, but has not yet received yet appropriate attention in environmental disciplines. This approach of decomposition of system dynamics to a "smooth" averaged motion and first-order Fourier harmonics we shall call subsequently first-order Fourier decomposition. It can be used while modelling any processes described by a quasilinear system of differential equations, if this system is subject to periodic disturbances (for example, due to the diurnal character of the processes modelled).

An idea of fourier decomposition

Let us consider an ordinary first-order differential quasilinear equation of the form

$$\frac{dA(t)}{dt} = K(A,t)A(t) + b(t),$$ (1)

where $A(t)$ is a dependent variable, and t is the independent variable representing time. Assume that $K(A,t)$ and $b(t)$ are subject to periodic changes in time (with unit period for simplicity). We are aware that the solution of (1) bears a definite periodic character. Therefore the following substitution can be made:

$$A(t) = A_0(t) + A_1(t)\sin(2\pi t) + A_2(t)\cos(2\pi t),$$ (2)

where $A_0(t)$ is a "slow motion" component or "averaged" dynamics; $A_1(t)\sin(2\pi t)$ and $A_2(t)\cos(2\pi t)$ are first-order Fourier harmonics of periodic movement.

Naturally we should assume that the components $A_0(t)$, $A_1(t)$ and $A_2(t)$ do not change significantly during one period of the process cycle; furthemore, we assume that $A_1(t)$, $A_2(t) \ll A_0(t)$, so we could disregard higher order terms of the decomposition in subsequent derivations. The term dA_0/dt we will treat as having the same order as $A_1(t)$, $A_2(t)$.

For a derivation of a set of decomposed equations, first let us substitute (2) into (1). The result is

$$\frac{dA_0}{dt} + 2\pi A_1\cos(2\pi t) - 2\pi A_2\sin(2\pi t) + ... = KA(t) + b,$$ (3)

where ... denotes terms of higher order than zero and first with respect to $A(t)$ and which will be disregarded in the following considerations. Secondly, let us integrate (3) three times over a period, i.e. interval from t to $t+1$, multiplied by 1, $\cos(2\pi t)$ and $\sin(2\pi t)$, respectively. We will get three new equations for our three new dependent unknowns, $A_0(t)$, $A_1(t)$ and $A_2(t)$. These equations can be sufficiently simplified if we treat them as constants during the derivation of the governing equations, utilising the basic assumptions above. We also suppose the following Taylor series decomposition to be valid:

$$K(A,t) = K(A_0,t) + \frac{\partial K}{\partial A}(A_0,t)[A_1(\sin 2\pi t) + A_2\cos(2\pi t)], \tag{4}$$

since the assumption was that $A_1(t)$, $A_2(t) \ll A_0(t)$. We get:

$$\pi A_1 = \int_0^1 \cos(2\pi t)[(A_0,t)dt + \frac{\partial K}{\partial A}(A_0,t)\{A_1\sin(2\pi t) + A_2\cos(2\pi t)\}] \times$$
$$\times [A_0 + A_1\sin(2\pi t) + A_2\cos(2\pi t)dt + \int_0^1 \cos(2\pi t)b(t)dt; \tag{5}$$

$$-\pi A_2 = \int_0^1 \sin(2\pi t)[K(A_0,t)dt + \frac{\partial K}{\partial A}(A_0,t)\{A_1\sin(2\pi t) + A_2\cos(2\pi t)\}] \times$$
$$\times [A_0 + A_1\sin(2\pi t) + A_2\cos(2\pi t)]dt + \int_0^1 \sin(2\pi t)b(t)dt; \tag{6}$$

$$\frac{dA_0}{dt} = \int_0^1 [K(A_0,t)dt + \frac{\partial K}{\partial A}(A_0,t)\{A_1\sin(2\pi t) + A_2\cos(2\pi t)\}] \times$$
$$\times [A_0 + A_1\sin(2\pi t) + A_2\cos(2\pi t)]dt + \int_1^0 b(t)dt . \tag{7}$$

Performing multiplication of terms in square parenthesis and skipping resultant terms of second and higher order with respect to the new dependent variables, it could be noted that, under the assumptions made, simplified equations (5) and (6) are linear with respect to A_1 and A_2. Furthemore, they are independent from equation (7) and therefore could be solved for A_1 and A_2 provided that we know A_0 and t. After (5)-(6) are resolved with respect to first-order components A_1 and A_2, we can substitute them into (7) and find dA_0/dt, completing integration of the "averaged" system.

This method easily extends with respect to increase in the number of dependent variables in the system. The dimension of the linear system equivalent to (5)-(6) in our example is always twice the number of independent variables; for example, for a three-equation system we must invert a 6x6 matrix to obtain first-order Fourier components. The only notable requirement is that the system under question should be aperiodic, i.e. does not exhibit oscillatory behavior. This is ascertained by the absense of eigenvalues of the corresponding characteristic matrix with positive real part. If this requirement does not hold, then techniques other than proposed here must be applied, since parametric resonances could require another substitution instead of (2). By the way, most mathematical models of environmental systems do not allow oscillatory solutions.

Let us reiterate the basic assumptions we made during derivation of the decomposed equations:

• The system of ordinary differential equations is quasilinear and aperiodic;

• The system is subject to periodic disturbances in coefficients and/or right-hand sides;

• Resulting periodical disturbances in dependent variables are much less than the average values.

2. Evaluation of the method

For demonstration and evaluation, the equation for phytoplankton growth without nutrients limitations was chosen. The same equation was used in the work of L.Somlyódy and L. Koncsos (1991) for studying the influence of sediment resuspension on algal growth in Lake Balaton (Hungary).

The equation for algae growth without limitation by nutrients reads:

$$\frac{dA}{dt} = k_g f_l f_T A - k_d \theta^{T - T_o} A, \tag{8}$$

where A is algal biomass in mg/l of chlorophyll-a,

f_l is light factor (defined with Steele function),
f_T is temperature reduction factor,
k_g is maximal growth rate,
k_d is mortality rate coefficient at temperature T_0,
T is actual water temperature,
Θ is mortality temperature coefficient,
and t is time.

Steele's function for light dependance factor f_l reads:

$$f_l = \frac{I}{I_S} \exp(1 - \frac{I}{I_S});$$
$$I(z) = I(0)\exp(-\varepsilon z); \tag{9}$$
$$\varepsilon = \varepsilon_0 + \alpha A,$$

where $I(0)$ is incident light intensity;
$I(z)$ is light intensity at a depth of z meters;
ε is the light attenuation factor in Lambert's formula;
I_S, α and ε_0 are empirical parameters.
Coefficient f_l was integrated from $z = 0$ to $z = H$ (waterbody depth) to represent average growth rate for the whole algae community. A triangular light pattern was used to represent diurnal changes in incident light intensity, with daily amount of radiation denoted as R.

The following values of parameters were used during simulations:

- A (t=0) = 0.03 mg/l of chlorophyll-a;
- k_g = 7.2 1/day;
- k_d = 1.8 1/day;
- T = 20°C.

Parameters of Steele function were set as follows:

- water depth H was set to 2.5 meters;
- phytoplankton self-shading constant α was set to 0.019 m² mg⁻³;
- $R/\lambda_T S$ ratio was set to 7.0;
- and background light attenuation coefficient ε_0 varied
from 0.5 to 2.0 1/m.

This set of parameters follows Somlyódy and Koncsos (1991).

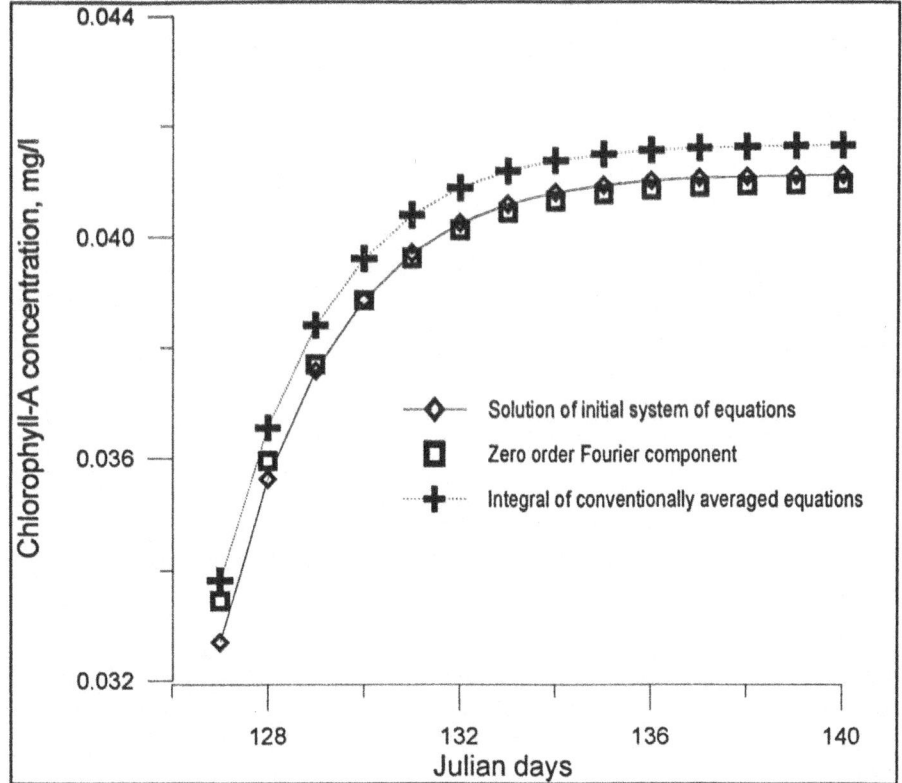

Figure 1. Integrals of the equation for phytoplancton growth. Background light attenuation factor 2.0 m⁻¹.

3.Simulation results

The results of the integration of equation (8) and its Fourier decomposition over a time of

15 days are presented on plots Fig. 1 - 6. In Figs. 1-3 the daily averaged values of the integral of eq. (8) (solved by Runge-Kutta fourth order method with an hourly time step) are plotted against the zero-order component of the decomposed system (integrated with daily time steps). Background light attenuation coefficient ε_0 was set subsequently to values of 2.0, 1.0 and 0.5 1/day (fig. 1, 2 and 3 respectively). The trajectory of the conventionally averaged equation (8) integrated with daily time step is also shown.

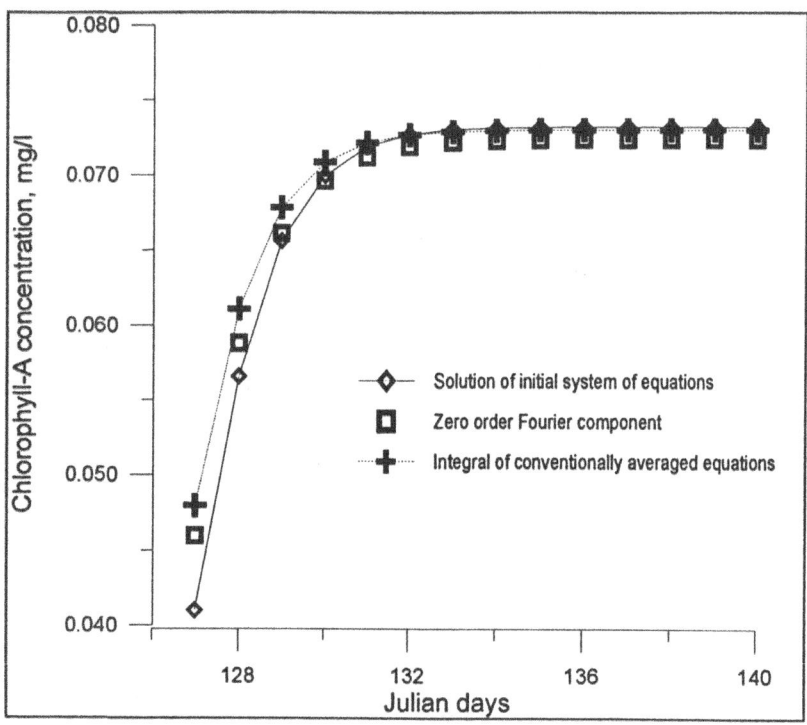

Figure 2. Integrals of the equation for phytoplancton growth. Background light attenu-ation factor 1.0 m^{-1}.

The plots demonstrate that the overall difference between all three trajectories in the case considered is small and negligible for any practical purpose (but this of course does not mean that this difference will always remain small for other cases/equations). However, the zero-order component of decomposed system tends to better represent the daily averaged values.

Fig. 4 shows the amplitudes of the first-order Fourier components as computed from the integral of eq. (8) from the set of linear equations (5)-(6). The overall accordance is suitable for practical purposes, although in the initial period the estimate of cosinusoidal the component is not pretty good. But since the cosinusoidal component is much less than the sinusoidal one, this discrepancy actually means only a slight shift in the phase of the solution.

Fig. 5 and 6 demonstrate evaluation of the diurnal effects with the use of the Fourier decomposition for two extreme values of the parameter ε_0. In this case diurnal components

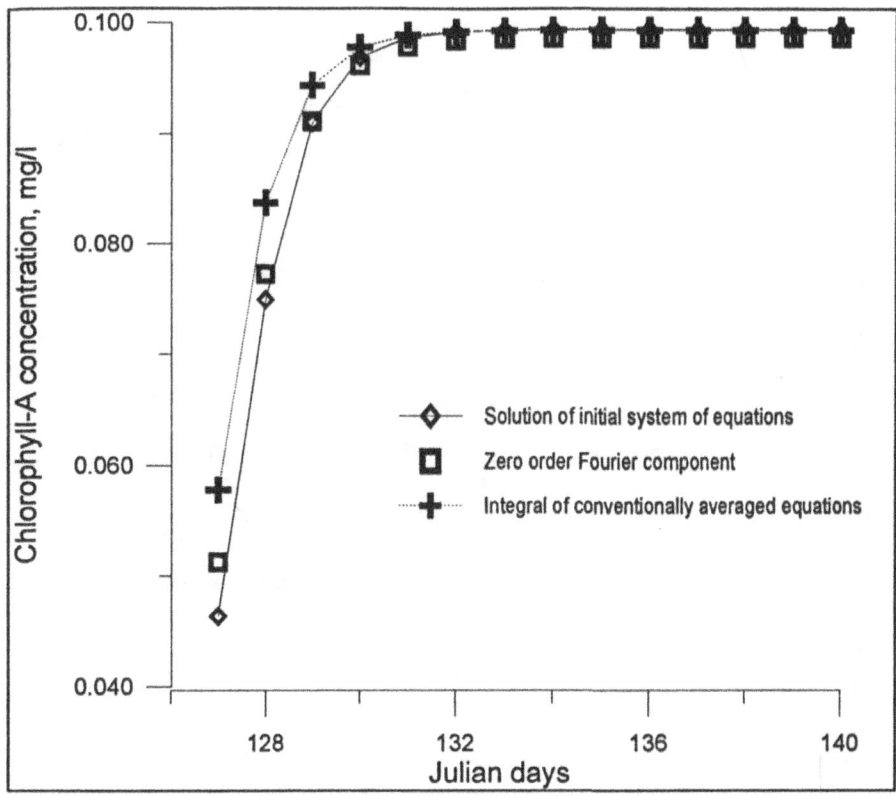

Figure 3. Integrals of the equation for phytoplancton growth. Background light attenuation factor 0.5 m⁻¹.

in algae biomass is considerable. The decomposition method is resolving the diurnal dynamics of algae growth with reasonable accuracy.

4. Conclusion

The application of a first-order Fourier decomposition method was demonstrated using the algae growth equation. From a practical point of view, the diurnal component of 30% in algae biomass dynamics is not exceedingly high, considering the existence of large measurement errors for this parameter. But if we take dissolved oxygen dynamics, then considerable diurnal changes can result in low night concentrations and even in periodical oxygen depletion which are highly unfavourable for fish survival. Other examples where diurnal changes in the system state variables are important could be found as well.

Precise computational gains from this method will vary significantly from one implementation to the other. In particular, they will depend upon the desired accuracy of the solution, the number of dependant variables in the model (which influence the dimension of the matrix to be reverted, see above), and on how efficiently the right-hand sides of equations (5)-(7) can be evaluated. Since the purpose of presenting an example of phytoplankton growth equation was primarily to illustrate usage of the Fourier decomposition method and its accuracy, it is felt that "benchmarking" computational comparisons of

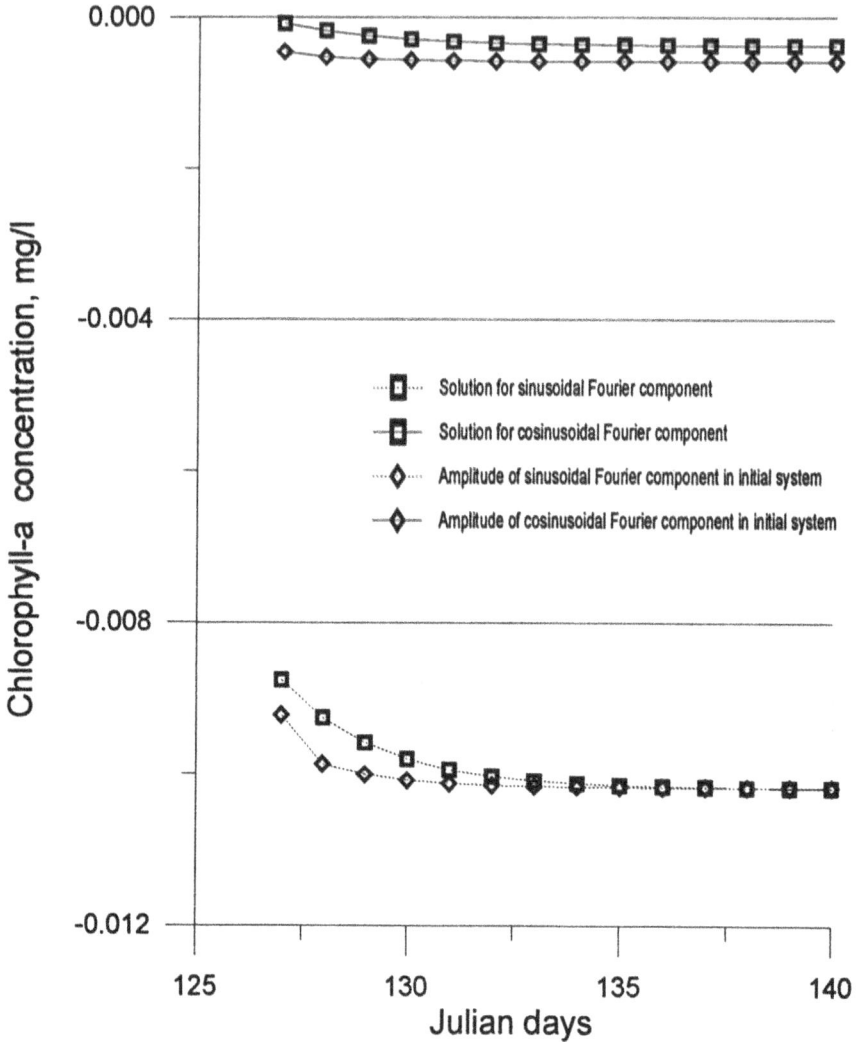

Figure 4. First-order Fourier amplitudes computed from the decomposed equations and from the integral of the initial equation.

this method with any of the conventional integration techniques would have little meaning in the scope of this methodological report.

The proposed method of decomposition of the dynamics of the system in question into "slow" and "fast" motion components is by no means new. Its usage often was succesful in other branches of science such as catchment hydrology, where Jakeman et al (1989) separated the streamflow responce to rainfall excess into a quick and slow flow compo-nent.. Hopefully it would be useful in the field of modeling nonlinear complex ecological and economical systems as well.

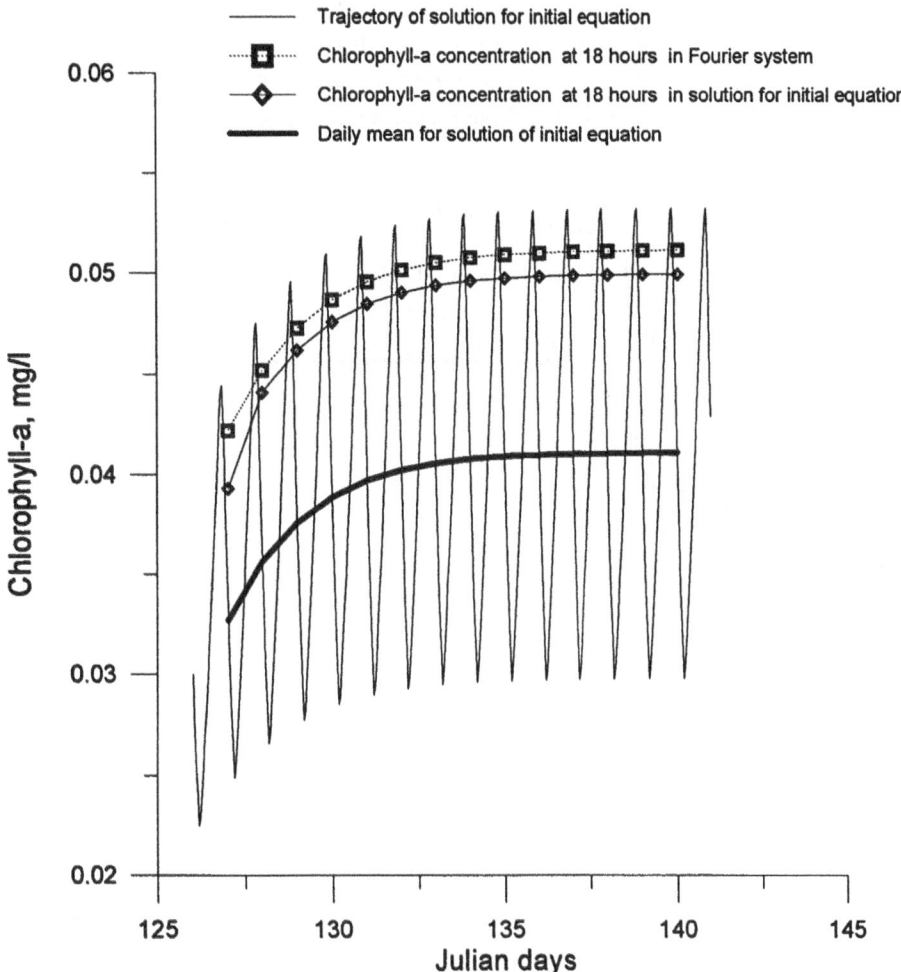

Figure 5. Diurnal effects and their simulation with decomposed equations. Background light attenuation factor 2.0 m⁻¹.

5. Acknowledgments

Author would like to thank Prof. O.F.Vasiliev and Prof. L.Somlyódy for encouraging, valuable suggestions and useful discussions of the material.

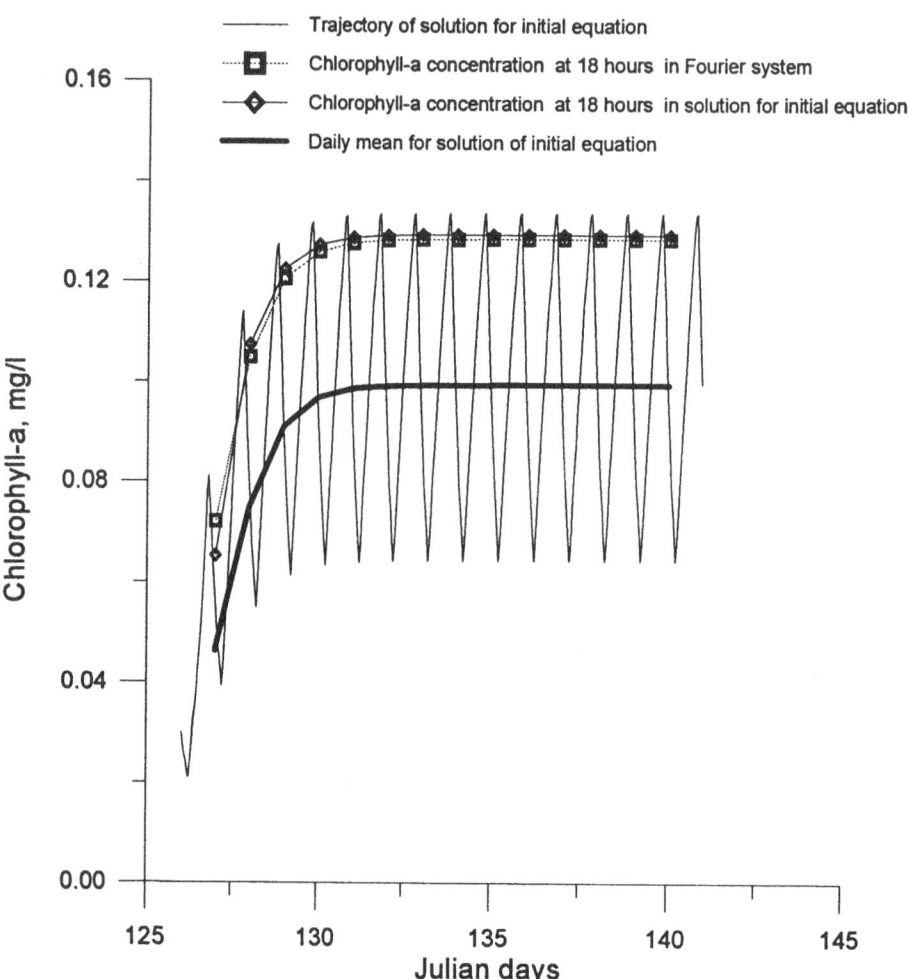

Figure 6. Diurnal effects and their simulation with decomposed equations. Background light attenuation factor 0.5 m⁻¹.

References

Bogolyubov N.N., Krylov N.M. (1937), 'Introduction to Nonlinear Mechanics', Kiev (in Russian).

M.B. Beck (1987), 'Water Quality Modelling: A Review of the Analysis of Uncertainty' *Water Res. Res.*, **23**, pp. 1393-1442.

L. Somlyódy, G. van Straten (eds), 'Modeling and Managing Shallow Lake Eutrophication' (1986), Springer-Verlag, Berlin.

L. Somlyódy and L.Koncsos(1991), 'Influence of sediment resuspension on the light conditions and algal growth in Lake Balaton', *Ecological Modelling*, **57**, pp. 173-192.

Thomann, R.V. and Mueller, J.A. (1987), 'Principles of Surface Water Quality Modeling and Control', Harper and Row, New York.

Jakeman A.J., Littlewood I.G., Whitehead, P.G. (1990), 'Computation of the instantaneous unit hydrohraph and identifiable component flows with application to two small upland catchments', *Journal of Hydrology*, **117**, pp.275-300.

MULTIOBJECTIVE INVERSE PROBLEMS WITH ECOLOGICAL AND ECONOMICAL MOTIVATIONS

OLEG I. NIKONOV

Dept. of Optimal Control, Institute of Mathematics and Mechanics,
Russian Acad. Sci., S.Kovalevskaya, 16, Ekatherinburg, Russia
e-mail: noi@imm.e-burg.su

Summary

The paper is devoted to some problems of multiobjective analysis which can be briefly outlined as follows. Two groups of restrictions on a variable x are given. The first one is treated as preassigned constraints and depends on the vector parameter ν, while the second group corresponds to controllable restrictions depending on the vector parameter μ. The following reciprocal problems are studied: find the set of controllable parameters m (the inputs of the system) which ensure the preassigned constraints on the variable x with n being given, and vice versa, specify the set of guaranteed values ν (the outputs) for the fixed μ. The paper deals with the structure, description and "extremal" elements of the above sets. Computerized implementation is also discussed. The questions under consideration are motivated by ecological and economical problems and closely related to those investigated in Kurzhanski (1986), Konstantinov (1983), Nikonov (1988), (1992).

1. Motivation

We begin with an example to motivate the mathematical problem formulation given below in section 2. Let $x = (x_1, ..., x_n)$ be a vector describing the emission of hazardous substances ejected by an industrial plant which causes an air (water) pollution. The components x_i, $i = 1, ..., n$ may stand e.g. for the concentrations of pollutants at the emission point. Assume there are several areas M_j $j = 1, ... M$, where the impact of pollution is measured. The ecological damage is described by a vector function $g(x) = (g_1(x),...,g_M(x))$, where $g_j(x)$ characterizes the level of toxic substances in the area M_j. For a fixed j the function $g_j(x)$ depends on various parameters (the distance between M_j and the ejection point, relief, climatic and meteorological conditions, hazardous substances etc.).This function may express the mean value of pollution for the given domain M_j or corresponding expenses for it's neutralization. Admissible level of the ecological damage is determined by the vector $\nu \in \mathbf{R}^M$ or, more precisely, by the inequalities

$$g_j(x) \le \nu_j , \qquad j = 1, ..., M. \tag{1.1}$$

Assume also, that there is a possibility to control the quantity of the fallouts e.g. by changing the production intensity, improving the cleaning technique, using other technologies etc. Let N be the number of the above "ecological" measures, and the measure k ($k=1,...,N$) results in restriction on the variable x of the form

$$\phi_k(x) \le \mu_k , \tag{1.2}$$

where $\phi_k(\cdot)$ is a characterization of the implemented action and the parameter μ_k corresponds to the intensity of the latter. (Here we suppose that decrease of the value μ_k implies reduction of the domain of all possible values x of hazardous parameters in the emission point area).

The following problem then arises: for a fixed level v of preassigned restrictions (1.1) to find the set of intensity (controllable) parameters $\mu = (\mu_1, ..., \mu_N)$ such that the fulfillment of corresponding restrictions (1.2) would guarantee the level v of ecological state determined by (1.1). The reciprocal problem can also be considered: with μ being fixed to specify the set of the level vectors v which are ensured by restrictions (1.2).

One can give an economical interpretation for the relations (1.1), (1.2) and the above problems assuming that the variable x characterizes the quantity of starting material for an industrial process. Relations (1.1) are then given constraints and the goal is to estimate (and maximize) the resulting product described by the vector function

$$\phi(x) = \phi_1(x),...,\phi_N(x)).$$

An additional point to emphasize is that the variable x may be not only a finite dimensional vector, but e.g. a state variable $x = x(t)$ governed by the differential equation

$$\frac{dx}{dt} = f(t,x),$$

with $x \in \mathbf{R}^n$ and $t \in [t_0, \Theta]$. For the details of this particular situation see Kurzhanski (1986), Nikonov (1988).

2. Mathematical Formalization

Analyzing the examples of the previous section we come to the following mathematical formalization of the problems considered above. Two groups of restrictions on the variable $x \in \mathbf{R}^n$ depending on the vector-valued parameters $\mu \in \mathbf{R}^N$ and $v \in \mathbf{R}^M$ are given

$$\phi_i(x) \le \mu_i , \quad i = 1, 2, ..., N \tag{2.1}$$
$$g_j(x) \le v_j , \quad j = 1, 2, ..., M \tag{2.2}$$

The first one is interpreted as the regulating constraints with regulating parameter μ (input of the system). The second group defines the quality of regulation, and vector v determines corresponding quality level (output). The following reciprocal problems are considered.

Problem 1 (inverse). With the quality level $v \in \mathbf{R}^M$ being given to specify set $M(v) \subseteq \mathbf{R}^N$ consisting of vectors μ with the properties:

a) system (2.1) has nonempty set of solutions
b) for any solution x to (2.1) the relations (2.2) are true.

Problem 2 (direct). For a given $\mu \in \mathbf{R}^N$ to specify the set $N(\mu)$ of the vectors $v \in \mathbf{R}^M$ such that for every solution x to (2.1) the relations (2.2) are true.

Denote the set of Pareto maximal and weak Pareto maximal (see e.g. Sawaragy *et al.* (1985)) points of $M(v)$ by $M^P(v)$ and $M^{WP}(v)$ respectively (with respect to ordering introduced by $\mathbf{R}_+^N = \mu \in \mathbf{R} : \mu_i \geq 0, i = 1, 2, ...,N\}$). Analogously $N^P(\mu)$ and $N^{WP}(\mu)$ will stand for Pareto minimal and weak Pareto minimal elements of the set $N(\mu)$.

Problem 3. Specify the sets $M^P(v)$ and $M^{WP}(v)$.

Problem 4. Specify the sets $N^P(\mu)$ and $N^{WP}(\mu)$.

We suppose the following assumption to be fulfilled.

Assumption 2.1 The functions $\phi_i(\cdot)$, $i = 1, ..., N$ and $g_j(\cdot)$, $j = 1,...,M$: $\mathbf{R}^n \rightarrow (-\infty, +\infty]$ are proper, convex and closed. The conditions: $\phi_i(x) \rightarrow +\infty$, $g_j(x) \rightarrow +\infty$ with $\|x\| \rightarrow +\infty$ hold.

Using the notations

$$A(\mu) = \{x \in \mathbf{R}^n: \phi(x) \leq \mu\} \tag{2.3}$$

$$B(\mu) = \{x \in \mathbf{R}^n: g(x) \leq v\} \tag{2.4}$$

where $\phi(x) = (\phi_1(x),...,\phi_N(x))$, $g(x) = (g_1(x),...,g_M(x))$, we can define the sets to be specified by the relations

$$M(v) = \{\mu \in \mathbf{R}^N: \varnothing \neq A(\mu) \subseteq B(v)\} , \tag{2.5}$$

$$M^P(v) = \{\mu^P \in M(v): \mu \geq \mu^P \ \& \ \mu \neq \mu^P \Rightarrow \mu \bar{\in} M(v)\} , \tag{2.6}$$

$$M^{WP}(v) = \{\mu^{WP} \in M(v): \mu > \mu^{wp} \Rightarrow \mu \bar{\in} M(v)\} , \tag{2.7}$$

$$N(\mu) = \{v \in \mathbf{R}^M: B(v) \supseteq A(\mu)\} , \tag{2.8}$$

$$N^P(\mu) = \{v^P \in N(\mu): v \leq v^P \ \& \ v \neq v^P \Rightarrow v \bar{\in} N(\mu)\} , \tag{2.9}$$

$$N^{WP}(\mu) = \{v^{wp} \in N(\mu): v < v^{wp} \Rightarrow v \bar{\in} N(\mu)\} . \tag{2.10}$$

Here the symbols "$\leq$", "$<$" are used in the following sense: for $y,z \in \mathbf{R}$

$$y \leq z \leftrightarrow y_i \leq z_i \ \forall i \in \overline{1,p}; \ y < z \leftrightarrow y_i < z_i \ \forall i \in \overline{1,p}.$$

Thus the problems 1-4 can be restated in terms of relations (2.5)-(2.10).

3. Structure of solutions

We start with the set $F = \{\mu \in \mathbf{R}^N: A(\mu) \neq \varnothing\}$. Let F^P and F^{WP} be the sets of Pareto minimal and weak Pareto minimal elements of F. Under conditions of Assumption 2.1 the following assertions are true.

Proposition 3.1.
i) The set F is nonempty, convex and closed.
ii) The equality holds:

$$F = F^P + \mathbf{R}_+^N = F^{WP} + \mathbf{R}_+^N \tag{3.1}$$

From the equality (2.11) we conclude that the set F is determined by it's Pareto minimal points. The next proposition gives a parametric description of the latter.

Proposition 3.2. The inclusions are true:

$$F^P \subseteq \{\mu \in \mathbf{R}^N: \mu = \phi(x) | x \in X^P\} \subseteq F^{WP} ,$$

where X^P is the set of solution to the inclusion

$$O \in co \bigcup_{i=1}^{N} \partial\phi_i(x) .$$

Here $\partial\phi_i(x)$ is a subdifferential (Rockafellar (1970)) of the function $\phi_i(\cdot)$, symbol co stands for the convex hull.

Remark. If $\phi_i(x) = \dfrac{1}{2}(x - a_i)^T A_i(x - a_i)$ with positive definite symmetric matrices A_i, then

$$X^P = \{x \in \mathbf{R}^n: x = \sum_{i=1}^{N} \alpha_i A^{-1}(a) A_i a_i | \Sigma a_i = 1\} ,$$

where

$$A(a) = \sum_{i=1}^{N} a_i A_i, \ a \in \mathbf{R}_+^N .$$

In particular, if $A_i = E$, $i = 1, ..., N$, then $X^P = co \{a_i \mid i = 1, ..., N\}$.

Coming to the problem of description the sets $\mu(v)$ we restrict ourselves by the strictly convex case.

Assumption 3.1. Functions $\phi_i(\cdot)$, $i = 1, ..., N$; $g_j(\cdot)$, $j = 1, ..., M$ are strictly convex and finite-valued.

We shall use the following notations: $L_f(\omega) = \{x \in \mathbf{R}^n: f(x) \leq \omega\}$ stands for the level set of a scalar or vector function $f(\cdot)$; for $\mu \in \mathbf{R}^N$ denote

$$\mu \backslash k = (\mu_1, ..., \mu_{k-1}, \mu_{k+1}, ..., \mu_N) \in \mathbf{R}^{N-1} ,$$

$$A^k(\mu \backslash k) = \{x \in \mathbf{R}^n : \phi_i(x) \le \mu_i \ \forall \ i \in \overline{1,N}, \ i \ne k\}.$$

A set $M \subseteq \mathbf{R}^N$ will be called $\mathbf{R}^N_+$ - convex, if for any pair $\mu^{(1)}$, $\mu^{(2)} \in M$ such that $\mu^{(1)} \le \mu^{(2)}$ from $\mu^{(1)} \le \mu \le \mu^{(2)}$ it follows that $\mu \in M$.

Proposition 3.1. The sets $M(\nu)$ are $\mathbf{R}^N_+$ -convex and closed.

Remark that $M(\nu)$ may be empty, nonconvex and unbounded.

Theorem 3.1. Under assumptions 2.1 and 3.1 the set $M(\nu)$ is nonempty iff $\nu \in Y^P + \mathbf{R}^N_+$, where where $Y^P = \{v \in \mathbf{R}^M : v = g(x) \mid x \in X^P\}$ and X^P is defined in Proposition 3.2.

For $\mu \in \mathbf{R}^N$ and $\nu \in \mathbf{R}$ define the functions

$$\Psi_k(\mu \backslash k, \nu) = sup\{\mu_k \mid \phi \ne L_{\phi_k}(\mu_k) \cap A^k(\mu \ k) \subseteq B(\nu)\}; \ k = 1,...,N.$$

These functions may have $+\infty$ and $-\infty$ as the values. By definition we put $\Psi_k = -\infty$ if supremum is calculated over empty set.

In the following theorems the assumptions 2.1, 3.1 are supposed to be fulfilled.

Theorem 3.2. The following representation is true

$$M(\nu) = \{v \in F : \mu_k \le \Psi_k (\mu \backslash k, \nu)\}$$

whichever $k,=,1,2,...,N$ is taken.

Denote by $\Omega_k(\nu)$ the graph of the function $\Psi_k(\mu \backslash k, \nu)$:

$$\Omega_k(\nu) = \{\mu \in \mathbf{R}^N : \mu_k = \Psi_k (\mu \backslash k, \nu)\}.$$

Theorem 3.3. Only one of the cases may occur

either (1) $\Omega_k(\nu) = \phi \ \forall k = 1,...,N$ and $M(\nu) = \phi$

or   (2) $\Omega_k(\nu) \ne \phi \ \forall k = 1,...,N$ and $M(\nu) \ne \phi$.

In terms of the graphs $\omega_k(\nu)$ a description of the sets $M^P(\nu)$ can be done.

Theorem 3.4. The following equalities hold

$$M^{WP}(\nu) = cl \ \bigcup_{k=1}^{N} \Omega_k(\nu) \ ; \ M^P(\nu) = \bigcap_{k=1}^{N} \Omega_k(\nu).$$

The last theorem deals with description of the sets $N(\mu)$, $N^P(\mu)$, $N^{WP}(\mu)$ which have more simple structure.

Theorem 3.5. If for a given $\mu \in \mathbf{R}^N$ the set $A(\mu)$ is nonempty, then $N^P(\mu) = \{v^0\}$, where $v_j^0 = \min\{g_j(x) \mid x \in A(\mu)\}$. Moreover, in this case we have

$$N(\mu) = v^0 + \mathbf{R}_+^M; \qquad N^{WP}(\mu) = v^0 + (\mathbf{R}_+^M \backslash \mathrm{int}\ \mathbf{R}_+^M).$$

Otherwise, if

$$A(\mu) = \phi\ , \text{ then } N(\mu) = \mathbf{R}^M.$$

For the proofs of the above theorems see Nikonov (1992).

4. Illustrations

First we consider a very simple example to illustrate the constructions of the previous section. All the sets and functions defined above can be determined in this case by explicit formulae. Let us have

$$N = 2,\ \phi_i(x) = \|x - a_i\|^2,\ x \in \mathbf{R}^2\ ,\ M = 1,\ g(x) = \|x - a_g\|^2.$$

Then solution to the problems 1, 3 for this particular case depends on the relative positions of the points a_1, a_2, a_g and on the value v. The complete solution can be obtained by consideration of several variants. We consider one of them. Denote

$$r_1 = \|a_1 - a_g\|^2,\ r_2 = \|a_2 - a_g\|^2,\ r = \|a_1 - a_2\|^2$$

and suppose the inequalities $r_1 < v < r_2$ to be fulfilled. Then we have

$$X^P = [a_1, a_2\backslash = \{x \in \mathbf{R}^2 : x = \lambda a_1 + (1 - \lambda)a_2 \mid 0 \le \lambda \le 1\},$$

$$F^P = \{\mu \in \mathbf{R}^2 : \sqrt{\mu_1} + \sqrt{\mu_2} = \sqrt{r}\ ,\ \mu_1 \ge 0,\ \mu_2 \ge 0\},$$

$$F = \{\mu \in \mathbf{R}^2 : \sqrt{\mu_1} + \sqrt{\mu_2} \ge \sqrt{r}\ ,\ \mu_1 \ge 0,\ \mu_2 \ge 0\}.$$

Functions $\Psi_1(\mu_1,v)$ and $\Psi_2(\mu_2,v)$ can be constructed by definition:

$$\Psi_2(\mu_1, v) = \begin{cases} +\infty, & 0 \le \mu_1 \le (\sqrt{v} - \sqrt{r_1})^2, \\ g(\mu_1), & (\sqrt{v} - \sqrt{r_1})^2 < \mu_1 \le \|q - a_1\|^2, \\ -\infty, & \mu_1 > \|q - a_1\|^2 \text{ or } \mu_1 < 0, \end{cases}$$

$$\Psi_1(\mu_2, v) = \begin{cases} -\infty, & 0 \le \mu_2 < \|a_2 - q\|^2 \text{ or } \mu_2 < 0, \\ q^{-1}(\mu_2), & \|a_2 - q\|^2 \le q\|^2 \le \mu_2 \le a_2 - p\|^2, \\ (\sqrt{v} - \sqrt{r_1})^2, & \mu_2 > \|a_2 - p\|^2. \end{cases}$$

Here p and q are the points of the cercumference $\|x - a_g\|^2 = v$ which belong to $[a_g, a_1]$ and $[a_1, a_2]$ respectively, $g(\mu_1) = \|z_{\mu_1} - a_2\|^2$, where z_{μ_1} is the point of intersection the above cercumference and that defined by $\|x - a_1\| = \mu_1$ lying in the angle pa_1q. The set $M(v)$ determined for the this case by relations

$$\sqrt{r} - \sqrt{\mu_1} \le \mu_2 \le \xi_2(\mu_1, v), \mu_1 \ge 0 \qquad (4.1a)$$

or

$$\sqrt{r} - \sqrt{\mu_2} \le \mu_1 \le \xi_1(\mu_2, v) \mu_2 \ge 0 \qquad (4.1b)$$

is unbounded and nonconvex (see Fig.1).

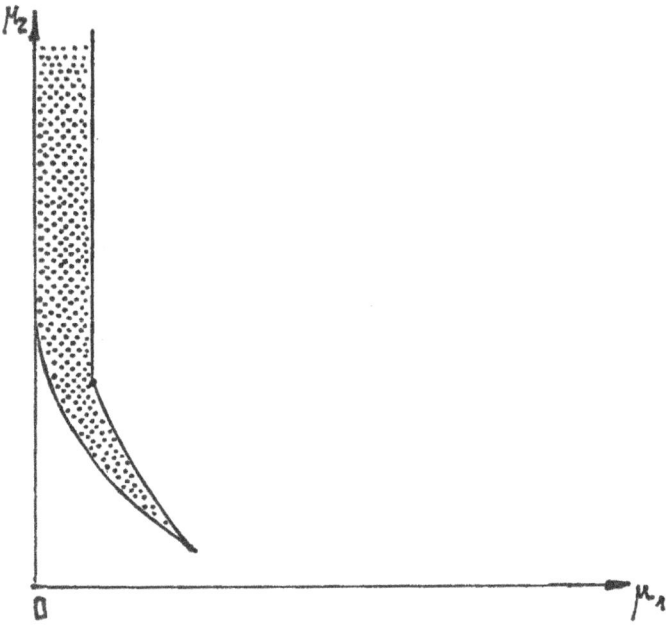

Figure 1. The set $M(v)$ for relations (4.1ab)

Theorems 3.2-3.4 concerning the structure and description of the sets $M(v)$, $M^P(v)$, $M^{WP}(v)$ provide a principal possibility of algorithmization the procedure of construction the mentioned sets. Such an algorithm has been developed and realized. Details are given in (Nikonov, 1988 and 1992). Figure 4 demonstrates the level sets of the functions ϕ_1, ϕ_2, g corresponding to the point μ^*, see Fig. 2.

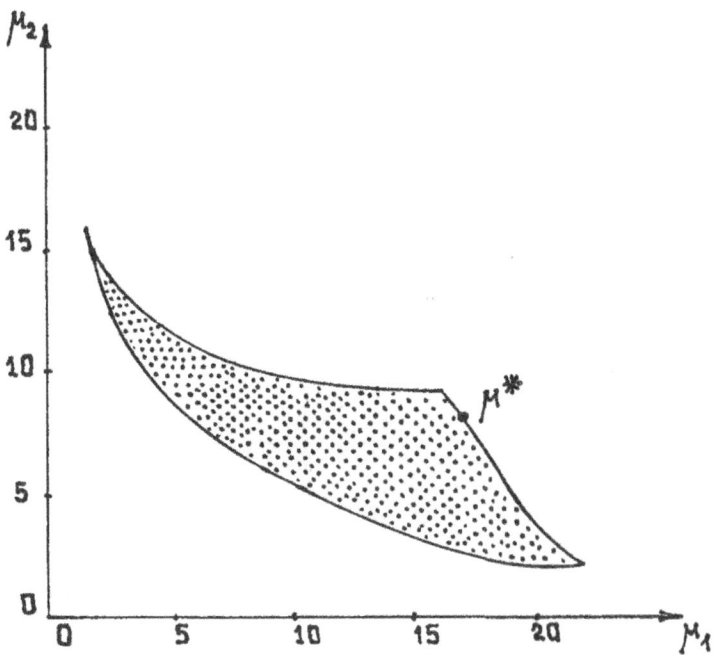

Figure 2. The set $M(v)$ for $N = 2$, $\phi_1(x) = 3x_1^2 + x_2^2$, $\phi_2(x) = (x_1 - 4)^2 + 2(x_2 + 3)^2$, $M = 1$, $d(x) = 8(x_1 - 1.5)^2 + (x_2 + 2.3)^2$, $v = 9.0$.

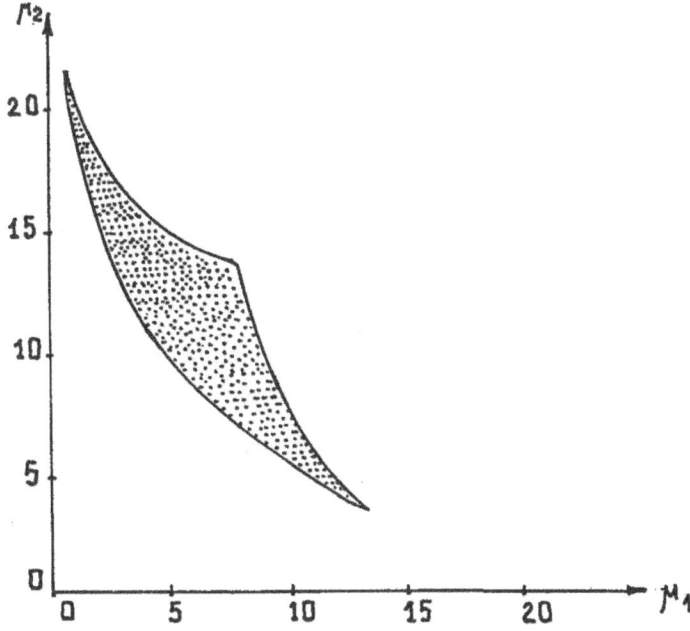

Figure 3. The set $M(v)$ for $N = 2$, $\phi_1(x) = 2x_1^2 + x_2^2$, $\phi_2(x) = (x_1 - 4)^2 + 2(x_2 + 3)^2$, $M = 1$, $g(x) = 8(x_1 - 1) + (x_2 + 1.8)^2$, $v = 4.5$.

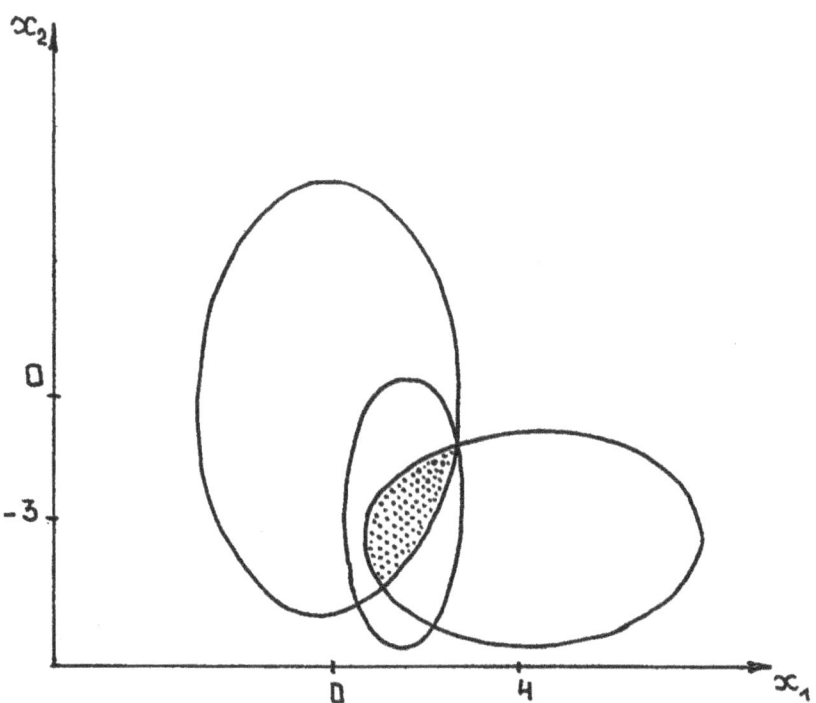

Figure 4. Level sets of the functions ϕ_1, ϕ_2 and g.

Futher applications of the presented technique can be done for uncertain dynamic systems using the methods developed in Krasovski (1985), Kurzhanski *et al.* (1977), (1979), (1990).

References

Konstantinov, G.N. (1993), 'The normalizing of the actions on dynamic system' Irkutsk. (Russian).

Krasovski, N.N. (1985), 'The control of a dynamic system', Moscow, Nauka (Russian).

Kurzhanski, A.B. (1977), 'Control and estimation under uncertainty', Moscow, Nauka. (Russian).

Kurzhanski, A.B. (1986), 'Inverse problems in multiobjective dynamic optimization: Proc VII Intern. Conf. on Multiple Criteria Decision Making, Kyoto', *Lect. Notes Econ. and Math. Syst.*, **285**, 374-382.

Kurzhanski, A.B., Filippova, T.F. (1989), 'On the set valued calculus in problems of viability and control for dynamic process: the evolution equation', *Ann. Inst. H.Poincare. Anal. non-lineaire*, 339-364.

Kurzhanski, A.B., Nikonov, O.I. (1990), 'Funnel equation and multivalued integration for control synthesis', *Perspectives in Control Theory: Proc. Sielpia Conf.*, Boston, Birkhauser, 143-153.

Nikonov, O.I. (1988), 'On the problem of guaranteed control with ector-valued performance criterion: Proc. Intern. Conf. on Multiobjective Problems of Math. Programming,
Yalta', *Lect. Notes Econ. and Math. Syst.*, **351**, 51-58.

Nikonov, O.I. (1992), 'On the structure and algorithms of multicriteria inverse problems',
Estimation and Identification of uncertain systems, Ekatherinburg, 167-187. (Russian).

Rockafellar, R.T. (1970), 'Convex analysis', Princeton Univ. Press.

Sawaragi, Y., Nakayama, H., Tanino, T. (1985), 'Theory of multiobjective optimization',
New York, Acad.Press.

AN EXPERT-OPINION APPROACH TO THE PREDICTION PROBLEM IN COMPLEX SYSTEMS

GERBRAND J. KOMEN

Royal Netherlands Meteorological Institute (KNMI)
P.O. Box 201, 3730 AE De Bilt, The Netherlands
e-mail: komen@knmi.nl

Summary

The use of model forecasts for decision making should be optimized. With this in mind, the concept of modelling the future is discussed from an epistemological point of view and on the basis of a stochastic model interpretation. Traditional definitions of model statistics make reference to an ensemble of systems. Since this does not work for a complex system with a unique state, an alternative approach, based on the subjective (Delphi) opinion of a group of experts, is also considered. This approach is then generalized to the situation in which a set of competing models is available. With a Delphi method a certain likelihood can be assigned to each model. Once the statistics is defined, one may face the issue of predictability. In hindsight (in a 'hindcasting mode') models can be validated by checking how accurate they have been describing observations and they can be falsified when their predictions differ in an unlikely way from the observations. 'Forecasting' is different, because models can never be proven. Therefore, exact prediction of the future is impossible. Definitions of predictability (two examples will be given) necessarily refer to the range of modelled possibilities. It is argued that all model predictions - also those resulting from physical models - should be considered as scenarios. To make rational decisions the likelihood of all possible model forecasts has to be taken into account. In case of complex systems and difficult decisions it appears useful to consider a large variety of models. Experts need not strive for consensus, because a diversity of opinions could lead to better decisions. It is recommended that more attention is paid to Delphi aspects of forecast likelihoods.

Keywords Predictability, complex models, likelihood, expert-opinion.

1. Motivation

The present discussion - although rather general and applicable to many different systems - was inspired by questions related with climate modelling. As is well known the atmospheric concentration of CO_2 and other Greenhouse gases increases due to human activities. It is expected that this will lead to a disturbance of the natural climate. This has led to political discussions, and to an increase of interest in (numerical) climate models.

In the development of these models two trends can be seen. The most advanced physical models of the coupled atmosphere-ocean system are still inadequate in describing (details of) the present climate. Therefore, one seeks improvement to obtain better description of the actual climate and more reliable 'predictions' of climate change. Improvements are expected to come from an enhancement of the spatial resolution, and from the use of more realistic sub-grid scale parametrizations, such as describing for example the cloud-radiation interaction or the effect of ocean waves on air/sea exchange. An example of this approach is given by Washington and Meehl (1989). A summary of similar approaches is given in the IPCC report (1990, 1992).

However, it is realized that chemical, biological and socio-economical factors are also crucial for a correct description of the anthropogenic effects on climate. To model these, simple physical models have been coupled with chemical, biological and socio-economic models (Rotmans, 1990).

Opponents of these latter models criticize by pointing out that the physical subsystem is modelled rather inaccurately, whereas the accuracy of the other subsystems is even less well known. Proponents argue that the interaction of the physical subsystem with the rest cannot be ignored.

This note tries to sort out the underlying assumptions, in an attempt to take away the prevailing confusion. The ideas are not new, but it is hoped that presenting them here may help stimulate the discussion.

Forecast models are often interpreted as stochastic models, predicting probabilities. Therefore, we begin with a discussion of these models. Next we will discuss the concept of probability, which is essential for understanding predictability and the meaning of model forecasts.

2. Stochastic models

Consider first a closed model system, which can be described by n prognostic variables X, and *for which we know the law of evolution M* apart from the value of (a set of) parameters and forcing variables A:

$$M(A): X(t_0) \rightarrow X(t). \tag{1}$$

The evolution operator M is nonlinear, in general, and acts on the state vector X to compute the state at a later time. In physical models it usually results from the discretization of a set of (integro-)differential equations. To be specific X may be thought of as the positions and velocities of interacting point particles; A would be their masses and M would be given by classical mechanics.

In this approach X and A are random variables (X is a 'stochastic process', see for example, Doob, 1953), so they define probability distributions f_X and f_A with the property that

$$f_X(x)dx \tag{2}$$

is the probability that X has a value between x and $x + dx$, and similarly for f_A.

In practice, we are often dealing with very complex systems with many degrees of freedom *for which we have only limited knowledge about the laws of evolution*. An example is an atmospheric model, in which case X would represent air pressure, velocity, density, temperature, etc. of the atmosphere on a finite difference grid on the globe at specified levels. But X can also be much larger: it could include ocean and sea-ice variables, the chemical composition of the atmosphere, emission rates, oil price, inflation rate, deforestation rate, etcetera. Typically X has $10^6 - 10^8$ components or more.

One of the central problems in modelling is the correct choice of the (high dimensional vector) space S, in which to describe the phenomena of interest. The choice of parameters

A and dynamical variables *X* requires a careful analysis of the system and the objectives of the study. In one approach the CO_2 emission rate could be prescribed as a parameter; in another model it could be treated as a dynamical variable depending on energy price, population growth and other variables. Often, the distinction between dynamic variables and parameters is somewhat artificial. Therefore, it is interesting to compare the perform-ance of models in different spaces *S*. This is most easily formulated by realizing that (1), for a given choice of dynamical variables, can also be interpreted as a collection of mappings (models) $\{M_a\}$ labelled with the possible values *a* of *A*, each with their given probability distribution. The generalization to the case of variable size of *X* and *A* is obtained by considering the collection of (all) proposed models $\{S_\alpha, M_\alpha\}$ specified by the mappings

$$M_\alpha : X(t_0) \rightarrow X(t), \ X \in S_\alpha . \tag{3}$$

This collection comprises (1), but it is more general because it simultaneously includes models in which the system is represented by state vectors in different spaces S_α. The label α is not a scalar. It has one component labelling the different models (i.e. sets of differential equations); the other components label the possible values of *A*. One may think of atmospheric grid point models with different spatial (and temporal) resolution but also of more complex biosphere models which do or do not include certain variables that may or may not be marginal for understanding the system (blue algae or dimethylsulfide concentration, for example).

3. The definition of probability

The definition of probability comes with a conceptual difficulty. Traditionally, one considers the abstract concept of an ensemble of realizations of the system, each with different values for the random components. When predicting the trajectory of a billiard-ball one may specify the uncertainty in its initial position by performing many independent measurements. This defines the ensemble - many billiard-tables with the balls initially in slightly different positions - and the required probability distribution. In weather prediction the distribution of *A* of (1) can be obtained by considering many *similar* situations in the past, and by determining the corresponding realizations *a* of *A*.

However, this procedure cannot be followed in very complex models of the world, because for all we know our world is quite unique at a given time, and therefore it is simply not possible to find *similar* situations. In these cases it is still possible to use statistical concepts, albeit on a rather different basis.

First, for simplicity, consider models of the type (1) and assume that the initial conditions are accurately known, so that the uncertainty is in *A*. To obtain the desired probability distribution one could then organize a simple Delphi-like procedure (see remark below) in which the opinions of experts are used. This can be done in infinitely many ways, but to be specific consider the case where each expert is asked to give the value of *a* that he finds most likely. From this then follows a probability distribution of *A*, not based on a direct objective analysis of past events, but expressing the subjective assessment of the expert panel. For a given panel this specifies the model uniquely. Of course, the width of the distribution of *A* contains interesting information, a narrow distribution indicating consensus, a broad one meaning disagreement. Also, one could form different panels and compare the results along the rules of test theory (see for example,

Crocker and Algine, 1986).

The same procedure may be applied to models of the type (3). Again there is an infinity of possible methods from which one must select and again the specific questions asked to the experts would be an integral part of the definition. An example? Suppose one considers six 'models', consisting of two different (sets of) differential equations, each with one undetermined constant which may assume three different values. Then one can ask the experts to rank them in order of likelihood of describing correctly the change in mean global sea level in the second half of the 21st century. The probabilities so defined specify the likelihood of the models. They are a (subjective) measure expressing the confidence experts have in the correctness of a particular model out of a given set. In this case it is even more interesting than before to know the width of the distribution and to apply test theory to the results of different expert groups.

In my opinion the traditional physical approach should be used whenever possible, in particular for the modelling of physical subsystems, because the results of this approach are much more likely to be correct. The gravitational acceleration at the surface of the earth will be approximately 9.8 m/s^2 also in 2050. However, to estimate the interest rates in that year the Delphi procedure could be helpful, because it quantifies the subjective estimate of the uncertainty in this parameter and allows for numerical modelling of the consequences of its uncertainty.

Remark In its original meaning the name Delphi method (see e.g. Linstone and Turoff, 1975) refers to a technique developed for obtaining judgements from a group of experts. Characteristics are feedback, anonimity and statistical presentation. Here we use the term in a loose way. Most conventional applications strive for consensus. The present application leaves room for feedback, but consensus is not necessary.

4. The use of models

The models can be used in two ways *independently from how the input probabilities have been defined.*

In the first type of model application ('hindcasting') one simulates the past and compares model results with observations. These observations - let us denote them by O - are themselves stochastic variables because of measurement errors. The so-called model counterparts of the observations O^m are functions of the model state vector X. Model I is usually considered to be better than model II in describing a particular observation O_i when the mean and variance (one must choose how to weight) of a time (or spatial) series $O_i - O_i^m(X)$ are smaller for I than for II. If the difference between O_i and $O_i^m(X)$ is unlikely large the model is falsified. One may then attempt to construct a better model and, in fact, it is along these lines that the understanding of the system is enlarged. In practice, one usually compares time series of particular realizations of O and O^m. In true stochastic models one could also attempt to validate the predicted probability distributions by comparing them with the distribution of observations obtained by averaging over analogous situations, but this is not usually done, and, in fact, as discussed in section 3, this would be impossible in a unique complex system.

In the second type of application ('forecasting') one forecasts the future. The justification is most easily expressed for traditional deterministic systems: (A) a system that evolves according to model M will lead to a state $x(t)$; (B) suppose that reality behaves as such a system; then (C) in reality we expect state $x(t)$ to occur at time t. Obviously, (B) is an assumption and for this reason at the moment the prediction is made there is no way of establishing its future correctness. One may hit a billiard-ball in the right direction, expecting with 100% certainty that the desired collision will occur, but the actual collision may never come, because unexpectedly the billiard-table collapses due to woodworm.

For a stochastic model the argument would go as follows: (A) a system that evolves according to the set of models $\{M_\alpha\}$ will lead to a state $X(t)$ (a set of states x_α each with a certain likelihood); (B) suppose that reality behaves as such a system; then (C) in reality we may expect to find state $x_\alpha(t)$ at time t with corresponding likelihood. (The meaning of this expectation in unique complex systems would differ from the conventional physical meaning). But, again, at the moment the prediction is made there is no way of establishing its future correctness. Our sophisticated stochastic climate predictions may be jeopardized by the unexpected penetration of a large meteorite through the ocean bottom.

5. Predictability

Let us first consider models of the type (1). We cannot, in general, compute a definite value for $X(t)$, because neither $X(t_0)$ nor A is known exactly. However, on the basis of statistical information on $X(t_0)$ and A, we can make a statement about the probability of finding a certain realization $x(t)$. Given the probability characteristics of A and X the probability distribution of $X(t)$ follows:

$$M: (f_A(a), f_{X(t_0)}(x)) \to f_{X(t)}(x) . \tag{4}$$

Equation (3) is realized most easily with the help of a Monte Carlo simulation. In such a simulation one generates values for A and $X(t_0)$ on the basis of their probability distributions. For each set a, $x(t_0)$ one solves (1) which then automatically generates the probability distribution of $X(t)$.

In this context predictability can be defined as the inverse width of the probability distribution of $X(t)$, or more precisely, for a particular observable O_i, as the inverse width of $f_{O_i(X_t)}(o_i)$. In chaotic systems this width gets quickly very large, even if the initial condition and the values A are known rather accurately. With this definition of predictability, it is quite possible for certain observables to be more predictable than others. Note that this concept of predictability necessarily refers to the range of *modelled* possibilities.

For models of the type (3) the same arguments can be given. Once a likelihood has beens assigned to each M_α, it is, in principle, straightforward to compute the corresponding likelihood distribution of a predicted observable O_i^m.

It is fascinating to speculate about something more profound. To this end consider a hierarchy of stochastic models $\{S_n, M_n\}$ (now labelled with a single integer n), more and more refined with ever more dynamical variables. One might order them according to complexity, say

$$\text{dimension } S_{n_1} \geq \text{dimension } S_{n_2} \quad \text{if} \quad n_1 > n_2 \tag{5}$$

Each model will predict a random state $X_n(t)$ from which the corresponding probability distribution of some observable $O_i^m[X_n(t)]$ is readily computed. We can then consider a sequence of model predictions $\{O_i^m[X_n(t)]\}$ and we would call an observable predictable if

$$\lim_{n \to n_{max}} O_i^m(X_n) \tag{6}$$

exists, with n_{max} the label of the most complex model considered. The basic idea is that some observables are sensitive to additional complexities of the system, whereas others are stable when you make the model more complex. To my knowledge such sequences have not been studied. I expect that for a given sequence of models some observables will be predictable and others will not, whereas for each observable one can construct a set of models in which this observable is not predictable. I have not attempted to prove these conjectures. They seem to be related to our inability to know the future and could perhaps explain why some observables are much harder to predict than others.

6. The psychology of decision making

In daily life, as in physics, one always makes 'models' of the future. These do not need to take the form (1) or (3), but they have in common with (1) and (3) that they are representations of the system in the neural network of the human brain, and that their predictions of the future need not come true. For instance, you want to go out and you expect that it will rain, so you pick up an umbrella. But then the model turns out to be inadequate, because there may be so much wind that you cannot use the umbrella or another unexpected event (not 'modelled') may prevent you from leaving the house.

Often people make decisions with a more or less explicit concept of the desired future situation in mind. They then select from possible courses of action by using (numerical) models to estimate the effect of these decisions on the future. From the foregoing discussion it will be clear that these models do not 'know' the future, but they can be used to generate possible future states with a certain likelihood. The argument then goes: if we do this, then there is a probability that such and so. After scanning all possible decisions they make that decision that brings them with the highest likelihood closest to the desired situation (see for example Lindley, 1985). It is interesting to note that decisions are always made on the basis of models, never on the basis of knowledge of the future state.

7. Conclusions

- A Delphi approach makes it possible to use stochastic mathematical models for forecasting, even for a complex system in which certain parameters are not known from experience.

- We have seen that it is not possible to base decisions on *knowledge* of the future. In fact, one always uses 'models' to generate scenarios with an attached likelihood. If the model predictions do not come true, decisions may have taken us away from the desired state, rather than that they have brought us closer to it.

- Physical models of simple, (nearly) closed systems have a very large likelihood.

- In case of complex systems, and difficult decisions, it appears useful to consider a large variety of models.

- Experts need not strive for consensus. A diversity of opinions could lead to better decision making, because it allows one to consider a larger number of possible futures, which reduces the risk of overlooking an important scenario with a small likelihood (a bifurcation in the thermohaline circulation of the world ocean, for example).

- It might be useful to pay more attention to details of Delphi procedures, such as the selection of (groups of) experts and the formulation of the questions asked.

8. Concluding remark

This note sketched a rationalistic approach to decision making, which pretends that one can control the system, to some extent at least. Of course, there are other approaches as well. For example, some people - they may be called ethicists - act according to certain principles ('thou shalt not lie') which they follow, irrespective of where it takes them. True modellers can not be stopped by this. On the contrary, they will be inspired to add two kinds of human actors to their models - rationalists and ethicists - and they would attempt to model the consequences of the interaction between these actors and the rest of the system.

Acknowledgements

My interest in the subject was stimulated by discussions with Henk Tennekes, who read a draft of this paper, as did Arie Kattenberg, Theo Opsteegh and Wieger Fransen. I would like to thank them all, for their suggestions and encouragement. The idea of the expert-opinion approach to likelihood came up during a discussion with Klaus Hasselmann, when we drove from Tallinn to Tartu.

References

Crocker, L. and J. Algine, (1986), 'Introduction to classical and modern test theory', *Holt, Rinehart and Windston, Inc.*
Doob, J.L. (1953), 'Stochastic processes', Wiley, New York.
IPCC, (1990), 'Climate change'. The IPCC scientific assessment, Report prepared for IPCC by working group 1; Cambridge University Press.
IPCC, (1992), 'Climate change 1992'. The supplementary report to the IPCC scientific assessment; Cambridge University Press.
Lindley, D.V. (1985), 'Making decisions', Wiley, New York.
Linstone, H.A. and M. Turoff (Eds) (1975), 'The Delphi method', *Techniques and Applications*, Reading, Mass.
Rotmans, J. (1990), 'Image: an integrated model to assess the Greenhouse effect'. Proefschrift Rijksuniversiteit Limburg.
Washington, W.M. and G.A. Meehl, (1989), 'Climate sensitivity due to increased CO_2: Experiments with a coupled atmosphere and ocean general circulation model', *Climate Dyn.* **4**, 1-38.

CRITICAL LOADS AND A DYNAMIC ASSESSMENT OF ECOSYSTEM RECOVERY

J.-P. HETTELINGH[1] and M. POSCH[2]

[1]National Institute of Public Health and Environmental Protection,
Coordination Center for Effects, P.O.Box 1,
NL-3720 BA Bilthoven, The Netherlands
[2]Water and Environment Research Institute, P.O.Box 250,
FIN-00101 Helsinki, Finland

Abstract

A dynamic soil acidification model (SMART) is used to assess the impact of the so-called 50%-gap-closure scenario on the state of forest soils in Europe. This scenario aims to reduce the excess deposition of sulfur over the critical loads by at least 50% everywhere in Europe by the year 2000 at minimal emission reduction costs and is currently under discussion as a basis for a new UN/ECE sulfur protocol. The concentration of aluminum in soil solution is used as an indicator for potential damage due to acidifying deposition. The time required for soils to recover, i.e. to reach an aluminum concentration of less than 0.2 eq/m3, is computed and mapped on a 150x150 km2 grid covering Europe. Results show that the implementation of the 50%-gap-closure scenario will "protect" an additional 10% of the European forest soils; however, they also indicate that deposition of acidifying nitrogen compounds have to be reduced as well in order not to create new areas with an elevated risk to ecosystem damage.

Keywords Acidification, critical loads, dynamic modeling, soil chemistry

1. Introduction

Revised protocols on the reduction of sulfur dioxide and nitrogen oxides are being prepared within the framework of the Convention on the Long-Range Transboundary Air Pollution (LRTAP) of the United Nations Economic Commission for Europe (UN/ECE). Reductions are being negotiated on the basis of the so-called critical load approach. The definition of a critical load is "a quantitative estimate of an exposure to one or more pollutants below which significant harmful effects on specified sensitive elements of the environment do not occur according to present knowledge" (Nilsson and Grennfelt, 1988). Critical loads have been mapped for the whole of Europe. The aim of the critical load approach is to achieve protection of ecosystems e.g., forest soils and surface waters, by sufficiently reducing emissions of acidifying compounds such as sulphur dioxide (SO_2), nitrogen oxide (NO_x) and ammonia (NH_3). Emission reductions are considered sufficient when acidic deposition does not exceed critical loads. The requirement to protect all European ecosystems in Europe, however, is not likely to be met for economical, political and technical reasons. In order to assess the influence of excess deposition, knowledge is needed of the dynamics of the geo-chemical processes in forest soils and surface waters.

The policy which is currently under debate within the framework of UN/ECE-LRTAP is to reduce sulfur emissions such that the excess of sulfur deposition over critical loads is diminished by 50% by the year 2000 and 2005 in Western and Eastern Europe, respectively. This reduction scenario is called the "50%-gap-closure" scenario. This paper analyzes the effects of the 50%-gap-closure scenario on European forest soils with the aid of a dynamic soil acidification model.

2. Methods and data

Modeling of acidification of forest soils described in this chapter addresses three questions: (1) what is a cost optimal distribution of sulfur emission reductions in European countries when a particular distribution of depositions over European forest soils is targeted (e.g. the 50%-gap-closure scenario); (2) are the emission reductions sufficient for long-term protection of forest soils in Europe, and (3) if not, when and where is damage likely to occur or, conversely, when and where will forest soils recover. The methodology used for these three research issues is addressed below.

2.1 Optimal reductions of sulfur emissions

The Regional Acidification, INformation and Simulation model (RAINS; Alcamo et al., 1989) is used to compute national cost- effective emission reductions in comparison to sulfur emissions in 1990. RAINS consists of: (1) an energy emission module which computes national emissions as a function of energy combustion (SO_2, NO_x) and agricultural practice (NH_3); (2) a deposition module computing acidic depositions in 150x150 km^2 grid cells using the EMEP transport model (Eliassen and Saltbones, 1983); (3) an effects module comparing acidic depositions to critical loads (Hettelingh et al., 1991; 1992a). RAINS allows for scenario analysis (computing deposition levels given an energy/emission pattern) or optimization (Amann, 1989; 1991). In the optimization mode minimum cost emission reductions over Europe as a whole can be computed subject to any particular set of deposition levels prescribed over 150x150 km^2 cells covering Europe (EMEP-grid). The optimization mode was used to find cost optimal sulfur emission reductions resulting in a 50% reduction of the present excess of critical loads in 2000. The levels of NO_x and NH_3 emissions were assumed not to change in comparison to 1990 emission levels, to allow for the analysis of the effects of sulfur reductions only. Due to the long-range transport of air pollutants the result of the optimization implies a reduction of sulfur deposition by more than 50% in some areas and even reductions in parts of Europe where the critical loads are presently not exceeded. Therefore forest soils might recover in certain areas due to the 50%-gap-closure scenario, and this will be investigated below.

2.2 The excess of critical loads by acid deposition

The potential detrimental effects of the deposition of acidifying compounds on forest soils in Europe is currently assessed by a static modeling approach. In this approach, which is usually referred to as the Steady-State Mass Balance (SSMB) method, a steady state is assumed in soil chemical processes such that the ability of soils to buffer acidity is not impaired. In practice, the objective is to avoid a selected chemical soil value being exceeded at critical load. A detailed description of SSMB can be found elsewhere (see Sverdrup et al.,1990; Hettelingh and de Vries, 1992; Hettelingh et al., 1992a). In general terms the SSMB model can be written as:

$$Y_{CL} = f(X_{crit})$$ (1)

where,

Y_{CL} =   critical load
X_{crit} =   critical chemical value of soil variable X.

and the function f describes the balance between acidifying and neutralizing ions. Different critical chemical values have been suggested such as a pH of 4, a molar base cation to aluminum ratio of 1 and an aluminum concentration of 0.2 eq m^{-3}. Only the latter one is used in this paper for characterizing the status of European forest soils. Critical loads have been computed for forest soils and surface waters on the EMEP grid. Data are either based on national contributions (see Hettelingh et al., 1991) or, when national data are lacking, on a European data base of forest soils reported in de Vries et al. (1993a).

2.3 Dynamic simulation of future acidification

The time development of the damage to or recovery of ecosystems due to different deposition scenarios on large spatial and temporal scales is generally assessed with dynamic models (see Hettelingh et al., 1992b). With this approach the time development of values of soil variables is simulated as a function of acid deposition and the previous state of the soil. In general terms this can be expressed as:

$$X_t = g(Y_t, Y_{t-1}, ..., X_{t-1}, X_{t-2},...)$$ (2)

where,

X_t = value of soil variable at time t
Y_t = deposition at time t

and g is a non-linear functional relationship describing the physical and chemical processes in the soil solution. The model used in this paper is the Simulation Model for Acidification's Regional Trends (SMART) by de Vries et al. (1989) and de Vries (1991). SMART is a simple, dynamic, process-oriented model based on the charge balance principle. The model is designed dynamic and process-oriented to allow for the analysis of long-term behavior of soils, and simple in order to minimize input data requirements for applications on a regional scale. The soil solution chemistry in SMART depends solely on the net element input from the atmosphere (deposition minus net uptake minus immobiliz-ation) and the geochemical interaction in the soil (CO_2 equilibria, weathering of carbon-ates, silicates and Al-hydroxides and cation exchange). The solute transport is described by assuming complete mixing of the element input within one homogeneous soil layer with a constant density and a fixed depth. The annual water flux percolating from this layer is taken to be equal to the annual precipitation surplus (precipitation minus interception minus evapotranspiration) by assuming that forests obtain all their water from the top layer. Since SMART is a single layer model neglecting the vertical heterogeneity, it predicts the concentration of the soil water leaving the root zone. A detailed model description can be found in Posch *et al.* (1993).

For the analysis described in this paper a number of preparations were made to obtain the initial conditions for SMART in 1990. Between 1840 and 1990 SMART was initialized as follows: Historical sulfur depositions were linearly interpolated between 1840 and 1900 assuming a background sulfur deposition of 0.01 g m^{-2} yr^{-1} for 1840. Historical sulfur depositions between 1900 and 1960 were estimated on the basis of (1) historical emissions (Fjeld, 1976), (2) the RAINS source receptor relationship, and (3) the assumption that the distribution over European countries of emissions is the same as in 1960. Historical depositions due to NO_x and NH_3 emissions have been linearly interpolated between 1840 and 1960 making similar assumptions as for sulfur. Acid depositions between 1960 and 1990 have been computed using national emission data for reference years and interpolating in between. Finally, the initial conditions reflecting the state of the soil (i.e. carbon to nitrogen ratio, base saturation) in 1840 were obtained from the SSMB model using actual critical loads.

The sensitivity of the SMART output, i.e. the concentration of aluminum in soil solution, to the initial conditions was tested by running SMART from 1840 to 1900 using different low (non ecosystem endangering) depositions, i.e. (1) actual critical loads and (2) the 1840 background deposition, and they were kept constant between 1840 and 1900. The resulting aluminum concentration in 1900 was similar in both cases, and this shows that the predictions of future aluminum concentration are independent of the details of the conditions in pre-industrialized times.

2.4 Ecosystem recovery under the 50%-gap-closure scenario

RAINS, SSMB and SMART were used to analyze the impacts on European forest soils of the 50%-gap-closure scenario which is assumed to become effective in 2000 with emissions linearly interpolated between 1990 and 2000. Nitrogen deposition was assumed to follow current reduction plans to the year 2000 and stay constant thereafter. The time horizon of the simulation is 80 years (1990-2070). Here we are interested in the lag between the time at which deposition changes and the time at which the aluminum concentration in the soil solution attains a 'safe' value (critical chemical value). This time lag - during which a risk of damage to ecosystems occurs due to the chemical condition of soils - is denoted as Damage Time Lag (DTL). In formula:

$$DTL = \min \{t\text{-}1990 : [Al]_t < [Al]_{crit} = 0.2, \ t = 1990,...,2070\} \qquad (3)$$

If the aluminum concentration still exceeds 0.2 eq m^{-3} in 2070 then the DTL is said to be undetermined (longer than 80 years).

3. Results

The result of the optimization of national cost subject to the deposition pattern of the 50%-gap-closure scenario is shown in Figure 1. It shows that the excess of critical loads for sulfur by sulfur deposition is restricted to northern, western and central Europe.

The largest excess occurs in the border area of Poland, Czech and Slovak republic (more than 1000 eq *ha*$^{-1}$ *yr*$^{-1}$). Note that the excess in the Nordic countries, in particular in Norway, largely reflects the sensitivity of surface waters. The result of the cost- minimal emission reduction leads to deposition which is lower than critical loads for sulfur in 81 % of the European area, which therefore may be considered "protected" against sulfur damage. However, the area of critical load excess increases when nitrogen deposition is included, and the magnitude of the excess in the area at risk is increased. An assessment of DTLs reveals to what extent the 50%-gap-closure contributes to recovery in the "protected" area and how important nitrogen reductions are. A map of DTLs related to the 50%-gap-closure scenario including nitrogen deposition is shown in Figure 2.

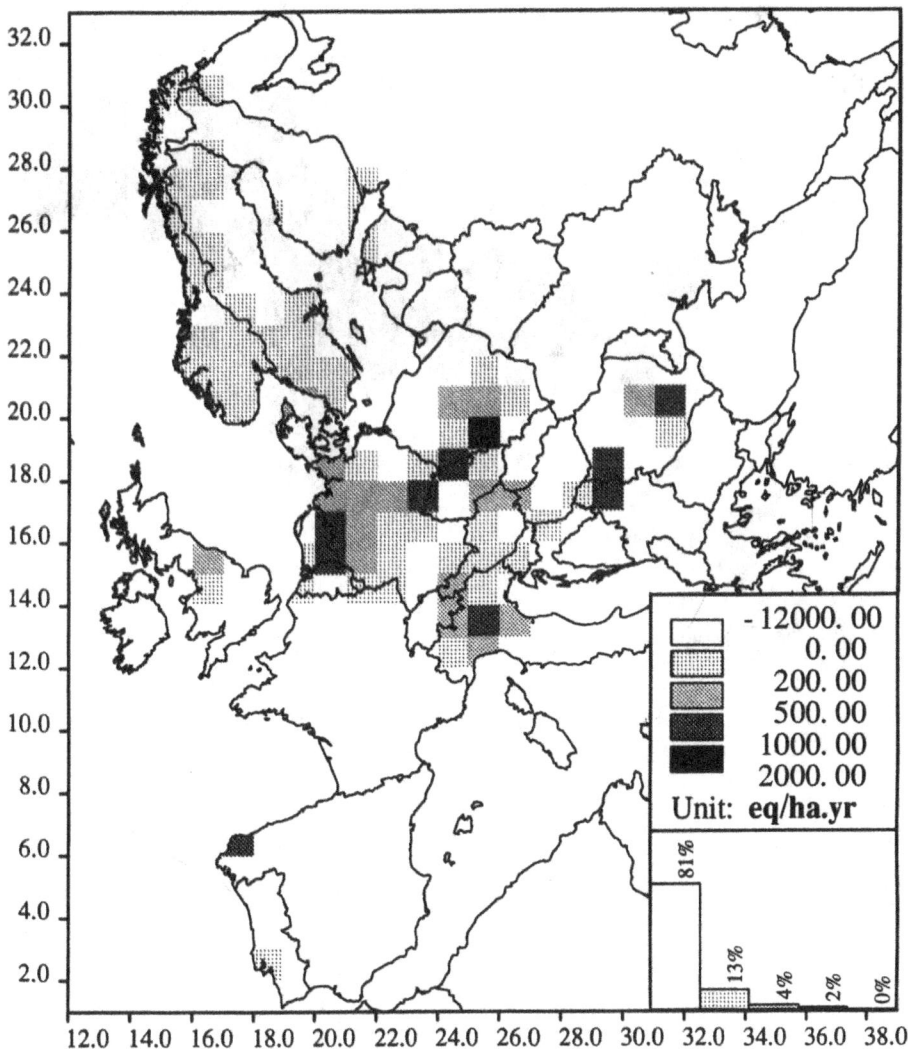

Figure 1. The excess of sulfur deposition over critical loads in 2000 resulting from the minimization of European costs for national reductions of sulfur emissions according to the 50%-gap-closure scenario.

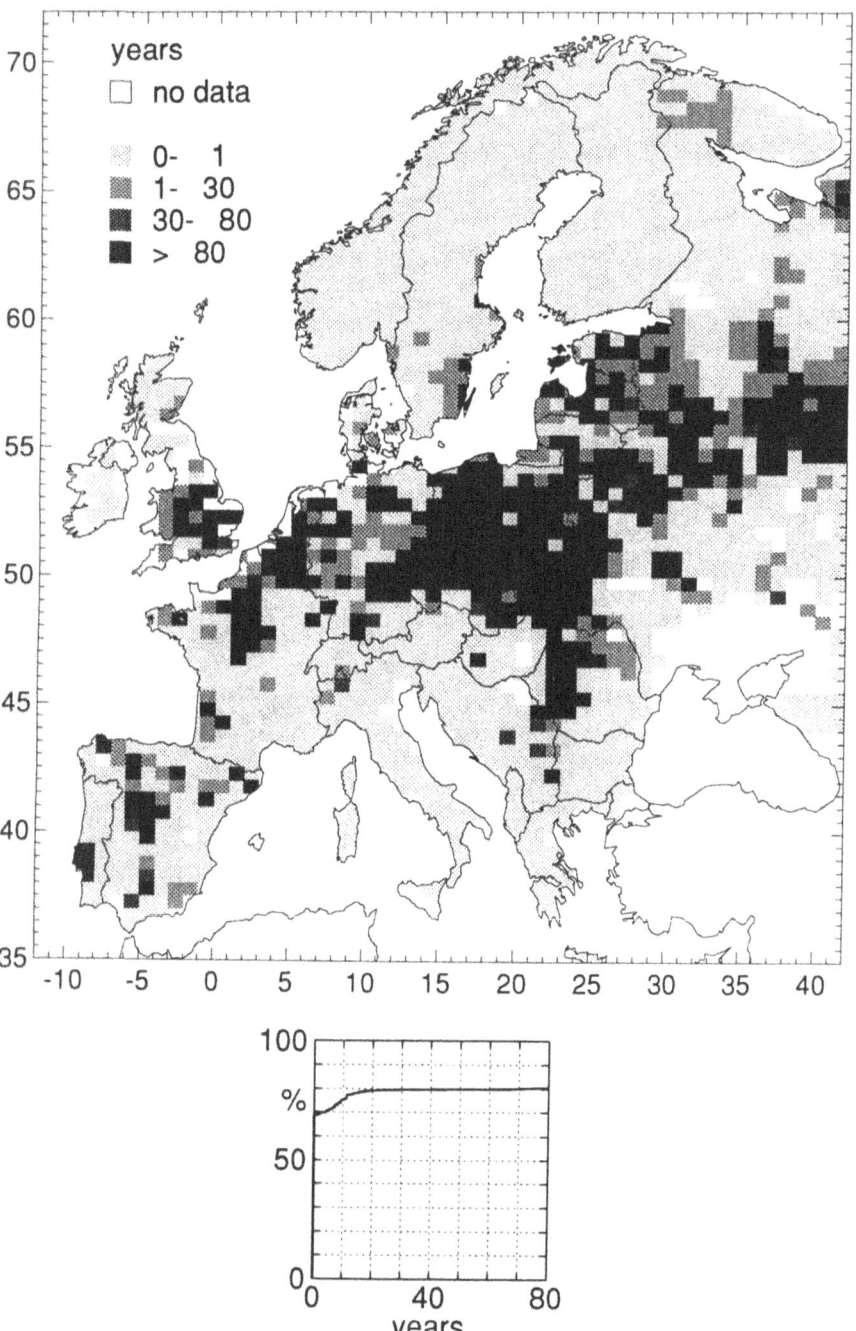

Figure 2. Damage Time Lags (DTLs) in the period 1990 to 2070 as a result of the 50%-gap-closure scenario becoming effective in 2000, assuming 1990 nitrogen deposition not to change till 2070.

The cumulative graph in figure 2 shows that 70% of the forest soils are not at risk or immediately recovered, i.e. DTL < 1 year, and an additional 10% of the area recovers within 20 years. However, the map of Figure 2 shows that parts of the "protected" area shown in Figure 1, i.e. in Spain, France and Russia will be at risk, because the DTL is undetermined (DTL > 80 years). The increase of DTLs in these regions is a result of the nitrogen deposition which was assumed constant between 2000 and 2070. DTLs may change as emissions of sulfur and nitrogen change over time. The time over which emission reductions are phased in and the individual effects of different pollutants become important in addition to the amount reduced. The conclusion is that it is not sufficient to restrict the evaluation of abatement policies to a comparison of deposition and critical loads alone. Knowledge of DTLs is required in addition to critical loads to prescribe also the timing of emission reduction strategies.

4. Conclusions and final remarks

This paper describes the assessment of emission reductions by comparing depositions to critical loads in combination with an estimate of the Damage Time Lag (DTL), i.e. the lag during which the risk of damage to ecosystems remains despite deposition reductions. It is shown that it is not sufficient to restrict the evaluation of emission reductions merely by investigating whether the magnitude of the reduction leads to depositions which are lower than critical loads. Knowledge of DTLs allows an assessment of the rate at which reductions should be made effective. This is especially true when more pollutants are involved and sufficient and timely reductions can not simultaneously be realized.

Note that the analysis described in this paper should be repeated for other critical parameters, other receptors such as surface waters, and should also include uncertainty analysis (see also the sensitivity analysis of SMART in de Vries *et al.*, 1993b).

Finally, the analysis of potential damage by investigating DTLs and critical loads will increasingly become important as the effects of combined stress (e.g., more pollutants, direct effects, climate change, acidification) become subject to policy measures. The reason is that the contributors to combined stress have varying time delays over which the effects and potential damage becomes imminent. This will be subject of future studies.

Acknowledgements

The authors are grateful to two anonymous referees of this chapter for their helpful comments.

References

Alcamo J., R. Shaw, L. Hordijk (eds.). (1989), *'The RAINS Model of Acidification: Science and Strategies in Europe'*, Kluwer Academic Publishers, Dordrecht.

Amann, M. (1989), 'Energy use, emissions, and abatement costs', In: Alcamo et al. *The RAINS Model of Acidification: Science and Strategies in Europe*, Kluwer Academic Publishers, Dordrecht.

Amann, M. (1991), 'Zur effizienten multinationalen Allokation von Emissions-minderungsmaßnahmen zur Verringerung der sauren Deposition - Anwendungsbeispiel Österreich', *Academic Dissertation*, University of Karlsruhe.

Eliassen, A. and J. Saltbones (1983), 'Modelling of long range transport of sulphur over Europe: A two year run and some model experiments', *Atmos. Environ.* **17**, 1457-1473.

Fjeld, B. (1976), 'Forbruk av fossilt brensel i Europa og utslipp av SO_2 i perioden 1900-1972', NILU Teksnik Notat 1/76 (in Norwegian).

Hettelingh, J.-P. and W. de Vries (1992), 'Mapping Vademecum', *report*, National Institute of Public Health and Environmental, Protection, Coordination Center for Effects, Bilthoven, The Netherlands.

Hettelingh, J.-P., R.J. Downing, P.A.M. de Smet (eds.). (1991), 'Mapping Critical Loads for Europe', *report*, National Institute of Public Health and Environmental Protection, CCE technical report no. 1, The Netherlands.

Hettelingh, J.-P., R.J. Downing, P.A.M. de Smet (1992a), 'The critical load concept for the control of acidification', In: T.Schneider (ed.), *Acidification research: evaluation and policy applications*, Elsevier, Studies in Environmental Science 50, pp. 161-174, Amsterdam.

Hettelingh, J.-P., R.H. Gardner, L. Hordijk (1992b), 'A Statistical Approach to the Regional Use of Critical Loads', *Environmental Pollution*, **77**, 177-183.

Nilsson, J. and P. Grennfelt (eds.). (1988), Critical Loads for Sulphur and Nitrogen, *Report from a workshop held at Skokloster*, 19-24 March 1988, Nordic Council of Ministers, Miljørapport 1988:15, Copenhagen.

Sverdrup, H., W. de Vries, and A. Henriksen (1990), 'Mapping Critical Loads: A Guidance Manual to Criteria, Calculation, Data Collection and Mapping', Nordic Council of Ministers, *Miljørapport 1990:14*, Copenhagen.

Vries, W. de, M. Posch, J. Kämäri (1989), 'Simulation of the long term soil response to acid deposition in various buffer ranges', *Water, Air and Soil Pollution* **48**, 349-390.

Vries, W. de (1991), 'Methodologies for the assessment and mapping of critical loads and of the impact of abatement strategies on forest soils', The Winand Staring Centre for Integrated Land, Soil and Water Research, *report 46*, Wageningen, The Netherlands.

Vries, W. de, G.J.Reinds and M.Posch (1993a), 'Assessment of critical loads and their exceedance on European forest using a one-layer steady state model', *Water Air and Soil Pollution*, (in press).

Vries W. de, M.Posch, G.J.Reinds, J.Kamari (1993b), 'Long term soil response to acidic deposition in Europe', submitted.

Posch M., G.J.Reinds, W.de Vries (1993), 'SMART-A Simulation Model for Acidification's Regional Trends: Model Description and User Manual', *Mimeograph Series of the National Board of Waters and the Environment 477*, Helsinki, Finland.

UNCERTAINTY ANALYSIS ON CRITICAL LOADS FOR FOREST SOILS IN FINLAND

MATTI P. JOHANSSON[1] and PETER H.M. JANSSEN[2]

[1] Water and Environment Research Institute
P.O.Box 250, 00101 Helsinki, Finland
e-mail: johanssonm@vyh.fi

[2] National Institute of Public Health and Environmental Protection (RIVM)
P.O.Box 1, 3720 BA Bilthoven, The Netherlands
e-mail: cwmpj@rivm.nl

Summary

The aim of the study is to present a comprehensive and quantitative estimation on the uncertainty of critical load values and their exceedances in a regional study for Finland. The critical loads are used to set goals for future deposition rates of acidifying compounds such that the environment is protected. In this study the critical loads for forest soils are determined using a steady-state mass balance approach. The critical load for a particular receptor varies from site to site, depending on its inherent sensitivity; its allocation for acidifying sulphur and nitrogen also depends on the deposition patterns. A software package UNCSAM (UNCertainty analysis by Monte Carlo SAMpling techniques) developed in RIVM has been used as a flexible tool for the analysis. The analysis presented here focuses on the estimation and effect of input parameter uncertainties. The study covers all relevant input parameters without preceding screening. The uncertainties are due to measurement errors or difficulties in the interpretation of measurement results. The effects of the uncertainties of the model structure and dose response assumptions are not included in this study. The uncertainties are calculated both for different areas in Finland and aggregated for the whole country. The uncertainties discovered are reasonable compared to the largest uncertainties of input parameters. The most influential parameters are shortly described both for the whole country and in spatial distribution.

1. Introduction

The flat rate reduction schemes of acidifying deposition in Europe are gradually yielding to abatement strategies, which are based on targeted cost-effective reductions. The concept of critical loads has been developed to serve as a potential basis for these policies. Critical loads describe the inherent tolerance of an ecosystem against acidifying deposition. They are usually based on steady state mass balance methods (Hettelingh et al., 1991). The method requires relatively modest input data, thus allowing its convenient regional application in many countries. The critical loads are calculated separately for each ecosystem considered, e.g. surface waters and forest soils, and then combined to the chosen map grid.

Many studies on uncertainty analysis of acidification models often concentrate on specific sites (examples in e.g. Beck, 1987). Analyses on the regional variability are still rare (De Vries, 1991) and deal mostly with water quality (e.g. Hettelingh et al., 1992;

Posch *et al.*, 1993; Kämäri *et al.*, 1993). On a regional level it is often hard to intuitively estimate the most influential parameters. The importance of various parameters also may change depending on the subregion e.g. due to changing geological properties, however, methods to overcome this problem have been suggested (Hettelingh, 1990; Hettelingh *et al.*, 1992).

This paper aims to assess the regional uncertainty of the critical loads and their exceedances for forest soils in Finland. An attempt has been made to take into account the uncertainties of all input parameters. The uncertainty analysis is first carried out on one site using the ranges of absolute input parameter values found in Finland. Then a regional analysis is carried out using over thousand plots with estimated uncertainty ranges for input parameters in each plot separately.

The potential sources of uncertainty in the assessment of the critical load calculations are the chemical criteria, the model structure and the estimation of the parameters. The chemical criterion of molar aluminium to base cation ratio is used to describe the potential detrimental effects of acidic deposition on the ecosystem (Ulrich, 1983; Tamminen and Starr, 1990; Hettelingh *et al.*, 1991). The value of this criterion is taken as given and not included in the uncertainty study. The model structure is also assumed to sufficiently describe the observed system and is therefore not considered in the uncertainty analysis. However, there may be some processes or mechanisms, which are not included in the current model version but could affect the results. The specification of the individual input parameters for the model and their uncertainties is the central part of this study. Depending of the examined parameter, the uncertainty can arise from poor measurement quality (e.g. total potassium content in biomass), erroneous interpretation of the phenomenon (e.g. base cation deposition) or lack of knowledge (e.g. spatial variability of forest filtering factors).

The distributions for parameter errors or uncertainties should preferably be based on statistical data. However, for many of the parameters an extensive measurement programme producing such data can be difficult in practice. An often used assumption of e.g. normally distributed errors for a parameter is in many cases difficult to justify due to lack of supporting data. Therefore, throughout this study symmetric triangular distributions have been used to portray the distribution of parameter value errors. The uncertainty in the parameter values is expressed in terms of the coefficient of variation. The min-max range for the assumed symmetric triangular distribution can then be derived from it.

The lack of sufficient data has also inhibited the quantitative correlation of some input parameters, e.g. the co-deposition of sulphate and ammonium. When the input parameter values are taken independently of each other, the resulting uncertainty distribution produces a more pessimistic result than in the case of correlations.

2. Critical loads

The critical load concept has been gradually developing into an operational tool, which is used to support decision-making in targetting emission reductions of potentially acidifying pollutants. The critical load for forest soils represents the highest deposition of acidifying compounds that will not cause chemical changes in soil leading to long-term harmful effects on ecosystem structure and function (Nilsson and Grennfelt, 1988). The most common way to quantify critical loads is to apply the simple mass balance method. In the following, a short description of the employed equations in Finland based on Hettelingh *et al.* (1991) and Kämäri *et al.* (1992) is given. The advantage of this specific approach is

that these critical load maps can be compared directly with acid deposition maps.
The critical loads of sulphur and nitrogen for forest soils are

$$CL\ (S) = \sigma \cdot BC_{\text{le, crit}} \ , \tag{1a}$$
$$CL\ (N) = N_u + (1 - \sigma) \cdot BC_{\text{le, crit}} \ , \tag{1b}$$
where
N_u = nitrogen net uptake in meq $m^{-2}\ a^{-1}$.

The neutralization factor σ is

$$\sigma = SO_{2,\ d} \ / \ (SO_{2,\ d} + N_d - N_u) \ , \tag{2}$$
where
$SO_{2,\ d}$ = total sulphur deposition to forest in meq $m^{-2}\ a^{-1}$,
$N_d = NO_{x,\ d} + NH_{4,\ d}$ = total nitrogen deposition to forest in meq $m^{-2}\ a^{-1}$.

The critical base cation leaching for acid forest soils is

$$\begin{aligned}
BC_{\text{le, crit}} &= BC_d^* + BC_w - BC_u - Alk_{\text{le, crit}} \\
&= BC_d^* + BC_w - BC_u - H_{\text{le}} - Al_{\text{le}} \ ,
\end{aligned} \tag{3}$$
where
BC_d^* = seasalt corrected base cation deposition to forests in meq $m^{-2}\ a^{-1}$,
BC_w = base cation production from the mineral weathering in meq $m^{-2}\ a^{-1}$,
BC_u = base cation net uptake by tree growth in meq $m^{-2}\ a^{-1}$,
H_{le} = hydrogen ion leaching in meq $m^{-2}\ a^{-1}$,
Al_{le} = aluminum ion leaching in meq $m^{-2}\ a^{-1}$.

The aluminum leaching is defined with the molar ratio of aluminum to divalent base cations, and calculated by

$$Al_{\text{le}} = R(Al/BC)_{\text{crit}} \cdot (BC_d^* + BC_w - BC_u) \ . \tag{4}$$
where
$R(Al/BC)_{\text{crit}}$ = critical molar $Al/(Ca+Mg)$ ratio of 1.0.

The gibbsite coefficient is used to operationally depict the aluminium concentration from an assumed equilibrium with a solid phase. An average value for K_{gibbs} of $10^{8.3}$ $(mol/l)^{-2}$ was chosen for the calculations of Finnish soils (Sverdrup *et al.*, 1990). The hydrogen leaching is

$$Al_{\text{le}} \ / \ Q = K_{\text{gibbs}} \cdot (H_{\text{le}} \ / \ Q)^3 \ , \tag{5}$$
where
Q = runoff in $m\ a^{-1}$.

The exceedance of the critical loads by the acid deposition is then simply

$$Ex\ (S) = SO_{2,\ d} - CL\ (S) \tag{6a}$$
$$Ex\ (N) = N_d - CL\ (N) \ . \tag{6b}$$

All nitrogen deposition, independent of its form, is assumed to acidify the forest soil. The net uptake of nitrogen is not allowed to be higher than total nitrogen deposition N_d. If necessary, N_u is reduced and a similar correction is made to BC_u to preserve their relative uptake ratio.

A minimum base cation leaching from forest soils, i.e. the sum of base cation deposition and weathering minus net uptake, is assumed to be 5 meq m^{-2} a^{-1}. The constant was derived from a minimum concentration threshold of 15 µeq l^{-1} for the tree root uptake (Hettelingh et al., 1991) with an average runoff of 0.3 m a^{-1}. The weathering rate is corrected to meet this requirement, if necessary.

3. Methodology of the analysis

Mathematical models have proven to be useful tools for studying environmental problems. To enable a reliable development and application of these models, it is often necessary to perform a thorough model evaluation consisting of sensitivity and uncertainty analyses. The results of the uncertainty analyses display the origins and effects of uncertainty in the model, and provide useful suggestions for further model development and data acquisition.

Since a complete assessment of the uncertainty (evaluating e.g. distributions) cannot be readily achieved by analytical manipulations, even not for simple models like the critical load model (Eqn. 1-6), the use of Monte Carlo oriented approaches is common for uncertainty analyses. The recently developed Latin Hypercube Sampling technique enhances the sampling efficiency. Regression and/or correlation analysis is applied to determine the important uncertainty sources (e.g. parameters). The software package UNCSAM (UNCertainty analysis by Monte Carlo SAMpling techniques) is designed for these purposes (see Heuberger and Janssen (1993)), and has been used in this study.

Conventional statistical information such as standard deviation, percentiles and cumulative distributions of model outputs, here the critical loads and their exceedances, are calculated. The contribution of the input parameters to the uncertainty of the model results has been assessed by (linear) regression and correlation analysis of the model outputs on the corresponding input parameters. If linear regression is appropriate, i.e. if the well-known R^2 of regression (coefficient of determination; COD) is near to 1, then the standardized regression coefficients (SRC) measure the fraction of the uncertainty in the model output which is contributed by the various input parameters (see Heuberger and Janssen (1993)). The SRC is especially suited for uncorrelated input parameters. When some input parameters are correlated, the use of the 'partial uncertainty contribution' PUC has been suggested instead by Kros et al. (1993). The PUC expresses the relative change in the uncertainty of a model outputs due to a relative change in the uncertainty of an input parameter, accounting for the fact that correlated input parameters should change accordingly. If correlations are weak, the PUC will be approximately equal to SRC^2. This will be the case in our study, where no correlations between input parameters are incorporated.

4. Specification of input parameters and their uncertainties

The effective temperature sum (ETS) denotes the amount of degree days, with the threshold level of + 5 °C. ETS is used in the estimation of both the weathering rate and the forest growth. The values, which depend on the latitude, height from the sea level and distance to sea, were calculated with the computer programme of the Finnish Forest Research Institute (Ojansuu and Henttonen, 1983) using the weather measurements 1951-

80 from the stations of the Finnish Meteorological Institute. The relative errors from the test material on the error estimation of the method were roughly under ± 10 %.

The weathering rate estimation is based on the method by Olsson and Melkerud (1990). The losses of *Ca* and *Mg* since deglaciation were found to correlate to effective temperature sum and the total element content in the soil *C*-horizon. The total analysis data for Finland are provided by the Geological Survey (Koljonen, 1992). The uncertainty analysis on the weathering rate was carried out using triangular distribution for the input variables as shown in Table 1. The resulting uncertainty for the weathering rate is characterised with the coefficient of variation of 24 %.

Table 1. The mean values and coefficients of variation (CV) from Finland for the input parameters used in the weathering rate estimation.

	mean	CV
		%
Ca content in mass-%	1.80	5
Mg content in mass-%	0.99	5
ETS in degree days	1070.0	10

The term net uptake refers to the net export of nutrients in the biomass of stem and bark via tree harvesting. The net uptake is estimated with annual forest growth and the average nutrient content in biomass. The growth curves of logistic type were estimated for all tree species with a non-linear regression method (Press *et al.*, 1986) using national forest inventory data (Kuusela, 1977) and ETS. The estimates for biomass density and element contents and their uncertainties are based on field measurements (Mälkönen, 1975; Rosén, unpublished). The values used in estimating the net uptake uncertainty are given in Table 2. The resulting coefficients of variation for the base cation net uptake were 37, 48 and 40 % for birch, spruce and pine, and for nitrogen 37, 55 and 44 % respectively.

Table 2. The input parameter values and their coefficients of variation (CV) for the uncertainty analysis on the net uptake.

	birch		spruce		pine	
		CV		CV		CV
biomass density in kg m^{-3}	500	6 %	410	6 %	410	6 %
content in stem and bark in mass-%						
Ca	0.160	28 %	0.209	16 %	0.115	29 %
Mg	0.040	28 %	0.019	16 %	0.018	36 %
K	0.080	28 %	0.064	22 %	0.044	36 %
N	0.170	28 %	0.104	32 %	0.079	29 %

Base cation deposition emerges from both natural sources, e.g. soil dust, and anthropogenic activities, e.g. fly ash from energy production. The deposition is estimated using the results of the years 1986 - 1988 from a nationwide network of stations measuring monthly bulk deposition (Järvinen and Vänni, 1990). An intercomparison of measurement methods and laboratory analysis (HELCOM, 1992) resulted in errors less than ± 20 % of the base cation deposition values for the measurement network used. However, other monitoring results in Finland (e.g. Leinonen and Juntto, 1991) suggest lower base cation deposition levels. Since the collection periods and methods vary from one network to the other, the overall coefficient of variation for the base cation deposition was increased to 30 %. This range then accounts for not only measurement errors but also the uncertainty of the phenomenon assessed.

The long-range atmospheric transport of the acidifying compounds is described by source-receptor transfer matrices based on EMEP models (Sandnes and Styve, 1992). The matrices for sulphur were manipulated to allow the use of sulphur mesoscale model to the Estonian and Russian nearby areas. The European emissions are from the databases of the RAINS-model by IIASA (Alcamo et al., 1990). For more detailed Finnish emissions of SO_2, NO_x and NH_4 mesoscale models were applied (Johansson et al., 1990). The different depositions were directly assigned uncertainty ranges, since it was not possible to carry out an uncertainty analysis on the emission and atmospheric transport models considered here. Not many relevant sources are available for the error quantification of the long-range models (Batterman and Amann, 1991). Therefore the sulphur deposition is assigned a roughly estimated coefficient of variation of 20 %, 30 % is assigned for NO_x and 35 % for NH_4.

The acid deposition fluxes to forests are estimated from a throughfall experiment (Hyvärinen, 1990). The throughfall values were compared to modelled deposition. Spruce was found to filter sulphur deposition 1.2 times compared to modelled deposition, but pine and especially deciduous forest showed filtering factors less than one. Since the amount of internal nitrogen cycling could not be separated from the throughfall measurements, the nitrogen deposition to forests was assumed to be equal to the modelled nitrogen depositi-on. An exact uncertainty was difficult to derive from these scarce data, and a small coefficient of variation of 5 % was assigned to all filtering factors.

The runoff map for Finland was prepared by the Hydrological office of the National Board of Waters and Environment by averaging values from the years 1961 - 75. A coefficient of variation of only 5 % was assigned to the runoff, since the measurement errors are very small. Due to the nature of the critical load approach, the annual variation is not included in the uncertainty estimate.

The critical load value for a plot is an area-weighed combination of values calculated for each forest type considered. The errors in the forest areas represented by the different tree species are quite small and therefore treated with a modest coefficient of variation 5 %.

5. Uncertainty studies on different model runs

First, an uncertainty analysis was performed to one plot using the range of possible input values that are found in Finland. The combination of the input data does not originate from one specific site in reality. However, it characterises some real sites in Finland, whose parameter values are within a ten-percent range of the average values in Table 3, which presents the ranges used for input parameter triangular distributions. Thus the use of this specific combination of values as an example is justified. The resulting distributions are presented in Fig. 1. These distributions describe the feasible range of critical loads and their exceedances, since they include the actual ranges of all input parameters. Here the critical loads have a smaller range of values than their exceedances, because the latter also includes the acid deposition compounds. The COD is high for all output variables.

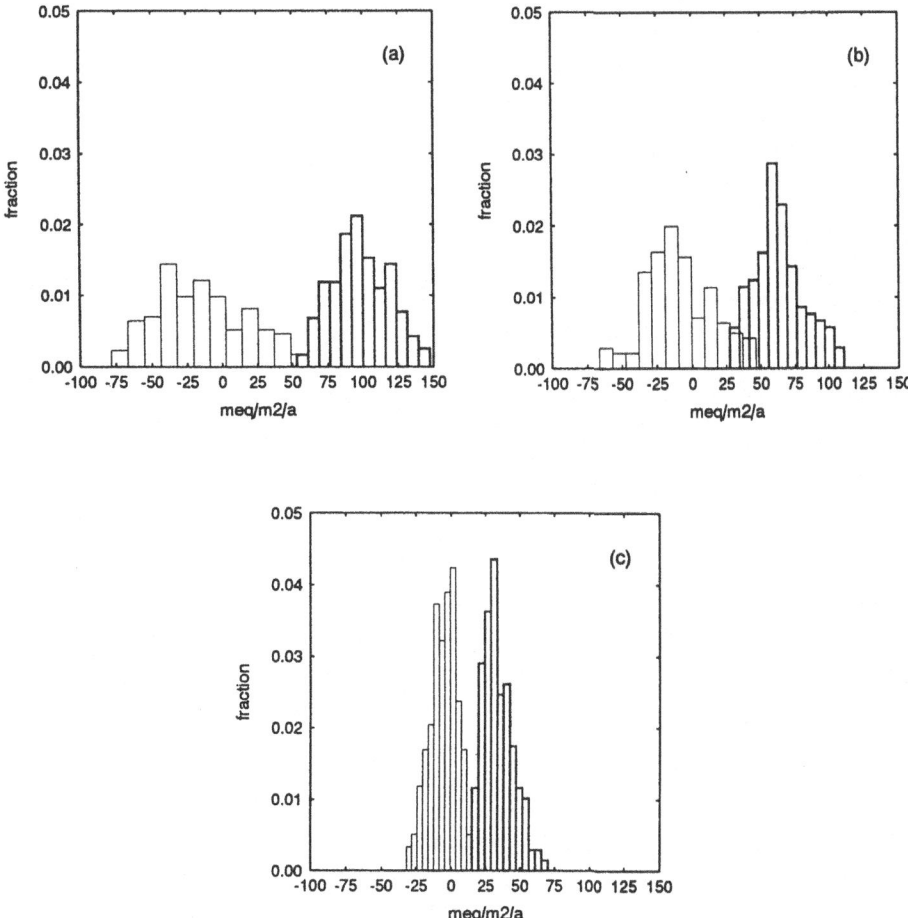

Figure 1. The uncertainty distributions of the critical loads (thick lines) and their exceedances (thin lines) of total acidity (a), sulphur (b) and nitrogen (c) using real ranges for input parameter values found in Finland.

Therefore, the PUC can be used to find out the most influencing parameters. The relative ranking according to PUC of some input parameters is shown in Table 4. The most important value explains on the average from 40 to 50 % of the resulting uncertainty, the second about 20 %, the third 10 % and the rest under ten percent. The three most important variables are the sulphur deposition, the weathering rate and the base cation deposition.

Next, the uncertainty analysis was carried out in more than thousand plots in Finland. Each plot has its own set of absolute input values derived from data sets and it may include all or only some of the three tree species considered. Different random seeds are used to get an independent random sampling for input parameters for each plot. The uncertainties of the input parameters are shown in Table 5. The model is then run 150

Table 3. Specifications of symmetric triangular distributions for the uncertainty analysis of one site.

parameter	min	midpoint	max
runoff Q in mm	150.	316.	490.
$SO_{2,d}$ in meq $m^{-2}\,a^{-1}$	13.2	32.3	111.9
$NO_{x,d}$ in meq $m^{-2}\,a^{-1}$	7.0	14.1	28.0
$NH_{4,d}$ in meq $m^{-2}\,a^{-1}$	2.4	12.5	22.1
BC_d^* in meq $m^{-2}\,a^{-1}$	7.3	15.8	25.0
BC_w in meq $m^{-2}\,a^{-1}$	3.6	13.0	41.1
filtering factor in fraction			
birch	0.878	1.000	1.122
spruce	1.054	1.200	1.346
pine	0.878	1.000	1.122
BC_u in meq $m^{-2}\,a^{-1}$			
birch	1.6	19.1	38.9
spruce	1.5	12.6	28.9
pine	1.0	6.4	10.9
N_u in meq $m^{-2}\,a^{-1}$			
birch	1.5	17.4	35.4
spruce	0.8	7.4	18.2
pine	0.7	4.8	8.8
percentage of grid area in %			
birch	0.	5.	66.
spruce	0.	15.	50.
pine	0.	43.	87.

Table 4. The relative ranking of some input parameters according to PUC. BC_u and N_u stands for whichever tree species that was most influential.

	CL	Ex	CL(S)	Ex(S)	CL(N)	Ex(N)
BC_w	1	2	1	2	3	2
BC_d^*	2	3	2	3	5	3
BC_u	3	5	5	5	6	4
N_u	4	7	7	7	7	7
$SO_{2,d}$	6	1	3	1	1	1
$NO_{x,d}$	5	4	4	4	4	5
$NH_{4,d}$	7	6	6	6	2	6

times with input parameter values varied randomly within their uncertainty ranges. Statistical measures for the simulation results are derived as well as the most influencing parameters separately for each plot. The results for the regional study are presented in Fig. 2, which gives the cumulative distribution functions both in a so-called EMEP grid and compiled for the whole country. The ranges are expressed as standard deviation around mean and a 90 % confidence interval with median. The critical loads do not seem to differ much spatially, but the standard deviation seems to be smaller in the north than in the south. The 5th percentile border shows the effect of the assumption concerning the minimum base cation leaching. It always leads to a minimum critical load of 12.5 meq $m^{-2}a^{-1}$, which is further increased by the proton leaching depending on the runoff. The exceedance uncertainties are highest in the south, where the acidifying deposition is also highest. Even within one grid element the uncertainty can vary significantly.

Table 5. Coefficients of variation for symmetric triangular distributions of input parameters, in addition to BC_d^*, BC_u and N_u which are explained in text.

BC_d^*	30 %
$SO_{2, d}$	20 %
$NO_{x, d}$	30 %
$NH_{4, d}$	35 %
filtering factors for acidifying deposition	5 %
forest area, all tree species	5 %
runoff	5 %

Some statistics on the most important parameters to different critical loads and their exceedances were compiled based on the ranking of three most influencing parameters in each site according to PUC. The COD's were calculated to judge for the use of PUC's.

The COD's tended to improve towards the north. For the whole country, the mean COD for the critical load of sulphur and nitrogen was 0.900 and the minimum was 0.613. For the exceedance the corresponding values are 0.902 and 0.623. Though the COD for some sites was thus somewhat low, the PUC was used throughout the country for consistency. The most influential parameter in the whole country was the base cation deposition. The next ones were the base cation and nitrogen net uptakes by birch. The growth rate of birch is high and the nutrient contents in the dense biomass are relatively high. Therefore, even a small fraction of birch forest may affect the results. The fourth most important parameter was the weathering rate. The importance of the base cation variables could also be judged to some extent from the employed model (Eqns. 1-6). The critical loads and exceedances for nitrogen were also influenced by the filtering factor of the deciduous forest.

The spatial distribution pattern of the most influential factors was shortly examined. For the critical load of sulphur and nitrogen the most important factor in some southern parts of the country was nitrogen net uptake, in most of the southern and central Finland the base cation uptake, and in Lapland the base cation deposition along with the weathering rate.

6. Discussion and conclusions

In a study involving input data from many disciplines, the interpretation of the values to be used may bring in considerable uncertainty, as e.g. in quantifying the base cation deposition not emerging from the internal forest cycle. Therefore, this study concentrated mostly on uncertainty emerging from measurement errors of input parameters.

The uncertainties of the base cation weathering and the base cation and nitrogen net uptakes were calculated separately before the regional application. This made it possible to estimate the effect of these individual parameters, because the summing of the uncertainty contributions of the parameters from which they were derived is not straightforward. However, their uncertainties were estimated as a countrywide average and not for each plot separately, which may have partially affected the results.

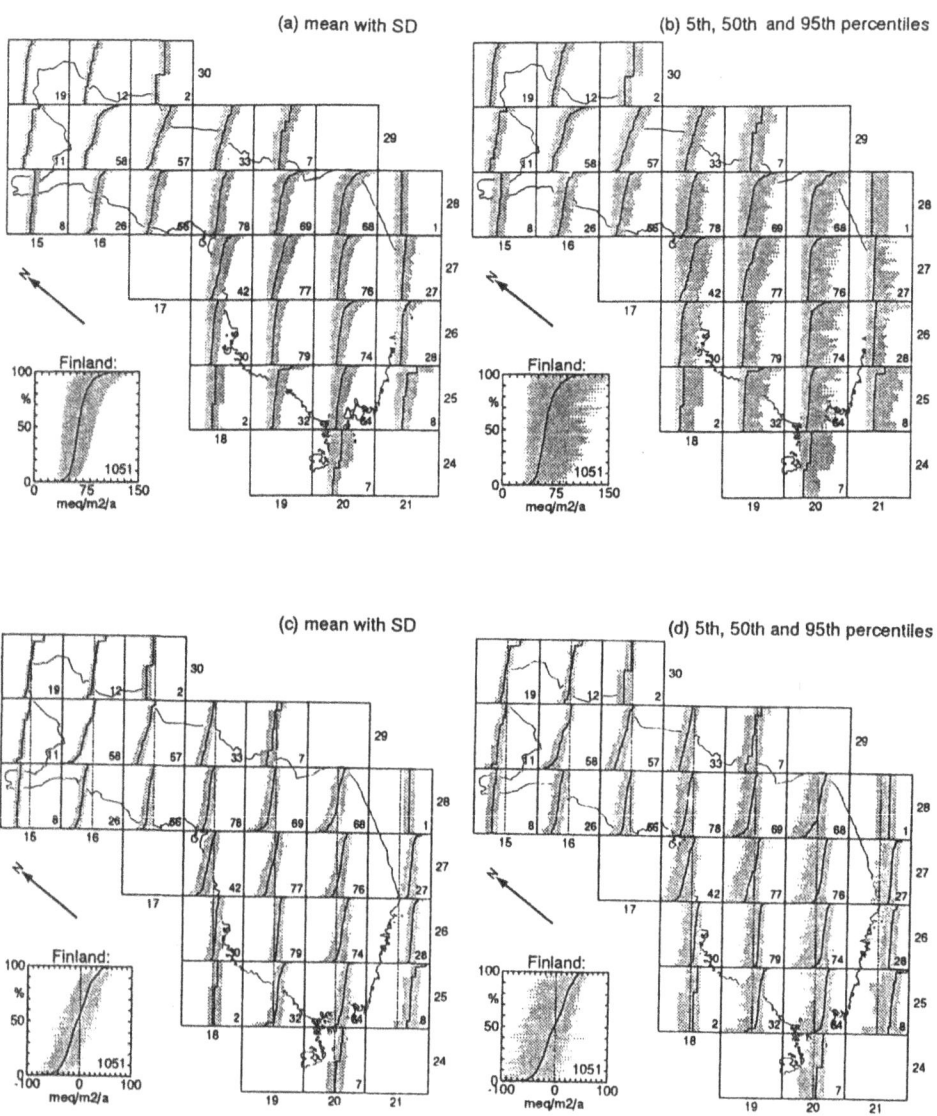

Figure 2. The uncertainty bands for the critical loads of sulphur and nitrogen (a, b) and their present exceedances (c, d). The results are presented both by mean with standard deviation (a, c) and by the 5th, 50th and 95th percentiles (b, d). The number of calculation points in each grid cell is denoted on the lower right corner of the grid square.

None of the input parameters were correlated due to lack of sufficient quantitative data. This has lead to a possible overestimation of the final uncertainties. E.g. the uncertainty of the net uptake values represents an upper limit, since the input parameters were varied independently of each other. This could even be the case in some instances of nutrient deficiencies. However, considering the nature of data available the approach of no correlations was judged to be suitable for this exercise.

The spatial variability of some parameters was not taken into account, mostly due to lack of data. E.g. the gibbsite coefficient was constant throughout the regional application and was not assigned any uncertainty, although it would have clearly affected the results. The effect of this constant could be assessed when the model structure and functioning are chosen and fixed.

Same type of reasoning is applied to the critical chemical Al/BC ratio. Although the effect of this ratio value of 1.0 to actual forest growth can be questioned, the value widely agreed upon provides a systematic approach and comparable results for international exercises. The involvement of other criteria, e.g. a critical limit for aluminium concentration, would have shown sensitivity to other input parameters like runoff, however, presently they were not considered relevant for forest soils in Finland.

When a critical load value is assigned to a map grid square, usually the 5th percentile value of the cumulative distribution function of the values from that square is shown. Therefore, rather than seeking most influential parameters for critical loads in general, the interest could also be focused on low critical load values in a grid cell and their influential parameters.

The results of the regional uncertainty analysis on the simple mass balance model seem to suggest, that the resulting uncertainties of the critical loads are not too big compared to the uncertainties of the input parameters. However, the results cannot be generalized, since this study was based only on data for forest soils in Finland. The inclusion of additional chemical criteria or lake ecosystems would already have given another view on the final uncertainties.

Acknowledgements

Dr. Maximilian Posch is gratefully acknowledged for his valuable comments on the work and the use of the graphical display programs.

References

Hettelingh, J.-P., R. Downing and P. de Smet (eds.) (1991), 'Mapping critical loads for Europe', UN/ECE, Convention on Long-Range Transboundary Air Pollution. CCE Technical report No. 1, RIVM, Bilthoven, The Netherlands.

Beck, M. (1987), 'Water quality modeling: a review of the analysis of uncertainty', *Water Resources Research*, **23(8)**, 1393-1442.

De Vries, W. (1991), 'Methodologies for the assessment and mapping of critical loads and of the impact of abatement strategies on forest soils', Report 46, DLO The Winand Staring Centre, Wageningen, The Netherlands.

Hettelingh, J.-P., R. Gardner and L. Hordijk (1992), 'A statistical approach to the regional use of critical loads', *Environmental Pollution*, **77**, 177-183.

Posch, M., M. Forsius and J. Kämäri (1993), 'Critical loads of sulphur and nitrogen for lakes I: model description and estimation of uncertainty', *Water, Air and Soil Pollution*, **66**, 173-192.

Kämäri, J., M. Forsius and M. Posch (1993), 'Critical loads of sulphur and nitrogen for lakes II: regional extent and variability in Finland', *Water, Air and Soil Pollution*, **66**, 77-96.

Hettelingh, J.-P. (1990), 'Uncertainty in modeling regional environmental system; The generalization of a watershed acidification model for predicting broad scale effects', Academic dissertation, RR-90-3, IIASA, Laxenburg, Austria.

Ulrich, B. (1983), 'Soil acidity and its relations to acid deposition', In: B. Ulrich and J. Pankrath (eds.) *Effects of accumulation of air pollutants in forest ecosystems*, D. Reidel Publ. Co., Dordrecht, The Netherlands, pp. 127-146.

Tamminen, P. and M. Starr (1990), 'A survey of forest soil properties related to soil acidification in Southern Finland', In: P. Kauppi, P. Anttila and K. Kenttämies (eds.) *Acidification in Finland*, Springer-Verlag, Berlin Heidelberg, pp. 235-251.

Nilsson, J. and P. Grennfelt (eds.) (1988), 'Critical loads for sulphur and nitrogen', Miljørapport 1988, 15, Nordic Council of Ministers.

Kämäri, J., M. Forsius, M. Johansson and M. Posch (1992), 'Critical loads for acidifying deposition in Finland', Report 111, Ministry of the Environment, Helsinki, Finland (in Finnish with English summary).

Sverdrup, H., W. de Vries and A. Henriksen (1990), 'Mapping critical loads', Miljørapport 1990, 98, Nordic Council of Ministers.

Heuberger P.S.C. and P.H.M. Janssen (1993), 'UNCSAM: a software tool for sensitivity and uncertainty analysis of mathematical models', Proceedings of the conference on predictability and non-linear modelling in natural sciences and economics, 5-7 Apr 93, Wageningen, The Netherlands.

Kros, J., W. de Vries, P. Janssen and C. Bak (1993), 'The uncertainty in forecasting trends of forest soil acidification', *Water, Air and Soil Pollution*, **66**, 19-58.

Ojansuu, R. and H. Henttonen (1983), 'Estimation of local values of monthly mean temperature, effective temperature sum and precipitation sum from the measurements made by the Finnish Meteorological Office', *Silva Fennica*, **17**(2), 143-160 (in Finnish with English summary).

Olsson, M. and P.-A. Melkerud (1991), 'Determination of weathering rates based on geochemical properties of the soil', In: E. Pulkkinen (ed.) Environmental geochemistry in northern Europe. *Geological Survey of Finland, Special Paper*, **9**, 69-78.

Koljonen, T. (ed.) (1992), 'The Geochemical Atlas of Finland, Part 2: Till', Geological Survey of Finland, Espoo, Finland.

Press, W.H., B.P. Flannery, S.A. Teukolsky and W.T. Vetterling (1986), 'Numerical recipes; The art of scientific computing', Cambridge University Press, New York, USA.

Kuusela, K. (1977), 'Increment and timber assortment structure and their regionality of the forests of Finland in 1970 - 76', *Folia Forestalia*, **320**. Finnish Forest Research Institute, Helsinki (in Finnish with English summary).

Mälkönen, E. (1975), 'Annual primary production and nutrient cycle in some scots pine stands', In: *Communicationes Instituti Forestalis Fenniae*, **84**.

Rosén, K. (unpublished), 'Data from recent Swedish national forest survey'.

Järvinen, O. and T. Vänni (1990), 'Bulk deposition chemistry in Finland', In: P. Kauppi, P. Anttila and K. Kenttämies (eds.) *Acidification in Finland*, Springer-Verlag, Berlin Heidelberg, pp. 151-165.

HELCOM (1992), 'Intercalibrations and intercomparisons of measurement methods for airborne pollutants', *Baltic Sea Environment Proceedings*, **41**.

Leinonen, L. and S. Juntto (eds.) (1991), 'Results on air quality at background stations, Jun-Dec 1990', Finnish Meteorological Institute, Air Quality Department, Helsinki.

Sandnes, H. and H. Styve (1992), 'Calculating budgets for airborne acidifying components in Europe 1985, 1987, 1988, 1989, 1990 and 1991', EMEP/MSC-W Report 1/92.

Alcamo, J., R. Shaw and L. Hordijk (eds.) (1990), 'The RAINS model of acidification; Science and strategies in Europe', Kluwer Academic Publishers, Dordrecht, The Netherlands.

Johansson, M., J. Kämäri, R. Pipatti, I. Savolainen, M. Tähtinen and J.-P. Tuovinen (1990), 'Development of an integrated model for the assessment of acidification in Finland', In: P. Kauppi, P. Anttila and K. Kenttämies (eds.) *Acidification in Finland*, Springer-Verlag, Berlin Heidelberg, pp. 1171-1193.

Batterman, S. and M. Amann (1991), 'Targeted acid rain strategies including uncertainty', *Journal of Environmental Management*, **32**, 57-72.

Hyvärinen, A. (1990), 'Deposition on forest soils', In: P. Kauppi, P. Anttila and K. Kenttämies (eds.) *Acidification in Finland*, Springer-Verlag, Berlin Heidelberg, pp. 199-213.

MONTE-CARLO SIMULATIONS IN ECOLOGICAL RISK ASSESSMENT

U. HOMMEN, U. DÜLMER, H.T. RATTE

Department of Biology V, Technical University Aachen,
Worringerweg 1, 52056 Aachen, Germany

Summary

Ecological risk assessment is usually based on two processes: exposure and effect assessment. Both have to deal with different kinds of uncertainties. Thus, predictions can only be probability statements about the expected hazard. In the paper we shall present two stochastic approaches of simulation models to analyse and predict toxicant effects in freshwater plankton communities using Monte-Carlo techniques. The first is an individual-based model which uses detailed measurements of life-table data to simulate community dynamics in laboratory systems. The natural variability of individuals is reflected in the model by describing life courses of every individual according to the means and probability distributions of the measured life-table data. The second model is a compartment model of plankton communities in outdoor microcosms, where taxa are modelled via differential equations. In this approach uncertainties due to parameter estimation are incorporated conducting many simulation runs with parameter values chosen from estimated probability distributions. We shall present application examples of both models and discuss some benefits and problems of using simulation models in Ecological Risk Assessment.

Keywords individual-based model, compartment model, freshwater plankton communities

1. Introduction

The US Environmental Protection Agency (EPA) defines Ecological Risk Assessment as 'a process that evaluates the likelihood that adverse effects may occur or are occurring as a result of exposure to one or more stressors' (EPA 1992). In the following we shall only consider chemicals as stressors.

Ecological Risk Assessment can be seen as a step by step procedure, a detailed description can be found in the framework of the EPA (1992) or in the text book of Suter (1993). In the beginning the problem has to be formulated and the potential hazard has to be defined. The ecosystem potentially at risk is described and first characteristics of the stressor and the expected effects are given. An important task during this phase is the formulation of endpoints, which may be for example the protection of a species or the maintenance of gamefish production in a lake. The following analysis phase usually consists of two parts: The exposure assessment investigates the expected fate of a toxicant as a result of the possible quantified inputs to the system, transportation within the system and different degradation processes. Ecosystem compartments and their populations with high probability of exposure to the stressor are identified. For the effects assessment usually toxicity tests are conducted in the laboratory to obtain NOECs (No Observed Effect Concentrations) or dose response relations. Field tests are often conducted to investigate fate and effects under more natural but controlled conditions (e.g. Heimbach *et al.* 1992).

The results of these assessments have to be integrated to characterise the risk which can be done with three different approaches (EPA 1992). The Quotient Method compares single effect and exposure values (e. g. Honeycutt and Ballantine 1983). This method usually results in safety factors. Another approach considers the probability distributions of effects and exposure (e.g. van Straalen and Dennemann 1989) and the risk is calculated from the overlap of the two distributions. An endpoint then may be the protection of 95% of all species which is done in the Netherlands today (HCN 1989). The third possibility is the use of simulation models. Depending on the specific endpoints, models of different complexity can be used. For example, single-species models to extrapolate from the individual to the population may be individual-based or matrix models (e.g. Hallam *et al.* 1990; Kooijman and Metz 1984; Ferson *et al.* 1989), while most models to study indirect effects of a stressor in an ecosystem are compartment models which handle populations or functional groups (e.g. Bartell et all. 1983; Poethke *et al.* 1991; Swartzman and Rose 1984).

As given in the definition above, the results of Ecological Risk Assessment are probabilistic statements. The uncertainties stem from stochasticity (natural variability such as weather conditions), ignorance (insufficient knowledge, e.g. hypotheses incorporated in the conceptual model, unknown parameters etc.), and errors (failures in measurements or observations) (Suter 1993). A common approach to consider uncertainties in simulation models is the Monte-Carlo technique: Before a simulation the model parameters are sampled randomly from measured (or estimated) distributions and the output of many of such simulations is analysed. The general principle to calculate ecological risks using Monte-Carlo techniques is shown in Fig. 1. Simulations without stressor are conducted to calculate the natural risk. For the stress scenarios, the results of the exposure assessment, e.g. as toxicant concentration time series, and of the effects assessment, e.g. as dose-response functions, are incorporated. Thus, the calculated risk under stress conditions can be compared with the natural risk.

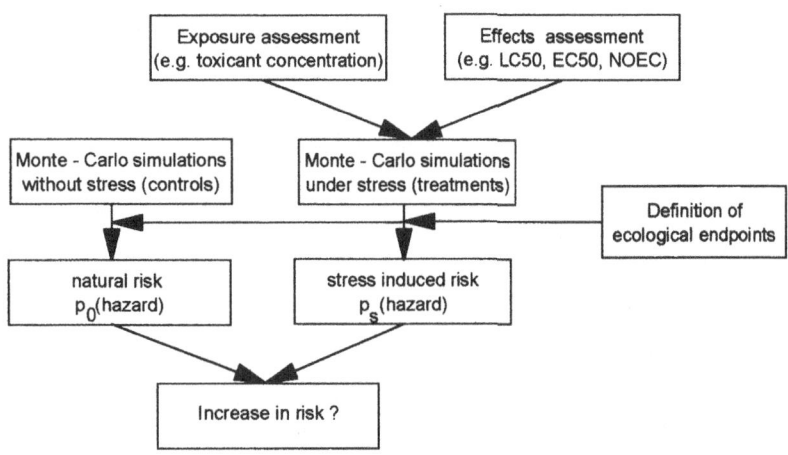

Figure 1. Monte-Carlo Simulations in Ecological Risk Assessment

In this paper we present two approaches. The first example is an individual-based model of water flea populations. The purpose of this model is to extrapolate from measured individual life-table data to the population level. As endpoint for the example given here we have chosen extinction of a population. The second example is a simple compartment model of a plankton community in outdoor microcosms to show prediction of indirect effects in a food web. Here the occurrence of algae blooms after application of an insecticide is the selected endpoint.

2. An individual-based model of water flea populations

Life-table experiments with the large waterflea *Daphnia magna* and the smaller species *Ceriodaphnia quadrangula* are conducted in the laboratory under different food levels, toxicant concentrations and temperatures where many individuals are observed during their whole life. The results are functional relationships of life-table parameters such as growth, juvenile development time, maximum age, reproduction etc. from the investigated factors. In addition, information about the natural variability of the life-table parameters under the given conditions are obtained. These life-table results serve as input for the simulation model. The model describes the life course of an individual animal (Fig. 2): e.g. a *Daphnia* is born, it feeds and grows until it has finished juvenile development and comes into the adult loop whilst feeding, growing, developing eggs and releasing juveniles until it dies. Many of this simulated individual life courses build the simulated population, where the life-table parameters for each individual are taken randomly from the measured statistical distributions. Therefore, the variability of simulated individuals reflects the variability of natural populations (Fitsch and Kaiser 1987). A detailed description of the modelling procedure is given in Fitsch (1990).

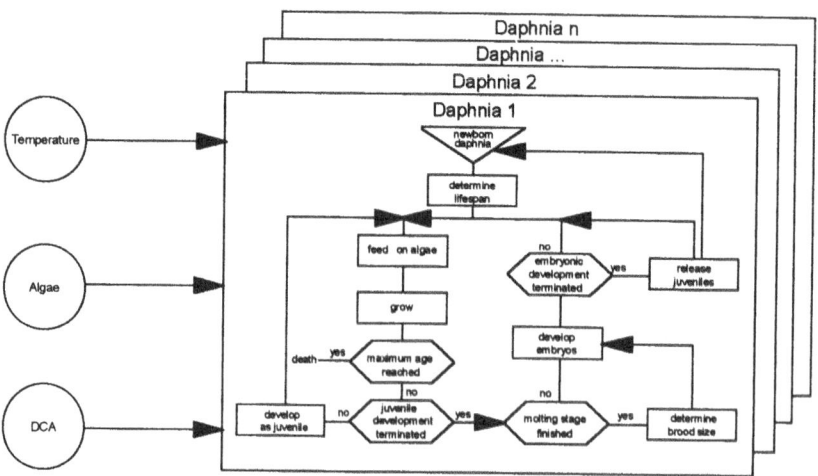

Figure 2. Life course of a simulated *Daphnia* individual (for explanations see text)

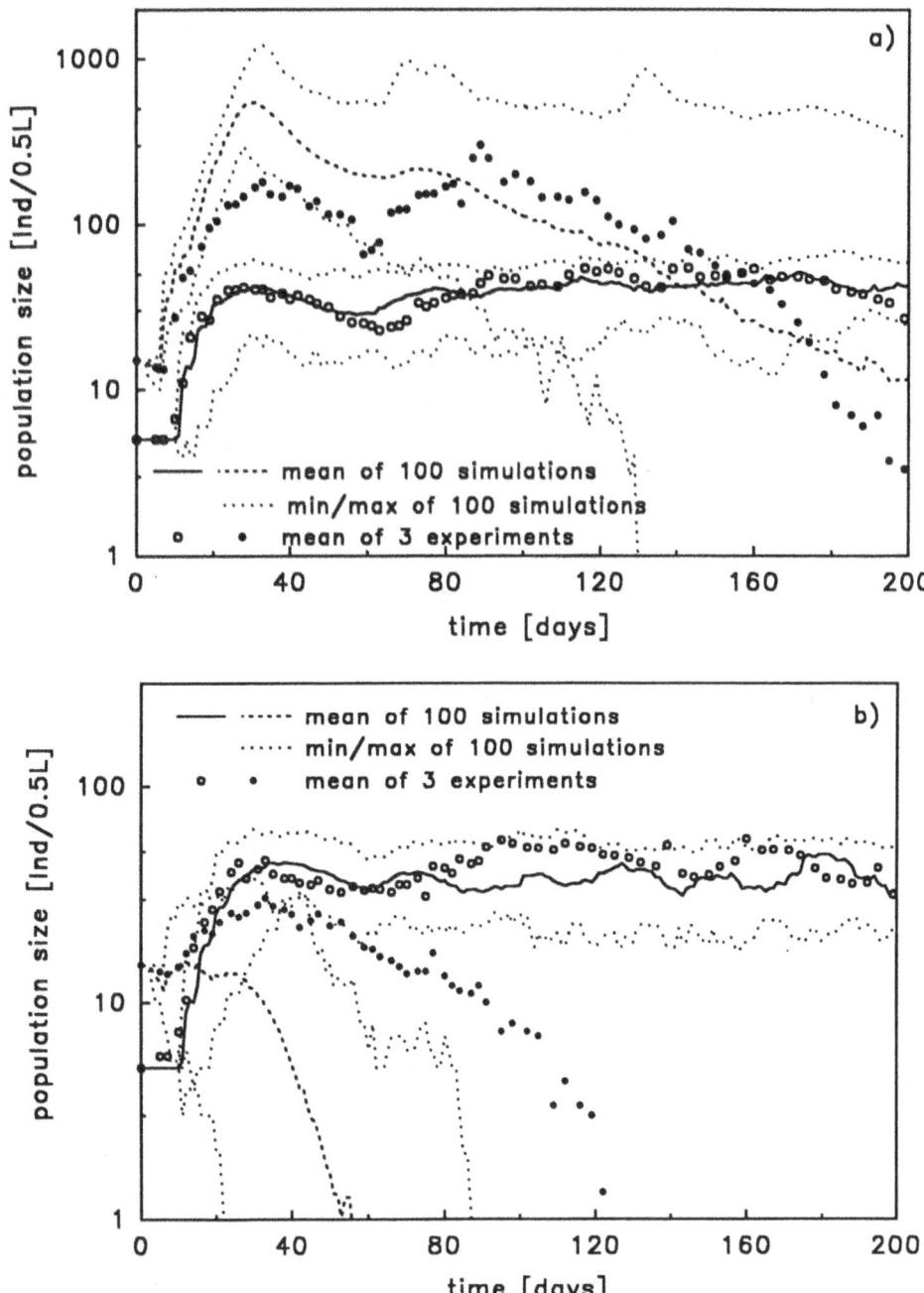

Figure 3. Simulation results and corresponding data for competition experiments a) control b) 10 μg *l*¹ DCA open symbols and solid line = Daphnia, closed symbols and dashed line = Ceriodaphnia

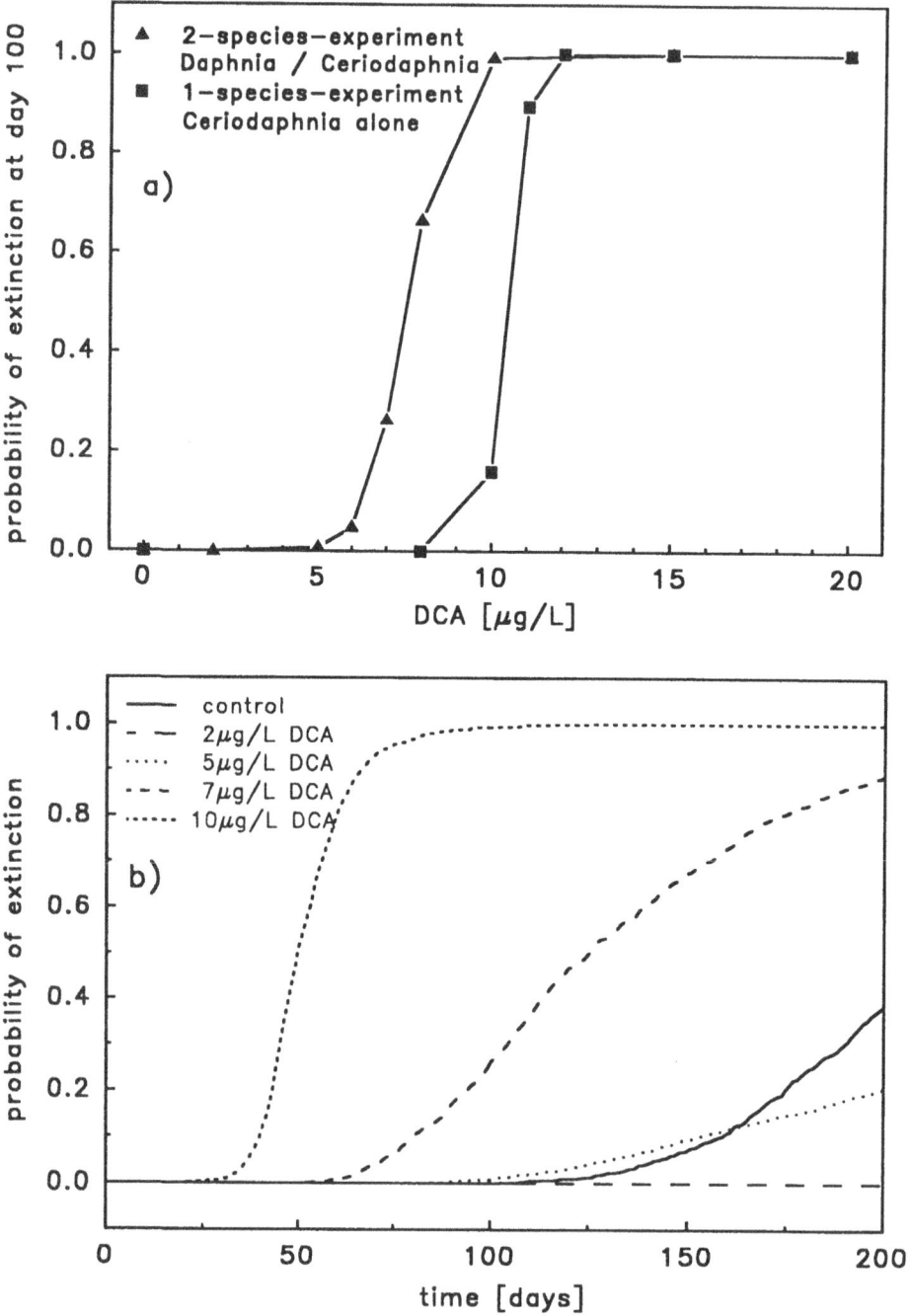

Figure 4. Extinction risk for *Ceriodaphnia* a) within 100 days, alone and in competition with *Daphnia* b) cumulative until day 200, in competition with *Daphnia*

The predicted population dynamics can be tested comparing simulation outputs with means of corresponding experiments as is shown in Fig. 3 for a competition experiment with *Ceriodaphnia quadrangula* and *Daphnia magna*. The simulations for the control experiment show a close agreement for *Daphnia* but population densities of *Ceriodaphnia* are overestimated at their maximum (Fig. 3a). The model predicts the extinction of *Cerioda-phnia* in the first 200 days from day 130 on, so experimental results are covered by the simulation runs. At 10 μg l-1 DCA again the predictions for *Daphnia* closely agree with the experimental results, whereas the abundance of *Ceriodaphnia* is underestimated (Fig. 3b). *Ceriodaphnia* is more sensitive to 3,4-Dichloroaniline (DCA) than Daphnia (Seitz and Ratte 1991) resulting in an earlier extinction under DCA stress. Time of extinction is predicted too early in the simulations, but in principle the results are in agreement with the experiments. We hope to improve the precision of the *Ceriodaphnia* model by additional life-table data at low food levels.

The results of 1000 Monte-Carlo runs for different toxicant levels are summarised in Fig. 4. Competition with *Daphnia* increases the extinction risk for *Ceriodaphnia* until day 100 (Fig. 4a) but these typical dose-response curves are changed if longer time periods are considered (Fig. 4b). The 'natural risk' under control conditions then exceeds the risk at low DCA concentrations (< 5 μg l^1 DCA). For 2 μg l^1 DCA the Monte-Carlo simulations predict no risk of extinction. This stimulation of DCA can be explained with the results of the life-table experiments where an increase of offspring production was observed.

3. A compartment model of an aquatic food web

The second example is a simplification of some of the classical aquatic food web models such as CLEANER or SWACOM (Park *et al.* 1974; O'Neill *et al.* 1982). It describes the dynamics of a plankton community in 5000 l outdoor microcosms (Hommen and Ratte, in press). Light, temperature and a toxicant, e.g. an insecticide, are the forcing functions driving the state variables population densities and nutrients (Fig. 5). For the following example the food web consists of one phytoplankton compartment, three zooplankton taxa - Cladocera, Copepoda, and Rotifers - and the nutrients Nitrogen and Phosphorus. The parameters in the differential equation system are e.g. maximum photosynthesis rate, respiration rates, optimum temperatures, maximum consumption rates etc. Only lethal effects are considered using saturation functions of the Monod type (LC50 values correspond to Monod's half saturation constants).

In the following we will concentrate on the indirect effect of an insecticide on phytoplankton based on an experiment with a pyrethroid which is toxic to waterfleas and fish while less or not toxic to rotifers and phytoplankton. The observed effects have been analysed by a former deterministic version of the model (Hommen *et al.* 1991). Here the simulation of the control microcosm shall serve as the base for a hypothetical risk assessment example.

Results of phytoplankton dynamics in 1000 Monte-Carlo runs under untreated conditions are shown to demonstrate the influence of different kind of uncertainties. This amount of simulations has been tested to be sufficient to stabilise mean and variance of phytoplankton peak density. In Fig. 6a) all parameters associated with population dynamics and nutrient cycling were kept constant but mean and amplitude of the assumed sinusoidal shape of the forcing functions temperature and light were varied with coefficients of variation of 10%. The thus calculated values were randomly varied ±20% per day (white noise). For Fig. 6b) the opposite was done: no variation in forcing function time series was employed but random sampling of all biological parameters was carried out from normal distributions with assumed coefficients of variation of 10 %. These magnitudes of variation are only rough estimates of uncertainties to demonstrate a simple kind of sensitivity analysis. Under these conditions (model structure and parameter values) uncertainty of biological parameters seems to have more effect on variability of model output. A more detailed sensitivity analysis can be carried out to test the importance of single parameters using multiple regression (e.g. Hommen *et al.* in press).

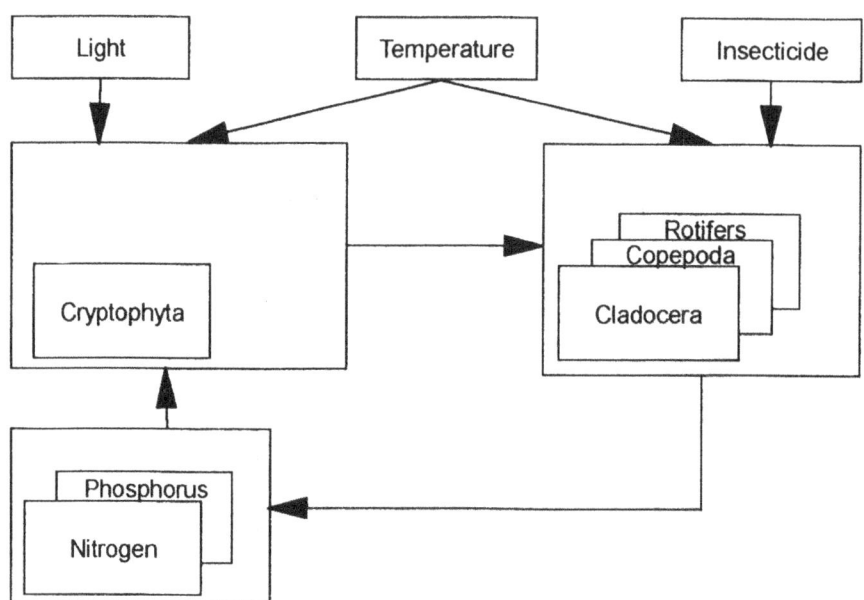

Figure 5. Conceptual diagram of the microcosm model (for explanations see text)

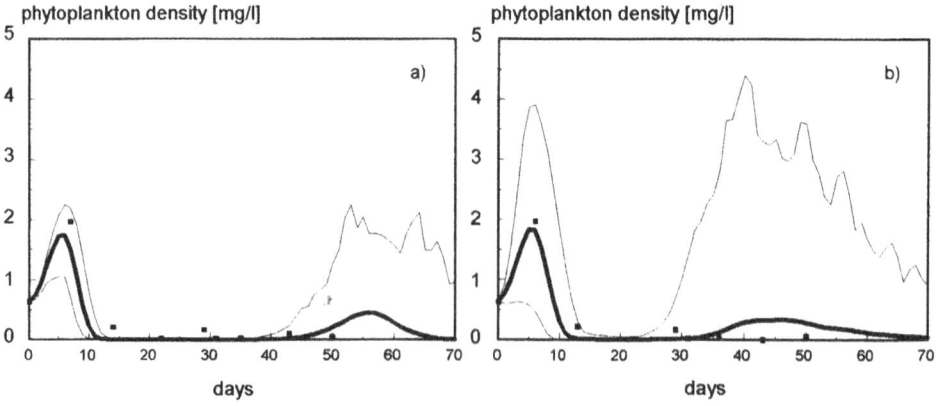

Figure 6. Phytoplankton dynamics in 1000 Monte-Carlo simulations, means, minima and maxima, a) variation of temperature and light, b) variation of biological parameters. Symbols are data (Hommen et al. 1991)

To demonstrate possible analysis of Monte-Carlo runs for Ecological Risk Assessments we conducted simulations for different application rates on an insecticide and calculated the risk of algae blooms measured as total net primary production. The insecticide is assumed to be highly toxic to Cladocera and Copepoda and not toxic to rotifers and phytoplankton which is typical for pyrethroids (Hommen et al. 1991). The simplest way is to compare the means of the chosen endpoint in the Monte-Carlo runs given in Fig. 7a). The next step may be to calculate the probability of a specific hazard, in our example that primary production exceeds a certain level. This was done for Fig. 7b). The saturation characteristics of both curves can be interpreted as a shift from grazer controlled primary production at low insecticide rates to nutrient limited production at higher toxicant concentrations where most Cladocerans and Copepods are killed.

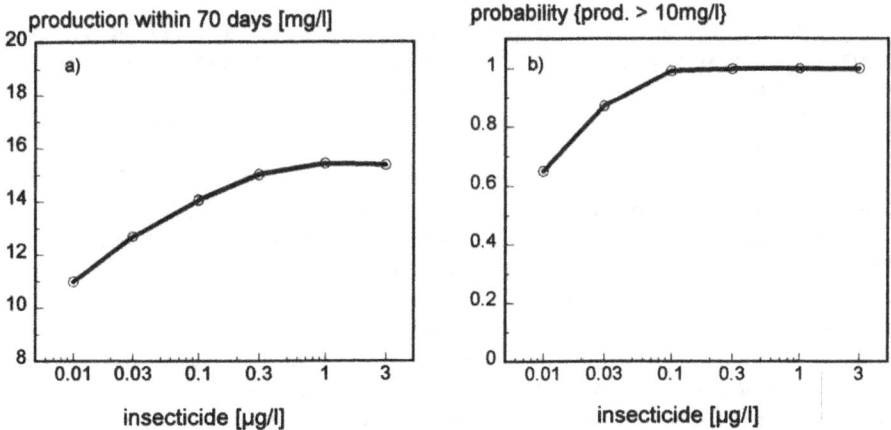

Figure 7. Analysis of Monte-Carlo results, primary production as endpoint a) comparison of means, b) comparison of risks to exceed a production of 10mg l^{-1} within 70 days

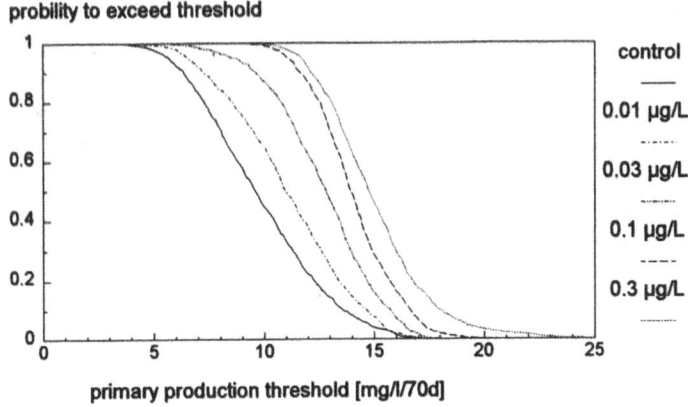

Figure 8. Risk of phytoplankton blooms depending on threshold levels for different insecticide doses

Instead of choosing one certain threshold level for the chosen endpoint, the risk can also be plotted in so called 'quasi-explosion' risk curves depending on threshold values of the endpoint (Ferson *et al.* 1989; Ginzburg and Akcakya 1990). In the example of Fig. 8 the risk curves are shown for different insecticide application rates. The shift of the risk curves to the right with increasing dose indicates the dose-response relationship as given in Fig. 7 and the risk in the different scenarios can be compared by vertical lines at the considered threshold. So, in this example, there is a natural risk of less then 5% exceeding a production of 15 mg l^{-1}, but the risk would be 50% after an application of 0.3 µg l^{-1} of the assumed insecticide.

4. Conclusions

As compared with the two other mentioned approaches to integrate exposure and effects assessment, problems of using Monte-Carlo simulations in ecological risk assessment are seen in the following:
- Simulation models often need a large amount of input data and parameters which are not routinely measured in laboratory toxicity tests. However, simulation models for risk assessment can use 'classical' ecological models and parameter values can often be obtained from ecological research. For example, ecotoxicological models for freshwater micro- or mesocosms can often be derived from lake models (Ratte *et al.*, in press).
- There is usually a large uncertainty in algorithms, parameter values and their distributions. Different approaches exist for example to describe the dependency of photosynthesis from light and parameters found in the literature can easily vary by a factor of 10 (Poethke *et al.*, in press). Known uncertainties are relatively easy to handle in Monte-Carlo simulations, but unknown parameters or in other words large parameter variances can have enormous effects on model predictions (Poethke and Seitz, in press).
- Experimental validation of the obtained probability statements may be possible in principle but it would require a lot of experiments, thus, generally will not be practicable. This problem also exists for other approaches giving risk estimates, but we assume that simulation model results may be assigned credibility if the predicted dynamics can be shown to correspond with results from several case studies as demonstrated for the water flea model. However, validation of models for risk assessment is based usually on deterministic model runs and compares simulated and observed effects instead of risks (e.g. Bartell *et al.* 1992).

The most important benefit of simulation models in our opinion is that they are 'mechanistic instead of statistic', allowing a deeper insight into the function of the system (Suter 1993). Knowledge and hypotheses can be incorporated into a model and then be tested comparing model predictions and experimental results. Due to the interactions between the model compartments it is also possible to investigate indirect effects due to competition and predator-prey relations as shown in principle with the microcosm model. This is seen as a promising alternative to commonly used approaches considering only direct effects. Moreover, simulation models allow to predict the behaviour of the system in a varying physico-chemical environment and thus, to make statements about seasonal sensitivity which is not included in the static approaches.

References

Bartell, S.M., R.V. O'Neill, and R.H. Gardner (1983), 'Aquatic ecosystem models for risk assessment', In: Lauenroth, W.K., G.V. Skogerboe, M. Flug (eds.), *Analysis of ecological systems: State-of-the-art in ecological modelling*, Elsevier, USA.

Bartell, S.M., R.H. Gardner and R.V. O'Neill (1992), 'Ecological risk estimation', Lewis Publishers, USA.

Ferson, S., L.R. Ginzburg and A. Silver (1989), 'Extreme event risk analysis for age-structured populations ', *Ecological Modelling*, **47**, 175-187.

Fitsch, V. (1990), 'Laborversuche und Simulationen zur kausalen Analyse der Populationsdynamik von *Daphnia magna*', PhD thesis, Technical University Aachen, Germany.

Fitsch, V. and H. Kaiser (1987), 'Population dynamics of Daphnia magna - simulations using the individual's approach', In: D.P.F. Möller (ed.), *Advances in Systems Analysis* Vol. 2, Vieweg Verlag, Braunschweig, Germany.

Ginzburg, L.R. and H.R. Akcakya (1990), 'Ecological risk assessment for single and multiple populations', In: A. Seitz and V. Loeschke (eds.), *Conservation: A population biology approach*, Birkhäuser, Basel, Switzerland.

Hallam, T.G., R.R. Lassiter, J. Li and W. Mckinney (1990), 'Toxicant-induced mortality in models of Daphnia populations', *Environmental Toxicology and Chemistry*, **9**, 597-621.

Health Council of the Netherlands (1989), 'Assessing the risk of toxic chemicals for ecosystems'. Report No. 1988/28E, The Hague, Netherlands.

Heimbach, F., H.T. Ratte and W. Pflueger (1992), 'Use of small artificial ponds for assessment of hazards to aquatic ecosystems', *Environmental Toxicology and Chemistry*, **11**, 27-34.

Hommen, U. and H.T. Ratte (in press), 'Application of a plankton simulation model on outdoor-microcosm case studies', In: Hill I.R.; F. Heimbach; P. Leeuwangh and P. Matthiessen (eds.), *Freshwater field tests for hazard assessment of chemicals*, Lewis Publishers, USA.

Hommen, U., H.T. Ratte and H.J. Poethke (1991), 'Modelling a mesocosm plankton community after insecticide application: a first approach', *System Analysis - Modelling - Simulation*, **11/12**, 821-828.

Honeycutt, R.C. and L.G. Ballantine (1983), 'Mathematical modeling application to environmental risk assessments', In: Swann, R.L. and A. Eschenroeder (Eds.), *Fate of chemicals in the environment: compartmental and multimedia models for predictions*, Symposium series 225, American Chemical Society, Washington DC, USA, 249-262.

Kooijman, S.A.L.M. and J.A.J. Metz (1984) 'On the dynamics of chemically stressed populations: The deduction of population consequences from effects on individuals' *Ecotoxicol. Environ. Safety*, **8**, 254-274.

O'Neill, R.V., R.H. Gardner, L.W. Barnthouse, G.W. Suter, S.G. Hildebrand and C.W. Gehrs (1982), 'Ecosystem risk analysis: a new methodology', *Environmental Toxicology and Chemistry*, **1**, 167-177.

Park, R.A. and 24 coauthors (1974), 'A generalized model for simulating lake ecosystems', *Simulation*, **23**, 33-50.

Poethke, H.J., D. Oertel and A. Seitz (1991), 'Risk assessment of toxicants to pelagic food-webs: a simulation study' In: D.P.F. Moeller and O . Richter (eds.), *Analyse dynamischer Systeme in Medizin, Biologie und Ökologie*, Springer Verlag, Berlin, Germany, 192-199.

Poethke, H.J., D. Oertel and A. Seitz (in press), 'Modelling effects of toxicants on pelagic food-webs: Many problems - some solutions', In: *Proceedings of the 8th ISEM-Conference*, Kiel, Germany.

Poethke, H.J. and A. Seitz (in press), 'Models of hazard assessment - problems and perspectives'. In: Hill I.R.; F. Heimbach; P. Leeuwangh and P. Matthiessen (eds.), *Freshwater field tests for hazard assessment of chemicals*, Lewis Publishers, USA.

Ratte, H.T., H.J. Poethke, U. Dülmer and U. Hommen (in press), 'Modelling aquatic field tests for hazard assessment' In: Hill I.R.; F. Heimbach; P. Leeuwangh and
P. Matthiessen (eds.), *Freshwater field tests for hazard assessment of chemicals*, Lewis Publishers, USA.

Seitz, A. and H.T. Ratte (1991), 'Aquatic ecotoxicology: on the problems of extrapolation from laboratory experiments with individuals and populations to community effects in the field', *Comp. Biochem. Physiol.*, **100**, 301-304.

Suter II, G.W. 1993. *'Ecological Risk Assessment'*, Lewis Publishers, USA.

Swartzman, G.L. and K.A. Rose (1984), 'Simulating the biological effects of toxicants in aquatic microcosm systems', *Ecological Modelling*, **22**, 123-134.

U.S. Environmental Protection Agency (1992), 'Framework for Ecological Risk Assessment', EPA /630/R-92/001, Risk Assessment Forum, Washington, D.C., USA

Van Straalen, N.M. and C.A.J. Denneman (1989), 'Ecotoxicological evaluation of soil quality criteria', *Ecotoxicology and Environmental Safety*, **18**, 241-251.

SENSITIVITY ANALYSIS OF A MODEL FOR PESTICIDE LEACHING AND ACCUMULATION

A. TIKTAK, F.A. SWARTJES, R. SANDERS and P.H.M. JANSSEN

National Institute of Public Health and Environmental Protection
PO BOX 1, 3720 BA BILTHOVEN, The Netherlands

e-mail: lbgat@rivm.nl

Summary

The sensitivity of pesticide leaching and accumulation to variations in pesticide properties, soil temperatures, soil water fluxes, and transport parameters was investigated with the general solute transport model, SOTRAS. Pesticide interactions include non-linear Freundlich equilibrium sorption, temperature and pressure head dependent first-order transformation kinetics, and plant uptake. For a number of pesticides with different mobility and half-lives, Monte Carlo simulations were carried out with Latin Hypercube Samples in a preset range of the input parameter domain. The sensitivity of model inputs to model outputs was quantified by statistics of linear regression. The time evolution of model sensitivity and the contribution of various model inputs to the total sensitivity were quantified as well. The standardized analysis gives rapid quantitative information about model behaviour. The results from the analysis are used to determine which parameters should be measured in greater detail and which need further calibration. Results are also used to set up sampling strategies. In general, the accumulation of pesticides in the plough layer was very sensitive to model inputs influencing the transformation rate of the pesticide (soil temperature and half-life) and almost insensitive to sorption characteristics and soil water fluxes. Only in the case of very persistent and mobile pesticides, accumulation was most sensitive to soil water fluxes. The concentration of pesticide in ground water was most sensitive to the Freundlich concentration exponent, and, to a lesser extent, to the Freundlich coefficient, except for some pesticides which are hardly sorbed. The leaching of these pesticides was most sensitive to half-life and soil temperature. The linear regression model could not be used for pesticides with high sorption coefficients, even if variation of the input was kept as low as 1%, but good results were obtained after logarithmic data transformation.

Keywords   Monte Carlo simulation, nonlinear model, pesticide leaching, regression analysis, sensitivity analysis, standardization.

1. Introduction

The most important environmental aspects of pesticide use are leaching into groundwater and accumulation in the upper soil layer. Assessment of the leaching and accumulation potential of a pesticide are now an important part of the pesticide admission procedure. In the Netherlands, these properties are assessed with the PESTicide Leaching and Accumulation (PESTLA) model (Boesten and Van der Linden, 1991). The most important pesticide interactions included are transport, sorption, transformation kinetics and plant uptake.

Simulations are carried out for a sandy soil cropped with maize, under Dutch weather conditions for the year 1980 and a fixed groundwater table at 1 m depth. The PESTLA model is primarily meant for *classifying* pesticides with respect to leaching and accumulation potential.

To assess the accuracy of the predictions for *actual* situations, the model results can be compared with field studies. However, as the number of model inputs is high and sampling is laborious (Boesten and Van der Linden, 1991), it is necessary to know which of these model inputs needs to be quantified at a high level of accuracy. This insight may be gained by carrying out sensitivity and uncertainty analyses on the basis of a Monte Carlo oriented approach. *Sensitivity analysis* is the study of the influence of variations in model parameters, boundary conditions and initial conditions on selected model outputs. In this study, variations in model inputs are described by normal distributions with a coefficient of variation of 1%. *Uncertainty analysis* deals with the influence of *real* distributions of all these model inputs. A sensitivity analysis gives insight into the behaviour of the model itself and can be carried out without a priori knowledge about real distributions; an uncertainty analysis provides insight into the combined effect of real distributions and model behaviour. This paper deals with sensitivity analysis. As the results are dependent on the choice of the nominal values of model inputs (*local* sensitivity analysis), the analysis was repeated for various nominal values (*global* sensitivity analysis). In this way, model sensitivity was analyzed for a broader range of model inputs.

This paper presents the results of a sensitivity analysis of a mathematical model for pesticide leaching and accumulation. First the model and the sensitivity analysis procedure will be described. Second, the model inputs and outputs and the sensitivity to various model inputs are discussed using objective, reproducible quantities. Finally, the importance of the major pesticide transport and transformation processes will be discussed. This paper will show that both the modelling and the sensitivity analysis are standardized.

2. Materials and methods

2.1 Description of the model

A framework for standardizing transport modelling
The general SOTRAS (SOlute TRansport ASsessment) model is used to describe transient flow, hydrodynamic dispersion, irreversible interactions or transformations, plant uptake and equilibria reactions for one or more components in an unsaturated column. The model consists of a number of modules, each module describing an individual process. For some processes, there are different modules available. Initialization and input of required (process specific) parameters are carried out within these modules. In this way SOTRAS offers a framework for standardizing the simulations and guarantees modules being used as often as possible, thus avoiding redundant programming. Standard modules are available for the calculation of soil water fluxes (Tiktak and Bouten, 1992; Van Grinsven and Makaske, 1993), soil heat fluxes and temperatures (Van Grinsven and Makaske, 1993), as well as convective and dispersive fluxes, sorption, integration, and input/output. Interactions or transformations, either through equilibria reactions or (irreversible) kinetic reactions, are often solute dependent and must be defined by whoever constructs a new transport program. Although general algorithms exist for the calculation of chemical equilibria, they are not used within SOTRAS. Instead, a procedure has been developed for standardizing the derivation of 'dedicated' chemical equilibria routines. These routines are

faster when the number of components is limited, thus making SOTRAS also suitable for multiple simulations such as Monte Carlo analyses, or regional applications.

At present, SOTRAS applications are available for pesticides, and under development for heavy metals (for cadmium in particular). This paper, will deal with the pesticide case.

Transport equations
The model provides a solution to the Richards equation for water transport and the convection-dispersion equation for solute transport:

$$C(h) \cdot \frac{\partial H}{\partial t} = \frac{\partial}{\partial z} \left[K(h) \cdot \left(\frac{\partial h}{\partial z} + 1 \right) \right] - S_w(h) \tag{1}$$

and

$$\frac{\partial \rho X}{\partial t} + \frac{\partial \theta c}{\partial t} = \frac{\partial}{\partial z} \left(\theta D \frac{\partial c}{\partial z} \right) - \frac{\partial \theta v c}{\partial z} - S_s - R_s \, , \tag{2}$$

where $C(m^{-1})$ is differential water capacity, $t(d)$ time, $z(m)$ vertical position, $h(m)$ soil water pressure head, $K(m \ d^{-1})$ unsaturated hydraulic conductivity, $S_w(d^{-1})$ sink term accounting for root water uptake, ρ (kg m^{-3}) dry bulk density of the soil, X(kg kg^{-1}) solid phase mass content, c(kg m^{-3}) mass concentration in the liquid phase, $D(m^2 \ d^{-1})$ hydrodynamic dispersion coefficient, $v(m \ d^{-1})$ rate of flow of the pore water, S_s (kg $m^{-3} \ d^{-1}$) rate of plant uptake of solute, and R_s(kg $m^{-3} \ d^{-1}$) rate of irreversible transformation of solute. The hydrodynamic dispersion coefficient is assumed to be equal to the sum of the molecular diffusion coefficient (D_m) and the hydrodynamic dispersion coefficient (D_d):

$$D = D_m + D_d = \lambda D_0 + a |v| \, , \tag{3}$$

where λ (-) is tortuosity factor, $D_o(m^{-2} \ d^{-1})$ molecular diffusion coefficient in water and α (m) dispersivity.

Case-dependent process formulations
For pesticide transport the most important processes are transport, sorption, transformation and plant uptake. Process formulations are derived from Boesten and Van der Linden (1991).

Sorption is described by the Freundlich equation:

$$X = K_F c^n \, , \tag{4}$$

where K_F ($m^3 \ kg^{-1}$) is the sorption coefficient and n (-) the Freundlich exponent. K_F is estimated from $f_{om} K_{om}$, where f_{om} (kg kg^{-1}) is mass fraction of organic matter and $K_{om}(m^3 \ kg^{-1})$ the coefficient for partitioning between organic matter and soil water.

Transformation of a pesticide in soil, R_s, is described with a first-order rate equation:

$$R_s = kc^* \, , \tag{5}$$

in which k (-) is the rate coefficient for transformation, and c^* (kg m^{-3}) the total content of pesticide in the soil system. The rate coefficient is calculated from the rate coefficient for reference conditions (k_{ref}):

$$k = f_T f_\theta f_z k_{ref} ,$$ (6)

where f_T, f_θ and f_z (-) are factors accounting for the influence of temperature, water content and depth in soil. The reference conditions are defined as those in fresh soil, sampled from the plough layer, kept at 20°C and at a soil water pressure head of -100 cm. The half-life time under reference conditions can be calculated from:

$$DT_{50} = \ln2/k_{ref} .$$ (7)

The reduction factors are described by:

$$f_T = e^{(\gamma(T - T_{ref}))}$$ (8)

and

$$f_\theta = (\theta/\theta_{ref})^B ,$$ (8)

where γ (-) is empirical parameter, T (°C) prevailing soil temperature, and T_{ref} (°C) reference temperature (20°C), θ_{ref} (m^3 m^{-3}) volumetric water content at reference conditions (i.e. water content at $h = -100$ cm) and B (-) empirical parameter. The depth-in-soil function, accounting for depth distribution of microbial activity, is described by a numerical function.

The rate of uptake of a pesticide by plant roots from soil, S_s, is described by:

$$S_s = FS_w c ,$$ (10)

where F (-) is the transpiration stream concentration factor.

3. Sensitivity analysis

In this paper sensitivity analysis is defined as the study of the influence of *variations* in model parameters, initial conditions, etc. on model outputs. The results of the sensitivity analysis are used to gain insight into the behaviour of the model, to evaluate the required accuracy of the model inputs and to assess the relevance of major processes when using pesticide models.

The sensitivity analysis is carried out with the UNCSAM software package (Janssen *et al.*, 1993). This package provides a standardized method for performing model analyses on a large variety of mathematical models on the basis of a Monte Carlo oriented approach. Monte Carlo based methods for sensitivity analysis rely on the fact that variations in model inputs can be described by specifying probability distributions, which reflect the variation of a model input around a 'nominal' value. Sampling is performed from these distributions, resulting in a set of values for the various model inputs. A disadvantage of the Monte Carlo approach is the large number of samples required. In order to reduce this, the efficient Latin Hypercube Sampling (LHS) technique is used. When using LHS, the parameter space is sampled in a representative fashion with only a few samples ($N > 5p$, where p is the number of model inputs to be sampled and N is the number of samples). The sampled model inputs are used to simulate model outputs. The further analysis consists of computing basic statistical information (means, percentiles and variances) and performing regression analysis to obtain insight into the contribution of the various model inputs to sensitivity. The coefficients of this regression analysis are used as sensitivity measurements. Suppose that least-squares linear regression of the model output $y(k)$ in the k-th simulation run ($k = 1, ..., N$) on the associated sampled model inputs $x_1(k), ..., x_p(k)$ results in the regression equation:

$$y(k) = \beta_0 + \sum_{i=1}^{p} \beta_i \, x_i(k) + \hat{e}(k) \qquad (k = 1,...,N) \, , \qquad (11)$$

where $\beta_0, ..., \beta_p$ denote the ordinary regression coefficients (*ORC*), and $\hat{e}(k)$ denotes the regression residual. β_i can be considered as absolute sensitivity measurements for the absolute change of Δy of y, if x_i changes with amount Δx_i, while the other model inputs remain constant. The coefficient of determination, *COD*, expressing the validity of the linear regression model, $\hat{y}(k)$, to approximate the original model output, $y(k)$, is equal to:

$$COD = \frac{S_{\hat{y}}^2}{S_y^2} \, , \qquad (12)$$

where S^2 denotes the sample variance of the associated quantity. *COD* has a value between 0 and 1; *COD* values close to 1 indicate a good approximation. In this paper the regression coefficients are standardized by dividing the absolute change by the average values, and considering the change $\Delta y/\bar{y}$ of y due to a relative change $\Delta x_i/\bar{x}_i$ of x_i. The associated regression coefficients are called the normalized regression coefficients (*NRC*), and, in fact, measure the relative changes with respect to the average values (i.e. relative sensitivity measurement). The *NRC* can be positive or negative. Notice that the regression coefficients are valid only if the linear regression model applies (i.e. *COD* approaches 1), and if no correlations exist between the model inputs.

As solute transport models often show a strong non-linear behaviour, the linear model only applies when the model inputs are given *small* variations around their nominal values. Therefore the model inputs are sampled from normal distributions with a mean μ equal to the nominal value of the model input and standard deviation $\sigma = 0.01\mu$. The sensitivity analysis thus performed is a *local* sensitivity analysis and must be repeated several times with different nominal values. Model inputs variable with time and/or depth are varied by multiplying them by a unique factor for all times and/or depths. These factors have a nominal value of 1. This implies that the variations of these model inputs are fully correlated in time and depth.

Strategy for pesticides
The sensitivity analysis was carried out for a vulnerable Dutch situation, i.e. a sandy soil cropped with maize, with moderate organic matter content, and with a fixed, hypothetical groundwater table at a 1m depth. Meteorological conditions are those recorded in De Bilt

in 1980, a 75% wet year. A pesticide dose of 1 kg ha^{-1} was applied once on 25 May.

Table 1. Summary of nominal parameter values and weather data used in the simulations. The nominal values are according to Boesten and Van der Linden (1991). The Freundlich coefficient (K_{om}) and the half-life (DT_{50}) are variable (cf. table 2)

Name and symbol			Value(s)	Unit
Mass fraction of organic	0	$< z < 0.3$ m	0.047	kg kg^{-1}
matter (f_{om})	0.3	$< z < 0.5$ m	0.008	kg kg^{-1}
	0.5	$< z < 0.6$ m	0.002	kg kg^{-1}
	0.6	$< z < 1.0$ m	0.002	kg kg^{-1}
		$z > 1.0$ m	0.001	kg kg^{-1}
Dry bulk density of	0	$< z < 0.3$ m	1310	kg m^{-3}
the soil (ρ)	0.3	$< z < 0.5$ m	1540	kg m^{-3}
		$z > 0.5$ m	1640	kg m^{-3}
Dispersivity (α)			0.05	m
Freundlich exponent (n)			0.9	(-)
Molecular diffusion coefficient in water (D_0)			4.10^{-3}	m^2 d^{-1}
Tortuosity factor for	$\theta \leq 0.035$		≤ 0.002	(-)
diffusion in the	$\theta = 0.070$		0.010	(-)
liquid phase (λ)	$\theta = 0.100$		0.030	(-)
	$\theta = 0.150$		0.060	(-)
	$\theta = 0.200$		0.100	(-)
	$\theta = 0.300$		0.200	(-)
	$\theta = 0.400$		0.340	(-)
Transpiration stream concentration factor (F)			0.5	(-)
Parameter in temperature reduction function (γ)			0.08	K^{-1}
Parameter in water content reduction function (B)			0.25	(-)
Function describing depth	$z < 0.3$ m		1.0	(-)
dependence of	$z = 0.5$ m		0.9	(-)
transformation (f_z)	$z = 1.0$ m		0.0	(-)
Annual rainfall			860	mm
Annual potential evapotranspiration			480	mm
Annual average soil temperature at 5 cm depth			11	°C

In the sensitivity analysis several model inputs were taken into account. A summary of the inputs is found in Table 1. The model inputs were grouped according to the process in which they play a role (i.e. transport, sorption, transformation or plant uptake). Sensitivity analyses were carried out for three hypothetical pesticides, characterized by different sorption coefficients and half-lives, and for nine existing pesticides (Table 2). As the nominal values of all other model inputs were constant, the sensitivity analyses were local with respect to soil parameters and meteorology dependent parameters (such as soil water fluxes and temperatures).

Two model outputs are considered: (1) the fraction of pesticide left in the plough layer (fr_{pl}) and (2) maximum concentration of pesticide from a depth of 1 to 2 m (c_{max}). The first model output is considered a measurement of pesticide accumulation in the top soil and the second a measurement of pesticide leaching.

Table 2. Nominal value of half-life (DT_{50}) and Freundlich coefficient (K_{om}) of the investigated pesticides.

Pesticide	K_{om} (dm^3 kg^{-1})	DT_{50} (d)
hypothetical; case 1	200	200
hypothetical; case 2	1	10
hypothetical; case 3	1	200
metalochloor	103	101
pirimifos-methyl	200	12
thiram	4	18
atrazin	70	50
metamitron	43	28
cyanazin	55	16
chloorthalonil	5031	10
isophalonitril	15	387
bentazon	0.4	48

4. Results and discussion

4.1 Fraction of pesticide left in the plough layer

Figures 1A-C show the percentiles of fr_{pl} as a function of time for the three hypothetical pesticides. Note that for each of the three pesticides considered, the relative variation (coefficient of variation) is small ($\approx$ 1%). The non-mobile pesticide (case 1) is still present in the plough layer five years after application. The other two pesticides are removed from the plough layer within one year of application, either by advection or by transformation. Sensitivity of model outputs to model inputs is time-dependent. As sensitivity analysis at times after the pesticide has been completely removed from the plough layer is meaningless, the point in time at which the analysis is carried out must be chosen carefully. We decided to evaluate the analysis when about half of the pesticide is still present in the plough layer.

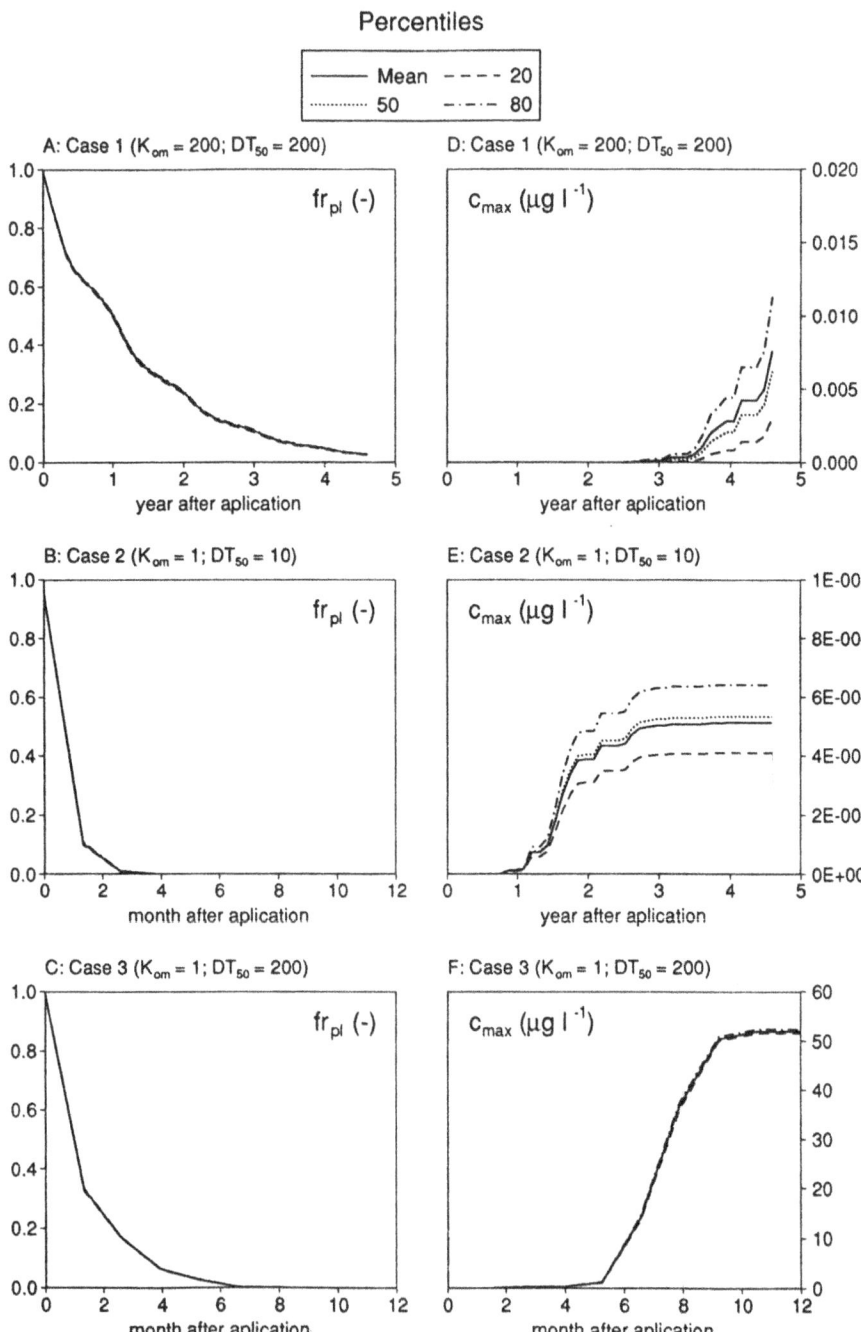

Figure 1. Percentiles and means for three hypothetical pesticides as a function of time. A-C: Fraction of pesticide, fr_{pl} (-), left in the plough layer. D-F: Maximum concentration of pesticide in the groundwater, c_{max} (μ l^{-1}).

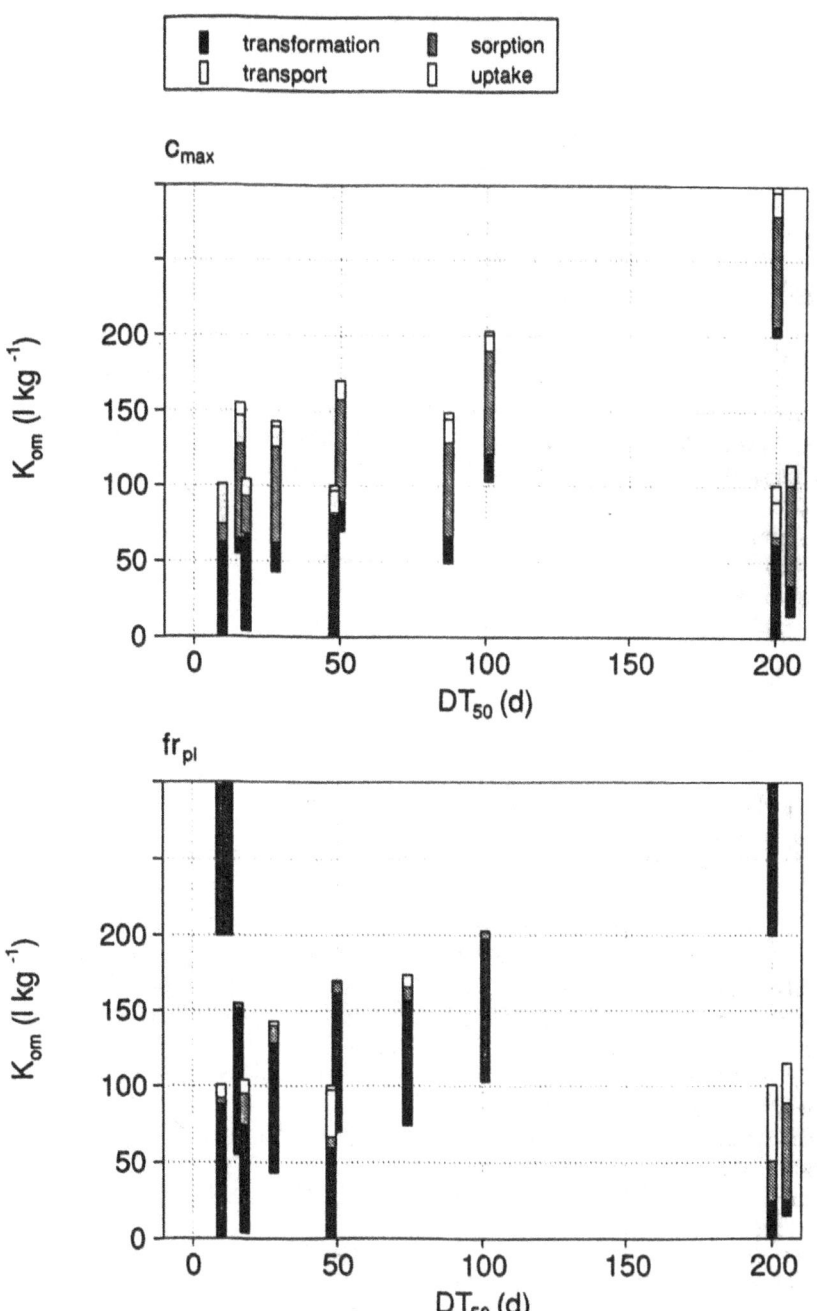

Figure 2. Contribution of the four major processes to model sensitivity for the three hypothetical pesticides and for the nine existing pesticides. The 12 pesticides are characterized by different K_{om} and DT_{50}. The length of the bars is proportional to the contribution of the major processes to the sensitivity of the model outputs considered.

Table 3. Sensitivity of the pesticide fraction left in the plough layer (fr_{pl}) to various model inputs. Analysis is carried out at the point in time when half the dose is still present in the plough layer.

		Analysis 1 $K_{om} = 200$ $DT_{50} = 200$ $COD = 1.000$		Analysis 2 $K_{om} = 1$ $DT_{50} = 10$ $COD = 0.998$		Analysis 3 $K_{om} = 1$ $DT_{50} = 200$ $COD = 0.999$	
Process		NRC	rank	NRC	rank	NRC	rank
T	transformation	−0.244	1	−5.514	1	−0.257	2
DT_{50}	transformation	+0.227	2	+4.490	2	+0.212	4
f_z	transformation	−0.226	3	−4.439	3	+0.210	5
γ	transformation	+0.040	4	·+1.083	4	+0.054	10
B	transformation	+0.011	5	+0.183	7	<0.001	>10
n	sorption	−0.003	6	−0.157	8	−0.064	9
f_{om}	sorption	+0.002	7	+0.110	11	+0.199	6
ρ	sorption	+0.002	8	0.156	9	+0.196	7
K_{om}	sorption	<0.001	> 10	+0.136	10	+0.192	8
v	transport	<0.001	> 10	−0.916	5	−0.993	1
α	transport	<0.001	> 10	+0.402	6	+0.248	3
D_0	transport	<0.001	> 10	<0.001	> 11	<0.001	>10
S	uptake	−0.001	9	<0.001	> 11	<0.001	>10
F	uptake	−0.001	10	<0.001	> 11	<0.001	>10

Table 3 shows the results for the three pesticides. The linear regression model gives a good fit $(COD \approx 1)$ for all cases. The following conclusions can be drawn from the table:
- In analysis 1, the model inputs to which fr_{pl} is most sensitive are: $f_z \approx DT_{50} \approx T > \gamma \approx B$. All model inputs concern transformation. Despite the high DT_{50}, transformation processes appear to control the pesticide fraction left, due to high sorption, the pesticide remains in the upper soil layer and can be transformed. This explains the importance of the depth dependence of transformation (f_z).
- In analysis 2, the ·inputs are ranked as follows: $T \approx DT_{50} \approx f_z > \gamma > v$. The model inputs ranking highest concern transformation.
- In analysis 3, the order is: $v > T \approx \alpha \approx DT_{50} \approx f_z$. Here, both transport and transformation parameters are relevant.

All *NRC*s are low in the first analysis, implying that here fr_{pl} is not very sensitive to the model inputs. In the second analysis the *NRC*s are high, implying high sensitivity. For example, increasing the soil temperatures by 1% decreases the fraction of pesticide in the plough layer by more than 5%. In none of the analysis, do plant uptake parameters play an important role.

Both the absolute value of the *NRC*s and the order in which the model inputs are ranked are dependent on the point in time at which the evaluation is carried out. For example, in the third case, the model inputs are ranked as follows six months after application:

$v \approx n \approx \alpha > f_{om} \approx \rho$. This implies that transport and sorption have become more important than transport and transformation.

Table 4. Sensitivity of the maximum concentration of pesticide in the upper groundwater (c_{max}) to various model inputs. The analysis is carried out at the point of time that the maximum concentration was reached.

		Analysis 1 $K_{om} = 200$ $DT_{50} = 200$ $COD = 0.925$		Analysis 2 $K_{om} = 1$ $DT_{50} = 10$ $COD = 0.951$		Analysis 3 $K_{om} = 1$ $DT_{50} = 200$ $COD = 0.960$	
Par	Process	NRC	rank	NRC	rank	NRC	rank
T	transformation	−2.678	9	−5.679	1	−0.436	1
DT_{50}	transformation	+1.235	14	+4.947	2	+0.390	2
f_z	transformation	+2.182	11	−4.752	3	−0.380	3
γ	transformation	−3.493	8	+1.635	6	+0.164	6
B	transformation	−1.352	13	>0.001	> 10	>0.001	>10
n	sorption	+62.89	1	+1.014	7	+0.057	7
f_{om}	sorption	−10.67	3	−0.582	8	+0.199	6
ρ	sorption	−7.731	5	−0.543	9	+0.196	7
K_{om}	sorption	−11.30	2	−0.455	10	−0.038	10
v	transport	+10.66	4	+4.027	4	+0.184	5
α	transport	+5.719	6	+1.824	5	−0.235	4
D_0	transport	−3.657	7	>0.001	> 10	>0.001	>10
S	uptake	+2.588	10	>0.001	> 10	−0.048	9
F	uptake	>0.001	> 14	>0.001	> 10	−0.05	8

In general, the sensitivity to the Freundlich exponent (n) increases with time. This is due to the non-linear sorption isotherm ($n < 1$) resulting in a slower decrease in the amount of pesticide sorbed than in the amount of pesticide in the liquid phase. Remarkable is that in none of the three analyses is the value of the Freundlich coefficient (K_{om}) important.

Note the importance of the dispersivity (α) in the third analysis. This is due to the fact that pesticide application is carried out during a dry period with low soil water fluxes and relatively high concentration gradients. If the application had been carried out during a wet period, the influence of the dispersivity would certainly have been negligible. This demonstrates the local character of the sensitivity analysis. However, the choice of application in dry periods is realistic, as no farmer would apply pesticides during wet periods.

4.2 Concentration of pesticide in groundwater

The percentiles of the pesticide concentration in the upper groundwater are presented in Figures 1D-F. It is obvious that the relative variation (coefficient of variation) of this model output is higher than for the fraction of pesticide in the plough layer. The maximum concentration of pesticide in the groundwater is reached between 2.5 and 5 years after application for case 1, between 1 and 3 years for case 2 and within 1 year for case 3.

Table 4 shows the sensitivity of the maximum concentration of pesticide in the upper groundwater. The analyses were carried out at five years after pesticide application. However, as we look at the *maximum* concentration of pesticide in the groundwater, the real point in time to which the analyses apply coincides with the time when the maximum concentration was reached. The *COD*s are high enough to allow linear regression. The following can be concluded:

- In analysis 1, the model inputs are ranked as follows: $n \gg K_{om} \approx f_{om} \approx v \approx \rho$. Four sorption parameters are found at the highest sensitivity level. The sensitivity to the Freundlich exponent (n) strongly dominates other model inputs.
- In analysis 2, the model inputs are ranked as follows: $T \approx DT_{50} \approx f_z \approx v > \alpha$. The maximum concentration in groundwater is sensitive to almost the same model inputs as the fraction left in the plough layer.
- In analysis 3, the order is: $T \approx DT_{50} \approx f_z > \alpha \approx v$. Again, transformation parameters are important.

From the sorption parameters, the Freundlich exponent dominates model sensitivity. Notice the very high NRC for this model input. With respect to transport parameters, the high sensitivity for the dispersivity (α) is remarkable. The importance of both model inputs has already been discussed.

4.3 Contribution of processes to model sensitivity

Figure 2 shows the contribution of the four major processes to model sensitivity for the 3 hypothetical pesticides described above and for 9 existing pesticides. As pesticides are characterized by different K_{om} and DT_{50}, their position can be plotted in a plot of K_{om} versus DT_{50}. The length of the bars is proportional to the contribution of the major processes to the sensitivity of the model outputs considered (i.e. fr_{pl} at the point of time that half the dose is present in the plough layer and c_{max} at $t = 5$ years), which is calculated as follows:

$$NRC_{process} = \sum_{i=1}^{m} NRC_i / m , \qquad (13)$$

where m is the number of evaluated model inputs for each process.

Model sensitivity for fr_{pl} is highest at high transformation rates (low DT_{50}). Except for the lower right hand corner where transport parameters control fr_{pl}, transformation parameters are the most important. This is also true for the upper right hand corner of the diagram. The most important conclusion which can be drawn from these findings is that model inputs concerning transformation should be measured with relatively high accuracy.

For c_{max}, model sensitivity is to a large extent controlled by sorption processes. Sensitivity is highest in the upper part of the diagram (strong sorption). The Freundlich exponent (n) is the most important sorption parameter. In cases of very high sorption, the *COD*s were too low for applying linear regression. These cases are not plotted in the diagram. In cases of high sorption, a clear exponential relation was found between n and c_{max} (Figure

3). This finding was verified by performing linear regression analysis between n and the logarithmic-transformed model output c_{max}. In such cases, high values of COD (0.916) were found and n strongly dominated model sensitivity.

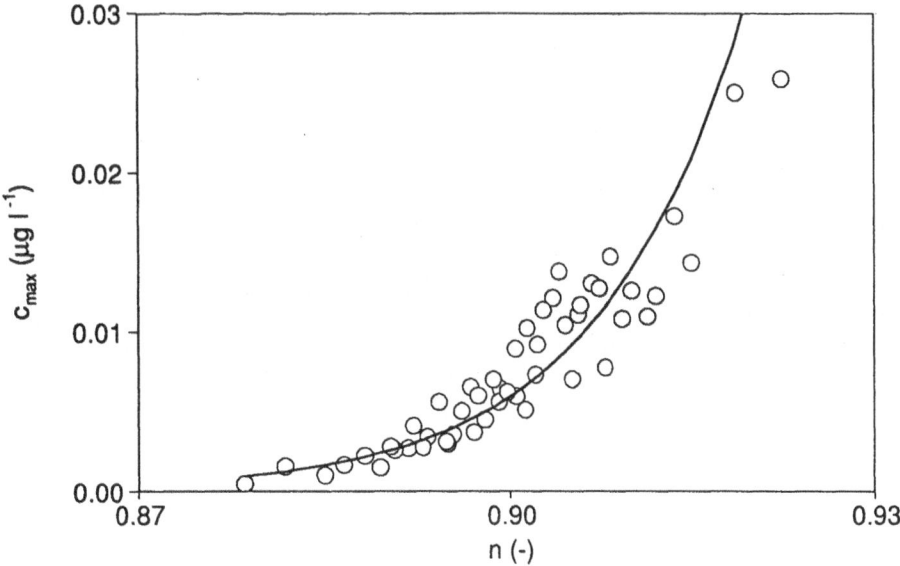

Figure 3. Scatterplot of the Freundlich exponent (n) versus the maximum concentration of pesticide in the groundwater (c_{max}) for analysis 1 (K_{om} = 200 dm^3 kg^{-1}; DT_{50} = 200 d).

5. Conclusions

The UNCSAM software provides a convenient and adequate tool for performing Monte Carlo based sensitivity analysis. The efficient Latin Hypercube Sampling technique allowed the number of samples to be limited to five times the number of investigated model inputs. In most cases, linear relations were found between the investigated model inputs and outputs if variations of the model inputs were kept low (standard deviation 1%). Only for pesticides with high sorption coefficients, the linear regression model did not apply. However, in these cases a good fit could be found after logarithmic data transformation.

Except for persistent pesticides with low sorption coefficients, the fraction of pesticide left in the plough layer (fr_{pl}) is most sensitive to transformation parameters. Therefore, for accurate predictions of accumulation in the plough layer, transformation parameters should be determined with high level of accuracy. The maximum concentration of pesticide in the groundwater (c_{max}) is very sensitive to sorption parameters, the most important parameter being the Freundlich exponent (n). In some cases, changing the Freundlich exponent by 1% results in a change of the maximum concentration of pesticide in the groundwater by 65%. It is obvious that for accurate predictions of the maximum concentration, the Freundlich exponent must be measured with high accuracy. We can also conclude that linear sorption models, which are often used in pesticide leaching studies, may yield incorrect results.

Because of the non-linear behaviour of the transport model, the linear regression model can only be applied if small variations in the model inputs were considered (standard deviation $\sigma = 0.01\mu$). Thus, the results of the analysis are strongly dependent on the value of model inputs (*local* sensitivity analysis). However, in this study the analysis has been successfully converted into a *global* sensitivity analysis with respect to K_{om} and DT_{50} by carrying out sensitivity analyses for a number of pesticides with different K_{om} and DT_{50}. The analysis is nevertheless still local with respect to other parameters. For example, rainfall distribution after pesticide application will strongly influence the results of the sensitivity analysis.

It should be mentioned that the sensitivity of model outputs will be different if the real distributions of the model inputs are considered ('uncertainty analysis'). For example, if we consider that the real uncertainty of the sorption coefficient K_{om} is much higher than the uncertainty of the sorption exponent (n), the influence of K_{om} will be higher than the influence of n (see also Boesten, 1991). In any case, both sensitivity analysis and uncertainty analysis provide useful tools in evaluating the model. A sensitivity analysis gives a clear insight into the behaviour of the model itself and can be carried out without a priori knowledge about real distributions; an uncertainty analysis provides insight into the combined effect of real distributions and model behaviour.

Acknowledgements

The authors would like to thank J.J.M. van Grinsven for critically reviewing the paper and R.E. Wijs-Christensen for editing the text.

References

Boesten, J.J.T.I. (1991), 'Sensitivity analysis of a mathematical model for pesticide leaching to groundwater', *Pestic. Sci.*, **31**, 375-388.

Boesten, J.J.T.I. and A.M.A. van der Linden (1991), 'Modelling the influence of sorption and transformation on pesticide leaching and persistence', *J. Environ. Qual.*, **20**, 425-435.

Janssen, P.H.M., P.S.C. Heuberger and R. Sanders (1993), 'UNCSAM: a useful tool for sensitivity and uncertainty analysis of mathematical models'. Submitted for publication to *Env. Software*.

Tiktak, A. and W. Bouten (1992), 'Modelling soil water dynamics in a forested ecosystem. III: Model description and evaluation of discretization', *Hydrol. Proc.*, **6**, 455-465.

Van Grinsven, J.J.M. and G.B. Makaske (1993), 'A one dimensional model for transport and accumulation of water and nitrogen based on the Swedish model SOILN', *National Institute of Public Health and Environmental Protection report 714908001*, Bilthoven, The Netherlands. (The water flow model on pp. 9-21).

BAYESIAN UNCERTAINTY ANALYSIS IN WATER QUALITY MODELLING

P.R.G. KRAMER, A.C.M. DE NIJS AND T. ALDENBERG

Laboratory for Water and Drinking Water Research, R.I.V.M.,
P.O. Box 1, 3720 BA Bilthoven, The Netherlands

Abstract

To analyse the influence of pollution and the impact of historical and future measures in river basins, an integrated modelling approach comprising the causality chain of emission, distribution and effects of toxicants was chosen. An integrated approach is also essential to study the interactions between abiotic and biotic components, which are considered vital to ecosystem functioning. After quantifying the sources of pollution, dynamic water quality models are used to determine the distribution of nutrients and toxicants over various spatial compartments. The models contain a large number of unknown parameters and therefore a statistical model analysis is needed. A combined uncertainty and sensitivity analysis procedure based on Bayesian inference is applied. This method leads to probability distributions for the uncertain parameters as well as for selected output variables, as shown in the results from the national water quality model. The inorganic matter in the water phase is analysed and calibrated on the sedimentation and resuspension parameters.

Keywords Water quality modelling; Bayesian inference; uncertainty analysis; probability distributions.

1. Introduction

Integrated studies are performed in order to analyse the influence of pollutant input into river basins. With this integrated approach it is possible to assess the impact of sanitary actions on the water quality and on the accumulation in ecosystems.

For the Netherlands an integrated modelling effort is performed in a national water quality model (de Nijs et al., in prep.), to determine and predict the combined effects of eutrophicating and toxic compounds in the main Dutch waterways. The project is an activity of the National Institute of Public Health and Environmental Protection (RIVM); the input and calibration data were supplied by the Institute of Inland Water Management and Waste Water Treatment (RIZA). The model offers the possibility to investigate national and international sanitary actions on a local scale. Results are shown for the analysis of the sedimentation and resuspension processes controlling the inorganic matter in the water phase.

Contaminant accumulation models are interfaced to the water quality models in order to assess the ecological effects of the compounds in specific areas (Fig. 1). The conceptual framework of CATS, Contaminants in Aquatic and Terrestrial ecoSystems, was developed to evaluate the impact of measures taken to reduce the emissions of chemicals (Traas & Aldenberg, 1992). Distributions of toxicant concentrations in organisms are used to evaluate the probability of exceedence of NOEC levels or other environmental standards. Also, the predicted effect of different loading scenarios can be estimated. A CATS model developed for the Rhine sedimentation areas, showed the response of cadmium concentration in phytoplankton and small molluscs to different loading scenarios (Kramer *et al.*, submitted).

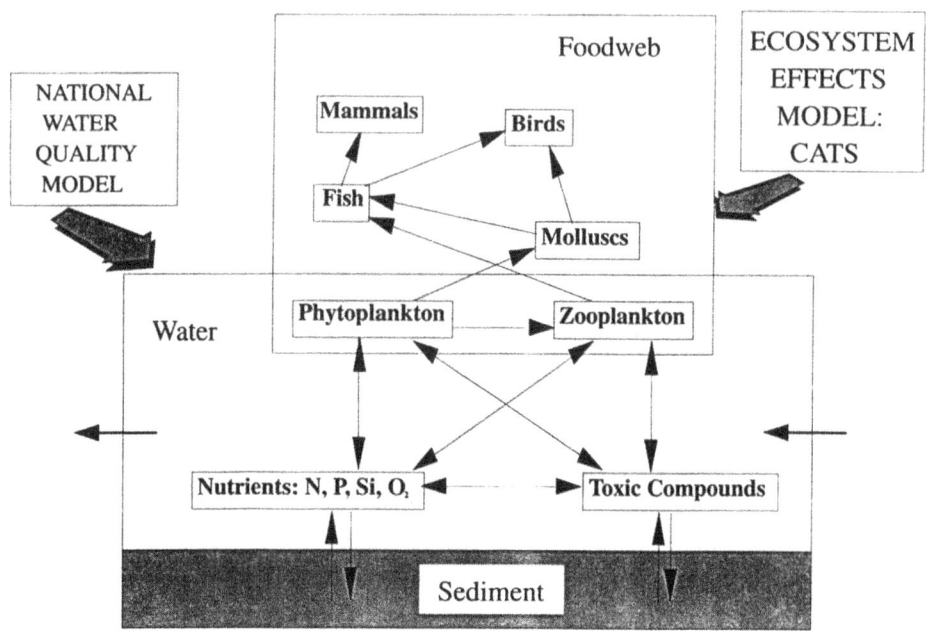

Figure 1. Schematic overview of the integrated risk assessment modelling approach.

The water quality model and the accumulation model will be linked in order to determine the consequences and benefits of managerial measures. However, these large scale models contain various sources of uncertainty. Sources of uncertainty that will never disappear are the environmental influences on the system. However, other sources can be assessed and their uncertainty reduced. First of all, the model structure itself with its assumptions and simplified processes is a cause of uncertainty. Initial conditions are not always well known. Finally, uncertain and/or unknown parameters contribute to the model uncertainty. A model analysis based on Bayesian inference enables a combined uncertainty analysis, calibration, sensitivity analysis as well as a tool for risk analysis for all types of uncertainty mentioned. In this paper the principles of this analysis are briefly described and results are shown from the application on the part of the WATNAT water quality model, describing the fate of the inorganic matter.

2. Water quality model

The WATNAT model consists of a hydrodynamic transport module of the network of the main Dutch water ways. This network is subdivided into 109 segments connected to each other by 385 links. Within each segment various water quality processes are incorporated (Fig. 2). A set of differential equations describes the behaviour of nutrients, inorganic and organic matter as well as primary producers through the whole system including the sediment. The cycles are modelled separately and the mass balances are completely closed within each segment (Janse *et al.*, 1992). Phytoplankton is divided into three functional groups: green algae, Cyanobacteria and diatoms. Output variables as dissolved and particulate nitrogen and phosphorus, chlorophyll-a and dissolved oxygen are included.

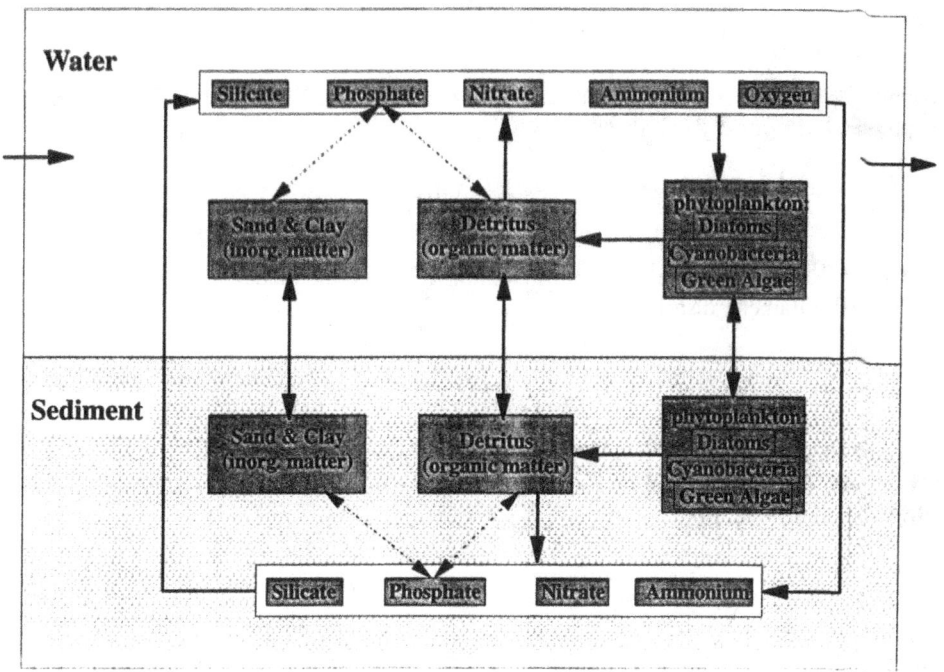

Figure 2. Schematic representation of the WATNAT water quality model segment constituents and processes.

At this moment model descriptions for heavy metals and organic xenobiotic compounds are incorporated to obtain a complete model including nutrients as well as toxicants. The model thus contains a large number of variables (28 per segment) and parameters (127), many of which are uncertain. For a reliable application the following model analysis method was developed.

3. Bayesian model analysis

A model analysis approach is used, based on Bayesian inference as described by Box & Tiao (1973) and Box (1971), which allows a combined uncertainty and sensitivity analysis, calibration of parameters as well as a tool for risk analysis (Aldenberg *et al.*, submitted). Given a model described as

$$\eta = M(x,\theta) \ , \tag{1}$$

where η is a predictive output variable to be compared with observed values, M is the model formulation, which is a function of x, a set of fixed parameters and θ, the uncertain parameters to be analysed. Let there be n observed values for the predictive output variable, y_i (with $i = 1,...,n$), then a comparison of model output variables with measurements leads to

$$y_i = M(x_i,\theta) + \varepsilon_i \ , \tag{2}$$

where the error term ε reflects the discrepancy between model and data. This error term is supposed to be normally distributed

$$\varepsilon \sim N(0,\sigma^2) \tag{3}$$

in which σ, like θ, is uncertain and to be examined. Before the model is confronted with the data a prior parameter distribution

$$p(\theta,\sigma) \tag{4}$$

is given, with θ and $\log(\sigma)$ independent and uniformly distributed (Box & Tiao, p. 186). Then, given the measurements a posterior parameter distribution can be obtained through Bayes' theorem:

$$p(\theta,\sigma|y) \propto p(y|\theta,\sigma) \cdot p(\theta,\sigma) \tag{5}$$

with

$$p(y|\theta,\sigma) \tag{6}$$

the likelihood.

This theory leads to the following application according to Box & Tiao (1973). With the prior given above the sum of squared differences between model output and observations can be calculated

$$S = \sum_i \left(y_i - M(x_i,\theta) \right)^2 \tag{7}$$

and the posterior parameter distribution is then given by:

$$p(\theta,\sigma|y) \propto \frac{1}{\sigma^{n+1}} \cdot e^{-\frac{1}{2\sigma^2}S} \ . \tag{8}$$

When the σ is not taken into account, it can be integrated out and the posterior parameter distribution will become:

$$p(\theta|y) \;\propto\; S^{-\frac{n}{2}}. \tag{9}$$

4. Application

This method was applied to the WATNAT model, in which the section describing the inorganic suspended matter was calibrated. Three uncertain parameters, θ, determining the sedimentation and resuspension of inorganic matter were calibrated. Other parameters are not supposed to affect these processes, so the parameters given in table 1 form a submodel. In the near future other parts of the model will be analysed, involving more than 20 parameters influencing a wide range of processes.

Table 1. Sedimentation and resuspension parameters, steering the inorganic matter concentrations in water and sediment, to be calibrated.

cVCritSedIM	critical shear stress velocity for sedimentation of inorganic matter	$[m \cdot s^{-1}]$
cVeloSedIM	sedimentation velocity in stagnant water at 20 °C	$[m \cdot d^{-1}]$
cFDResusMin	minimum resuspension flux (if present)	$[gDW \cdot m^{-2} \cdot d^{-1}]$

In this case $M(x, \theta)$ represents the model or better, the numerical solution to the set of differential equations describing the behaviour of nutrients, inorganic and organic matter and phytoplankton. The fixed parameters, x, may vary per case and in this case x denotes the parameters time and segment. The predictive output variable y is given by the observed concentrations of inorganic matter at specific times in specific segments; x therefore is also input for y. Hence, the predictive output variable y_i is linked to x_i and thus to $M(x_i, \theta)$ as given in eq. 2. For example, at day 2 in segment 45 (lake IJsselmeer) a measured value for the inorganic matter is available. Then this value is compared to the model output variable for inorganic matter at the same time in the segment mentioned, adding up to the sum of squared differences as given in eq. 7. For all segments these comparisons are made each time a measurement in a specific segment is present, finally leading to the posterior parameter distribution as given by eq. 8 or eq. 9. Prior parameter distributions are chosen uniform between a lower (θ_{min}) and an upper boundary (θ_{max}), based upon prior knowledge. The discrete parameter values are generated from:

$$\theta_j \;=\; \theta_{min} + \left(\frac{j-0.5}{nBins}\right) \cdot \left(\theta_{max} - \theta_{min}\right) \tag{10}$$

with $j = 1, \ldots, n$Bins and nBins the number of discrete values (Bins) in between the boundaries (Aldenberg et al., submitted). The parameters are combined to form a rectangular equally spaced grid and simulation runs are performed with each combination. Prior all parameter combinations have the same probability. At first a relatively coarse grid sample was generated, to search the area where the higher probabilities are to be found, and simulations were performed with the WATNAT model using the simulation software ACSL (Mitchell & Gauthier, 1991). With σ not to be estimated and therefore integrated out, probability distributions were generated as described in eq. 9. A new grid based on these results was chosen. Thus, in an iterative way the appropriate ranges of the parameters were determined. This iterative manner of working was necessary to avoid computational problems with this large scale model. Eventually a sufficiently detailed grid offering a good picture of the distribution was obtained. Then this grid was used in runs including σ, analysed as a fourth parameter (eq. 8). Table 2 shows the discrete grid values with which the final simulation runs were performed.

Table 2. Discrete grid values for the parameters in the final simulation runs.

parameter	parameter values					nBins
cVCritSedIM	0.095	0.105	0.115	0.125	0.135	5
cVeloSedIM	0.08	0.09	0.1	0.11	0.12	5
cFDResusMi	0.2	0.25	0.3	0.35	0.4	5
σ_{DIMW}	12.5	15.0	17.5	20.0	22.5	5

The year 1985 was used for the calibration of the inorganic matter parameters. Results are shown for two segments: the river IJssel and lake IJsselmeer. For 1985 there were 25 measurements of inorganic matter in the river IJssel and 44 observations for the lake IJsselmeer.

5. Results

The method described resulted in a distribution in which all probability was found at grid samples in which the value of σ_{DIMW} was 15.0 [$gDW\,m^{-2}$]. At this value, 125 combinations of the remaining parameters resulted in the distribution shown in Fig. 3. The maximum probability is 0.076, while 50% of the probability can be assigned to only 9 combinations; 44 parameter combinations account for 95% of the probability. In the first grid samples generated, almost all weight was found in one combination and consequently the maximum probability was close to 1.0.

The marginal posterior two-parameter distributions show the areas where the most probable combinations are found. Between cVeloSedIM and cFDResusMin (Fig. 4a) and between cVCrit and cVeloSedIM (Fig. 4b) no correlations are observed. However, in case of cVCrit and cFDResusMin (Fig. 4c) it is shown that the lower the treshold value for resuspension, the lower the minimum resuspension flux should be in order to match the data. The value of

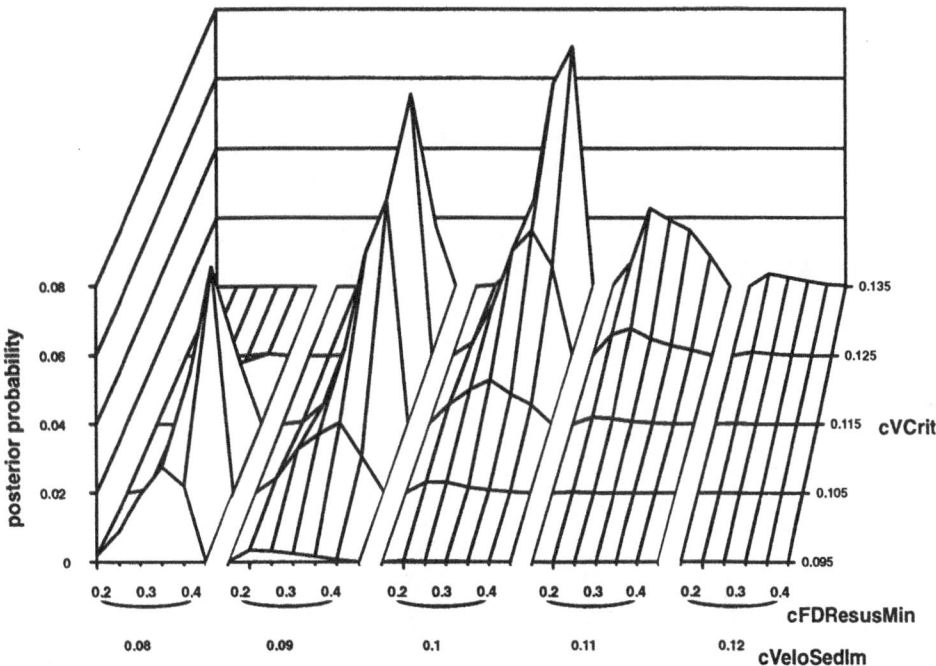

Figure 3. Posterior probability distribution of the final grid sample of 125 parameter combinations. Mark the 2 parameters shown in combination on the *x*-axis.

cVeloSedIM ≈ 0.12 [$m·d^{-1}$] has the highest probability, whereas cFDResusMin should be between 0.3 - 0.4 [$gDW m^{-2}·d^{-1}$] and cVCrit between 0.105 - 0.115 [$m·s^{-1}$]. The parameter combination with the highest probability is to be found in this area.

To obtain 95% of the total posterior probability, the 44 parameter combinations with the highest probability were selected. In the IJssel, a branch of the Rhine, stream velocity caused the shear stress to be too high. Thus, sedimentation played no role in this segment and the model was not sensitive to the parameters (Fig. 5a). On the other hand, sedimentation and resuspension processes are main forces in the lake IJsselmeer, a large freshwater lake into which the IJssel discharges. Stream velocity is relatively low and the shear stress is mainly wind induced. As a consequence the parameter combinations lead to a much wider range of predicted concentrations inorg. matter (Fig. 5b). The lines represent the expected value for the model, $M(x,θ)$. Not drawn in Fig. 5 is the deviation due to the σ = 15.0 [$gDW m^{-2}$]. Every single simulation run has ± 15.0 [$gDW m^{-2}$] bounds, which represents the area of expected values for the observations. Still, in case of sedimentation areas, the model needs refinement.

A distinction between coarse and fine inorganic matter is likely to improve the results, as well as distinct parametervalues for river segments with high stream velocities and for segments where sedimentation is the main process.

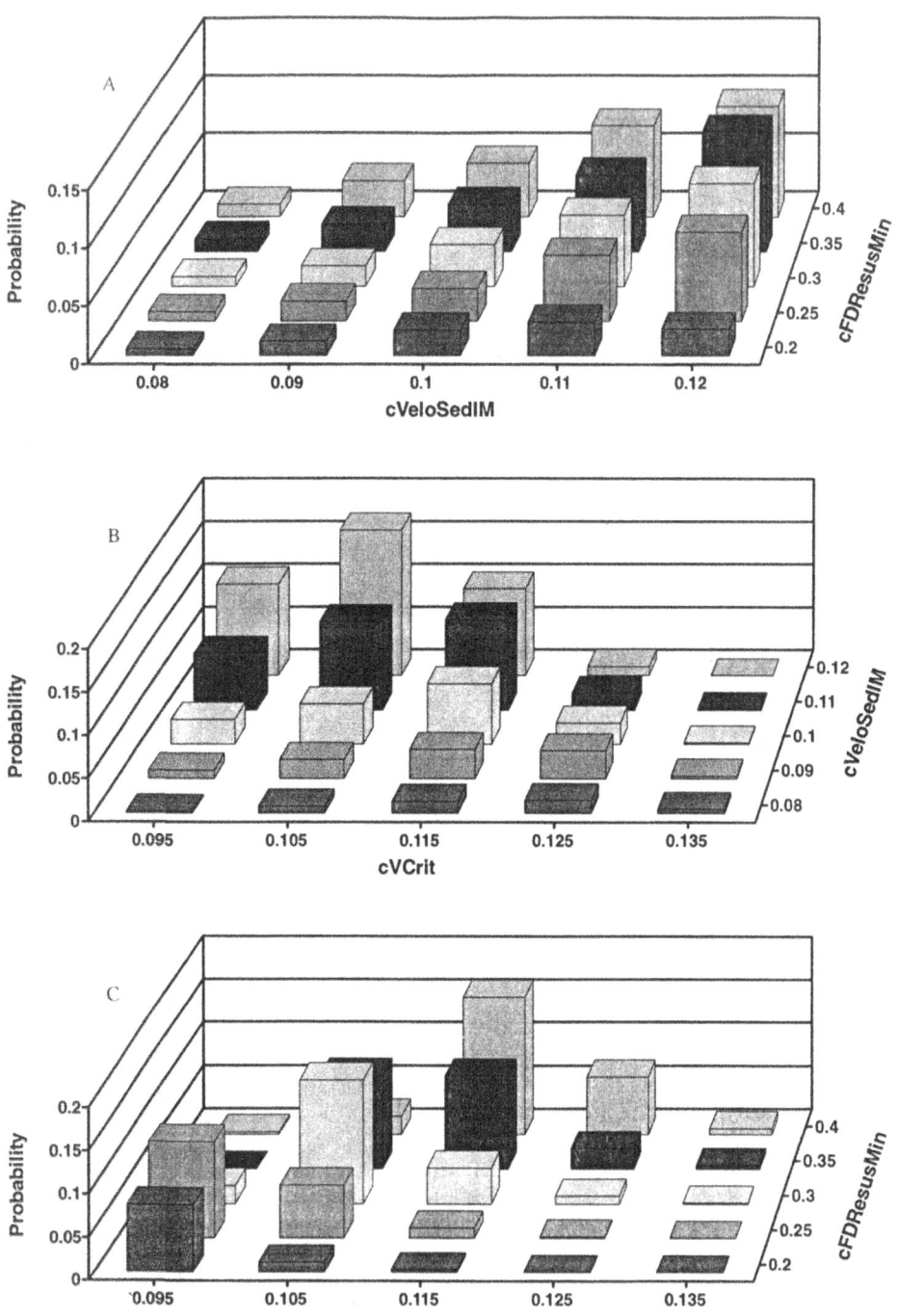

Figure 4. Marginal posterior parameter distributions for combinations of a) cVeloSedIM and cFDResusMin, b) cVCrit and cVeloSedIM and c) cVCrit and cFDResusMin.

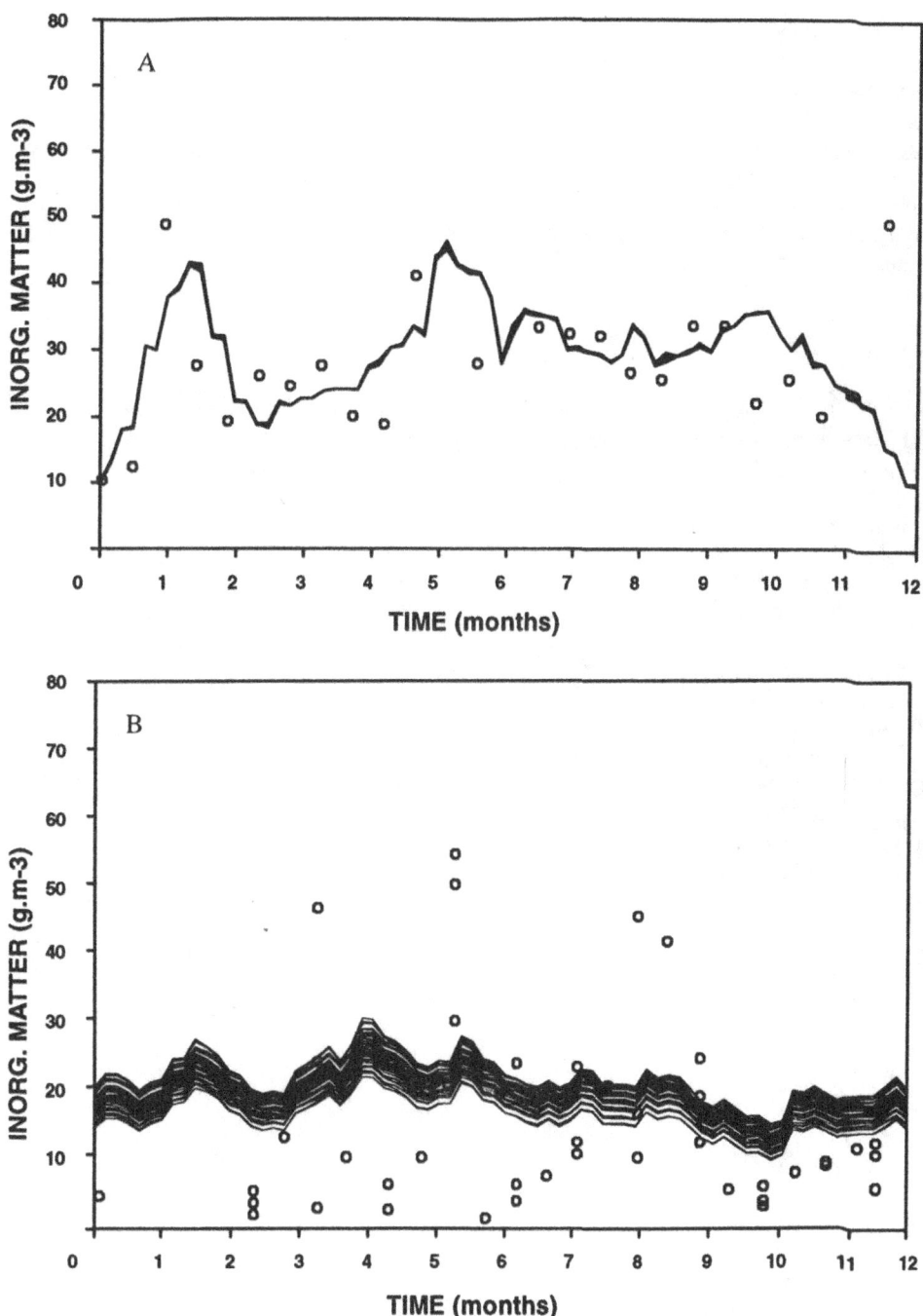

Figure 5. Simulation runs with the 44 parameter combinations defining 95% of the total posterior probability in a) the river IJssel and b) the freshwater lake IJsselmeer. The observed concentrations are represented by the circles.

6. Conclusions

Bayesian statistics in model analysis enables a combined uncertainty analysis, calibration of parameters as well as a tool for risk analysis. This way uncertainty in water quality models and risk assessment models can be distinguished and diminished. The uncertainty of the parameters determining the sedimentation and resuspension of inorganic suspended matter was reduced by means of this Bayesian model analysis. A subdivision of the parameters into a part with more resuspension in case of rivers and into a sedimentation part for more stagnant waters has to be considered and analysed.

References

Aldenberg, T., Janse, J.H. & Kramer, P.R.G. (submitted), 'Fitting the dynamic model PCLAKE to a multi-lake survey through Bayesian statistics', *Ecological Modelling*.
Box, G.E.P. & Tiao, G.C. (1973), 'Bayesian inference in statistical analysis', Addison Wesley.
Box, M.J. (1971), 'A parameter estimation criterion for multiresponse models applicable when some observations are missing', *Applied Statistics*, **20(1)**, 1-7.
Janse, J.H., Aldenberg, T. and Kramer, P.R.G. (1992), 'A mathematical model of the phosphorus cycle in Lake Loosdrecht and simulation of additional measures', *Hydrobiologia*, **233**, 119-136.
Kramer, P.R.G., Traas, Th.P., Aldenberg, T. & de Vries, M.B. (submitted), 'Modelling foodweb accumulation of toxicants in sedimentation areas of the Rhine delta', *Water Science and Technology*.
Mitchell & Gauthier Associates (MGA) Inc. (1991), 'Advanced Continuous Simulation Language (ACSL)', *Reference manual level 10.0*.
De Nijs, A.C.M., Janse, J.H., Wortelboer, F.G., Kramer, P.R.G. & Aldenberg, T. (in preparation)', *WATNAT. Model documentation, Version 1.0, RIVM report*, RIVM, Bilthoven.
Traas, Th.P. & Aldenberg, T. (1992), 'CATS-1: a model for predicting contaminant accumulation in a meadow ecosystem. The case of cadmium'. *RIVM report no. 719103001*, RIVM, Bilthoven.

MODELLING DYNAMICS OF AIR POLLUTION DISPERSION IN MESOSCALE

PIOTR HOLNICKI

Systems Research Institute, Polish Academy of Sciences
6, Newelska str, 01-447 Warszawa, Poland

e-mail: holnicki@ibspan.waw.pl

Summary

In the paper a multilayer, dynamical model of air quality analysis is presented. The model has been designed for predicting dispersion and deposition of air pollution in urban and industrial areas and for evaluation of emission control strategies. The computer implementation of pollutant dispersion is based on numerical solving advection-diffusion equations. The wind field structure and dynamics are preprocessed by a specialized generator. The real data computational examples are presented.

Keywords air quality control, mathematical model, computer simulation

1. Introduction

Air quality forecasting and analysis are especially important in major urban and industrial areas, where the emission field is characterized by a variety of polluting species, spatial complexity and high emission intensity. Emissions are mainly due to central or domestic heating, industrial sources and urban transportation system. Moreover, in many urban areas, unfavorable topographical conditions and stress meteorological episodes, "heat island" effect and complex chemical processes that occure in the surface layer of the atmosphere - lead to spatial and temporal cumulation of polluting factors. As a consequence, the resulting pollution concentrations often reach very high values and exceed the admissible air quality standards. High population density additionally enlarges the environmental vulnerability of those areas.

The forecasting computer models of air pollution dispersion in regional scale are powerful tools for analysis of ecological impact. The difficulty in mathematical and computer modelling of urbanized and industrial areas is due to structural complexity (topography, aerodynamical and thermal effects) as well as to day-variability of meteorological conditions (stratification, temperature inversion). An important factor that should be considered in mathematical description is dynamics of dispersion process. Its meaning is evident in short-term forecasts of air pollution, simulation of stress meteo scenarios and alarm emission episodes. The dynamics aspect can also be important in analysis of long-term scenarios. In such a case, the process of transition between the consecutive meteo situations can be continuously simulated and reflected in the resulting, time-averaged pollution patterns. Some examples of such an analysis are presented in Secion 4.

2. General model's structure and the input data

In the paper, the implementation of a multilayer version of air pollution forecasting model for mesoscale is presented. The spatial dimensions of the rectangle area considered are limited to about 100km x 100km. The dispersion process is analysed in a cuboid bounded by the terrain elevation and the mixing height - HM. Vertical stratification of this domain consists of three horizontal layers:

1-st, the surface layer: $H_1 = 0 - 50m$,
2-nd, the middle layer: $H_2 = 50 - 150m$,
3-rd, the upper layer: $H_3 = 150 - HM$.

This approximation of three-dimensional process allows to consider vertical structure of the wind (vertical profile) and that of the pollution concentration field. All the ground level, area and linear sources as well as small pointwise sources are located in the first layer. The strong influence of the aerodynamic roughness, turbulence effects and the wind shear are observed at this level. In the second layer the most of emissions of industrial and intermediate energy installations are located. The effective emission points of the major power and heating plants are placed in the upper layer. The depth of this layer depends on the mixing height and can vary in time.

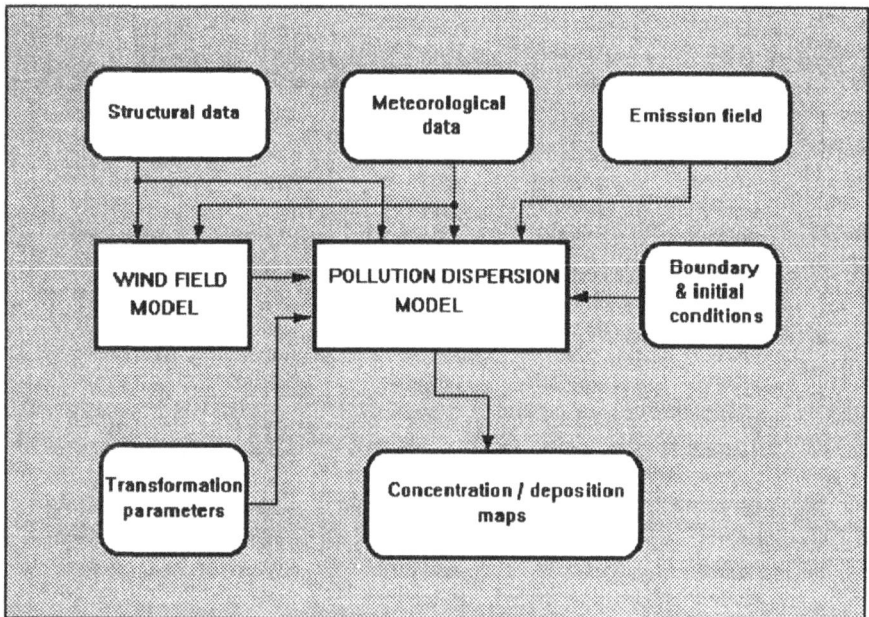

Figure 1. General block diagram of the model

The main input data used in the model are presented in Fig. 1. They can be arranged in the form of the following basic groups:

-       structural characteristics of the domain (geometry of the area, topography, aerody-

namic properties, ecological parameters),
- meteorological forecast (mixing depth, wind components, atmospheric stability, precipitation intensity, difference of temperaures between urban and rural areas),
- emission field characteristics (pointwise sources and areal emission flux),
- physical and chemical transformation parameters (dry and wet deposition coefficients, chemical transformation rate),
- initial and boundary conditions.

In general case, meteorological and emission data are time and space dependent functions.

This version of the model is mainly sulfur-oriented, but it can be extended to analyse other types of air pollution. Time scale of the forecast can be optionally selected, and ranges from 24hr (short-term forecast) up to a season or a year period (long-term forecast). The dynamics of the dispersion process is reflected in both scales of analysis. The forecast is generated by a sequence of the consecutive meteorological and emissions episodes, entered with the interval - DT (the default value DT = 6hr). The same time resolution is applied for the resulting concentration fields. The interval DT is next discretized with the step of numerical procedure - τ; the length of this step is automatically set up to minimize the computing time and to satisfy numerical stability conditions. The space discretization step depends on the area dimensions and is about 0.5 - 1km in urban scale and 2 - 5km in regional scale, respectively.

3. The mathematical model

In the course of the atmospheric transport, the emitted species are transformed in chemical and physical reactions. In case of sulphur-type pollution, these processes can be approximated by two-component model (Dernwent, 1988; Eliassen and Saltbones, 1983; Holnicki and Zochowski, 1990). A general view of the sulfur life cycle is shown in Fig. 2.

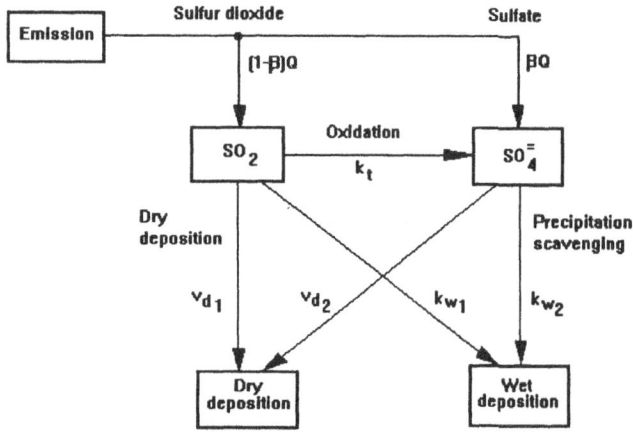

Figure 2. The sulfur life cycle and chemical transformations

The full mathematical representation of the dispersion process integrates the wind induced transport, turbulent diffusion and the chemistry. The same form of the model is applied at each vertical layer. It consists of the set of two advection-diffusion equations, for the primary and secondary type pollutants, respectively, considered in the rectangle domain
$\Omega = L_x \times L_y$ and the time interval $(0,T)$

$$\frac{\partial c_{1,i}}{\partial t} + \vec{w}_i \cdot \nabla c_{1,i} - K_{H_i} \Delta c_{1,i} - \frac{\partial}{\partial z} K_{V_i} \frac{\partial c_{1,i}}{\partial z} + (\frac{v_{d,i}}{H_i} + k_{w_1} + k_i) c_{1,i} = (1-\beta)\bar{Q}_i , \tag{1}$$

$$\frac{\partial c_{2,i}}{\partial t} + \vec{w}_i \cdot \nabla c_{2,i} - K_{H_i} \Delta c_{2,i} - \frac{\partial}{\partial z} K_{V_i} \frac{\partial c_{2,i}}{\partial z} + (\frac{v_{d,i}}{H_i} + k_{w_2}) c_{2,i} - k_i c_{1,i} = \beta \bar{Q}_i \tag{2}$$

with the boundary conditions $(k = 1,2, i = 1,2,3)$

$$c_{k,i} = c_{k,i}^b \quad on \quad S^- = \{\partial\Omega \times (0,T) | \vec{w}_i \cdot \vec{n} < 0\}, \tag{3}$$

$$\frac{\partial c_{k,i}}{\partial \vec{n}} = 0 \quad on \quad S^+ = \{\partial\Omega \times (0,T) | \vec{w}_i \cdot \vec{n} \geq 0\} \tag{4}$$

and the initial condition $(k = 1,2, i = 1,2,3)$

$$c_{k,i}(0) = c_{k,i}^0 \quad in \quad \Omega \times \{t = 0\}. \tag{5}$$

The following notation is applied:
T -- time horizon of the forecast,
$i = 1,2,3$ -- the layer index,
$k = 1,2$ -- pollutant type index,
$c_1(x,t)$, $c_2(x,t)$ -- primary (SO_2) and secondary (SO_4) pollutant concentrations,
$H_i(x,t)$ -- the i-th layer height,
$K_{H_i}(x,t)$, $K_{V_i}(x,t)$ -- horizontal and vertical diffusion coefficients in the i-th layer,
$Q_i(x,t)$ -- emission field of the i-th layer,
$v_{dk,i}(x)$ -- dry deposition velocity of SO_2 (or SO_4) in the i-th layer,
$k_{wk}(\alpha)$ -- wet removal factor of SO_2 (or SO_4),
$\alpha(x,t)$ -- precipitation intensity,
$k_i(t)$ -- chemical transformation rate $SO_2 \Rightarrow SO_4$,
β -- relative share of sulfate in emission.

The direct environmental effect of air pollution is due to people and other biota exposure to sulfur dioxide and sulfate aerosol. The current concentrations are strongly related to dynamics of the dispersion process. Temporal and spatial variability of the pollution field depends on the current meteorological episode, influence of the area stuctural features (topography, aerodynamical roughness) and emission characteristics. These factors are mainly relevant to short-term forecasts and are directly reflected in current concentration distributions.

Dynamics of the system is also important in case of long-term forecasts, since the season- or year-averaged concentration patterns actually reflect not only the steady meteorological episodes, but also the transition process between the consecutive situations. The real trajectories of pollution puff transport can be considered in such an approach. In the model presented, this type of analysis is performed as a sequence of meteorological episodes that cover the forecast interval. In general case, the emission field intensity can also be time dependent function.

The total sulfur deposition is another polluting factor that mainly affects soil and groundwater quality. Its influence is important in long-term perspective and can be calculated as a sum of dry and wet depositions, according to the formula

$$D = \sum_{j=1}^{N} [D_d(j) + D_w(j)] \, , \qquad (6)$$

where the components are as follows:

$$D_d(j) = [v_{d_1,1} c_{1,1}(j) + v_{d_2,1} c_{2,1}(j)] \cdot \Delta t,$$

$$D_w(j) = \sum_{i=1}^{3} [k_{w_1} c_{1,i}(j) + k_{w_2} c_{2,i}(j)] \cdot H_i \cdot \Delta t.$$

It can be observed that dry deposition is determined by the current concentration and deposition velocity in the surface layer. We denote here $\Delta t = T/N$, and $j = 1,...,N$ is an index of the current time interval.

The vertical transfer between layers and the horizontal diffusion coefficient are parametrized according to meteorological conditions, as well as to the time and space discretization parameters (Holnicki et al., 1992, 1993). Dry deposition velocities and chemical transformation rate in (1), (2) are adopted from the literature (Dernwent, 1988; Eliassen and Saltbones, 1983), but in general they can be space and time dependent functions, respectively. The wet removal factors are defined as nonlinear functions of precipitation intensity, as suggested in Holnicki and Zochowski (1990).

Horizontal transport of air pollution is mainly due to advection process, thus the calculation of wind components in (1) -- (4) is an essential step of simulation. In the model presented, the wind field is preprocessed by a specialized, linked submodel (compare Holnicki et al., 1992, 1993). The main effects, taken into account in the algorithm applied, are as follows:

- vertical wind profile (wind shear),
- influence of topography and aerodynamic roughness,
- urbanized area "heat island".

Dynamics of the input data is mainly related to the time-dependent meteorological forecast. It consists of the mixing height, atmospheric stability conditions, geostrophic and anemometric wind components, precipitation intensity and the difference of temperatures between the urban and rural areas. Structural features of the area are characterized by the space-dependent matrices of topography, terrain aerodynamical roughness and the urbanized area borders.

Table 1. Selected results of the model URFOR3 validation (concentrations in µgm^{-3})

Observ. point	Episode 1			Episode 2		
	Measured	URFOR	Pasquill	Measured	URFOR	Pasquill
1	40.00	94.99	50.00	35.00	13.31	6.00
2	92.00	99.39	19.00	54.00	36.82	1.00
3	97.00	108.16	61.00	24.00	36.46	23.00
4	131.00	73.39	29.00	35.00	26.22	3.00
5	130.00	138.43	110.00	75.00	64.36	102.00
6	63.00	141.06	32.00	73.00	41.36	11.00
7	98.00	100.82	42.00	71.00	30.58	7.00
8	112.00	189.74	106.00	30.00	61.99	4.00
9	79.00	55.92	29.00	18.00	4.83	1.00
10	51.00	133.60	46.00	63.00	36.45	36.00
11	24.00	74.62	231.00	26.00	32.65	2.00
12	92.00	134.15	98.00	64.00	85.10	92.00
13	10.00	4.73	2.00	8.00	7.04	60.00
14	11.00	6.79	13.00	27.00	8.17	23.00
15	25.00	4.48	1.00	12.00	71.38	4.00
16	34.00	41.01	8.00	18.00	18.75	1.00
17	41.00	33.13	12.00	18.00	6.76	1.00
18	38.00	7.01	3.00	22.00	28.04	5.00
19	43.00	46.84	9.00	23.00	32.53	6.00
20	65.00	60.14	43.00	64.00	108.95	47.00
21	56.00	79.59	43.00	36.00	43.40	19.00
22	4.00	4.68	1.00	13.00	5.59	36.00
23	10.00	6.00	1.00	21.00	7.52	3.00
24	79.00	19.86	45.00	25.00	7.10	2.00
relative error	-0.16	0.27	-	0.05	0.42
correlation coef.	0.70	0.30	-	0.56	0.44

For computational efficiency of the final algorithm, the linearity of the system is assumed. That means, the listed above effects are considered separately and then the results are superimposed on each other to form the final wind field in the area. Calculation of the mean wind components in three horizontal layers and calculation of the topographical and thermal corrections constitute the main steps of the algorithm (see Holnicki and Zochowski, 1990; Holnicki *et al.*, 1992, 1993; for details). The procedure is repeated in the subsequent intervals of the main forecasting routine.

The resulting wind fields are obtained for the beginning and the end of the subsequent intervals (compare computational examples presented in Section 4). In regional scale, an additional information on spatial distribution of the anemometric wind is utilized. In this case, the available measurement data of the surface wind are interpolated and used in the wind-field generation procedure.

The numerical method used for solving the advective part of (1) - (4) is based on combination of the method of characteristics and the finite element technique. The positivity and monotonicity conditions of approximation scheme are satisfied. The turbulent diffusion process is parametrized according to the discretization parameters and the current meteorological conditions (wind velocity, atmospheric stability). Some details can be found in Holnicki *at al.* (1993).

Selected results of test computations are shown in Table 1. They present validation of URFOR3 model in case of the short-term concentration forecast. The experiment has been performed for Krakow urban area (24km x 18km, South-East corner of the region shown in Fig. 4). The 30-min SO_2 concentrations, measured at 24 observation sites are compared with the model forecast; the results are also favourably compared with the Pasquill-type model prediction. The URFOR3 model predictability is respectively higher in long-term forecast case.

4. The computational examples

In this section some computational examples of the model applications are presented. The implementation of the system consists of the main forecasting model -- REGFOR3 (URFOR3 in urban scale, respectively), the wind-field preprocessor -- WIND3, and the graphics routine -- REGMAP3 that presents the results of computation in a form of isoline, concentration/deposition maps (compare Fig. 4). This version of the system is designed for a microcomputer configuration, compatible with IBM PC/486 and SVGA graphics. In Fig. 3 examples of short-term simulation of the wind field dynamics and the resulting concentration field are presented. The area considered is a rectangle region 90km x 50km, with complex topography. The meteo scenario is composed of five sbsequent episodes, (the basic time interval DT = 6hr) related to clockwise rotating inflow geostrophic wind (ranging from S-W to N-E). The respective wind fields and the vertical wind profile, preprocessed by WIND3 generator, reflect the influence of topography and the spatial interpolation of the surface wind measurements at meteo stations. The emission field consists of 3 point sources. The effective emission points of two of them are placed in the second layer and the last one is in the third layer. The pollution concentration fields in the subsequent intervals (the upper layer), calculated by the forecasting model, show dynamics of the transport and dispersion processes. They also reflect the wind-field evolution and the vertical exchange between layers.

An example of the long-term environmental analysis is presented in Fig. 4. The season-averaged, ground level SO_2 concentrations represent environmental effect of one of the major power plants in Silesia-Krakow region (110km x 74km). The maps represent winter and summer concentrations, respectively. The simulation was performed as a real sequence

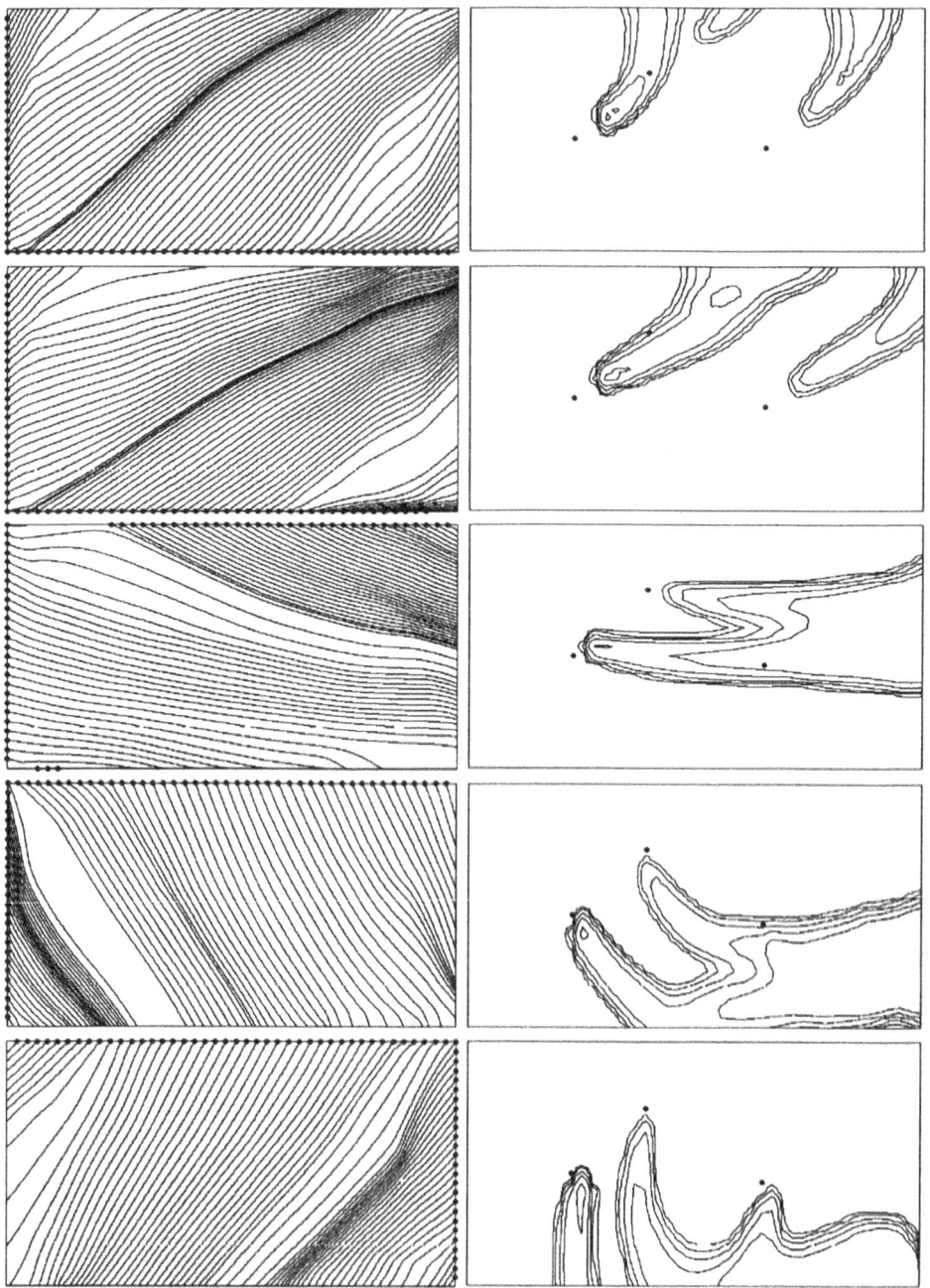

Figure 3. Left - streamlines evolution for clockwise rotation of the inflow wind (dots indicate the inflow border). Right - the respective concentration dynamics from three industrial point sources)

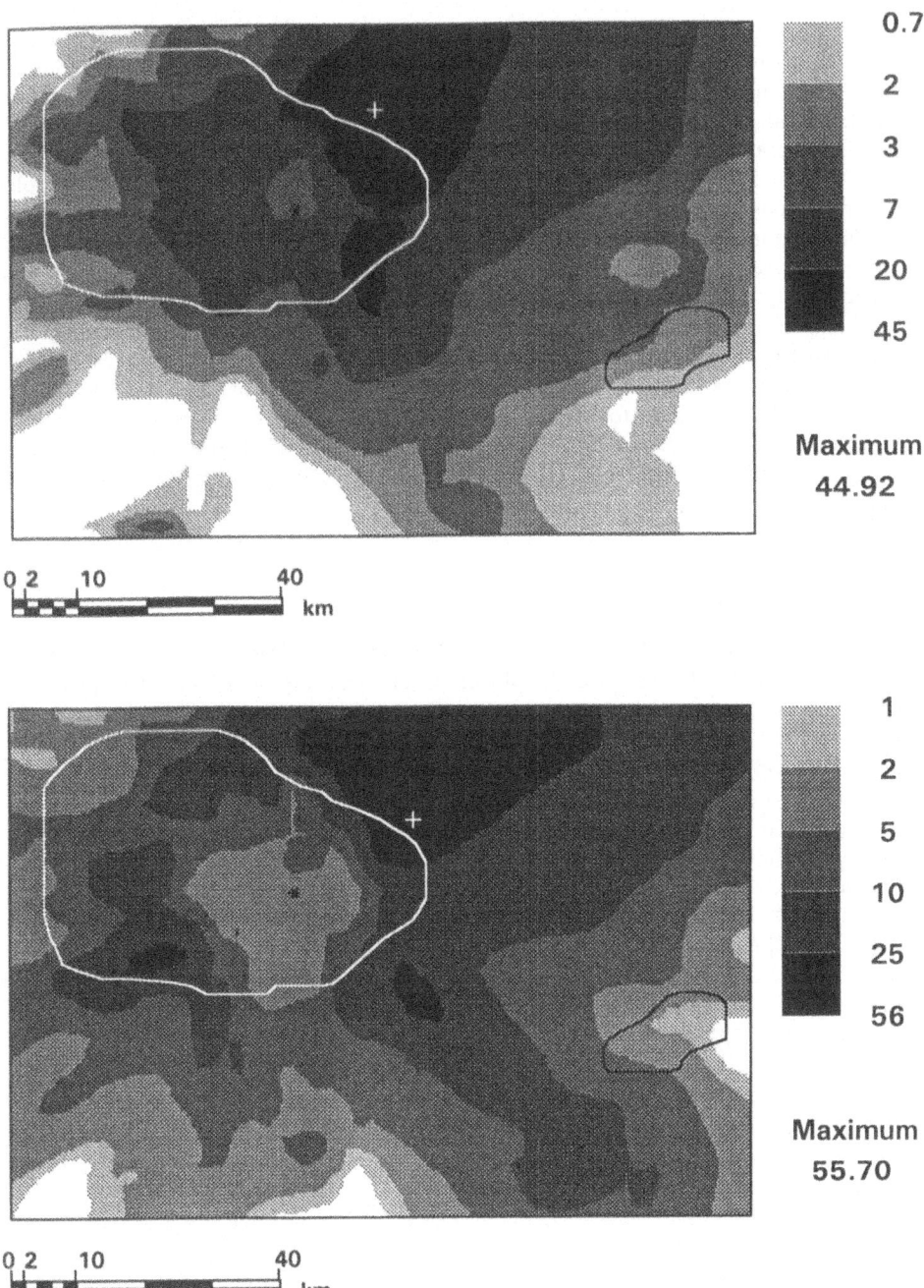

Figure 4. Katowice Industrial Region and Krakow. The averaged surface-layer SO₂ concentration from a major power plant, for the summer (top) and the winter (bottom) seasons.

of the consecutive 6-hr meteorological situations that cover the total period of simulation. In such an approach, the transition processes between the subsequent episodes (compare Fig. 3) are simulated. In general, emission intensity can also be time-dependent function, but in this example it is assumed constnt within one season.

The maps presented in Fig. 4 refer to the winter (182 episodes) and summer (183 episodes) ground-level averaged concentrations. The isoline maps, besides the meteorological and aerodynamical conditions, reflect the influence of the source parameters (the stack height, diameter, the outlet gasses temperature and velocity). One can see, for example, relatively low concentrations in the neighbourhood of the source; the maximum concentration area indicates the dominating wind directions in the season considered.

The computer system presented in the paper, can be applied as a decision support tool for regional planning, evaluating environmental damage of industrial sources, optimal location of new installations, or optimal strategy selection of emission abatement. Dynamics of the short-term forecasting model can also be utilized in real time emission control, especially in local or urban scale. Some examples of such applications can be found in Holnicki and Zochowski (1990), Holnicki and Kałuszko (1991), Holnicki et al. (1993).

References

Demwent, R.G. (1988), 'A better way to control air pollution', *Nature*, **331**, 575 - 578.

Eliassen, A. and J. Saltbones (1983), 'Modelling of long-range transport of sulphur over Europe: A two-year model run and some model experiments', *Atmos. Environment*, **17**, 1457 - 1473.

Holnicki, P. and A. Zochowski (1990), '*Selected Mathematical Methods in Air Quality Analysis*' (In Polish), Polish Scientific Publishers (PWN), Warszawa.

Holnicki, P. and A. Kałuszko (1991), 'Decision support algorithm for air quality planning by emission abatement', In Proc. *The 15th IFIP Conference on System Modelling and Optimization*, Zurich.

Holnicki, P., A. Kałuszko and A. Zochowski (1992), 'Regional/urban air quality management system', *Report ZTS-2/M181/92*, Systems Research Institute, Warszawa.

Holnicki P., A. Kałuszko and A. Zochowski A (1993), 'A multilayer computer model for air quality forecasting in urban/regional scale', *Control and Cybernetics* (to appear).

UNCERTAINTY FACTORS ANALYSIS IN LINEAR WATER QUALITY MODELS

ANDRZEJ KRASZEWSKI
Institute of Environmental Engineering Systems,
Warsaw University of Technology, Warsaw, Poland
e-mail: akk@saturn.iis.pw.edu.pl

RODOLFO SONCINI-SESSA
Dipartimento di Elettronica e Informazione,
Politecnico di Milano, Milano, Italy
e-mail: soncini@ipmel2.elet.polimi.it

Summary

The parameter uncertainties that result from the calibration of Biochemical Oxygen Demand (BOD) - Dissolved Oxygen (DO) river quality models is quantitatively analyzed as a function of the number of sampling points, of the parameter "true" values, and of the model complexity.

Keywords Water quality, parameter estimation, uncertainty

1. Introduction

The common approach in the development of water quality models is as follows: the system analyst selects the structure of the water quality model, instream water quality measurements are collected, and then, based on these measurements, the model is finally calibrated. In the last twenty years, many papers have been devoted to the definition of calibration algorithms, but only seldom was attention paid to the estimated parameter uncertainties (as measured by the variances of the estimate) induced by different sources of errors. Among the first who analyzed the influence of measurement errors in the calibration of water quality models were Koivo and Philips (1971), O'Neil (1973) and Argentesi and Olivi (1976). Significant progress resulted from the work of Whitehead and Young (1979), Rinaldi and Soncini-Sessa (1978), Beck (1978, 1979, 1981), and van Straten (1983). An exhaustive review of the uncertainty issues surrounding water quality models was undertaken by Beck (1987).

All of these papers consider how to determine the uncertainties of the estimated parameters. However, none addresses how these uncertainties are affected by
- the number of sampling points
- the "true" values of the parameters
- the complexity of the model.

The knowledge of these dependencies would give a qualitative idea of how many measurements would be necessary to obtain a desired level of uncertainty for a model with a given structure (complexity). Unfortunately, the results reported in this paper cannot be more than qualitative since the parameters' uncertainty depends not only on the three items previously mentioned, but also on the particular case at hand; i.e., on how the considered river stretch is structured (location of inflow and discharge points, positions of the sampling points, etc.). However, once a particular system is at hand and a model structure has been selected, a quantitative analysis can be obtained by repeating the procedure described in the paper.

The parameter uncertainty is produced by three different errors:
a)   the input measurement error;
b)   the process error (i.e., the error induced by a simplified description of the real processes);
c)   the output (or state) measurement error.

Only the last two have been considered in the literature, and error b) only when the calibration tool was the Kalman filter. The first error has never been considered, probably due to the extreme difficulty in analyzing its effects. Unfortunately, even if this is a common practice, it makes no sense to neglect error a) is intrinsically a non-sense because the input error is surely of the same relevance as the output error since the measurement instruments are the same for both input and output variables. However, since this paper represents the first attempt to develop a quantitative analysis of parameter uncertainty, we considered output measurement errors only. Moreover, the analysis is restricted to the sub-class of linear (in the state variables) BOD-DO models where the temperature influence on the process rates is not taken into account. We chose the linear BOD-DO model because it is implemented in a commercially available Modeling Support System (WODA, see Kraszewski and Soncini-Sessa (1989)), that includes a calibration facility. We restricted the analysis to models with temperature independent rates after observing that these models are the ones actually considered by WODA users in the great majority of applications.

2. Model calibration

2.1 The BOD-DO water quality model

When written along the characteristic line (Rinaldi et al., 1979), the model we consider (Kraszewski and Soncini-Sessa, 1989) is described by the following two equations

$$\frac{db}{dt} = -\frac{S_q(t)}{A} b - (k_1 + k_2)b + \frac{1}{A} u_b(t) , \tag{1a}$$

$$\frac{dc}{dt} = -\frac{S_q(t)}{A} c - k_1 b + k_r(t)(c_s(T) - c) + k_5 \Psi(t) + k_6 + \frac{1}{A} u_c(t) , \tag{1b}$$

where t is time (d), b and c are the mean sectional BOD and DO concentrations respectively (g m^{-3}), $c_s(T)$ is the DO saturation level (g m^{-3}) as a function of the water temperature T (°C), u_b and u_c are the BOD and DO distributed load patterns (g d^{-1} m^{-1}), $S_q(t)$ is the pattern of the distributed water inflow rate (m^{3}d^{-1}m) so that the first terms in Eqs (1a) and (1b) represent the dilution induced by the water inflows. The term

$$\Psi(t) = \begin{cases} \sin\left[\pi\dfrac{t \bmod(24) - t_r}{t_s - t_r}\right] & \text{if } t_r \le t \bmod(24) \le t_s \,, \\ 0 & \text{otherwise} \end{cases} \tag{1c}$$

is the light function governing the net photosynthetic oxygen production rate (O'Connor and Di Toro, 1970) (t_r and t_s denotes the sunrise and sunset times respectively).

The parameters and rates that appears in Eq. (1) have the following meaning:

k_1 is the BOD decay parameter (d^{-1});

k_2 is the sedimentation parameter (d^{-1});

$k_r(t)$ is the water reaeration parameter (d^{-1}), as a function of the flow rate $Q(t)$ (m^3s^{-1}), given by

$$k_r = k_3 Q(t)^{k_4} \,, \tag{1d}$$

where k_3 is in $d^{-1}(m^3s^{-1})^{-k_3}$ and k_4 is a dimensionless exponent;

k_5 is the maximum rate of photosynthetic oxygen production ($g\ m^{-3}d^{-1}$);

k_6 is the oxygen consumption rate due to sediment or plant respiration ($g\ m^{-3}d^{-1}$).

By setting to zero some of these parameters, one can obtain particular well known models. The simple Streeter-Phelps (1925) model corresponds to setting $k_2=k_4=k_5=k_6=0$, and the O'Connor-Di Toro model (1970) is given by $k_4=k_6=0$.

2.2 The measurement model

We assume that, for calibrating model (1), n measurements of BOD and DO concentrations are collected along the characteristic line in n sampling points at time t ($i=1,...,n$). We assume that these measurements are corrupted by errors; more precisely the values B_i and C_i measured at time t, the "true" concentrations $b(t_i)$ and $c(t_i)$, and the errors β_i and γ_i, all are related by the following output equations

$$\begin{aligned} B_i &= b(t_i) + \beta_i \,, \\ C_i &= c(t_i) + \gamma_i \,. \end{aligned} \tag{2}$$

These measurement errors are made up of the sum of three different errors: the sampling, handling and analytic errors. The sampling error occurs because water is collected at one or more points of the river cross section, while a one-dimensional model, such as model (1), logically requires that B_i and C_i represent the cross sectional mean concentrations. The handling error arises when the chemical analysis of the sample is not coincident with its collection. It is generally negligible for DO, while it can be reduced, but never completely avoided, in the case of BOD. In fact, in the period between collecting the sample and analyzing it, the biodegradation process cannot be completely stopped, thus causing a difference between the real BOD concentration in the stream and the value later determined in the sample. Finally, the analytic error results from all the uncontrolled perturbation factors that act in the laboratory analysis. No one of these three errors can be completely eliminated, but it is surprising how much attention has been paid to studying and controlling only the analytic errors; until now little or no attention has been given to the sampling or handling errors.

The errors β and γ are related to the reproducibility of the measurement. It is generally assumed that they are normally distributed, with zero mean. Little is known of their standard deviations, since only the standard deviations σ_b^a and σ_c^a of the analytic error have been studied. Standard Methods (1981) reports that, in the case of DO, σ_c^a is 0.05 g m^{-3} with the Winkler method and 0.1 g m^{-3} with the tests based on membrane sensors. The standard deviation σ_b^a of BOD (g m^{-3}), however, depends on the module of the measured value B:

$$\sigma_b^a = 0.18B + 1.26 \ .$$

It must be stressed that these values only refer to the analytic errors; the standard deviations of β and γ are definitively larger, but, unfortunately, unknown. To overcome this ignorance, we have made an approximate statistical analysis of the residuals of models that we calibrated in the past, and we estimated that σ_c is of the order of 1 g m^{-3} , while σ_b is roughly two times σ_b^a. Therefore, in the following study we will adopt these latter values, being conscious that it must be considered only a rough guess since model residuals result from all sources of errors listed in Section 1.

2.3 Parameter estimation

Let's denote with the symbol ϑ the vector of the model parameters to be estimated. These parameters are not always all the parameters k_j ($j = 1,...,6$) of model (1), but only those parameters that appear in the particular model we consider. For example, when dealing with the classical Streeter-Phelps model, ϑ is the vector $|k_1 \ k_3|^T$ (the upper script T denotes transposition).

Given the upstream condition (B_0, C_0) and a value $\overline{\vartheta}$ of the parameter vector ϑ, the system (1) can be integrated for computing the BOD and DO concentrations $b(t_i|\overline{\vartheta})$ and $c(t_i|\overline{\vartheta})$ in the n sampling points. Then the vector

$$\varepsilon_i(\overline{\theta}) = |(B_i - b(t_i|\overline{\theta}) \quad (C_i - c(t_i|\overline{\theta})|^T \tag{3}$$

of BOD and DO deviations is an estimate of the measurement error vector $|\beta_i \ \gamma_i|$ in sampling point i.

Since we have assumed in the previous paragraph that the errors β and γ are normally distributed, it is rational to assume as an estimate of ϑ, the value $\hat{\vartheta}$ that minimizes the following function

$$J(\overline{\theta}) = \sum_{i=1}^{n} \varepsilon_i^T(\overline{\theta})W^{-1}\varepsilon_i(\overline{\theta}) , \tag{4a}$$

where the matrix

$$W = \text{diag}(\sigma_b^2, \sigma_c^2) \qquad (4b)$$

is the variance-covariance matrix of the measurement errors (its form implies that the measurement errors of BOD and DO are not correlated). This estimation criterion is equivalent to the maximum likelihood model, i.e. that model for which the probability of obtaining the n collected measurements is maximum.

For determining the minimum of $J(\vartheta)$, a Gauss-Newton algorithm had been used; given an a priori estimate ϑ^0, it recursively computes successive approximations to $\hat{\vartheta}$ by means of the following equation

$$\theta^{k+1} = \theta^k + \left[\sum_{i=1}^{n} \Phi^T(t_i,\theta^k) W^{-1}\Phi(t_i,\theta^k) \right]^{-1} \times \sum_{i=1}^{n} \Phi^T(t_i,\theta^k) W^{-1}\varepsilon_i(\theta^k) , \qquad (5a)$$

where $\Phi^T(t_i,\vartheta^k)$ is the Jacobian of the vector $|b(t_i|\vartheta^k) \ c(t_i|\vartheta^k)|$ with respect to the vector ϑ^k. In our case this Jacobian is easily determined since the system equations (1) are linear in the state. Once $\hat{\vartheta}$ has been calculated, the member of Eq. (5a)

$$P(\hat{\theta}) = \left[\sum_{i=1}^{n} \Phi^T(t_i,\hat{\theta}) W^{-1}\Phi(t_i,\hat{\theta}) \right]^{-1} \qquad (5b)$$

approximates the parameters variance-covariance matrix. The formula in unfortunately correct only asymptotically, i.e. when $n \to \infty$.

3. Uncertainty evaluation

Our goal is to evaluate how the uncertainty (variance) of the parameter estimates depends upon three factors:
 a) the number of sampling points;
 b) the "true" values of the parameters;
 c) the complexity of the model.
As far as factor c) is concerned we will consider three types of models: the Streeter-Phelps model with 2 parameters, the O'Connor-Di Toro model with 4 parameters, and the WODA model which is a full model (1) with 6 parameters. For each of these models we compute the sample variances of the parameter estimates obtained in a large number of calibration experiments. Since the experiments must differ only in the realizations of the measurement errors, the measurement set $\{(B_i^h, C_i^h), i=1,...,n\}$ to be used in the h-th experiment must be artificially produced according to Eq. (2). More precisely, for each sampling point i, the measurements B_i^h and C_i^h are given by

$$B_i^h = b(t_i|\theta^*) + \beta_i^h ,$$
$$C_i^h = c(t_i|\theta^*) + \gamma_i^h ,$$

5. Environmental sciences

where the errors β_i^h and γ_i^h are randomly extracted from the normal distributions discussed in Section 2.2 and the "true" concentrations $b(t_i|\vartheta^*)$ and $c(t_i|\vartheta^*)$ are generated by a model, parameterized with ϑ^*. The model adopted in this generation is the same model under calibration. The rationale is twofold. First, as stated in the introduction, we want to analyze only the uncertainty induced by measurement errors, thereby excluding process errors so that the "reality" must belong to the class of the considered models. Second, by doing so we know the "true" values ϑ^* of the parameters. Since it is a priori known that the parameter variances do depend upon the case study at hand, we decided to consider the simplest case we could imagine: a stretch of river with no discharges or tributaries. An exception had to be made when the full model was considered, because in this case, at least one tributary inflow is compulsory to produce the variation of flow rate necessary for making k_3 and k_4 separately identifiable (see Eq. (1d)).

More precisely, we considered a stretch (4 days flow time in length) of a lowland stream with a flow-rate of 25 m^3s^{-1} in spring conditions ($T=15°C$ with a daylight period of 12 hours); the upstream DO concentration is assumed to be at saturation level (10 g m^{-3}) and the BOD concentration, due to untreated dairy waste, of 40.0 g m^{-3}. When the full model is considered, a tributary (BOD = 30 g m^{-3}; DO= 6 g m^{-3}) with an inflow rate of 5 m^3s^{-1} is assumed to be located in the middle of the stretch. As a consequence of this scenario the nominal values ϑ^* of the parameters for the three considered models are reported in Table 1.

Table 1. The parameter nominal values for the three considered models.

	k_1^* (d^{-1})	k_2^* (d^{-1})	k_3^* (d^{-1})	k_4^* -	k_5^* (gm^{-3}d^{-1})	k_6^* (gm^{-3}d^{-1})
Streeter-Phelps	0.70	-	0.50	-	-	-
O'Connor-Di Toro	0.60	0.15	0.50	-	1.50	-
WODA	0.60	0.15	0.15$^{*)}$	0.30	1.50	-0.2

*) For the WODA model, units for parameter k_3 are (d^{-1}(m^{3}s^{-1})$^{-k_3}$)

Thus, to study the effects of factor c), how the variance $\sigma_{k_i}^2$ of the estimate $\hat{k}_i$ depends upon the model complexity, we computed it for each one of the three models in Table 1. To investigate factor b), how the variance $\sigma_{k_i}^2$ of the estimate $\hat{k}_i$ depends on the "true" value $\hat{k}_i$, we computed the variance for increasing values of $\hat{k}_i$, from a very small value up to a value such that $\sigma_{k_i}^2$ turns out to be independent of it. In these experiments, the remaining parameters k_j $(j\neq i)$ are kept at the nominal values reported in Table 1. Finally, to explore factor a), how the variance $\sigma_{k_i}^2$ depends on the number of sampling points n, the previous computations were repeated for $n=5, 15, 20, 30, 50, 100$. In each case, a sample estimate of the variance $\sigma_{k_i}^2$ was computed on the basis of 6000 calibration experiments.

3.1 Results

In all the cases the parameter estimates turned out to be unpolarized. As far as the variances of the estimates are concerned, instead of reporting their computed values $\hat{\sigma}_{k_i}^2$, we prefer to show the percent standard deviations

$$s_{k_i} = 100\sqrt{\sigma^2_{k_i}}/k_i^*$$ (6)

since their values are more directly comprehensive then the ones of the variances. In Figs. 1-3 these deviations are reported for Streeter-Phelps, O'Connor-Di Toro and WODA models.

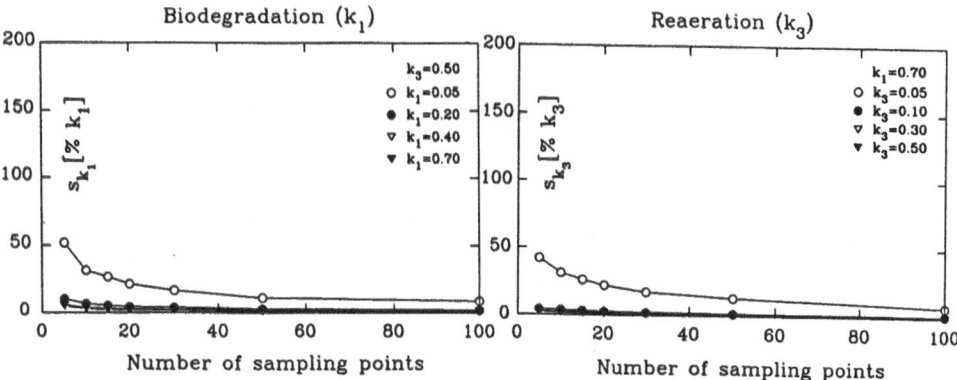

Figure 1. Percent standard deviations s_k of the parameters of Streeter Phelps model.

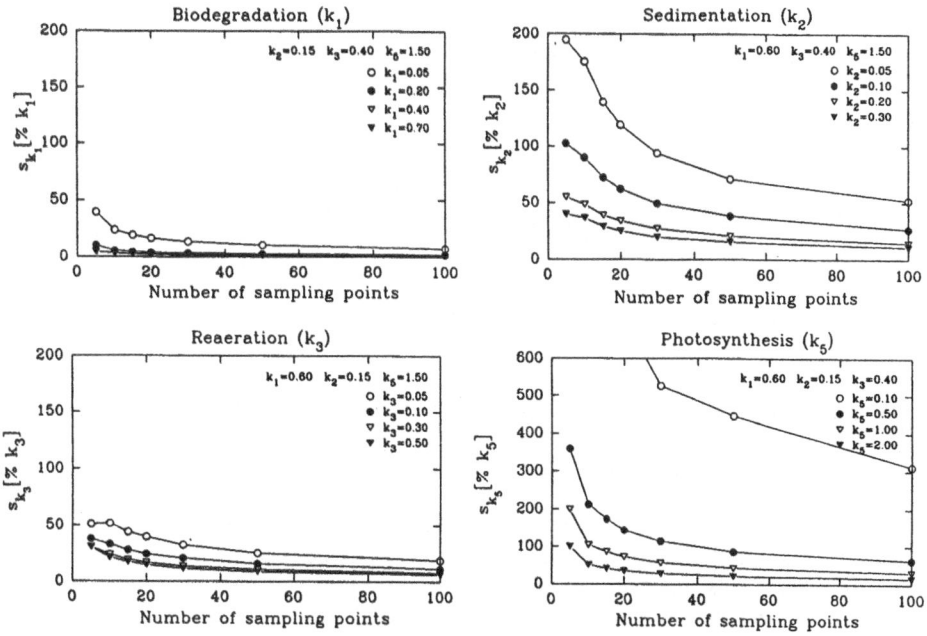

Figure 2. Percent standard deviations s_k of the parameters of O'Connor Di Toro model.

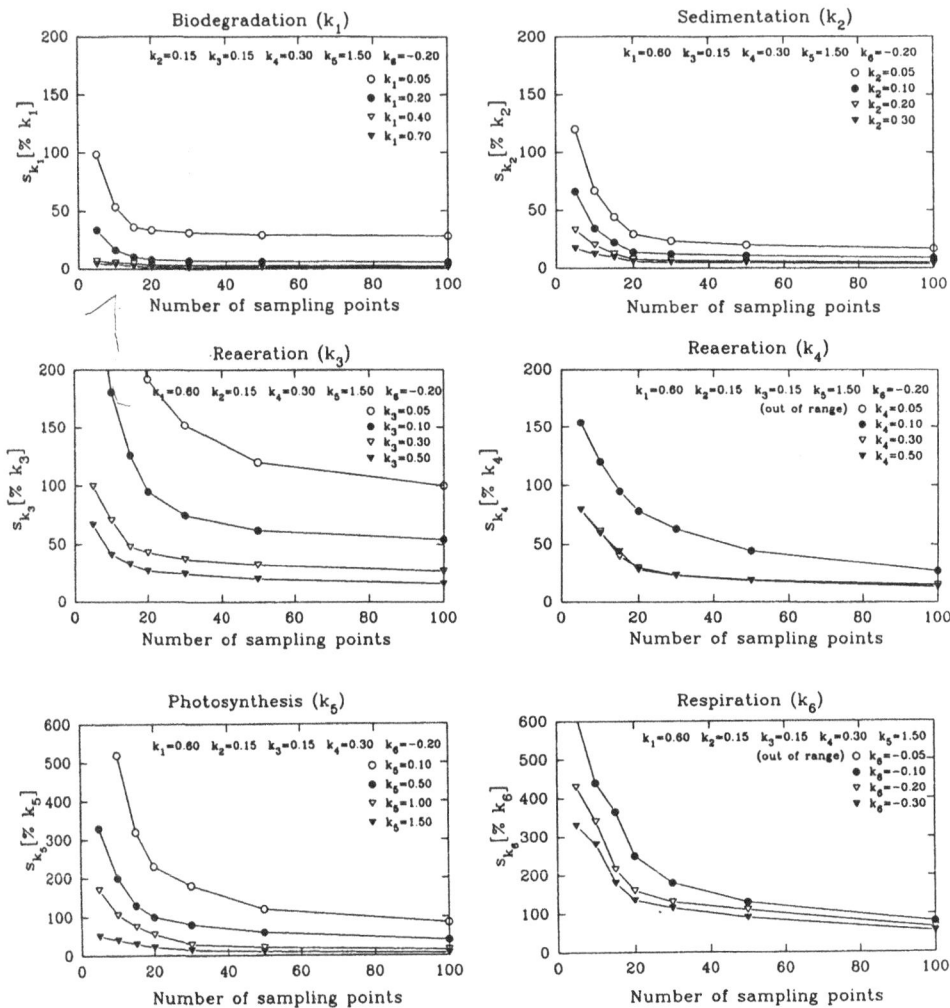

Figure 3. Percent standard deviations s_k of the parameters of WODA model.

At first glance, the parameter uncertainties in the majority of cases are larger than one would intuitively expect from such a simple scenario. This observation does not hold for the BOD decay rate and the reaeration rate (this latter only when the Streeter-Phelps or the O'Connor-Di Toro model is adopted), where the percent standard deviation s_k is of the order of 100%, if not larger. We think that this surprising result is due mainly to the absence of "excitation" of the model (remember that discharges and tributaries are nearly absent). In other words, the "simple" case is indeed a very hard one when one wants to discriminate between the several parameters.

Let's now analyze in detail the effects of the factors a)-c), as listed at the beginning of Section 3.

Factor a): in all the cases, s_{k_i} significantly decreases, as the number of sampling points increases. Fifty measurement points are required and sufficient in the majority of cases to make s_k independent of the number of sampling points (the few exceptions occur when the "true" value of the parameter is extremely low). In some case (e. g. for the BOD decay rate or the reaeration rate in the case of Streeter-Phelps model) this number can be reduced, but it can never be as low as five to ten, the range generally assumed to be sufficient.

Factor b): the value of s_k significantly reduces as the "true" value of the parameter increases; moreover there exist a threshold value, over which s_k becomes nearly independent on the parameter "true" value. To obtain a clearer perception of this threshold consider Fig. 4, where s_k is plotted as a function of the k_1 true value for the case of Streeter-Phelps model and different numbers of sampling points: it can be noticed that, from a practical point of view the value of s_k does not vary for sufficiently high values of k_1, this value (threshold) is decreasing with increasing number of measurements.

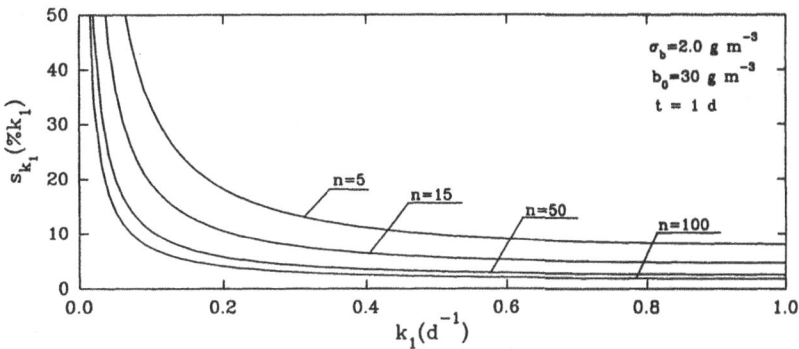

Figure 4. Treshold values of the s_{k_1} of Streeter-Phelps model.

Factor c): one would a priori expect that for a given number of sampling points the uncertainty will increase with an increase in model complexity. The results obtained don't seem to confirm completely such expectations. In fact, a small increase for the BOD decay parameter k_1 can be observed when moving from Streeter-Phelps to WODA, but not to the O'Connor-Di Toro. Sedimentation and photosynthesis are estimated from the WODA model with even less uncertainty than from the O'Connor-Di Toro model. The only significant increase in the uncertainty of k_3 can be related to an increase in model complexity. However, we must note that in the case of WODA, the parameter k_3 is not directly the reaction parameter, and that it synergically interacts with k_4 to determine the reoxygenation rate; thus, the increase in the uncertainty is both a cause and a consequence of the uncertainty of k_4. For the case study considered the uncertainties for the sedimentation, photosynthetic oxygen production rate and respiration rate are so large (from two to six times the parameter "true" value) that they make the estimate meaningless. Only when the associated process is significantly relevant is the parameter value estimate sufficiently accurate.

Finally, it is worth noting that the uncertainty of the sedimentation rate is a few times larger with the O'Connor-Di Toro model than with the WODA one. The reason is not yet clear. We may suggest that it is a consequence of the presence of a tributary in the case dealt with by the WODA model, since this might positively "excite" the model but a definitive conclusion requires further study.

4. Conclusion

The uncertainty (as measured by the variance) of the parameter estimates of BOD-DO river quality models has been quantitatively analyzed as a function of three main factors: the number of sampling points, the parameter "true" values and the complexity of the model. The experiments have been organized to explore the role of instream measurement errors only; input measurements errors and process errors have been intentionally set to zero. Therefore, in real world applications one has to expect higher uncertainty for a given case study.

The reader must be aware that the results of the present study cannot easily be extrapolated, since they strongly depend on the case at hand. In particular, the "simple" case we dealt with in this study is probably a "difficult" case, since the system is poorly excited and therefore the state variable pattern is unimodal (i.e. include only one extreme); but further studies are required to transform our intuition into a scientific statement.

The methodology used in this study may be usefully employed to design a data gathering-campaign. In fact, an uncertainty analysis based on the a priori knowledge of the river to be modeled may be of significant help in choosing the right number of measurement points, given the complexity of the model to be applied.

References

Argentesi, F. and L. Olivi (1976). Statistical sensitivity analysis of a simulatil for the biomass nutrient dynamics in aquatic ecosystems. *Proc., 4-th Summer Computer Simulation Conference*, 389-393, Simulation Council, La Jolia, Cal.

Beck, M. B. (1978). Random signal analysis in an environmental sciences problem. *Appl. Math. Model.*, **2**, 23-29.

Beck, M. B. (1979). System identification, estimation, and forecasting of water quality, 1. Theory. *Working Pap. WP-79-31*. Int. Inst. for Appl. Syst. Anal., Laxenburg, Austria.

Beck, M. B. (1981). Operational estimation and prediction of nitrification dynamics in an activated sludge process, *Water Resour. Res.*, **15**, 1313-1330.

Beck, M. B. (1987) Water Quality Modeling: A review of the Analysis of Uncertainty, *Water Resour. Res.*, **23**, 1393-1442.

Koivo, A.J. and G.R. Philips (1971). Identification of mathematical models for DO and BOD concentrations in polluted streams from noise corrupted measurements, *Water Resour. Res.*, **7**, 853-862.

Kraszewski, A.K. and R. Soncini-Sessa (1989). WODA. *Un modello di qualita' fluviale (BOD e DO)* (in Italian), CLUP, Milano.

O'Connor, D.J. and D.M. Di Toro (1970). Photosynthesis and oxygen balance in streams, *J. San. Eng. Div., Proc. ASCE*, **96**, 547-571.

O'Neil, R,V (1973). Error analysis of ecological models. *Radionuclides in ecosystems, Conf. 710501*, 898-908, Natl. Tech. Inf. Serv., Springfield, Va., 1973.

Rinaldi, S. and R. Soncini-Sessa (1978). Sensitivity analysis of generalized Streeter-Phelps models, *Advances in Water Res.*, **1**, 141-146.

Rinaldi, S., R. Soncini-Sessa, H. Stehfest, H. Tamura (1979). *Modeling and control of river quality*, Mc Graw-Hill, New York.

Standard Methods for the examination of water and waste water (1981), American Public Health Ass., New York.

Streeter, H.W. and E.B. Phelps, (1925). A study on the pollution and natural purification of the Ohio River, vol. III, *Public Health Bulletin*, **146**, United States Public Health Service, Reprinted by U.S. Department of Health, Education and Welfare.

Whitehead, P.G and P.C. Young (1979). Water quality in river systems: Monte Carlo analysis. *Water Resour. Res.*, **15**, 1305-1312.

van Straten, G. (1983). Maximum likelihood estimation of parameters and uncertainty in phytoplankton models, in *Uncertainty and forecasting of water quality*, edited by M.B. Beck and G. van Straten, 157-171, Springer Verlag, New York.

UNCERTAINTY ANALYSIS AND RISK ASSESSMENT COMBINED: APPLICATION TO A BIOACCUMULATION MODEL

THEO P. TRAAS and TOM ALDENBERG

National Institute of Public Health and Environmental Protection (RIVM)
PO Box 1, 3720 BA Bilthoven, The Netherlands

Summary

A bioaccumulation model has been formulated for contaminant accumulation in meadow-ecosystems. This type of model generally is parameter-rich, which poses problems for risk prediction due to parameter uncertainty. A procedure is needed to deal with model uncertainty and risk-assessment simultaneously. The model was subjected to uncertainty analysis, leading to probability distributions of all model output variables. Uncertainty measures were calculated using a linear regression model. The probability that environmental standards or No Observed Effect Concentrations are exceeded, was derived from the same distributions, used for the uncertainty analysis. Effects of different toxicant loading scenarios on these probabilities were calculated. The procedure discussed here facilitates the use of complex ecosystem models for risk-assessment.

Keywords Ecotoxicological models, risk assessment, uncertainty analysis.

1. Introduction

Specific chemicals are known to accumulate in terrestrial ecosystems, such as heavy metals. Near metal smelter works, cadmium has been shown to accumulate in soil invertebrates and their predators (Ma et al., 1991). In agricultural areas, cadmium loading is caused mainly by contamination of artificial fertilizer, organic manure and dry deposition (RIVM, 1990). To prevent cadmium accumulation in relatively clean agro-ecosystems, the risk of cadmium accumulation and the effect of load-reduction measures on cadmium accumulation need to be calculated. In the present study, the objective was to predict the risk of cadmium accumulation at different trophic levels within meadow ecosystems. Therefore, all major flows of biomass and cadmium had to be modelled and integrated into a comprehensive ecosystem model.

Ecosystem models are generally highly uncertain, due to inherent biological variation, and knowledge gaps. Furthermore, ecotoxicological data are often collected at different locations, extrapolated from other species or estimated, e.g. by means of body size relationships (Peters, 1983). Model components are often ill-defined and model behaviour can be considered valid within quite wide bounds. These arguments call for a probabilistic treatment of deterministic ecotoxicological models (Douben and Aldenberg, 1991). Consequently, model inputs (or components *sensu* Janssen (1993)) are random variables with a specific distribution, from which samples can be drawn. The model can then be run many times, Monte Carlo style, with different input combinations, and model output is also a random variable with a certain distribution (Janssen, 1993). Uncertainty analysis based on a regression model (Janssen et al., 1990) can now be performed, which should lead to a better understanding of the model, and subsequently lead to model improvement and suggestions for data collection. Model output distributions can also be used for the calculation of environmental risks, according to environmental standards derived from toxicity testing. In this paper, the combined use of uncertainty analysis and risk assessment is discussed.

2. Agro-ecosystem model

The model presented here is CATS-1, an acronym for Contaminants in Aquatic and Terrestrial ecoSystems, version 2.31, fully documented in a technical report including the uncertainty analysis (Traas and Aldenberg, 1992). It is a comprehensive ecotoxicological model, including the most important biotic and abiotic compartments and processes that determine bioaccumulation of persistent toxicants. Species in an ecosystem can be combined into functional groups (Cummins, 1974), regarding the fact that different species fulfill almost the same role with respect to nutrient cycling, food preferences, metabolic rates etc. The working hypothesis is, that bioaccumulation in ecosystems can be modelled at the level of functional groups.

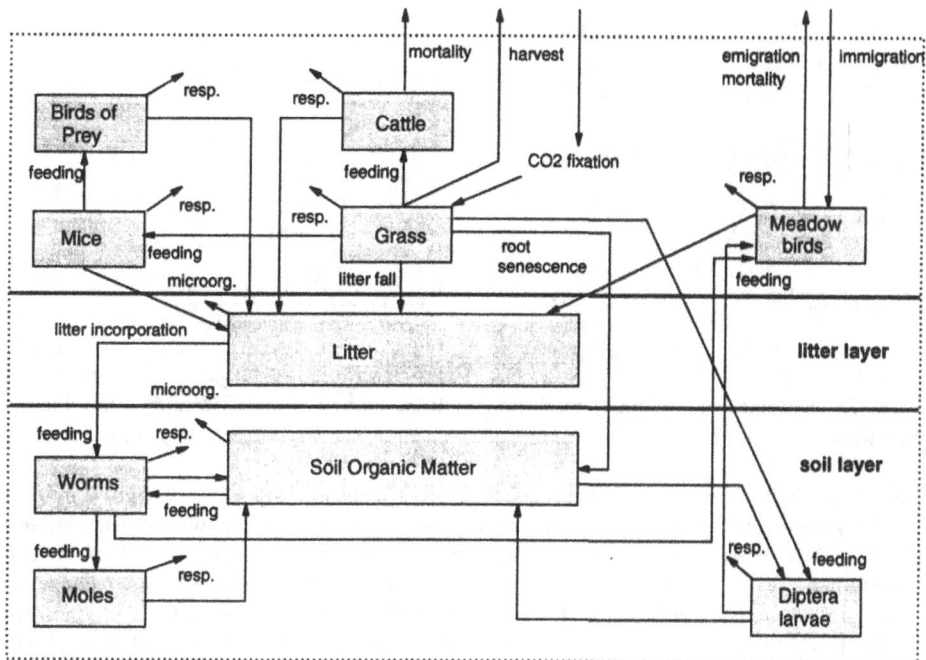

Figure 1. Diagram of biomass (rectangles) and biomass fluxes (arrows) in CATS-1. Arrows for egestion and mortality are unlabeled, respiration is abbreviated 'resp'.

The present model consists of animals and plants living on top of the soil and in the soil layer. In the meadow ecosystem we find grass, mice, cattle, meadow birds and birds of prey. Soil fauna consists of Diptera larvae, earthworms and moles (Fig. 1). Biomass passes through several stages of decomposition. First, decaying plant and animal biomass is collected in the litter layer, which may be very thin in meadows. Second, earthworm feeding activity results in the transfer of the litter to the soil organic matter (SOM). Below ground, SOM is formed because of plant root decay, egestion and mortality fluxes, and SOM is recycled through invertebrate feeding activity. All biomass compartments mentioned are present as state variables in dry weight ($gDW.m^{-2}$). They form the ecological backbone of the model, obeying the law of mass conservation.

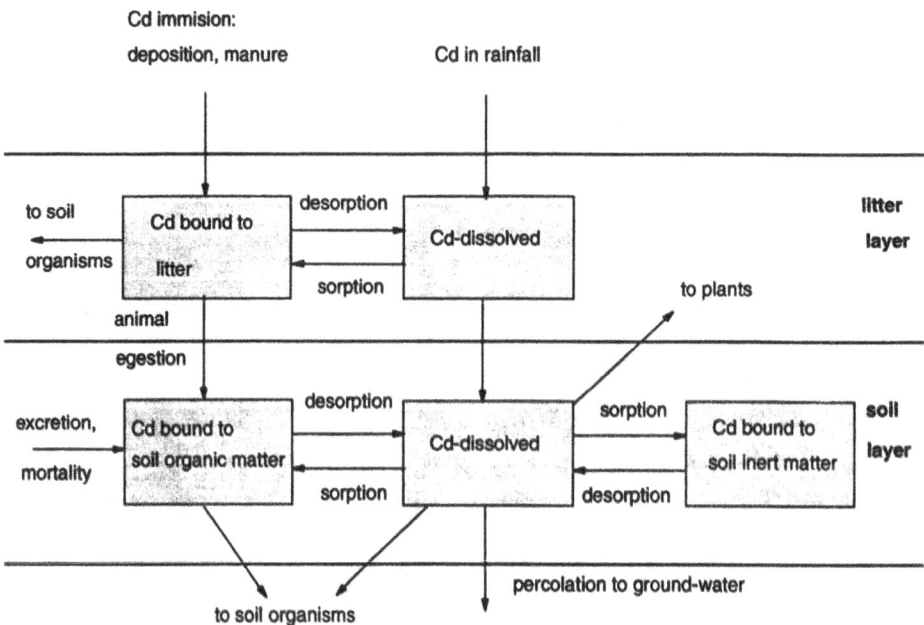

Figure 2. Diagram of cadmium equilibria in litter and in soil and connections to other compartments.

All organic compartments are also modelled for toxicant mass ($gX.m^{-2}$), effectively doubling the number of state variables. Additional state variables are the pools of dissolved and sorbed toxicant (Fig. 2). Cadmium equilibria are considered fast reactions relative to other ecosystem processes (Clasen, 1967). In the litter layer, equilibrium between sorbed and dissolved cadmium is modelled. In the soil layer, equilibrium is presumed between dissolved cadmium and two sorbing fractions with different sorption characteristics; palatable SOM and unpalatable soil fractions, such as strongly humified fractions, and soil inorganic matter, abbreviated SIM. The dissolved cadmium can percolate to deeper soil layers outside the model boundary.

Processes acting on the biomass and toxicant mass of a functional group, and the corresponding fluxes, are depicted in Fig. 3. When food is consumed (DCons), part of it is assimilated (DAss) and part is egested (DEges). The same holds true for the contaminant present in the food: it is consumed (XCons), partly assimilated (XAss) and partly egested (XEges). Biomass leaves the compartment by way of respiration (DResp), predation and mortality (DPred and DMort). Many toxicant fluxes are proportional to biomass fluxes, such as consumption (XCons) and mortality (XMort). Because of a different assimilation efficiency and excretion of toxicant, other fluxes are not proportional: assimilation of toxicant versus biomass and respiration versus excretion. These differences dynamically determine the state of biomass and toxicant of all compartments, thereby determining the concentration of the toxicant, which is the ratio of these two states ($gX.gDW^{-1}$). Since the model is expressed in mass density and not in concentration, concentrations are output variables that can be calculated any time.

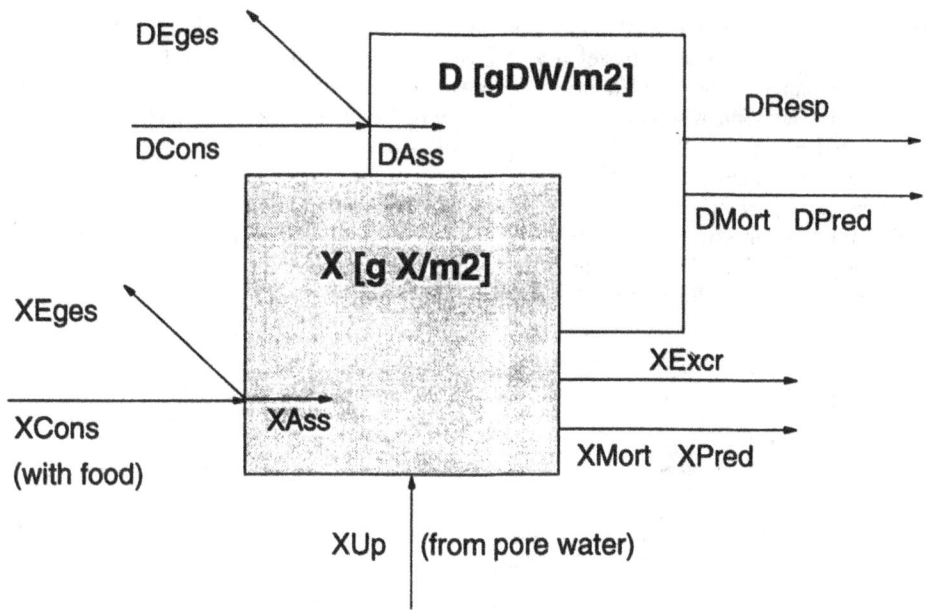

Figure 3. Diagram of biotic (D, $gDW.m^{-2}$) and pollutant (X, $gX.m^{-2}$) fluxes going to and coming from a functional group. See text for explanation.

3. Uncertainty analysis

Various techniques have been proposed for uncertainty analysis (Janssen *et al.*, 1990). The model CATS-1 was subjected to uncertainty analysis with the tool UNCSAM (Heuberger and Janssen, 1993). Uncertainty of all model components was specified using data, extrapolated data and expert judgement (Traas and Aldenberg, 1992). In most cases, only uniform distributions could be specified due to data gaps. One positive correlation was specified, between the initial cadmium concentration of top soil and the sorption constant of SOM, assuming high initial cadmium concentrations in soils with high sorption constants.

Efficient Latin Hypercube Sampling was performed to generate parameter combinations from input distributions. Model calculations were compared with ranges of acceptable model output, using the concept of Hornberger & Spear (1981). The ranges of model output such as biomass and bioconcentration factors of toxicants were extracted from the open literature. If the simulation yielded values outside the defined bounds, the simulation was stopped and a new parameter combination was used for the next model run (Janse *et al.*, 1992). This procedure was used to rule out invalid (random) parameter combinations. All model output within acceptable bounds was collected for uncertainty analysis.

Uncertainty measures were calculated with UNCSAM, using linear regression analysis of model input versus model output. To assess the uncertainty contributions of model components, two uncertainty measures have been used. The Standardized Regression Coefficient (SRC) measures the fraction of the uncertainty in a model output which is contributed by a certain model input. The SRC is especially suited when model components are uncorrelated (Janssen, 1993). If (some) model components are correlated however, the use of the Root of the Partial Uncertainty Contribution (RTU) has been suggested (Kros *et al.*, 1993). The RTU expresses the relative change in the uncertainty of a model output due to a relative change in the uncertainty of a model input, accounting for the fact that correlated model components should change accordingly (Heuberger and Janssen, 1993). If correlations are weak, the RTU is almost equal to the SRC. By comparing these two measures, the influence of correlations on uncertainty measures can be identified.

The most important model output is cadmium concentration of the top soil, since all biotic compartments follow the same qualitative accumulation pattern as top soil (Traas and Aldenberg, 1992). Distributions of total soil concentrations are calculated at the end of the simulation period, for three different cadmium loading scenarios (Fig. 4).

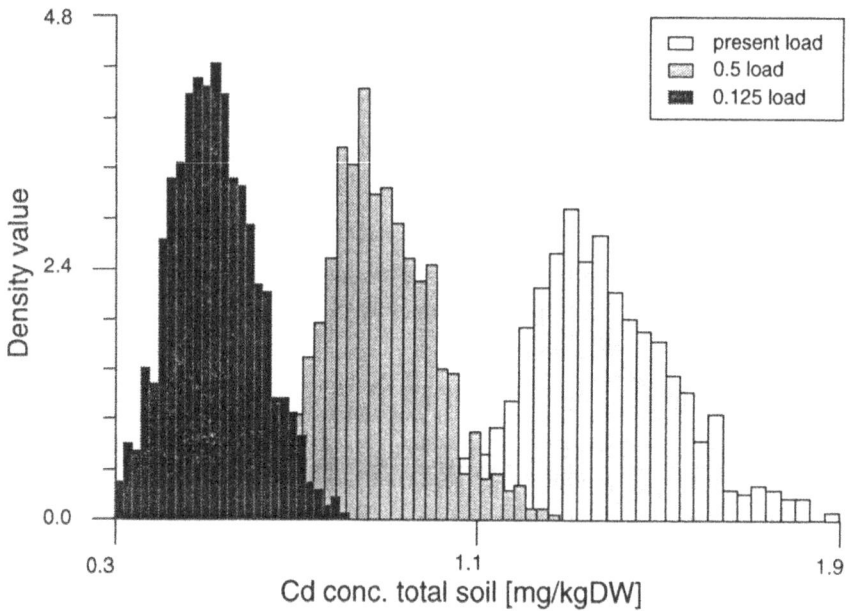

Figure 4. Probability distributions of Cd concentration in total soil in 2015, for different Cd loading scenario's.

Table 1. Ranking (rank between brackets) of the 10 most important sources of uncertainty, for Cd conc. in total soil at different loading scenarios in 2015. Uncertainty measures RTU and SRC and parameters are explained in the text.

Scenario	Present load		0.5 load		0.125 load	
uncert. measure	RTU	SRC	RTU	SRC	RTU	SRC
parameter						
RhoSIM	0.70 (1)	-0.70 (1)	0.57 (2)	-0.58 (2)	0.45 (2)	-0.45 (2)
cXTotSoilIn	0.66 (2)	0.67 (2)	0.74 (1)	0.75 (1)	0.72 (1)	0.71 (1)
cXAdsSOMIn	0.40 (3)	0.00(45)	0.44 (3)	-0.02(18)	0.44 (3)	0.01(76)
cXAdsSIMIn	0.18 (4)	0.20 (3)	0.17 (4)	0.19 (3)	0.13 (4)	0.14 (3)
hXUpCr	0.13 (5)	0.10 (5)	0.12 (6)	0.10 (5)	0.07 (9)	0.05 (7)
kXUpCr	0.12 (6)	-0.13 (4)	0.12 (5)	-0.13 (4)	0.11 (5)	-0.11 (4)
kXUpW	0.07 (7)	0.00(44)	<0.06	<|0.03|	<0.07	<|0.04|
cXDMiceIn	0.07 (8)	0.00(89)	0.07 (7)	0.00(75)	0.08 (6)	0.01(73)
fPreSoilDeep	0.06 (9)	-0.06 (6)	0.06 (8)	-0.06 (7)	<0.07	<|0.04|
hDPrB	0.06(10)	-0.01(25)	0.06(10)	-0.01(59)	0.05(19)	0.02(42)
R² or COD	0.99		0.94		0.79	

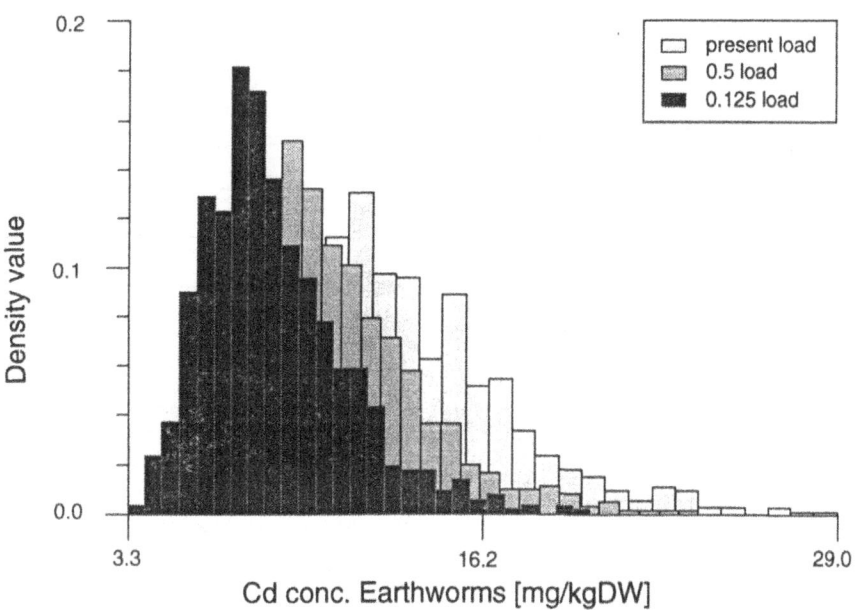

Figure 5. Probability distributions of Cd concentration in earthworms in 2015, for different Cd loading scenario's.

Table 1 shows that RTU value and ranking are very different from the SRC for the cadmium sorption constant for soil organic matter *cXAdsSOMIn*. The SRC ranks this parameter as 45, i.e. 44 sources have a greater influence on uncertainty. According to the RTU, *cXAdsSOMIn* is a most important source (with a high value). This is a typical case of correlation of a weak source with a strong source: *cXAdsSOMIn* was correlated on purpose with high initial soil concentrations. This makes sense, since low initial soil concentrations would generally be found in soils with a low sorption constant for soil organic matter (*cXAdsSOMIn*). Here, the prediction of the SRC should be used, and the RTU overestimates the importance of this source.

The density of the soil *RhoSIM* is a very strong source of uncertainty, together with initial soil concentration that becomes the primary source of uncertainty if Cd load decreases. Ranking of most other parameters does not change much when loads are reduced. It is no surprise that the sorption constant for soil inert matter is important, since it influences cadmium equilibrium directly. Parameters governing cadmium uptake of the grass (*hXUpCr* and *kXUpCr*) are weaker sources of uncertainty. Because of its high biomass, the grass takes up much cadmium from the soil, influencing cadmium concentration in the soil.

A comparison of uncertainty measures for different load-reduction scenarios shows, that the only significant change is the growing importance of the initial soil concentration (*cXTotSoilIn*). A comparison of the Coefficient of Determination (COD) shows that the greater the load reduction, the lower the COD (or R^2). R^2 is a number between 0 and 1, indicating the validity of the regression model to approximate the orginal model output. A high R^2 indicates a good approximation (Heuberger and Janssen, 1993). Fig. 4 shows that the distribution is skewed to the right, suggesting that a log transformation could improve the fit.

Table 2. Ranking (rank between brackets) of the 10 most important sources of uncertainty, for Cd conc. in earthworms at different loading scenarios in 2015. Uncertainty measures RTU and SRC and parameters are explained in the text

Scenario	Present load		0.5 load		0.125 load	
uncert. measure	RTU	SRC	RTU	SRC	RTU	SRC
parameter						
hXUpW	0.49 (1)	-0.51 (1)	0.46 (1)	-0.47 (1)	0.39 (2)	-0.40 (2)
kXExcrW	0.39 (2)	-0.41 (2)	0.36 (3)	-0.38 (3)	0.30 (3)	-0.31 (3)
RhoSIM	0.37 (3)	-0.27 (3)	0.34 (4)	-0.34 (4)	0.27 (4)	-0.27 (4)
cXTotSoilIn	0.34 (4)	0.34 (4)	0.41 (2)	0.41 (2)	0.44 (1)	0.43 (1)
cXAdsSIMIn	0.30 (5)	-0.29 (6)	0.29 (5)	-0.27 (6)	0.21 (7)	-0.20 (6)
kDMortW	0.30 (6)	-0.32 (5)	0.29 (6)	-0.30 (5)	0.23 (6)	-0.25 (5)
kXUpW	0.23 (7)	0.27 (7)	0.22 (8)	0.25 (7)	0.16 (8)	0.19 (7)
cXAdsSOMIn	0.22 (8)	0.01(33)	0.25 (7)	0.00(90)	0.27 (5)	0.00(90)
hDEatW	0.17 (9)	-0.22 (8)	0.17 (9)	-0.21 (8)	0.14(10)	-0.18
kDAssW	0.16(10)	-0.15 (9)	0.16(10)	-0.15 (9)	0.16 (9)	-0.13 (9)
R² or COD	0.96		0.92		0.72	

The uncertainty in the prediction of the Cd concentration of worms is studied because worms are an important source of cadmium for meadow birds and especially moles. Probability distributions for cadmium concentrations in earthworms are also skewed to the right (fig. 5). RTU and SRC usually have the same ranking of parameters, except for the sorption constant *cXAdsSOMIn*, which shows the same problem as for Cd concentration in top soil (Table 2). The half saturation constant for cadmium uptake (*hXUpW*) from pore water is a very strong source. This suggests that the shape of the Monod saturation function used here is important with respect to uncertainty. Top soil density (*RhoSIM*) is important in the calculation of the dissolved concentration, which is shown in the ranking of sources. Excretion, mortality and uptake rates are strong sources of uncertainty, the more important since data on these rates are sparse. The uncertainty in the soil sorption constant (*cXAdsSIMIn*), though quite a wide range has been used, is a weaker source than soil density (*RhoSIM*). This reflects that important sources of uncertainty in the calculation of soil solution concentration are also important for uncertainty of earthworm concentration.

4. Risk assessment

Probability distributions of model output have been generated for uncertainty analysis. These distributions were also used to calculate the probability that No Observed Effect Concentrations (NOECs) or other environmental standards are exceeded, by calculating the right tail probabilities (Fig. 6). These probabilities were calculated at the end of the simulation period (2015), for several cadmium load reduction scenarios.

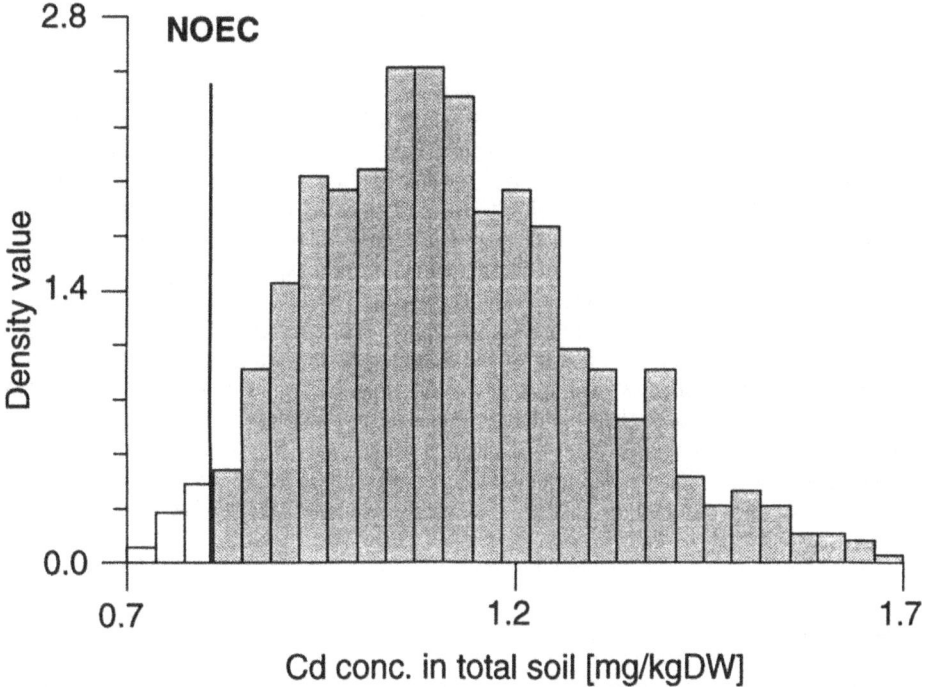

Figure 6. The right tail probability that a NOEC is exceeded, is calculated from the shaded area.

An extrapolation procedure has been used to calculate a soil cadmium concentration at which 95% of all species are protected (Aldenberg and Slob, 1993). The Maximum Permissible Concentration (MPC) is calculated from a set of NOEC values (Van der Meent *et al.*, 1990), at a 95 % species protection level with 50% confidence. The present

Cd load leads to exceeding the MPC with 100% probability (Table 3), saying that more than 5% of the species are unprotected. Load reductions of up to one quarter of the present load do not lead to a lower probability. Only the strongest load reduction of one eighth offers an improvement but we must look beyond 2050 before the MPC can be achieved (results not shown).

Table 3. Probabilities for exceeding selected environmental standards in 2015 for different Cd loading scenarios, calculated with model CATS-1.

Scenario	Present load	0.5 load	0.250 load	0.125 load
Soil standard (MPC)	100%	100%	100%	100%
Water quality standard	0%	0%	0%	0%
Food standard meadow birds	100%	100%	100%	100%
Food standard raptorial birds	5%	0%	0%	0%

Cadmium concentration of interstitial water is used to assess whether water that seeps down to deeper layers, conforms to drinking water quality standards. The present quality standard for the preparation of drinking water from surface water is 1.5 µg/l (CCRX, 1985) and the probability that it is exceeded for any scenario up to 2050 is zero (Table 3).

Romijn et al. (1991) calculated an NOEC for cadmium in food of birds and mammals, extrapolated from existing toxicological data. The NOEC in food, only valid for birds and mammals, is defined such that 95% of these species do not experience sublethal effects of cadmium. The model calculates the probability that the cadmium concentration in food of mammals or birds is exceeded, using average food concentrations. The food of meadow birds contains much cadmium, because of the earthworms in their diet, and the probability that more than 5% of meadow bird species are unprotected, is 100% for all scenarios and times studied. This warrants field validation, since meadow birds are considered important indicators of ecosystem quality. Raptorial birds, however, which feed on herbivorous mice, are not expected to experience sublethal effects of cadmium. At the present load, the probability that the standard is exceeded is very low, and becomes zero when load is reduced.

5. Discussion

Uncertainty analysis allows the evaluation of the importance of certain mechanisms incorporated in the model, given the fact that many parameters are ill-defined. Bio-availability of cadmium to earthworms is such a mechanism, supposedly governed by the dissolved cadmium concentration (Van Gestel and Ma, 1988). This warrants model details on cadmium sorption equilibria and uptake functions of earthworms. Much model uncertainty is caused by the cadmium sorption constant of the parent soil material, which represents the bulk of cadmium in the system. The dissolved concentration calculated from the cadmium sorption equilibrium, turns out to be very sensitive to the density of the soil. Measurements of dissolved cadmium concentrations were not available, limiting calibration. Uncertainty of dissolved cadmium concentrations is transmitted to all plants and animals that are directly exposed to soil pore water. Therefore, more attention needs to be paid to formulation of sorption equilibria because of these consequences for bio-availability, if we want to predict bio-availability with less uncertainty than at present.

Other important parameters that largely determine model uncertainty are rates of uptake, assimilation and excretion of cadmium, immediately followed by respiration and mortality rates. Due to the conceptual division of the model in a biomass cycle and a toxicant cycle, uncertainty in both cycles influences our prediction of bioaccumulation. With this type of model, it is easy to find out which cycle contributes most to prediction uncertainty.

The present analysis also showed that elaborate details of the model, like multiple cadmium sorption equilibria (Fig. 2), and earthworms feeding from different layers, do not always lead to large uncertainty of model outputs. This does not necessarily mean that these model details are superfluous, if we want to preserve a model structure applicable to different ecosystems on different types of soil. Attention to generic processes makes the model complex, but also more scientifically credible (Hornberger and Cosby, 1985). At the same time, the data might not warrant such elaborate detail (Beck, 1981).

Although individual model outputs can be studied in great detail, it is very hard to get a clear picture of how well the model fits the available data. With this type of uncertainty analysis, output distributions are basically independent of output data. They can only be compared by testing whether the data could be generated by these distributions. Through Bayesian maximum likelihood estimation of parameters based on output data, or Bayesian statistics, input uncertainty can be calibrated to match the data. This in its turn will lead to reduced prediction uncertainty (Aldenberg et al., 1992, Kramer et al., 1993).

Probability distributions of model outputs like concentrations allow the simultaneous evaluation of environmental standards or noels. The authors feel that it is much better to speak of a 10% probability that a drinking water standard is exceeded, than to put faith in deterministic scenarios for worst case or best case guesses. On the other hand, it is not easy to interpret the significance of a 20% probability that a NOEC extrapolated from laboratory toxicity tests is exceeded. A fundamental problem in this interpretation is the uncertainty associated with the estimation of parameters calibrated on data, such as uptake rates of soil invertebrates from the soil solution. Since data are scarce, a very wide range of uptake rates must be assumed, but how wide is often hard to establish. Again, Bayesian calibration (Aldenberg et al., 1992) could provide us with better techniques for calibration, and thus for reducing uncertainty in our prediction of the probability of exceeding environmental standards.

Acknowledgements

This study was supported by the Ministry of VROM, Directorate of Chemicals and Risk Management, within the framework of the Project Ecological Sustainability of the Use of Chemicals (PEIS). Thanks are due to Mr R. Lammers for support on computer and network facilities.

References

Aldenberg, T. and W. Slob (1993), 'Confidence limits for hazardous concentrations based on logistically distributed NOEC toxicity data', Ecotox. Environ. Saf., 25, 48-63.
Aldenberg, T., J.H. Janse and P.R.G. Kramer (1992), 'Fitting the dynamic model PCLake

to a multi-lake survey through Bayesian statistics', *International Conference on Mathe-matical Modelling in Limnology*, Innsbruck, Austria, 4-7 November, submitted for publication.

Beck, M.B. (1981), 'Hard or soft environmental systems?'. *Ecol. Model.*, **11**, 233-251.

CCRX (1985), 'Cadmium loading of the environment in the Netherlands' (in Dutch), *CCRX evaluatierapport*, Ministerie van VROM, Leidschendam, The Netherlands.

Clasen, R.J. (1967), 'The numerical integration of kinetic equations for chemical systems having both slow and fast reactions', *Technical Report nr. P-3547*, The RAND Corporation, Santa Monica, California.

Cummins, K.W. (1974), 'Trophic relations of aquatic insects', *Bioscience*, **24**, 631-641.

Douben, P.E.T. and T. Aldenberg (1991), 'Ecosystem recovery and efficiency of load reduction measures' (In Dutch), in: *Flora en Fauna chemisch onder druk*, Eds: Hekstra, G.P. and F.J.M. Van Linden, Pudoc, Wageningen, pp. 213-230.

Heuberger, P.S.C. and P.H.M. Janssen (1993), 'UNCSAM: a software tool for sensitivity and uncertainty analysis of mathematical models', *This congress*.

Hornberger, G.M. and B.J. Cosby (1985), 'Selection of parameter values in environmental models using sparse data: a case study', *Appl. Math. Comp.*, **17**, 335-355.

Hornberger, G.M. and R.C. Spear (1981), 'An approach to the preliminary analysis of environmental systems', *J. Environ. Mgmt.*, **12**, 7-18.

Janse, J.H., T. Aldenberg and P.R.G. Kramer (1992), 'A mathematical model of the phosphorous cycle in Lake Loosdrecht and simulation of additional measures', *Hydro-biologia*, **233**, 119-136.

Janssen, P.H.M. (1993), 'Assessing sensitivities and uncertainties in models: a critical evaluation', *This congress*.

Janssen, P.H.M., W. Slob and J. Rotmans (1990), 'Sensitivity analysis and uncertainty analysis, an inventory of ideas, methods and techniques' (in Dutch), *RIVM rapport nr. 958805001*, RIVM Bilthoven, The Netherlands.

Kramer, P.R.G., A.C.M. de Nijs and T. Aldenberg (1993), 'Bayesian uncertainty analysis in water quality modelling', *This Congress*.

Kros, J., W. de Vries, P.H.M. Janssen and C.I. Bak (1993), 'The uncertainty in forecasting trends of forest soil acidification', *Water, Air, and Soil Pollut.*, **66**, 29-58.

Ma, W.C., W. Denneman and J. Faber (1991), 'Hazardous Exposure of Ground-Living Small Mammals to Cadmium and Lead in Contaminated Terrestrial Ecosystems', *Arch. Environ. Contam. Toxicol.*, **20**, 266-270.

Peters, R.H. (1983), 'The ecological implications of body size', *Cambridge studies in ecology*, Cambridge University Press, Cambridge.

RIVM (1990), 'Concern for tomorrow', Samson Tjeenk Willink, Alphen aan de Rijn.

Romijn, C.A.F.M., R. Luttik, W. Slooff and J.H. Canton (1991), 'Presentation of a general algorithm for effect-assessment on secondary poisoning. II: terrestrial foodchains', *RIVM report nr. 679102007*, RIVM Bilthoven, The Netherlands.

Traas, Th.P. and T. Aldenberg (1992), 'CATS-1: a model for predicting contaminant accumulation in a meadow ecosystem, the case of cadmium', *RIVM report nr. 719103001*, RIVM Bilthoven, The Netherlands.

Van der Meent D., T. Aldenberg, J.H. Canton, C.A.M. van Gestel and W. Slooff (1990), 'Desire for levels. Background study for the policy document "Setting environmental quality standards for water and soil"', *RIVM report nr. 670101002*, RIVM Bilthoven, The Netherlands.

Van Gestel, C.A.M. and W. Ma (1988), 'Toxicity and bioaccumulation of chlorophenols in earthworms, in relation to bioavailability in soil', *Ecotox. Environ. Safety*, **15**, 289-297.

DIAGNOSIS OF MODEL APPLICABILITY BY IDENTIFICATION OF INCOMPATIBLE DATA SETS ILLUSTRATED ON A PHARMACOKINETIC MODEL FOR DIOXINS IN MAMMALS

OLIVIER KLEPPER and WOUT SLOB

Centre for Mathematical Methods
National Institute of Public Health and Environmental Protection
P.O. Box 1, 3720 BA Bilthoven, The Netherlands

Abstract

When calibrating a model the problem often arises that different data sets (e.g. different outputs of the same system; data obtained under different conditions) are incompatible: parameter values that give an acceptable fit to one data set provide an unacceptable fit to another set. Occasionally, a previously calibrated model fits poorly to a new data set, but it may be possible to re-calibrate the model to both data sets. However, if data sets are in fact incompatible, calibration of the model to all available data may simply result in a poor compromise-fit without any insight into the cause of the problem. This paper discusses how the situation can be analysed as a multi-objective optimization problem, with the goodness of fit values to the different data sets as optimization goals. It is shown that the set of Pareto-optimal solutions (i.e. where an increase in fit to a particular data set must necessarily lead to a decrease elsewhere) provides an efficient way to analyse the situation. If there is only a single Pareto-optimal point, the same set of parameter values can be used for all applications. If this is not the case, it will be shown how trade-offs between goodness of fit values can be indicated, and clusters of mutually compatible data sets (i.e. those that can be fitted by single set of parameter values) can be identified. Analysis of parameter values corresponding to the different clusters provides insight to arrive at a more generally applicable model. The method proposed in this paper is illustrated on a simple growth model with artificial data sets as well as on a pharmacokinetic model for dioxins using data from a number of independent experiments using mice, rats and cows. The present analysis provides valuable insight for the further development of generic toxicokinetic models that can be used for interspecies extrapolations.

Keywords model identification, Pareto-optimal calibration, model error diagnosis

1. Introduction

Model construction generally proceeds in an iterative fashion, starting with some initial model, calibrating it to the available data, and subsequently either expanding the model (in case of an insufficient fit) or simplifying it (in case of unidentifiable model parameters). In this process, we may distinguish three main activities: *model formulation* (selecting variables and mathematically formulating the relations between them) *parameter identification* (calibrating parameter values, indicating superfluous parameters) and *model assessment* (judging the fit of the model to the data). Of these three activities, the latter two have received a great deal of attention in the literature (e.g., Beck, 1987 and Tarantola, 1987 on model identification, and Draper and Smith, 1981 and Bates and Watts, 1988 on model assessment); and formal methods have replaced subjective procedures ("fitting by eye" and judging the model as "sufficiently accurate"). If the calibration procedure indicates superfluous parameters it is generally obvious where to simplify the model. However, it is very common to have an insufficient model fit (i.e. residuals are considerably larger than measurement errors and the desired accuracy), even in the presence of superfluous parameters. This indicates that model formulations have to be changed. This part of the model construction process seems a rather neglected area, and is perhaps the reason why modelling is occasionally regarded more as an art than a science. If the model fails, there are almost no tools or formal methods that help the modeller in formulating an improved model. This paper aims at developing a diagnostic tool that uses the calibration results to direct the search towards an improved model.

In a series of papers by Beck, Young and coworkers (Beck and Young, 1976; Scavia, 1980; Whitehead, 1983; Young, 1984; Beck, 1987) the extended Kalman filter (EKF) has been used as a diagnostic tool to improve model formulation. The EKF provides recursive (time-dependent) estimates of the model parameters based on a sequential analysis of the available data. EKF estimates varying in time (for example, showing a seasonal pattern) indicate that the associated model parameters are in fact variables. The time-pattern of the parameter value and its correlation with other variables may help to improve model structure. Examples are a variable re-aeration coefficient which correlates with algal biomass (conclusion: primary production contributes significantly to oxygen input) and a variable carbon to chlorophyl ratio which correlates with species composition (conclusion: ratio is species-dependent). Despite numerous successful applications, there are several situations where the EKF approach can not be used. It is only suited for situations where parameters are not constant through time; often, one is also interested in situations where different locations (e.g., applying a eutrophication model to different lakes) or different systems (e.g., applying a toxicokinetic model to different species) yield different parameter estimates. Furthermore, the EKF linearizes the model and assumes Normally distributed errors. The linearization results in numerical instabilities for large and highly nonlinear models, especially if uncertainties are relatively large. The assumption of Normality for the errors may be restrictive also: occasionally, a criterion that is more robust than the standard least-squares is preferable (e.g., the L_1-norm (Tarantola, 1987), or -after some data smoothing- a L_∞-norm is preferred for its clarity of interpretation (Hornberger and Spear, 1981).

In this paper we want to formulate a tool that is in some sense similar to the EKF approach, but is also applicable in some of the situations where EKF fails. For this purpose, we will first state some definitions. We will assume that we have a model (with output that is generally multivariate and time-dependent) that is intended to be applied to a number of similar systems. We further have available several *data sets*: these may relate to a single system's output through time (e.g., winter, summer data), to several variables in the same system (e.g., blood pressure, heart beat) or to the same variables in different systems (rat, mouse). We will further assume that we have some measure of *lack of fit* (functions $L_i(p)$, with $i=1...n$ the data sets and p the parameter vector) with an *acceptable threshold* ($L_i(p)<c_i$). This may be acceptable in the "standard" sense (e.g., lack-of-fit function based on likelihood and threshold based on a likelihood-ratio) or in a set-theoretic sense (e.g., define set of acceptable model-outcomes subjectively; Hornberger and Spear, 1981). A model can be termed overall acceptable if we can find a single set of parameters that gives an acceptable fit to all available data sets. We will assume that the model can always be made acceptable for a single data set; this may be achieved by splitting the available data into sufficiently small sets, possibly consisting of a single point. The problem of a model that is not overall

acceptable can now be viewed in two complementary ways. In parameter space: which parameters have to be adjusted to achieve acceptable fits for the different data sets? And in system space: which data sets are *compatible* (do give a mutually acceptable fit for a single vector of parameter values) and which not? An answer to the first question provides a diagnostic tool towards the formulation of a more generally applicable model; the second question helps us to define the range of applicability of the model. Note that compatability as defined above is a property of the data *in combination* with the model.

The paper first introduces the concept of *Pareto-optimal calibration* as an aid to investigate trade-offs between the lack-of-fit functions L. It is shown that in case of compatible data sets Pareto-optimal calibration is equivalent to standard calibration techniques. A method of clustering the data sets into compatible clusters is proposed, which is the basis to indicate parameters that are significantly different between clusters. The procedure is illustrated both on a simple model and an actual case study using a toxicokinetic model of dioxin (TCDD) in mammals. The method is particularly relevant for the latter model because extrapolation from test animals to untested species such as man (i.e., having a wide range of applicability) is the main purpose of such models. The results indicate that the procedure is useful as a framework to define model applicability and as a tool to indicate a direction towards improved model formulations.

2. Methods

The problem of finding compatible data sets can be formulated by defining for data set i a lack of fit function L_i and a threshold c_i a set in parameter space S_i which corresponds to all parameter values p so that $L_i(p)<c_i$. For convenience we scale each L_i by c_i so that our threshold is 1 throughout. A group of n data sets is compatible if the intersection is not empty: $S_1 \cap S_2...\cap S_n \neq \emptyset$. Theoretically, one could check this by minimizing the L_∞ norm on the scaled L_i's:

$$L^*(p) = \max \left\{ \frac{L_1(p)}{c_1}, \frac{L_2(p)}{c_2}, ..., \frac{L_n(p)}{c_n} \right\} \qquad (1)$$

with a minimum value $L^*<1$ indicating compatability of the sets 1...n with respect to the model. In practice, this may not be a feasible approach however. In the case study treated below, there are 19 data sets. If these are not compatible as a single group, one would be interested in finding smaller groups that *are* mutually compatible. Applying equation (1) directly to all possible subdivisions of the 19 data sets would be very laborious: checking all pairs on compatibility requires $\binom{19}{2}$ optimizations; checking all groups of 3 another $\binom{19}{3}$ etc. Clearly it is necessary to use a more efficient method both to check on compatibility and to group compatible data sets.

A central issue in the investigation of compatibility is *trade-off*: how much does L_i increase with a decrease in L_j. For this purpose we first restrict ourselves to the subset of parameter space D where there *is* such trade-off. Clearly, a parameter value which is dominated in the sense that we can find another one that is better for all lack-of-fit criteria L_i (i.e.: no trade-off) can be discarded (figure 1). We are interested only in the set of non-dominated or Pareto-optimal parameter values:

$$P = \{\ p \in D \ | \ \forall \ p' \in D \ \exists \ i: \ L_i(p) \leq L_i(p')\ \} \tag{2}$$

i.e., those parameter values showing trade-offs in lack of fit between the data sets.

Although in general there will be *some* trade-off (two data sets are seldomly perfectly compatible), we are mainly concerned with the situation where this trade-off is relatively large. In statistical analysis, it is usually assumed that the observations are realizations of some stochastic process with zero mean around the model-output for the true parameter value p_o. Under this assumption, trade-offs only result from random errors in the realizations of the data. For a sufficiently large sample trade-offs become negligible and the set shrinks to (almost) a single point which is similar to the usual maximum likelihood parameter value (i.e., the point which minimizes ΣL_i, see fig. 1).

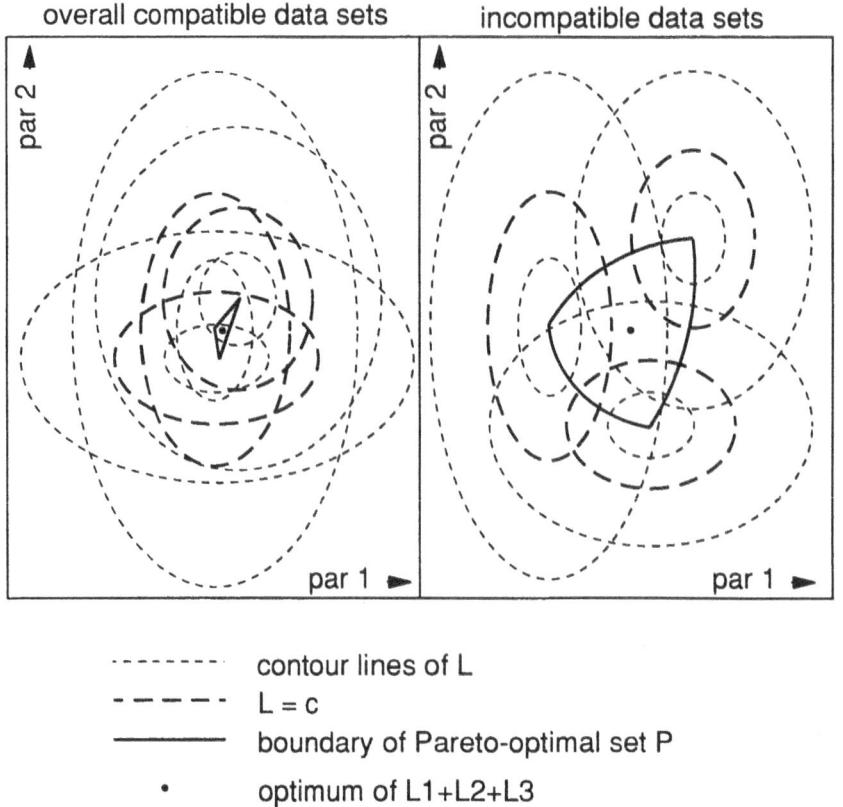

Figure 1. A hypothetical example showing the Pareto set in 2 dimensions (parameters) with 3 data sets (L-functions). The dot indicates the overall optimum of $L_1+L_2+L_3$. When trade-offs in lack of fit functions lare large compared to uncertainty in the data (indicated by the contours $L=c$), the data sets are termed incompatible.

A practical method to find points in the Pareto set is to minimize linear combinations of L_i (Yu, 1989):

let   - w be a vector ($w_i \geq 0$, $i=1..n$) with w_i not all equal to 0
and - p^* minimizes $\Sigma w_i L_i$
then: - $p^* \in P$.

The reverse may not be true: there may be points in P that do not correspond to any vector w. An example of the latter is when P lies along the edges of a polygon in L-space, but only vertices of the polygon can be reached by minimizing linear combinations of L_i's. In this case, linear interpolation between the vertices also yields points in P. However, if the functions L_i are not convex (for example, have multiple local minima) the linear interpolation along its edges may fail (P may not be connected). We will assume that all L_i's *are* convex.

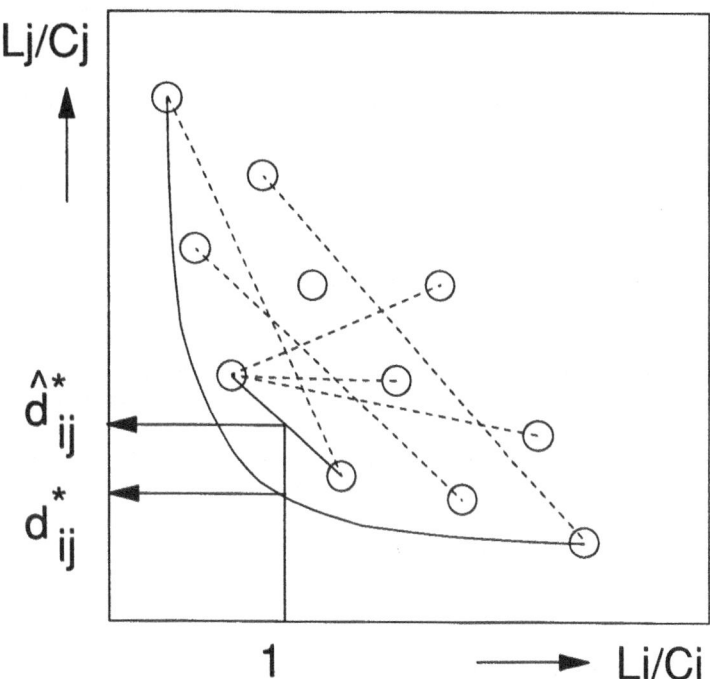

Figure 2. Definition of d^*_{ij} (the lowest value of L_j/c_j that can be achieved for $L_i \leq c_i$) and calculation of $\hat{d}^*_{ij}$ by linear interpolation. The circles represent the 2-dimensional projection of the sample of N Pareto-optimal points. (Note that because only two dimensions are shown, some of the points appear to be dominated). The dotted lines represent some of the possible linear interpolations between the N points; the drawn line yields the lowest value and gives the estimate $\hat{d}^*_{ij}$.

We can now define for each pair (i,j) an asymmetric distance d^*_{ij} as the minimum value of L_j that can be achieved under the constraint $L_i/c_i < 1$. An algorithm to estimate d^*_{ij} as $\hat{d}^*_{ij}$ was constructed as follows. First we us the n unit vectors for w, i.e. we minimize the individual L_i's. (Note that by assumption min$\{L_i\} < c_i$: the model fit can always be made acceptable for a single data set). By randomly choosing $N-n$ additional vectors w and minimizing $\Sigma w_i L_i$ we are able to generate a total of N points in P, in particular: N pairs (L_i^k, L_j^k) ($k=1...N$). In order to estimate $\hat{d}^*_{ij}$ we have to distinguish three cases: all N points have a value $L_i < c_i$, there is a point with $L_i=c_i$, or some L_i are below c_i and some above the threshold. In the first case, we simply estimate $\hat{d}^*_{ij}$ as the minimum of the N L_j/c_j values. In the second case, we use the L_j/c_j value corresponding to $L_i=c_i$. In the third case we calculate values L_j^{kl} by linear interpolation between two combinations (L_i^k,L_j^k) and (L_i^l,L_j^l) $(k,l \in 1...N)$, with $L_i^k<c_i$ and $L_i^k>c_i$. The minimum of all possible interpolations L_j^{kl}/c_j gives us the estimate $\hat{d}^*_{ij}$: see figure 2. Linear interpolation of a convex function L has the properties that: $\hat{d}^*_{ij}\geq d^*_{ij}$ and in the limit $N\rightarrow\infty$ $\hat{d}^*_{ij}=d^*_{ij}$. Clearly, in general $d^*_{ij}\neq d^*_{ji}$; a symmetric distance d_{ij} is now defined as $\min[d^*_{ij},d^*_{ji}]$.

On the basis of the distances d_{ij} we can group the individual L_i's by hierarchical clustering (Pielou, 1984). In hierarchical cluster analysis, the two nearest objects (in this case, L_i's) are linked and subsequently treated as a single new object. Although the lumping inevitably introduces some distortions, it avoids the problem of actually having to check all groups of 3, 4 and more data sets. In our variant of clustering we define the distance between a new cluster (the combination of two existing objects) and another (third) object as the maximum distance (L_∞, equation 1) between each of the two components of the cluster and the third object (instead of the usual average). The clustering leads to a so-called dendrogram which links the objects at increasing mutual distances. One can distinguish clusters of objects by choosing a particular threshold: all objects which are linked at a level below this threshold form a cluster. In our case we scaled the threshold to unity so that a cluster consists of compatible data sets. Note that the clustering is not unique: in the first place, the sample of N points in P is random, and another sample could yield different estimates d_{ij}. In addition, the fact that some L_i's may be redundant (i.e. there may be some $d_{ij}=0$) makes the inclusion of these L_i into a particular cluster arbitrary.

The grouping of L_i's solves part of our problem (which data sets are compatible with a single model formulation). The next step is to look at parameter values corresponding to a particular cluster. For example, if data sets $\{i, j, k\}$ and $\{l, m\}$ are grouped in two clusters one looks at the sets $\{p| L_i(p)<c_i \wedge L_j(p)<c_j \wedge L_k(p)<c_k\}$ and $\{p| L_l(p)<c_l \wedge L_m(p)<c_m\}$. In this way one can distinguish which parameters discriminate between the clusters. Parameters that are non-overlapping among the clusters are apparently variables rather than constants. This may provide diagnostic information towards model improvement.

3. Case studies

The issues in this paper are illustrated by two case studies. The first one is designed to identify the discriminating power of the approach, the second is intended to illustrate the viability of the approach in a real life case study. The first case study is a numerical experiment illustrated in figure 3. The data were generated by adding random Normal noise to the logistic growth model. A total of 6 data sets was generated, in three pairs of two compatible sets. One parameter was kept constant in the three pairs, a second varied between two of the three pairs, and a third was different for all three pairs (see figure 3). In two pairs the data sets were obtained by randomly splitting 44 generated observations, in one pair by splitting the first and the last 22 observations in time. As a measure of the lack of fit the mean absolute deviation was used. As a threshold a value of 0.1 was used ($c_i=0.1$, $i=1...6$).

The second case study concerns the fate of dioxins (in particular, 2,3,7,8-tetrachlorodibenzo-p-dioxin, TCDD) in mammals. This "physiologically based pharmacokinetic" (PBPK) model has been formulated by Leung (1991) and has been tested on various animals (mice, rats, cows). It is hoped that the realism of the model (describing actual organs and physiological processes rather than abstract linear compartments) helps in extrapolation to untested animals, in particular man. Actual measurements on human beings are scarce and difficult to interpret (usually only a single measurement -at death- is possible while exposure is unknown). If it is possible to develop a model that is able to explain all experiments ranging from mice to cows this would considerably increase confidence in (cancer) risk estimates for humans based on this model.

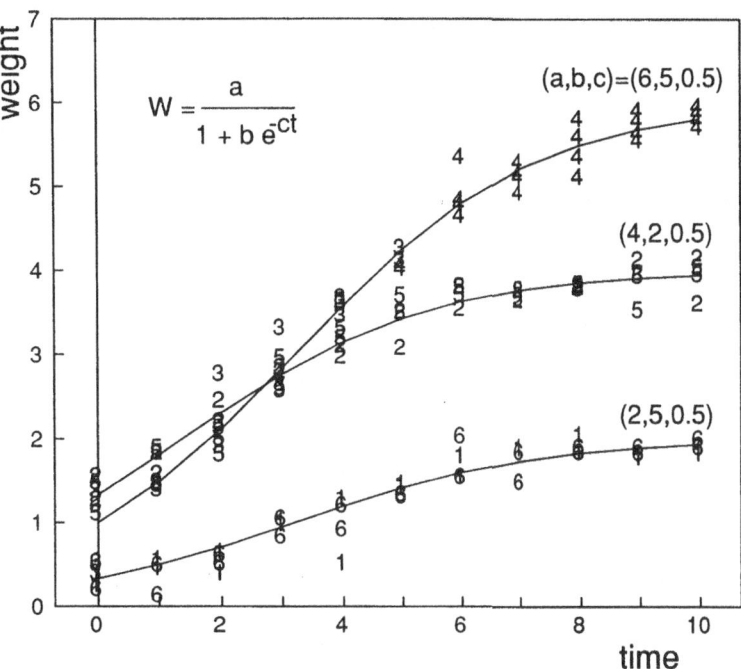

Figure 3. An artificial data set generated around the three curves shown. The data are split into 6 sets as indicated by the labels.

Table 1. Measurements used for calibration of PBPK model. A full description of the experiments and literature references are given by Klepper (1992).

experiment number:	code	1	2	3	4	5	6	7
organ: fat	F	+	+	+	+	+	+	+
liver	L	+	+			+	+	+
blood	B						+	+
milk (fat)	M			+				
milk (bulk)	MR						+	
richly perfused	R					+		+
slowly perfused	S							+

experimental animals:	1, 2	rats
	3, 6	lactating cows
	4, 5	non-lactating cows
	7	mice

A detailed description of the model with literature references for model and data is given by Klepper (1992). There are a total of 19 data sets, all concerning the ^{10}log of the concentrations, see table 1. Again the average absolute deviation was used for the lack of fit. As a threshold either a value of 0.1 (i.e.: an average relative error of $10^{0.1}-1=25\%$) or (for the data set B7 with a relatively high scatter) a value of 1.5 times the minimum lack of fit was used. The model contains a total of 14 uncertain parameters; after an initial uncertainty analysis, 8 of these were selected for further analysis (see table 2). The range of values for each parameter included the literature values, but was deliberately set considerably wider.

Table 2. Selected parameters in the PBPK model and estimated uncertainty ranges. A detailed discussion of model formulation and literature sources for the estimated range of parameter values is given by Klepper, 1992.

parameter name	dimension	meaning	range
ALLOCOEF	-	allometric scaling	0.6 - 1.0
BM20	pg.(g liver)$^{-1}$	"naive" TCDD binding capacity to microsomal proteins in liver	$2\ 10^5$ - $2\ 10^6$
KB2	pg.g^{-1}	TCDD binding affinity to microsomal proteins in liver	10^2 - 10^5
KFC	day^{-1}(g bw)$^{(\text{allocoef-1})}$	rate of TCDD metabolism	100-1000
MTS1	day^{-1}(g bw)$^{(\text{allocoef-1})}$	exchange rate between deep and superficial fat	0 - 0.5
PF	(-)	partition coefficient in fat	10^2 - 10^4
PR	(-)	part. coeff. in richly perfused organs	5 - 50
PS	(-)	part. coeff. in slowly perfused organs	1 - 200

4. Results and discussion

The results for the first case study are given in figures 4 and 5. The clustering algorithm links the compatible data sets at a low level; at the threshold level of 1, three clusters corresponding to the original pairs of data sets are recovered. Note that the data in cluster 3 (data sets 3 and 4) which were split over seperate periods are closer than the randomly split data sets in clusters 1 and 2. This is related to the fact that the data lie on different parts of a curve resulting in less trade-off.

In parameter space there is a clear separation between the clusters for the parameter representing asymptotic weight (a) only. The preset parameter values 2, 4 and to a lesser degree, 6 are recovered. For the other two parameters the (1-dimensional) recovery is not as clear. This is a result of the fact that the parameters b and c have negatively correlated effects on model fit: a high value of b *and* c gives a similar fit as low values for both parameters. As a result, we cannot determine both parameters independently. Setting one of them at a predetermined value, we *would* get a separation in values for the other. For example, if we set c at its true value of 0.5, we obtain $b \approx 2$ in cluster 2 and a common range of values $4<b<8$ in the clusters 1 and 3. This can be compared to the true values of 2 and 5, respectively. Alternatively, we might choose $b=5$, and obtain $c \approx 0.6$ for clusters 1 and 3, and $c \approx 0.9$ for cluster 2. This illustrates that separation of the right parameters is difficult in case of correlated parameters.

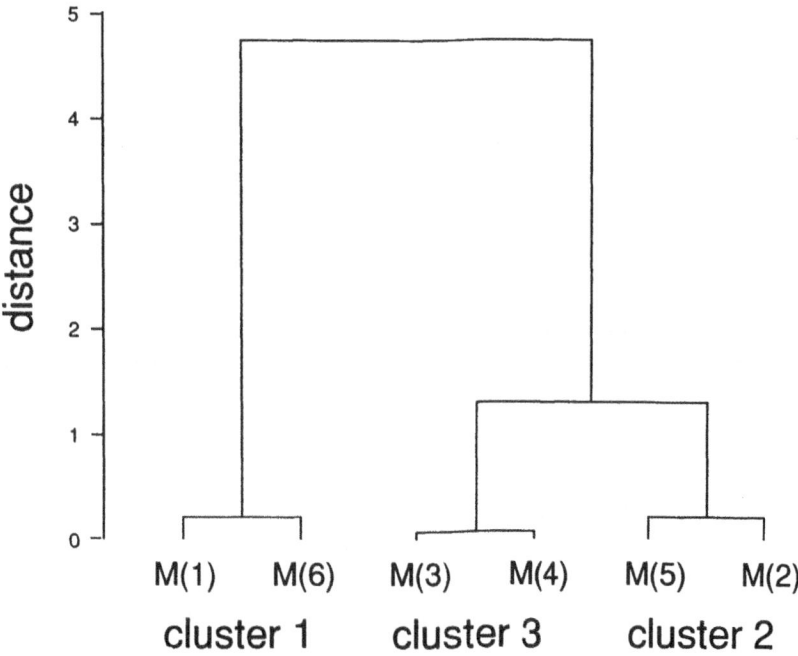

Figure 4. Clustering of the data sets from figure 2. A distance of unity indicates an acceptable level; data sets linked at a higher levels in the dendrogram are incompatible.

Table 3. Symmetric distances d_{ij} between datasets in the PBPK model. For codes see table 1.

	L1	L2	L5	L6	L7	F1	F2	F3	F4	F5	F6	F7	B6	B7	M3	MR6	R5	R7	S7
L1	0.1																		
L2	1.4	0.1																	
L5	2.0	0.0	0.0																
L6	4.4	1.5	0.0	0.0															
L7	0.4	0.1	0.3	2.2	0.2														
F1	0.3	0.2	0.0	0.0	0.2	0.2													
F2	0.2	0.1	0.0	0.0	0.1	0.1	0.1												
F3	0.3	0.1	0.0	0.0	0.1	0.1	0.2	0.1											
F4	0.1	0.1	0.0	0.0	0.1	0.1	0.1	0.4	0.1										
F5	0.1	0.1	0.0	0.0	0.1	0.1	0.1	0.2	0.2	0.1									
F6	0.2	0.0	0.0	0.0	0.0	0.0	0.0	0.1	0.0	0.0	0.0								
F7	0.4	0.2	0.0	0.4	0.2	0.3	0.1	0.1	0.1	0.1	0.6	0.5							
B6	0.2	0.1	0.0	0.0	0.2	0.1	0.1	0.1	0.1	0.1	0.0	0.3	0.1						
B7	0.7	0.2	0.5	1.7	0.3	0.3	0.1	0.4	0.2	0.2	0.0	0.5	1.1	0.7					
M3	0.4	0.2	0.0	0.0	0.2	0.3	0.1	0.1	0.1	0.1	0.0	0.7	0.2	0.8	0.5				
MR6	0.1	0.2	0.0	0.0	0.2	0.2	0.1	0.2	0.1	0.1	0.0	0.4	0.1	0.7	0.4	0.1			
R5	0.0	0.0	0.0	0.0	0.0	0.0	0.0	0.0	0.0	0.0	0.0	0.0	0.0	0.0	0.0	0.0	0.0		
R7	0.2	0.2	0.0	0.0	0.2	0.2	0.1	0.1	0.1	0.1	0.0	0.2	0.1	0.2	0.2	0.2	0.0	0.2	
S7	0.4	0.1	0.0	0.5	0.2	0.4	0.2	0.1	0.1	0.1	1.0	0.5	0.1	0.4	0.5	0.3	0.0	0.2	0.4

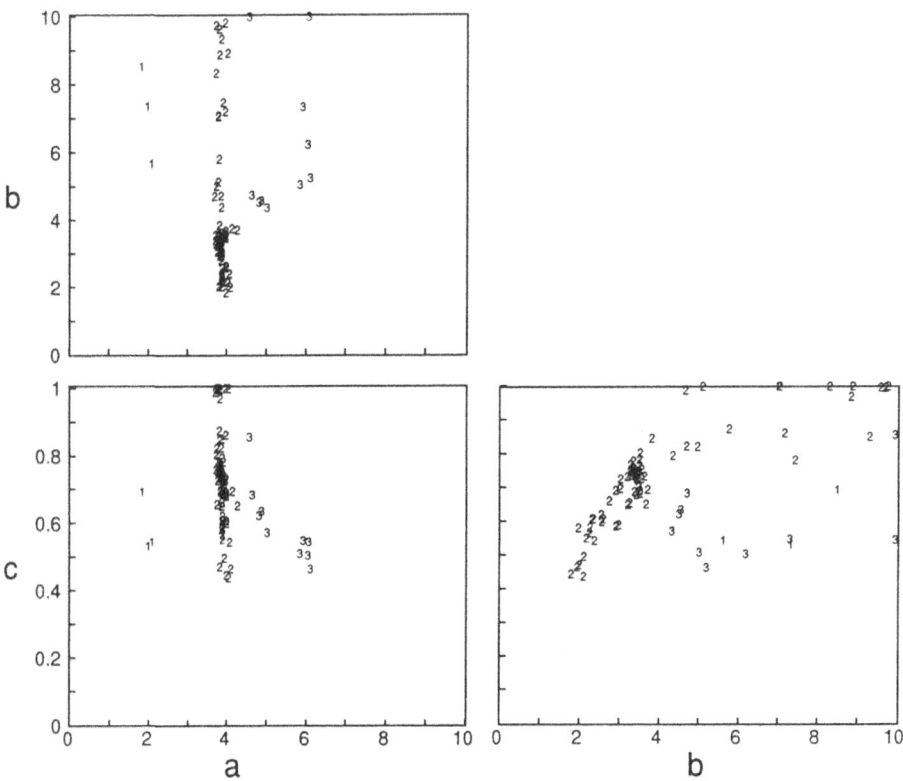

Figure 5. Parameter values in the Pareto set, labeled according to the clusters in figure 3.

Of course, in this artificial case study it is immediately obvious that the data are incompatible, and also that asymptotic weights (*a*) are different. It should be stressed however that this is a consequence of the deliberately simple model that was chosen. Often, a certain bias *is* evident in the model results, but its diagnosis is not immediately obvious. It may be that the bias is caused by underestimated inputs, by processes that were overlooked or (as in the present case) from trying to compromise on different data sets that are in fact incompatible.

For the PBPK model the distance matrix is given in table 3. Note that the distances can not be interpreted in the usual (Euclidean or equivalent) sense: for example, the distance d_{ii} onto itself is normally zero, but in the present case represents the best (unconstrained) fit that can be achieved. Furthermore, the fact that both the pairs (L_i, L_j) and (L_i, L_k) are close does not imply that the pair (L_j, L_k) is close. This is illustrated on the dataset R5 (consisting of a single point), which has distances 0 to all other data sets, which would normally imply that *all* distances are zero. An attempt to visualise the situation by projection into Euclidean space (by principal coordinate analysis; Pielou, 1984) is given in figure 6. The clustering algorithm produces the dendrogram in figure 7. At the threshold level (1) three clusters are distinguished. Although it may be noted that the clustering in this figure is to some extent arbitrary (for example, R5 may be included in *any* cluster), the fact that we need three clusters to include all data sets is *not* arbitrary. This is also evident in figure 6: one needs at least three circles with diameter 1 to cover all data sets.

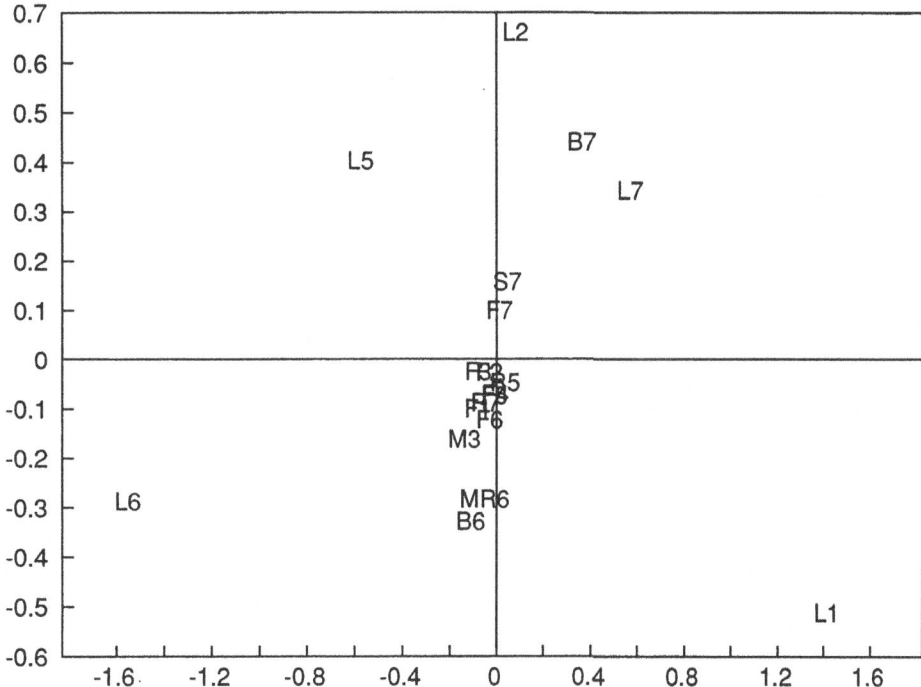

Figure 6. Results of a principal coordinate analysis on the distances between the 19 data sets of the PBPK model. This results in the best-fitting 2-dimensional representation of the data sets. For abbreviations of the data sets: see table 1.

The parameter values clearly distinguish between the three clusters (figure 8). In most cases there is a good separation in parameter values already for a single dimension and it is not necessary to consider correlations between parameters. In some cases a particular combination of parameters does improve the discrimination: for example, PS distinguishes between clusters 2 (low PS) and 1+3 (higher values); KB2 further splits the 1+3 group (see figure 8).

It is of particular interest that the present analysis points to a number of rather "lumped" parameters (the partition coefficients PR, PS, PF) as the main cause of the incompatability of the data sets with respect to the model. In contrast, the more "physiologically based" parameters BM20 and KFC show overlapping ranges between the clusters. A first step towards model improvement would be to reconsider the partition coefficients and make them species-dependent. A possible explanation for the differences between the species would be differences in composition of the various organs (fat, protein content, type of fat, etc.). Including these properties in the model (i.e., the partitioning is a result of specific properties of the organ rather than a fixed coefficient) would probably increase model applicability. This would be an obvious advantage in extrapolation from test animals to other species (man, wild species).

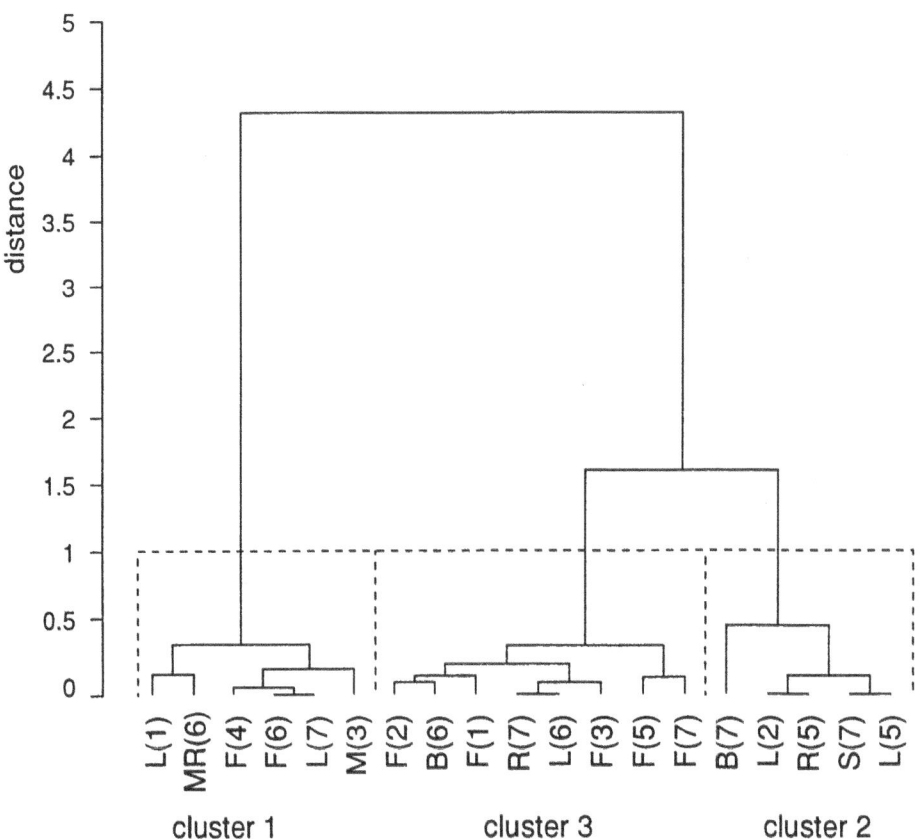

Figure 7. Clustering of the 19 PBPK data sets from table 1. A distance of unity indicates an acceptable level; data sets linked at a higher level in the dendrogram are incompatible.

5. Conclusion

The method presented in this paper presents a framework to analyse situations where different data sets are incompatible with a single model formulation. It can be used both to delineate model applicability more clearly (which data sets *are* compatible with a single model formulation) and to improve the model (how should the model be modified so that it *does* fit all the data?). The case studies show that the methods works well both for a test case (it recovers the known inputs) and for a complex real-life model (it identifies incompatabilities and points towards model improvement).

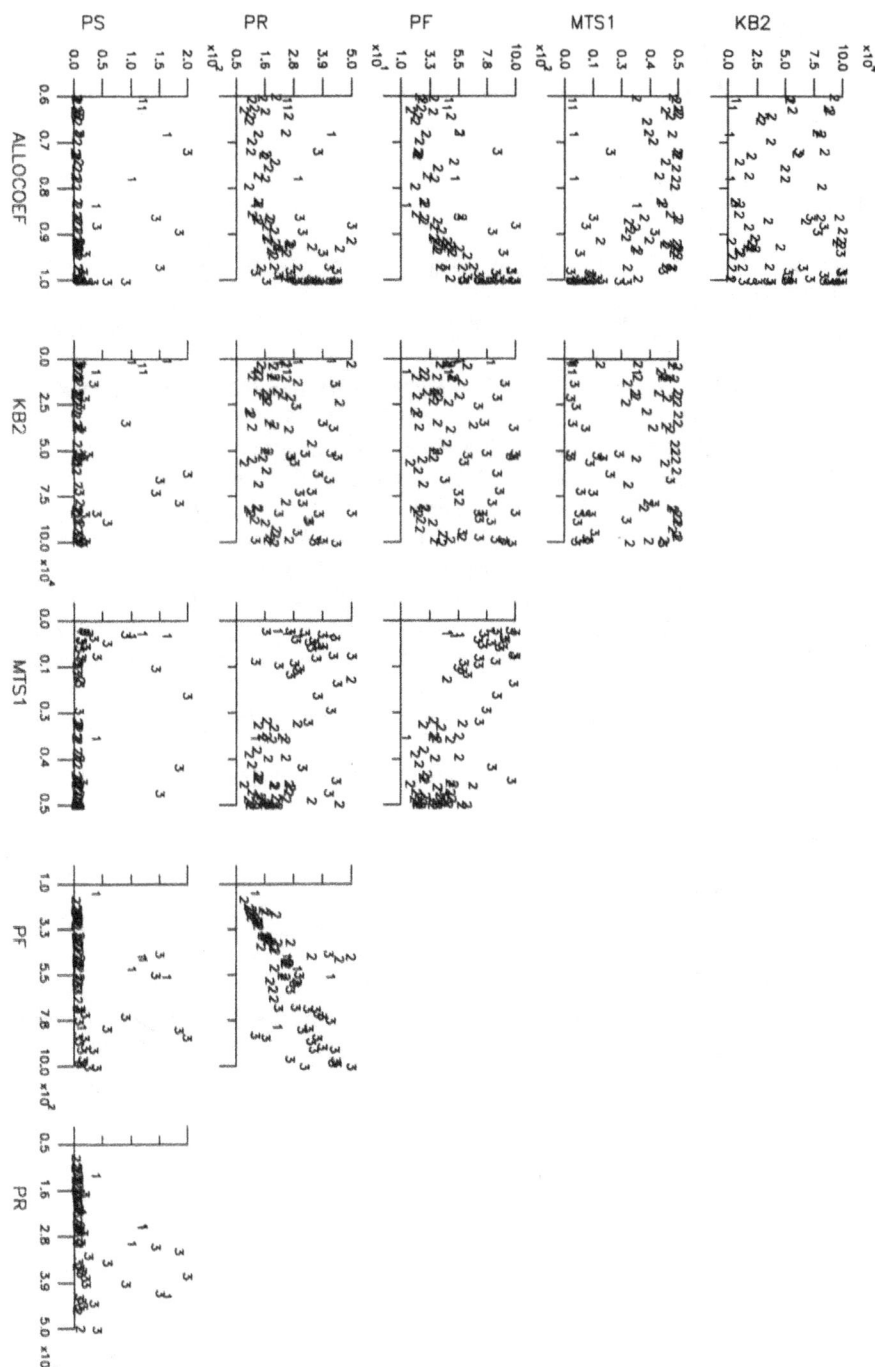

Figure 8. 2-dimensional projections of the parameter values corresponding to the three clusters from figure 6. Parameters are explained in table 2. Parameters BM20 and KFC did not show any discrimination between the clusters and were left out of the graph.

Acknowledgements

The authors wish to thank Drs. J. Van der Meer and Dr. Ir. M.J. Zeilmaker for their stimulating comments.

References

Bates, D.M. and D.G. Watts, (1988), 'Nonlinear regression analysis and its applications', Wiley, New York.

Beck, M.B. (1987), 'Water quality modelling: a review of the analysis of uncertainty', *Water Res. Res.* **23(8)**, 1393-1442.

Beck, M.B. and P.C. Young (1976), 'Systematic identification of DO-BOD model structure', *Proc. Am. Soc. Civ. Eng. J. Environ. Eng. Div.* **102(EE5)**, 902-927.

Draper, N.R. and H. Smith (1981), 'Applied regression analysis', Wiley, New York.

Hornberger, G.M. and R.C. Spear (1981), 'An approach to the preliminary analysis of environmental systems', *J. Environ. Manage.* **12**, 7-18.

Klepper, O. (1992), 'Analysis of a physiologically based pharmacokinetic model for dioxins in mammals', *Note RIVM-CWM*, Bilthoven, The Netherlands.

Leung, H.W. (1991), 'Development and utilization of physiologically based pharmacokinetic models for toxicological applications', *Toxicol. Appl. pharmacol.* **103**, 411-419.

Pielou, E.C. (1984), 'The interpretation of ecological data, a primer on classification and ordination' Wiley, New York.

Scavia, D. (1980), 'Uncertainty analysis of a lake eutrophication model', *Ph.D. dissertation*, Univ. of Michigan, Ann Arbor.

Tarantola, A. (1987), 'Inverse problem theory. Methods for data fitting and model parameter estimation', Elsevier, Amsterdam.

Whitehead, P.G. (1983), 'Modelling and forecasting water quality in non-tidal rivers: the Bedford Ouse study'. In: *'Uncertainty and forecasting of water quality'*, M.B. Beck and G. Van Straten (eds.), 321-337, Springer, New York.

Young, P.C. (1984) 'Recursive estimation and time-series analysis: an introduction', Springer, New York.

Yu, P.L. (1989), 'Multiple criteria decision making: five basic concepts'. In: G.L. Nemhauser, A.H.G. Rinnooy Kan and M.J. Todd (eds.) *'Optimization'*, *Handbooks in operations research and management science*, **1**, 663-700, North Holland, Amsterdam.

REGIONAL CALIBRATION OF A STEADY-STATE MODEL TO ASSESS CRITICAL ACID LOADS

J. KROS[1], P.S.C. HEUBERGER[2], P.H.M. JANSSEN[2] & W. DE VRIES[1]

[1] DLO Winand Staring Centre, P.O. Box 125, 6700 AC Wageningen, Netherlands;
[2] National Institute of Public Health and Environmental Protection, P.O. Box, 3720 BA Bilthoven, Netherlands

Abstract

The Model to Assess Critical Acid Loads (MACAL) has been developed for assessing and mapping critical acid loads on a national scale. MACAL simulates soil solution concentrations of major ions in a forest soil at any given depth at steady state for a given deposition level. The critical acid load is calculated from defined critical values for the Al^{3+} concentration and the Al^{3+}/Ca^{2+} ratio by inverse modelling. In order to minimize the uncertainty in the critical load computations, which is due to insufficient knowledge of parameter values, a multi-signal calibration of poorly defined important model parameters was performed using a data set on soil solution concentrations of 150 forest stands in the Netherlands. Since no detailed data was available on site scale (i.e. individual forest stands), a regional calibration was preferred. The cumulative distribution functions (CDF) of the model outputs for the 150 forest stands where fitted to those of the associated measurements. All model parameters could be identified with the objective function used except for forest filtering factors for nitrogen deposition. The calibration showed to be useful to reduce parameter ranges for some of the important model parameters, resulting in a lower uncertainty in model predictions.

Keywords Calibration, critical load, soil acidification model

1. Introduction

The multi-layer Model to Assess Critical Acid Loads (MACAL) has been developed for assessing and mapping critical acid loads on a national scale (De Vries, 1991). For given critical Al^{3+} concentrations or Al^{3+}/Ca^{2+} ratios, MACAL can derive a critical acid load related to any given depth. Data required by MACAL are inputs (deposition data and hydrological data), variables (nutrient contents in various tree compartments) and parameters that describe the various interactions occurring in the forest canopy and in the soil. Unlike model inputs and variables, which can be derived directly from available literature or measurements, most parameters can only be derived in an indirect way. Consequently, thorough knowledge of the parameter values to be used is generally lacking. This seriously affects the credibility of the model results.

In order to reduce the uncertainty in the critical load computations poorly defined important model parameters were calibrated, using a data set on soil solution concentrations of H^+, Al^{3+}, Ca^{2+}, Mg^{2+}, K^+, Na^+, NH_4^+, NO_3^-, SO_4^{2-}, Cl^- and HCO_3^- in the topsoil (0-30 cm) and subsoil (60-100 cm) of 150 forest stands. The calibration aimed at obtaining better estimates for poorly defined model parameters through the comparison of model outputs and measurements, i.e. the concentration of each individual ion including concentrations of the sum of H^+ and Al^{3+}, the sum of Ca^{2+}, Mg^{2+}, K^+ and Na^+, and the sum of NH_4^+ and NO_3^-, and concentration ratios of Al^{3+}/Ca^{2+}, NH_4^+/K^+ and NH_4^+/Mg^{2+}. The ultimate aim of this study was (1) to obtain a smaller range of parameter values, i.e. reduction of parameter uncertainty, and (2) to reduce the number of parameters in case of overparameterization, or obtain guidelines for gathering additional data for further calibration experiments.

2. The Macal model

MACAL is a steady-state soil acidification model (De Vries, 1991). For a given deposition level, it simulates the soil solution concentration of major ions in a forest soil at any given depth at steady state. It only includes processes influencing acid production and consumption during infinite time. Instead of defining discrete soil layers, ionic concentrations can be calculated at any given depth using continuous functions for the various processes occurring in the rooting zone. Ions included are H^+, Al^{3+}, Ca^{2+}, Mg^{2+}, K^+, Na^+, NH_4^+, NO_3^-, SO_4^{2-}, Cl^- and HCO_3^-. Model outputs of major importance are the concentrations of Al^{3+} and Ca^{2+}, since the Al^{3+} concentration and the Al^{3+}/Ca^{2+} ratio are considered important parameters with respect to forest decline, for which critical values have been defined (De Vries, 1993). Other important model outputs are the concentrations of NH_4^+, K^+ and Mg^{2+}, since the NH_4^+/K^+ and NH_4^+/Mg^{2+} ratios are also related to effect on forests. The critical acid load is calculated by inverse modelling using the given critical values for the Al^{3+} concentration and Al^{3+}/Ca^{2+} ratio.

MACAL is based on the assumption that dynamic processes such as cation exchange, adsorption/desorption of SO_4^{2-} and NH_4^+ and mineralization/immobilization dynamics of N, S and base cations (BC) are unimportant for the assessment of a long-term critical load. On the long-term these relative fast processes are considered to be in equilibrium. Further assumptions are negligible nitrogen fixation and negligible reduction and precipitation of SO_4^{2-}. A justification of these assumptions is given by De Vries (1993). An overview of the processes and process formulations included in MACAL is given in Table 1.

3. Methodology

3.1 Measurements Used for Calibration

Measured soil solution concentrations were available for the mineral topsoil (0-30 cm) and the mineral subsoil (60-100 cm) of 150 forest stands (De Vries and Leeters, 1993). Composite soil samples of these layers, consisting of 20 subsamples, were taken from February to May 1990. During this period (i.e. early spring) the composition of the soil solution reasonably corresponds with the flux-weighted annual average soil solution concentration. The soil solution was extracted by centrifuging a soil sample. The locations were restricted to non-calcareous soils throughout the country. The included tree species were Scotch pine, black pine, Douglas fir, Norway spruce, Japanese larch, oak and beech.

MACAL simulates yearly averaged values, whereas the data set represents the concentration of ions in early spring (February to May). This affects the quality of the calibration, because one field observation in early spring does not necessarily reflect the flux weighted annual average value as predicted by MACAL.

For the calibration 134 of the 150 Dutch forest stands were used. The excluded sites

appeared to be calcareous, influenced by groundwater or extremely dry, resulting in non-representative concentrations. Because the numbers of observations per soil and forest type were too small for a calibration on each type alone, the tree and soil types were lumped into classes. The seven tree species were lumped into three forest type classes:
- 31 spruce stands (Douglas fir and Norway spruce), evergreen trees with high forest filtering capacity, growth rate and transpiration rate;
- 58 pine stands (Scotch pine and black pine), evergreen trees with moderate forest filtering capacity, growth rate and transpiration rate;
- 45 deciduous stands (Japanese larch, oak and beech), needle or leave sheddy trees with low forest filtering capacity, growth rate and transpiration rate.

A distinction per soil type was not considered, because the differences in soil characteristics are relatively small and, even more important, there is no correlation between soil type and measured soil solution concentration (De Vries and Leeters, 1993). For this reason we lumped all the distinguished soil types into one class, i.e. non-calcareous sandy soils.

Table 1. Processes and process formulations included in MACAL

Processes	Description
Hydrological processes:	
Water flow	Variable flow with depth
Biogeochemical processes:	
Foliar uptake	Proportional to total deposition
Foliar exudation	Proportional to H^+ and NH_4^+ deposition
Litterfall	First order reaction
Mineralization	Equals litterfall
Immobilization	Proportional to the total N deposition minus N root uptake
Growth uptake	Constant growth
Maintenance uptake	Forcing function[1]
Nitrification	Proportional to NH_4^+ flux
Denitrification	Proportional to NO_3^- flux
Geochemical processes:	
CO_2 dissociation	Equilibrium
Silicate weathering	Zero order reaction
Al-hydroxide weathering	Gibbsite equilibrium

[1] Equals litterfall plus foliar exudation minus foliar uptake.

3.2 Selected parameters for calibration

An overview of the values and their origin of the model inputs, variables and parameters for the national application of MACAL is presented by De Vries *et al.* (1994). Here we restrict ourselves to an enumeration and specification of the required model parameters only (Table 2).

Table 2. Model parameters

Name	Unit	Explanation	Dependent on[1]	Derivation
$ffNH_{4\,dt}$	-	Filtering factor of NH_4^+	F	Throughfall monitoring[2]
$ffNO_{3\,dt}$	-	Filtering factor of NO_3^-	F	Throughfall monitoring[2]
$ffSO_{4\,dt}$	-	Filtering factor of SO_4^{2-}	F	Throughfall monitoring[2]
$f_{dd\,di}$	-	Dry deposition factor for Ca^{2+}, Mg^{2+}	F	Throughfall monitoring[3]
$f_{dd\,mo}$	-	Dry deposition factor for Na^+, K^+, Cl^-	F	Throughfall monitoring[3]
fr_{ic}	-	Interception fraction	F	Hiege (1985)
f_{re}	-	N reallocation factor	F	Turner (1975)
$fr_{ni\,in}$	-	Nitrification fraction in humus layer	S	Tietema (1992)
$fr_{ni\,rz}$	-	Nitrification fraction in mineral soil	S	Tietema (1992)
fr_{de}	-	Denitrification fraction	S	Literature[4]
$fr_{ru\,in}$	-	Root uptake fraction in the humus layer	S	Literature[5]
frN_{im}	-	N immobilization fraction	S	Calibration
$frCa_{fe}$	yr^{-1}	Foliar exudation fraction for Ca^{2+}	F	De Vries *et al.* (1994)
$frMg_{fe}$	yr^{-1}	Foliar exudation fraction for Mg^{2+}	F	De Vries *et al.* (1994)
frK_{fe}	yr^{-1}	Foliar exudation fraction for K^+	F	De Vries *et al.* (1994)
$frNH_{4\,fu}$	yr^{-1}	Foliar uptake fraction for NH_4^+	F	De Vries *et al.* (1994)
$frNO_{3\,fu}$	yr^{-1}	Foliar uptake fraction for NO_3^-	F	De Vries *et al.* (1994)
$frSO_{4\,fu}$	yr^{-1}	Foliar uptake fraction for SO_4^{2-}	F	De Vries *et al.* (1994)
Am_{lv}	$kg\ ha^{-1}$	Amount of leaves	F	Kimmins *et al.* (1985)
kr_{gc}	$m^3ha^{-1}yr^{-1}$	Net growth rate	F, S	De Vries *et al.* (1994)
rho_{st}	$kg\ m^{-3}$	Wood density	F	Kimmins *et al.* (1985)
$alpha$	-	Dissolution constant of Al hydroxides	S	Klein *et al.* (1989)
$beta$	-	Depth dependency of Al hydroxides constant	S	Klein *et al.* (1989)
pCO_2	HPa	partial CO_2 pressure	G	De Vries (1991)
D_{ni}	cm	Depth of the nitrification-zone	F, S	De Vries *et al.* (1994)
D_{de}	cm	Depth of the denitrification-zone	S	De Vries *et al.* (1994)
D_{rz}	cm	Depth of the rooting zone	F	De Vries *et al.* (1994)
ni_{exp}	-	Nitrification parameter	F	De Vries *et al.* (1994)
de_{exp}	-	Denitrification parameter	S	De Vries *et al.* (1994)
ru_{exp}	-	Water and nutrient uptake parameter	F, S	De Vries *et al.* (1994)
we_{exp}	-	Weathering parameter	S	De Vries *et al.* (1994)

[1] The parameters were derived as a function of soil type (S), forest type (F) or forest and soil type (F, S) with the expectation of the global data (G).

[2] Values for the fitting factors were derived from a comparison of throughfall data below spruce, pine and deciduous forests at 42 sites in the Netherlands (Erisman, 1990; Houdijk, 1990) and total deposition estimates with the TREND model using 1985 emission data.

[3] Values were derived from a comparison of Na in throughfall and bulk deposition (cf Bredemeier, 1988) for the 42 sites mentioned above.

[4] The denitrification fraction was related to soil type using data from Breeuwsma *et al.*(1991) for agricultural soils while correcting them for the more acid circumstances in forerst soils.

[5] The root uptake fraction in het humus layer was set to zero except for NH_4^+. For this ion, we used a value of 0.3 for coniferous forests and 0.5 for deciduous forests based on data from Tietema and Verstraten (1991).

The MACAL version used here contains 31 parameters, of which 30 depend either on soil type or forest type or on both soil and forest type, and 1 parameter is global (c.f. Table 2). In this study only the dependency on forest type was taken into account. Considering three forest type classes (c.f. Section 3.1), the number of parameters became 71. It was clear that a calibration of such an excessive amount of parameters would cause identification problems. A preliminary reduction of the parameters was based on expert judgement; those parameters which are certain were omitted from the calibration and taken at their nominal values.

Further reduction was based on a sensitivity analysis, using the linear correlation coefficient as a sensitivity measure. The results of this sensitivity analysis were used to determine which parameters are most important in an optimization procedure. It is reasonable to fix those parameters to which the model output is insensitive. Next, we excluded those parameters from calibration, which we considered well known (e.g. the depth of the rooting zone, D_{rz} and the net growth rate, kr_{gc}). Finally, we decided to fix those parameters which were almost impossible to identify with this data set in combination with parameters having the same overall effect on the soil solution concentration (e.g. $ffNH_{4\ dt}$ and $ffNO_{3\ dt}$ together with frN_{im}). A higher filtering factor for NH_4^+ or a lower NH_4^+ immobilization, in both cases results in an increase in the net NH_4^+ input. By fixing one of these parameters we avoided identifiability problems.

However, by doing this the ability to assign a realistic meaning to the finally obtained parameter values is more or less lost. Selected parameters to be calibrated and their uncertainty range, expressed as minimal and maximal feasible value is given in Table 3. The total number of parameters to be calibrated was 25.

Table 3. Selected parameters for the calibration and their ranges of values and units for the distinguished forest classes and soil class

Forest dependent parameters

Forest type	$ffSO_{4\ dt}$		$f_{dd\ di}$		$f_{dd\ mo}$		$frNH_{4\ fu}$		Am_{tv} (kg ha^{-1})		$fr_{ru\ in}$		ru_{exp}	
	min	max	min	max	min	max	min	max	min	max	min	max	min	max
Deciduous	0.8	1.4	1.0	6.0	1.0	6.0	0.1	0.4	2000	5000	0.1	0.7	1.0	4.0
Pine	1.0	1.8	1.5	8.0	1.5	8.0	0.1	0.4	4000	10000	0.1	0.7	1.0	4.0
Spruce	1.2	2.0	2.0	10.0	2.0	10.0	0.1	0.4	6000	18000	0.1	0.7	1.0	4.0

Soil dependent parameters

$fr_{ni\ in}$		$fr_{ni\ rz}$		frN_{im}		$alpha$	
min	max	min	max	min	max	min	max
0.2	0.6	0.5	1.0	0.0	1.0	6.0	10.0

With the selected set of parameters to be calibrated, the calibration method was first tested using a synthetic data set followed by calibration on the real data set (c.f. Section 1).

3.3 Calculation method

So-called deposition areas of 10×10 km^2 were defined for which the model inputs (deposition and precipitation) were available, which were assumed to be constant within this grid cell (De Vries *et al.*, 1994). For each sampled forest site MACAL was evaluated with the deposition and precipitation of the corresponding grid cell. For each forest site the model was parameterized for the involved forest-soil combination (there were 7 tree species and 10 soil types included, see De Vries and Leeters (1993)) as far as the non-adjustable parameters and variables were concerned, whereas the adjustable parameters were either lumped into 3 forest classes or considered global (c.f. Section 3.1).

Although MACAL calculates the concentrations (and critical loads) for a specific soil-forest type combination at a specific geographical location (site scale), we decided to calibrate MACAL on a regional scale. The empirical cumulative distribution functions (CDFs; c.f. Conover, 1971) of model outputs for the 134 forest stands were fitted to those of the associated measurements independent of the geographical location. Regional calibration was preferred to a comparison on every distinguished soil/forest type combination because of a lack of observed deposition data on a site scale (i.e. individual forest stands). The estimated uncertainty in the annual potential acid deposition is 15% for the Netherlands as a whole, 45% for a 5×5 km^2 grid. On an individual forest stand this will be even much more (Erisman, 1992). This implies that a calibration on site level makes no sense, because this is extremely biased by the uncertainty in deposition for which the model is very sensitive. However, for a calibration on a regional scale the uncertainty in deposition has little or no effect (Kros *et al.*, in prep.). In addition, a regional calibration is also more in line with the purpose for which the model is developed: regional/national critical load mapping.

In order to compare the model output with the measurements, the inspected model output was calculated as the arithmetic mean of the values for each centimetre, for 0 to 30 cm for the topsoil and 60 to 100 cm for the subsoil.

3.4 Calibration

Calibration method
Model parameters were calibrated by simultaneously, minimizing differences between various model outputs and measurements (multiple signals) using certain parameter ranges of *a priori* values. Although a number of objective functions can be proposed for this multisignal calibration problem, we chose the sum of the weighted absolute error of the considered signals.

In general terms the model can be written as:

$$\hat{Y}_{r,s,l} = f_s(x_l; \theta) \ , \tag{1}$$

where $\hat{Y}_{r,s,l}$ is the model output for signal r, in soil layer s at location l, x_l is the location dependent model input, and θ a vector of m adjustable parameters $(\theta_1, ..., \theta_m)$. The model parameters θ are dependent on either the forest type or global (see Sections 3.2). The observations for each signal, layer and location were defined as: $Y_{r,s,l}$.

The calibration was based on the minimization of differences between several percentile values of the distribution of measured and modelled signals over all sites. Each considered output signal $\hat{Y}_{r,s,l}$ and measured signal $Y_{r,s,l}$ was converted to cumulative distribution functions $\hat{F}_{r,s}$ and $F_{r,s}$ respectively:

$$\{\hat{Y}_{r,s,l}\}_{l=1,\dots,k} : \qquad\qquad \hat{F}_{r,s},$$

$$\{Y_{r,s,l}\}_{l=1,\dots,k} : \qquad\qquad F_{r,s},$$

where k is the number of the considered locations. From the model output and measured distribution function various percentile values were calculated from the inverse distribution functions:

$$\hat{Y}_{r,s}^{(p_i)}(\theta) = \hat{F}_{r,s}^{-1}(p_i; \theta) \qquad\qquad (i = 1,\dots,n), \tag{2}$$

$$Y_{r,s}^{(p_i)} = F_{r,s}^{-1}(p_i) \qquad\qquad (i = 1,\dots,n), \tag{3}$$

where n is the number of considered percentile values. Here we used 9 percentiles: 0.1, 0.2, ... , 0.8, 0.9. The weighted model residual (output error) for the considered percentiles, $e_{r,s}$ was defined as:

$$e_{r,s}(\theta) = \sum_{i=1}^{n} |\hat{Y}_{r,s}^{(p_i)}(\theta) - Y_{r,s}^{(p_i)}| \, \omega_i \, , \tag{4}$$

where ω_i represents a percentile-dependent weighting factor. Which was defined as:

$$\omega_i = \frac{1}{l_i} \, , \tag{5}$$

where l_i is the length of the estimated 80% confidence interval (c.f. Conover, 1971) of the ith percentile. This weighting factor was introduced to ensure that the uncertainty in the estimated percentile values was equally distributed. For example, a lower weight will be assigned to percentile values in the neighbourhood of 0 and 1. For the overall criterion the sum of the weighted residuals of the various outputs in the considered soil layers was used:

$$C(\theta) = \sum_{s=1}^{j} \sum_{r=1}^{i} \omega_{r,s} \, e_{r,s}(\theta), \tag{6}$$

where $\omega_{r,s}$ is a signal and layer dependent weighting factor. The optimal values for θ were derived by minimizing $C(\theta)$, with θ within the predefined parameter space (c.f. Section 3.2).

Optimization method
The actual computations for the optimization of the calibration criterion were carried out with the constraint minimization function of the Matlab Optimization Toolbox (Anonymous, 1992; Grace, 1990). The algorithm uses a sequential quadratic programming method. In this method, a quadratic programming subproblem is solved at each iteration. At each iteration an estimate of the Hessian of the Lagrangian is updated and a line search is performed. For more information on the method, see Grace (1990) and the references therein. On average a calibration run took about 30-40 iterations to let the (normalized) criterion converge within

4 decimals. Each iteration requires p function evaluations, where p is the number of parameters. The computations were carried out on a SUN SPARC workstation, where the run time of the model is approximately 15 seconds (processing all 134 sites, see Section 3.1), hence a typical calibration run took 3-4 hours.

Model outputs used for the calibration
The model outputs used for the calibration were (c.f. Section 1): (i) concentrations of each individual ion, (ii) concentrations of the sum of H^+ and Al^{3+}, the sum of Ca^{2+}, Mg^{2+}, K^+ and Na^+ and the sum of NH_4^+ and NO_3^-, and (iii) ratios of Al^{3+}/Ca^{2+}, NH_4^+/K^+ and NH_4^+/Mg^{2+} in the soil solution. The calibration was carried out simultaneously for the topsoil ($s = 1$) and the subsoil ($s = 2$). Based on the importance of the considered output, the following weighting factors $\omega_{r,s}$ values were used:
$\omega_{r,s} = 2$ for $r = Al^{3+}$, Ca^{2+}, NH_4^+, NO_3^-, Al^{3+}/Ca^{2+}, NH_4^+/K^+, NH_4^+/Mg^{2+}
$\omega_{r,s} = 1$ for $r = $ all other considered model outputs
The weighting factors in the topsoil and subsoil were set equal.
It must be noticed that nearly all considered output signals were correlated, resulting in an implicit weighing, which in turn caused bias. However, this effect is likely to be small because of the complex structure of the model and criterion.

4. Results

4.1 Calibration

In order to check the calibration procedure and the objective function, a calibration was carried out on a synthetic data set generated by MACAL with nominal parameter values. The algorithm appeared to be adequate in finding the original CDFs and the original parameter values underlying the synthetic data set. However, complications raised due to the existence of various local optima of the objective function. This was reflected by the fact that different starting values for the parameters in the optimization of the objective function can lead to different final values. In order to cope with this problem for the calibration of the real data set, we decided to perform 50 calibration runs with randomly selected initial parameter values. This resulted in 50 different estimated parameters and 50 different values for the objective function. The resulting CDFs over all sites of the calibration with the best and the worst fit are presented in Figure 1, together with the CDFs of the real data. In general, the fit to the real data was good for the subsoil but only moderate for the topsoil. This is a general problem in this study, which will be discussed later. The differences between the calibration runs with the best and worst fits were generally small, except for pH and the Al^{3+}/Ca^{2+} ratio in the topsoil. It is remarkable that the calibration run with the lowest objective criterion (best fit) does not always lead to the best approximation of the data (in terms of the data) for all model outputs. See for instance the pH in the topsoil, where the CDF for the worst fit approximates the data remarkably better than the best fit.

The set of the estimated parameter values, obtained from the 50 calibration experiments with random starting values were considered as indicative for the optimal parameter range (Table 4). A comparison between the initial parameter ranges (see Table 3) and the optimal ranges (Table 4), showed that only in some cases a serious reduction of parameter uncertainty has been obtained: ru_{exp}, frN_{im} and *alpha*. For $f_{dd\,di}$, $f_{dd\,mo}$, $fr_{ni\,in}$ and $fr_{ni\,rz}$ the reduction in parameter uncertainty is rather small, while in all other cases the optimal ranges equal the initial ranges.

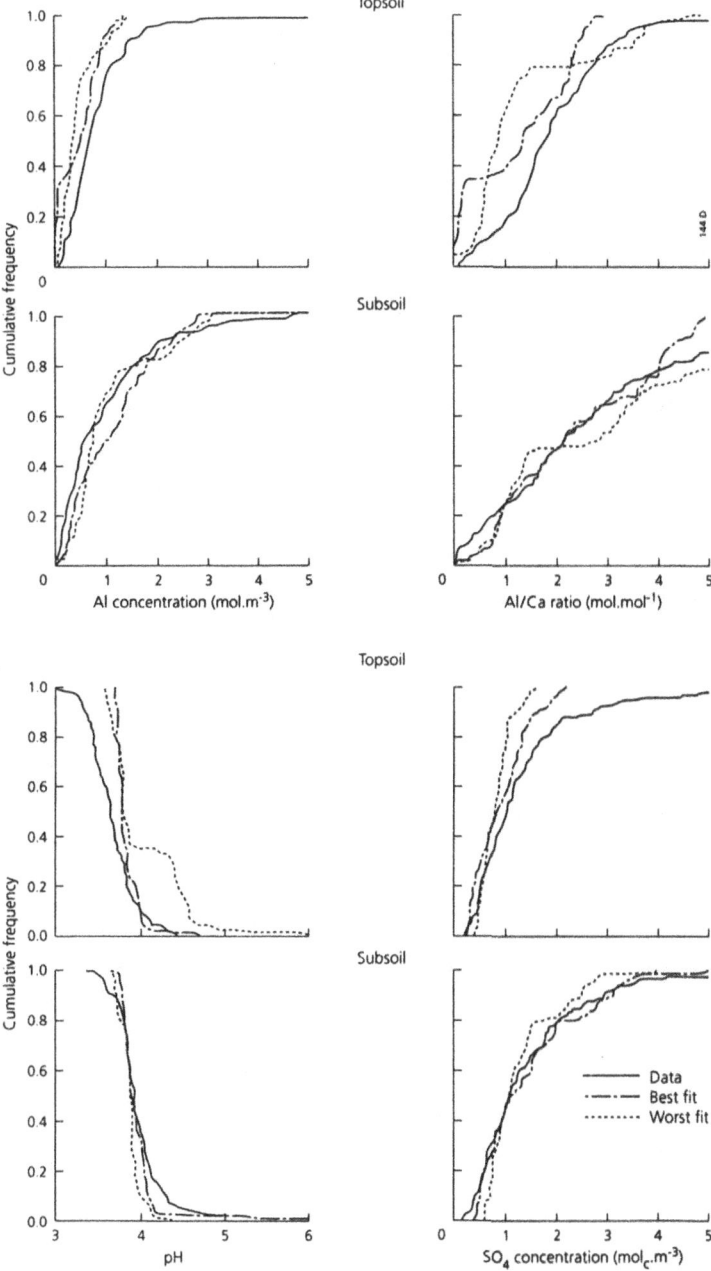

Figure 1. Comparison of the CDF's of all location for the best and worst calibration runs and the observed data for the Al^{3+} concentration, Al^{3+}/Ca^{2+} ratio, pH and SO_4^{2-} concentration in the topsoil and subsoil.

Table 4. Nominal parameter values, and 5, 50 and 95 percentile values of the calibrated parameters for the 50 calibration runs

Vegetation dependent parameters

Forest type	$ff_{SO_4,d}$ Nominal	Calibrated 5-perc.	Median	95-perc.	$f_{d,d}$ Nominal	Calibrated 5-perc.	Median	95-perc.	$f_{d,no}$ Nominal	Calibrated 5-perc.	median	95-perc.	$fr_{NH_4,fs}$ Nominal	Calibrated 5-perc.	median	95-perc.
Deciduous	1.2	0.8	1.0	1.4	2.0	2.6	3.6	6.0	2.0	1.9	2.9	4.2	0.3	0.2	0.2	0.4
Pine	1.4	1.1	1.7	1.8	2.5	4.1	7.6	8.0	2.5	2.4	3.2	4.5	0.3	0.1	0.1	0.4
Spruce	1.6	1.5	1.8	2.0	3.0	2.1	2.8	4.0	3.0	2.1	2.8	4.0	0.3	0.1	0.2	0.3

Vegetation dependent parameters (continued)

Forest type	AM_b ($kg\ ha^{-1}$) Nominal	Calibrated 5-perc.	Median	95-perc.	$fr_{m,rt}$ Nominal	Calibrated 5-perc.	Median	95-perc.	fr_{Nm} Nominal	Calibrated 5-perc.	Median	95-perc.	ru_{arg} Nominal	Calibrated 5-perc.	Median	95-perc.
Deciduous	3375	2000	2557	4946	0.5	0.1	0.4	0.7	2.0	2.5	4.0	4.0	4.0	4.0	4.0	4.0
Pine	7375	4000	6854	10000	0.3	0.1	0.4	0.7	2.5	2.9	3.9	4.0	4.0	4.0	4.0	4.0
Spruce	13725	6213	12927	18000	0.3	0.2	0.5	0.7	2.0	2.1	3.7	4.0	4.0	4.0	4.0	4.0

Soil dependent parameters

$fr_{m,in}$ Nominal	Calibrated 5-perc.	Median	95-perc.	$fr_{m,rt}$ Nominal	Calibrated 5-perc.	Median	95-perc.	$alpha$ Nominal	Calibrated 5-perc.	median	95-perc.
0.4	0.3	0.5	0.6	1.0	0.5	0.6	0.9	8.0	8.0	8.1	8.2

Table 5. Comparison of the median observed values and the median values of the 50 calibration runs for various model outputs in the topsoil and subsoil

Forest type		pH Obs.	Cal.	Al^{3+} conc. ($mol_c\ m^{-3}$) Obs.	Cal.	Al^{3+}/Ca^{2+} ratio ($mol\ mol^{-1}$) Obs.	Cal.	SO_4^{2-} conc. ($mol_c\ m^{-3}$) Obs.	Cal.
Deciduous	topsoil	3.66	3.97	0.47	0.21	1.49	0.90	0.78	0.48
	subsoil	4.00	3.92	0.43	0.61	1.65	2.90	0.94	0.66
Pine	topsoil	3.64	3.72	0.66	0.52	1.76	1.09	0.91	0.89
	subsoil	3.89	3.85	0.58	1.00	2.02	1.61	0.98	1.27
Spruce	topsoil	3.42	3.64	1.30	0.97	2.04	1.61	2.02	1.35
	subsoil	3.80	3.73	1.60	2.35	3.98	2.57	1.82	2.70

In general, the nominal parameter values, which were used for the national application, (see De Vries *et al.* 1994), were within the calibrated parameter ranges, except for: $f_{dd\ di}$, ru_{exp} and $fr_{ni\ rz}$. The high calibrated values of $f_{dd\ di}$ in relation to the nominal values indicate that the input of divalent base cations was under estimated in the national application of MACAL as described by the De Vries *et al.* (1994). Furthermore, the centrifugation method may lead to an overestimation of the base cation concentration in the soil solution (Verhagen and Diederen, 1991). The relatively high values of ru_{exp} indicate that water and nutrient uptake by roots, which is the main cause of concentration differences with depth, occurs at a shallower depth as assumed. A value of 4 resulted in an uptake of almost 100% of the water and nutrient demand within the first 25 cm of the soil profile. As a consequence, MACAL calculated a relatively small difference in soil solution concentrations between the topsoil and the subsoil. But even a value of 4 for ru_{exp} seems too small to fit the observed concentrations for both considered depths. The relatively low value of $fr_{ni\ rz}$ indicates that the assumption that no NH_4^+ leaches from forest soils in the Netherlands is not justified. Values of $fr_{ni\ rz}$ below 1 (a value of 1 results in nitrification of all the remaining NH_4^+) will cause NH_4^+ leaching.

4.2 Validation

The calibration criterion does not distinguish the output values per forest class, since the percentiles of all output values were lumped (c.f. Section 3.3). To study the potential of the employed procedure for making statements on the level of specific forest classes, we compared the observed and simulated median concentrations of the 50 median values of the calibration runs per forest class (Table 5). The comparison in Table 5 shows that the fit of the pH in both the topsoil and the subsoil was good (a deviation of less than 10%) for all the forest classes. The Al^{3+} and SO_4^{2-} concentrations and the Al^{3+}/Ca^{2+} ratio were fitted reasonably (a deviation of 10 to 50%) for all the classes and layers, except for the Al^{3+} concentration in the topsoil of the deciduous forest. Again, it was not possible to realize a good fit in the topsoil and subsoil at the same time. In general, modelled concentrations in the topsoil were higher (pH lower) than observed concentrations, whereas the opposite was true for the subsoil.

5. Evaluation and conclusions

It has been shown that the calibration procedure described is a useful tool to find parameter ranges with a lower uncertainty. However, the number of parameter with a serious reduction in parameter uncertainty is rather small (i.e. ru_{exp}, frN_{im} and *alpha*). The present data set of field observations is apparently insufficient for estimating unequivocal values of the parameters, although the resulting optimal ranges are relatively small and the nominal values, which are derived independently, generally are within this range. A serious problem exists, however, it was not possible to identify the highly uncertain and sensitive forest filtering factors for nitrogen deposition ($ffNH_{4\ dt}$ and $ffNO_{3\ dt}$). Regarding this, it is undoubted that any model prediction of MACAL should be accompanied with an uncertainty analysis.

The CDFs for all sites together were generally well fitted in the subsoil, but the fit in topsoil was less convincing. The field observations used here hardly showed any differences between topsoil and subsoil concentrations. Probably, this is an artefact of data collection procedure. Because sampling was carried out in early spring, the soil solution concentrations are not affected by water and nutrient uptake, which is the main cause of concentration differences with depth. It is therefor plausible that there are hardly any differences between topsoil and subsoil in this period. The similarity between the calibrated model results and the observations per forest class was moderate. This is probably due to the calibration on forest classes not being included in the criterion.

References

Anonymous (1992), *'MATLAB User's Guide'*, The MathWorks Inc., Natick MA.

Bredemeier, M. (1988), 'Forest canopy transformation of atmospheric deposition', *Water Air and Soil Poll.*, **40**, 121-138.

Breeuwsma, A., J.P. Chardon, J.F. Kragt and W. de Vries (1991), 'Pedotransfer functions for denitrification', In: *ECE 1991, Soil and Groundwater Research Report II "Nitrate in Soils"*, Commission of the European Community, Luxembourg, pp. 207-215.

Conover, W.J. (1971), *'Practical Nonparametric Statistics'*, John Wiley & Sons Inc., New York.

De Vries, W. (1991), *'Methodologies for the assessment and mapping of critical loads and of the impact of abatement strategies on forest soils'* DLO-Winand Staring Centre, Wageningen, Netherlands, Report 46.

De Vries, W. (1993), 'Average critical loads for nitrogen and sulfur and its use in acidification abatement policy in the Netherlands', *Water, Air, and Soil Pollution*, **68**, in press.

De Vries, W. and E. Leeters (1993), *'Effects of acid deposition on 150 forest stands in the Netherlands. 1. Chemical composition of the humus layer, mineral soil and soil solution'*, DLO-Winand Staring Centre, Wageningen, the Netherlands, Report 69.1.

De Vries, W., J. Kros, and J.C.H. Voogd, (1994). 'Assessment of Critical Loads and their Exceedance on Dutch Forest using a Multi-Layer Steady-State Model', *Water, Air, and Soil Poll.*, accepted for publication.

Erisman, J.W. (1990), *'Atmospheric deposition of acidifying compounds onto forests in the Netherlands: throughfall measurements compared to deposition estimates from inference'*, National Institute of Public Health and Environmental Protection, Bilthoven, Netherlands, Report nr. 723001001.

Erisman, J.W. (1992), *'Atmospheric deposition of acidifying compounds in the Netherlands'*, Ph.D. Thesis, Rijksuniversiteit Utrecht, Utrecht, Netherlands.

Grace, A. (1990), *'Optimization Toolbox for use with MATLAB User's Guide'*, The MathWorks, Inc., Natick MA.

Hiege, W. (1985), *'Wasserhaushalt von Forsten und Wälder und der Einfluss der Wassers auf Wachstrum und Gesundheit von Forsten und Wälder: eine Literaturstudie'*, Studiecommissie Waterbeheer, Natuur, Bos en Landschap Utrecht, Netherlands, Rapport 7a.

Houdijk, A.L.F.M. (1990), *'Effecten van zwavel- en stikstof depositie op bos- en heide vegetaties'*, Katholieke Universiteit Nijmegen, Nijmegen, Netherlands.

Kimmins, J.P., D. Binkley, L. Chatarpaul and J. de Catanzaro (1985), *'Biogeochemistry of temperate forest ecosystems: Literature on inventories and dynamics of biomass and nu-*

trients', Petawawa National Forestry Institute, Information Report PI-X-47E/F.

Kleijn, C.E., G. Zuidema and W. de Vries (1989), '*De indirecte effecten van atmosferische depositie op de vitaliteit van Nederlandse bossen. 2. Depositie, bodemeigenschappen en bodemvochtsamenstelling van acht Douglas opstanden*', Stichting voor Bodemkartering, Wageningen, Netherlands, Rapport 2050.

Kros, J., P.S.C. Heuberger, P.H.M. Janssen and W. de Vries (in prep), 'Reduction of parameter uncertainty for the assessment of critical loads using a multi-layer steady-state model'.

Tietema, A. and J.M. Verstraten (1991), 'Nitrogen cycling in an acid forest ecosystem in the Netherlands at increased atmospheric nitrogen input: the nitrogen budget and the effects of nitrogen transformations on the proton budget', *Biogeochem.* **15**, 21-46.

Tietema, A., W. de Boer, L. Riemer and J.M. Verstraten (1992), 'Nitrate production in nitrogen saturated acid forest soils: vertical distributions and characteristics', *Soil Biol. and Biochem.* **24**, 235.

Turner, J.P. (1975), '*Nutrient cycling in a douglas-fir ecosystem with respect to age and nutrient status*', Ph.D. Thesis, University of Washington, Seattle.

Verhagen, H.L.M. and H.S.M.A. Diederen (1991), '*Vergelijkingsmetingen van de analyse- en monsternemingsmethode van de vaste en vloeibare fase van bodemmonsters*', TNO/IMW, Delft, Netherlands, R91/171.

UNCERTAINTY ANALYSIS FOR THE COMPUTATION OF GREENHOUSE GAS CONCENTRATIONS IN IMAGE

MAARTEN S. KROL

Global Change Department
National Institute of Public Health and Environmental Protection
P.O. Box 1, 3720 BA, Bilthoven, the Netherlands

e-mail: mobimart@rivm.nl

Summary

Uncertainties in simulations of greenhouse gas concentrations are analyzed. Greenhouse gas concentrations are simulated using the Atmospheric Composition model of IMAGE. Uncertainties arise from, amongst others, uncertainties in greenhouse gas emissions and in parameters in the description of atmospheric processes. The total uncertainty in the greenhouse gas concentrations is quantified as well as the contributions of the individual sources of uncertainty to this total uncertainty, by making a multi-dimensional Monte Carlo analysis using Latin Hypercube sampling. Focus is on the non CO_2 gases like methane and ozone, for which atmospheric processes play a key role in determining the changes in concentration.

Keywords Uncertainty analysis, greenhouse gas, Monte Carlo method

1. Introduction

Over the last century concentrations of gases like carbon dioxide, methane, CFCs and nitrous oxide have risen very fast. The changes in the concentrations of these gases are mainly due to increased anthropogenic emissions. In the atmosphere, these gases can effectively absorb solar or infra-red radiation. In that way the atmosphere is heated; this is the greenhouse effect. Increased anthropogenic emissions of nitrogen oxides and hydrocarbons have led to increased levels of tropospheric ozone, another important contributor to the greenhouse effect.

In the stratosphere halogenated species like CFCs have led to increased levels of chlorine radicals. These radicals are the main cause of ozone depletion. Ozone depletion in the lower stratosphere cools the atmosphere, but leads to higher UV radiation levels. Enhanced levels of UV radiation can have effects on for instance public health or plant growth.

Both the climatic impacts and the UV impacts due to anthropogenic emissions are of serious concern to mankind. The 'ultimate aim' of the Climate Convention in Rio 1992 is ".. *to achieve stabilization of greenhouse gas concentrations in the atmosphere at a level that would prevent dangerous anthropogenic interference with the climate system. Such level should be achieved within a time frame sufficient to allow ecosystems to adapt naturally to climate change, to ensure that food production is not threatened and to enable economic development to proceed in a sustainable manner*". Modelling of atmospheric processes is a necessity to translate this aim into quantitative policy goals.

In this paper we investigate the uncertainties in the computation of greenhouse gas concentrations in IMAGE. We quantified the uncertainties and identified the contributions of the individual sources of uncertainty to the total uncertainty. For the analysis we used the atmospheric chemistry and climatic impact modules of IMAGE 1.5, an update of the Integrated Model to Assess the Greenhouse Effect, IMAGE, Rotmans (1990), Rotmans *et al.* (1992). Carbon cycle calculations were replaced by emission data from the IPCC IS92a scenario and a fixed airborne fraction for emitted CO_2.

We concentrated on the uncertainties due to uncertainties in emissions and uncertainties in parameters in the description of atmospheric processes. We did not try to quantify the uncertainties resulting from the type of modelling. Furthermore we constrained the joint variations of the parameters to those situations where the present day situation is represented well by the model, ie. we only varied the parameters over possible calibrations of the model. This gives rise to a constraint on the parameters appearing in the descriptions of processes with a short time scale. We took this approach since the model results, concentrations, are much better established than the model forcing or model parameters, emissions and process parameters.

Uncertainties in reaction kinetics and their impact on the atmospheric chemical equilibrium are significant, Thompson and Stewart (1991): standard deviations of reaction rates average 30% of their means and photodissociation rates are even more uncertain; as a result standard deviations of atmospheric compounds average over 25% of their means in model simulation. Emission estimates show uncertainties ranging up to 30%, IPCC (1992). Sensitivities of concentrations of oxidants to emissions in chemical coherent regions tend to amplify those uncertainties slightly, Thompson and Cicerone (1986), Thompson *et al.* (1989).

2. Calculation of greenhouse gas concentrations in IMAGE

The atmospheric chemical model used in IMAGE 1.5 was a one box model of the troposphere with some stratospheric data added. The model aims at calculating the concentrations of greenhouse gases with an accuracy satisfactory for impact calculations in situations not differing too much from the present day situation. The concentrations of greenhouse gases were represented by their tropospheric average. This representation is at least satisfactory for long lived greenhouse gases that are distributed homogeneously, like carbon dioxide (CO_2), methane (CH_4), nitrous oxide (N_2O) and CFCs. Tropospheric ozone (trop O_3) and compounds involved in the CH_4 chemistry, like hydroxyl (OH), nitrogen oxides (NO_x), carbon monoxide (CO) and non methane hydrocarbons (NMHC), are distributed much more heterogeneously.

The model was set up similar to AMAC, Prather (1989). Global averages of reaction kinetics were evaluated using distributions of hydroxyl and temperature following Prather and Spivakovsky (1990). Linear representations of the sensitivity of the atmospheric chemistry to changes in the NO_x emissions, the column ozone and methane and carbon monoxide were calculated using regional sensitivity results from Thompson *et al.* (1989) for chemically coherent regions. The regional sensitivities were weighted taking into account the present day distributions of the compounds and the present day regional trends in the emissions or concentrations of the compounds.

CH_4: Methane is emitted into the troposphere. The emission was converted linearly into a global increase term for the concentration in the troposphere. The sinks of methane include transport to the stratosphere and deposition/soil uptake. These loss terms were modelled to be proportional to the methane concentration. The main sink is caused by the oxidation of methane by hydroxyl. This loss term is proportional to both the methane and the hydroxyl concentrations. The change in atmospheric concentration was described by formula (1) with $X = CH_4$ and parameter values as given in table 1:

$$dpX = f_X \cdot Em_X - \left(l_X + k_{XOH} \cdot pOH\right) \cdot pX \ . \tag{1}$$

CO: Carbon monoxide competes with methane for the consumption of hydroxyl. The resulting oxidation is the main sink of carbon monoxide. The loss is proportional to the concentrations of carbon monoxide and hydroxyl. Another sink is soil uptake and transport to the stratosphere; these sinks are assumed to be proportional to the concentration of carbon monoxide. The sources for carbon monoxide included are threefold: direct emissions, atmospheric oxidation of NMHCs and atmospheric oxidation of methane. The source from NMHC is included in the direct emissions of carbon monoxide in this model, taking into account an average yield YNMHC of 0.651 (mass/mass), as calculated by the 2-D TNO-Isaksen model, Roemer (1991). For the source from methane we assume that 80% of the oxidized methane ends up as carbon monoxide, following Logan et al. (1981). This leads to:

$$dpCO = f_{CO}\{Em_{CO} + YNMHC \cdot Em_{NMHC}\} + 0.8 \cdot k_{CH_4OH} \cdot pOH \cdot pCH_4$$

$$-\left(l_{CO} + k_{COOH} \cdot pOH\right) \cdot pCO \ . \tag{2}$$

Table 1. Values for the parameters in the evolution equations for the concentrations of the compounds assumed not to be in equilibrium. f_X denotes the conversion factor from emissions to concentrations (in their respective units), k_{X+OH} the reaction rate for the oxidation of X by OH in $cm^{-3} \cdot year^{-1}$, l_X stands for the transport losses to both stratosphere and biosphere (deposition) in $year^{-1}$ and lft_X for the average atmospheric residence time (lifetime) in years, p_X and Em_X denote the concentration and emission of the compound, and are both given in the base year 1990.

compound X	p X(1990)	Em_X(1990)	f_X	lft_X	k_{X+OH}	l_X
CH$_4$	1.72 ppmv	506 Tg	$3.90 \cdot 10^{-4}$	-	$1.23 \cdot 10^{-7}$	0.0091
CO	0.090 ppmv	1164 Tg	$2.22 \cdot 10^{-4}$	-	$5.93 \cdot 10^{-6}$	0.713
NMHC	-	608 Tg	-	-	-	-
N$_2$O	308 ppbv	12.9 TgN	0.212	150	-	-
CFC-11	272 pptv	298 Gg	0.0413	55	-	-
HCFC-22	113 pptv	138 Gg	0.0723	-	$9.05 \cdot 10^{-8}$	0.0042

OH: Hydroxyl has a very short atmospheric lifetime It was approximated by an equilibrium concentration. The production of hydroxyl was used as a separate quantity in the model. The loss of hydroxyl is caused by both methane and carbon monoxide. The steady state assumption used reads:

$$pOH = \frac{prodOH}{k_{CH_4OH} \cdot pCH_4 + k_{COOH} \cdot pCO} \ . \qquad (3)$$

The production of hydroxyl depends on the concentrations of its precursors, excited oxygen atoms and tropospheric water vapour. The concentration of excited oxygen atoms is represented by the concentration of tropospheric ozone and by the column ozone (determining the radiation levels required for the photodissociation of ozone in the troposphere). Furthermore nitrogen oxides have their impact, recycling hydroxyl in the oxidation chain of for instance carbon monoxide. The equation used reads:

$$dlnprodOH = LDOHTRO3 \cdot dlntropO_3 - 0.7 \cdot dlncolO_3$$
$$+ 0.5 \cdot dlnpH_2O + LDOHNOX \cdot dlnEm_{NO_x} \ , \qquad (4)$$

where LDOHTRO3 = 0.5, LDOHNOX = 0.368.

O_3:   Ozone is the most important greenhouse gas which is not homogeneously distributed over the troposphere. In the model changes in the concentration of tropospheric ozone were being related to changes in the concentrations of its precursors and to changes in the levels of the radiation responsible for its dissociation. The precursors included were methane and carbon monoxide and nitrogen oxides (only as emissions); the radiation level was parameterized the column ozone:

$$dlntropO_3 = LDO3CH4 \cdot dlnpCH_4 + LDO3CO \cdot dlnpCO$$
$$+ LDO3NOX \cdot dlnEm_{NO_x} + LDO3CMO3 \cdot dlncolO_3 \ , \qquad (5)$$

where LDO3CH4 = 0.13, LDO3CO = 0.15, LDO3NOX = 0.064, LDO3CMO3 = 0.2.
The column ozone was parameterized by the stratospheric concentration of chlorine, fitting the TOMS data. A threshold level PSTRCLTH of 1.3 ppbv of active chlorine was used and an ozone depletion rate LDCMO3CL of 3.4 % ppbv^{-1} for chlorine levels exceeding the threshold level:

$$colO_3 = min\left(1 \ , 1 - LDCMO3CL \cdot \left(pCL_{strat} - PSTRCLTH\right)\right) \ . \qquad (6)$$

H_2O: The tropospheric water vapour mixing ratio was parameterized by the average tropospheric temperature T_{trop}, following Prather (1989):

$$dlnpH_2O = 0.062 \cdot dt_{trop} \ . \qquad (7)$$

T_{trop}:   The increase in mean tropospheric temperature was assumed to be similar to the increase in surface temperature and linearly related to changes in radiative forcing. Changes in the concentrations of greenhouse gases were translated into changes in radiative forcing (dF) following IPCC (1990). The temperature increase due to CO_2 doubling T2XCO2 is assumed to be 2.5 °C.

$$dT_{trop} = \frac{T2XCO2 \cdot dF}{4.35} \ . \tag{8}$$

N_2O: Nitrous oxide is emitted from the surface and depleted by photodissociation. The emissions, in units of mass, were converted into concentrations assuming that nitrous oxide is homogeneously distributed over the troposphere. The lifetime lft_{N2O} is estimated to be 150 years, IPCC (1990). Formula (9), with $Y = N_2O$, describes the evolution of the concentration of N_2O; parameter values are listed in table 1:

$$dpY = f_Y \cdot Em_Y \ - \ \frac{1}{lft_Y} \cdot pY \ . \tag{9}$$

CFCs: CFC 11 was modelled in the same way as N_2O. The lifetime lft_Y was taken from WMO (1992): 60 years. Other CFCs were included similarly.

HCFCs: HCFC-22 was treated similar to methane. The changing concentrations follow from formula (1) with $X = HCFC\text{-}22$ and the data in table 1. Other HCFCs, methylchloroform and methylchlorine were included similarly.

Cl: The concentration of stratospheric free chlorine was calculated from the tropospheric concentrations of chloride containing compounds. A timelag of 4 years was included to represent the transportation time from the troposphere to the stratosphere. The chlorine yield of the various compounds was taken to be 2.95 for CFC-11 and 0.6 for HCFC-22, WMO (1992). Other halogenated species were included similarly.

The box model has been used in a scenario study examining control options leading to stabilization of atmospheric methane concentrations, Rotmans *et al.* (1992). The model has been validated by comparing model results to historical methane data and to results of other scenario studies, IPCC (1990). The historical methane trend is represented well by the model, while model results for scenario studies lie within the range of the state of the art models.

3. Methodology of the uncertainty analysis.

In the uncertainty analysis we followed the methodology described in Janssen *et al.* (1990) and using the software package UNCSAM, Janssen *et al.* (1992).
Our main points of interest are
   ♦ a quantification of the uncertainties in the simulation of greenhouse gas concentrations

♦ an identification of the main individual sources of the uncertainties

Uncertainties in model simulations can, in general, be due to a number of sources, Janssen *et al.* (1992):

♦ uncertainty due to model structure
♦ uncertainty in external factors, forcing
♦ uncertainty in initial conditions, boundary conditions
♦ uncertainty in model parameters
♦ uncertainty due to model operation

Model intercomparison studies give insight in the first source of uncertainty. For the simulation of methane concentrations Guthrie and Yarwood (1991) found that this contribution to the uncertainty can be significant. In that study the results of 5 state of the art atmospheric chemistry models are compared. The models each simulate the methane concentration in 2100 resulting from the IPCC Business as Usual scenario, IPCC (1991). The results vary between 3.2 ppmv and 5.1 ppmv, averaging 4.0 ppmv. This large variation is due to not only the uncertainty in model structure but also to different external factors, initial conditions and model parameters used. Therefore it is hard to obtain an estimate of just the first source of uncertainty.

We concentrated on the uncertainties due to uncertainties in the forcing of the model and uncertainty in model parameters. For our model the forcing were the emissions of the greenhouse gases and the emissions of the source gases for ozone.. The model parameters were the reaction rates, the chemical sensitivity coefficients, the coefficients for the radiative forcing and the climate sensitivity. Table 2 lists the parameters that were varied in the analysis, their actual value in the calibrated model and the range used in the uncertainty analysis. The ranges of the uncertainties in the global sources and sinks and the climate sensitivity were taken from IPCC (1992), radiative forcing data from IPCC (1990), lifetimes and parameters on ozone depletion from WMO (1992), reaction kinetics from Thompson and Stewart (1991) and the chemical sensitivities from Thompson *et al.* (1989).

In the reference case, the UNCSAM package sampled values for all parameters, except for LCO, from independent uniform distributions using a Latin Hypercube algorithm. For each set of sampled parameters the value for LCO was set under the assumption of an equilibrium for the carbon monoxide concentration and the observed trend for the methane concentration in 1990. Both the carbon monoxide concentration and the methane concentration trend are of course only known within an uncertainty range. If the value calculated for LCO was outside its uncertainty range the set of parameters was dropped. In this way 150 sets of parameter values were constructed. Due to the constraints on LCO these sets show correlations between the parameters; the strongest correlations are shown in Table 3.

The 150 sets of parameter values were used for model simulations for the period 1990 to 2100 using the IS92a emissions scenario, IPCC (1992). The global emissions are scaled with the sampled values representing the uncertainties in these emissions (CO2EMUNC, CH4EMUNC, ...). The model parameters are used in a straightforward way. For each of the model simulations UNCSAM performed statistical calculations for percentiles, means, variations and regressions. The regressions resulted in estimates for the contributions of individual sources of uncertainty to the total uncertainty.

Two alternative approaches were taken. The first alternative approach was to drop the assumption of simulating the methane trend in 1990. As a consequence the system showed a transition to a new equilibrium for carbon monoxide, methane and hydroxyl on a timescale of 10 year. This new equilibrium was often considerably different from the

Table 2. Parameters that were varied in the uncertainty analysis.

parameter	unit	actual value	minimum	maximum
CO2EMUNC		1.0	0.8	1.2
CH4EMUNC		1.0	0.7	1.6
COEMUNC		1.0	0.6	1.8
NOXEMUNC		1.0	0.75	1.2
N2OEMUNC		1.0	0.6	1.4
KCH4OH	cm^3yr^{-1}	$1.228 \cdot 10^{-7}$	$0.863 \cdot 10^{-7}$	$1.603 \cdot 10^{-7}$
KCOOH	cm^3yr^{-1}	$5.934 \cdot 10^{-6}$	$3.56 \cdot 10^{-6}$	$8.31 \cdot 10^{-6}$
KHC22OH	cm^3yr^{-1}	$9.051 \cdot 10^{-8}$	$7.7 \cdot 10^{-8}$	$10.4 \cdot^{-8}$
LCH4	yr^{-1}	0.0091	0.0045	0.0136
LCO	yr^{-1}	0.713	0.35	1.15
YNMHC		0.65	0.35	1.0
LDOHNOX		0.368	0.2	0.4
LDOHTRO3		0.5	0.1	1.0
LDO3CH4		0.13	0.08	0.18
LDO3CO		0.15	0.1	0.2
LDO3NOX		0.064	0.044	0.084
LDO3CMO3		0.2	0.12	0.28
LDCMO3CL		0.034	0.029	0.039
PSTRCLTH	ppbv	1.3	1.0	1.6
LFTN2O	yr	150	110	168
LFTCFC11	yr	55	42	66
PCO90	ppmv	0.09	0.085	0.095
CH4INCR	$Tg\ yr^{-1}$	45	40	50
AIRBORFR		0.55	0.4	0.6
T2XCO2	°C	2.5	1.5	4.5
RADNCL		1.0	0.85	1.15
RADCL		1.0	0.8	1.2

observed present day situation. The uncertainties found were much larger. A second alternative approach also avoided that sets of samples were dropped. Here the distributions were assumed to be correlated from the beginning. Correlations were chosen in such a way that the consistent value for LCO automatically complied with the constraints on the uncertainty range for LCO. We found however that the correlations necessary to obtain this were very high, a group of 6 parameters had to be correlated with correlations all above 0.9. These strong correlations tend to reduce uncertainty considerably.

4. Results

4.1 Comparison of the sampling methods

The percentiles for the simulated methane concentrations for the three approaches are shown in figure 1. For each approach the mean, median and the 10 and 90 percentiles of the simulated concentrations are shown as a function of time. We see that for the reference case the variations grow steadily in time to 15% in 2050 and 25% in 2100. (For the methane increase since 1990 the coefficients of variation are 35% in 2050 and 40% in 2100.) These variations are similar to the variations between the results for different state of the art models for such a scenario, Guthrie and Yarwood (1991). For the unbalanced case variations grow much more rapidly to 45% in 2050 and 50% in 2100. Variations for the correlated case are similar as for the reference case: 10% in 2050 and 20% in 2100.

Table 3. Correlations between the sampled parameters after dropping parameter sets leading to inadmissible values for LCO.

	CH4EMUNC	COEMUNC	KCH4OH	KCOOH	LCH4	YNMHC
CH4EMUNC	1.00	0.25	0.34	-0.32	-0.06	0.08
COEMUNC	0.25	1.00	-0.13	0.50	-0.18	0.11
KCH4OH	0.34	-0.13	1.00	0.31	-0.15	-0.13
KCOOH	-0.32	0.50	0.31	1.00	-0.07	0.18
LCH4	-0.06	-0.18	-0.15	-0.07	1.00	0.09
YNMHC	0.08	0.11	-0.13	0.18	0.09	1.00

In the reference case sets of samples were dropped when they led to values for LCO outside its uncertainty range. Nevertheless, the distributions of samples in the reference case were close to the originally chosen distributions, see the upper panel of figure 2. The distributions of the simulated concentrations were often skew, see the middle panel of figure 2, but the linear regressions of the uncertainties in terms of the parameters always had a coefficient of determination above 0.9 after the year 2000. The uncertainty in pCH_4 can be attributed to different sources. On the short longer time scale the uncertainty of LDOHTRO3 dominates (50%); other important sources are the uncertainty in the methane emission and the effects of NO_x emissions and temperature increase. The unbalanced case shows a similar picture, but KCOOH and COEMUNC are of strong influence here; for the correlated case the relative importance of all parameters in the CH_4-CO-OH equilibrium is similar since all these parameters are strongly correlated.

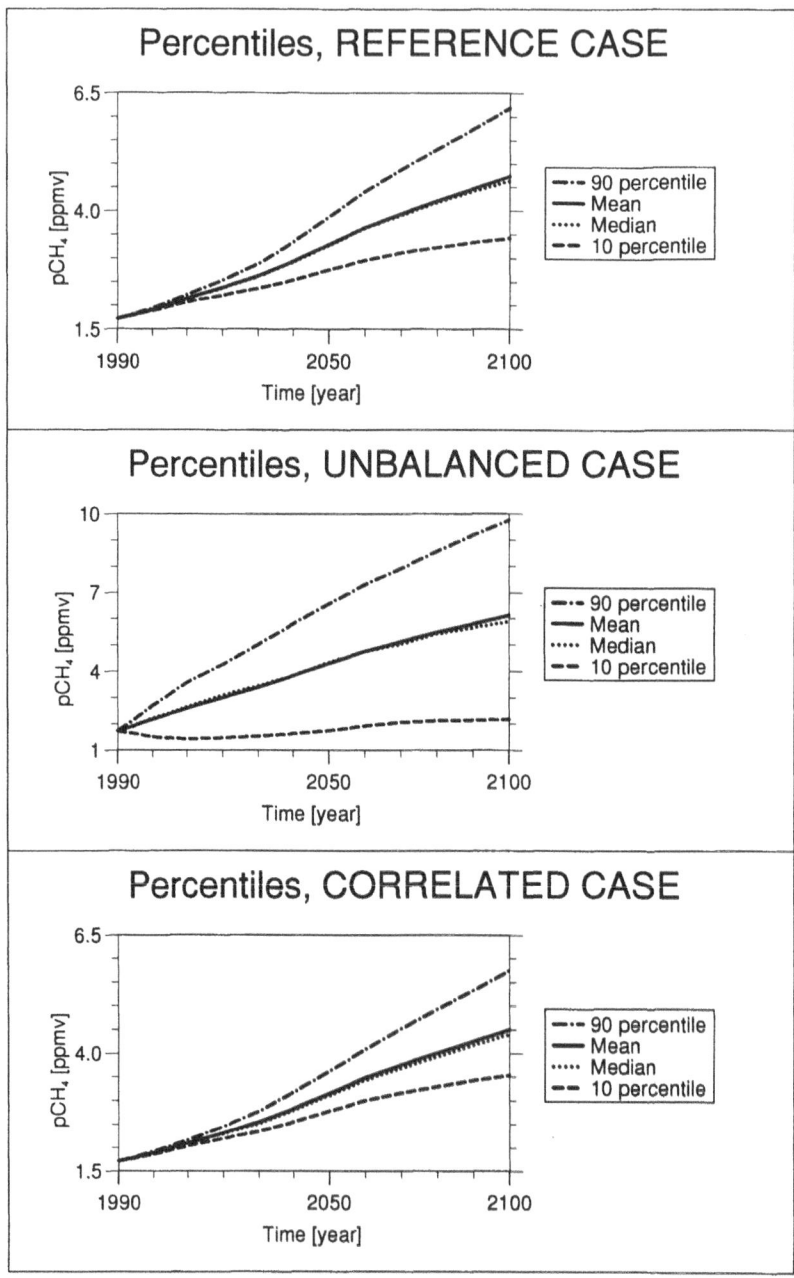

Figure 1. Percentiles for the simulated methane concentrations as a function of time for the three cases considered.

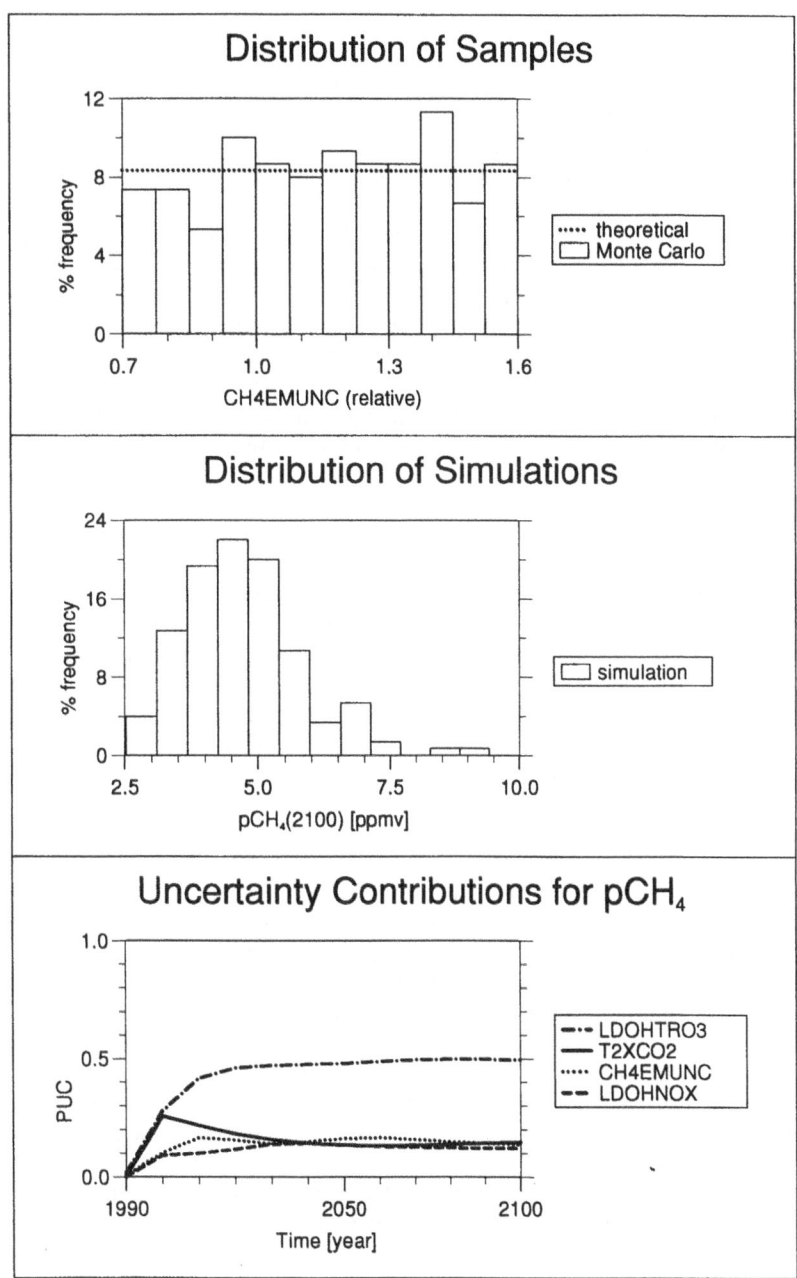

Figure 2. Distributions for the reference case of the samples of CH4EMUNC and of the simulated methane concentrations in 2100. The total uncertainty in the methane concentration is split up into sources of the uncertainty, as a function of time.

4.2 Uncertainties in the reference case

Apart from methane, a number of greenhouse gases were analyzed for their uncertainty. For tropospheric ozone the coefficient of variation for the ozone increase since 1990 is 25% in 2050 and 30% in 2100. On the short time scales LDO3CH4 and LDO3CO are important sources of uncertainty, on the longer timescale again LDOHTRO3 dominates (figure 3). Emissions in the source gases and temperature related uncertainty both account for some 10 % of the uncertainty.

For nitrous oxide no constraint on the parameters is assumed. As a result the present day trend is not necessarily simulated and the uncertainty is large: the variation in the increase of nitrous oxide is 80% in 2050 and 75% in 2100. The two sources of uncertainty, N2OEMUNC and LFTN2O account for 80% and 20% of the uncertainty (figure 3).

For HCFC 22 the situation is quite different. Here the uncertainty is mainly determined by uncertainties in the concentration of hydroxyl. These are caused by uncertainties in emissions of hydrocarbons, oxidation rates for hydrocarbons and atmospheric feedback processes. The oxidation rate of HCFC 22 itself plays a minor role (5% in 2100). The coefficient of variation of the increase in HCFC 22 since 1990 is 35% in 2050 and 30% in 2100.

No serious uncertainty analysis is done for CO_2. The uncertainty range for CO_2 in figure 4 is indicative and just meant to compare it to the uncertainty in non CO_2 contributions to the equivalent CO_2 concentration. Here we see that the uncertainty contributions of CO_2 uncertainty and of non CO_2 uncertainty are of the same order of magnitude. Uncertainties are in large part due to uncertainties in the radiative properties of the greenhouse gases, but on the long timescale LDOHTRO3 is the largest source of uncertainty (figure 4). The uncertainties in the equivalent CO_2 concentration are significant. For impact studies however they are small compared to the uncertainty in the climate sensitivity.

5. Discussion

Three methods to sample parameters were used. In the unbalanced case parameters were simply sampled from independent distributions. As a result, methane concentrations immediately started to drift away from the presently observed concentrations. The uncertainties thus obtained should be interpreted as a representation of the sensitivity of the atmospheric equilibrium rather than as a realistic estimate of the uncertainty in the simulation of methane concentrations. In the correlated case the initial distributions were correlated such that the present day trends were simulated. The correlations required were so large however, that regressions of the uncertainty cannot be expected to provide us with any information. In the reference case correlations came in automatically by dropping inadmissible sets of sampled variables. This method appears to give a reasonable representation of the uncertainties. For the simulation of methane concentrations the uncertainties found are comparable to the variations of simulations by the different state of the art models. Individual sources of uncertainty can be distinguished; the effects of parameters that are directly related to each other by the constraints cannot be separated however. This approach does not seem to be appropriate for parameters in processes involving only two or three parameters. In that case the distributions will be seriously affected by the procedure of dropping sets of sampled parameters.

The uncertainty in LDOHTRO3 is found to be particularly important for not only methane and tropospheric ozone concentrations, but even for the equivalent CO_2 concentration on

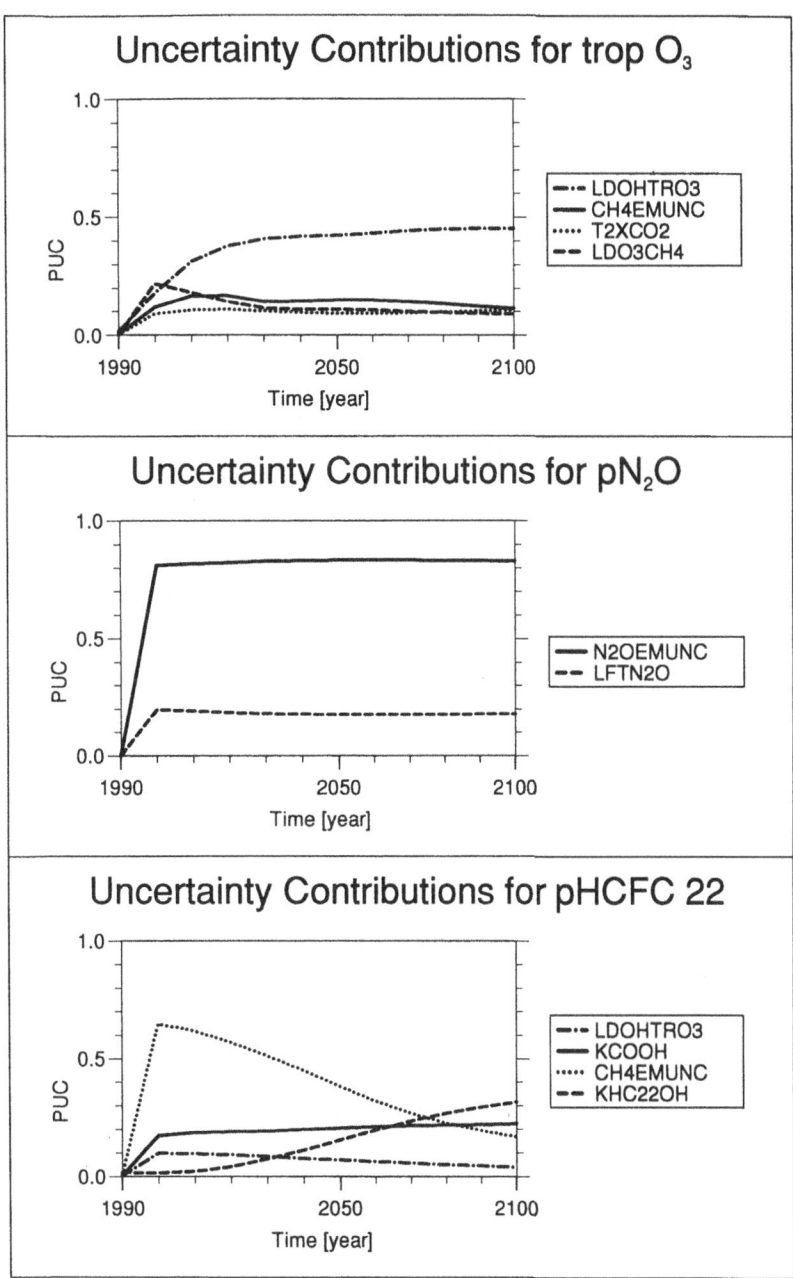

Figure 3. Uncertainty contributions for the concentrations of tropospheric ozone, nitrous oxide and HCFC 22 in the reference case.

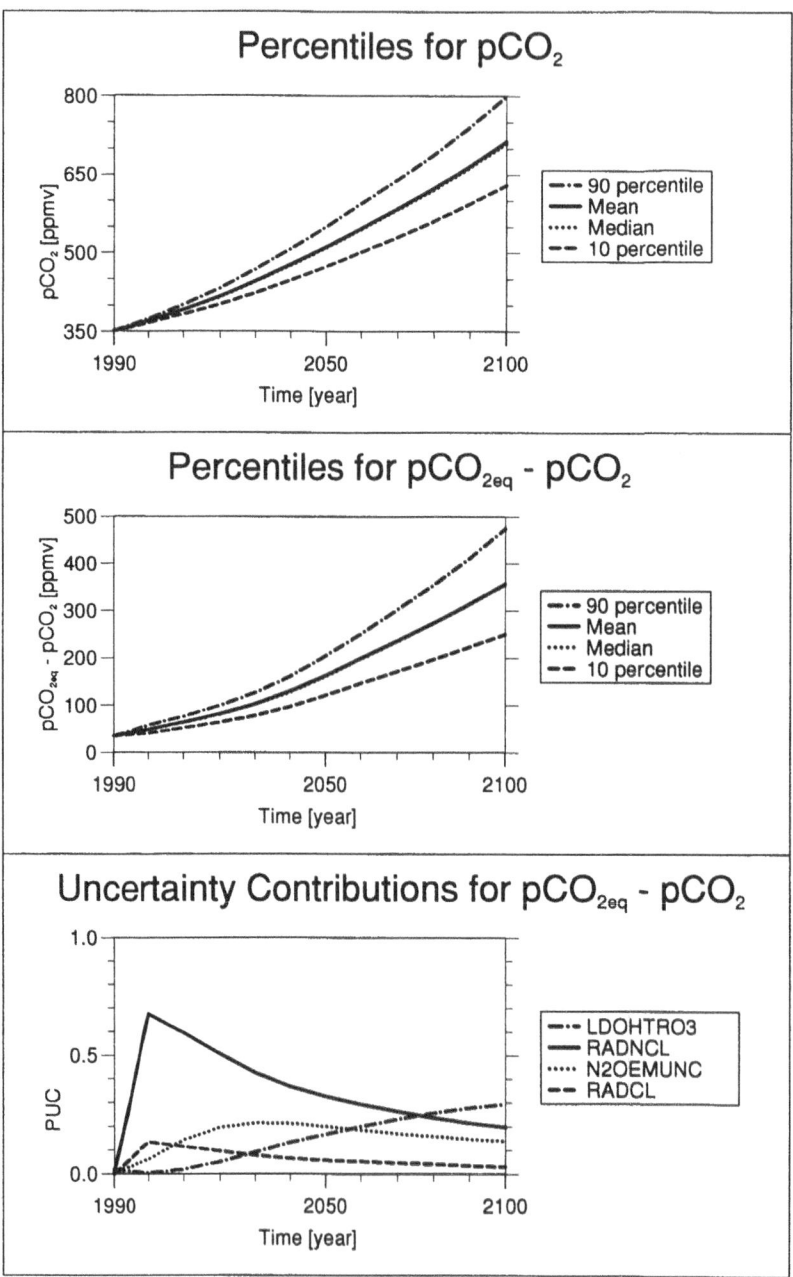

Figure 4. Percentiles for the simulations of the carbon dioxide concentration and the contribution of non CO_2 greenhouse gases to the equivalent CO_2 concentration. The uncertainty in this contribution is split up into the individual sources of uncertainty.

the longer timescales. Uncertainties in HCFC concentrations are dominated by uncertainties in the CH_4-CO-OH equilibrium rather than by uncertainties in the oxidation rate for the HCFCs itself. Uncertainties in the non CO_2 greenhouse gases are equally important for climate impacts as the uncertainties in CO_2. Both sources of uncertainty are significant but small compared to the uncertainty in the climate sensitivity.

Acknowledgements

The work described herein was supported by grant number 851042 of the Dutch National Research Program on Global Air Pollution and Climate Change.

References

Guthrie, P.D., and Yarwood, G. (1991), 'Analysis of the intergovernmental panel of climate change (IPCC) future methane emissions' (SYS-APP-91/114), Systems applications International.

IPCC. (1991), 'Climate Change: The IPCC Response Strategies' , Washington, D.C.: Island Press.

IPCC, Houghton, J.T., Callander, B.A. and Varney, S.K. (Eds.). (1992), 'Climate Change 1992. The Supplementary Report to the IPCC Scientific Assessment', Cambridge Univ. Press.

IPCC, Houghton, J.T., Jenkins, G.J. and Ephraums, J.J. (Eds.). (1990), 'Climate Change. The IPCC Scientific Assessment', Cambridge: Cambridge Univ. Press.

Janssen, P.H.M., Heuberger, P.S.C. and Sanders, R. (1992), 'UNCSAM 1.1: a Software Package for Sensitivity and Uncertainty Analysis Manual' (959101004), RIVM.

Janssen, P.H.M., Slob, W. and Rotmans, J. (1990), 'Gevoeligheidsanalyse en Onzekerheidsanalyse: een Inventarisatie van Ideeën, Methoden en Technieken' (958805001), RIVM.

Logan, J.A., Prather, M.J., Wofsy, S.C. and McElroy, M.S. (1981), 'Tropospheric chemistry: a global perspective', *J. Geophys. Res.*, **86**, 7210-7254.

Prather, M. and Spivakovsky, C.M. (1990), 'Tropospheric OH and the lifetimes of hydrochlorofluorocarbons', *J. Geophys. Res.*, **95**, 18723-18729.

Prather, M.J. (Eds.). (1989), 'An assessment model for atmospheric composition' (NASA Conf. Pub. 3023), New York: NASA.

Roemer, M.G.M. (1991), 'Ozone and the greenhouse effect' (R 91/227). IMW-TNO.

Rotmans, J. (1990), 'IMAGE: an Integrated Model to Assess the Greenhouse Effect' Dordrecht: Kluwer Academic Publishers.

Rotmans, J., Elzen, M.G.J.D., Krol, M.S., Swart, R.J. and Van de Woerd, H. (1992), 'Stabilizing atmospheric concentrations: towards international methane control', *Ambio*, **21**(6), 404-413.

Thompson, A.M. and Cicerone, R.J. (1986), 'Possible perturbations to atmospheric CO, CH4, and OH', *J. Geophys. Res.*, **91**(D10), 10853-10864.

Thompson, A.M. and Stewart, R.W. (1991), 'Effect of chemical kinetics uncertainties on calculated constituents in a tropospheric photochemical model', *J. Geophys. Res.*, **96**(D7), 13089-13108.

Thompson, A.M., Stewart, R.W., Owens, M.A. and Herwehe, J.A. (1989), 'Sensitivity of tropospheric oxidants to global chemical and climate change', *Atm. Env.*, **23**(3), 519-532.

WMO. (1992), 'Scientific Assessment of Ozone Depletion- 1991', (Global Ozone Research and Monitoring Project Rep. No. 25), WMO.

FORECAST UNCERTAINTY IN ECONOMICS

F.J. HENK DON

Central Planning Bureau and University of Amsterdam
Van Stolkweg 14, 2585 JR The Hague, The Netherlands

Summary

Forecasting and policy analysis in (macro-)economics has been the core business of the Dutch Central Planning Bureau ever since its inception in 1945. Econometric models are used to organize knowledge and to ensure consistency of the many variables in the forecast. The track record shows large forecast errors, both at a one year and a four year horizon. Forecasts are always conditional on declared government policies, hence forecast errors can.be partly traced to changes in economic policy. The paper reports on a Monte-Carlo based study into the relative importance of four different sources of forecast uncertainty; especially for a somewhat longer horizon, uncertainty originating from error terms in the model and errors in exogenous variables dominates uncertainty originating from preliminary data and uncertain model parameters.

It is unlikely that forecast errors can be significantly reduced; indeed, continuous efforts are required to keep them at their current level. The main purpose of CPB's forecasting activities is to provide a benchmark for economic policy preparation. The uncertainty that comes with the forecast must of course also be communicated to the policy makers, so they can develop contingency plans. By using different scenarios, 'no-regrets'-policies can be separated from strategic policy decisions.

Keywords Economic forecasts, uncertainty, Monte Carlo methods, policy feedback.

1. Introduction

Forecasting and policy analysis in (macro-)economics has been the core business of the Dutch Central Planning Bureau (CPB) ever since its inception in 1945. Econometric models are used to organize knowledge and to ensure consistency of the many variables in the forecast. Because of the Bureau's role in the preparation of economic policy, its forecasts are conditional on current government policy, see Don and Van den Berg (1990). Often some policy alternatives are presented along with the baseline forecast. In addition, alternative scenarios are usually included to stress the forecast uncertainty in some areas.

There are different ways to study the uncertainty that applies to economic forecasts. The common approach is to study the track record of past forecasts, and compute statistics for observed forecast errors. For the track record of CPB forecasts a recent reference is Westerhout (1990). This approach cannot do justice to the conditional character of the forecasts. Additional insight is gained when the observed forecast errors are decomposed, with the help of a model, into components that can be ascribed to different sources of uncertainty. In particular, for each forecast it is helpful to see what part of the *ex post* errors resulted from forecast errors in the various (policy and non-policy) exogenous variables of the model. Recent studies for CPB forecasts along these lines are Van den Berg (1986), CPB (1988) and Hers (forthcoming).

While these studies of the track record are important for understanding past errors, it is not straightforward to infer uncertainty in new forecasts from them. Indeed, the models used in economic forecasting are continuously being revised, both in reaction to the observed forecast errors and in reaction to observed or expected changes in the real world. To assess forecast uncertainty for a particular model, one should work through the impact of different sources of uncertainty on the model forecast. This approach has been followed in a series of recent papers by Van Vlimmeren, Don and Okker (1991a,b,c,d) for a simplified and annualized version of the large quarterly model FK'85 of the Central Planning Bureau. Though simplified, this model is still nonlinear. Therefore Monte Carlo techniques were used to study how the different sources of uncertainty affect the forecasts produced by the model. A pilot study was published as Gallo and Don (1991).

This paper brings together the main results of the Monte Carlo research project and offers an evaluation. Numerical results in the present paper may differ from those reported earlier because of changes in definitions and correction of errors.

Section 2 discusses the different sources of forecast uncertainty and how the relevant covariance matrices have been determined. In section 3 the contributions from the different sources to the forecast uncertainty are evaluated. In section 4, total conditional forecast uncertainty as derived along those lines is compared with the results of Westerhout (1990) for the CPB forecasting track record. This section also discusses the impact of policy reactions on forecast errors. Section 5 concludes and offers some advice on how to cope with forecast uncertainty in economics.

2. The sources of forecast uncertainty

2.1 The forecasting process

Figure 1 gives a stylized picture of the forecasting process with an econometric model. Data for the past are both input for determining parameter values of the model (possibly by formal estimation procedures) and feed directly into the model as data for lagged variables. Over the forecast horizon, the model requires input for the exogenous variables and for the error terms (disturbances in behavioural equations). The latter are usually set at zero, but autocorrelation processes or non-model information may lead to nonzero values.

Expert opinion is an important ingredient of any forecasting exercise. It brings in non-model information, e.g. on variables or relations that were neglected at the modelling stage but are considered to be relevant in the forecasting period. In Figure 1, expert opinion is pictured as a filter for the model forecasts, with possible feedback on parameter values and error terms.

The different types of information feeding into the model define as many sources of forecast uncertainty. It is the character and relative importance of these sources of uncertainty that is the focus of this paper.

For lack and inaccessibility of data on expert interference, we will ignore the influence of expert opinion on the forecasting process. This is likely to imply an upward bias in our estimates of forecasting uncertainty, because generally expert opinion is found to reduce forecast errors (e.g. Klein (1981), p. 43). Hence the computed values for forecast uncertainty presented in this paper should be discounted by some unknown factor; it seems reasonable to assume, however, that our conclusions on the character and relative importance of the four remaining sources of forecast uncertainty are not affected by ignoring expert interference in the forecasting process. In section 4 we return to the impact of expert interference, when comparing computed forecast uncertainty with observed forecast errors.

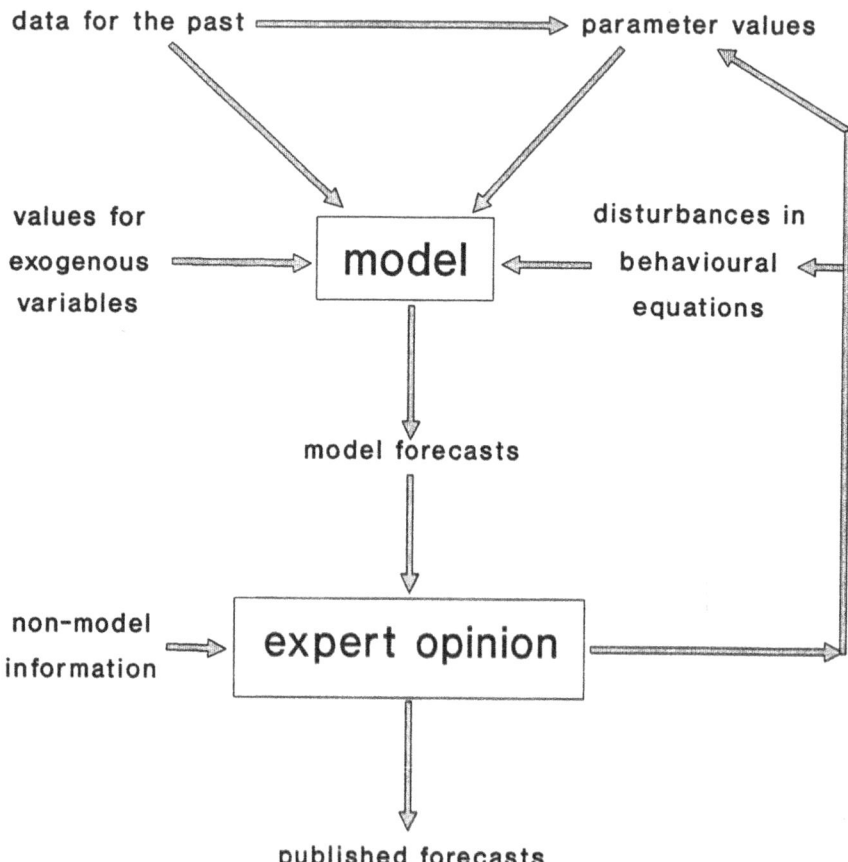

Figure 1. The forecasting process

   In computing forecast uncertainty, we will also ignore any interdependencies between the different sources. In particular, when discussing the impact of unreliable data for the past, we will assume that data errors have not affected the parameter values used in our model. There are two grounds for this simplifying assumption. First, data errors are particularly important for the most recent years of the past, on which only provisional and preliminary data are available for most macro-economic variables. Such provisional data often were not yet available or ignored at the estimation stage of the model. More pervasive data errors related to measurement problems and the like, will often not harm the observed forecasting errors because they affect data for past and future in a largely systematic way. Second, at the model building stage expert opinion is also an important factor in determining the parameter values to be used in the model. At least in the Dutch Central Planning Bureau, model parameters are hardly ever the result of a mechanical estimation procedure and through the intervention of expert opinion are likely to be more robust with respect to possible data errors.

We have studied forecast uncertainty for the ZOEM model (Gelauff and Okker, 1989), which has 295 equations. This is a simplified and annualized version of the large quarterly macro-econometric model FK'85 (Van den Berg et al., 1988), which has some 1300 equations. The latter has been used routinely for several years as an aid in macro-economic forecasting and policy analysis at the Dutch Central Planning Bureau. Reflecting the perpetual process of updating and improving an operational econometric model, FK'85 was renamed FK'89 in 1989 and in the course of 1991 it was replaced by FKSEC (CPB, 1992), which distinguishes exposed (manufacturing, transport, agriculture) and sheltered (non-transport services, building industry) sectors in the supply side of the model. We believe that the ZOEM model used for this study is still adequate for the purpose of studying the major sources of model forecast uncertainty and their relative importance in the Dutch situation. For an exposition of the activities of the Dutch Central Planning Bureau and the role played by its different models, the interested reader is referred to Don and Van den Berg (1990).

2.2 Specifying the uncertainty in model input

To study the effects on the model forecasts due to uncertainty in the model input variables (past data, parameters, exogenous variables and error terms), we had to establish the numerical magnitudes of standard errors related to the latter. To this end, we have relied upon available data sources and economic expertise.

The error structure of the *data revision* process was studied extensively in Van Vlimmeren et al. (1991a, 1991b), using published preliminary and final data for 1960-1984. The data revision process is modelled as

$$w_{1i} = a_{1i} + b_{1i}w_{3i} + u_{1i}$$
$$w_{2i} = a_{2i} + b_{2i}w_{3i} + u_{2i}$$

where w_{1i}, w_{2i}, w_{3i} are the first estimate, the revised estimate and the final observation for variable i. The errors u_{1i} and u_{2i} are allowed to exhibit autocorrelation and intratemporal cross-correlations.

It was found that preliminary data in the Dutch national accounts have shown a bias, tending to systematically underestimate gross national product and private consumption. This experience is shared in other countries, see Lützel (1989). However, we decided to ignore any such bias in preliminary data when assessing forecast uncertainty. Experienced forecasters are likely to correct for estimated or perceived bias when they apply preliminary data in their forecasting exercise. Hence our study of the effects of uncertainty in preliminary data refers to zero mean error processes describing the non-systematic part of data revisions. Standard errors, auto-correlations and cross-correlations for preliminary data are taken from Van Vlimmeren et al. (1991b, appendix B). They refer to the properties of preliminary data on 40 different variables. Through definitions and accounting identities, the uncertainty in those data spreads to all lagged variables in the ZOEM model. The methodology for handling uncertainty in initial data in the presence of accounting identities was described in Gallo and Don (1991).

The uncertainty about *model parameters* has not been directly linked to econometric estimation results. Although estimated standard errors for regression coefficients have served as a guideline in some cases, expert information has been the primary source in assigning standard errors to uncertain parameters. It should be remembered that incompletely specified a priori information usually plays a role in the model building process. Estimated standard errors in a particular regression are likely to overstate parameter uncertainty when, as often is the case, formally or informally, other datasets have also been brought to bear on the determination of the parameter values used in the model.

For the numerical specification of parameter uncertainty used in this study, we refer to Van Vlimmeren *et al.* (1991c). Largely on *a priori* grounds, uncertainty was specified for the 47 parameters in the behavioural equations of the ZOEM model. In some cases, a nonzero cross-correlation is assigned for parameters within the same equation. Parameters from different equations are always assumed to be independent. Parameter constancy is assumed throughout. Any effect of changing parameters is caught in the error terms of the behavioural equations.

Apart from demographic variables, the *exogenous variables* in the ZOEM model relate either to the international environment (potential growth of export markets, foreign prices, foreign interest rates) or to domestic policy variables (real government expenditures, tax rates, controlled housing rents). Forecasts for the short and medium run published by the Central Planning Bureau are always conditional on stated domestic policies, hence in our study of forecast uncertainty we start from the assumption that policy instruments are given (and free of errors). The issue of uncertainty in policy instruments is taken up again in section 4.

As much as possible, we have used published data to assess the uncertainty in the international and demographic exogenous variables. However, we had to mix in expert opinion for several variables because of the very small sample size available, especially for errors in exogenous variables more than one year into the future. The numerical specification of the uncertainty in the nine non-policy exogenous variables of the model is described in Van Vlimmeren *et al.* (1991d). The correlation structure includes several auto-correlations and some cross-correlations.

It is important to note here that the sample period used to establish uncertainty in the non-policy exogenous variables, 1972-1989, includes several oil price shocks. The estimated standard forecast error for the price of imports one year ahead is over 8%, and that for (reweighted) world trade is almost 4%. Though oil crises may be considered outliers, we felt it was only fair to include their surprises in our sample of forecast errors.

The numerical specification of the variances and covariances for the *error terms* of the 17 behavioural equations is also given in Van Vlimmeren *et al.* (1991d). Standard errors were based on mean squared residuals for the period 1976-1987. Most estimated autocorrelations proved statistically insignificant and for less than half of the error terms a nonzero autocorrelation parameter was implemented. Estimated cross-correlations were highly implausible while the estimated cross-correlation matrix was deficient in rank because the number of behavioural equations exceeded the number of years in the database. We decided to proceed on the assumption of zero cross-correlations of the error terms, though in some instances a case could be made for plausible nonzero values.

3. Decomposition of forecast uncertainty

Before we discuss the standard forecast errors that were computed in the Monte Carlo experiments, we assess in this section what the relative importance of the different sources of uncertainty is. For some important forecast variables, Table 1 shows the contributions of the different sources to the total error variance. Clearly, the uncertainty in exogenous variables and the error terms in the behavioural equations are the two dominant sources of forecast uncertainty in our model for these variables.

Table 1. Contributions to total forecast error variance

	preliminary data	model parameters	non-policy exog. vars	error terms	total
% share in forecast error variance					
A One year ahead forecast					
real growth rates					
real growth rates	3	5	33	59	100
private consumption	10	13	4	72	100
business investment	2	2	83	13	100
non-energy exports production enterprises	3	5	49	42	100
nominal growth rates					
consumption price	2	2	60	36	100
wage rate enterprises	3	20	27	50	100
other indicators					
government surplus	14	1	70	15	100
current account balance	5	3	67	25	100
B Four years ahead forecast					
real growth rates					
real growth rates	4	12	48	36	100
private consumption	4	11	34	51	100
business investment	2	6	62	30	100
non-energy exports production enterprises	6	11	45	38	100
nominal growth rates					
consumption price	0	1	91	8	100
wage rate enterprises	1	10	75	13	100
other indicators (end-year)					
government surplus	1	2	93	4	100
current account balance	1	2	89	9	100

The impact of errors in *preliminary data* is not entirely negligible in the one year ahead forecast for business investment and the government surplus. In Van Vlimmeren *et al.* (1991b) we found that this type of uncertainty is largely accounted for by data errors on capacity utilisation, inventory formation and consumption. For short term forecasting of the government surplus, the quality of preliminary data on interest flows and labour market variables is more important.

Most behavioural equations in the model are written in growth rates, because in forecasting practice this has been found to provide a good default treatment of the last recorded residuals. As a result, forecasts of growth rates more than one year ahead are less sensitive to errors in initial data than are forecasts of levels for the same horizon. Indeed, for all variables presented in Table 1, the contribution of uncertainty in preliminary data to the forecast error variance four years ahead is 6% or less.

The contribution of *parameter uncertainty* to forecast error variance shows a more interesting pattern. In a one year ahead forecast, parameter uncertainty contributes some 13% to the variance of the investment growth forecast and about 5% to that of the consumption growth forecast. In a four year ahead forecast, the contribution of parameter uncertainty to forecast errors in average investment growth has decreased somewhat, while its contribution to the forecast error variance of average consumption growth has more than doubled. The wage rate forecast is (relatively) most affected by parameter uncertainty; more than 20% of its one year ahead error variance stems from parameter uncertainty. While the contribution of parameter uncertainty to forecast errors may be relatively small, for policy evaluation it is very important, cf. Rutten (1984), p. 92-93.

In Van Vlimmeren *et al.* (1991c), we found that different groups of parameters are responsible for short term versus longer term forecast uncertainty. Uncertainty about the parameters in the investment equations is the most important source of short term forecast errors, but their contribution is more than halved on a longer horizon. The parameters in the labour market block of the model, in particular those of the wage rate equation, are already an important source of uncertainty in the one year ahead forecast, but their share in the contribution of parameter uncertainty still increases for a longer horizon. These general statements are based upon an analysis of contributions to average forecast error variance of a set of 32 relevant forecast variables. Of course, for any particular individual variable the relative importance of different groups of parameters may be different. For forecasting production of enterprises, for instance, the parameters of the trade equations (exports and imports) are more important.

Because the model is nonlinear, the assessment of the composition of forecast errors is not independent of the baseline used in our numerical experiments. This is particularly true for the role of parameter uncertainty, because parameter errors are transmitted through multiplicative rather than additive channels. For the same reason, one might expect serious interaction effects from uncertainty in parameters and uncertainty in exogenous variables. When a parameter α and an exogenous variable x are multiplied to obtain a forecast for y (i.e. $y = \alpha.x$), the uncertainty in y measured by its variance exceeds the sum of the variances obtained from the two sources separately:

$$var(y) = \bar{\alpha}^2 \, var(x) + \bar{x}^2 \, var(\alpha) + var(x)var(\alpha),$$

where $\bar{\alpha}$ and $\bar{x}$ indicate mean values. Yet a column with such interaction contributions is absent from Table 1, because they proved to be of negligible numerical importance. To see how this can be realistic, consider $y = \alpha.x$ with y for percentage growth in domestic exports, x for percentage growth in world trade and α for the parameter relating the two. With $\bar{\alpha} =$

1, $\bar{x} = 5.5$, $var(\alpha) = (.09)^2$ and $var(x) = (4)^2$ the interaction term contributes only 0.8% to the total variance in y.

The bulk of forecast uncertainty, in both the short and medium run, stems from uncertainty in *exogenous variables* and *error terms* in behavioural equations. Considering that most (non-policy) exogenous variables relate to the international environment, it does not come as a surprise that exports and the current account are predominantly affected by uncertainty in exogenous variables. More interesting is the shift in the pattern of contributions to forecast error variance between the short and the medium term. While the errors in the behavioural equations are responsible for an important share of one year ahead forecast error for almost all variables of table 1, the four years ahead forecasts for wage and price inflation, the current account balance and the government surplus are hardly affected by such errors. Four years ahead, uncertainty in the (non-policy) exogenous variables is responsible for about 90% of the forecast error variance in inflation, the current account balance and the government surplus; probably the impact of observed uncertainty in oil prices is an important contributor to all three.

The autocorrelations imposed on errors in foreign prices are somewhat larger than those imposed on error terms of domestic price equations, which helps to explain the shift in emphasis to exogenous uncertainty for inflation. Probably more important are feedback mechanisms in the model which tend to stabilize the forecasts vis-à-vis errors in behavioural equations, while changes in the international environment have a more permanent effect. Also at one year ahead, the forecast for the government surplus is not very sensitive to the error terms of the model, the latter accounting for only 15% of the forecast error variance. The high sensitivity of the government budget to the external environment was also apparent in Brandsma *et al.* (1988).

4. Total forecast uncertainty: an evaluation

4.1 Computed versus observed forecast errors

Of course the Monte Carlo experiments not only allow us to study the relative importance of the different sources of uncertainty. They also provide standard forecast errors as computed on the basis of the four sources of uncertainty taken together. In this section we evaluate these computed standard errors and compare them with observed forecast errors. In addition, the impact of policy reactions is discussed.

The left-hand part of Table 2 gives standard errors of forecasts one year ahead and four years ahead. The first column reports standard errors as observed in the CPB track record by Westerhout (1990) and Hers (forthcoming). For the one year ahead forecast, Westerhout studied the September forecasts over the period 1962-1988. For the four year ahead forecast, Westerhout used nine medium term forecasts published between 1966 and 1986; Hers used nine medium term forecasts published between 1976 and 1990. To assess the 1990 forecast, the most recent short term outlook was used as 'preliminary realization'.

The second column of Table 2 gives the standard forecast errors as computed from our Monte Carlo experiments on the basis of the four sources of uncertainty. When comparing these with the observed standard errors in the first column, we should remember three reasons why they are different: (i) the computed standard errors are likely to overstate the true forecasting uncertainty of the model-cum-expert forecasting system; (ii) the computed standard errors are valid for conditional forecasts with given policy instrument values, while the observed errors include the effects of policy changes; and (iii) the computed standard errors refer to one particular model built in 1988, while observed errors relate to actual published forecasts made with different (and changing) tools.

Table 2. Forecast errors and impact of policy feedback

| | standard forecast error | | policy feedback |
	observed[a]	computed[b]	factor[c]
A One year ahead			
** forecast**			
real growth rates	*% per annum*		*ratios*
private consumption	2.3	2.3	1.43
business investment	6.4	9.6	1.11
non-energy exports	4.5	5.0	0.98
production enterprises	2.2	2.5	1.51
nominal growth rates	*% per annum*		
consumption price	1.6	2.5	1.22
wage rate enterprises	2.4	2.1	1.46
other indicators	*% NNP*		
government surplus	na	1.3	0
current account balance	na	2.6	0.82
B Four years ahead			
** forecast**			
real growth rates	*% per annum*		*ratios*
private consumption	1.7 (1.9)	1.5	2.79
business investment	3.9 (3.7)	4.0	2.22
non-energy exports	3.0 (1.9)	3.2	0.91
production enterprises	1.1 (1.6)	1.4	2.24
nominal growth rates	*% per annum*		
consumption price	1.7 (1.9)	3.5	1.48
wage rate enterprises	3.4 (1.8)	3.6	2.07
other indicators	*% NNP (end year)*		
government surplus	na (1.9)	6.0	0
current account balance	na (na)	7.0	0.54

[a] Based on Westerhout (1990); numbers between parentheses based on Hers (forthcoming).
[b] Based on Monte-Carlo exercises with four sources of uncertainty.
[c] Standard error of forecast with policy feedback, divided by standard error of forecast without policy feedback (based on uncertainty from exogenous variables and residuals only).

To shed some light on point (ii), the impact of policy changes, a 'policy feedback factor' is reported in the right-hand part of Table 2. As will be explained below, this number indicates the multiplication factor that can be applied to the computed standard errors of the second column, to obtain standard forecast errors for when the model is extended with two policy feedback rules that give a stylized impression of Dutch economic policy in the eighties. In practice, policy feedback is much more relevant for the medium term forecasts. The short term forecasts published in September are based on the final budget proposals which tend to be changed only little afterwards.

The one year ahead forecasts for real growth rates prove to carry larger margins of uncertainty than the four years ahead (average) real growth rates. For inflation, it is the other way around: the short term wage and price increases carry smaller margins than those at medium term range. Expert interference appears to improve in particular the short term forecast for investment and inflation. For the medium term forecast on inflation and wages, model changes that follow real world changes in wage determination and in cost shares should be quoted as a reason why observed forecast errors are smaller than those computed with the ZOEM model. The fact that Westerhout finds a larger medium term standard error on wages than Hers, relates to the different sample period.

4.2 The impact of policy reactions

The government surplus and the current account balance, both as a percentage of net national product, show a considerable increase in forecast uncertainty between the first and fourth forecast year. Because the government surplus is an important (intermediate) target of economic policy, the relatively large and increasing uncertainty on that variable is particularly disturbing. Precisely because it is an important target of economic policy, observed forecast errors on the government surplus are much smaller than the numbers given in the left hand part of Table 2: by adjusting economic policy when the target is in danger, the government acts to stabilize the result. More generally, one might expect that adding policy reaction functions to the model would reduce forecasting errors. In other words, the uncertainty in policy instruments is not independent of that from other sources. At least for policy targets, its contribution to forecast errors is likely to be negatively correlated with the other contributions.

Anderson and Enzler (1986) set out to design stabilizing policy reaction functions for their model of the US economy. In contrast, Van Vlimmeren et al. (1991d) studied the effects of a *policy feedback* mechanism which gives a stylized description of actual policy. In the eighties, economic policy in the Netherlands put strong emphasis on numerical targets for the government surplus and the burden of taxes and social security premiums. For our analysis of forecast errors this means that policy instruments, so far assumed to be known exogenously without error, in fact observe feedback rules that stabilize the result for these policy targets.

The right hand part of Table 2 reports on the effects on forecast error of incorporating two policy feedback rules in the model. The (stylized) rules fix the government surplus and the collective burden by adjusting government expenditure and tax rates; also the allowed housing rent increases are made to move with inflation. The effects of policy feedback were studied only in the context of forecast errors due to uncertainty in exogenous variables and error terms. As these two are the dominant sources of forecast errors, the reported policy feedback factors on standard forecast errors may be considered representative. The policy feedback factors reported in Table 2 indicate by how much the standard forecast error is multiplied as a result of policy feedback. Clearly the forecast error on the government surplus is annihilated, but most variables suffer an increase in forecast uncertainty rather than a decrease. The sensitivity of the wage rate and consumption forecast errors to policy reaction is confirmed in the *ex post* evaluation of the 1983-1986 forecast of the Central Planning Bureau, see Van den Berg (1986).

More generally, the destabilizing effect of these particular feedback rules are readily understood from the Keynesian mechanisms in the model. A fixed target for the government budget of course generates procyclical policies. However, it should be noted that the results are somewhat extreme, because our stylized policy rules abstract from any flexibility on the part of the policy maker as regards his targets and from his ingenuity in contriving policies that have optical effects on the budget but hardly any real effects on the economy.

5. Conclusion

The forecast errors of the Dutch Central Planning Bureau are similar in size to those reported for other forecasting institutions in economics, see e.g. Klein (1981), Wallis *et al.* (1984-1987) and NIESR (1992). The relatively large role of uncertainty in (non-policy) exogenous variables reflects the importance of the international environment for the very open Dutch economy. Standard errors of the model forecast are reduced somewhat by expert interference. Also, our sample included some hectic periods, with several surprises in the oil price. While our research gives no indications of chaotic behaviour in the national economy, realistic models for exchange rates and oil prices can generate chaos. See De Grauwe and Dewachter (1990) for a plausible chaos model of the exchange rate. In addition, the error terms in the behavioural equations might represent the outcome of chaotic processes. Surprises in exogenous variables and error terms will continue to dominate forecast uncertainty. At the same time, continuous efforts are required to keep the econometric models up to date and the contribution of error terms at current levels. In my opinion, therefore, it is unlikely that forecast errors can be significantly reduced.

While an objective yardstick is not available, it seems fair to say that, at least for Dutch economic policy makers, the forecast errors are uncomfortably large. The main purpose of CPB's forecasting activities is to provide a benchmark for economic policy preparation. The forecast uncertainty is such that the policy goals are strongly affected by normal forecast errors, hence the politicians must find a way to cope with the uncertainty. To help them, the recent medium term outlook (CPB, 1993) of the Planning Bureau contains two scenarios, one cautious and one more optimistic. The future course of the Dutch economy is likely to lie within these bounds, barring major unexpected events. The international economic environment is the major source of difference between the two scenarios. Policy assumptions are the same for both scenarios, defining a common 'no policy change' starting point for the policy debate. Policy options should be evaluated in both scenarios, to allow a systematic treatment of the most important risks and to develop contingency plans to deal with the future as it unfolds. 'No regrets' policies are those that prove useful in both scenarios. If the success of a policy option depends on the choice of scenario, the decision is best postponed until more information is available. If that is not possible, a strategic choice must be made which weighs the costs and benefits in both alternatives. Considering the ease with which pleasant surprises tend to be handled and the large difficulties triggered by bad luck, it seems wise to focus on the cautious scenario when drafting economic policy for the medium term.

Acknowledgements

The research project reported upon here was supported by the Foundation for the

Advancement of Research in Economic Science (ECOZOEK), a subsidiary of the Netherlands Organisation for the Advancement of Research (NWO).

This is an improved and extended version of a paper which was presented at the Macromodels'91 conference in Warsaw, December 1991, at the econometrics seminar of the University of Groningen, March 1992 and at the economics seminar of the University of Aarhus, November 1992. Comments of participants at those meetings and of colleagues at CPB are gratefully acknowledged. I am indebted to John Blokdijk for his skilful research assistance and his determination to get all the numbers right.

References

Anderson, R. and J.J. Enzler (1986), 'Toward realistic policy design: policy reaction functions that rely on economic forecasts', FRB research paper, Washington DC.

Berg, P.J.C.M. van den (1986), 'De betrouwbaarheid van macro-economische voorspellingen', *Economisch Statistische Berichten* **71**, 1153-1157.

Berg, P.J.C.M. van den, G.M.M. Gelauff and V.R. Okker (1988), 'The FREIA-KOMPAS model for the Netherlands: A quarterly macroeonomic model for the short and medium term', *Economic Modelling* **5**, 170-236.

Brandsma, A.S., G.M.M. Gelauff, B. Hanzon and A.M.A. Schrijver (1988), 'Retracing the preferences behind macroeconomic policy: the Dutch experience', *De Economist* **136**, 468-490.

CPB (1988), 'Waarom het miljoen werklozen er niet kwam'. Een "What-if"-simulatie van de middellange-termijnverkenning 1984-1987, CPB Werkdocument 24, Den Haag.

CPB (1992), '*FKSEC, a macro-econometric model for the Netherlands*', Stenfert Kroese, Leiden/Antwerpen.

CPB (1993), '*Centraal Economisch Plan 1993*', Sdu, Den Haag.

De Grauwe, P. and H. Dewachter (1990), 'A chaotic monetary model of the exchange rate', unpublished paper, University of Leuven.

Don, F.J.H., and P.J.C.M. van den Berg (1990), 'The Dutch Central Planning Bureau - its role in the preparation of economic policy', unpublished paper, CPB, Den Haag.

Gallo G.M. and F.J.H. Don (1991), 'Forecast uncertainty due to unreliable data', *Economic and Financial Computing* **1**, 49-69.

Gelauff, G.M.M., and V.R. Okker (1989), 'ZOEM: a condensed version of the FREIA-KOMPAS model of the Dutch economy', CPB Research memorandum 49, Den Haag.

Hers, J.F.P. (forthcoming), 'De kwaliteit van de middellange-termijnramingen van het Centraal Planbureau'.

Klein, L.R. (1981), 'Econometric Models as a guide to decision making', *The Charles C. Moskowitz Memorial Lectures no. XXII*, The Free Press, New York.

Lützel, H. (1989), 'The impact of GDP revisions on growth rates; an international comparison', paper presented at the twenty-first general conference of the International Association for Research in Income and Wealth.

NIESR (1992), 'Forecast error margins', *National Institute Economic Review*, November, p.16.

Rutten F.W. (1984), 'De betekenis van macro-econometrische modellen bij de beleidsvoorbereiding'. In: H. Den Hartog en J. Weitenberg (eds), *Toegepaste economie, grenzen en mogelijkheden*, CPB, Den Haag.

Vlimmeren, J.C.G. van, F.J.H. Don and V.R. Okker (1991a), 'Modelling data uncertainty due to revisions', CPB Research memorandum 75, Den Haag.

Vlimmeren, J.C.G. van, F.J.H. Don and V.R. Okker (1991b), 'Composition and pattern of the forecast uncertainty due to unreliable data: further results', CPB Research memorandum 81, Den Haag.

Vlimmeren, J.C.G. van, F.J.H. Don and V.R. Okker (1991c), 'Uncertain parameters and forecast errors: an empirical study of parameter sensitivity', unpublished paper, CPB, Den Haag.

Vlimmeren, J.C.G. van, F.J.H. Don and V.R. Okker (1991d), 'Policy feedback and forecast uncertainty', unpublished paper, CPB, Den Haag.

Wallis, K.F. (ed.) et al. (1984-1987), 'Models of the UK Economy'. *Reviews by the ESRC Macroeconomic Modelling Bureau*, Oxford University Press.

Westerhout, E.W.M.T. (1990), 'Voorspelling en realisatie', Discussienota 9001, Ministerie van Economische Zaken, Den Haag.

SOME ASPECTS OF NONLINEAR DISCRETE-TIME DESCRIPTOR SYSTEMS IN ECONOMICS

TH. FLIEGNER[1], H.NIJMEIJER[1] & Ü. KOTTA[2]

[1]Department of Applied Mathematics, University of Twente
P.O.Box 217, 7500 AE Enschede, The Netherlands
Fax:+31-53-340733
E-mail: fliegner@math.utwente.nl or twhenk@math.utwente.nl

[2]Institute of Cybernetics,~Estonian Academy of Sciences
Akademia tee 21, Tallinn, Estonia
Fax:3722-527-901

Summary

In this paper we study nonlinear discrete-time descriptor (or singular, implicit, general) systems. Some of the problems connected with such systems, which arise frequently in modelling certain classes of economic relationships, are questions concerning existence and uniqueness of solutions. By means of a step by step procedure we will give, under generic conditions, a local reduction mechanism yielding a possibly lower dimensional system in standard state space form.

Keywords Nonlinear systems, descriptor variables, discrete-time systems, economic modelling.

1. Introduction

Descriptor systems, wether linear or nonlinear, in continuous or discrete time, arise in a variety of situations and have, therefore, received considerable attention in the last decades. Among others, one encounters such systems in the study of large-scale interconnected systems, noncausal systems, networks, and economic systems. Especially in the latter case, the so-called *descriptor approach to modelling*, leads to systems of this form. The crucialpoint of this approach is that, first, all variables are determined which are considered to have some importance to the description of the system and which are, therefore, referred to as *descriptor variables*, a concept introduced by Luenberger (see Luenberger, 1977). In terms of (macro)economic models, such variables could be the gross domestic product, consumption, investments, imports and exports of goods and services, inflation rate etc.. The process of modelling then consists of relating these variablesaccording to the system laws. Usually, this leads to a combination of dynamic and static equations in the model of the system. In macroeconomic modelling, one in general divides the descriptor variables into so-called *endogenous* and *exogenous variables*. Endogenous are those variables for which the values are simultaneously determined by the model and for which the model is designed to explain. The exogenous variables are variables for which the values are determined outside the model but which influence the model. In general the exogenous variables are either historically given, policy variables (or *instruments*, or determined by some separate mechanism (see e.g. Intriligator, 1978). A usual structural form of the model is then

$$F(x(k + 1), x(k), q_{np}(k), q_p(k)) = 0 ,$$ (1)

where x denotes the vector of endogenous variables, q_{np} and q_p are the non-policy and policy (exogenous) variables, respectively, and k denotes time. A control aspect enters in that a *policy maker* can make use of the instruments to achieve certain goals for (part of) the endogenous variables or combinations of them referred to as *targets*. We will denote them by y. In literature, a frequently considered case is where one tries to influence all of the endogenous variables simultaneously in a desired way, although this situation is not of great relevance in practice as argued in Wohltmann and Krömer (1984). A more realistic situation for the target variables y is described by the following equation:

$$H(y(k), x(k), q_{np}(k), q_p(k)) = 0 .$$ (2)

The above mentioned case then corresponds to

$$H(y(k), x(k), q_{np}(k), q_p(k)) = y(k) - x(k) = 0 .$$ (3)

We would like to emphasize that the procedure of modelling described here, differs in its methodology from the now in system theory generally accepted so-called *behavioural approach* to modelling (see e.g. Willems, 1989; Willems, 1991). In particular, (1,2) is not necessarily what is called an *input-output (instrument-target) system*. It is, however, the by now usual one as far as economics is concerned. Clearly, given a set of time functions $(\bar{x},(k), \bar{q}_{np}(k), \bar{q}_p(k), \bar{y}(k))$, the equations (1,2) can be used to check wether or not this set satisfies the system laws. For control purposes, however, the model equations should preferrably have an evolutionary character as it is expressed in a *standard state space model* (see Aoki, 1976). But also (1,2) itself can be thought of as a rule producing $x(k+1)$ and $y(k)$ iteratively. As such, despite of all carefullness in the modelling process, one can come up with a model which is not well-behaved in that it may for instance not be possible to determine the endogenous variables at time $k + 1$ given the endogenous and the exogenous variables at time k. This is another reason why it would be desirable to have sufficient conditions under which a descriptor system is equivalent to a system in standard state space form. By now, there exist numerous results on the well-posedness of linear descriptor systems (see e.g. Luenberger, 1977; Luenberger, 1978; Dai, 1978) which also provide methods to actually obtain a solution. To our best knowledge, Luenberger was also the only author who considered nonlinear descriptor systems, restricted to the finite horizon case (see Luenberger, 1979). In connection with the representation of linear systems in descriptor form from a behavioural point of view, Kuijper (1992) is an interesting reference (see also Nieuwenhuis and Willems, 1991). Here, conditions can be found under which a linear descriptor system is equivalent to a system in standard state space form. It turns out that our conditions can be viewed as the nonlinear extension of them. A feature of all those contributions is that they rely in one way or another on rank conditions of Jacobian matrices. We believe that considerations of the following type can be helpful in understanding and analyzing economic models.

2. Definitions and Problem formulation

We consider nonlinear discrete-time descriptor systems of the form

$$f(x(k + 1), x(k), q(k)) = 0 , \qquad (4)$$

$$h(y(k), x(k), q(k)) = 0 , \qquad (5)$$

where the endogenous variables $x(\cdot)$ belong to an open part X of $\mathbb{R}^n$, the instruments q $(\cdot)$ are in an open part Q of $\mathbb{R}^m$ and the targets y belong to an open set Y of $\mathbb{R}^p$. Observe that we assume all exogenous variables to be policy variables. We shall come back to the more general situation later. The mappings $f\colon X \times X \times Q \to \mathbb{R}^n$ times and: $Y \times X \times Q \to \mathbb{R}^p$ are supposed to be smooth, although everything remains valid for $f, h \in C^l$ with l sufficiently large. Moreover, we shall work in a neighbourhood of an equilibrium point (x_e, q_e, y_e) of the system (4,5), that is $f(x_e, x_e, q_e) = 0$ and $h(h_e, x_e, q_e) = 0$ (and so we implicitly assume the existence of an equilibrium solution). In what follows, we list some situations in which the system (4,5) is equivalent to a system in standard state space form

$$\tilde{x}(k + 1) = \tilde{f}(\tilde{x}(k), q(k)) , \qquad (6)$$

$$\tilde{y}(k) = \tilde{h}(\tilde{x}(k), q(k)) \qquad (7)$$

possibly with $\dim \tilde{X} < \dim X$ and $\dim \tilde{X} < \dim Y$ where $\tilde{X}$ and $\tilde{Y}$ are the spaces where $\tilde{x}$ and $\tilde{x}$ take their values. In equation (7), $\tilde{y}(k)$ denotes some part of the original (coordinate transformed) target variables. The missing part of $y(k)$ in $\tilde{y}(k)$ is functionally depending on $\tilde{y}(k)$. The importance of this equivalence is due to the fact that assuming only that $\tilde{f}$ and $\tilde{h}$ are well defined, we have an unique evolution of $\tilde{x}$ (which is also simple to determine) and $\tilde{y}$ for each set of initial conditions and sequence of instruments.

3. Sufficient Conditions for the Equivalence between Descriptor Systems and Standard State Space Systems

In what follows, we are concerned with some cases for which the aforementioned equivalence can be shown.

3.1 Case 1

Consider the following special case of system (4,5)

$$f(x(k + 1), x(k), q(k)) = 0, \quad f(x_e, x_e, q_e) = 0 , \qquad (8)$$

$$y(k) = h(x(k), q(k)) ,$$ (9)

so (5) is in this situation already in the desired form (7). Let us assume that in a neighbourhood of the equilibrium point (x_e, q_e, y_e), $(y_e = h(x_e, q_e))$, the $n \times n$ Jacobian matrix $D_1f(x(k + 1), x(k), q(k))$ has constant rank $r \leq n$. In case $r = n$, f can locally be solved for $x(k+1)$ reducing (8,9) to a system in state space form with state space X. If $r < n$, permute if necessary, the components of $f(\cdot)$ and $x(k+1)$ such that the $r \times r$ matrix

$$D_1f_1(x_1(k+1), x_2(k + 1), x_1(k), x_2(k), q(k))$$

has rank r around (x_e, x_e, q_e) where f_1 and x_1 denote the first r components of f and x respectively after permutation (correspondingly, f_2 and x_2 denote the remaining $n - r$ components). As usual, D_jg denotes the derivative of a function g with respect to its j-th argument. Note, that all transformations of $x(k + 1)$ have to be performed with respect to $x(k)$ in an analogous way. So, f can be partitioned into the following two equations

$$f_1(x_1(k + 1), x_2(k + 1), x_1(k), x_2(k), q(k)) = 0 ,$$ (10)

$$f_2(x_1(k + 1), x_2(k + 1), x_1(k), x_2(k), q(k)) = 0$$ (11)

with $rank\ D_1f(x(k + 1), x(k), q(k)) = rank,\ D_1f_1(x_1(k + 1), x_2(k), q(k)) = r$ locally around (x_e, x_e, q_e). The latter implies that the last $n - r$ components f_2 of f are functionally dependent on the first r components f_1 of f and can, therefore, be expressed as a function of $f_1(\cdot)$, that is

$$f_2(z,x,q) = \xi(f_1(z,x,q),x,q) .$$ (12)

Since $f_1(\cdot) = 0$ along solutions of (8), we finally obtain for solutions along (8)

$$f_2(x(k + 1), x(k), q(k)) = \phi(x(k), q(k))$$ (13)

and (10,11) transform into

$$f_1(x_1(k + 1), x_2(k + 1), x_1(k), x_2(k)) = 0 ,$$ (14)

$$\phi(x_1(k), x_2(k), q(k)) = 0 .$$ (15)

Now, the assumption $rank\ D_1f_1(x_1(k + 1), x_2(k + 1)\ ,\ x_1(k), x_2(k), q(k)) = r$ ensures the existence of a coordinate transformation $\tilde{x} = S(x)$ such that in the new coordinates (14,15) take the form

$$\bar{f}_1(\tilde{x}_1(k + 1), \tilde{x}_1(k), \tilde{x}_2(k), q(k)) = 0 , \tag{16}$$

$$\bar{\phi}(\tilde{x}_1(k), \tilde{x}_2(k), q(k)) = 0 . \tag{17}$$

For more detailed information about the existence of a coordinate transformation with the required properties, see e.g. Spivak (1970). The main feature here is that, after transformation, f_1 is only depending upon the first r components of $\tilde{x}(k + 1)$. If we now assume that $D_2\bar{\phi}(\tilde{x}_1(k), \tilde{x}_2(k), q(k))$ has rank $n - r$ in $(\tilde{x}_e, q_e)$, we can solve (17) for $\tilde{x}_2(k)$ yielding

$$\tilde{x}_2(k) = \tilde{\phi}(\tilde{x}_1(k), q(k)) . \tag{18}$$

Observe that in case (17) is solvable for $\tilde{x}_2(k)$, at least locally no restrictions are imposed on the instruments $q(k)$. Substituting (18) into (16), we obtain

$$\bar{f}_1(\tilde{x}_1(k + 1), \tilde{x}_1(k), \tilde{\phi}(\tilde{x}_1(k), q(k)), q(k)) = 0 , \tag{19}$$

which can finally be solved for $\tilde{x}_1(k + 1)$ yielding a reduced system in standard state space form

$$\tilde{x}_1(k + 1) = \hat{f}_1(\tilde{x}(k), q(k)) , \tag{20}$$

$$y(k) = \bar{h}(\tilde{x}_1(k), q(k)) . \tag{21}$$

Remarks
(i) For linear systems, the assumption of an explicit output equation is not very restrictive because it can be shown that every linear system which allows for an *autoregressive* (AR) representation can be expressed in the form

$$Ex(k + 1) = Ax(k) + Bu(k) , \tag{22}$$

$$y(k) = Cx(k) + Du(k) , \tag{23}$$

when inputs and outputs are selected at the outset (see Kuijper, 1992).
(ii) In a linear setting, the assumptions used in the previous section are strongly related to the input-output structure of the system.
(iii) The assumption that all exogenous variables are instruments is certainly not very realistic. The question arises what consequences the existence of (time-varying) non-policy exogenous variables has for the procedure just described. First of all, this implies that we can no longer restrict ourselves to a neighbourhood of an equilibrium point. A second problem is caused by the fact that it is often necessary to consider time as exogenous variable as well. This way, time enters the model equations explicitly. A well known model of that type is for instance the *Klein-Goldberger Model* (see e.g. Adelman and Adelman,

1959). The explicit time dependence can be removed by introducing an additional equation $x_{n+1}(k + 1) - x_{n+1}(k) - 1 = 0$ in (1) treating time as extra endogenous variable. With this manipulation, (1,2) become time-invariant. In order to overcome the nonexistence of equilibrium points in case of the presence of (time-varying) non-policy variables, one can perform the procedure above around some specific trajectory of (1,2), that is some specific set of time functions $(\bar{x}(k), \bar{q}_{np}(k), \bar{q}_p(k), \bar{y}(k))$ that satisfies these equations; this procedure has been used successfully in a slightly different context in Maas and Nijmeijer (1993).

(iv) The procedure uses coordinate transformations. From a practical point of view, this is usually not desired because it disguises the meaning of the descriptor variables and their relations among each other. If one is not going to accept this drawback, the test of the constant rank conditions, possibly numerically, can still be used to check the behaviour of the model. Once this is done, one can continue with the original variables. On the other hand, performing the reduction (if possible!) offers the possibility to apply all the results of (non)linear control theory. Hence, in a concrete situation, there has to be made a decision which aspect is preferred.

3.2 Example

Consider the following example.

$$f(x(k + 1),x(k),q(k)) = \begin{pmatrix} f_1(x(k + 1),x(k),q(k)) \\ f_2(x(k + 1),x(k),1(k)) \end{pmatrix} =$$

$$\begin{pmatrix} e^{x1(k + 1)} + x_2(k + 1) - q(k)x_2(k) - 1 \\ 2e^{x_1(k + 1)} + 2x_2(k + 1) + e^{x_1(k)} + q(k) - 3 \end{pmatrix} = 0, f(0,0,0) = 0, \tag{24}$$

where $x = (x_1, x_2)$. Since an output equation does not play any role in our considerations so far, we omit it. We decided to use a rather artificial example because here one can easily see what is happening. Now, we obviously have

$$rank\ D_y f(x(k + 1),x(k),q(k)) = rank\ D_y f_1(x_1(k + 1),x_2(k + 1),x_1(k),x_2(k)q(k)) = 1 \tag{25}$$

in a neighbourhood of $(0,0,0)$ that means, no permutations are necessary. Moreover, we can express f_2 in terms of f_1 in the following way

$$f_2(z,x,q) = 2f_1(z,x,q) + 2qx_2 + e^{x1} - x_2 + q - 1 . \tag{26}$$

Restricting ourselves to solutions of (8), we obtain

$$f_2(x(k + 1),x(k),q(k)) = (2q(k) - 1)x_2(k) + e^{x_1(k)} + q(k) - 1 = \phi(x(k),q(k)) = 0 . \tag{27}$$

We apply the following coordinate transformation $\tilde{x} = S(x)$ given by

$$\begin{aligned}\tilde{x}_1 &= e^{x_1} + x_2 \\ \tilde{x}_2 &= x_2\end{aligned} \quad \text{with inverse} \quad \begin{aligned} x_1 &= \ln(\tilde{x}_1 - \tilde{x}_2) \\ x_2 &= \tilde{x}_2 \end{aligned} \tag{28}$$

In the new coordinates, we are working around the transformed equilibrium point $(\tilde{x}_e, \tilde{x}_e, q_e)$ $= (S(0),S(0),0) = ((1,0),(1,0),0)$. Performing the transformation, we obtain

$$\bar{f}_1(\tilde{x}_1(k+1),\ \tilde{x}(k),q(k)) = \tilde{x}_1(k+1) - q(k)\tilde{x}_2(k) - 1 = 0 , \tag{29}$$

$$\bar{\phi}(\tilde{x}(k),\ q(k)) = (2q(k) - 2)\tilde{x}_2(k) + \tilde{x}_1(k) + q(k) - 1 = 0 . \tag{30}$$

It is easily seen that $rank\ D_2\bar{\phi}(\tilde{x}_1(k),\ \tilde{x}_2(k),\ q(k)) = 1$ in $((1,0),0)$ so that we can locally solve for $\bar{\phi}$ for $\tilde{x}_2(k)$ to obtain

$$\tilde{x}_2(k) = \frac{-(q(k) + \tilde{x}_1(k) - 1}{2q(k) - 1} . \tag{31}$$

Replacing $\tilde{x}_2(k)$ in (29) by (31) we finally get

$$\tilde{x}_1(k+1) = \bar{f}(\tilde{x}_1(k),1(k)) = 1 - \frac{q(k)q(k) + \tilde{x}_1(k) - 1}{2q(k) - 2} . \tag{32}$$

$\square$

Remark. Some typical properties for descriptor systems which satisfy the assumptions above can immediately be recognized by this example. Firstly, the system can not be initialized arbitrarily but $x(k_0)$ and $q(k_0)$ have to be chosen in such a way that (27) is satisfied, where k_0 denotes the initial time. Secondly, once this is done, (32) ensures the unique evolution of some combination (given by the coordinate transformation (S) of the original endogenous variables. They can be reconstructed using (28) and (30) if desired. Since the methods used in the sequel parallel very much those in *Case 1*, the following will not be in full detail.

3.3 Case 2

In this section we consider descriptor systems where also the output map is given implicitly.

$$f(x(k+1),\ x(k),\ q(k)) = 0, \quad f(x_e,x_e,q_e) = 0 , \tag{33}$$

$$h(y(k),\ x(k),\ q(k)) = 0, \quad h(y_e,x_e,q_e) = 0 . \tag{34}$$

Again all exogenous variables are supposed to be instruments. In addition to $rank\ D_1f(x(k+1),\ x(k),\ q(k)) = r_1$ locally around $(x_e,\ x_e,\ q_e)$, we assume $rank\ D_1h(y(k),x(k),q(k)) = r_2$

locally around (y_e, x_e, q_e). As before, $r_1 = n$ and $r_2 = p$ implies solvability of f and h for $x(k + 1)$ and $y(k)$ respectively. If only $r_2 = p$ holds, we are in the situation of *Case 1* after applying the *Implicit Function Theorem* to (34).

Case 2.1

We therefore assume first that $r_1 = n$ and $r_2 < p$. Then, f can be solved for $x(k + 1)$

$$x(k + 1) = \tilde{f}(x(k)) \tag{35}$$

and we are only left with an implicit output equation. This equation can now be manipulated in a similar way as f in *Case 1*. That means, we split up h into

$$h_1(y_1(k), y_2(k), x(k), q(k)) = 0 , \tag{36}$$

$$\psi(x(k), q(k)) = 0 \tag{37}$$

by performing analogous steps leading to (14,15). Moreover, we can again find a coordinate transformation $\tilde{y} = T(y)$ such that in the new coordinates (36) turns to

$$\bar{h}_1(\tilde{y}_1(k), x(k), q(k)) = 0 , \tag{38}$$

which can be solved for $\tilde{y}_1(k)$ yielding

$$\tilde{y}_1(k) = \tilde{h}_1(x(k), q(k)) . \tag{39}$$

We are now in the position to incorporate the constraints (37) into (35) and (39). This can be done if there are $p - r_2$ variables $x_1, ..., x_{p-r2}$ (possibly after permutation) such that

$$\frac{\partial \psi(x,q)}{\partial(x_1, ..., x_{(p-r_2)})} \Big|_{(x,q)}$$

has rank $p - r_2$. Then $x_1, ..., x_{p-r2}$ can be replaced in (35) and (39) via solving (37), so completing the reduction.

Case 2.2

The last and most involved situation we want to deal with arises from $r_1 < n$ and $r_2 < p$. Here we cannot make (in general) statements with respect to the solvability of neither equation (33) nor (34), except for the equilibrium. It is therefore not possible to apply *Case 1* or *Case 2.1* directly. What we can do is using a combination of them. Under the assumptions made and on the analogy of *Case 1* and *Case 2.1* we can find coordinate transformations $\tilde{x} = S(x)$ and $y = T(y)$ such that (33,34) turn into

$$\bar{f}(\tilde{x}_1(k + 1), \tilde{x}_1(k), \tilde{x}_2(k), q(k)) = 0 , \tag{40}$$

$$\bar{\phi}(\tilde{x}_1(k), \tilde{x}_2(k), q(k)) = 0 , \tag{41}$$

$$\bar{\psi}(\tilde{x}_1(1), \tilde{x}_2(k), q(k)) = 0 , \tag{42}$$

$$\bar{h}_1(\tilde{y}_1(k), \tilde{X}_1(k), \tilde{x}_2(k), q(k)) = 0 \tag{43}$$

after transformation. Now, the question is again to what extent $\tilde{x}_2(k)$ can be replaced in (40) by solving (41,42) for $x_2(k)$ and substituting into (40). Note, that in this case we have more equations to our disposal than necessary in order to solve for $x_2(k)$. This implies that by solving (41,42) only for $x_2(k)$ (if possible) not all the constraints (41,42) can be incorporated into (40,43). A reduction of (33,34) to the form (6,7) is therefore only possible in case we can select $(n - r_1) + (p - r_2)$ components of $\tilde{x}(k)$ (which have to include $\tilde{x}_2(k)$ for which (41,42) can be solved. Substituting these components into (40,43) and solving (40) for $\tilde{x}_1(k + 1)$ and (43) for $\tilde{x}(k)$ completes the reduction.

4. Final Remarks

Implicit system equations have been studied for continuous time systems. For linear implicit systems, the theory for continuous time and discrete time systems is more or less the same (see e.g. Kuijper, 1992). The problem becomes quite more involved in the nonlinear setting and in particular a specific (general) case has been treated in some detail. These are the so-called nonholonomic systems in which a set of differential-algebraic equations describes the evolution of the system, see e.g. Neimark and Fufaev (1972). A discrete time analogon of this occurs if the Jacobian matrix of for instance (17) with respect to $\tilde{x}_2(k)$ has rank less than $n - r$ in $(\tilde{x}_e, q_e)$ or similarly if not all of the constraints in *Cases 2.1,2.2* can be incorporated into input equation and output equation, respectively. This situation deserves some further investigation in the future following Reyhanoglu (1992) because it is this situation which is in our opinion the most restrictive point of our approach. We expect that regulation of economic systems (as for instance in Maas and Nijmeijer, 1993) can be done for the *Cases 1, 2.1 and 2.2*. This will be a topic for further research.

References

Adelman, I. and F.L. Adelman (1959), 'The Dynamic Properties of the Klein-Goldberger Model', *Econometrica*, **27(4)**, 596-625.

Aoki, M. (1976), 'Optimal Control and System Theory in Dynamic Economic Analysis', American Elsevier Publishing Co., New York.

Dai, L. (1989), 'Singular Control Systems', *Lecture Notes in Control and Information Science*, **118**, Springer.

Intriligator, M.D. (1978), 'Econometric Models, Techniques and Applications', *Advanced Textbooks in Economics*, C.J.Bliss and M.D.Intriligator, eds., Prentice-Hall.

Kuijper, M. (1992), 'First-order Representations of Linear Systems', Ph.D. Thesis, University of Brabant, Tilburg.

Luenberger, D.G. (1977), 'Dynamic Equations in Descriptor Form', *IEEE Trans. Automat. Control*, **22(3)**, 312-321.

Luenberger, D.G. (1978), 'Time-Invariant Descriptor Systems', *Automatica*, **14**, 473-480.

Luenberger, D.G. (1979), 'Non-Linear Descriptor Systems', *Journal of Economic Dynamics and Control*, **1**, 219-242.

Maas, W.C.A. and H. Nijmeijer (1993), 'Dynamic path controllability in economic models: from linearity to nonlinearity', to appear in Journal of Economic Dynamics and Control,

Neimark, J.I. and F.A. Fufaev (1972), 'Dynamics of Nonholonomic Systems', *AMS Translations of Mathematical Monographs*, **33**.

Nieuwenhuis, J.W. and J.C. Willems (1991), 'On the Nature of Descriptor Systems', *Kybernetica*, **27(3)**, 282-288.

Reyhanoglu, M. (1992), 'Control and Stabilization of Nonholonomic Dynamic Systems', Ph.D. Thesis, University of Michigan, Ann Arbor, USA.

Spivak, M. (1970), 'A comprehensive introduction to differential geometry', **1**, Publish or Perish, Boston.

Willems, J.C. (1989), 'Models for Dynamics', In: *Dynamics Reported*, **2**, U. Kirchgraber and H.O. Walther, eds., Teubner and Wiley & Sons, Stuttgart.

Willems, J.C. (1991), 'Puzzels and Paradigms in the Theory of Dynamical Systems', *IEEE Trans. Automat. Control*, **36(3)**, 259-294.

Wohltmann, H.W. and W. Krömer (1984), 'Sufficient Conditions for Dynamic Path Controllability of Economic Systems', *Journal of Economic Dynamics and Control*, **7**, 315-330.

QUASI-PERIODIC AND STRANGE, CHAOTIC ATTRACTORS IN HICK'S NONLINEAR TRADE CYCLE MODEL

CARS H. HOMMES

Dept. Economic Statistics, University of Amsterdam,
Roetersstraat 11, NL-1018 WB Amsterdam, The Netherlands.

Abstract

Hicks' nonlinear trade cycle model is an unstable multiplier-accelerator model together with an 'income-ceiling' and an 'investment-floor'. We show that the simplest, 2-D version of the model can have a quasi-periodic attractor. When consumption and/or investment is distributed over several time periods higher dimensional versions of the model are obtained. We present numerical evidence that the 3-D Hicks-model can have strange, chaotic attractors.

keywords: nonlinear business cycle model, chaos, attractors.

1. Introduction

In the last decade in economics there has been a rapidly growing interest in nonlinear dynamic models exhibiting chaotic dynamical behaviour. For a survey and examples see e.g. Lorenz (1989) and Medio (1992). The use of nonlinear dynamic economic models however is not new and dates back to Kaldor (1940), Hicks (1950) and Goodwin (1951). In the fifties economists had not learned about chaos yet and the 'classical' nonlinear economic models focussed on regular, periodic behaviour rather than irregularity and chaos. However, in this paper we present numerical evidence that the 'classical' Hicks-model can have a strange, chaotic attractor.

Hicks' nonlinear trade cycle model, as introduced by Hicks (1950), is an unstable linear multiplier-accelerator model together with an income 'ceiling' (full employment upper bound) and an investment 'floor' (investment lower bound). The Hicks-model is a piecewise linear dynamic model generating cycles. The simplest version of the Hicks-model is a 2-dimensional model, which we will consider in section 2. Extra time lags in the consumption and/or the investment equation of the model lead to higher dimensional versions. In section 3 we will investigate the 3-D Hicks-model. We concentrate on the following question: *does each time path in Hicks' nonlinear trade cycle model converge to a periodic time path?*

2. The elementary Hicks-model

Hicks' nonlinear trade cycle model is perhaps the simplest nonlinear model of the business cycle, and can be found in many textbooks on economic dynamics, see e.g. Allen (1965) or Blatt (1983). The simplest version of the model is given by the following 4 equations:

consumption $\qquad\qquad\qquad C_t = mY_{t\text{-}1},$ $\qquad\qquad\qquad\qquad\qquad$ (1)

total investment $\qquad\qquad\quad I_t = I_t^{ind} + I^{aut},$ $\qquad\qquad\qquad\qquad$ (2)

induced investment $\qquad\quad I_t^{ind} = \max\{a(Y_{t\text{-}1} - Y_{t\text{-}2}), -I^f\},$ $\qquad\qquad$ (3)

income $\qquad\qquad\qquad\qquad Y_t = \min\{C_t + I_t, Y^c\}.$ $\qquad\qquad\qquad\qquad$ (4)

According to the first equation current consumption C_t is proportional to previous income $Y_{t\text{-}1}$, with m the marginal propensity to consume, $0 < m < 1$. Equation (2) states that total investment I_t equals constant autonomous investment I^{aut} plus induced investment I_t^{ind}.

Equations (3) and (4) are the two nonlinear, or more precisely the piecewise linear equations of the model. According to (3) induced investment is proportional to the growth in national income, as long as it is larger than the *investment floor* $-I^f$. The negative net investment $-I^f$ corresponds to zero gross investment. The parameter a is the accelerator. Based on economic data Hicks observed that the accelerator $a > 1$. Finally, according to equation (4) income Y_t equals consumption plus investment, as long as it is smaller than the (full employment) *income ceiling* Y^c. We call the model given by (1-4) *the elementary Hicks-model*. Substituting (1-3) into (4) a piecewise linear second order difference equation, describing the dynamics of income is obtained:

$$Y_t = \min\{mY_{t\text{-}1} + \max\{a(Y_{t\text{-}1} - Y_{t\text{-}2}),-I^f\} + I^{aut}, Y^c\}. \qquad\qquad (5)$$

Equation (5) has a unique equilibrium $Y^e = I^{aut}/(1-m)$ and by assumption $Y^e < Y^c$. A simple computation shows that the equilibrium Y^e is unstable, since the accelerator $a > 1$. Writing $x_t = Y_t$ and $y_t = Y_{t\text{-}1}$ (5) is transformed into:

$$\begin{aligned} x_{t+1} &= \min\{mx_t + \max\{a(x_t - y_t),-I^f\}, Y^c\} \\ y_{t+1} &= x_t. \end{aligned} \qquad\qquad (6)$$

Consequently, the elementary Hicks-model is given by a difference equation

$$(x_{t+1},y_{t+1}) = H(x_t,y_t),$$

where H is the 2-dimensional piecewise linear map

$$H(x,y) = (\min\{mx + \max\{a(x - y),-I^f\} + I^{aut},Y^c\}, x) \qquad\qquad (7)$$

The map H has an unstable equilibrium $E = (Y^e,Y^e)$, and one can easily show that all time paths are bounded. We investigate the following question: *Does a time path in the elementary Hicks' model always converge to a periodic time path?* Although the Hicks-model can be found in many textbooks on economic dynamics, it seems that this question has not been answered previously.

The long term behaviour of a dynamical system is determined by the attractors of that system. In order to define an attractor we first introduce the notion of an attracting set. A compact set A is called an *attracting set* of a map f if (1) A is f-invariant, that is $f(A) \subset A$, and (2) there exists a neighbourhood U of A, such that for all x in U, the distance $d(f^n(x),A)$ tends to zero as n tends to infinity. An attracting set A is called an *attractor* if there exists a point x in A such that the orbit of x is dense in A. Simple examples of attractors are a stable equilibrium and a stable periodic orbit. More complicated examples are a quasi-periodic, a chaotic or a strange attractor.

Before we state the main result, we recall the definition of a quasiperiodic attractor. Let $F: \mathbf{R}^m \rightarrow \mathbf{R}^m$ be a map, and A an attracting set of F. The set A is called a *quasi-periodic attractor* of the map F, if $F: A \rightarrow A$ is topologically conjugate to an irrational rotation R_α, that is, there exists a homeomorphism h from A to the circle, such that $R_\alpha oh = hoF$.

The main result concerning the dynamics of the elementary Hicks-model is:

Theorem *For $a > 1$ the map H in (7) has an attracting set A, which is homeomorphic to a circle. All orbits, except for the unstable equilibrium E, are attracted to the attracting set A. For the dynamics on the attracting set A, one of the following two possibilities occurs:*
(1)   The map $H: A \rightarrow A$ is topologically conjugate to a piecewise linear (orientation preserving) circle homeomorphism, or
(2)   The map $H: A \rightarrow A$ is topologically conjugate to a piecewise linear, non-decreasing circle map f of degree 1. There exists a circle arc I where f is constant, while f is strictly increasing on the complement of I.

A proof of the theorem can be found in Hommes (1991). Here we only give a brief intuitive explanation of the result to the mathematically less informed reader. The theorem states that all time paths are attracted to a piecewise linear closed curve (the attracting set A). The unstable equilibrium lies inside A. The time paths can be described as a cyclic movement around the unstable equilibrium, along the attracting set. According to the theorem this cyclic movement is equivalent (topologically conjugate) to the dynamics of an orientation preserving circle map. This means that the dynamics on the attracting set is regular. In particular, all time paths in the elementary Hicks-model rotate with the same average rotation around the unstable equilibrium. We call this unique average rotation the rotation number $\rho(H)$ of the elementary Hicks-model. When the rotation number is rational, all time paths converge to a (stable) periodic orbit. However, the rotation number may also be irrational. In that case the time paths (except for the equilibrium) are aperiodic, that is, they are not periodic and they do not converge to a periodic time path. In particular we have:

Corollary *The elementary Hicks-model can have a quasi-periodic attractor.*

An example of a quasi-periodic attractor (or a periodic attractor with a very long period) is shown in figure 1. The quasi-periodic attractor is a piecewise linear closed curve with the equilibrium lying inside. All orbits, except for the equilibrium, are dense in the attractor.

Although aperiodic time paths can occur, the above theorem implies that the dynamics of the elementary Hicks-model is always regular. The time paths are periodic or almost periodic and all time paths rotate with the same average rotation around the unstable equilibrium. Chaotic dynamics does not occur in the simplest version of Hicks' nonlinear trade cycle model.

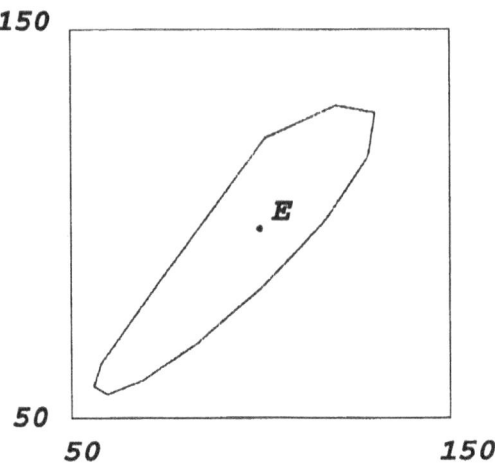

Figure 1. A quasi-periodic attractor in the 2-D Hicks-model. The parameters are $Y^c = 150$, $I^f = 10$, $I^{aut} = 20$, $m = 0.75$ and $a = 1.25$. The figure shows 5000 points of the orbit of $(x_0, y_0) = (Y^c, Y^c)$ after a transient of 10 time periods.

3. The 3-D Hicks-model

In the elementary Hicks-model all consumption lags one period behind income, while all induced investment outlays are concentrated in the time period immediately following the originating change in income. Hicks (1950) already posed the following two questions: (1) *what difference does it make when some consumption is lagged more than one period behind income?* (2) *what difference does it make if induced investment is postponed or distributed over a number of periods?* With distributed consumption and/or investment, higher dimensional versions of the Hicks-model are obtained. When consumption is distributed over the three time periods following income and induced investment is distributed over the two time periods following the change in income, the Hicks-model is a 3-dimensional piecewise linear model. The 3-D Hicks-model is in fact the simplest version of the model, after the elementary Hicks-model. In this section we investigate whether or not the regularity in the dynamics of the elementary Hicks-model also holds for the 3-D version. In particular we investigate the following fundamental problem: *Can Hicks' nonlinear trade cycle model with distributed consumption- and/or investment-lags, generate chaotic time paths?*

The 3-D *Hicks-model* is given by the following 4 equations:

$$C_t = m_1 Y_{t-1} + m_2 Y_{t-2} + m_3 Y_{t-3}, \tag{8}$$

$$I_t = I_t^{ind} + I^{aut}, \tag{9}$$

$$I_t^{ind} = \max\{a_1(Y_{t-1} - Y_{t-2} + a_2(Y_{t-2} - Y_{t-3}), -I^f\}, \tag{10}$$

$$Y_t = \min\{C_t + I_t, Y^c\}. \tag{11}$$

The parameters m_1, m_2 and m_3 are called the partial consumption coefficients, while a_1 and a_2 are called the partial investment coefficients. Substituting (8-10) into (11) and writing $x_t = Y_t$, $y_t = Y_{t-1}$ and $z_t = Y_{t-2}$, we get:

$$
\begin{aligned}
x_{t+1} &= \min\{m_1 x_t + m_2 y_t + m_3 z_t + \max\{a_1(x_t - y_t) + a_2(y_2 - z_t), -I^f\} + I^{aut}, Y^c\}, \\
y_{t+1} &= x_t, \\
z_{t+1} &= y_t,
\end{aligned}
\tag{12}
$$

Hence, the 3-D Hicks-model is given by a difference equation
$(x_{t+1}, y_{t+1}, z_{t+1}) = H(x_t, y_t, z_t),$

where H is the 3-dimensional piecewise linear map given by

$$H(x,y,z) = (\min\{m_1 x + m_2 y + m_3 z + \max\{a_1(x-y) + a_2(y-z), -I^f\} + I^{aut}, Y^c\}, x, y). \tag{13}$$

Let $m = m_1 + m_2 + m_3$, $0 < m < 1$, be the overall marginal propensity to consume. We write $Y^e = I^{aut}/(1-m)$ and we assume that $Y^e < Y^c$. The map H has a unique equilibrium given by $E = (Y^e, Y^e, Y^e)$. Define the income floor-level $Y^f = (I^{aut} - I^f)/(1-m)$. Note that $Y^f < Y^t < Y^c$. Let D be the set $D = \{(x,y,z) \mid Y^f \le x,y,z \le Y^c\}$. A simple computation shows that H maps D into itself. Therefore, all time-paths in the 3-D Hicks-model are bounded. *What can be said about the dynamics of the 3-D Hicks-model when the equilibrium is unstable?*

There seems to be some confusion in the literature about the notions of chaotic and strange attractors. The definition of chaos which we use is based on the so-called Lyapunov exponents. Chaotic behaviour is characterized by sensitive dependence on initial conditions: nearby initial states diverge at an exponential rate. Lyapunov exponents measure the average rate at which initial states separate. An attractor A is called a *chaotic attractor* if there exists a set S of positive Lebesgue measure, such that for all x in S the orbit of x is attracted to A, and the corresponding largest Lyapunov exponent is positive. For mathematical details see e.g. Guckenheimer and Holmes (1986) or Eckmann and Ruelle (1985). Next we would like to describe the notion of *strange attractor*. We will call an attractor strange, if the attractor has a "Cantor-like structure". A strange attractor can be defined as an attractor having a non-integer fractal dimension; for a definition of different fractal dimension see e.g. Falconer (1990). Summarizing, a chaotic attractor is characterized by sensitivity to initial states, while a strange attractor is characterized by a complicated structure. Hence, the word chaotic refers to the dynamics on the attractor, while the word strange refers to the geometry of the attractor.

In figures 2 and 3 we present pictures of the (x,y)-projections of attractors of the 3-D Hicks-model. The corresponding parameter values are given in the captions of the figures. Figures 2a and 3a show two examples of strange, chaotic attractors. The pictures suggest a Cantor-like structure (see also the enlargements in figures 2c, 2d, 3b, 3c and 3d), the corresponding (Lyapunov) fractal dimensions are $D \approx 1.75$ and $D \approx 2.02$, and the corresponding largest Lyapunov exponents are positive. The fractal dimension and Lyapunov exponents have been computed by using the DYNAMICS-program (Yorke, 1990).

In the first example (figure 2), in addition to the strange, chaotic attractor, a stable period 20 orbit occurs, see figure 2b. Hence, in this example we have coexistence of (at least) two different attractors. It depends on the initial state whether regular (periodic) or irregular (chaotic) behaviour occurs. In the second example (figure 3) it seems that the model has a unique strange, chaotic attractor; the orbits of all initial states that we tried converged to this strange, chaotic attractor.

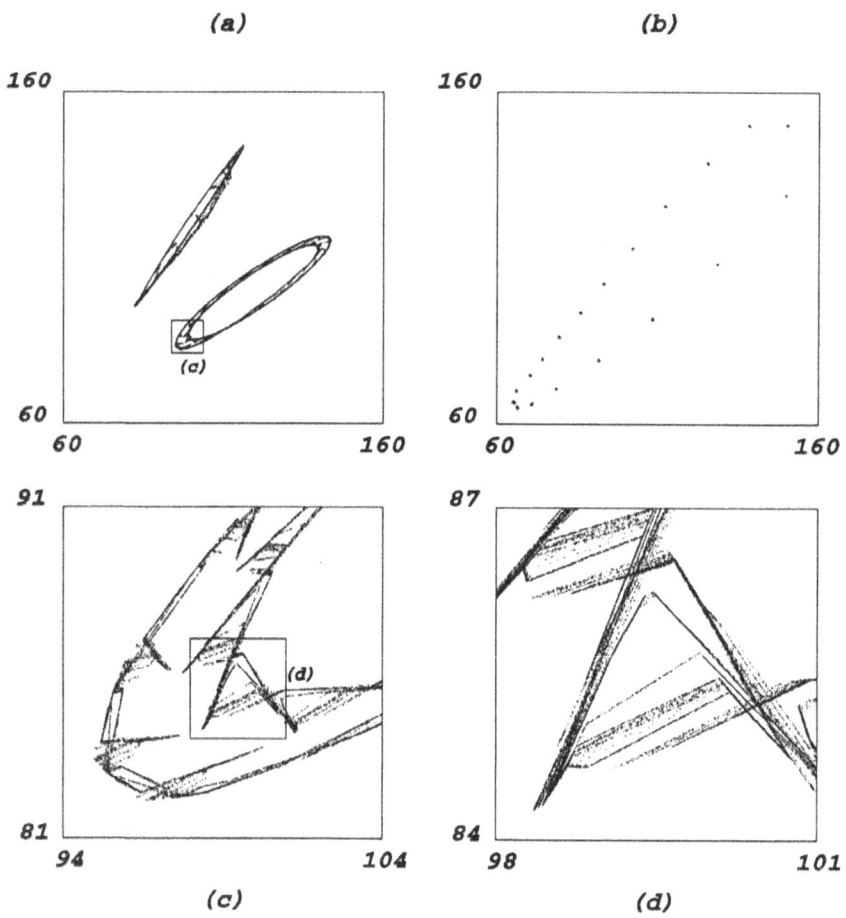

Figure 2. Coexistence of a strange, chaotic attractor and a stable period 20 orbit. The parameters are: $Y^c = 150$, $I^f = 10$, $I^{aut} = 20$, $m_1 = 0.54$, $m_2 = 0.25$, $m_3 = 0$, $a_1 = 0.6$ and $a_2 = 1.4$.   (a) The orbit of $(x_0,y_0,z_0) = (120,100,150)$ converges to a strange, chaotic attractor, with (Lyapunov) fractal dimension $D \approx 1.75$ and largest Lyapunov exponent $\lambda \approx 0.013$. (b) The orbit of $(x_0,y_0,z_0) = (Y^c,Y^c,Y^c)$ converges to a stable period 20 orbit. Only 19 points are visible, since two points of the period 20 orbit are projected to the point $(x,y) = (Y^c,Y^c)$. (c-d) Enlargements of the strange chaotic attractor in (a).

Notice that in both examples we have $m_1 > m_2 > m_3 > 0$, so that the highest fraction of income is consumed with a delay of one period, while the lowest fraction of income is consumed with a delay of three periods. Furthermore, in both examples $a_2 > a_1$ so that most of the induced investment takes place in the second time period following the originating change in income. As for example pointed out by Allen (1965), such an investment pattern may be the rule rather than the exception. Usually, the time needed for investment-decisions and investment outlays is longer than the shortest consumption lag. Therefore, in the 3-D Hicks-model it seems that chaos may occur for reasonably realistic parameter values.

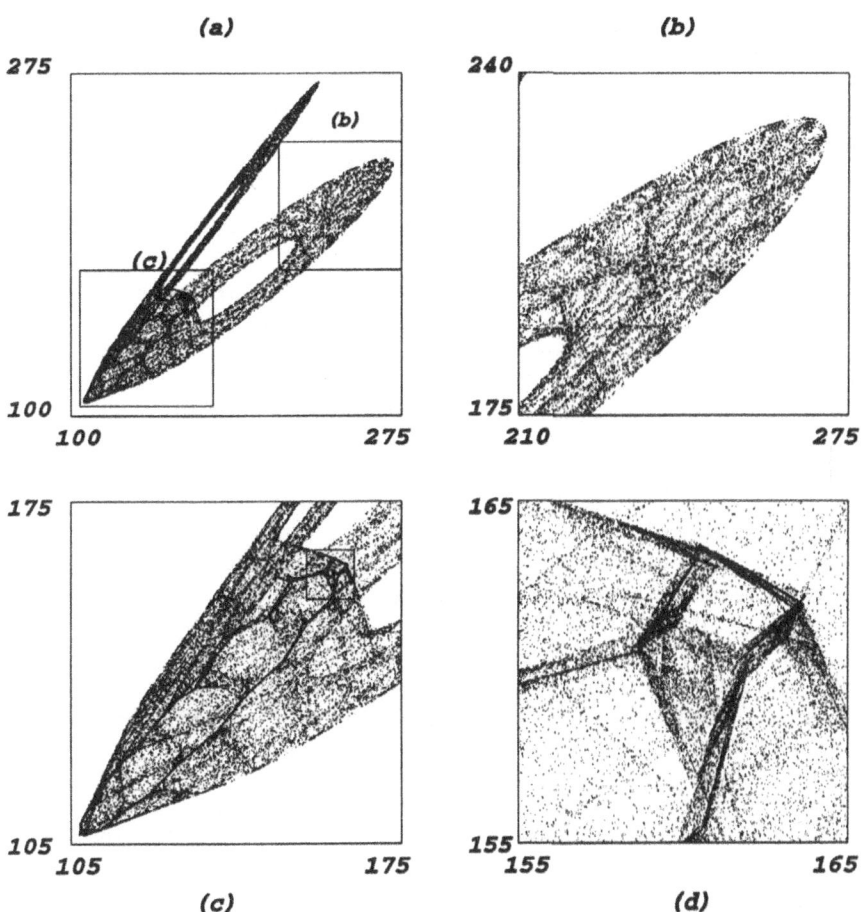

Figure 3. A strange, chaotic attractor with (Lyapunov) fractal dimension D ≈ 2.02 and largest Lyapunov exponent λ ≈ 0.022. The parameters are: $Y^c = 275$, $I^f = 10$, $I^{aut} = 20$, $m_1 = 0.55$, $m_2 = 0.2$, $m_3 = 0.1$, $a_1 = 0.25$ and $a_2 = 1.27$. Figures 3b-d show enlargements of (a). Almost all time paths seem to converge to the strange, chaotic attractor.

4. Concluding Remarks

We have investigated the dynamics of Hicks' nonlinear trade cycle model. Even in the simplest version of the model a time path does not always converge to a periodic time path. In addition to stable periodic behaviour the elementary Hicks-model also exhibits quasi-periodic dynamics. On the other hand, chaos does not occur in the elementary Hicks-model. However, extra time lags in the consumption or the investment equation may cause erratic, chaotic dynamics. Our numerical results indicate that even in the one but simplest version of the model, the 3-D Hicks model, strange chaotic attractors occur.

The Hicks-model is a prototype of a model with a very simple type of non-linearities: natural 'ceilings' (upper bounds) and 'floors' (lower bounds) imposed on an (unstable) linear system. Within the bounds the equations of the model are linear, while the bounds prevent that the time paths tend to infinity. In particular for economics this type of nonlinearities may be important. The importance of such nonlinear models for describing economic data has e.g. been pointed out by Brock (1988). As we have seen, this simple type of nonlinearities can produce complex dynamical behaviour.

References

Lorenz, H.-W. (1989), 'Nonlinear dynamical economics and chaotic motions, *Lect. Notes in Economics and Math. Systems* 334, Springer-Verlag, Berlin.

Medio, A. (1992), 'Chaotic dynamics. Theory and applications to economics', Cambridge University Press, Cambridge.

Kaldor, N. (1940), 'A model of the trade cycle', *Economic Journal* **50**, 78-92.

Hicks, J.R. (1950), 'A contribution to the theory of the trade cycle', Clarendon Press, Oxford.

Goodwin, R.M.(1951), 'The nonlinear accelerator and the persistence of business cycles', *Econometrica* **19**, 1-17.

Allen, R.G.D. (1965), 'Mathematical economics', MacMillan & Co., London.

Blatt, J.M. (1983), 'Dynamic economic systems. A post-Keynesian approach', M.E. Sharpe, Inc., Armonk, New York.

Hommes, C.H. (1991), 'Chaotic dynamics in economic models. Some simple case-studies', Thesis University of Groningen, Wolters-Noordhoff Groningen.

Guckenheimer, J. and Holmes, P. (1986), 'Nonlinear oscillations, dynamical systems and bifurcations of vector fields', Springer Verlag, New York.

Eckmann, J.P. and Ruelle, D. (1985), 'Ergodic theory of chaos and strange attractors', *Reviews of Modern Physics* **57**, 617-656.

Falconer, K. (1990), 'Fractal geometry. Mathematical foundations and applications', John Wiley & Sons, Chichester.

Yorke, J.A. (1990), 'DYNAMICS. An interactive program for IBM PC-clones', IPST, University of Maryland, College Park.

Brock, W.A. (1988), 'Hicksian nonlinearity', Paper presented at the *International Economic Association International Conference "Value and Capital fifty years later"*, Bologna, Italy, September 3-5, 1988.

MONTE CARLO EXPERIMENTATION
FOR LARGE SCALE FORWARD-LOOKING
ECONOMIC MODELS

RAOUF BOUCEKKINE

CREST-ENSAE, UNIVERSIDAD CARLOS III de Madrid and CEPREMAP

142, rue de Chevaleret, 75013 Paris, France

Summary

In this paper, a Monte Carlo experimentation scheme is developed for rational expectations large scale models with a special attention to the theoretical foundations of the underlying deterministic algorithm and to the *a posteriori* statistical validation of the experimentation. The base-deterministic algorithm is of the Newton-Raphson type. The Monte Carlo experimentation uses a perfect foresight approximation and then, requires *a posteriori* validation. Numerical exercises are proposed in order to show clearly the adequacy of our methodology, by evaluating either its purely numerical bias or the goodness of its perfect foresight approximation, on a canonical growth model.

Keywords Rational expectations, Large scale models, Monte Carlo methods, Statistical validation.

1. Introduction

Throughout the two last decades, the simulation of rational expectations economic models has been one of the most debated topics in empirical economics. Because the rational expectation assumption leads to a particular theoretical framework, it requires a special numerical treatment. To get some important outlines of the induced theoretical framework, consider the following univariate model :

$$x_t = E_t \{f(x_t, x_{t+1}, x_{t-1}, z_t)\}, \quad t \geq 1, \quad x_0 \text{ given,}$$

where x_t is the endogenous variable of the model, z_t the unique exogenous innovation affecting the model at each period t, $f(.)$ is a real -valued function and $E_t \{.\}$ is the expectation operator conditionally to the information set available at the beginning of the period t. Given this simple model, we can see the theoretical implications of the rational expectation assumption as essentially twofold :

i) For any period t, the solution for x_t depends on the complete sequence of the future innovations $(z_{t+i})_{i \geq 0}$: in trivial terms, this means that the "length" of uncertainty is infinite when the expectations are rational.

ii) Even for the particular case of perfect foresight, the model may admit an infinite number of stable solutions.

Property i) implies that the usual Monte Carlo experimentation can not reproduce exactly all the possible out-turns of the simulated economy, as the rational expectation assumption implies that the "length" of uncertainty is infinite at any period. As a consequence, the Monte Carlo estimates can only be seen as approximative solutions, and then, the practitioners have to check *a posteriori* their accuracy. Few years ago, a number of "exact" stochastic simulation methods were proposed to handle directly the infinite "length" property of uncertainty, without using the Monte Carlo approximation. Based on various mathematical principles (see Taylor and Uhlig (1990) and Boucekkine (1992-a) for some exhaustive surveys), these "exact" methods follow actually the same methodological scheme: they primarily compute the *stochastic* steady states of the models using some distributional assumptions, then they replicate the obtained steady states using some stochastic generators, in order to get the desired pseudo-samples.

Property ii) refers to the "saddlepoint" models characterizations : before solving rational expectations models, the practitioners must primarily ensure that these models admit a unique stable solution. An earlier contribution in this field is due to Blanchard and Kahn (1980) on finite difference linear models : they provide an efficient *a priori* test, based on the spectra of some well-defined matrices, which allows to identify the "saddlepoint" finite difference linear models. Stokey and Lucas (1989) proved that the so-called Blanchard-Kahn saddlepoint conditions are relevant in the nonlinear case, when adopting a local approach around the deterministic steady states of the models. Assuming, of course, that the linearized models are of a finite difference configuration.

The developments given above may suggest that conducting "exact" stochastic simulation methods and identifying the saddlepoint solutions became more and more tractable. Unfortunately, this is not the case at all, when *dealing with large scale models*:

* *"Exact" stochastic simulation methods are usually intractable when dealing with large scale models*: that is because the computational cost of such methods grows exponentially with the number of state variables, which is typically the case of large scale models. Indeed, as the majority of the "exact" methods use discretization techniques on well-specified state spaces, their usefulness is very limited in practice as the discretization costs grow exponentially with the number of state variables. Given this undeniable feature, Taylor and Uhlig (1990) concluded that ".. It might be quite impossible to compute the solution for a model with 15 state variables, say, using some grid methods.." (p. 16).

** *Identifying the "saddlepoint" large scale models with a priori tests is almost impossible*: indeed, as Blanchard-Kahn saddlepoint conditions have to be checked on finite difference linear (or linearized in the nonlinear case) models, large scale models almost never exhibit such a (local) configuration because of the typical numerous *redundancies* existing between their variables.

These important remarks lead us to argue that approximation methods are unavoidable for the purpose of stochastic simulation of large scale forward-looking models. Furthermore, large scale "saddlepoint" models have to be identified using some specific *a posteriori* simulation procedures. An earlier contribution fulfilling these two characteristics is due to Gagnon (1990). Using the "extended-path" algorithm suggested by Fair and Taylor (1983), Gagnon developed a Monte Carlo experimentation type and applied it on a standard growth model. However, his framework suffers from two major disadvantages, arising from the use of the "extended-path" technique. First, as this algorithm requires three Gauss-Seidel loops, each of them with its own specific iteration type, it is suspected to be highly time consuming when dealing with large scale models. This point is made by Taylor and Uhlig themselves when they state ".. that is, grid methods and the extended-path method are computationally quite involved.." (p. 16). More crucially, the use of the "extended-path" algorithm could be largely misleading regarding to the saddlepoint paths extraction problem. By construction, the latter algorithms exclusively selects the solution paths which remain insensitive to the solution time horizon; now, given some recent results (mainly due to the ESRC Macroeconomic Modelling Bureau, see Wallis *et alii* (1986)), it appears that this selection criterium could lead to some non-saddlepoint solutions, which ultimately disqualifies the "extended-path" technique from the crucial point of view of the saddlepoint paths extraction problem.

In this paper, we provide an alternative methodology which allows to solve the previous puzzles. Section 2 presents a stochastic simulation method of a Monte Carlo type, thus using a perfect foresight approximation. As this ultimately leads us to a deterministic setting, we need to a deterministic solver and we choose the one suggested by Laffargue (1990). This choice is motivated by the existence of a saddlepoint path extraction strategy associated to this algorithm, and developed by Boucekkine (1993). Section 3 is an application of the presented method on a canonical growth model. We focuse deliberately on the investigation of the numerical bias of the algorithm and especially on the *a posteriori statistical validation* of the method, as it uses a perfect foresight approximation. Some indicative computing times are also provided.

2. The method

As announced before, the proposed solution method is of a Monte Carlo type : given a rational expectations model, the method consists in replicating the simulated economy using a *finite* number of draws of the model's innovations at each replication, which provokes a deviation from the rational expectation hypothesis. As a consequence of the induced perfect foresight approximation, we need to a deterministic algorithm in order to compute the solutions of each Monte Carlo replication. Our experimental scheme is then constituted by two different algorithms: the first one is devoted to compute the required deterministic solutions for each replication, the second one provides the disposition of the Monte Carlo replications. The presentation is conducted on the following canonical model:

$$E_t \{f(X_t, X_{t+1}, X_{t-1}, Z_t) \} = 0 \text{ for } t \geq 1,$$

where X_t is the vector of the endogenous variables, Z_t the vector of the exogenous variables, $f(.)$ a well dimensioned vectorial function and X_0 given.

2.1 The base-deterministic algorithm : a saddlepoint path extraction strategy

Contrary to the existing deterministic algorithms which exclusively use Gauss-Seidel iterative schemes, such as Fair and Taylor's algorithm (1983), we shall use a relaxation algorithm of the Newton-Raphson type, suggested by Laffargue (1990). Our choice is motivated by several theoretical and numerical reasons, we summarize them in the following points :
* *The applicability to large scale models*: As the chosen algorithm is of a Newton--Raphson type, it requires the inversion of the global jacobian of the system, which is highly time consuming for large scale models. Fortunately, as described in Laffargue (1990), instead of directly inverting the jacobians, a Gaussian triangulation, taking advantage of the jacobians' sparse configuration, is introduced, which greatly lowers the numerical cost, including the storage cost. Finally, the algorithm can be used without any problem for many large scale economic models, until about some hundred equations. What makes the algorithm very tractable is that the adopted approach is definitely local : focusing the simulations around the steady states, which is the most common exercise in economics, and given the local weak nonlinearities of the economic models, a very low

number (usually between three and five) of Newton-Raphson iterations is needed to achieve convergence

** *The saddlepoint paths extraction problem*: Dwelling on the weaknesses of Fair and Taylor's algorithm concerning the identification of the "saddlepoint" models, Boucekkine (1993) develops an alternative methodology based on Laffargue's Newton-Raphson algorithm: using the particular structure of this algorithm, the author provides *theoretical* characterizations of the algorithm's convergence in terms of the saddlepoint conditions : In particular, he demonstrates that the algorithm always converges when the Blanchard-Kahn saddlepoint conditions are (locally) checked; on the contrary, it is explosive if the system is itself (locally) explosive. However, if it exists an infinity of (local) stable solutions and when the relaxation is initialized exclusively with the deterministic steady state values, then the algorithm may converge, which is an inefficiency regarding to the saddlepoint path extraction problem. Fortunately, the author shows that when the relaxation is not exclusively initialized with the long run values, the algorithm is explosive if the saddlepoint conditions are not (locally) checked, which is a very great property.

This suggests a rigorous deterministic algorithm : to present its main principles, we consider a deterministic version of the model given before: $f(X_t, X_{t+1}, X_{t-1}, Z_t) = 0$ with $X_0 = X_s$ where X_s is a selected deterministic steady state of the model consistent with the value Z_s of the environment. We consider a small deviation ΔZ_s of this environment at the first period and we aim at solving the induced system. Thus, we only consider transitory shocks for simplification. Permanent shocks are also allowed by the Newton-Raphson algorithm. The proposed algorithm takes the following steps (implicitly assuming that the model does not generate a unit root):

i) Linearize the model around the steady state X_s. If it is possible (linearized model of a finite difference form), compute the eigenvalues of the linearized system : if the Blanchard-Kahn saddlepoint conditions are not checked, the algorithm is stopped, otherwise, go to the next step. If it is not possible to compute the eigenvalues, go to the next step. Return after this preliminary step to the *nonlinear* model.

ii) Choose a simulation time horizon T consistent with the values of the eigenvalues or a deliberately high horizon if they are not available. Set the terminal condition : $X_{T+1}=X_s$. Choose the trajectory base equal at each period to X_s.

iii) Solve the induced stacked nonlinear system with Laffargue's algorithm. If it is (explosively) divergent, the algorithm is interrupted. If not , go to step iv). (Newton-Raphson divergence is almost always explosive. This is a great advantage of this method.)

iv) Check the adequacy of the chosen simulation time horizon by comparing X_T to X_s. If T is not adequate, increase it and conduct a step iii) until X_T is sufficiently close to X_s . If this property can not be achieved (see remark at the end of this paragraph), the algorithm is interrupted. Otherwise go to step v).

v) Choose now a relaxation initialization which slightly deviates from the long run values. Then, conduct steps iii) and vi). If the solution path is insensitive to these initializations, accept it; otherwise, reject it.

As one can see, this deterministic scheme includes three different components : the first one (step i)) is a preliminary investigation of the local properties of the model. The second axis (step iii)) consists in applying the Newton-Raphson algorithm for a given simulation time horizon and a considered relaxation initialization. The final component (steps iv) and v)), the most important one, is our saddlepoint path extraction strategy, and it is entirely based on the theoretical considerations given above. This algorithm is denoted BDA (Base-Deterministic Algorithm) hereafter.

Remark In relation with step iv) it is remarked that if this property is not achieved, this is due to two possible situations : first, no solution exists nearby and the algorithm diverges explosively for a simulation horizon sufficiently high. Otherwise, another steady state is reached, which is very unlikely to occur, given the small magnitude of the deviation considered and if no unit root is allowed, as it is assumed in our setting.

2.2 The Monte Carlo algorithm

Given the base-deterministic algorithm, it is now possible to describe the Monte Carlo setting. To this end, we have to specify the law of motion of the exogenous variables. To be consistent with the usual economic exercises, we assume that the exogenous variables of the model follow a stationary p-order linear autoregressive process:

$$Z_t - Z_s = \Phi(L)(Z_{t-1} - Z_s) + \varepsilon_t \, , \tag{ST}$$

where L is the lag operator, $\Phi(.)$ is a polynome of degree $p-1$ with $\Phi(1) \neq 1$ for the stationarity requirement and ε_t is drawn from the normal law $N(\mu, \Sigma)$. The linearity assumption is not necessary ; however, as it will appear in the presentation of the stochastic algorithm, this significantly simplifies the computations. In order to generate an artificial time series $(s_1, s_2,....., s_r)$, $r > 2$, for the vector of endogenous variables X_t, we use the following algorithm (denoted MCA, for Monte Carlo Algorithm, hereafter):
i) Given initial conditions for the predetermined endogenous variables, say $X_0 = X^0$, where X^0 is in the neighborhood of X_s, and initial conditions for Z_t, $Z^0 = (Z^0_{1-p} , Z^0_{2-p} ,..., Z^0_0)$, draw ε_1 from $N(\mu, \Sigma)$ and assume that $\varepsilon_t = 0$ for $t > 1$: *thus, we approximate the rational expectation hypothesis by a perfect foresight one, which generates a bias.*
ii) Using equation (ST) and step i), compute the complete exogenous sequence $(Z_1, Z_2,.., Z_{Tz})$ where Z_{Tz} is close to Z_s. Solve the model with the simulation horizon initially set to Tz, using the algorithm BDA. Especially, compute the vector X_1. Set $s_1 = X_1 = r$. Store r. Store Z_1. Actually, we do not need to perform all the steps of the algorithm BDA at all the replications : except for the first replication, as we need to know if the model is "saddle-point", only steps ii), iii) and iv) have to be performed.
iii) Set $X_0 = r$ and $Z^0 = (Z^0_{2-p} , Z^0_{3-p} ,..., Z^0_0 , Z_1)$. Draw ε_1 from $N(\mu, \Sigma)$ and assume that $\varepsilon_t = 0$ for $t > 1$. Repeat step ii) and compute the vector X_1. Set $s_2 = X_1 = r$. Store X_1 and Z_1.
iv) Repeat the step iii) until reaching the length needed for the time series.
    Because of the perfect foresight approximation, the algorithm MCA is potentially biased. It is well known that this bias depends on *the nonlinearities* of the models. Hence, *a posteriori* statistical validation is needed to evaluate this bias : the most rigorous and tractable way to do this is to analyze whether the generated series of the residuals of the nonlinear Euler equations included in the models check the martingale difference property, inherent in the rational expectation scheme. In any case, it is usually possible to take advantage of the local weak nonlinearities of the economic models to minimize this bias, for example by considering "small" variance matrices for the models' innovations, which is consistent with the usual economic practice. These points are developed in the application example.

3. Indicative example

The model used in this working example is of a small dimension, to give a simple and significative numerical framework. However, as it is detailed in the previous sections, the method is essentially devoted to solve large scale models, as exact methods dominate the Monte Carlo approximation for such small models.

3.1 The model

The model considered here is a version of the Ramsey growth model in order to refer to Gagnon's conclusions (1990). Gagnon uses the same stochastic generator but Fair and Taylor's algorithm as a base-deterministic algorithm. As the model is very known, we describe it very briefly. Let us consider a representative agent whose intertemporal optimization problem is :

$$Max \ E_0 \left\{ \sum_{t \geq 0} \delta^t u(c_t) \right\}$$

under the constraints :

$$c_t + I_t = Q_t = Az_t k_{t-1}^\alpha ,$$ (1)
$$I_t = k_t - \mu k_{t-1} ,$$ (2)
$$Ln(z_t) = \rho Ln(z_{t-1}) + \varepsilon_t ,$$ (3)

where ε_t is a normal law $N(0; \sigma^2)$.

Table 1. The model's parameterizations

Parameters	α	γ	μ	δ	A	ρ	σ
1	0.9	1	0	0.95	1.5	0.9	
2	0.5	0.5	1	0.9	1	0.8	0.01
3	0.5	0.5	1	0.9	1	0.5	0.09
4	0.5	0.5	1	0.9	1	0.8	0.12
5	0.8	0.25	1	0.9	0.15	0.5	0.01
6	0.33	0.5	1	0.95	1	0.95	0.1

The agent maximizes his expected discounted intertemporal consumption utility (i.e. $u(c_t)$) under the budgetary constraint (equation (1)) which stipulates that consumption, plus investment (I_t) must be equal to the Cobb-Douglas production function $Q_t = Az_t k_{t-1}^\alpha$ where k_{t-1} is the lagged capital stock. z_t, the unique stochastic variable of the model, is the productivity shock. Equation (2) is the usual capital accumulation law and the stochastic evolution of z_t is given in equation (3). All the other figures are economic parameters. We also set: $u(c)=c^{1-\gamma}/(1-\gamma)$ and $u(c) = Ln(c)$ if $\gamma = 1$. The optimization problem gives the following Euler equation:

$$c_t^{-\gamma} = \delta E_t \{ c_{t+1}^{-\gamma} (A\alpha z_{t+1} k_t^{\alpha-1} + \mu) \}.$$ (4)

Given the equations (1), (2), (3) and (4), it is easy to establish that the model admits an explicit and unique deterministic steady state.

Now, we present the results of some simulation exercises we perform on six parameterizations of the model, given in the table 1 (each line of it referring to a particular parameterization), additional exercises can be found in Boucekkine (1992-b).

3.2 The purely numerical bias of the stochastic algorithm MCA

To evaluate the numerical performances of our experimentation, we perform some steps ii) computations of the algorithm MCA in a case where an explicit solution exists. As steps ii) of the Monte Carlo algorithm scheme given in section 2.2 use in particular the algorithm BDA, the proposed investigation relies also on this algorithm. Indeed, parameterization 1 provides an explicit time invariant decision rule for the capital stock: $k_t = A\alpha z_t k_{t-1}^{\alpha}$. Given k_{t-1} and z_t, it is then possible to compute the exact capital decision. Setting $z_0 = 1$, we compute the values of k_1 for different values of k_0 and when $z_1=0.7$, for a tolerance level equal to $5\cdot10^{-4}$ for the base-deterministic algorithm, in order to conform to Gagnon's framework (1990). Table 2 gives the obtained results:

Table 2. Capital decision rules for $z_1=0.7$, convergence tolerance level $4\cdot10^{-5}$

k_0	MCA	Gagnon	Exact
5	3.81979	3.82	3.82145
10	7.13103	7.12	7.13108
15	10.27334	10.26	10.27158
20	13.30932	13.29	13.30706
25	16.26880	16.25	16.26677

The purely numerical bias is then very satisfactory, better than Gagnon's one. Actually, when adopting stricter tolerance levels, this bias vanishes! Table 3 provides the computed decision rules k_1 for various values of the initial stock of capital k_0, for $z_1 = 1.3$, and when setting the tolerance level for the base-deterministic algorithm equal to 10^{-5}.

Table 3. Capital decision rules for $z_1=1.3$, convergence tolerance level 10^{-5}

k_0	MCA	Gagnon	Exact
5	7.09698	7.09	7.09698
10	13.24343	13.23	13.24343
15	19.07580	19.06	19.07580
20	24.71312	24.69	24.71312
25	30.20972	30.19	30.20972

However, despite these very valuable performances, this is not sufficient to assess the goodness of our perfect foresight approximation and further statistical validation is needed.

3.3 Statistical validation

A common way to test if the generated series satisfy to the rational expectation property, is to see if the residuals of the Euler equation (4) check the martingale difference property, which is specific to the rational expectations. Denote by η_t these residuals, that is:

$$\eta_{t+1} = \frac{\delta c_{t+1}^{-\gamma} [A \, \alpha \, z_{t+1} \, k_t^{\alpha-1} + \mu]}{c_t^{-\gamma}} - 1 \; .$$

The martingale difference property stipulates simply that $E_t\{\eta_{t+1}\}=0$. A convenient statistic to test this property is the so-called m-statistic (see the mathematical details in Taylor and Uhlig (1990)). This statistic is given by :

$$m = \hat{a}'(\Sigma_t \, w_t'w_t)(\Sigma_t \, w_t'w_t \, \eta_t^2)^{-1}(\Sigma_t \, w_t'w_t)\hat{a},$$

where w_t is a vector of instruments taken in the information set available at the period t, and $\hat{a}$ is the least squares regressor of the generated residuals η_t on the vector of instruments, w_t. With our instruments' choices (five lags for both the consumption variable and the technology shock, plus a constant), this statistic follows asymptotically a $\chi^2(11)$ law. To accept the martingale difference property using a two-sided test with a significance level of 2.5% for each side, the m-statistic values obtained with the generated time series must lay between 3.82 and 21.92. 1000 replications were needed to stabilize the results. Table 4 gives the m-statistic values for the parameterizations 2 to 5, additionally to the empirical means of the generated residuals :

Table 4. m-statistic values and residuals' empirical means

Parameters	Empirical means	m-statistic
2	$2 \; 10^{-5}$	7.782
3	0.00010	9.76
4	0.00012	13.18
5	0.566	244.96

The reported results prove clearly that the generated series generally check the martingale difference property. In fact, only parameterization 5 does not satisfy to the m-statistic test: this is actually the case for all the parameterizations which include high values for the parameters α and γ, which ultimately signifies that increasing α or γ increases the local nonlinearities. Fortunately, these values are not realistic : for example, as parameter α can be seen economically seen as the complementary to one of the ratio "wages over GNP", realistic values are about 0.5 (see King *et alii* (1988) for the US economy case). For the realistic parameterizations, the m-statistic test is clearly satisfied, as they include "sufficiently small" values for the innovation variance, σ, which keeps the problem local and minimizes the effects of nonlinearities. This justifies *a posteriori* our Monte Carlo approximation. Such a treatment, if theoretically unsatisfactory as the statistical goodness of the Monte Carlo experimentation ultimately depends on the nonlinearities of the models, allows to perform rigorously the usual stochastic exercises in economics. Of course, to get statistically relevant results, one has to choose "sufficiently small" variances for the models' innovations; actually, this limitation may not be really restrictive in practice : in our example, for some realistic parameterizations of the model, the approximation is acceptable, even for $\sigma=0.12$, which is actually a "high" value regarding to the usual numerical exercises in economics.

3.4 Computing times

In table 5, we report the computing times required by our method and Gagnon's one, for two typical exercises conducted on the parameterization 6 of the model. The first exercise consists in computing the decision rules for the endogenous variables, exactly as in subsection 3.2. To compare our results to Gagnon's ones (reported in Taylor and Uhlig (1990), p. 16), we consider the following grid of values for the state variables: $K_0 = 5, 10,$ 15, 20 and 25, $z_1 = 0.4, 0.7, 1, 1.3$ and 1.6. The second exercise consists in computing a simulation of 2000 data points. In table 5, Time 1 (Resp. Time 2) gives the CPU time in seconds, needed to perform the first (Resp. second) exercise.

Table 5. Computing times

Method	Machine	Megahertz	Software	Time1	Time2
Gagnon	Amdahl 5850		Troll 13.0	396	5320
MCA	Dell 486P	66	Gauss-386i 3.01	183	2815

Although we can not rigorously compare the numbers reported in the table because they depend strongly on the machine and the software used, and given that our computational environment does not appear to be clearly better than Gagnon's one, it is worth pointing out that time consuming comparison is undeniably favorable to our algorithm. Observe that this outcome is not surprising at all as the latter is essentially based on Newton-Raphson iterations whereas the "extended-path" technique, used by Gagnon, includes three Gauss-Seidel loops.

4. Conclusion

The presented deterministic and stochastic simulation strategies adopt deliberately a local approach around the steady states. This approach has two important advantages: it allows to develop a saddlepoint extraction strategy with some theoretical justifications and finally, it allows to validate *a posteriori* a Monte Carlo scheme in a rational expectation context. Given the local weak nonlinearities of the existing large scale economic forecasting models, the approach is likely to be adequate. We do not deny, of course, its inefficiency in the presence of local strong nonlinearities, but this occurrence is, as said before, very unexpected in practice ; otherwise, handling it for large scale forward-looking models is too problematic, given the current computational capacities.

References

Blanchard, O and C. Kahn (1980), 'The solution of linear difference models under rational expectations', *Econometrica*, **48(5)**, 1305-1311.

Boucekkine, R (1992-a), 'Simulations stochastiques et anticipations rationnelles', *Discussion Paper*, N° **9212**, Catholic University of Louvain, Belgium.

Boucekkine, R (1992-b), 'Quelques idées simples pour la simulation stochastique des modèles non-linéaires à anticipations rationnelles et méthodes de validation' , *Discussion Paper*, N° **9215**, CEPREMAP, Paris.

Boucekkine, R (1993), 'Some new developments on the analysis of the numerical solutions of consistent expectations models', Pre-accepted for publication in *Journal of Economic Dynamics and Control*.

Fair, R and J-B.Taylor (1983), 'Solution and maximum likelihood estimation of dynamic rational expectations models', *Econometrica*, **51 (4)**, 1169-1186.

Gagnon, J (1990), 'Solving the stochastic growth model by deterministic extended path', *Journal of Business and Economic Statistics*, **8 (1)**, 35-36.

King, R, C.Plosser and J.Rebelo (1988), 'Production, growth and business cycles I', *Journal of Monetary Economics*, **21 (2)**, 196-232.

Laffargue, J-P (1990), 'Résolution d'un modèle macroéconomique à anticipations rationnelles', *Annales d'Economie et Statistique*, **17**, 97-119.

Stokey, N and R.Lucas (1989), Recursive methods in economic dynamics, *Harvard University Press*, Cambridge, pp 148-156.

Taylor, J-B and H.Uhlig (1990), 'Solving nonlinear stochastic growth models : a comparison of alternative solution methods', *Journal of Business and Economic Statistics*, **8 (1)**, 1-18.

Wallis, K (ed.), M.Andrews, D.Bell, P.Fisher and J.Whitley (1986), Models of the U.K economy: a third review by the ESRC Macroeconomic Modelling Bureau, *Oxford University Press*, Oxford.

ERRATIC DYNAMICS IN A RESTRICTED TATONNEMENT PROCESS WITH TWO AND THREE GOODS

CLAUS WEDDEPOHL

Department of Actuarial Science an Econometrics,
Faculty of Economics and Econometrics,
University of Amsterdam, Roetersstraat 11, 1011 WB Amsterdam, The Netherlands

Summary

It is well known that adjustment processes like tatonnement can show erratic dynamic behavior. The path of prices generated in such a process generally shows big jumps and varies over a wide range. This is not acceptable. A discrete tatonnement process in simple two and three goods exchange economies is studied. It is shown that in the examples studied, the region in which the prices can vary can be restricted to a neighborhood of an equilibrium, if the relative price adjustment is restricted by a maximal rate of increase or decrease of the price. Within the interval any type of erratic dynamics remains possible.

Keywords: price adjustment, chaos, nonlinear dynamics.

1. Introduction

The tatonnement process is usually considered to be a first approximation of the price mechanism, although it is generally agreed that it is not really a good representation of it. The basic idea is that the price of a good rises if demand exceeds supply and falls if supply exceeds demand. It is still widely believed among economists that this guarantees that finally a set of equilibrium prices is attained. It has been known for a long time that this is not generally true (see Arrow and Hahn, 1971, ch. 12), but it has been realized only quite recently that such a process could generate erratic price paths (Saari, 1985, Day and Piagiani, 1991, Bela and Majumdar, 1992). Simulations of non converging paths show that prices tend to vary over a wide range and that prices make big jumps up and down in absolute or relative terms. This is not reasonable: neither an auctioneer nor "the market", whatever that may be, would continue to do that, but would rather reduce the adjustment. In the present paper it is shown that, at least in some simple economies, setting bounds to the rate of price adjustment, not only rules out big jumps, but also ensures that prices converge to a neighborhood of an equilibrium price vector, while the path may show erratic behavior within this neighborhood.

2. Tatonnement

Demand and supply in an n-goods economy are expressed by excess demand functions $Z_i(p_1,p_2,...,p_n)$. $Z_i(p) > 0$ means that demand for good i is larger than supply and $Z_i(p) < 0$ means that supply exceeds demand. An equilibrium occurs at p^* if $Z_i(p^*) = 0$ for all $i = 1,2,...,n$. Tatonnement is a process to generate an equilibrium price with the help of an *auctioneer*, who is assumed to set a price vector $p(t)$ in stage t and get information on

$Z_i(p(t))$. Then he adjusts the prices in such a way that $p_i(t)$ is increased if $Z_i(p(t)) > 0$ and decreased if $Z_i(p(t)) < 0$. The new price vector $p(t+1)$ is announced and new information on excess demands is received, leading again to a new price. The process continues until an equilibrium price is obtained. Only then the agents are assumed to realize trade. This latter condition makes the concept less realistic, because agents may have to wait very long. Usually the process is specified (with the provision that p_i cannot be negative) either by the difference equation

$$p_i(t+1) = p_i(t) + \lambda_i Z_i(p(t) \tag{2.1}$$

or by the differential equation

$$dp_i/dt = \lambda_i Z_i(p(t)) \tag{2.2}$$

with constants λ_i, though any sign preserving function $\varphi_i(Z^i)$ could replace $\lambda_i Z_i$. Although the continuous time process (2.2) has been best studied, the discrete time process (2.1) seems to be more realistic (compare Saari, 1985, p.1119). In this note I only deal with discrete processes.

In (2.1) which I call the *additive* process, the price adjustment equals the level of the excess times a constant. An alternative is the *multiplicative* process

$$p_i(t+1) = p_i(t)(1 + \lambda_i Z_i(p(t))), \tag{2.3}$$

where the adjustment $p_i(t+1)-p_i(t) = \lambda_i p_i(t) Z_i(p(t))$ is multiplicative. (2.2) is known to converge (Arrow and Hahn, 1971, p.283) with two goods and isolated equilibria, or (with more goods) if Z_i satisfy the (not very plausible) condition of *gross substitutability*, that is if

$$\text{for all } i,j: \partial Z_i/\partial p_i < 0 \text{ and } \partial Z_i/\partial p_j > 0 \tag{GS}$$

(GS) also ensures the uniqueness of the equilibrium. Under a similar condition the continuous version of (2.3) converges (see Arrow and Hahn, p.293). Proving the convergence of (2.2) is very simple and for that reason probably continuous tatonnement is usually expressed by (2.2) rather than by the continuous version of (2.3). (2.1) however needs not to converge even under (GS). Arrow and Hahn (1971, p.308) note: "in the finite time adjustment rule stability depends on the fine properties of the adjustment rule itself even when so convenient a hypothesis as GS is made". I think that the multiplicative process is more realistic than the additive process. In the real world usually price adjustments are announced as percentage increases and decreases. Apart from that the multiplicative process is homogeneous of degree 1 in prices, hence if an initial price $p(0)$ generates a path $p(t)$, then the initial price $\alpha p(0)$ generates a path $\alpha p(t)$. It may be argued that the adjustment should rather depend on the relative excess, such that it becomes $\lambda_i(D_i/S_i - 1)$, D_i and S_i being obtained by adding individual demands and supplies separately. This however becomes quite complicated and probably does not affect general results. Below I study process (2.3) for a special case.

Day and Pianigiani (1991) give an example of discrete tatonnement in a two goods economy where (2.1) becomes chaotic, (although (2.2) converges). Saari (1985) and Bela and Majumdar (1992) give conditions on the occurrence of chaos in a two goods economy with (2.1). (2.3) also can be chaotic, as will be shown below. Typically in these non converging processes prices make big jumps up and down. I consider this to be unrealistic: the agents who adjust the price will generally adjust a price only by a small percentage. Therefore I replace (2.3) by a process where the price change is bounded by a maximum multiplicative adjustment. Given the numbers $R > 1$ and $r \in (0,1)$ it is required that $rp(t) \leq p(t+1) \leq Rp(t)$, and (2.3) is replaced by

$$p_i(t+1) = \begin{cases} p_i(t) + \min \{\lambda_i Z_i(p(t)),(R-1)\}p_i(t) & \text{if } Z_i(p(t) \geq 0 \\ p_i(t) + \max \{\lambda_i Z_i(p(t)),(r-1)\}p_i(t) & \text{if } Z_i(p(t) < 0 \end{cases} \tag{2.4}$$

hence maximal upwards or downwards adjustments are $(R-1)p_i(t)$ and $(1-r)p_i(t)$ respectively. I shall study this process in a Cobb-Douglas exchange economy with two and three goods (where (GS) is satisfied). In an equilibrium only the ratios between prices are determined, that is, if p^* is an equilibrium price, so is μp^* for any $\mu > 0$. One price, usually p_n, is chosen as a *numéraire* and fixed a priori. Then the tatonnement process only adjusts n-1 prices. The numéraire determines the price level. If the price of all n goods is adjusted, then it may happen that all prices go to zero or to infinite. This tends to happen in the cases studied below, so p_n will be fixed.

3. Tatonnement with two goods

First a two goods economy with Cobb-Douglas utility functions (which is a very regular case) is studied, as in Day and Pianigiani (1991). The demand functions satisfy (GS). Consumers $h = 1,2,...,n$ have resources $w^h = (w_1^h, w_2^h)$ and Cobb-Douglas utility functions ($0 < \alpha^h < 1$)

$$u^h = (x_1^h)^{\alpha^h}(x_2^h)^{1-\alpha^h}. \tag{3.1}$$

The excess demand function of h for good 1 follows from maximizing u^h under the budget constraint $p_1 x_1^h + p_2 x_2^h \leq p_1 w_1^h + p_2 w_2^h$ and, since $z_i^h = x_i^h - w_i^h$, it becomes

$$z_1^h(p_1,p_2) = -(1-\alpha^h)w_1^h + \alpha^h p_2 w_2^h/p_1 \equiv -A^h + B^h p_2/p_1$$

and by Walras law $z_2^h = -p_1 z_1^h/p_2$. Aggregate excess demand becomes

$$Z_1(p_1,p_2) = -\Sigma A^h + (p_2/p_1)\Sigma B^h \equiv -A + Bp_2/p_1 \tag{3.2}$$

with the equilibrium price vector satisfying $Z_1(p_1^*,p_2^*) = 0$, hence $p_1^* = (B/A)p_2^*$. Let $p_2 = 1$ be the numéraire, and denote the price of good 1 by p and excess demand of good 1 by $Z(p)$. The three discrete adjustment rules mentioned in section 2, are specified.
(i)   Additive price adjustment (2.1) gives

$$p(t+1) = p(t) + \lambda Z(p(t)) = p(t) - \lambda[A-B/p(t)] \tag{3.3}$$

with derivative at p^* equal to $1 - \lambda B/p^{*2} = 1 - \lambda A^2/B$, which is smaller than -1 at p^* if

$$\lambda > 2B/A^2, \tag{3.4}$$

which is possible for normal configurations of the parameters (see example below).
(ii)   Multiplicative adjustment (2.3) gives

$$p(t+1) = (1+\lambda Z(p(t)))p(t) = (1 + \lambda[-A + B/p(t)])p(t) = (1 - \lambda A)p(t) + B \tag{3.5}$$

with derivative smaller than -1 at p^* if

$$\lambda > 2/A, \tag{3.6}$$

which also may occur at normal parameter values.
(iii)   The ratio of excess demand D over excess supply S, with $D = \Sigma_h \max\{z^h,0\}$ and
$S = -\Sigma_h \min \{z^h,0\}$, gives

$$p(t+1) = [1 + \lambda(D/S -1)]p(t). \tag{3.7}$$

By summing over agents in excess demand and supply respectively, we obtain for any given p

$$D(p) = -A^+ + B^+/p \quad S(p) = A^- - B^-/p$$

with $A^+ + A^- = A$ and $B^+ + B^- = B$. Which agents are in excess supply depends on p, so A^- and A^+ depend on p. In p^*

$$D(p^*)/S(p^*) = (-A^+p^* + B^+)/(A^-p^* - B^-) = 1.$$

If in a neighborhood of p^*, the partition of the set of consumers is constant, then at p^*

$$\partial p(t+1)/\partial p(t) = 1 + \lambda[\partial(D/S)/\partial p]p^* = 1 + \lambda[A^+/A - B^+/B]^{-1} = 1 + \lambda[B/(A^+p^* - B^+)] < 0.$$

The process is not stable if

$$\lambda > 2(-A^+p^* + B^+)/B. \tag{3.8}$$

Example

Two consumers: $\alpha^1 = 0,2$, $\alpha^2 = 0,8$, $w^1 = (1,0)$, $w^2 = (0,1)$, giving $A = 0.8$, $B = 0.8$, $A^+ = 0$, $A^- = 0.8$, $B^+ = 0.8$;$B^- = 0$. Then by (3.4), (3.6) and (3.8) instability appears at $\lambda_1 > 2.5$, $\lambda_2 > 2.5$ and $\lambda_3 > 2$ respectively. If there are $2n$ consumers, n of each type, then $A = 0.8n$, $B = 0.8n$, $A^+ = 0$, $A^- = 0.8n$, $B^+ = 0.8n$, $B^- = 0$. Now instability obtains if $\lambda_1 > 2.5/n$, $\lambda_2 > 2.5/n$, $\lambda_3 > 2$. Hence in the first two cases, with high n, instability occurs already at low λ. Note however that nothing can be said a-priori on the value of λ.

I apply the restricted multiplicative process (2.4), for reasons given above. Although the last scheme may be still better, it is more complex, because at each p consumers have to be separated into two groups for each i. However, the results for the three cases seem not to be essentially different, since in each case λ can be found that make the difference equation unstable. If we insert (3.2) in (2.4) we get, with $p^* = B/A$ (compare (3.5))

$$p(t+1) \equiv \varphi(p(t)) = \begin{cases} \min\{(1-\lambda A)p(t)+\lambda B, Rp(t)\} & \text{if } p(t) < p^*, \\ \max\{(1-\lambda A)p(t)+\lambda B, rp(t)\} & \text{if } p(t) \geq p^*. \end{cases} \qquad (3.9)$$

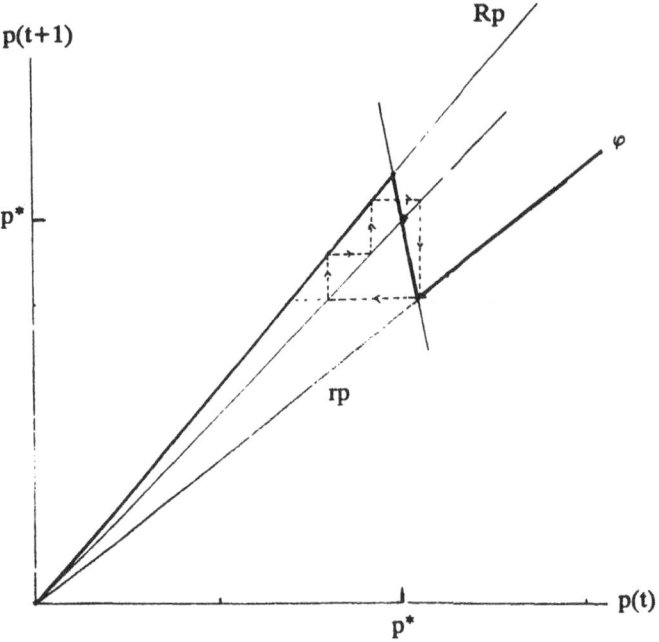

Figure 1. Function φ with 3-cycle (for $h = 1,2$, $\alpha^1 = 0.1$, $\alpha^2 = 0.9$, $w_1^1 = w_2^2 = 0$, $w_2^1 = w_1^2 = 1$, $r = 0.755$, $R = 1.15$, $\lambda = 1.48$).

φ, as specified by (3.9), is a continuous and piecewise linear function: $p+\lambda pZ(p)$ is linear and the upper and lower bounds $Rp(t)$ and $rp(t)$ are linear (see Fig. 1). p^* is the point where the 45° line intersects $\varphi(p) = p+\lambda pZ(p)$. For low values of λ, the slope at $\varphi(p^*)$ is larger than -1 and the process converges to p^*. For λ high enough, in the neighborhood of p^* the process diverges, but this divergence is stopped by the increasing parts of φ, given by Rp and rp. If $p(0)$ is low or high, initially the path will move in the direction of p^* and then it is trapped within a neighborhood $N = [p^-, p^+]$ of p^*, which depends on r and R and becomes smaller if $R-1$ and $1-r$ decrease. Inside N, φ is a kind of (possibly inverse) tent map. Inside N the path cannot converge, so it must be periodic or chaotic: in our numerical simulation a 3-cycle is the stable solution for some value of λ (see Figure 1), which by the Li-York theorem implies existence of paths of all periods and chaotic paths, for different values of λ. A bifurcation diagram is given in Fig. 3, which suggests chaos for sufficiently high λ.

Theorem 1: For φ given by (3.9), with $\lambda > 2/A$, there exist p^- and p^+, with $p^- \leq p^* \leq p^+$ such that

(i) $p^- \leq \varphi(p) \leq p^+$, if $p^- \leq p \leq p^+$;

(ii) $\varphi(p) > p$, if $p < p^-$ and $\varphi(p) < p$, if $p > p^+$;
(iii)   if $p(0) > 0$ then for some t, $p^- \leq p(t) \leq p^+$.

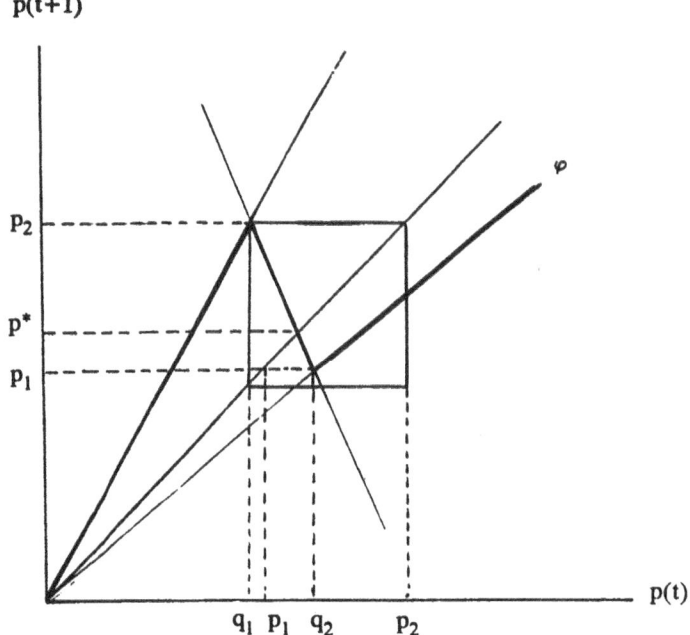

Figure 2. Construction of N.

Proof: Choose $p_2 = \max_{0<p<p^*}\varphi(p) = \varphi(q_1)$ and $p_1 = \min_{p>p^*}\varphi(p) = \varphi(q_2)$. By definition $q_1 < p^* < q_2$ and $p_1 < p^* < p_2$. Choose $p^- = \min\{p_1,q_1\}$ and $p^+ = \max\{p_2,q_2\}$ (see Fig. 2).

(i)   If $p < p^-$, then $p < q_1$, hence $\varphi(p) = Rp > p$. If $p > p^+$, $\varphi(p) = rp < p$.
(ii)   If $p \in [p^-,p^+]$, then $\varphi(p) \in [q_1,q_2] \subset [p^-,p^+]$.
(iii)   If $p(0) < p^-$, then $p(0) < p(1) = Rp(0) \leq p^+$, hence after a finite number of steps $p^- < p(t) \leq p^+$. Similarly for $p(0) > p^+$.   ∎

4. Tatonnement with three goods

The resources are $w^h = (w_1^h,w_2^h,w_3^h)$ and the utility functions become, with $\alpha^h+\beta^h+\gamma^h = 1$, $\alpha^h,\beta^h,\gamma^h > 0$

$$u^h = (x_1^h)^{\alpha^h}(x_2^h)^{\beta^h}(x_3^h)^{\gamma^h} .$$ (4.1)

Excess demands of h of goods 1 and 2, as functions of (p_1,p_2,p_3) are obtained by maximizing u^h under the constraint $p_1x_1^h+p_2x_2^h+p_3x_3^h \leq p_1w_1^h+p_2w_2^h+p_3w_3^h$, giving

$$z_1^h(p_1,p_2,p_3) = -(1-\alpha^h)w_1^h + \alpha^h p_2 w_2^h/p_1 + \alpha^h w_3^h p_3/p_1 \equiv -A_1^h + B_1^h p_2/p_1 + C_1^h p_3/p_1,$$

$$z_2^h(p_1,p_2,p_3) = -(1-\beta^h)w_2^h + \beta^h p_1 w_1^h/p_2 + \beta^h w_3^h p_3/p_2 \equiv -A_2^h + B_2^h p_1/p_2 + C_2^h p_3/p_2$$

and by Walras law $z_3^h = -(p_1 z_1^h + p_2 z_2^h)/p_3$. Aggregate excess demand is, for $p_3 = 1$ and $p = (p_1,p_2)$

$$Z_1(p) = -\Sigma A_1^h + (p_2/p_1)\Sigma B_1^h + \Sigma C_1^h/p_1 \equiv -A_1 + B_1 p_2/p_1 + C_1/p_1,$$

$$Z_2(p) = -A_2 + B_2 p_1/p_2 + C_2/p_2$$

(4.2)

with the unique equilibrium price p^* computed from $Z_1(p^*) = Z_2(p^*) = 0$, giving

$$p_1^* = (A_2 C_1 + B_1 C_2)(A_1 A_2 - B_1 B_2)^{-1}, \qquad p_2^* = (-B_2 C_1 + A_1 C_2)(A_1 A_2 - B_1 B_2)^{-1}.$$

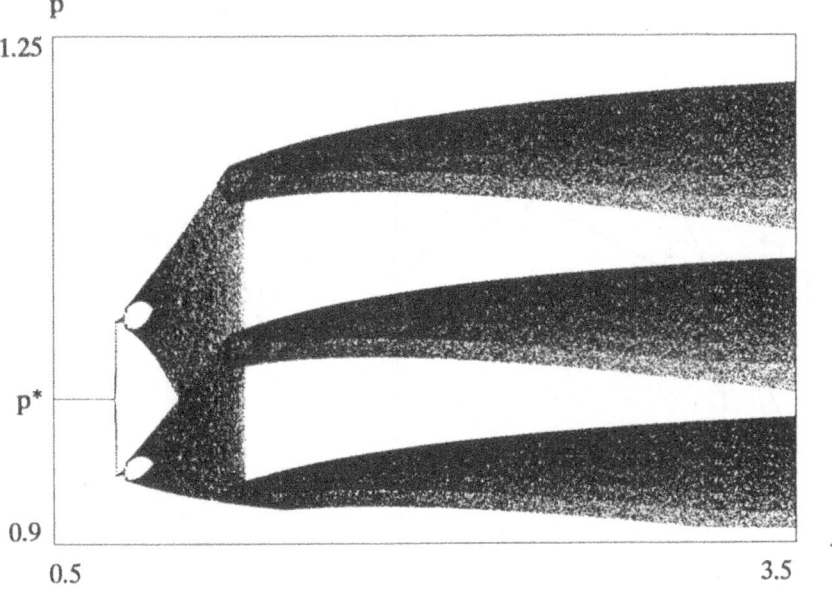

Figure 3. Bifurcation diagram for φ with increasing λ (with $r = 0.9$, $R = 1.25$; other parameters as in Figure 1).

Note that $p^* > 0$, since $A_1 A_2 > B_1 B_2$, by the definitions of these parameters. Applying (2.4), the price adjustments are given by the functions $\varphi_i:\mathbb{R}^2 \to \mathbb{R}$ ($i,j = 1,2$, $i \neq j$), with $\lambda = \lambda_1 = \lambda_2$

$$p_i(t+1) = \varphi_i(p(t)) = \begin{cases} \min\{(1-\lambda A_i)p_i(t)+\lambda B_i p_j(t)+\lambda C_i), Rp_i(t)\}, & \text{if } Z_i(p(t)) > 0, \\ \max\{(1-\lambda A_i)p_i(t)+\lambda B_i p_j(t)+\lambda C_i), rp_i(t)\}, & \text{if } Z_i(p(t)) \leq 0. \end{cases}$$

(4.3)

For $p(t)$ such that $r < 1+\lambda Z_i(p(t)) < R$, which is certainly holds near p^*,

$$p_i(t+1) = (1-\lambda A_i)p_i + \lambda B_i p_j + \lambda C_i$$

hence

$$(\partial\varphi_i/\partial p_j) = \begin{bmatrix} 1 - \lambda A_1 & \lambda B_1 \\ \lambda B_2 & 1 - \lambda A_2 \end{bmatrix} \equiv M.$$

The eigenvalues of M are

$$\mu = \tfrac{1}{2}[(1-\lambda A_1)+(1-\lambda A_2)] +/- \tfrac{1}{2}\{[(1-\lambda A_1)-(1-\lambda A_2)]^2 + 4\lambda^2 B_1 B_2\}^{\frac{1}{2}},$$

hence both roots are real. Values of the parameters with one or two roots smaller than -1 appear for sufficiently high values of λ and then the equilibrium p^* will be unstable.

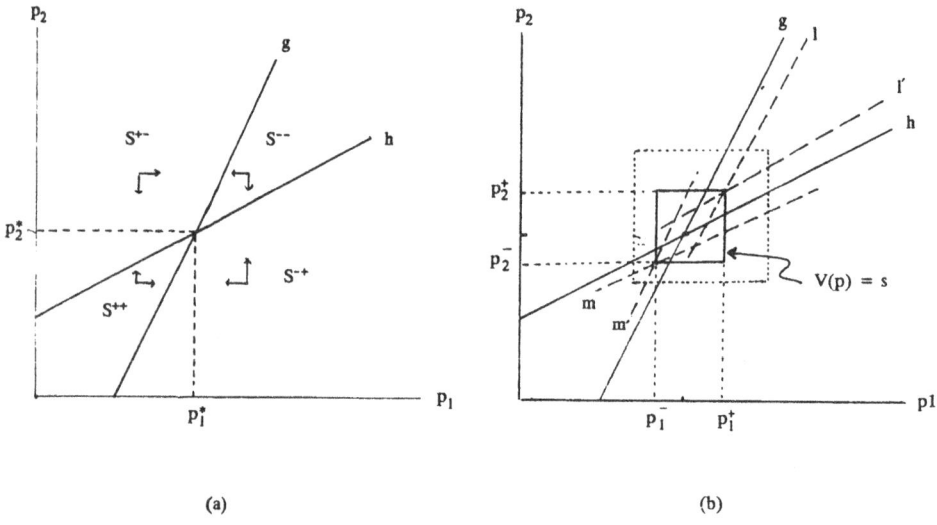

(a)   (b)

Figure 4   Regions S (a) and construction of N and $V(p) = s$ (b), (for $h = 1,2,3$; all CD parameters equal to $1/3$ and all $w_i^h = 1$, $r = 0.8$, $R = 1.2$).

It will be shown that, as in the two goods model, the path of prices generated by (4.3) will converge to a neighborhood of the equilibrium, and may show erratic behavior near p^*. Let the lines g and h be defined by

$$p_1 Z_1(p) = -A_1 p_1 + B_1 p_2 + C_1 = 0, \tag{g}$$

$$p_2 Z_2(p) = -A_2 p_2 + B_2 p_1 + C_2 = 0, \tag{h}$$

then $p*$ is the point where g and h intersect. g and h divide $\mathbb{R}^2_+$ into the following four regions

$$S^{+-} = \{p > 0 \mid Z_1(p) \geq 0,\ Z_2(p) \leq 0\}, \qquad\qquad S^{++} = \{p > 0 \mid Z_1(p) \geq 0,\ Z_2(p) \geq 0\},$$
$$\tag{4.5}$$
$$S^{--} = \{p > 0 \mid Z_1(p) \leq 0,\ Z_2(p) \leq 0\}, \qquad\qquad S^{-+} = \{p > 0 \mid Z_1(p) \leq 0,\ Z_2(p) \geq 0\}.$$

(see Fig. 4a). (Clearly $Z_1(p_1,p_2) \geq 0$ if and only if $p_1Z(p_1,p_2) \geq 0$, etc.) The directions of price adjustment in the regions differ. The set S^{+-} is bounded by g and h. If $p(t) \in S^{+-}$, then p_1 will increase and p_2 will decrease; $p(t+1)$ needs not lie in S^{+-}, but, because of the maximum adjustment, it cannot pass the lines l and m, defined by (see Fig. 4b)

$$Z_1(p_1/R, p_2/r) = 0, \tag{l}$$

$$Z_2(p_1/R, p_2/r) = 0, \tag{m}$$

since $p_1(t+1) \leq Rp_1(t)$ and $p_2(t+1) \geq rp_2(t)$. Similarly, for $p(t) \in S^{-+}$, $p(t+1)$ cannot pass the lines l' and m', defined by

$$Z_1(p_1/r, p_2/R) = 0, \tag{l'}$$

$$Z_2(p_1/r, p_2/R) = 0. \tag{m'}$$

Let p^+ and p^- be the intersections of l and l' and m and m' respectively and define

$$N = \{p \mid p^- \leq p \leq p^+\}. \tag{4.6}$$

Clearly $p^- \in S^{++}$ and $p^+ \in S^{--}$, hence $p_i^- < p_i^* < p_i^+$. $p* \in N$, hence N is a neighborhood of $p*$. N will shrink if $(R-1)$ and $(1-r)$ are reduced and it can be made as small as desired.

Theorem 2:

(i)   if $p(t) \in N$ then $p(t+1) \in N$,
(ii)   if $p(0) \notin N$, then there exist t such that $p(t) \in N$.

Proof:

(i) It is first proved that

$$p^- < rp^*,\quad Rp^* < p^+ \tag{*}$$

Let $q = (p_1^-/r, p_2^-/R)$, then $Z_2(q) = 0$ and $q < p^*$. Therefore $p_1^- = rq_1 < rp_1^*$. Similarly, for $q' = (p_1^-/R, p_2^-/r)$ and $Z_1(q') = 0$, it follows $p_2^- = rq'_2 < rp_2^*$. If $p(t) \in S^{++} \cap N$, then $p(t) < p^*$, hence by (*) $p(t) < p(t+1) \leq Rp(t) < Rp^* < p^+$. Similarly, if $p(t) \in S^{--} \cap N$, then $p(t) > p(t+1) > p^-$; hence $p(t+1) \in N$. If $p(t) \in S^{+-} \cap N$, then $p_1(t) > q_1$, hence $rp_1(t) \geq p_1^-$. Also $p_2(t) < p_2^*/R$, hence $Rp_2(t) < p_2^+$. Hence $p(t+1) \in N$. Similarly for $p(t) \in S^{-+}$. This proves (i).

(ii) A Lyapunov function $V(p)$ is constructed in the following way (se Fig. 4b)

$$V(p) = \min_{p' \in N} \max\{s \mid s(p'-p^*) \leq p-p^*\}, \qquad\qquad \text{if } p \notin N,$$

$$V(p) = 1, \qquad\qquad\qquad \text{if } p \in N,$$

hence for some p' on the boundary of N, $p-p^* = s(p'-p^*)$ and the set $\{p \mid V(p) = s\}$ is a square around N as in Fig. 4b. If $V(p) = s$ then either

$$p_i = sp_i^+ \text{ and } p_j^- \leq p_j \leq p_j^+, \qquad i \neq j,$$

or

$$p_i = sp_i^- \text{ and } p_j^- \leq p_j \leq p_j^+, \qquad i \neq j.$$

It remains to prove that $p(t) \notin N$, implies $V(p(t+1)) < V(p(t)) = s$.
(a) Assume that $p(t) \in S^{+-}W$. Then $p_1(t+1) > p_1(t)$ and $p_2(t+1) < p_2(t)$. If $p(t+1) \in N$, then $V(p(t+1)) = 1 < s$. So let $p(t+1) \notin N$. We either have

$$p_1(t)-p_1^* < s(p_1^+ - p_1^*) \text{ and } p_2(t)-p_2^* = s(p_2^+ - p_2^*) \qquad\qquad (i)$$

(because $p^*+s(p^+-p^*) \notin S^{+-}$), or

$$p_1(t)-p_1^* = s(p_1^- - p_1^*) < 0 \text{ and } p_2(t)-p_2^* > s(p_2^- - p_2^*). \qquad\qquad (ii)$$

If (i), then $p_2(t+1) < p_2(t)$, hence $p_2(t+1) - p_2^* \equiv s'(p_2^+ - p_2^*)$, with $s' < s$; $p_1(t+1) > p_1(t)$, but $p_1(t+1)-p_1^* < s'(p_1^+ - p_1^*)$, since $p^*+s'(p^+-p^*)$ lies to the left of l. Hence $V(p(t+1)) = s' < s$. Similarly, if (ii) $p_1(t+1) > p_1(t)$, hence $-(p_1(t+1)-p_1^*) \equiv -s'(p_1^- - p_1^*) < -s(p_1^- - p_1^*)$ with $s' < s$, while $p_2(t+1) < p_2^*) + s'(p_2^- - p_2^*)$, since $p^*+s'(p^--p^*)$ lies below l'. Hence $V(p(t+1)) = s' < s$.
An analogous argument holds if $p(t) \in S^{-+}W$.
(b) Assume $p(t) \in S^{--}W$. Then $p_1(t) - p_1^* \leq s(p_1^+ - p_1^*)$ and $p_2(t) - p_2^* \leq s(p_2^+ - p_2^*)$, with at least one equality. Since $p(t+1) < p(t)$, $V(p(t+1)) < s$. An analogous argument holds if $p(t) \in S^{++}W$. ∎

Outside N $p(t)$ can only jump in one step from S^{+-} to S^{++} or S^{--}, but not to S^{-+}. This enforces that it comes closer to N. But inside N it can (and will) jump from S^{+-} to S^{-+} and back, without approaching p^*, following an erratic path, which may be cyclic, but usually looks chaotic, as simulations suggest where $p(t)$ mainly alternates between S^{+-} and S^{-+} in an irregular way, sometimes passing also through S^{--} or S^{++}. In Fig. 5(a) and (b) the attractors are depicted for two different values of λ. Note that both pictures 5(a) and 5(b) lie completely inside N, which is defined by $p_1^- = p_2^- = 0.76$ and $p_1^+ = p_2^+ = 1.82$.

5. Conclusion

It has been shown that for the Cobb-Douglas economy with two or three goods, though in general the discrete tatonnement process does not converge to the unique equilibrium, a restriction on the rate of price adjustment enforces convergence to the neighborhood of the equilibrium. Within that neighborhood erratic dynamics occur. It seems that due to the restriction the discrete system behaves "more similar" to the continuous process, which, as noted in the introduction, converges in the cases studied above. The equations applied above are piecewise linear. But the behavior of the system will be similar, if the system is smoothened (see Weddepohl, 1992).

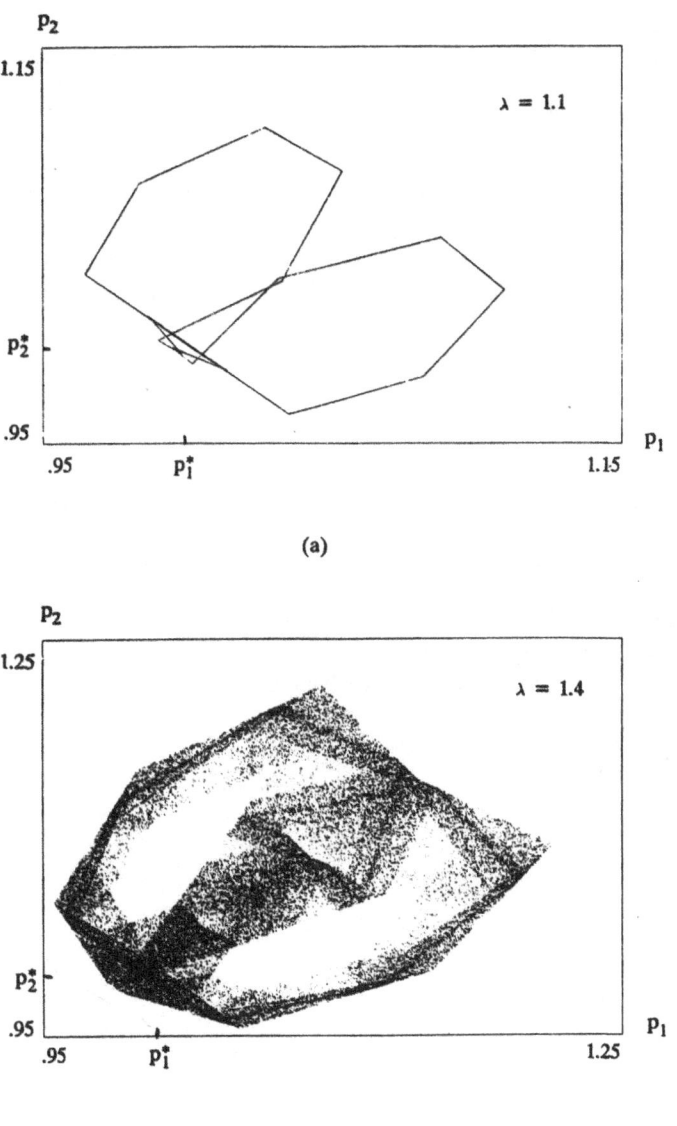

(a)

(b)

Figure 5   Attractors of φ for λ = 1,1 (a) and λ = 1.4 (b), (for $r = 0,95$ and $R = 1.25$, other parameters as in Figure 4).

Theorem 1 for the two goods economy can be generalized to an economy with several isolated equilibria (see Weddepohl, 1992). It is conjectured that also theorem 2 can be applied to economies where (GS) is satisfied, since the precise structure of demand is not used in the proof. A proof that in the three goods economy chaos can obtain, is still lacking, but simulations suggest that this is indeed the case: Fig. 5a suggests some kind of periodicity, whereas Fig. 5b suggests chaos. This and the n-goods economy will be subjects of further research.

Finally it must be noted that, since the system does not converge, the auctioneer has to decide when to announce the final (non-equilibrium) price at which trade is to be concluded. Some rationing will then be necessary.

Acknowledgement

The author thanks Cars Hommes for valuable discussions and for help with computations and with the construction of the figures 3 and 5. He also thanks an unknown referee for his comments.

References

Arrow, K.J. and F. Hahn (1971), 'General competitive analysis', Holden Day, San Francisco.

Bela, V and M. Majumdar (1992), 'Chaotic tatonnement', *Economic Theory*, **2**, 437-445.

Day, R.H. and G. Pianigiani (1991), 'Statistical dynamics and Economics', *Journal of Economic Behavior and Organization*, **16**, 37-83.

Saari, D.G. (1985), 'Iterative price mechanisms', *Econometrica*, **53**, 1117-1131.

Weddepohl, C. (1992), 'Restricting erratic dynamics in a tatonnement process', A.E. Report, 12/92.

CHAOTIC DYNAMICS IN A TWO-DIMENSIONAL OVERLAPPING GENERATION MODEL: A NUMERICAL INVESTIGATION

CARS H. HOMMES[1]

SEBASTIAN J. VAN STRIEN[2]

ROBIN G. DE VILDER[1,3]

[1] Dept. Economic Statistics, University of Amsterdam,
Roetersstraat 11, NL-1018 WB Amsterdam, The Netherlands.

[2] Dept. Mathematics, University of Amsterdam,
Plantage Muidergracht 24, NL-1018 TV Amsterdam,The Netherlands.

[3] The Tinbergen Institute, University of Amsterdam,
Roetersstraat 11, NL-1018 WB Amsterdam, The Netherlands.

Abstract

Grandmont (1985) showed that in a 1-dimensional overlapping generations (OLG-) model chaotic fluctuations can occur, when the traders offer curve is highly nonlinear due to a very strong income effect. We present a numerical investigation of the global dynamics of a 2-dimensional OLG-model as introduced by Grandmont (1992). Chaotic output fluctuations already arise when the income effect is not too strong.

keywords: nonlinear business cycles, Hopf bifurcation, chaos.

1. Introduction

There are two different ways of explaining observed economic fluctuations. According to the first business cycles are caused by repeated exogenous, random shocks to the fundamentals of the economic system. In the absence of these exogenous forces, the economy would converge to a stable equilibrium time path. According to the second, endogenous explanation a significant part of the fluctuations are caused by nonlinear deterministic "economic laws". Even in the absence of exogenous shocks, economic variables could exhibit oscillatory behaviour.

Early attempts to explain economic fluctuations by nonlinear models are due to Kaldor (1940), Hicks (1950) and Goodwin (1951). These models possessed several shortcomings however. For example there was no optimizing behaviour, and systematic forecasting errors were made by the agents. Thereafter macroeconomics was dominated by (linear) models with a globally stable equilibrium, including optimizing behaviour and self-fulfilling expectations, and subject to exogenous random shocks. In the last decade however, the discovery of chaotic, random looking behaviour in deterministic nonlinear models, has renewed the interest in an endogenous explanation of business cycles, see e.g. Lorenz (1989), Hommes (1991) and Medio (1992).

An important class of economic models are the overlapping generation (OLG) models. In OLG models the standard assumptions that traders optimize, expectations are self-fulfilling and markets clear at every date are satisfied. It has been shown by Benhabib and Day (1982) and Grandmont (1985) that periodic as well as chaotic equilibrium time paths can occur in a simple version of the OLG-model, where the output time paths are descri-bed by a 1-dimensional difference equation $y_{t+1} = \chi(y_t)$ with χ the trader's offer curve. Periodic and chaotic dynamics arise, because of the non-monotonic shape of the offer curve, due to a conflict between the substitution and the income effects. Although this result seems to be of fundamental importance to economic theory, one should add that chaos can only arise when the income effect is very strong, perhaps too strong to be empirically plausible.

Recently Grandmont (1992) proposed a two-dimensional OLG-model with productive investment as in Reichlin (1986). The output time paths are described by a second order, nonlinear difference equation. Grandmont (1992) showed that periodic and quasi-periodic behaviour may occur, even when the substitution effect dominates the income effect everywhere, i.e. when the trader's offer curve is monotonic. The occurrence of oscillatory behaviour follows from a local analysis of the dynamics near the stationary state. A Hopf-bifurcation occurs when a parameter (the concavity of the old trader's utility function) is increased.

The motivation for the present paper is to investigate the global dynamics of the 2-D OLG-model in Grandmont (1992). At this point we would like to refer to Medio (1992, chapter 12) who investigates the dynamics of a related, but different 2-D OLG-model introduced by Reichlin (1986). In particular, we focus on the question whether or not in addition to periodic and quasi-periodic output time paths, chaotic fluctuations can occur when the income effect is not too strong. In the 1-D OLG-model the question whether or not chaos occurs can be answered by investigating the shape of the offer curve. In the 2-D OLG-model the analysis of the global dynamics is much more difficult. Therefore, in this paper, as a first step, we present a numerical analysis of the global dynamics. The paper is organized as follows. In section 2 we briefly discuss the results for the 1-D OLG-model. Section 3 deals with the 2-D version and presents a stability analysis of the steady state and numerical results concerning the global dynamics. Finally some concluding remarks are given in section 4.

2. A 1-D OLG-model

In order to be selfcontained we briefly recall a simple version of the OLG-model following the lines of Grandmont (1985, 1992). Assume that people participate in the market for two periods, work and save wages when young and consume when old. The total money stock is assumed to be constant and no bequests are allowed. One unit of output (a perishable consumption good) is produced from one unit of labour in the same period. The house-hold born at time t supplies the quantity of labour $0 \leq l_t \leq l_1^*$, where $l_1^* > 0$ denotes the labour endowment of the young generation, saves the amount of money $m_t \geq 0$, and consumes $c_{t+1} \geq 0$ of the perishable consumption good at the next time $t+1$. Markets for consumption goods, labour and money are competitive, so that the price of consumption goods and the money wage can be identified. The household born at time t will choose his current labour supply l_t, his current demand for money m_t and plan his future consumption c_{t+1}, in order to maximize his utility function under the budget constraint $p_t l_t = m_t = p_{t+1} c_{t+1}$, where p_{t+1} denotes the expected price for period $t+1$ made at

t. The trader's maximization problem yields the first order condition

$$l_t \, V_1'(l_1^* - l_t) = c_{t+1} \, V_2'(c_{t+1}), \tag{1}$$

where V_1 and V_2 are the utility function of the young and the old household respectively, $V_1'(l_1^*-l_t)$ is the marginal utility of leisure and $V_2'(c_{t+1})$ is the marginal utility of future consumption which are both positive. Writing $v_1(l_t)$ and $v_2(c_{t+1})$ for the left and right hand side of (1) yields

$$v_1(l_t) = v_2(c_{t+1}). \tag{2}$$

From the usual assumption that the marginal utility function is decreasing it follows easily that v_1 is increasing so that v_1 has an inverse v_1^{-1}. Therefore (2) can be rewritten as

$$l_t \equiv v_1^{-1} \, [v_2(c_{t+1})] \equiv \chi[c_{t+1}]. \tag{3}$$

The trader's offer curve χ consists of all optimal pairs (l_t, c_{t+1}) of current labour and future consumption. The shape of the offer curve χ turns out to be very important for the dynamical behaviour of the model. When the price of future consumption increases this will give rise to both substitution and income effects. In terms of labour supply the two effects work in the opposite direction: the substitution effect generates less labour supply while the income effect generates more labour supply. In terms of the offer curve (3) this means that the function χ increases when the substitution effect dominates and decreases when the income effect dominates.

In equilibrium at time t, $l_t = y_t = c_t$. It is assumed that agents have perfect foresight, that is the consumption by the old traders is what they planned to consume when they were young, so that $c_{t+1} = y_{t+1}$. A deterministic equilibrium time path with perfect foresight is given by a sequence $y_t \geq 0$ satisfying

$$y_t = \chi(y_{t+1}), \qquad \forall \, t \geq 0. \tag{4}$$

Notice that in (4) the present output y_t is a function of the future output y_{t+1}. This may seem awkward, but it is due to the assumption of perfect foresight. We say that (4) describes the backward perfect foresight (bpf) dynamics. Of course we are not interested in predicting the past, but we are interested in the forward perfect foresight (fpf) dynamics, that is all possible sequences of future outputs satisfying (4). The fpf dynamics can be described by the difference equation $y_{t+1} = \chi^{-1}(y_t)$. However, as already noted, the map χ can be non-monotonic, and in that case the inverse χ^{-1} is not uniquely defined. In order to obtain information about the fpf dynamics, it is useful to investigate the bpf dynamics first, and translate the results to the fpf dynamics afterwards.

Whether the map χ is increasing or decreasing is in fact determined by the "concavity" of the old trader's utility function, which can be measured by the so called Arrow-Pratt "relative degree of risk aversion" $R_2(c) = -cV_2''(c)/V_2'(c)$. Straightforward computation shows that χ is increasing when $R_2(c) < 1$, while χ is non-monotonic if $R_2(c) > 1$. In particular, under the assumption that $R_2(c)$ is increasing and $\sup\{R_2(c)\} > 1$, the map χ has a unique critical point y^* where χ has a maximum, χ is increasing for $0 < y < y^*$, while χ is decreasing for $y > y^*$. Chaos arises when the map χ has a "sufficiently large hump". Grandmont (1985) gives sufficient conditions, in terms of the trader's utility function, for the occurrence of chaos. We would like to end this section by discussing the numerical results of Grandmont (1985) with the trader's utility functions given by

$$V_1(l_1^* - l) = \frac{(l_1^* - l)^{1-\alpha_1}}{1-\alpha_1}, \qquad V_2(c) = \frac{(c + l_2^*)^{1-\alpha_2}}{1-\alpha_2}. \qquad (5)$$

The parameter l_2^* is the labour endowment of the old generation (cf. Grandmont (1985)) and it is important for the shape of the offer curve. For $l_2^* = 0$, the offer curve is increasing when $\alpha_2 < 1$ and decreasing when $\alpha_2 > 1$. For $l_2^* > 0$ the offer curve is increasing when $\alpha_2 < 1$ and non-monotonic with a maximum at $y = l_2^*/(\alpha_2 - 1)$ when $\alpha_2 > 1$. We will use the same utility functions in our numerical work concerning the 2-D OLG-model in the next section, and compare the results. Notice that the relative degree of risk aversion is $R_2(c) = \alpha_2 c/(c + l_2^*)$, so for $l_2^* > 0$, $R_2(c)$ increases from 0 to α_2 when c increases from 0 to ∞. We fix the parameters $\alpha_1 = 0.5$, $l_1^* = 2$, $l_2^* = 0.5$. The bifurcation diagram in figure 1 shows how the long run dynamical behaviour (in the bpf dynamics) of output changes, as the parameter α_2 is increased from 2 to 16. The bifurcation diagram shows the familiar period doubling route to chaos, arising at $\alpha_2 \approx 9$. The conclusion of this experiment is that when the relative degree of risk aversion of the old traders is increased the bpf dynamics (and therefore also the fpf dynamics) becomes more and more complicated. However, chaos only occurs for large values of α_2, when the graph of the map χ displays a large 'hump'. In the 1-D OLG-model chaos may arise, but only when the income effect is very strong.

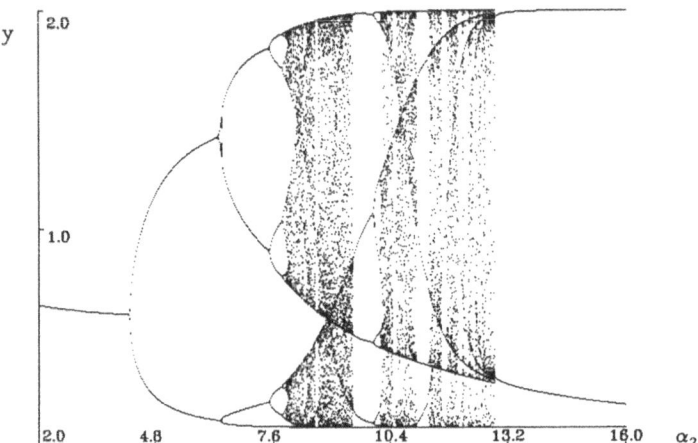

Figure 1. Bifurcation diagram of the 1-D OLG-model, with respect to the parameter α_2, with $\alpha_1 = 0.5$, $l_1^* = 2$ and $l_2^* = 0.5$.

3. A 2-D OLG-model with productive investment

Recently Grandmont (1992) proposed a 2-dimensional version of the OLG-model with productive investment as in Reichlin (1986). By a local analysis of the dynamics around the stationary state, Grandmont showed that a Hopf-bifurcation may occur, in which the steady state changes stability and an invariant closed curve, exhibiting periodic and quasi-periodic behaviour, is created. Our aim here is to investigate the global dynamics of the 2-D OLG-model. In particular we are interested in the question whether chaotic fluctuations can arise.

In the 2-D OLG-model it is assumed that output y_t in period t is produced from labour supply l_t by the young household and from the capital stock k_{t-1} that is available at the beginning of the period. Both inputs are used in fixed proportions, $y_t = \min\{l_t, k_{t-1}/a\}$, where a is the capital output ratio. In equilibrium in period t output consists of consumption c_t by the old generation and investment i_t, i.e. $y_t = c_t + i_t$. The capital stock available at the beginning of period $t+1$ is:

$$k_t = k_{t-1}(1-\delta) + i_t, \quad 0 < \delta \le 1, \tag{6}$$

where δ is the depreciation rate of capital. It is assumed that workers have no access to ownership of capital, while the production sector is operated by entrepreneurs who maximize profits. Households choices for current labour supply l_t and future consumption c_{t+1} are, as in the 1-D version, determined by the offer curve χ, that is $l_t = v_1^{-1}[v_2(c_{t+1})] = \chi(c_{t+1})$ In equilibrium the productive sector yields $y_t = l_t = k_{t-1}/a$. Under the assumption of perfect foresight expected consumption $c_{t+1} = y_{t+1} - i_{t+1}$. Using (6) future investment can be written as $i_{t+1} = k_{t+1} - k_t(1-\delta)$, and substituting $k_{t-1} = ay_t$ yields

$$i_{t+1} = ay_{t+2} - a(1-\delta)y_{t+1} = a[y_{t+2} - y_{t+1}(1-\delta)]. \tag{7}$$

Expected consumption is then given by

$$c_{t+1} = y_{t+1} - a[y_{t+2} - y_{t+1}(1-\delta)] = [1 + a(1-\delta)]y_{t+1} - ay_{t+2}. \tag{8}$$

By substituting (8) and $l_t = y_t$ into $l_t = \chi(c_{t+1})$ we conclude that intertemporal equilibria with perfect foresight are characterized by sequences of outputs $y_t \ge 0$ satisfying

$$y_t = \chi[(1 + a(1-\delta))y_{t+1} - ay_{t+2}]. \tag{9}$$

We have obtained a second order nonlinear difference equation describing the backward perfect foresight dynamics of output. When the map χ is not invertible, the forward perfect foresight dynamics is not uniquely defined. In order to obtain information about the fpf dynamics in that case, it is useful to investigate the bpf dynamics first, by substituting $z_t = y_{-t}$. If we take $n=-t$ and $p_n = z_{n-1}$ and $q_n = z_{n-2}$ we obtain a 2-D difference equation $(p_{n+1}, q_{n+1}) = F(p_n, q_n)$, where the map F is given by

$$F(p,q) = (\chi([1+a(1-\delta)]p - aq),\ p). \tag{10}$$

This 2-D difference equation is equivalent to the second order difference equation generating the bpf dynamics. Note that the χ is only defined for nonnegative values and $\chi(0)=0$. The function F is defined only for nonnegative values of p and q or if the argument of χ is positive, and $F(0,0)=0$. The point $(0,0)$ is called the autarkic steady state, where consumption and output are 0. We extend the definition of the map F in a natural way. When $[1+a(1-\delta)]p-aq < 0$ we define $F(p,q)=(0,p)$, and $F^2(p,q) =F(0,p)=(0,0)$. Hence, these points tend to the autarkic steady state. In the next section we investigate the stability of the steady state of F.

3.1 Stability analysis of the steady state

The positive steady state $y^s > 0$ of the model is determined by the equation $y^s = \chi\ (c^s)$, where $c^s=(1-a\delta)y^s$ is the steady state consumption. The steady state y^s exists and is unique if and only if $a\delta < 1$ and the slope of the offer curve at the origin is larger then $1/(1-a\delta)$. The stability of the equilibrium is determined by the eigenvalues of the Jacobian matrix $JF(y^s,y^s)$, where F is the 2-D map in (10) and (y^s,y^s) is the (positive) fixed point of F. The Jacobian matrix $JF(y^s,y^s)$ is

$$JF(y^s,y^s) = \begin{pmatrix} \chi'(c^s)(1+a(1-\delta)) & -a\chi'(c^s) \\ 1 & 0 \end{pmatrix} \tag{11}$$

Writing $\sigma = \chi'(c^s)$ the characteristic equation is given by

$$\lambda^2 - \lambda\sigma[1+a(1-\delta)] +a\sigma = 0. \tag{12}$$

Straightforward computation yields the stability results, of the steady state (y^s,y^s) as summarized in table I, under the assumption $a > 1/(1+\delta)$. Notice that table I summarizes the stability results for the bpf dynamics. The stability of the equilibrium in the fpf dynamics is obtained by reversing the results in table I. It can be seen from table I that a Hopf bifurcation may occur when $\sigma=\chi'(c^s)=\frac{1}{a}$ and a period doubling bifurcation when $\sigma = -1/(a(2-\delta)+1)$. Hence, a Hopf bifurcation can arise even when the offer curve is increasing, while a period doubling bifurcation can only arise when the offer curve is decreasing at the steady state consumption.

3.2 Some numerical results with the 2-D OLG-model

In order to investigate the global dynamics we have to make a choice for the trader's utility functions. As for the 1-D OLG-model, we choose the utility functions defined in (5). We fix the parameters $\alpha_1=0.5$, $l_1^*=2$, $l_2^*=0.5$, as in section 2, the capital output ratio $a=3$ and the depreciation rate of capital $\delta=0.05$. We investigate how the dynamics depends upon the parameter α_2, measuring the concavity of the old traders utility function. In figure 2 graphs of the offer curve χ are shown for different values of α_2. For $\alpha_2<1$ χ is increasing, while for

$\alpha_2>1$ χ is non-monotonic, having a unique critical point $x = l_2^{*}/(\alpha_2-1)$ where χ assumes its maximum. In figure 2 we have also drawn the line $y = x/(1-a\delta)$. The steady state y^s is the x-coordinate of the intersection point of the graph of χ with the line $y = x/(1-a\delta)$. It can be seen from figure 2 that the steady state y^s is almost constant ($y^s \approx 0.59$) for α_2 between 0.5 and 6.

Table 1. Stability results (bpf dynamics) of the steady state y^s, when $a > 1/(1+\delta)$.

$\sigma = \chi'(c^s)$	conclusion
$\sigma \geq \dfrac{4a}{[1+a(1-\delta)]^2}$	unstable with positive real eigenvalues
$\dfrac{1}{a} < \sigma < \dfrac{4a}{[1+a(1-\delta)]^2}$	unstable with complex eigenvalues
$0 < \sigma < \dfrac{1}{a}$	stable with complex eigenvalues
$\dfrac{-1}{a(2-\delta)+1} < \sigma < 0$	stable with real eigenvalues, $-1 < \lambda_1 < 0 < \lambda_2 < 1$
$\sigma < \dfrac{-1}{a(2-\delta)+1}$	unstable with real eigenvalues, $\lambda_1 < -1 < 0 < \lambda_2 < 1$

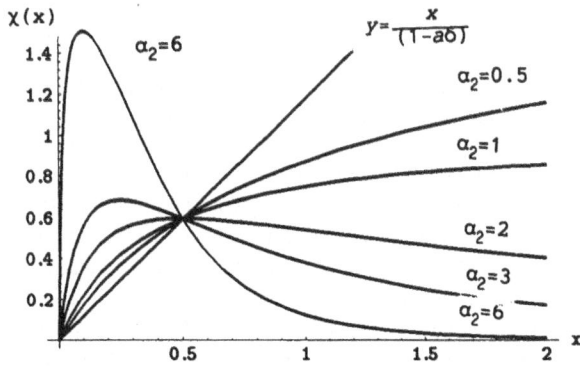

Figure 2. Graphs of the offer curve χ for different values of α_2.

Recall from table I that we have stated the stability results for the steady state, in terms of the slope $\sigma = \chi'(c^s)$. In figure 3 we have plotted σ as a function of the parameter α_2. It is easy to show that σ decreases as α_2 increases. The horizontal dotted lines $y=1/a$ and $y=-1/(a(2-\delta)+1)$ correspond to the σ-values where the steady state changes stability. The steady state is unstable for σ above the line $y=1/a$ or below the line $y=-1/(a(2-\delta)+1)$, and stable for σ between these two lines. For the α_2-value corresponding to the intersection point with the line $y=1/a$ ($\alpha_2 \approx 1.306$) two complex eigenvalues cross the unit circle and a Hopf-bifurcation occurs. With our choice of the utility functions the relative degree of risk aversion

$$R_2(c^s) = \alpha_2 \, c^s / (c^s - l_2^*) \approx 0.5 \,(1.306) \approx 0.65,$$

so that the Hopf bifurcation indeed occurs when $R_2(c^s)$ is increased from 0 to 1, as in Grandmont (1992). For the α_2-value corresponding to the intersection point with the line $y=-1/[a(2-\delta)]$ ($\alpha_2 \approx 2.292$) the smallest real eigenvalue is -1 and a period doubling bifurcation occurs.

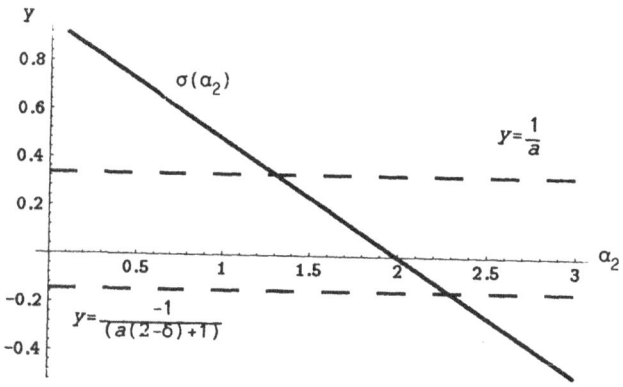

Figure 3. Graph of the slope $\sigma=\chi'(c^s)$ as a function of α_2.

First we concentrate on the Hopf-bifurcation. As α_2 is increased and crosses the bifurcation value $\alpha_2 \approx 1.306$, the steady state, which was unstable (bpf dynamics), becomes stable and an unstable invariant closed curve is created. In figure 4a and 4b two phase portraits in the (p,q)-plane are shown, one before and one after the Hopf bifurcation. In figure 4a the steady state (y^s,y^s) is unstable, while in figure 4b it is stable. The dark region in the phase space consists of all states (p,q) converging to the autarkic steady state $(0,0)$,while the white region in figure 4b consists of all states (p,q) that are attracted by the stable steady state (y^s,y^s).The boundary between the white and the dark regions is the unstable invariant closed curve created in the Hopf bifurcation. For α_2-values close to the bifurcation value, the dynamical behaviour on the invariant closed curve is either periodic or quasi-periodic. For larger α_2-values more complicated dynamical behaviour might occur.

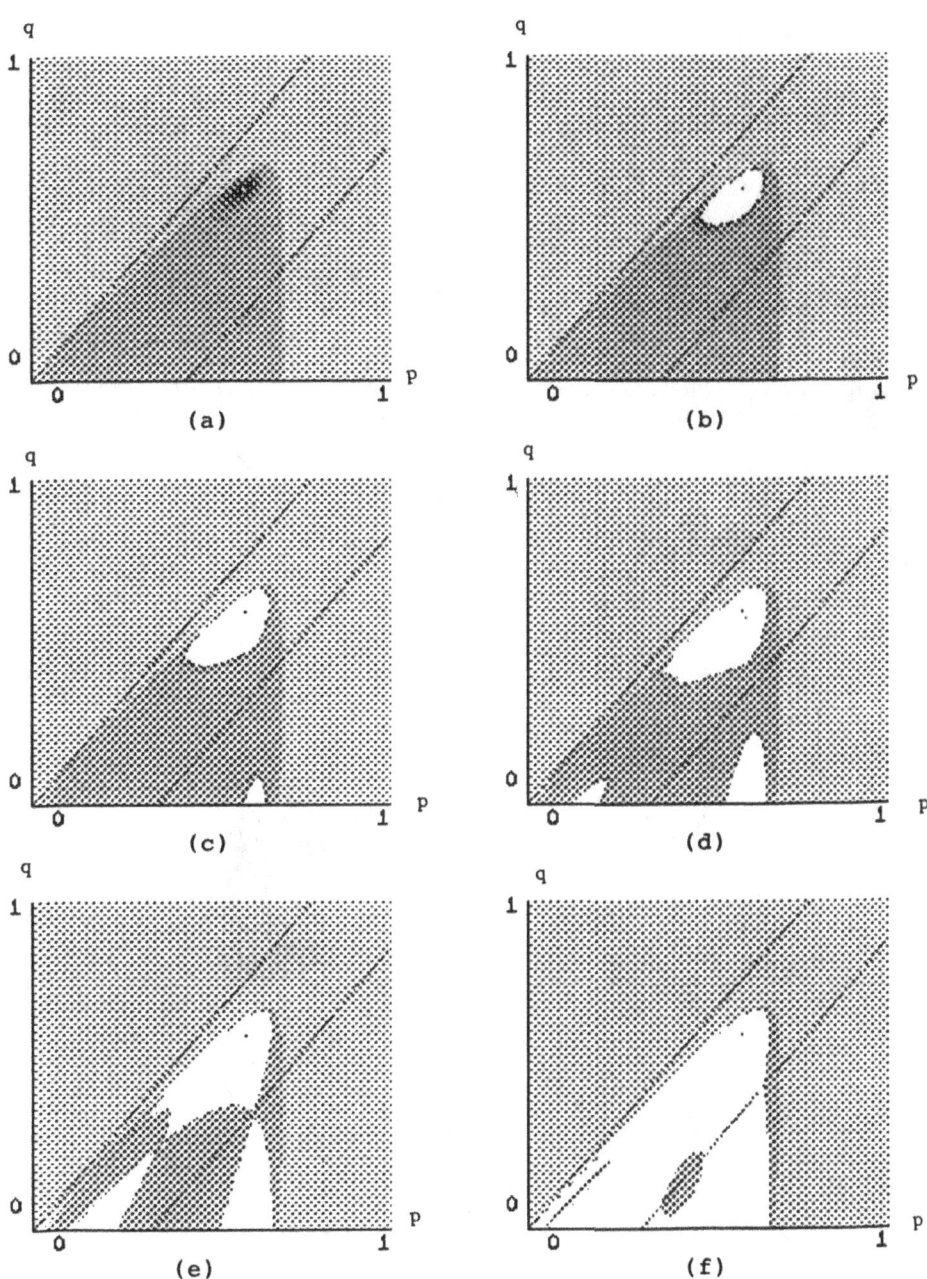

Figure 4. Phase portraits for different values of α_2.

(a) $\alpha_2 = 1.29$ (d) $\alpha_2 = 1.374$
(b) $\alpha_2 = 1.34$ (e) $\alpha_2 = 1.383$
(c) $\alpha_2 = 1.36$ (f) $\alpha_2 = 1.404$

We emphasize that the invariant closed curve, which is unstable in the bpf dynamics, is stable in the fpf dynamics. Therefore, it is important to investigate what happens to this invariant closed curve when α_2 is further increased. Figures 4c-f show 4 phase portraits for different values of α_2. All points lying above or on the upper dotted line $q=(5/a+1-\delta)p$ in figures 4c-f, are mapped onto the autarkic steady state in at most two time periods. For all points lying on the lower dotted line $q=(1/a+1-\delta)p - l_2^*/(a(\alpha_2-1))$ output in the next period reaches its maximum value $\chi(x^*)$, where $x^*=l_2^*/(\alpha_2-1)$ is the critical point of the offer curve. Figures 4c-f show that the invariant closed curve becomes larger and more irregular when α_2 is increased. In figure 4c at the left hand side, a second white region appears. This white region is mapped into the white region inside the invariant closed curve and is then attracted by the stable steady state. The boundary of the second white region is the preimage of the invariant curve, and appears because of the unimodal shape of the map χ for these α_2-values. Note that the lower dotted line where output reaches its maximum value is between the invariant closed curve and the presumed preimage of this curve. When α_2 is further increased (figure 4d) a third white region appears, which is mapped through the previous preimage inside the invariant closed curve. When α_2 is increased further the invariant closed curve "collides" with the presumed preimages (see figure 4e) and after the "collision" it seems that the invariant closed curve disappears, see figure 4f. There should still be some (unstable) invariant set, but apparently it is not a closed curve anymore in the plane (p,q). This observation certainly deserves a more detailed analysis in future research. When α_2 is further increased it seems that the dynamics generally stays the same for a while, i.e. the steady state is stable and the shape of the white region changes only slightly.

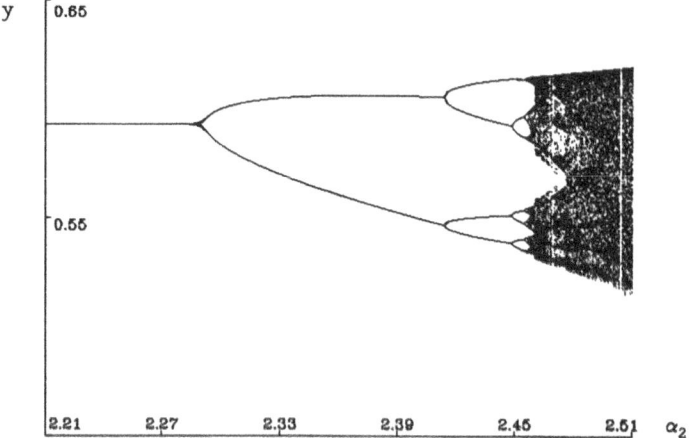

Figure 5. Bifurcation diagram of the 2-D OLG-model with respect to α_2, with $\alpha_1=0.5$, $l_1^*=2$, $l_2^*=0.5$, $a=3$ and $\delta=0.05$.

Next we investigate what happens after the period doubling bifurcation for $\alpha_2 \approx 2.292$. Figure 5 shows a bifurcation diagram for α_2 between 2.2 and 2.51. The bifurcation diagram shows the familiar period doubling route to chaos. The first accumulation point of period doubling bifurcations occurs for $\alpha_2 \approx 2.47$. Figure 6 shows the attractor of the system in the phase space, for α_2=2.5. It seems that a strange chaotic attractor occurs. For $\alpha_2 > 2.505$ the strange, chaotic attractor suddenly disappears, and it seems that almost all points are attracted to the autarkic steady state. A strange, chaotic attractor in the bpf dynamics corresponds to a strange, chaotic repellor in the fpf dynamics. Initial states close to this repellor will behave erratic, in the fpf dynamics, for a long time. Thereafter, the forward time paths might converge to a stable steady state or to a stable periodic orbit. On the other hand the fpf time paths also might converge to a strange, chaotic attractor (a repellor in the bpf dynamics). Which of these possibilities actually occurs remains to be seen.

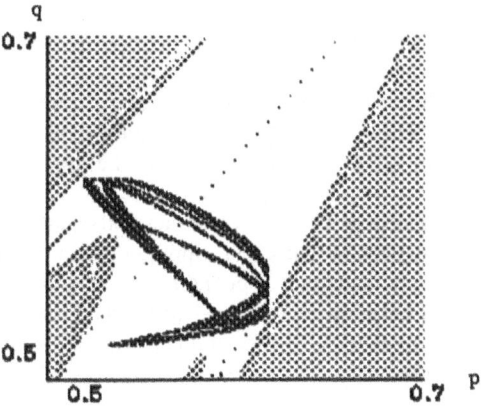

Figure 6. Strange, chaotic attractor of the 2-D OLG-model with $\alpha_2 = 2.5$.

We emphasize that the period doubling route to chaos occurs for much lower values of α_2 than in the 1-D OLG-model. In the 1-D OLG-model, for $\alpha_2 < 4$, the steady state is even globally stable. In the 2-D OLG-model, with the realistic extension of productive investment, complicated dynamical behaviour can occur for much more realistic parameter values.

4. Concluding remarks

We have investigated the dynamics of a 2-D OLG-model with productive investment, introduced by Grandmont (1992). We have seen that in addition to periodic and quasi-periodic dynamics arising after a Hopf-bifurcation, the model also exhibits complex, chaotic dynamical behaviour after a cascade of period doubling bifurcations. The parameter values for which complex dynamics occur are much more realistic than in the 1-D case. In particular, in the 1-D OLG-model chaos only arises with a very strong income effect. In the 2-D OLG-model chaos can already arise when the income effect is only slightly stronger than the substitution effect. The numerical evidence for complex dynamical behaviour in this paper will serve as a basis for a more detailed analysis of the global dynamics of the model.

Acknowledgements

We are indebted to Jean-Michel Grandmont for helpful comments on an earlier version of the paper. The comments of an anonymous referee are also gratefully acknowledged. Finallly we would like to thank Jan de Gooijer and Peter Molenaar for discussions and Erik Wassink for assistence with the numerical work.

References

Grandmont, J.M. (1985), 'On endogenous competitive business cycles', *Econometrica* **53**,995-1045.

Grandmont, J.M. (1992), 'Expectations driven nonlinear business cycles', to appear in: *Proc. FIEF conference on business cycle*, Stockholm, august 1991.

Kaldor, N. (1940), 'A model of the trade cycle', *Economic Journal* **50**, 78-92.

Hicks, J.R. (1950), 'A contribution to the theory of the trade cycle', Clarendon Press, Oxford.

Goodwin, R.M. (1951), 'The nonlinear accellerator and the persistence of business cycles', *Econometrica* **16**, 1-17.

Lorenz, H.-W. (1989), 'Nonlinear dynamical economics and chaotic motion', Lecture Notes in Economic and Mathematical Systems 334, Springer-Verlag, Berlin.

Hommes, C.H. (1991), 'Chaotic dynamics in economic models. Some simple case studies', Ph-D thesis University of Groningen, Wolters-Noordhoff, Groningen.

Medio, A. (1992), 'Chaotic dynamics. Theory and applications to economics', Cambridge University Press, Cambridge.

Benhabib, J. and Day, R. (1982), 'A characterization of erratic dynamics in the overlapping generations model', *Journal of Economic Dynamics and Control* **4**, 37-55.

Reichlin, P. (1986), 'Equilibrium cycles in an overlapping generations economy with production', *Journal of Economic Theory* **40**, 898-102.

NONLINEARITY AND FORECASTING ASPECTS
OF PERIODICALLY INTEGRATED AUTOREGRESSIONS

by

PHILIP HANS FRANSES

Econometric Institute, Erasmus University Rotterdam
P.O. Box 1738, NL-3000 DR Rotterdam, The Netherlands

Summary

This paper deals with forecasting and nonlinearity aspects of linear periodic models for seasonally observed time series which contain a single unit root. This unit root imposes a nonlinear restriction on the model parameters. Multi-step ahead forecasts differ from forecasts obtained from nonperiodic models in the sense that they can reflect slowly changing seasonal patterns observed within the estimation sample.

Keywords Seasonal time series, periodic models, unit roots

1. Introduction

Two often encountered characteristics of seasonally observed time series in economics is that they show trending behavior and slowly changing seasonal patterns. Traditionally, time series analysts describe such variables using linear autoregressive models for the appropriately transformed series. Usually these transformations are based on the application of the first order differencing filter or the seasonal differencing filter. These filters imply for log transformed time series that the resulting series correspond to seasonal and annual growth rates, respectively.

For some economic series these models can be considered to be only rough approximations. Seasonal growth rate models assume that the seasonal patterns can be described by constant seasonal dummies. However, recursive estimates of the corresponding parameters sometimes suggest that they may not be constant within the estimation sample, see, e.g., Canova and Ghysels (1992). On the other hand, the annual growth rate models assume that the annual series, which contain the observations in the separate seasons, each have a unit root. In theory, this implies that the observations in the different seasons may not have the tendency to move together, which may be unreasonable for many economic time series. This calls for a class of models that can cope with these drawbacks of the traditional models. An example of such a class is that of periodic autoregressions with a single unit root, which are also called periodically integrated autoregressions [PIAR]. Detailed discussions of these models are given in Osborn (1988), Franses (1991), and Boswijk and Franses (1992). In the present paper the focus is on the forecasting and nonlinearity aspects of this class.

First, in section 2, some notational issues are briefly discussed. Then, in section 3, the nonlinearity aspect is highlighted. Section 4 deals with forecasts from a PIAR. Section 5 concludes with some remarks.

2. Preliminaries

Consider a time series x_t which is observed s times per year during N years, where s usually is 2, 4 or 12. The index t runs from 1 through n, with $n = sN$. A commonly applied assumption for x_t is that it can be described by an autoregression of order p [AR(p)], i.e.

$$\phi_p(B)x_t = \mu + \varepsilon_t, \tag{1}$$

where $\phi_p(B) = 1 - \phi_1 B - \ldots - \phi_p B^p$ is a p-th order polynomial in the operator B which is defined by $B^k x_t = x_{t-k}$. The ε_t is a standard white noise process, and μ is a constant.

When $\phi_p(B)$ can be decomposed as $\phi_p(B) = \phi_{p-s}^*(B)(1 - B^s)$ model (1) becomes $\phi_{p-s}^*(B)y_t = \mu + \varepsilon_t$, where $y_t = \Delta_s x_t$ i.e., the annually differenced x_t series. The Δ_s filter is defined by $\Delta_s x_t = x_t - x_{t-j}$. Alternatively, when $\phi_p(B)$ in (1) can be decomposed as $\phi_{p-1}^{**}(B)(1 - B)$, then model (1) becomes $\phi_{p-1}^{**}(B)z_t = \mu + \varepsilon_t$, where $z_t = \Delta_1 x_t$ with Δ_1 is the first order differencing filter. In the latter case, the μ is usually replaced by seasonal dummies to account for seasonal variation.

Model (1) assumes autoregressive parameter constancy, i.e. the ϕ_i parameters do not vary with season s. A model that allows periodic parameters is the so-called periodic autoregression of order p [PAR(p)], see, e.g., Anderson and Vecchia (1993) and the references cited therein. There are several ways to represent a PAR process, of which a simple one is

$$\phi_{ps}(B)x_t = \mu_s + \varepsilon_t, \tag{2}$$

with $\phi_{ps}(B) = 1 - \phi_{1s}B - \ldots - \phi_{ps}B^p$, where the index s indicates that the parameters vary with the season. Of course not all parameters in $\phi_{ps}(B)$ have to be unequal to zero. Hence, also the lag lengths in each of the seasons may be different.

It is clear from (2) that a PAR model considers each season differently. Therefore, an alternative representation of (2) is given by a multivariate model for X_T which is the ($s \times 1$) vector of stacked seasonal observations, i.e. $X_T = (X_{1T}, X_{2T}, \ldots, X_{sT})'$, where X_{iT} is the observation in season i in year T, see, e.g., Lutkepohl (1991) and the references cited therein. The index T runs from 1 through N. A vector representation of (2) is

$$A_0 X_T = \delta + A_1 X_{T-1} + \ldots + A_m X_{T-m} + \varepsilon_T \tag{3}$$

where A_i, $i = 0, \ldots, m$ are ($s \times s$) parameter matrices, δ is the ($s \times 1$) vector of constants, and ε_T is vector white noise process. The order m is related to the order p in (2) by $m \leq sp$.

3. A Unit Root and Nonlinearity

The vector process X_T and the univariate process x_t do not contain unit roots if the solutions to the characteristic equation

$$|A_0 - A_1 z - ... - A_m z^m| = 0 \qquad (4)$$

are outside the unit circle. Franses and Paap (1993) in a yet unpublished paper find evidence of the presence of a single unit root in about twenty quarterly UK macroeconomic time series. The presence of a unit root can be checked by testing whether $z=1$ is the only unity solution of (4). It is clear from (1) that this unity solution implies a nonlinear parameter restriction. Hence, a periodically integrated autoregression is a linear model with a nonlinear parameter restriction. A simple example is given by a PAR(1) process

$$x_t = \phi_{1s} x_{t-1} + \varepsilon_t .$$

The multivariate model for the corresponding X_T vector process contains a unit root when $\Pi_{i=1}^s \phi_{1i} = 1$.

Pre-multiplying (3) with A_0^{-1}, and some rewriting yields

$$\Delta X_T = \delta + \Gamma_1 \Delta X_{T-1} + ... + \Gamma_{m-1} \Delta X_{T-m+1} + \Pi X_{t-m} + \omega_T , \qquad (5)$$

where the Γ_j and Π, $j = 1,...,m-1$, are functions of the A_i in (3), $i=0,..,m$, where Δ is the first order differencing filter for annual time series, and $\omega_T = A_0^{-1} \varepsilon_T$. Note that the Δ filter here corresponds to the Δ_s filter for the univariate quarterly series X_t.

The representation in (5) is convenient to test for cointegration between the elements of X_T, see Franses (1991). For example, if the rank of the matrix Π is equal to zero, the model reduces to a vector process for the ΔX_T series. Otherwise stated, when there are no cointegration relations between the elements of X_T, the transformed series $\Delta_s X_t$ can be used for further modeling. Then, each of the time series X_{sT} contains a stochastic trend, and there are no linear combinations that ensure that these series tie together.

A PIAR process of some order p which contains a single unit root has $s-1$ cointegrating relations between the X_{sT} series. Hence, the rank of Π in (5) is equal to $s-1$. If this is the case, it can easily be shown that these cointegration relations are given by

$$X_{iT} - \alpha_i X_{i-1,T}, \quad \text{with} \quad \prod_{j=1}^s \alpha_j = 1 \quad \text{for} \quad i = 1,2,...,s , \qquad (6)$$

where $X_{0T} = X_{s,T-1}$. Hence, a PIAR(p) process can also be written as

$$(1 - \alpha_i B)x_t = \beta_{1s}(1 - \alpha_{i-1} B)x_{t-1} + ... + \beta_{p-1,s}(1 - \alpha_{i-p+1} B)x_{t-p+1} + \mu_s + \varepsilon_t , \qquad (7)$$

where the α_i satisfy the restriction in (6), and where $\alpha_{-k} = \alpha_{s-k}$, for $k=0, 1,...$. The β_{js} are again periodic parameters. The filter $(1 - \alpha_i B)$ is called a periodic differencing filter. Obviously, when all α_i are equal to one, the conventional $(1-B)$ filter emerges.

Together with (4), the equation in (7) indicates a useful model selection strategy. A first step is to estimate the order p of the PAR. This can be done using familiar model selection criteria as the Akaike and Schwarz information criteria, or using F type tests for parameter redundancy. Once this order is determined, the next step is to test whether the autoregressive parameters are seasonally varying. Again an F type test can be constructed. In Franses and Paap (1993) it is shown that for both F tests standard asymptotic results apply. The third step is to check the number of unit roots in a PAR by checking the solutions of (4). When there is only a single unit root, which is usually the case in practice, one can reformulate the model as in (7) to test for adequacy of filters such as $(1-B)$, or even $(1 - B^4)$ if the order p is large enough.

The model selection strategy, discussed in the previous paragraph, is a general-to-simple-method. The crucial tests for periodicity and unit roots are performed within the context of a prespecified model. An alternative, since simple-to-general, method is given in Vecchia and Ballerini (1991). This approach checks for periodicity in the autocorrelation function. A drawback of this method however is that two distinct time series processes, like AR(2) and AR(3), can have similar estimated autocorrelation functions. Moreover, not rejecting the null hypothesis of no periodic autocorrelations does not automatically imply that the underlying process is nonperiodic.

4. Forecasting

Consider again the expressions in (6) and (7). In practice, the α_i are usually estimated to be close to, though not equal to, one. This means that the distance between the observations X_{iT} and $X_{i-1,T}$ is not constant over time, which implies that a PIAR process displays a slowly changing seasonal pattern. The changes in this pattern depend on the changes in the stochastic trend in the X_T process.

The cointegration relations in (6) also effect the pattern of the multistep ahead forecasts. In fact, these forecasts will display slowly changing patterns too. The magnitudes of these changes depend on the estimated values of α_i. On the other hand, the nonperiodic models for the Δ_4 and Δ_1 transformed time series do not generate such slowly changing patterns in the forecasts.

Of course, when the $(1 - \alpha_i B)$ filter is appropriate, one can expect some gain in one-step ahead forecasting of using PIAR models. This is because some nonperiodic models misspecify either the number or the form of the cointegration relations. Empirical evidence of such a gain is reported in Osborn and Smith (1989), Franses (1992), and Franses and Romijn (1992).

5. Remarks

Periodically integrated autoregressions can be useful for the description of time series with a stochastic trend and a slowly changing seasonal pattern. This seasonal pattern changes because of variations in the stochastic trend. In other words, seasonality, trend and possibly cycles may not be easily separable. Since this is the underlying assumption of seasonal adjustment methods as Census-X11, one may question the usefulness of these methods when applied to periodic time series with a single unit root. Future research will be directed to investigate the effects of seasonal correction methods.

Although Osborn (1988) derives a PIAR(1) process for nondurable quarterly consumption directly from an economic theory, for many economic time series it is unlikely that a univariate PIAR process is the underlying data generating process. A natural step may then be to consider periodic cointegration models. These models incorporate error correction mechanisms with periodic equilibrium and adjustment parameters.

Acknowledgements

Thanks are due to an anonymous referee for several constructive comments. The financial support from the Royal Netherlands Academy of Arts and Sciences is gratefully acknowledged.

References

Anderson, P.L. and A.V. Vechhia (1993), 'Asymptotic results for periodic autoregressive moving-average processes', *J. of Time Series Analysis*, **14**, 1-18.

Boswijk, H.P. and P.H. Franses (1992), 'Testing for periodic integration', Econometric Institute Report 9216, Erasmus University Rotterdam.

Canova, F. and E. Ghysels (1991), 'Changes in seasonal patterns: are they cyclical?', *J. of Economic Dynamics and Control*, to appear.

Franses, P.H. (1991), 'A multivariate approach to modeling univariate seasonal time series', Econometric Institute Report 9101, *J. of Econometrics*, to appear.

Franses, P.H. (1992), 'Periodically integrated subset autoregressions for Dutch industrial production and money stock', *J. of Forecasting*, to appear.

Franses, P.H. and R. Paap (1993), 'Model selection in periodic autoregressions', Econometric Institute Report 9213, Erasmus University Rotterdam.

Franses, P.H. and G. Romijn (1992), 'Periodic integration in UK macroeconomic time series', Econometric Institute Report 9209, *Int. J. of Forecasting*, to appear.

Lütkepohl, H. (1991), *Introduction to multiple time series analysis*, Berlin: Springer Verlag.

Osborn, D.R. (1988), 'Seasonality and habit persistence in a life-cycle model of consumption', *J. of Applied Econometrics*, **3**, 255–266.

Osborn, D.R. and J.P. Smith (1989), 'The performance of periodic autoregressive models in forecasting seasonal U.K. consumption', *J. of Business and Economic Statistics*, **7**, 117-127.

Vecchia, A.V. and R. Ballerini (1991), 'Testing for periodic autocorrelations in seasonal time series data', *Biometrika*, **78**, 53-63.

CLASSICAL AND MODIFIED RESCALED RANGE ANALYSIS: SOME EVIDENCE

BEN JACOBSEN

Financial Management Department, University of Amsterdam,
Roetersstraat 11, 1018 WB Amsterdam, The Netherlands.
e-mail: ben@james.fee.uva.nl

Abstract

In this paper it is shown that the 'modified' rescaled range statistic, as suggested by Lo (1991), is for practical purposes an unnecessary complication, since it is also possible to first correct the time series for its idiosyncratic short term dependence and then simply apply the 'classical' rescaled range test developed by Hurst (1951). These two methods are illustrated by applying them to an index of the Amsterdam Stock Exchange. A comparison between the empirical power of both methods is made, using a Monte Carlo simulation.

Key Words time series, stock returns, long term dependence, rescaled range analysis, R/S analysis, Hurst exponent.

1. Introduction

Stock prices are probably among the most intensively studied time series in the world. Since the beginning of the century evidence has been growing that these series follow a random walk. Recently however, several researchers have reported evidence of long term dependence in stock returns, see for instance, De Bondt and Thaler (1985), Fama and French (1988), Poterba and Summers (1988), Jegadeesh (1990) and Kim, Nelson and Startz (1991). Unfortunately, this evidence is not conclusive.

Using a technique called rescaled range analysis (R/S analysis) Greene and Fielitz (1977), Booth, Kaen and Koveos (1982) and Peters (1991) also report evidence of long term dependence in economic time series. We will use the definition of long term dependence suggested by Beran (1993, page 405): "correlations that decay hyperbolically, that is, like $|k|^a$ with $a \in (0,1)$", where k denotes the size of the lag. Intuitively one might think of an autocorrelation function that fails to dampen out, even when the lag becomes very large. Recently, R/S analysis has been criticised by Lo (1991). His basic conclusion is that, due to short term correlation, R/S analysis can give misleading results. Consensus among financial economists, that stock returns are also short term correlated, is now growing. See for instance, Lo and MacKinlay (1988). Lo suggests a different form of R/S analysis, introducing a measure he calls the 'modified' rescaled range. Using this measure, thus correcting for short term dependence, his results for stock returns of the United States do not indicate the existence of long term dependence. He concludes that the finding of long term dependence in economic time series, as suggested by other authors, might have been premature. These results are confirmed by Jacobsen (1993), who reports additional evidence for stock markets of several European countries, Japan and the United States. However, Cheung (1990), also applying the modified rescaled range technique to time series of exchange rates, still finds evidence of long term dependence in these series.

The purpose of this paper is to show that Lo's suggestion to modify the rescaled range, is for practical purposes an unnecessary complication, since it is also possible to first correct the time series for its idiosyncratic short term dependence and then simply apply the 'classical' rescaled range test. These two methods are illustrated by applying them to an index of the Amsterdam Stock Exchange. Moreover, the empirical power of both methods is compared, using a Monte Carlo simulation.

Several statistical procedures related to long term dependence in time series have currently gained interest, especially the research regarding the so-called ARFIMA (Autoregressive Fractionally Integrated Moving Average) models. This research, although closely related to the statistics used in this research, will not be discussed here. The interested reader can find an overview of current research in this area in Beran (1992). References of long term dependence in other economic time series can be found in Cheung (1990,1993).

This paper is organised as follows. In Section 2, we introduce the 'classical' rescaled range statistic and the 'modified' rescaled range statistic. In Section 3, we report the results of the rescaled range statistics applied to an index of the stock market of the Netherlands. Results of a Monte Carlo simulation, using fractional Brownian motion, indicating the empirical power of the methods used, are given in Section 4. Some conclusions are given at the end.

2. Rescaled range analysis

2.1 Introduction

In this section the classical rescaled range statistic, introduced by Hurst (1951), and a modification of the rescaled range statistic, suggested by Lo (1991), will be discussed. In the last part of this section, we will shortly address the graphical procedure to estimate the so-called 'Hurst exponent', which is closely related to the rescaled range.

2.2 The classical rescaled range statistic

The Rescaled Range or R/S is a statistic developed by Harold Edwin Hurst (1951). The R/S statistic is the range of partial sums of deviations of a time series from its mean, rescaled by its standard deviation. To be more precise; let r_j be the return on a stock over period j, then the average return over some period n for $r_1, r_2, ..., r_n$ is:

$$\overline{r}_n = \frac{1}{n} \sum_{j=1}^{n} r_j \tag{1}$$

The difference between the maximum and the minimum accumulated deviation from the mean over period n is called the Range (R_n), thus:

$$R_n = \max_{1 \leq k \leq n} \sum_{j=1}^{k} \{r_j - \overline{r}_n\} - \min_{1 \leq k \leq n} \sum_{j=1}^{k} \{r_j - \overline{r}_n\} \tag{2}$$

To rescale this range, or in other words to make this measure dimensionless, Hurst divided the range by the usual standard deviation over period n:

$$S_n = \left(\frac{1}{n} \sum_j \{ r_j - \overline{r_n} \}^2 \right)^{1/2} \tag{3}$$

This ratio R_n/S_n (or R/S) is called the classical rescaled range. In the case of i.i.d observations the asymptotic distribution function of the normalized (that is, divided by the square root of n) classical rescaled range statistic is given by (see also Lo (1991)):

$$F_v(v) = 1 + 2 \sum_{k=1}^{\infty} (1 - 4k^2v^2) \, e^{-2(kv)^2} \tag{4}$$

with $E[V] = \sqrt{(\pi/2)} \approx 1.25$ and $E[V^2] = \pi^2/6$. Tests of the null hypothesis of no long term dependence are performed (in the absence of short term dependence) by verifying (after calculating the confidence intervals with respect to some significance level), if the normalized classical R/S statistic (V_c from now on) lays in- or outside the desired interval (Asymptotic p-values are given in table 4). The main advantage of the R/S statistic is, that it still gives reliable results if the time series exhibits large skewness and kurtosis. However, as shown by Lo (1991) the V_c can give biased results in the case of short term dependence. Lo proves that if the time series exhibits short term dependence, this changes the limiting distribution of the rescaled range by a multiplicative constant, for which the expression depends on the short term dependence structure. The modified rescaled range corrects for the bias caused by short term dependence.

2.3 The modified rescaled range statistic

To take the shortcoming of the classical rescaled range into account, Lo (1991) suggests to modify the rescaled range. This correction means that we replace the term S_n by $\sigma_n(q)$ where:

$$\sigma_n^2(q) = \frac{1}{n} \sum_{j=1}^{n} \left(r_j - \overline{r_n} \right)^2 + \frac{2}{n} \sum_{j=1}^{q} w_j(q) \left\{ \sum_{i=j+1}^{n} \left(r_i - \overline{r_n} \right) \left(r_{i-j} - \overline{r_n} \right) \right\} \tag{5a}$$

$$= \sigma_r^2 + 2 \sum_{j=1}^{q} w_j(q)\gamma_j, \quad w_j(q) = 1 - \frac{j}{q+1}, \quad q<n \tag{5b}$$

and σ_r^2 and γ_j are the usual estimators of the sample variance and autocovariance of r. Lo calls the statistic adjusted this way, the 'modified' rescaled range. This 'modified' R/S can, when properly normalized (that is divided by $\sqrt{n}$), be used for detecting long term dependence. Thus, we can use the statistic:

$$V(q) = \frac{1}{\sqrt{n}} \cdot \frac{R}{\sigma(q)} \tag{6a}$$

instead of:

$$V_c = \frac{1}{\sqrt{n}} \cdot \frac{R}{S} \tag{6b}$$

where V_c is the normalized classical rescaled range and $V(q)$ is the modified normalized rescaled range with lag q. Tests of the null hypothesis of no long term dependence, for a 95 percent confidence level, for instance, can then be performed according to whether $V(q)$ is contained in the interval $[0.809, 1.862]$. (Asymptotic p-values are given in the appendix). The argument Lo gives for this correction is that if the series $\{r_j\}$ is subject to short range dependence, the variance of the partial sum is not simply the sum of the variance of the individual terms, but also includes the autocovariances. Therefore, the estimator $\sigma_n(q)$ involves not only sums of squared deviations of r_j, but also its weighted autocovariances up to lag q.

The main advantage of Lo's method is, that it is a very general method as it applies to many forms of short term dependent structures. However, the main drawback is that we have to choose a q. The choice of q is an arbitrary one. Lo concludes that little is known about how best to pick q in finite samples. This means that if we pick q too small we might still leave some short term dependence in the time series and the modified rescaled range statistic might give biased results. On the other hand if we pick q too large, the $V(q)$ might become insensitive to long memory in the data (see also Cheung (1990)). Thus, the price to pay for the ability of the modified rescaled range to adjust, for general forms of short term dependence, is rather high. In fact, it might be easier to first correct the time series for its specific short term dependence and then simply calculate the classical R/S statistic. Although in this case we must assume some short term dependence structure, for instance an AR(1)-process, we do not have to choose an arbitrary value for q. Lo acknowledges this possibility, but his attention focuses on the statistical aspects of the modified rescaled range. As we are interested in the application of the statistic, we can still use this procedure. Using proposition 3.1 (Lo (1991), page 1293) these calculations can even be done by hand, if we have calculated the classical rescaled range statistic. The procedure also enables us to correct in a 'tailor made' way for the dependence observed. Therefore, we will not only calculate the classical R/S and the modified R/S for the original time series, but we will also analyze the time series after fitting some AR(1)-processes. In section 4 we will also compare the empirical power of both procedures. Lo (1991) and Jacobsen (1993) compare the results of the classical rescaled range, V_c with the (normalized) modified R/S statistic, $V(q)$. Their main conclusion is that V_c gives misleading results for time series that exhibit short term dependence.

2.4 The graphical procedure

The graphical procedure of rescaled range analysis consists of first averaging the value of the rescaled range for a number of values up to n for a given value of n. Let $Q(n)$ denote this average. The limit of the ratio $log[Q(n)]/log\ n$ is often called the 'Hurst exponent'. Mandelbrot and Wallis (1969) suggest a plotting technique of $log[Q(n)]$ against $log\ n$ for different values of n. The slope of this plot can then be estimated using ordinary least squares and is an estimate of the Hurst exponent. It might be useful to get a grip on the Hurst exponent (H) by relating H to the parameter d in the class of ARFIMA processes. A time series $\{x_t\} = \{x_1,...,x_T\}$ follows an ARFIMA(p,d,q) process if $\Phi(B)(1-B)^d x_t = \Theta(B)\varepsilon_t$, where $\varepsilon_t\sim$i.i.d $(0,\sigma^2)$ and B is the Backward shift operator. ARFIMA processes with $d \in$ (0,0.5) display long memory. For $d = 0$ an ARFIMA process reduces to an ARMA process. In these models, H is equal to $d + \frac{1}{2}$, when $-\frac{1}{2} < d < \frac{1}{2}$.

It is shown in Feller (1951) that the range R_n for an independent Gaussian process becomes asymptotically proportional to $\sqrt{n}$. This means that the Hurst exponent or the ratio $log[Q(n)]/log\ n$ for an independent process approaches $\frac{1}{2}$ in the limit. However, if the time series exhibits long term dependence, this ratio should converge to values larger or less than $\frac{1}{2}$. Mandelbrot and Wallis (1969) also state that for any short term dependent stochastic process the Hurst exponent would converge to $\frac{1}{2}$ if n goes to infinity. It is common usage to call a time series with a Hurst exponent larger than $\frac{1}{2}$ persistent and a time series with a Hurst exponent less than $\frac{1}{2}$ anti-persistent. Mandelbrot has in several papers advocated the use of the R/S analysis, as it has some advantages over other conventional methods (as analyzing autocorrelations) of determining long range dependence. For instance, R/S analysis can still be used if the time series has large skewness and kurtosis. Mandelbrot (1972) notes that R/S can even be used for stochastic processes that have infinite variances. Although much is known about the empirical distribution of stock returns, the question of which theoretical distribution conforms best to the empirical evidence is still an open one. Therefore this is an important advantage because the R/S statistic would still hold if stock returns conform to a stable Paretian distribution with a characteristic exponent less than 2 and does not rule out the possibility of such a distribution in advance. This class of distributions has the property that the variance does not exist or is infinite and has been suggested by different authors in relation to stock returns (See for instance, Fama (1965)). This graphical procedure has some major drawbacks too. The most important one is that the asymptotic distribution of the Hurst exponent has not been determined and is in fact intractable (Cheung (1990)). Furthermore the Hurst exponent is also sensitive to short term dependence. Although it is true that the Hurst exponent is equal to $\frac{1}{2}$, in the limit, for short range dependent processes, the *estimated* Hurst coefficient might be biased for short term dependent processes. This shortcoming was pointed out by Davies and Harte (1987), and the results in Lo (1991) support this evidence. Even though the Hurst exponent for an AR(1)-process should be equal to $\frac{1}{2}$ in the limit. Davies and Harte (1987) show that the regression test specified above too often rejects this null hypothesis for finite samples. Another potential drawback, one that will not be considered here, has been suggested by Klemeš (1984). He claims that the Hurst phenomenon is due to nonstationarities in the underlying mean of the process. In Jacobsen (1993) however, it is shown that even after correcting for short term dependence, the graphical procedure, developed by Mandelbrot and Wallis (1969), still gives estimates for the Hurst exponent higher than $\frac{1}{2}$, indicating long term dependence.

3. R/S applied to an index of the Amsterdam Stock Exchange

In this section we will compare the modified R/S statistic, introduced by Lo (1991), with the procedure in which we first correct the time series for its idiosyncratic short term dependence and then use the classical statistic V_c. The data used in this research are the return series of the (value-weighted) CBS-index (Royal Dutch excluded, see the remark below) of the Amsterdam Stock Exchange over the period December 1952 - December 1990. This series consists of 456 monthly continuously compounded returns. The results and the basic characteristics of the CBS-index are reported in Table 1. If we look at the values of V_c and $V(q)$, neither the classical nor the modified statistic suggest significant long term dependence at a 5 percent significance level. This confirms the conclusion of no long term dependence in stock returns (Lo (1991) and Jacobsen (1993)). Indeed as indicated by Lo (1991) short term dependence might very well have biased the value of V_c upward, since the $V(q)$ is lower for all q.

Table 1. Basic characteristics; classical and modified R/S

number of obs.	456	$\rho(1)$	0.1774*	[0.061]	$\rho(7)$	0.0194	[0.0445]
mean	0.0050	$\rho(2)$	0.0071	[0.041]	$\rho(8)$	−0.0188	[0.0462]
var	0.0024	$\rho(3)$	0.0720	[0.048]	$\rho(9)$	0.1523*	[0.0447]
std	0.0491	$\rho(4)$	−0.0559	[0.052]	$\rho(10)$	0.0972*	[0.0463]
skewness	−0.6899	$\rho(5)$	−0.0376	[0.046]	$\rho(11)$	0.0747	[0.0454]
kurtosis	3.4926	$\rho(6)$	−0.0008	[0.041]	$\rho(12)$	0.0902*	[0.0427]

		CBS-index	AR(1) $\rho = 0.15$	AR(1) $\rho = 0.175$	AR(1) $\rho = 0.20$
V_c		1.66	1.42	1.38	1.34
$V(q)$	$q = 3$	1.60	1.42	1.39	1.36
	$q = 6$	1.59	1.45	1.36	1.35
	$q = 9$	1.54	1.34	1.34	1.33
	$q = 12$	1.35	1.28	1.27	1.26

- heteroscedastisty consistent standard errors in square brackets
- asterisks denote significance using a 5 percent significance level

We stated that instead of calculating the 'modified' rescaled range it is also possible to first correct for short term dependence and then use the classical rescaled range statistic. Therefore, based on the autocorrelation function, we fitted three different AR(1)-processes with correlation 0.15, 0.175 and 0.20 respectively. The evidence, that short term dependence might be the cause of the long term dependence found, using rescaled range analysis, gets even stronger if we look at these results. (Note, that after imposing a short term dependence structure we can use the same confidence interval for the V_c as for the $V(q)$.) The lowest value for the $V(q)$ in the first column is 1.35 ($q=12$), whereas the highest value for the V_c in the other columns, adjusting the time series for its observed dependence, has already dropped to 1.42. If we just take into account the first order correlation, the V_c is close to 1.25 (or $\sqrt{(\pi/2)}$), and this value is the expected value of V_c for a process with no long term dependence. The value of the V_c is slightly lower than this expected value but this might even be due to the fact that we only corrected for the positive short term dependence, as the negative serial correlations were high at lag four and five but not significant. The hypothesis of long term dependence is, by this tailor made procedure, also rejected. The values for V_c are even lower than the values found for the more general $V(q)$ in the first column (apart from $q=12$). It can also be seen that the differences between the V_c's and the $V(q)$'s of the corrected series are not large: the maximum of the difference between V_c and $V(q)$ is 1.42 minus 1.28, or 0.14.

Thus, if we first correct the time series for the specific short term dependence observed, we can still use the classical statistic. For practical purposes it therefore seems less complicated to first correct for the observed short term dependence and then applying the classical test instead of using the modified rescaled range as suggested by Lo (1991).

Remark On the Dutch stock market about half of the trading is in stock of the Royal Dutch. This means that the index including Royal Dutch is highly influenced by the price changes of this stock. Therefore, normal procedure is to exclude Royal Dutch from the index. The Hurst exponent, the V_c and the $V(q)$'s were also calculated for the index with Royal Dutch included. Since the results are basically the same as the results given here we do not report them here.

Although, using an AR(1)-process, we have imposed a structure on the form of short term dependence, the results do not differ much from the more complex procedure of calculating the modified rescaled range. Furthermore, another advantage of this procedure is, that the problem of choosing q is no longer there.

4. Empirical Power

4.1 Introduction

Although we know the asymptotic distribution of the rescaled range statistics, we do not know how working with a finite sample of 456 observations might give misleading results. To test the power of the classical and the modified rescaled range statistic, we performed a Monte Carlo simulation on fractional Brownian motion with different Hurst exponents. Moreover, we calculated the power of the procedure where we first fitted an AR(1)-process to these time series and then applied the classical rescaled range statistic.

4.2 Simulation

In our simulation we generated time series using a discrete version of fractional Brownian motion. We used the formula given in Feder (1988, page 174):

$$B_H(t) - B_H(t-1) = \frac{n^{-H}}{\Gamma(H+1/2)} \left\{ \sum_{i=1}^{n} (i)^{H-1/2} \xi_{(1+n(M+t)-i)} + \sum_{i=1}^{n(M-1)} ((n+i)^{H-1/2} - (i)^{H-1/2}) \xi_{(1+n(M-1+t)-i)} \right\} \quad (7)$$

where $\{\xi_i\}$, with $i = 1,2,...,M,...$, is a set of Normal random variables with unit variance and zero mean, H is the Hurst exponent and M and n are two constants. The choice of M influences the length of the long term dependence, whereas the n mainly influences the short term dependence. We used different values for M and n but these values did not seem to influence our results heavily. The results reported here were obtained by using M=200 and n=5. Using this formula we generated 1000 time series of 456 observations with Hurst exponents of 0.75 and 0.85. The classical and the modified rescaled range with truncation lags (q) of 3, 6, 9 and 12, were then calculated for these series. We also calculated the classical rescaled range after fitting an AR(1)-process. The power of those statistics were calculated using different α's (1, 5 and 10 percent, respectively). Furthermore, we calculated the mean, maximum, minimum and standard deviation of these statistics. The results of these simulations are reported in tables 2 and 3.

Table 2. The power of the classical and the modified rescaled range statistic, and the power of the classical rescaled range after correcting for short term dependence using an AR(1)-process. H=0.85, 1000 simulated time series of length 456.

	Power 10%	power 5%	Power 1%	Mean	Max.	Min.	Stddev.
$V(3)$	92.3%	87.9%	76.2%	2.630	4.579	1.034	0.634
$V(6)$	80.0%	74.3%	58.6%	2.237	3.667	1.105	0.497
$V(9)$	70.5%	62.6%	42.7%	2.021	3.177	1.054	0.423
$V(12)$	61.7%	51.7%	31.0%	1.880	2.876	1.012	0.374
V_c	99.6%	98.8%	96.1%	3.746	7.606	1.396	1.060
$V_{AR(1)}$	74.0%	67.0%	49.9%	2.134	3.384	0.994	0.504

Table 3. The power of the classical and the modified rescaled range statistic, and the power of the classical rescaled range after correcting for short term dependence using an AR(1)-process. H=0.75, 1000 simulated time series of length 456.

	Power 10%	Power 5%	Power 1%	Mean	Max.	Min.	Stddev.
$V(3)$	79.2%	69.0%	51.5%	1.757	3.934	1.017	0.526
$V(6)$	65.5%	54.9%	37.8%	1.655	3.306	0.943	0.438
$V(9)$	54.4%	43.3%	26.4%	1.593	2.946	0.906	0.384
$V(12)$	46.0%	36.8%	16.4%	1.550	2.716	0.909	0.346
V_c	93.3%	90.9%	80.4%	1.975	5.640	1.247	0.758
$V_{AR(1)}$	60.8%	52.2%	34.5%	1.924	3.553	0.878	0.494

4.3 Results

The simulation results reported here are well in line with the results found by Lo (1991) and Cheung (1990). In the case of strong long term dependence (high Hurst exponent, Table 2) we see that the power of all the statistics is reasonably high. Of course, the power of the V_c is the highest as we have simulated long term persistent time series that are likely to be positive short term dependent. This conclusion is verified if we look at the power of the $V_{AR(1)}$, which is lower for all significance levels. The power of the modified rescaled range is also lower for all q compared with the classical statistic. The power of the modified rescaled range statistic decreases with increasing q. This too is not surprising since in the case of positive short term dependence the denominator of $V(q)$ can only increase, while the range itself is not affected. If we compare the $V(q)$'s with the $V_{AR(1)}$ in table 2 we see that the power of the $V_{AR(1)}$ ranges between the power of the $V(6)$ and the $V(9)$. For the 5 percent level for instance, the power of the $V_{AR(1)}$ of 67.0% lies between the power of the $V(6)$ (74.3%) and $V(9)$ (62.6%). Similar results hold for the case where the Hurst exponent is equal to 0.75 (Table 3). Based on these results, one can argue that, for reasons of simplicity, it might be preferable to first correct the time series for short term dependence and then calculate the classical rescaled range statistic instead of using the modified rescaled range. Moreover, one could first calculate the classical rescaled range statistic and then use Lo's proposition 3.1 (Lo (1991), page 1293). If we compare the different tables we see that the power of all statistics decreases when the level of persistence (lower Hurst exponent) decreases.

Conclusion

The results indicate that instead of calculating the 'modified' rescaled range statistic, as suggested by Lo (1991), a much easier procedure is to first correct the time series at hand for its idiosyncratic short dependence structure, and then calculate the 'classical' rescaled range. The results of calculation of the 'classical' R/S statistic and the results of the 'modified' R/S statistic, as suggested by Lo (1991), do not indicate evidence of long term dependence for the index of the Amsterdam Stock Exchange. However, before drawing any strong conclusions about the existence of long term dependence based on the classical and the modified rescaled range statistics, the low power of the statistics in the case of weak long term dependence should be considered.

Table 4. Fractiles of the distribution $F_V(v)$

$P(V<v)$	v	$P(V<v)$	v
0.005	0.721	0.543	$\sqrt{(\pi/2)}$
0.025	0.809	0.600	1.294
0.050	0.861	0.700	1.374
0.100	0.927	0.800	1.473
0.200	1.018	0.900	1.620
0.300	1.090	0.950	1.747
0.400	1.157	0.975	1.862
0.500	1.223	0.995	2.098

Acknowledgements

The author thanks Pol Ankum, Arnoud Boot, Jan de Gooijer, Johan Grasman, Cars Hommes, Henk Koster, Theo Nijman and Paul Torfs for detailed comments. Programming assistance by Michael Blok and Angelo Welling is gratefully acknowledged. The usual disclaimer applies.

References

Beran, J. (1992), 'Statistical Methods for Data with Long-Range Dependence', *Statistical Science*, **7**, pp. 404-427.

Booth, C.G., Kaen, F.R., & Koveos, P.E. (1982), 'R/S Analysis of Foreign Exchange Rates Under Two International Monetary Regimes', *Journal of monetary Economics*, **10**, pp.407-415.

Cheung, Y. (1990), 'Long Memory in Foreign Exchange Rates and Sampling Properties of some Statistical Procedures related to Long Memory Models', Phd. Dissertation, University of Pennsylvania.

Cheung, Y. (1993), 'Long Memory in Foreign-Exchange Rates', *Journal of Business & Economic Statistics*, **11**, pp. 93-101.

Davies, R.B. and Harte, D.S. (1987), 'Tests for Hurst effect', *Biometrika*, **74**, pp. 95-101.

De Bondt, W.F.M., & Thaler, R. (1985), 'Does the Stock Market Overreact?' *Journal of Finance*, **XL**, pp. 793-808.

Fama, E.F. (1965), 'The behavior of stock market prices', *Journal of Business*, **38**, pp 34-105.

Fama, E.F. & French, K.R. (1988), 'Permanent and Temporary Components of Stock Prices', *Journal of Political Economy*, **96**, pp. 246-273.

Feder, J.(1988), 'Fractals', Plenum press, New York.

Feller, W.(1951), 'The asymptotic distribution of the range of sums of independent variables', *Ann. Math. Stat.*, **22**, pp. 427-432.

Greene, M. T. and Fielitz B.D. (1977), 'Long-Term Dependence in Common Stock Returns', *Journal of Financial Economics*, **4**, pp. 339-349.

Greene, M.T. & Fielitz, B.D. (1979), 'The Effect of Long Term Dependence on Risk Return Models of Common Stocks', *Operations Research*, **27**, pp. 944-951.

Harvey, A. (1990), 'The Econometric Analysis of Time Series', Philip Allan, New York.

Hurst H.E. (1951), 'Long-term storage capacity of reservoirs', *American Society of Civil Engineers* **116**, pp. 770-799.

Jacobsen, B. (1993), 'Long Term Dependence in Stock Returns', Working paper, Financial Management Department, University of Amsterdam.

Jegadeesh, N. (1990), 'Evidence of Predictable Behavior of Security Returns', *Journal of Finance*, **XLV**, pp. 881-899.

Kim, M.J., Nelson, C.R., & Startz, R. (1991), 'Mean Reversion in Stock Prices? A Reappraisal of the Empirical Evidence', *Review of Economic Studies*, **58**, pp. 515-528.

Klemeš, V. (1974), 'The Hurst Phenomenon: A Puzzle?' *Water Resources Research*, **10**, pp. 675-688.

Lo, A.W. and MacKinlay, A.C. (1988), 'Stock Market Prices Do Not Follow Random Walks: Evidence from a Simple Specification Test', *The Review of Financial Studies*, **1**, pp. 41-66.

Lo, A.W. (1991), 'Long-Term Memory in Stock Market Prices', *Econometrica*, **59**, pp. 1279-1313.

Mandelbrot, B. B. & Van Ness, J.W. (1968), 'Fractional Brownian Motions, Fractional Noises and Applications', *SIAM Review*, **10**, pp 422-437.

Mandelbrot B.B. & Wallis J.R. (1969), 'Robustness of the Rescaled Range R/S in the measurement of Noncyclic Long Run Statistical Dependence'. *Water Resources Research*, **5**, pp.967-988.

Mandelbrot, B.B. (1972), 'Statistical Methodology for Non-Periodic Cycles: From the Cova riance to R/S Analysis', *Annals of Economic and Social Measurement*, **1**, pp.259-290

Mandelbrot, B.B. & Taqqu M.S. (1979), 'Robust R/S Analysis of Long Run Serial Correlation'. *Invited paper of the 42nd session of the International Statistical Institute*, Manilla, 4-14 December.

Peters, E.E. (1991), 'Chaos and order in the capital markets', Wiley, New York.

Porter-Hudak, S. (1991), 'An Application of the Seasonal Fractionally Differenced Model to the Monetary Aggregates', *Journal of the American Statistical Association*, **85**, pp. 338-344.

Poterba, J.M. & Summers, L.H. (1988), 'Mean Reversion in Stock Prices, evidence and implications', *Journal of Financial Economics*, **22**, pp. 27-59.

SUBJECT INDEX

The manufacturer's authorised representative in the EU is Springer
Nature Customer Service Centre GmbH, Europaplatz 3, 69115 Heidelberg,
Germany. If you have any concerns regarding our products, please
contact ProductSafety@springernature.com

Printed and bound by CPI Group (UK) Ltd, Croydon, CR0 4YY
24/04/2026
02096348-0013